Student's Solutions Manual

Beverly Fusfield
James J. Ball

Finite Mathematics

Tenth Edition

Mathematics with Applications

Tenth Edition

Margaret Lial
American River College

Thomas Hungerford
St. Louis University

John Holcomb, Jr.
Cleveland State University

Addison-Wesley
is an imprint of

PEARSON

The author and publisher of this book have used their best efforts in preparing this book. These efforts include the development, research, and testing of the theories and programs to determine their effectiveness. The author and publisher make no warranty of any kind, expressed or implied, with regard to these programs or the documentation contained in this book. The author and publisher shall not be liable in any event for incidental or consequential damages in connection with, or arising out of, the furnishing, performance, or use of these programs.

Reproduced by Addison-Wesley from electronic files supplied by the author.

ISBN-13: 978-0-321-64582-1
ISBN-10: 0-321-64582-0

1 2 3 4 5 6 BRR 14 13 12 11 10

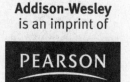

Addison-Wesley
is an imprint of

PEARSON

www.pearsonhighered.com

CONTENTS

1   **Algebra and Equations** 1

2   **Graphs, Lines, and Inequalities** 32

3   **Functions and Graphs** 62

4   **Exponential and Logarithmic Functions** 104

5   **Mathematics of Finance** 128

6   **Systems of Linear Equations and Matrices** 152

7   **Linear Programming** 202

8   **Sets and Probability** 262

9   **Counting, Probability Distributions, and Further Topics in Probability** 283

10 **Introduction to Statistics** 305

11 **Differential Calculus** 327

12 **Applications of the Derivative** 375

13 **Integral Calculus** 412

14 **Multivariate Calculus** 450

Chapter 1 Algebra and Equations

Section 1.1 The Real Numbers

1. True. This statement is true, since every integer can be written as the ratio of the integer and 1.

   For example, $5 = \dfrac{5}{1}$.

3. Answers vary with the calculator, but
   $\dfrac{2,508,429,787}{798,458,000}$ is the best.

5. $6(t + 4) = 6t + 6 \cdot 4$
   This illustrates the distributive property.

7. $0 + (-7) = -7 + 0$
   This illustrates the commutative property of addition.

9. Answers vary. One possible answer: The sum of a number and its additive inverse is the additive identity. The product of a number and its multiplicative inverse is the multiplicative identity.

For Exercises 11–13, let $p = -2$, $q = 3$ and $r = -5$.

11. $-3(p + 5q) = -3\big[-2 + 5(3)\big] = -3[-2 + 15]$
    $= -3(13) = -39$

13. $\dfrac{q + r}{q + p} = \dfrac{3 + (-5)}{3 + (-2)} = \dfrac{-2}{1} = -2$

15. Let $r = 1.5$.
    $APR = 12r = 12(1.5) = 18\%$

17. Let $APR = 9$.
    $APR = 12r$
    $9 = 12r$
    $\dfrac{9}{12} = \dfrac{3}{4} = r$
    $r = .75\%$

19. $3 - 4 \cdot 5 + 5 = 3 - 20 + 5 = -17 + 5 = -12$

21. $8 - 4^2 - (-12)$
    Take powers first.
    $8 - 16 - (-12)$
    Then add and subtract in order from left to right.
    $8 - 16 + 12 = -8 + 12 = 4$

23. $-(3 - 5) - \Big[2 - \big(3^2 - 13\big)\Big]$
    Take powers first.
    $-(3 - 5) - [2 - (9 - 13)]$
    Work inside brackets and parentheses.
    $-(-2) - [2 - (-4)] = 2 - [2 + 4]$
    $\qquad\qquad\qquad\quad = 2 - 6 = -4$

25. $\dfrac{2(-3) + \frac{3}{(-2)} - \frac{2}{\left(-\sqrt{16}\right)}}{\sqrt{64} - 1}$

    Work above and below fraction bar. Take roots.
    $\dfrac{2(-3) + \frac{3}{(-2)} - \frac{2}{(-4)}}{8 - 1}$

    Do multiplications and divisions.
    $\dfrac{-6 - \frac{3}{2} + \frac{1}{2}}{8 - 1}$

    Add and subtract.
    $\dfrac{-\frac{12}{2} - \frac{3}{2} + \frac{1}{2}}{7} = \dfrac{-\frac{14}{2}}{7} = \dfrac{-7}{7} = -1$

27. $\dfrac{2040}{523}, \dfrac{189}{37}, \sqrt{27}, \dfrac{4587}{691}, 6.735, \sqrt{47}$

29. 12 is less than 18.5.
    $12 < 18.5$

31. x is greater than or equal to 5.7.
    $x \geq 5.7$

33. z is at most 7.5.
    $z \leq 7.5$

35. $-6 < -2$

37. $3.14 < \pi$

39. a lies to the right of b or is equal to b.

41. $c < a < b$

43. $(-8, -1)$
    This represents all real numbers between -8 and -1, not including -8 and -1. Draw parentheses at -8 and -1 and a heavy line segment between them. The parentheses at -8 and -1 show that neither of these points belongs to the graph.

45. $[-2, 2)$

This represents all real numbers between -2 and 2, including -2, not including 2.
Draw a bracket at -2, a parenthesis at 2, and a heavy line segment between them.

47. $(-2, \infty)$

This represents all real numbers x such that $x > -2$. Start at -2 and draw a heavy line segment to the right. Use a parenthesis at -2 since it is part of the graph.

49. 3; 2000, 2006, and 2007

51. 7; 1998, 1999, 2001, 2002, 2003, 2004, 2005

53. 0

55. **a.** Steffi Graf's height in inches is $5(12) + 9 = 69$ in.
$$B = \frac{.455W}{(.0254H)^2} = \frac{.455(119)}{(.0254(69))^2}$$
$$= \frac{54.145}{(1.7526)^2} \approx 17.6$$

   b. No, Steffi Graf's body mass index falls below the desirable range.

57. **a.** Tiger Wood's height in inches is $6(12) + 2 = 74$ in.
$$B = \frac{.455W}{(.0254H)^2} = \frac{.455(180)}{(.0254(74))^2}$$
$$= \frac{81.9}{(1.8796)^2} \approx 23.2$$

   b. Yes, Tiger Wood's body mass index falls within the desirable range.

59. A wind at 10 miles per hour with a 30° temperature has a wind-chill factor of 21°. A wind at 30 miles per hour with a $-10°$ temperature has a wind-chill factor of $-39°$.
$$|21 - (-39)| = |21 + 39| = 60°$$

61. A wind at 25 miles per hour with a $-30°$ temperature has a wind-chill factor of $-64°$. A wind at 15 miles per hour with a $-30°$ temperature has a wind-chill factor of $-58°$.
$$|-64 - (-58)| = |-64 + 58| = |-6| = 6°$$

63. $|8| - |-4| = 8 - (4) = 4$

65. $-|-4| - |-1 - 14| = -(4) - |-15|$
$$= -(4) - 15 = -19$$

67. $|5| \underline{\quad} |-5|$
$$5 \underline{\quad} 5$$
$$5 = 5$$

69. $|10 - 3| \underline{\quad} |3 - 10|$
$$|7| \underline{\quad} |-7|$$
$$7 \underline{\quad} 7$$
$$7 = 7$$

71. $|-2 + 8| \underline{\quad} |2 - 8|$
$$|6| \underline{\quad} |-6|$$
$$6 \underline{\quad} 6$$
$$6 = 6$$

73. $|3 - 5| \underline{\quad} |3| - |5|$
$$|-2| \underline{\quad} 3 - 5$$
$$2 \underline{\quad} -2$$
$$2 > -2$$

75. When $a < 7$, $a - 7$ is negative.
So $|a - 7| = -(a - 7) = 7 - a$.

Answers will vary for exercises 77–79. Sample answers are given.

77. No, it is not always true that $|a + b| = |a| + |b|$. For example, let $a = 1$ and $b = -1$. Then,
$|a + b| = |1 + (-1)| = |0| = 0$, but
$|a| + |b| = |1| + |-1| = 1 + 1 = 2$.

79. $|2 - b| = |2 + b|$ only when $b = 0$. Then each side of the equation is equal to 2.

81. The statement is true for 2001, 2002, and 2007. For example, for the year 2001,
$|18,300,000 - 20,000,000| = |-1,700,000|$
$$= 1,700,000$$

Section 1.2 Polynomials

1. $11.2^6 \approx 1{,}973{,}822.685$

3. $\left(-\dfrac{18}{7}\right)^6 \approx 289.0991339$

5. -3^2 is negative, whereas $(-3)^2$ is positive. Both -3^3 and $(-3)^3$ are negative.

7. $4^2 \cdot 4^3 = 4^{2+3} = 4^5$

9. $(-6)^2 \cdot (-6)^5 = (-6)^{2+5} = (-6)^7$

11. $\left[(5u)^4\right]^7 = (5u)^{4 \cdot 7} = (5u)^{28}$

13. degree 4; coefficients: 6.2, –5, 4, –3, 3.7; constant term 3.7.

15. Since the highest power of x is 3, the degree is 3.

17. $\left(3x^3 + 2x^2 - 5x\right) + \left(-4x^3 - x^2 - 8x\right)$
$= \left(3x^3 - 4x^3\right) + \left(2x^2 - x^2\right) + (-5x - 8x)$
$= -x^3 + x^2 - 13x$

19. $\left(-4y^2 - 3y + 8\right) - \left(2y^2 - 6y + 2\right)$
$= \left(-4y^2 - 3y + 8\right) + \left(-2y^2 + 6y - 2\right)$
$= -4y^2 - 3y + 8 - 2y^2 + 6y - 2$
$= \left(-4y^2 - 2y^2\right) + (-3y + 6y) + (8 - 2)$
$= -6y^2 + 3y + 6$

21. $\left(2x^3 + 2x^2 + 4x - 3\right) - \left(2x^3 + 8x^2 + 1\right)$
$= \left(2x^3 + 2x^2 + 4x - 3\right) + \left(-2x^3 - 8x^2 - 1\right)$
$= 2x^3 + 2x^2 + 4x - 3 - 2x^3 - 8x^2 - 1$
$= \left(2x^3 - 2x^3\right) + \left(2x^2 - 8x^2\right) + (4x) + (-3 - 1)$
$= -6x^2 + 4x - 4$

23. $-9m\left(2m^2 + 6m - 1\right)$
$= (-9m)\left(2m^2\right) + (-9m)(6m) + (-9m)(-1)$
$= -18m^3 - 54m^2 + 9m$

25. $(3z + 5)\left(4z^2 - 2z + 1\right)$
$= (3z)\left(4z^2 - 2z + 1\right) + (5)\left(4z^2 - 2z + 1\right)$
$= 12z^3 - 6z^2 + 3z + 20z^2 - 10z + 5$
$= 12z^3 + 14z^2 - 7z + 5$

27. $(6k - 1)(2k + 3)$
$= (6k)(2k + 3) + (-1)(2k + 3)$
$= 12k^2 + 18k - 2k - 3$
$= 12k^2 + 16k - 3$

29. $(3y + 5)(2y + 1)$
Use FOIL.
$= 6y^2 + 3y + 10y + 5$
$= 6y^2 + 13y + 5$

31. $(9k + q)(2k - q)$
$= 18k^2 - 9kq + 2kq - q^2$
$= 18k^2 - 7kq - q^2$

33. $(6.2m - 3.4)(.7m + 1.3)$
$= 4.34m^2 + 8.06m - 2.38m - 4.42$
$= 4.34m^2 + 5.68m - 4.42$

35. $5k - [k + (-3 + 5k)]$
$= 5k - [6k - 3]$
$= 5k - 6k + 3$
$= -k + 3$

37. $R = 5(1000x) = 5000x$
$C = 200{,}000 + 1800x$
$P = (5000x) - (200{,}000 + 1800x)$
$\quad = 3200x - 200{,}000$

39. $R = 9.75(1000x) = 9750x$
$C = 260{,}000 + (-3x^2 + 3480x - 325)$
$\quad = -3x^2 + 3480x + 259{,}675$
$P = (9750x) - (-3x^2 + 3480x + 259{,}675)$
$\quad = 3x^2 + 6270x - 259{,}675$

41. a. According to the bar graph, the net earnings in 2001 were $179,000,000.

   b. Let $x = 1$.
   $$.875x^4 - 16.47x^3 + 110.35x^2 - 212.6x + 300$$
   $$= .875(1)^4 - 16.47(1)^3 + 110.35(1)^2$$
   $$-212.6(1) + 300$$
   $$= 182.155$$
   According to the polynomial, the net earnings in 2001 were $182,155,000.

43. a. According to the bar graph, the net earnings in 2005 were $494,000,000.

   b. Let $x = 5$.
   $$.875x^4 - 16.47x^3 + 110.35x^2 - 212.6x + 300$$
   $$= .875(5)^4 - 16.47(5)^3 + 110.35(5)^2$$
   $$-212.6(5) + 300$$
   $$= 483.875$$
   According to the polynomial, the net earnings in 2005 were $483,875,000.

45. Let $x = 8$.
   $$.875x^4 - 16.47x^3 + 110.35x^2 - 212.6x + 300$$
   $$= .875(8)^4 - 16.47(8)^3 + 110.35(8)^2$$
   $$-212.6(8) + 300$$
   $$= 812.96$$
   According to the polynomial, the net earnings in 2008 were $812,960,000.

47. Let $x = 10$.
   $$.875x^4 - 16.47x^3 + 110.35x^2 - 212.6x + 300$$
   $$= .875(10)^4 - 16.47(10)^3 + 110.35(10)^2$$
   $$-212.6(10) + 300$$
   $$= 1489$$
   According to the polynomial, the net earnings in 2010 were $1,489,000,000.

For exercises 49–51, we use the polynomial $.002722x^2 + .003x + .37$.

49. Let $x = 0$.
   $$.002722(0)^2 + .003(0) + .37 = .37$$
   Thus, there were $.37(1,000,000) = 370,000$ knee implants in 2000. The statement is true.

51. Let $x = 14$.
   $$.002722(14)^2 + .003(14) + .37 = .945512$$
   Thus, there will be $.945512(1,000,000) =$ 945,512 knee implants in 2014. The statement is false.

For exercises 53–55, we use the polynomial $1 - .0058x - .00076x^2$.

53. Let $x = 10$.
   $$1 - .0058x - .00076x^2$$
   $$= 1 - .0058(10) - .00076(10)^2 = .866$$

55. Let $x = 22$.
   $$1 - .0058x - .00076x^2$$
   $$= 1 - .0058(22) - .00076(22)^2 = .505$$

57. a. Calculate the volume of the Great Pyramid when $h = 200$ feet, $b = 756$ feet and $a = 314$ feet.
   $$V = \frac{1}{3}h\left(a^2 + ab + b^2\right)$$
   $$V = \frac{1}{3}(200)\left(314^2 + (314)(756) + 756^2\right)$$
   $$\approx 60,501,067 \text{ cubic feet}$$

   b. When $a = b$, the shape becomes a rectangular box with a square base, with volume b^2h.

   c. If we let $a = b$, then $\frac{1}{3}h\left(a^2 + ab + b^2\right)$ becomes $\frac{1}{3}h\left(b^2 + b(b) + b^2\right)$ which simplifies to hb^2. Yes, the Egyptian formula gives the same result.

59. a. Some or all of the terms may drop out of the sum, so the degree of the sum could be 0, 1, 2, or 3 or no degree (if one polynomial is the negative of the other).

   b. Some or all of the terms may drop out of the difference, so the degree of the difference could be 0, 1, 2, or 3 or no degree (if they are equal).

   c. Multiplying a degree 3 polynomial by a degree 3 polynomial results in a degree 6 polynomial.

Section 1.3 Factoring

1. $12x^2 - 24x = 12x \cdot x - 12x \cdot 2 = 12x(x - 2)$

3. $r^3 - 5r^2 + r = r\left(r^2\right) - r(5r) + r(1)$
$$= r\left(r^2 - 5r + 1\right)$$

5. $6z^3 - 12z^2 + 18z$
$$= 6z\left(z^2\right) - 6z(2z) + 6z(3)$$
$$= 6z\left(z^2 - 2z + 3\right)$$

7. $3(2y - 1)^2 + 7(2y - 1)^3$
$$= (2y - 1)^2(3) + (2y - 1)^2 \cdot 7(2y - 1)$$
$$= (2y - 1)^2[3 + 7(2y - 1)]$$
$$= (2y - 1)^2(3 + 14y - 7)$$
$$= (2y - 1)^2(14y - 4)$$
$$= 2(2y - 1)^2(7y - 2)$$

9. $3(x + 5)^4 + (x + 5)^6$
$$= (x + 5)^4 \cdot 3 + (x + 5)^4(x + 5)^2$$
$$= (x + 5)^4\left[3 + (x + 5)^2\right]$$
$$= (x + 5)^4\left(3 + x^2 + 10x + 25\right)$$
$$= (x + 5)^4\left(x^2 + 10x + 28\right)$$

11. $x^2 + 5x + 4 = (x + 1)(x + 4)$

13. $x^2 + 7x + 12 = (x + 3)(x + 4)$

15. $x^2 + x - 6 = (x + 3)(x - 2)$

17. $x^2 + 2x - 3 = (x + 3)(x - 1)$

19. $x^2 - 3x - 4 = (x + 1)(x - 4)$

21. $z^2 - 9z + 14 = (z - 2)(z - 7)$

23. $z^2 + 10z + 24 = (z + 4)(z + 6)$

25. $2x^2 - 9x + 4 = (2x - 1)(x - 4)$

27. $15p^2 - 23p + 4 = (3p - 4)(5p - 1)$

29. $4z^2 - 16z + 15 = (2z - 5)(2z - 3)$

31. $6x^2 - 5x - 4 = (2x + 1)(3x - 4)$

33. $10y^2 + 21y - 10 = (5y - 2)(2y + 5)$

35. $6x^2 + 5x - 4 = (2x - 1)(3x + 4)$

37. $3a^2 + 2a - 5 = (3a + 5)(a - 1)$

39. $x^2 - 81 = x^2 - (9)^2 = (x + 9)(x - 9)$

41. $9p^2 - 12p + 4 = (3p)^2 - 2(3p)(2) + 2^2$
$$= (3p - 2)^2$$

43. $r^2 + 3rt - 10t^2 = (r - 2t)(r + 5t)$.

45. $m^2 - 8mn + 16n^2 = (m)^2 - 2(m)(4n) + (4n)^2$
$$= (m - 4n)^2$$

47. $4u^2 + 12u + 9 = (2u + 3)^2$

49. $9p^2 - 16 = (3p)^2 - 4^2 = (3p - 4)(3p + 4)$

51. $4r^2 - 9v^2 = (2r + 3v)(2r - 3v)$

53. $x^2 + 4xy + 4y^2 = (x + 2y)^2$

55. $3a^2 - 13a - 30 = (3a + 5)(a - 6)$.

57. $21m^2 + 13mn + 2n^2 = (7m + 2n)(3m + n)$

59. $y^2 - 4yz - 21z^2 = (y - 7z)(y + 3z)$

61. $121x^2 - 64 = (11x + 8)(11x - 8)$

63. $a^3 - 64 = a^3 - (4)^3 = (a - 4)\left(a^2 + 4a + 16\right)$

65. $8r^3 - 27s^3$
$$= (2r)^3 - (3s)^3$$
$$= (2r - 3s)\left[(2r)^2 + (2r)(3s) + (3s)^2\right]$$
$$= (2r - 3s)\left(4r^2 + 6rs + 9s^2\right)$$

67. $64m^3 + 125$

$= (4m)^3 + (5)^3$

$= (4m+5)\left[(4m)^2 - (4m)(5) + (5)^2\right]$

$= (4m+5)\left(16m^2 - 20m + 25\right)$

69. $1000y^3 - z^3$

$= (10y)^3 - (z)^3$

$= (10y-z)\left[(10y)^2 + (10y)(z) + (z)^2\right]$

$= (10y-z)\left(100y^2 + 10yz + z^2\right)$

71. $x^4 + 5x^2 + 6 = \left(x^2 + 2\right)\left(x^2 + 3\right)$

73. $b^4 - b^2 = b^2\left(b^2 - 1\right) = b^2(b+1)(b-1)$

75. $x^4 - x^2 - 12 = \left(x^2 - 4\right)\left(x^2 + 3\right)$

$\qquad\qquad\quad = (x+2)(x-2)\left(x^2 + 3\right)$

77. $16a^4 - 81b^4 = \left(4a^2 - 9b^2\right)\left(4a^2 + 9b^2\right)$

$\qquad\qquad\quad = (2a+3b)(2a-3b)\left(4a^2 + 9b^2\right)$

79. $x^8 + 8x^2 = x^2\left(x^6 + 8\right) = x^2\left(\left(x^2\right)^3 + 2^3\right)$

$\qquad\qquad = x^2\left(x^2 + 2\right)\left(x^4 - 2x^2 + 4\right)$

81. $6x^4 - 3x^2 - 3 = \left(2x^2 + 1\right)\left(3x^2 - 3\right)$ is not the

correct complete factorization because $3x^2 - 3$ contains a common factor of 3. This common factor should be factored out as the first step. This will reveal a difference of two squares, which requires further factorization. The correct factorization is

$6x^4 - 3x^2 - 3 = 3\left(2x^4 - x^2 - 1\right)$

$\qquad\qquad\quad = 3\left(2x^2 + 1\right)\left(x^2 - 1\right)$

$\qquad\qquad\quad = 3\left(2x^2 + 1\right)(x+1)(x-1)$

83. $(x+2)^3 = (x+2)(x+2)^2$

$\qquad\quad = (x+2)(x^2 + 4x + 4)$

$\qquad\quad = x^3 + 4x^2 + 2x^2 + 8x + 4x + 8$

$\qquad\quad = x^3 + 6x^2 + 12x + 8,$

which is not equal to $x^3 + 8$. The correct factorization is $x^3 + 8 = (x+2)(x^2 - 2x + 4)$.

Section 1.4 Rational Expressions

1. $\dfrac{8x^2}{56x} = \dfrac{x \cdot 8x}{7 \cdot 8x} = \dfrac{x}{7}$

3. $\dfrac{25p^2}{35p^3} = \dfrac{5 \cdot 5p^2}{7p \cdot 5p^2} = \dfrac{5}{7p}$

5. $\dfrac{5m+15}{4m+12} = \dfrac{5(m+3)}{4(m+3)} = \dfrac{5}{4}$

7. $\dfrac{4(w-3)}{(w-3)(w+6)} = \dfrac{4}{w+6}$

9. $\dfrac{3y^2 - 12y}{9y^3} = \dfrac{3y(y-4)}{3y\left(3y^2\right)}$

$\qquad\quad = \dfrac{y-4}{3y^2}$

11. $\dfrac{m^2 - 4m + 4}{m^2 + m - 6} = \dfrac{(m-2)(m-2)}{(m+3)(m-2)} = \dfrac{m-2}{m+3}$

13. $\dfrac{x^2 + 2x - 3}{x^2 - 1} = \dfrac{(x+3)(x-1)}{(x+1)(x-1)} = \dfrac{x+3}{x+1}$

15. $\dfrac{3a^2}{64} \cdot \dfrac{8}{2a^3} = \dfrac{3a^2 \cdot 8}{64 \cdot 2a^3} = \dfrac{3}{16a}$

17. $\dfrac{7x}{11} \div \dfrac{14x^3}{66y} = \dfrac{7x}{11} \cdot \dfrac{66y}{14x^3} = \dfrac{7x \cdot 66y}{11 \cdot 14x^3} = \dfrac{3y}{x^2}$

19. $\dfrac{2a+b}{3c} \cdot \dfrac{15}{4(2a+b)} = \dfrac{(2a+b) \cdot 15}{(2a+b) \cdot 12c} = \dfrac{15}{12c} = \dfrac{5}{4c}$

21. $\dfrac{15p-3}{6} \div \dfrac{10p-2}{3} = \dfrac{15p-3}{6} \cdot \dfrac{3}{10p-2}$

$\qquad = \dfrac{3(5p-1) \cdot 3}{3 \cdot 2 \cdot 2 \cdot (5p-1)}$

$\qquad = \dfrac{3(5p-1) \cdot 3}{3(5p-1) \cdot 2 \cdot 2} = \dfrac{3}{4}$

23. $\dfrac{9y-18}{6y+12} \cdot \dfrac{3y+6}{15y-30} = \dfrac{9(y-2)}{6(y+2)} \cdot \dfrac{3(y+2)}{15(y-2)}$

$\qquad = \dfrac{27(y-2)(y+2)}{90(y+2)(y-2)} = \dfrac{27}{90} = \dfrac{3}{10}$

25. $\dfrac{4a+12}{2a-10} \div \dfrac{a^2-9}{a^2-a-20} = \dfrac{4a+12}{2a-10} \cdot \dfrac{a^2-a-20}{a^2-9}$

$\qquad = \dfrac{4(a+3)}{2(a-5)} \cdot \dfrac{(a-5)(a+4)}{(a+3)(a-3)}$

$\qquad = \dfrac{4(a+3)(a-5)(a+4)}{2(a-5)(a+3)(a-3)}$

$\qquad = \dfrac{2(a+4)}{a-3}$

27. $\dfrac{k^2-k-6}{k^2+k-12} \cdot \dfrac{k^2+3k-4}{k^2+2k-3}$

$\qquad = \dfrac{(k-3)(k+2)}{(k+4)(k-3)} \cdot \dfrac{(k+4)(k-1)}{(k+3)(k-1)}$

$\qquad = \dfrac{(k-3)(k+2)(k+4)(k-1)}{(k+4)(k-3)(k+3)(k-1)} = \dfrac{k+2}{k+3}$

29. Answers will vary. Sample answer: To find the least common denominator for two fractions, factor each denominator into prime factors, multiply all unique prime factors raising each factor to the highest frequency it occurred.

31. The common denominator is $35z$.

$\dfrac{2}{7z} - \dfrac{1}{5z} = \dfrac{2 \cdot 5}{7z \cdot 5} - \dfrac{1 \cdot 7}{5z \cdot 7} = \dfrac{10}{35z} - \dfrac{7}{35z} = \dfrac{3}{35z}$

33. $\dfrac{r+2}{3} - \dfrac{r-2}{3} = \dfrac{(r+2)-(r-2)}{3}$

$\qquad = \dfrac{r+2-r+2}{3} = \dfrac{4}{3}$

35. The common denominator is $5x$.

$\dfrac{4}{x} + \dfrac{1}{5} = \dfrac{4 \cdot 5}{x \cdot 5} + \dfrac{1 \cdot x}{5 \cdot x} = \dfrac{20}{5x} + \dfrac{x}{5x} = \dfrac{20+x}{5x}$

37. The common denominator is $m(m-1)$.

$\dfrac{1}{m-1} + \dfrac{2}{m} = \dfrac{m \cdot 1}{m \cdot (m-1)} + \dfrac{(m-1) \cdot 2}{(m-1) \cdot m}$

$\qquad = \dfrac{m}{m(m-1)} + \dfrac{2(m-1)}{m(m-1)}$

$\qquad = \dfrac{m+2(m-1)}{m(m-1)} = \dfrac{m+2m-2}{m(m-1)}$

$\qquad = \dfrac{3m-2}{m(m-1)}$

39. The common denominator is $5(b+2)$.

$\dfrac{7}{b+2} + \dfrac{2}{5(b+2)} = \dfrac{7 \cdot 5}{(b+2) \cdot 5} + \dfrac{2}{5(b+2)}$

$\qquad = \dfrac{35+2}{5(b+2)} = \dfrac{37}{5(b+2)}$

41. The common denominator is $20(k-2)$.

$\dfrac{2}{5(k-2)} + \dfrac{5}{4(k-2)} = \dfrac{8}{20(k-2)} + \dfrac{25}{20(k-2)}$

$\qquad = \dfrac{8+25}{20(k-2)} = \dfrac{33}{20(k-2)}$

43. First factor the denominators in order to find the common denominator.

$x^2 - 4x + 3 = (x-3)(x-1)$

$x^2 - x - 6 = (x-3)(x+2)$

The common denominator is $(x-3)(x-1)(x+2)$.

$\dfrac{2}{x^2-4x+3} + \dfrac{5}{x^2-x-6}$

$\qquad = \dfrac{2}{(x-3)(x-1)} + \dfrac{5}{(x-3)(x+2)}$

$\qquad = \dfrac{2(x+2)}{(x-3)(x-1)(x+2)} + \dfrac{5(x-1)}{(x-3)(x+2)(x-1)}$

$\qquad = \dfrac{2(x+2)+5(x-1)}{(x-3)(x+2)(x-1)} = \dfrac{2x+4+5x-5}{(x-3)(x-1)(x+2)}$

$\qquad = \dfrac{7x-1}{(x-3)(x-1)(x+2)}$

45. First factor the denominators in order to find the common denominator.

$$y^2 + 7y + 12 = (y+3)(y+4)$$
$$y^2 + 5y + 6 = (y+3)(y+2)$$

The common denominator is $(y+4)(y+3)(y+2)$.

$$\frac{2y}{y^2+7y+12} - \frac{y}{y^2+5y+6}$$

$$= \frac{2y}{(y+4)(y+3)} - \frac{y}{(y+3)(y+2)}$$

$$= \frac{2y(y+2)}{(y+4)(y+3)(y+2)} - \frac{y(y+4)}{(y+4)(y+3)(y+2)}$$

$$= \frac{2y(y+2)-y(y+4)}{(y+4)(y+3)(y+2)} = \frac{2y^2+4y-y^2-4y}{(y+4)(y+3)(y+2)}$$

$$= \frac{y^2}{(y+4)(y+3)(y+2)}$$

47. $\dfrac{1+\frac{1}{x}}{1-\frac{1}{x}}$

Multiply both numerator and denominator of this complex fraction by the common denominator, x.

$$\frac{1+\frac{1}{x}}{1-\frac{1}{x}} = \frac{x\left(1+\frac{1}{x}\right)}{x\left(1-\frac{1}{x}\right)} = \frac{x\cdot1+x\left(\frac{1}{x}\right)}{x\cdot1-x\left(\frac{1}{x}\right)} = \frac{x+1}{x-1}$$

49. $\dfrac{\frac{1}{x+h}-\frac{1}{x}}{h}$

The common denominator in the numerator is $x(x+h)$

$$\frac{\frac{1}{x+h}-\frac{1}{x}}{h} = \frac{\frac{x-(x+h)}{x(x+h)}}{h} = \frac{\frac{x-x-h}{x(x+h)}}{h} = \frac{\frac{-h}{x(x+h)}}{h}$$

$$= \frac{-h}{x(x+h)} \div h = \frac{-h}{x(x+h)} \cdot \frac{1}{h}$$

$$= \frac{-1}{x(x+h)} \text{ or } -\frac{1}{x(x+h)}$$

51. The length of each side of the dartboard is $2x$, so the area of the dartboard is $4x^2$. The area of the shaded region is πx^2.

a. The probability that a dart will land in the shaded region is $\dfrac{\pi x^2}{4x^2}$.

b. $\dfrac{\pi x^2}{4x^2} = \dfrac{\pi}{4}$

53. The length of each side of the dartboard is $5x$, so the area of the dartboard is $25x^2$. The area of the shaded region is x^2.

a. The probability that a dart will land in the shaded region is $\dfrac{x^2}{25x^2}$.

b. $\dfrac{x^2}{25x^2} = \dfrac{1}{25}$

55. a. $\dfrac{x^2-10x+25}{x^3-15x^2+50x} = \dfrac{(x-5)(x-5)}{x(x^2-15x+50)}$

$$= \frac{(x-5)(x-5)}{x(x-10)(x-5)}$$

$$= \frac{x-5}{x(x-10)} = \frac{x-5}{x^2-10x}$$

b. Let $x = 11$.

$$\frac{11-5}{(11)^2-10(11)} = \frac{6}{121-110} = \frac{6}{11}$$

$$\left(\frac{6}{11}\right)(60) \approx 32.7 \text{ sec}$$

Let $x = 15$.

$$\frac{15-5}{(15)^2-10(15)} = \frac{10}{225-150} = \frac{10}{75} = \frac{2}{15}$$

$$\left(\frac{2}{15}\right)(60) = 8 \text{ sec}$$

Let $x = 20$.

$$\frac{20-5}{(20)^2-10(20)} = \frac{15}{400-200} = \frac{15}{200} = \frac{3}{40}$$

$$\left(\frac{3}{40}\right)(60) = 4.5 \text{ sec}$$

57. a. Let $x = 25$. Then

$$\frac{.072(25)^2 + .744(25)+1.2}{25+2} = 2.4$$

The ad cost $2.4 million in 2005

b. Let $x = 40$. Then

$$\frac{.072(40)^2 + .744(40)+1.2}{40+2} = 3.48$$

The cost of an ad will not reach $4 million in 2020.

Section 1.5 Exponents and Radicals

1. $\dfrac{7^5}{7^3} = 7^{5-3} = 7^2 = 49$

3. $(4c)^2 = 4^2 c^2 = 16c^2$

5. $\left(\dfrac{2}{x}\right)^5 = \dfrac{2^5}{x^5} = \dfrac{32}{x^5}$

7. $\left(3u^2\right)^3 \left(2u^3\right)^2 = \left(27u^6\right)\left(4u^6\right) = 108u^{12}$

9. $7^{-1} = \dfrac{1}{7^1} = \dfrac{1}{7}$

11. $2^{-5} = \dfrac{1}{2^5} = \dfrac{1}{32}$

13. $-6^{-5} = -\dfrac{1}{6^5} = -\dfrac{1}{7776}$

15. $(-y)^{-3} = \dfrac{1}{(-y)^3} = -\dfrac{1}{y^3}$

17. $\left(\dfrac{1}{7}\right)^{-3} = \left(\dfrac{7}{1}\right)^3 = 7^3 = 343$

19. $\left(\dfrac{4}{3}\right)^{-2} = \left(\dfrac{3}{4}\right)^2 = \dfrac{9}{16}$

21. $\left(\dfrac{a}{b^3}\right)^{-1} = \left(\dfrac{b^3}{a}\right)^1 = \dfrac{b^3}{a}$

23. $49^{1/2} = 7$ because $7^2 = 49$.

25. $(5.71)^{1/4} = (5.71)^{.25} \approx 1.55$ Use a calculator.

27. $27^{2/3} = \left(27^{1/3}\right)^2 = 3^2 = 9$

29. $-64^{2/3} = -\left(64^{1/3}\right)^2 = -(4)^2 = -16$

31. $\left(\dfrac{8}{27}\right)^{-4/3} = \left(\dfrac{27^{1/3}}{8^{1/3}}\right)^4 = \left(\dfrac{3}{2}\right)^4 = \dfrac{3^4}{2^4} = \dfrac{81}{16}$

33. $\dfrac{5^{-3}}{4^{-2}} = \dfrac{4^2}{5^3} = \dfrac{16}{125}$

35. $4^{-3} \cdot 4^6 = 4^3 = 64$

37. $8^{2/3} \cdot 8^{-1/3} = 8^{1/3} = \left(2^3\right)^{1/3} = 2$

39. $\dfrac{4^{10} \cdot 4^{-6}}{4^{-4}} = 4^{10} \cdot 4^{-6} \cdot 4^4 = 4^8 = 65{,}536$

41. $\dfrac{9^{-5/3}}{9^{2/3} \cdot 9^{-1/5}} = 9^{-5/3} \cdot 9^{-2/3} \cdot 9^{1/5} = 9^{-7/3} \cdot 9^{1/5}$

$\qquad = 9^{-32/15} = \dfrac{1}{9^{32/15}}$

43. $\dfrac{z^6 \cdot z^2}{z^5} = \dfrac{z^8}{z^5} = z^{8-5} = z^3$

45. $\dfrac{3^{-1}\left(p^{-2}\right)^3}{3p^{-7}} = \dfrac{3^{-1}p^{-6}}{3^1 p^{-7}} = 3^{-1-1}p^{-6-(-7)}$

$\qquad = 3^{-2}p^1 = \dfrac{1}{3^2} \cdot p = \dfrac{p}{9}$

47. $\left(q^{-5}r^3\right)^{-1} = q^5 r^{-3} = q^5 \cdot \dfrac{1}{r^3} = \dfrac{q^5}{r^3}$

49. $\left(2p^{-1}\right)^3 \cdot \left(5p^2\right)^{-2} = 2^3\left(p^{-1}\right)^3 (5)^{-2}\left(p^2\right)^{-2}$

$\qquad = 2^3\left(p^{-3}\right)\left(\dfrac{1}{5^2}\right)\left(p^{-4}\right)$

$\qquad = 2^3\left(\dfrac{1}{p^3}\right)\left(\dfrac{1}{5^2}\right)\left(\dfrac{1}{p^4}\right)$

$\qquad = \dfrac{8}{25p^7}$

51. $(2p)^{1/2} \cdot \left(2p^3\right)^{1/3} = 2^{1/2}p^{1/2} \cdot 2^{1/3} \cdot \left(p^3\right)^{1/3}$

$\qquad = 2^{1/2}p^{1/2} \cdot 2^{1/3} \cdot p^1$

$\qquad = 2^{5/6}p^{3/2}$

53. $p^{2/3}\left(2p^{1/3} + 5p\right) = p^{2/3}\left(2p^{1/3}\right) + p^{2/3}(5p)$

$\qquad = 2p + 5p^{5/3}$

55. $\dfrac{\left(x^2\right)^{1/3}\left(y^2\right)^{2/3}}{3x^{2/3}y^2} = \dfrac{(x)^{2/3}(y)^{4/3}}{3x^{2/3}y^2}$

$\qquad = \dfrac{1}{3y^{2-4/3}} = \dfrac{1}{3y^{2/3}}$

57. $\dfrac{(7a)^2(5b)^{3/2}}{(5a)^{3/2}(7b)^4} = \dfrac{7^2a^2 5^{3/2}b^{3/2}}{5^{3/2}a^{3/2}7^4b^4} = \dfrac{a^{2-\frac{3}{2}}}{7^2 b^{4-\frac{3}{2}}}$

$\qquad = \dfrac{a^{1/2}}{49b^{5/2}}$

59. $x^{1/2}\left(x^{2/3} - x^{4/3}\right) = x^{1/2}x^{2/3} - x^{1/2}x^{4/3}$

$\qquad = x^{7/6} - x^{11/6}$

61. $\left(x^{1/2}+y^{1/2}\right)\left(x^{1/2}-y^{1/2}\right) = \left(x^{1/2}\right)^2 - \left(y^{1/2}\right)^2$

$\qquad = x - y$

63. $(-3x)^{1/3} = \sqrt[3]{-3x}$, (f)

65. $(-3x)^{-1/3} = \dfrac{1}{(-3x)^{1/3}} = \dfrac{1}{\sqrt[3]{-3x}}$, (h)

67. $(3x)^{1/3} = \sqrt[3]{3x}$, (g)

69. $(3x)^{-1/3} = \dfrac{1}{(3x)^{1/3}} = \dfrac{1}{\sqrt[3]{3x}}$, (c)

71. $\sqrt[3]{125} = 125^{1/3} = 5$

73. $\sqrt[4]{625} = 625^{1/4} = 5$

75. $\sqrt[7]{-128} = (-128)^{1/7} = -2$

77. $\sqrt[3]{81}\cdot\sqrt[3]{9} = \sqrt[3]{729} = 9$

79. $\sqrt{81-4} = \sqrt{77}$

81. $\sqrt{81}-\sqrt{4} = 9 - 2 = 7$

83. $\sqrt{8}\sqrt{96} = \sqrt{8}\sqrt{8\cdot12} = \sqrt{8}\sqrt{8}\sqrt{12} = 8\sqrt{4\cdot3}$

$\qquad = 8\sqrt{4}\sqrt{3} = 8\cdot2\sqrt{3} = 16\sqrt{3}$

85 $\sqrt{75}+\sqrt{192} = 5\sqrt{3}+8\sqrt{3} = 13\sqrt{3}$

87. $\left(\sqrt{3}+2\right)\left(\sqrt{3}-2\right) = \left(\sqrt{3}\right)^2 - 2^2 = 3 - 4 = -1$

89. $\left(\sqrt{3}+4\right)\left(\sqrt{5}-4\right) = \sqrt{15} - 4\sqrt{3} + 4\sqrt{5} - 16$

91. $\dfrac{3}{1-\sqrt{2}} = \dfrac{3}{1-\sqrt{2}}\cdot\dfrac{1+\sqrt{2}}{1+\sqrt{2}} = \dfrac{3\left(1+\sqrt{2}\right)}{(1)^2-\left(\sqrt{2}\right)^2}$

$\qquad = \dfrac{3\left(1+\sqrt{2}\right)}{1-2} = \dfrac{3\left(1+\sqrt{2}\right)}{-1}$

$\qquad = -3\left(1+\sqrt{2}\right) = -3 - 3\sqrt{2}$

93. $\dfrac{9-\sqrt{3}}{3-\sqrt{3}} = \dfrac{9-\sqrt{3}}{3-\sqrt{3}}\cdot\dfrac{3+\sqrt{3}}{3+\sqrt{3}} = \dfrac{27+9\sqrt{3}-3\sqrt{3}-3}{3^2-\left(\sqrt{3}\right)^2}$

$\qquad = \dfrac{24+6\sqrt{3}}{9-3} = \dfrac{24+6\sqrt{3}}{6} = 4+\sqrt{3}$

95. $\dfrac{3-\sqrt{2}}{3+\sqrt{2}} = \dfrac{3-\sqrt{2}}{3+\sqrt{2}}\cdot\dfrac{3+\sqrt{2}}{3+\sqrt{2}}$

$\qquad = \dfrac{9-2}{9+6\sqrt{2}+2} = \dfrac{7}{11+6\sqrt{2}}$

97. $x = \sqrt{\dfrac{kM}{f}}$

Note that because x represents the number of units to order, the value of x should be rounded to the nearest integer.

a. $k = \$1, f = \$500, M = 100{,}000$

$\qquad x = \sqrt{\dfrac{1\cdot100{,}000}{500}} = \sqrt{200} \approx 14.1$

The number of units to order is 14.

b. $k = \$3, f = \$7, M = 16{,}700$

$\qquad x = \sqrt{\dfrac{3\cdot16{,}700}{7}} \approx 84.6$

The number of units to order is 85.

c. $k = \$1, f = \$5, M = 16{,}800$

$\qquad x = \sqrt{\dfrac{1\cdot16{,}800}{5}} = \sqrt{3360} \approx 58.0$

The number of units to order is 58.

For exercises 99–101, we use the model
sales $= 11.68x^{.1954}$, $x \geq 3$, $x = 3$ corresponds to 2003.

99. Let $x = 5$. Then $11.68(5)^{.1954} \approx 15.996$

The domestic sales for 2005 were about $\$15{,}996{,}000{,}000$.

101. Let $x = 11$. Then $11.68(11)^{.1954} \approx 18.661$

The domestic sales for 2011 will be about $18,661,000,000.

For exercises 103–105, we use the model transplants $= 5274.7x^{.4159}$, $x \geq 7$, $x = 7$ corresponds to 1997.

103. Let $x = 11$. Then $5274.7(11)^{.4159} \approx 14,299$

The number of kidney transplants in 2001 was approximately 14,299.

105. Let $x = 16$. Then $5274.7(16)^{.4159} \approx 16,711$

The number of kidney transplants in 2006 was approximately 16,711.

107. Let $x = 18$. Then $5274.7(18)^{.4159} \approx 17,550$. So

approximately 17,550 people had kidney transplants and $76,300 - 17,550 = 58,750$ people on the waiting list did not.

For exercises 109–111, we use the model

life expectancy $= \dfrac{92.7}{1 + .237 \times 10^{.0189x}}$.

109. Let $x = 22$. Then $\dfrac{92.7}{1 + .237 \times 10^{.0189(22)}} \approx 57.3$

According to the model, a person who is currently 22 years old has life expectancy of 57.3 years.

111. Answers will vary.

Section 1.6 First-Degree Equations

1.
$$3x + 8 = 20$$
$$3x + 8 - 8 = 20 - 8$$
$$3x = 12$$
$$\frac{1}{3}(3x) = \frac{1}{3}(12)$$
$$x = 4$$

3.
$$.6k - .3 = .5k + .4$$
$$.6k - .5k - .3 = .5k - .5k + .4$$
$$.1k - .3 = .4$$
$$.1k - .3 + .3 = .4 + .3$$
$$.1k = .7$$
$$\frac{.1k}{.1} = \frac{.7}{.1} \Rightarrow k = 7$$

5.
$$2a - 1 = 4(a + 1) + 7a + 5$$
$$2a - 1 = 4a + 4 + 7a + 5$$
$$2a - 1 = 11a + 9$$
$$2a - 2a - 1 = 11a - 2a + 9$$
$$-1 = 9a + 9$$
$$-1 - 9 = 9a + 9 - 9$$
$$-10 = 9a$$
$$\frac{-10}{9} = \frac{9a}{9} \Rightarrow -\frac{10}{9} = a$$

7.
$$2[x - (3 + 2x) + 9] = 3x - 8$$
$$2(x - 3 - 2x + 9) = 3x - 8$$
$$2(-x + 6) = 3x - 8$$
$$-2x + 12 = 3x - 8$$
$$12 = 5x - 8$$
$$20 = 5x \Rightarrow 4 = x$$

9. $\dfrac{3x}{5} - \dfrac{4}{5}(x + 1) = 2 - \dfrac{3}{10}(3x - 4)$

Multiply both sides by the common denominator, 10.

$$10\left(\frac{3x}{5}\right) - 10\left(\frac{4}{5}\right)(x + 1)$$
$$= (10)(2) - (10)\left(\frac{3}{10}\right)(3x - 4)$$
$$2(3x) - 8(x + 1) = 20 - 3(3x - 4)$$
$$6x - 8x - 8 = 20 - 9x + 12$$
$$-2x - 8 = 32 - 9x$$
$$-2x + 9x = 32 + 8$$
$$7x = 40$$
$$\frac{1}{7}(7x) = \frac{1}{7}(40) \Rightarrow x = \frac{40}{7}$$

11. $\dfrac{5y}{6} - 8 = 5 - \dfrac{2y}{3}$

$$6\left(\frac{5y}{6} - 8\right) = 6\left(5 - \frac{2y}{3}\right)$$
$$6\left(\frac{5y}{6}\right) - 6(8) = 6(5) - 6\left(\frac{2y}{3}\right)$$
$$5y - 48 = 30 - 4y$$
$$9y - 48 = 30$$
$$9y = 78$$
$$y = \frac{78}{9} = \frac{26}{3}$$

13.
$$\frac{m}{2} - \frac{1}{m} = \frac{6m+5}{12}$$
$$12m\left(\frac{m}{2} - \frac{1}{m}\right) = 12m\left(\frac{6m+5}{12}\right)$$
$$(12m)\left(\frac{m}{2}\right) - (12m)\left(\frac{1}{m}\right) = m(6m) + m(5)$$
$$6m^2 - 12 = 6m^2 + 5m$$
$$-12 = 5m$$
$$\frac{1}{5}(-12) = \frac{1}{5}(5m) \Rightarrow -\frac{12}{5} = m$$

15.
$$\frac{4}{x-3} - \frac{8}{2x+5} + \frac{3}{x-3} = 0$$
$$\frac{4}{x-3} + \frac{3}{x-3} - \frac{8}{2x+5} = 0$$
$$\frac{7}{x-3} - \frac{8}{2x+5} = 0$$
Multiply each side by the common denominator, $(x-3)(2x+5)$.
$$(x-3)(2x+5)\left(\frac{7}{x-3}\right) - (x-3)(2x+5)\left(\frac{8}{2x+5}\right)$$
$$= (x-3)(2x+5)(0)$$
$$7(2x+5) - 8(x-3) = 0$$
$$14x + 35 - 8x + 24 = 0$$
$$6x + 59 = 0$$
$$6x = -59 \Rightarrow x = -\frac{59}{6}$$

17.
$$\frac{3}{2m+4} = \frac{1}{m+2} - 2$$
$$\frac{3}{2(m+2)} = \frac{1}{m+2} - 2$$
$$2(m+2)\left(\frac{3}{2(m+2)}\right)$$
$$= 2(m+2)\left(\frac{1}{m+2}\right) - 2(m+2)(2)$$
$$3 = 2 - 4(m+2)$$
$$3 = 2 - 4m - 8$$
$$3 = -6 - 4m \Rightarrow 9 = -4m \Rightarrow m = -\frac{9}{4}$$

19.
$$9.06x + 3.59(8x - 5) = 12.07x + .5612$$
$$9.06x + 28.72x - 17.95 = 12.07x + .5612$$
$$9.06x + 28.72x - 12.07x = 17.95 + .5612$$
$$25.71x = 18.5112$$
$$x = \frac{18.5112}{25.71} = .72$$

21.
$$\frac{2.63r - 8.99}{1.25} - \frac{3.90r - 1.77}{2.45} = r$$
Multiply by the common denominator $(1.25)(2.45)$ to eliminate the fractions.
$$(2.45)(2.63r - 8.99) - (1.25)(3.90r - 1.77)$$
$$= (2.45)(1.25)r$$
$$6.4435r - 22.0255 - 4.875r + 2.2125 = 3.0625r$$
$$1.5685r - 19.813 = 3.0625r$$
$$-19.813 = 1.494r$$
$$-\frac{19.813}{1.494} = \frac{1.494r}{1.494}$$
$$r \approx -13.26$$

23.
$$4(a + x) = b - a + 2x$$
$$4a + 4x = b - a + 2x$$
$$4a = b - a - 2x$$
$$5a - b = -2x$$
$$\frac{5a - b}{-2} = \frac{-2x}{-2}$$
$$-\frac{5a - b}{2} = x \text{ or } x = \frac{b - 5a}{2}$$

25.
$$5(b - x) = 2b + ax$$
$$5b - 5x = 2b + ax$$
$$5b = 2b + ax + 5x$$
$$3b = ax + 5x$$
$$3b = (a + 5)x$$
$$\frac{3b}{a+5} = \frac{(a+5)x}{a+5} \Rightarrow \frac{3b}{a+5} = x$$

27.
$$PV = k \text{ for } V$$
$$\frac{1}{P}(PV) = \frac{1}{P}(k) \Rightarrow V = \frac{k}{P}$$

29.
$$V = V_0 + gt \text{ for } g$$
$$V - V_0 = gt$$
$$\frac{V - V_0}{t} = \frac{gt}{t} \Rightarrow \frac{V - V_0}{t} = g$$

31.
$$A = \frac{1}{2}(B+b)h \text{ for } B$$

$$A = \frac{1}{2}Bh + \frac{1}{2}bh$$

$$2A = Bh + bh \quad \text{Multiply by } 2.$$

$$2A - bh = Bh$$

$$\frac{2A - bh}{h} = \frac{Bh}{h} \qquad \text{Multiply by } \frac{1}{h}.$$

$$\frac{2A - bh}{h} = B$$

33. $|2h - 1| = 5$

$2h - 1 = 5 \quad$ or $\quad 2h - 1 = -5$

$\quad 2h = 6 \quad$ or $\qquad 2h = -4$

$\qquad h = 3 \quad$ or $\qquad\quad h = -2$

35. $|6 + 2p| = 10$

$6 + 2p = 10 \quad$ or $\quad 6 + 2p = -10$

$\quad 2p = 4 \quad$ or $\qquad 2p = -16$

$\qquad p = 2 \quad$ or $\qquad\quad p = -8$

37. $\left|\dfrac{5}{r-3}\right| = 10$

$\dfrac{5}{r-3} = 10 \qquad$ or $\qquad \dfrac{5}{r-3} = -10$

$\quad 5 = 10(r-3) \quad$ or $\qquad 5 = -10(r-3)$

$\quad 5 = 10r - 30 \quad$ or $\qquad 5 = -10r + 30$

$\quad 35 = 10r \qquad$ or $\qquad -25 = -10r$

$\dfrac{35}{10} = \dfrac{7}{2} = r \quad$ or $\quad \dfrac{-25}{-10} = \dfrac{5}{2} = r$

39. $1.250 = \dfrac{x}{8} \Rightarrow x = 10$

The stroke lasted 10 hours.

41.
$$-5 = \frac{5}{9}(F - 32)$$

$$-5\left(\frac{9}{5}\right) = \left(\frac{9}{5}\right)\left(\frac{5}{9}\right)(F - 32)$$

$$-9 = F - 32 \Rightarrow 23 = F$$

The temperature $-5°C = 23°F$.

43.
$$22 = \frac{5}{9}(F - 32)$$

$$22\left(\frac{9}{5}\right) = \left(\frac{9}{5}\right)\left(\frac{5}{9}\right)(F - 32)$$

$$39.6 = F - 32 \Rightarrow 71.6 = F$$

The temperature $22°C = 71.6°F$.

45. $y = .5x + 5.33$

Substitute 8.83 for y.

$8.83 = .5x + 5.33$

$3.5 = .5x \Rightarrow 7 = x$

Therefore, the federal deficit will be $8.83 billion in 2007.

47. $y = .5x + 5.33$

Substitute 12.83 for y.

$12.83 = .5x + 5.33$

$7.5 = .5x \Rightarrow 15 = x$

Therefore, the federal deficit will be $12.83 billion in 2015.

49. $E = 86x + 1648$

Substitute $2250 in for E.

$2250 = 86x + 1648$

$602 = 86x \Rightarrow 7 = x$

The health care expenditures were $2250 billion in 2007.

51. $E = 86x + 1648$

Substitute $2680 in for E.

$2680 = 86x + 1648$

$1032 = 86x \Rightarrow x = 12$

The health care expenditures will be $2680 billion in 2012.

53. $-.12(x - 2006) = 14y - 2.8$

Substitute .26 for y and solve for x.

$-.12(x - 2006) = 14(.26) - 2.8$

$-.12x + 240.72 = .84$

$\qquad -.12x = -239.88$

$\qquad\qquad x = 1999$

26% of workers were covered in 1999.

55. $-.12(x - 2006) = 14y - 2.8$

Substitute .14 for y and solve for x.

$-.12(x - 2006) = 14(.14) - 2.8$

$-.12x + 240.72 = -.84$

$\qquad -.12x = -241.56$

$\qquad\qquad x = 2013$

14% of workers will be covered in 2013.

57. $P = 2.67x + 282.62$

Substitute 320 for P and solve for x.

$320 = 2.67x + 282.62$

$37.38 = 2.67x \Rightarrow x = 14$

The United States will have a population of 320,000,000 in 2014.

59. $P = 2.67x + 282.62$
Substitute 346.7 for P and solve for x.
$346.7 = 2.67x + 282.62$
$64.08 = 2.67x \Rightarrow x = 24$
The United States will have a population of 346,700,000 in 2024.

61. $f = 800$, $n = 18$, $q = 36$
$$u = f \cdot \frac{n(n+1)}{q(q+1)} = 800 \cdot \frac{18(19)}{36(37)}$$
$$= 800 \cdot \frac{342}{1332} \approx 205.41$$
The amount of unearned interest is \$205.41.

63. a. $k = .132\left(\dfrac{B}{W}\right) = .132\left(\dfrac{20}{75}\right) = .0352$

b. $R = kd = .0352(.42) \approx .015$ or 1.5%

c. Number of cases of cancer
$$= \frac{R \cdot 5000}{72} = \frac{.015 \cdot 5000}{72} \approx 1 \text{ case}$$

65. Let x represent the amount invested at 4%. Then $20,000 - x$ is the amount invested at 6%. Since the total interest is \$1040, we have
$$.04x + .06(20,000 - x) = 1040$$
$$.04x + 1200 - .06x = 1040$$
$$-.02x + 1200 = 1040$$
$$-.02x = -160$$
$$x = 8000$$
She invested \$8000 at 4%.

67. Let x represent the amount invested at 4%. \$20,000 invested at 5% (or .05) plus x dollars invested at 4% (or .04) must equal 4.8% (or .048) of the total investment (or $20,000 + x$). Solve this equation.
$$.05(20,000) + .04x = .048(20,000 + x)$$
$$1000 + .04x = 960 + .048x$$
$$1000 = 960 + .008x$$
$$40 = .008x \Rightarrow x = 5000$$
\$5000 should be invested at 4%.

69. Let x represent the distance between the two cities. Since $d = rt$, $t = \dfrac{d}{r}$. Make a table.

	Distance	Rate	Time
Going	x	50	$\dfrac{x}{50}$
Returning	x	55	$\dfrac{x}{55}$

Since the total traveling time was 32 hr,
$$\frac{x}{50} + \frac{x}{55} = 32$$
$$550\left(\frac{x}{50} + \frac{x}{55}\right) = 550(32)$$
$$11x + 10x = 17,600$$
$$21x = 17,600$$
$$x \approx 838.10$$
The distance between the two cities is approximately 838 mi.

71. Let x = the number of liters of 94 octane gas;
200 = the number of liters of 99 octane gas;
$200 + x$ = the number of liters of 97 octane gas.
$$94x + 99(200) = 97(200 + x)$$
$$94x + 19,800 = 19,400 + 97x$$
$$400 = 3x$$
$$\frac{400}{3} = x$$
Thus, $\dfrac{400}{3}$ liters of 94 octane gas are needed.

73. Let x = number of miles driven
$$20 + .18x = 45.56$$
$$.18x = 25.56$$
$$x = 142$$
You must drive 142 miles in a day for the costs to be equal.

75. $y = 10(x - 65) + 50$
Substitute 100 in for y.
$$100 = 10(x - 65) + 50$$
$$50 = 10(x - 65)$$
$$5 = x - 65 \Rightarrow 70 = x$$
Paul was driving 70 mph.

77. Let x represent the width. Since the length is 3 cm less than twice the width, the length is $2x - 3$. The perimeter of a rectangle is twice the length plus twice the width. The perimeter is 54 cm, so the equation is $2x + 2(2x - 3) = 54$.
Solve the equation.
$$2x + 2(2x - 3) = 54$$
$$2x + 4x - 6 = 54$$
$$6x - 6 = 54$$
$$6x = 60 \Rightarrow x = 10$$
The width is 10 cm. The length is $2(10) - 3 = 20 - 3 = 17$ cm.
Since $2(10) + 2(17) = 20 + 34 = 54$, as stated in the original problem, the solution checks.

79. Let x = the length of the shortest side, $2x$ = the length of the second side, and $x + 7$ = the length of the third side.
The perimeter, 27, is the sum of the lengths of the three sides.
$$x + 2x + x + 7 = 27$$
$$4x + 7 = 27$$
$$4x = 20 \Rightarrow x = 5$$
The length of the shortest side is 5 cm.

Section 1.7 Quadratic Equations

1. $(x + 4)(x - 14) = 0$
$\quad x + 4 = 0 \quad$ or $\quad x - 14 = 0$
$\quad\quad x = -4 \quad$ or $\quad\quad x = 14$
The solutions are -4 and 14.

3. $x(x + 6) = 0$
$\quad x = 0 \quad$ or $\quad x + 6 = 0$
$\quad\quad\quad\quad\quad\quad x = -6$
The solutions are 0 and -6.

5. $\quad 2z^2 = 4z$
$2z^2 - 2z = 0$
$2z(z - 2) = 0$
$2z = 0 \quad$ or $\quad z - 2 = 0$
$\quad z = 0 \quad$ or $\quad\quad z = 2$
The solutions are 0 and 2.

7. $y^2 + 15y + 56 = 0$
$(y + 7)(y + 8) = 0$
$y + 7 = 0 \quad$ or $\quad y + 8 = 0$
$y = -7 \quad$ or $\quad y = -8$
The solutions are -7 and -8.

9. $\quad 2x^2 = 7x - 3$
$2x^2 - 7x + 3 = 0$
$(2x - 1)(x - 3) = 0$
$2x - 1 = 0 \quad$ or $\quad x - 3 = 0$
$\quad x = \dfrac{1}{2} \quad$ or $\quad\quad x = 3$
The solutions are $\dfrac{1}{2}$ and 3.

11. $\quad 6r^2 + r = 1$
$6r^2 + r - 1 = 0$
$(3r - 1)(2r + 1) = 0$
$3r - 1 = 0 \quad$ or $\quad 2r + 1 = 0$
$\quad r = \dfrac{1}{3} \quad$ or $\quad\quad r = -\dfrac{1}{2}$
The solutions are $\dfrac{1}{3}$ and $-\dfrac{1}{2}$.

13. $\quad 2m^2 + 20 = 13m$
$2m^2 - 13m + 20 = 0$
$(2m - 5)(m - 4) = 0$
$2m - 5 = 0 \quad$ or $\quad m - 4 = 0$
$\quad m = \dfrac{5}{2} \quad$ or $\quad\quad m = 4$
The solutions are $\dfrac{5}{2}$ and 4.

15. $\quad m(m + 7) = -10$
$m^2 + 7m + 10 = 0$
$(m + 5)(m + 2) = 0$
$m + 5 = 0 \quad$ or $\quad m + 2 = 0$
$\quad m = -5 \quad$ or $\quad\quad m = -2$
The solutions are -5 and -2.

17. $\quad 9x^2 - 16 = 0$
$(3x + 4)(3x - 4) = 0$
$3x + 4 = 0 \quad$ or $\quad 3x - 4 = 0$
$\quad 3x = -4 \quad\quad\quad\quad 3x = 4$
$\quad x - \dfrac{4}{3} \quad$ or $\quad\quad x = \dfrac{4}{3}$
The solutions are $-\dfrac{4}{3}$ and $\dfrac{4}{3}$.

19. $16x^2 - 16x = 0$

$16x(x-1) = 0$

$16x = 0$ or $x - 1 = 0$

$x = 0$ or $x = 1$

The solutions are 0 and 1.

21. $(r-2)^2 = 7$

$r - 2 = \sqrt{7}$ or $r - 2 = -\sqrt{7}$

$r = 2 + \sqrt{7}$ or $r = 2 - \sqrt{7}$

We abbreviate the solutions as $2 \pm \sqrt{7}$.

23. $(4x-1)^2 = 20$

Use the square root property.

$4x - 1 = \sqrt{20}$ or $4x - 1 = -\sqrt{20}$

$4x - 1 = 2\sqrt{5}$ or $4x - 1 = -2\sqrt{5}$

$4x = 1 + 2\sqrt{5}$ or $4x = 1 - 2\sqrt{5}$

The solutions are $\dfrac{1 \pm 2\sqrt{5}}{4}$.

25. $2x^2 + 7x + 1 = 0$

Use the quadratic formula with $a = 2$, $b = 7$, and $c = 1$.

$x = \dfrac{-b \pm \sqrt{b^2 - 4ac}}{2a}$

$= \dfrac{-7 \pm \sqrt{7^2 - 4(2)(1)}}{2(2)}$

$= \dfrac{-7 \pm \sqrt{49 - 8}}{4} = \dfrac{-7 \pm \sqrt{41}}{4}$

The solutions are $\dfrac{-7 + \sqrt{41}}{4}$ and $\dfrac{-7 - \sqrt{41}}{4}$, which are approximately $-.1492$ and -3.3508.

27. $4k^2 + 2k = 1$

Rewrite the equation in standard form.

$4k^2 + 2k - 1 = 0$

Use the quadratic formula with $a = 4$, $b = 2$, and $c = -1$.

$k = \dfrac{-2 \pm \sqrt{2^2 - 4(4)(-1)}}{2(4)} = \dfrac{-2 \pm \sqrt{4 + 16}}{8}$

$= \dfrac{-2 \pm \sqrt{20}}{8} = \dfrac{-2 \pm 2\sqrt{5}}{8} = \dfrac{2\left(-1 \pm \sqrt{5}\right)}{2 \cdot 4}$

$k = \dfrac{-1 \pm \sqrt{5}}{4}$

The solutions are $\dfrac{-1 + \sqrt{5}}{4}$ and $\dfrac{-1 - \sqrt{5}}{4}$, which are approximately $.309$ and $-.809$.

29. $5y^2 + 5y = 2$

$5y^2 + 5y - 2 = 0$

$a = 5$, $b = 5$, $c = -2$

$y = \dfrac{-5 \pm \sqrt{5^2 - 4(5)(-2)}}{2(5)} = \dfrac{-5 \pm \sqrt{25 + 40}}{10}$

$= \dfrac{-5 \pm \sqrt{65}}{10} = \dfrac{-5 \pm \sqrt{65}}{10}$

The solutions are $\dfrac{-5 + \sqrt{65}}{10}$ and $\dfrac{-5 - \sqrt{65}}{10}$, which are approximately $.3062$ and -1.3062.

31. $6x^2 + 6x + 4 = 0$

$a = 6$, $b = 6$, $c = 4$

$x = \dfrac{-6 \pm \sqrt{6^2 - 4(6)(4)}}{2(6)} = \dfrac{-6 \pm \sqrt{36 - 96}}{12}$

$= \dfrac{-6 \pm \sqrt{-60}}{12}$

Because $\sqrt{-60}$ is not a real number, the given equation has no real number solutions.

33. $2r^2 + 3r - 5 = 0$

$a = 2$, $b = 3$, $c = -5$

$r = \dfrac{-(3) \pm \sqrt{9 - 4(2)(-5)}}{2(2)} = \dfrac{-3 \pm \sqrt{9 + 40}}{4}$

$= \dfrac{-3 \pm \sqrt{49}}{4} = \dfrac{-3 \pm 7}{4}$

$r = \dfrac{-3 + 7}{4} = \dfrac{4}{4} = 1$ or $r = \dfrac{-3 - 7}{4} = \dfrac{-10}{4} = \dfrac{-5}{2}$

The solutions are $-\dfrac{5}{2}$ and 1.

35. $2x^2 - 7x + 30 = 0$

$a = 2$, $b = -7$, $c = 30$

$x = \dfrac{-(-7) \pm \sqrt{49 - 4(2)(30)}}{2(2)} = \dfrac{7 \pm \sqrt{-191}}{4}$

Since $\sqrt{-191}$ is not a real number, there are no real solutions.

37. $1 + \dfrac{7}{2a} = \dfrac{15}{2a^2}$

To eliminate fractions, multiply both sides by the common denominator, $2a^2$.

$2a^2 + 7a = 15 \Rightarrow 2a^2 + 7a - 15 = 0$

$a = 2, b = 7, c = -15$

$a = \dfrac{-7 \pm \sqrt{7^2 - 4(2)(-15)}}{2(2)} = \dfrac{-7 \pm \sqrt{49 + 120}}{4}$

$= \dfrac{-7 \pm \sqrt{169}}{4} \Rightarrow a = \dfrac{-7 + 13}{4} = \dfrac{6}{4} = \dfrac{3}{2}$ or

$a = \dfrac{-7 - 13}{4} = \dfrac{-20}{4} = -5$

The solutions are $\dfrac{3}{2}$ and -5.

39. $25t^2 + 49 = 70t$

$25t^2 - 70t + 49 = 0$

$b^2 - 4ac = (-70)^2 - 4(25)(49)$

$\qquad = 4900 - 4900$

$\qquad = 0$

The discriminant is 0.
There is one real solution to the equation.

41. $13x^2 + 24x - 5 = 0$

$b^2 - 4ac = (24)^2 - 4(13)(-5)$

$\qquad = 576 + 260 = 836$

The discriminant is positive.
There are two real solutions to the equation.

For Exercises 43–45 use the quadratic formula:

$x = \dfrac{-b \pm \sqrt{b^2 - 4ac}}{2a}$.

43. $4.42x^2 - 10.14x + 3.79 = 0$

$x = \dfrac{-(-10.14) \pm \sqrt{(-10.14)^2 - 4(4.42)(3.79)}}{2(4.42)}$

$\approx \dfrac{10.14 \pm 5.9843}{8.84} \approx .4701$ or 1.8240

45. $7.63x^2 + 2.79x = 5.32$

$7.63x^2 + 2.79x - 5.32 = 0$

$x = \dfrac{-2.79 \pm \sqrt{(2.79)^2 - 4(7.63)(-5.32)}}{2(7.63)}$

$\approx \dfrac{-2.79 \pm 13.0442}{15.26} \approx -1.0376$ or $.6720$

47. a. Let $R = 450$ ft.

$450 = .5x^2 \Rightarrow 900 = x^2 \Rightarrow 30 = x$
The maximum taxiing speed is 30 mph.

b. Let $R = 615$ ft.

$615 = .5x^2 \Rightarrow 1230 = x^2 \Rightarrow 35 \approx x$
The maximum taxiing speed is about 35 mph.

c. Let $R = 970$ ft.

$970 = .5x^2 \Rightarrow 1940 = x^2 \Rightarrow 44 \approx x$
The maximum taxiing speed is about 44 mph.

49. a. Let $D = 1$.

$1 = .0031x^2 - .291x + 7.1$

$0 = .0031x^2 - .291x + 6.1$
Store

$\sqrt{b^2 - 4ac} = \sqrt{(-.291)^2 - 4(.0031)(6.1)}$

$\qquad\qquad \approx .0951$ in your calculator.

By the quadratic formula, $x \approx 62$ or $x \approx 32$. People of about 32 or 62 years of age have a driver fatality rate of about 1 death per 1000.

b. Let $D = 3$.

$3 = .0031x^2 - .291x + 7.1$

$0 = .0031x^2 - .291x + 4.1$
Store

$\sqrt{b^2 - 4ac} = \sqrt{(-.291)^2 - 4(.0031)(4.1)}$

$\qquad\qquad \approx .1840$ in your calculator.

By the quadratic formula, $x \approx 77$ or $x \approx 17$. People of about 77 or 17 years of age have a driver fatality rate of about 3 deaths per 1000.

c. $T = -.26x^2 + 3.62x + 30.18$

$0 = -.26x^2 + 3.62x + 30.18$

$x = \dfrac{-3.62 \pm \sqrt{(3.62)^2 - 4(-.26)(30.18)}}{2(-.26)}$

≈ -5.87 or 19.79

The Corporation will be out of money late in 2019. (Note that the negative solution is not applicable.)

51. $R = -.315x^2 + 8.44x + 14.27$

    a. Let $R = 58$ (in billions).

$$58 = -.315x^2 + 8.44x + 14.27$$
$$0 = -.315x^2 + 8.44x - 43.73$$

Using the quadratic formula, we have

$$x = \frac{-8.44 \pm \sqrt{8.44^2 - 4(-.315)(-43.73)}}{2(-.315)}$$

$$= \frac{-8.44 \pm \sqrt{16.1338}}{2(-.315)} \approx 7.021 \text{ or } 19.773$$

19.773 is not applicable because the formula is defined for $2 \le x \le 8$. Thus, revenues were about \$58 billion in 2007.

    b. Let $R = 67.170$ (in billions)

$$67.170 = -.315x^2 + 8.44x + 14.27$$
$$0 = -.315x^2 + 8.44x - 52.9$$

Using the quadratic formula, we have

$$x = \frac{-8.44 \pm \sqrt{8.44^2 - 4(-.315)(-52.9)}}{2(-.315)}$$

$$= \frac{-8.44 \pm \sqrt{4.5796}}{2(-.315)} = 10 \text{ or about } 16.8$$

$16.8 > 15$, so this is not applicable. If the formula remains valid until 2015, revenues reach \$67,170,000,000 in 2010.

53. Triangle *ABC* represents the original position of the ladder, while triangle *DEC* represents the position of the ladder after it was moved.

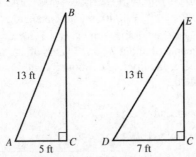

Use the Pythagorean theorem to find the distance from the top of the ladder to the ground.
In triangle *ABC*,

$$13^2 = 5^2 + BC^2 \Rightarrow 169 - 25 = BC^2 \Rightarrow$$
$$144 = BC^2 \Rightarrow 12 = BC$$

Thus, the top of the ladder was originally 12 feet from the ground.

In triangle *DEC*, $\quad 13^2 = 7^2 + EC^2 \Rightarrow 169 - 49 = EC^2 \Rightarrow$
$$120 = EC^2 \Rightarrow EC = \sqrt{120}$$

The top of the ladder was $\sqrt{120}$ feet from the ground after the ladder was moved. Therefore, the ladder moved down $12 - \sqrt{120} \approx 1.046$ feet.

55. **a.** The eastbound train travels at a speed of $x + 20$.

    b. The northbound train travels a distance of $5x$ in 5 hours.
The eastbound train travels a distance of $5(x + 20)$ in 5 hours.

    c.

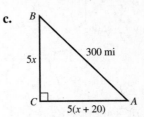

By the Pythagorean theorem,
$$(5x)^2 + (5x + 100)^2 = 300^2$$

    d. Expand and combine like terms.
$$25x^2 + 25x^2 + 1000x + 10,000 = 90,000$$
$$50x^2 + 1000x - 80,000 = 0$$

Factor out the common factor, 50, and divide both sides by 50.

$$50(x^2 + 20x - 1600) = 0$$
$$x^2 + 20x - 1600 = 0$$

Now use the quadratic formula to solve for x.

$$x = \frac{-20 \pm \sqrt{20^2 - 4(1)(-1600)}}{2(1)}$$
$$\approx -51.23 \text{ or } 31.23$$

Since x cannot be negative, the speed of the northbound train is $x \approx 31.23$ mph, and the speed of the eastbound train is $x + 20 \approx 51.23$ mph.

57. **a.** Let x represent the length. Then, $\dfrac{300 - 2x}{2}$ or $150 - x$ represents the width.

    b. Use the formula for the area of a rectangle.
$$LW = A \Rightarrow x(150 - x) = 5000$$

c. $150x - x^2 = 5000$

Write this quadratic equation in standard form and solve by factoring.

$0 = x^2 - 150x + 5000$

$x^2 - 150x + 5000 = 0$

$(x - 50)(x - 100) = 0$

$x - 50 = 0$ or $x - 100 = 0$

$x = 50$ or $x = 100$

Choose $x = 100$ because the length is the larger dimension. The length is 100 m and the width is $150 - 100 = 50$ m.

59. Let $x =$ the width of the uniform strip around the rug.

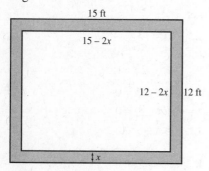

The dimensions of the rug are $15 - 2x$ and $12 - 2x$. The area, 108, is the length times the width.

Solve the equation.

$(15 - 2x)(12 - 2x) = 108$

$180 - 54x + 4x^2 = 108$

$4x^2 - 54x + 72 = 0$

$2x^2 - 27x + 36 = 0$

$(x - 12)(2x - 3) = 0$

$x - 12 = 0$ or $2x - 3 = 0$

$x = 12$ or $x = \dfrac{3}{2}$

Discard $x = 12$ since both $12 - 2x$ and $15 - 2x$ would be negative.

If $x = \dfrac{3}{2}$, then

$15 - 2x = 15 - 2\left(\dfrac{3}{2}\right) = 12$

and $12 - 2x = 12 - 2\left(\dfrac{3}{2}\right) = 9$.

The dimensions of the rug should be 9 ft by 12 ft.

For exercises 61–65, use the formula $h = -16t^2 + v_0 t + h_0$, where h_0 is the height of the object when $t = 0$, and v_0 is the initial velocity at time $t = 0$.

61. $v_0 = 0$, $h_0 = 625$, $h = 0$

$0 = -16t^2 + (0)t + 625 \Rightarrow 16t^2 - 625 = 0 \Rightarrow$

$(4t - 25)(4t + 25) = 0 \Rightarrow t = \dfrac{25}{4} = 6.25$ or

$t = -\dfrac{25}{4} = -6.25$

The negative solution is not applicable. It takes 6.25 seconds for the baseball to reach the ground.

63. a. $v_0 = 0$, $h_0 = 200$, $h = 0$

$0 = -16t^2 - (0)t + 200 \Rightarrow 16t^2 = 200 \Rightarrow$

$t^2 = \dfrac{200}{16} \Rightarrow t = \pm\dfrac{\sqrt{200}}{4} \approx \pm3.54$

The negative solution is not applicable. It will take about 3.54 seconds for the rock to reach the ground if it is dropped.

b. $v_0 = -40$, $h_0 = 200$, $h = 0$

$0 = -16t^2 - 40t + 200$

Using the quadratic formula, we have

$t = \dfrac{-(-40) \pm \sqrt{(-40)^2 - 4(-16)(200)}}{2(-16)}$

$= -5$ or 2.5

The negative solution is not applicable. It will take about 2.5 seconds for the rock to reach the ground if it is thrown with an initial velocity of 40 ft/sec.

c. $v_0 = -40$, $h_0 = 200$, $t = 2$

$h = -16(2)^2 - 40(2) + 200 = 56$

After 2 seconds, the rock is 56 feet above the ground. This means it has fallen $200 - 56 = 144$ feet.

65. a. $v_0 = 64$, $h_0 = 0$, $h = 64$

$64 = -16t^2 + 64t + 0 \Rightarrow$

$16t^2 - 64t + 64 = 0 \Rightarrow$

$t^2 - 4t + 4 = 0 \Rightarrow (t - 2)^2 = 0 \Rightarrow t = 2$

The ball will reach 64 feet after 2 seconds.

b. $v_0 = 64$, $h_0 = 0$, $h = 39$

$$39 = -16t^2 + 64t + 0 \Rightarrow$$

$$16t^2 - 64t + 39 = 0 \Rightarrow (4t - 13)(4t - 3) \Rightarrow$$

$$t = \frac{13}{4} = 3.25 \text{ or } t = \frac{3}{4} = .75$$

The ball will reach 39 feet after .75 seconds and after 3.25 seconds.

c. Two answers are possible because the ball reaches the given height twice, once on the way up and once on the way down.

In exercises 67–71, we discard negative roots since all variables represent positive real numbers.

67. $S = \frac{1}{2}gt^2$ for t

$$2S = gt^2$$

$$\frac{2S}{g} = t^2$$

$$\sqrt{\frac{2S}{g}} \cdot \frac{\sqrt{g}}{\sqrt{g}} = t$$

$$\frac{\sqrt{2Sg}}{g} = t$$

69. $L = \frac{d^4 k}{h^2}$ for h

$$Lh^2 = d^4 k$$

$$h^2 = \frac{d^4 k}{L}$$

$$h = \pm\sqrt{\frac{d^4 k}{L}} \cdot \frac{\sqrt{L}}{\sqrt{L}} = \pm\frac{\sqrt{d^4 kL}}{L}$$

$$h = \pm\frac{d^2 \sqrt{kL}}{L}$$

71. $P = \frac{E^2 R}{(r + R)^2}$ for R

$$P(r + R)^2 = E^2 R$$

$$P\left(r^2 + 2rR + R^2\right) = E^2 R$$

$$Pr^2 + 2PrR + PR^2 = E^2 R$$

$$PR^2 + \left(2Pr - E^2\right)R + Pr^2 = 0$$

Solve for R by using the quadratic formula with $a = P$, $b = 2Pr - E^2$, and $c = Pr^2$.

$$R = \frac{-\left(2Pr - E^2\right) \pm \sqrt{\left(2Pr - E^2\right)^2 - 4P \cdot Pr^2}}{2P}$$

$$= \frac{-2Pr + E^2 \pm \sqrt{4P^2 r^2 - 4PrE^2 + E^4 - 4P^2 r^2}}{2P}$$

$$= \frac{-2Pr + E^2 \pm \sqrt{E^4 - 4PrE^2}}{2P}$$

$$= \frac{-2Pr + E^2 \pm \sqrt{E^2\left(E^2 - 4Pr\right)}}{2P}$$

$$R = \frac{-2Pr + E^2 \pm E\sqrt{E^2 - 4Pr}}{2P}$$

73. **a.** Let $x = z^2$.

$$x^2 - 2x = 15$$

b.
$$x^2 - 2x = 15$$

$$x^2 - 2x - 15 = 0$$

$$(x - 5)(x + 3) = 0$$

$$x = 5 \text{ or } x = -3$$

c. Let $z^2 = 5$.

$$z = \pm\sqrt{5}$$

75. $2q^4 + 3q^2 - 9 = 0$

Let $u = q^2$; then $u^2 = q^4$.

$$2u^2 + 3u - 9 = 0 \Rightarrow (2u - 3)(u + 3) = 0$$

$$2u - 3 = 0 \quad \text{or} \quad u + 3 = 0$$

$$u = \frac{3}{2} \quad \text{or} \quad u = -3$$

Since $u = q^2$,

$$q^2 = \frac{3}{2} \quad \text{or} \quad q^2 = -3$$

$$q = \pm\sqrt{\frac{3}{2}} \quad \text{or} \quad q = \pm\sqrt{-3} \text{ (not real)}$$

$$q = \pm\frac{\sqrt{3}}{\sqrt{2}} \cdot \frac{\sqrt{2}}{\sqrt{2}} = \pm\frac{\sqrt{6}}{2}$$

The solutions are $\pm\dfrac{\sqrt{6}}{2}$.

77. $z^4 - 3z^2 - 1 = 0$

Let $x = z^2$; then $x^2 = z^4$.

$x^2 - 3x - 1 = 0$

By the quadratic formula, $x = \dfrac{3 \pm \sqrt{13}}{2}$.

$\dfrac{3 + \sqrt{13}}{2} = z^2$

$\pm\sqrt{\dfrac{3 + \sqrt{13}}{2}} = z$

Chapter 1 Review Exercises

1. 0 and 6 are whole numbers.

2. $-12, -6, -\sqrt{4}, 0$, and 6 are integers.

3. $-12, -6, -\dfrac{9}{10}, -\sqrt{4}, 0, \dfrac{1}{8}$, and 6 are rational numbers.

4. $-\sqrt{7}, \dfrac{\pi}{4}, \sqrt{11}$ are irrational numbers.

5. $9[(-3)4] = 9[4(-3)]$
Commutative property of multiplication

6. $7(4 + 5) = (4 + 5)7$
Commutative property of multiplication

7. $6(x + y - 3) = 6x + 6y + 6(-3)$
Distributive property

8. $11 + (5 + 3) = (11 + 5) + 3$
Associative property of addition

9. x is at least 9.
$x \geq 9$

10. x is negative.
$x < 0$

11. $-7, -3, -2, 0, \pi, 8$
(π is about 3.14.)

12. $-\dfrac{5}{4}, -\dfrac{2}{3}, -\dfrac{3}{8}, \dfrac{1}{2}, \dfrac{5}{6}$

13. $|6 - 4| = 2, -|-2| = -2, |8 + 1| = |9| = 9$,

$-|3 - (-2)| = -|3 + 2| = -|5| = -5$

Since $-5, -2, 2, 9$ are in order, then

$-|3 - (-2)|, -|-2|, |6 - 4|, |8 + 1|$ are in order.

14. $-\left|\sqrt{16}\right| = -4, -\sqrt{8}, \sqrt{7}, \left|-\sqrt{12}\right| = \sqrt{12}$

15. $-|-4| + |3| = -4 + 3 = -1$

16. $|-6| + |-9| = 6 + 9 = 15$

17. $7 - |-8| = 7 - 8 = -1$

18. $|-3| - |-9 + 6| = 3 - |-3| = 3 - 3 = 0$

19. $x \geq -3$
Start at -3 and draw a ray to the right. Use a bracket at -3 to show that -3 is a part of the graph.

20. $-4 < x \leq 6$
Put a parenthesis at -4 and a bracket at 6. Draw a line segment between these two endpoints.

21. $x < -2$
Start at -2 and draw a ray to the left. Use a parenthesis at -2 to show that -2 is not a part of the graph.

22. $x \leq 1$
Start at 1 and draw a ray to the left. Use a bracket to show that 1 is a part of the graph.

23. $(-6 + 3 \cdot 5)(-2) = (-6 + 15)(-2) = 9(-2) = -18$

24. $-4(-8 - 9 \div 3) = -4(-8 - 3) = -4(-11) = 44$

25. $\dfrac{-9 + (-6)(-3) \div 9}{6 - (-3)} = \dfrac{-9 + 18 \div 9}{6 + 3} = \dfrac{-9 + 2}{9} = -\dfrac{7}{9}$

26. $\dfrac{20 \div 4 \cdot 2 \div 5 - 1}{-9 - (-3) - 12 \div 3} = \dfrac{5 \cdot 2 \div 5 - 1}{-9 - (-3) - 4}$

$= \dfrac{10 \div 5 - 1}{-9 + 3 - 4} = \dfrac{2 - 1}{-6 - 4} = -\dfrac{1}{10}$

27. $\left(3x^4 - x^2 + 5x\right) - \left(-x^4 + 3x^2 - 6x\right)$

$= 3x^4 - x^2 + 5x + x^4 - 3x^2 + 6x$

$= \left(3x^4 + x^4\right) + \left(-x^2 - 3x^2\right) + (5x + 6x)$

$= 4x^4 - 4x^2 + 11x$

28. $\left(-8y^3 + 8y^2 - 5y\right) - \left(2y^3 + 4y^2 - 10\right)$

$= -8y^3 + 8y^2 - 5y - 2y^3 - 4y^2 + 10$

$= -8y^3 - 2y^3 + 8y^2 - 4y^2 - 5y + 10$

$= -10y^3 + 4y^2 - 5y + 10$

29. $-2\left(q^4 - 3q^3 + 4q^2\right) + 4\left(q^4 + 2q^3 + q^2\right)$

$= -2q^4 + 6q^3 - 8q^2 + 4q^4 + 8q^3 + 4q^2$

$= \left(-2q^4 + 4q^4\right) + \left(6q^3 + 8q^3\right) + \left(-8q^2 + 4q^2\right)$

$= 2q^4 + 14q^3 + \left(-4q^2\right) = 2q^4 + 14q^3 - 4q^2$

30. $5\left(3y^4 - 4y^5 + y^6\right) - 3\left(2y^4 + y^5 - 3y^6\right)$

$= 15y^4 - 20y^5 + 5y^6 - 6y^4 - 3y^5 + 9y^6$

$= 14y^6 - 23y^5 + 9y^4$

31. $(4z + 2)(3z - 2)$

$= 12z^2 - 8z + 6z - 4$

$= 12z^2 - 2z - 4$

32. $(8p - 4)(5p + 2)$

$= 40p^2 + 16p - 20p - 8$

$= 40p^2 - 4p - 8$

33. $\left(5k - 2h\right)\left(5k + 2h\right) = \left(5k\right)^2 - \left(2h\right)^2 = 25k^2 - 4h^2$

34. $(2r - 5y)(2r + 5y) = (2r)^2 - (5y)^2$

$\qquad\qquad\qquad = 4r^2 - 25y^2$

35. $\left(3x + 4y\right)^2 = \left(3x\right)^2 + 2(3x)(4y) + \left(4y^2\right)$

$\qquad\qquad\quad = 9x^2 + 24xy + 16y^2$

36. $(2a - 5b)^2 = (2a)^2 - 2(2a)(5b) + (5b)^2$

$\qquad\qquad\quad = 4a^2 - 20ab + 25b^2$

37. $2kh^2 - 4kh + 5k = k\left(2h^2 - 4h + 5\right)$

38. $2m^2n^2 + 6mn^2 + 16n^2 = 2n^2\left(m^2 + 3m + 8\right)$

39. $5a^4 + 12a^3 + 4a^2 = a^2\left(5a^2 + 12a + 4\right)$

$\qquad\qquad\qquad\quad = a^2\left(5a + 2\right)\left(a + 2\right)$

40. $24x^3 + 4x^2 - 4x = 4x\left(6x^2 + x - 1\right)$

$\qquad\qquad\qquad\quad = 4x(3x - 1)(2x + 1)$

41. $6y^2 - 13y + 6 = (3y - 2)(2y - 3)$

42. $8q^2 + 3m + 4qm + 6q$

$= \left(8q^2 + 4qm\right) + (3m + 6q)$

$= 4q(2q + m) + 3(m + 2q)$

$= 4q(2q + m) + 3(2q + m)$

$= (2q + m)(4q + 3)$

43. $25a^2 - 20a + 4 = \left(5a - 2\right)^2$

44. $36p^2 + 12p + 1 = (6p + 1)^2$

45. $144p^2 - 169q^2 = (12p)^2 - (13q)^2$

$\qquad\qquad\qquad\quad = (12p - 13q)(12p + 13q)$

46. $81z^2 - 25x^2 = (9z)^2 - (5x)^2$

$\qquad\qquad\qquad = (9z + 5x)(9z - 5x)$

47. $27y^3 - 1 = \left(3y\right)^3 - 1^3$

$\qquad\quad = (3y - 1)\left[\left(3y\right)^2 + (3y)(1) + 1^2\right]$

$\qquad\quad = (3y - 1)\left(9y^2 + 3y + 1\right)$

48. $125a^3 + 216 = (5a)^3 + (6)^3$

$\qquad\qquad\quad = (5a + 6)\left[(5a)^2 - 5a(6) + 6^2\right]$

$\qquad\qquad\quad = (5a + 6)\left(25a^2 - 30a + 36\right)$

49. $\dfrac{3x}{5} \cdot \dfrac{45x}{12} = \dfrac{3x \cdot 45x}{5 \cdot 12} = \dfrac{3 \cdot 5 \cdot 3 \cdot 3x^2}{4 \cdot 5 \cdot 3} = \dfrac{9x^2}{4}$

50. $\dfrac{5k^2}{24} - \dfrac{70k}{36} = \dfrac{5k^2 \cdot 3}{24 \cdot 3} - \dfrac{70k \cdot 2}{36 \cdot 2} = \dfrac{15k^2}{72} - \dfrac{140k}{72}$

$\qquad\qquad = \dfrac{15k^2 - 140k}{72} = \dfrac{5k(3k - 28)}{72}$

51. $\dfrac{c^2-3c+2}{2c(c-1)} \div \dfrac{c-2}{8c} = \dfrac{(c-1)(c-2)}{2c(c-1)} \cdot \dfrac{8c}{(c-2)}$

$\qquad\qquad = \dfrac{8c(c-1)(c-2)}{2c(c-1)(c-2)} = \dfrac{8}{2} = 4$

52. $\dfrac{p^3-2p^2-8p}{3p\left(p^2-16\right)} \div \dfrac{p^2+4p+4}{9p^2}$

$\qquad = \dfrac{p\left(p^2-2p-8\right)}{3p(p+4)(p-4)} \cdot \dfrac{9p^2}{(p+2)(p+2)}$

$\qquad = \dfrac{p(p-4)(p+2) \cdot 9p^2}{3p(p+4)(p-4)(p+2)(p+2)}$

$\qquad = \dfrac{3p(p-4)(p+2) \cdot 3p^2}{3p(p-4)(p+2) \cdot (p+4)(p+2)}$

$\qquad = \dfrac{3p^2}{(p+4)(p+2)}$

53. $\dfrac{2y-6}{5y} \cdot \dfrac{20y-15}{14} = \dfrac{2(y-3)}{5y} \cdot \dfrac{5(4y-3)}{14}$

$\qquad\qquad = \dfrac{(y-3)(4y-3)}{7y}$

54. $\dfrac{m^2-2m}{15m^3} \cdot \dfrac{5}{m^2-4} = \dfrac{m(m-2)}{5 \cdot 3m^3} \cdot \dfrac{5}{(m+2)(m-2)}$

$\qquad\qquad = \dfrac{5m(m-2) \cdot 1}{5m(m-2) \cdot 3m^2(m+2)}$

$\qquad\qquad = \dfrac{1}{3m^2(m+2)}$

55. $\dfrac{2m^2-4m+2}{m^2-1} \div \dfrac{6m+18}{m^2+2m-3}$

$\qquad = \dfrac{2\left(m^2-2m+1\right)}{(m+1)(m-1)} \cdot \dfrac{m^2+2m-3}{6m+18}$

$\qquad = \dfrac{2(m-1)^2}{(m+1)(m-1)} \cdot \dfrac{(m+3)(m-1)}{6(m+3)}$

$\qquad = \dfrac{2(m-1)(m+3) \cdot (m-1)^2}{2(m-1)(m+3) \cdot 3(m+1)} = \dfrac{(m-1)^2}{3(m+1)}$

56. $\dfrac{x^2+6x+5}{4\left(x^2+1\right)} \cdot \dfrac{2x(x+1)}{x^2-25}$

$\qquad = \dfrac{(x+5)(x+1) \cdot 2x(x+1)}{4\left(x^2+1\right) \cdot (x+5)(x-5)}$

$\qquad = \dfrac{2(x+5) \cdot x(x+1)^2}{2(x+5) \cdot 2\left(x^2+1\right)(x-5)} = \dfrac{x(x+1)^2}{2\left(x^2+1\right)(x-5)}$

57. $\dfrac{6}{15z} + \dfrac{2}{3z} - \dfrac{9}{10z}$

$\qquad = \dfrac{6 \cdot 2}{15z \cdot 2} + \dfrac{2 \cdot 10}{3z \cdot 10} - \dfrac{9 \cdot 3}{10z \cdot 3}$

$\qquad = \dfrac{12}{30z} + \dfrac{20}{30z} - \dfrac{27}{30z} = \dfrac{5}{30z} = \dfrac{1}{6z}$

58. $\dfrac{5}{y-3} - \dfrac{4}{y} = \dfrac{5(y)}{(y-3)(y)} - \dfrac{4(y-3)}{y(y-3)}$

$\qquad\qquad = \dfrac{5y-4(y-3)}{y(y-3)} = \dfrac{5y-4y+12}{y(y-3)}$

$\qquad\qquad = \dfrac{y+12}{y(y-3)}$

59. $\dfrac{2}{5q} + \dfrac{10}{7q} = \dfrac{2 \cdot 7}{5q \cdot 7} + \dfrac{10 \cdot 5}{7q \cdot 5} = \dfrac{14}{35q} + \dfrac{50}{35q} = \dfrac{64}{35q}$

60. Answers will vary. Sample answer:
$125^{2/3} = \left(125^2\right)^{1/3} = (15,625)^{1/3} = 25$

or $125^{2/3} = \left(125^{1/3}\right)^2 = (5)^2 = 25$

61. $5^{-3} = \dfrac{1}{5^3}$ or $\dfrac{1}{125}$

62. $10^{-2} = \dfrac{1}{10^2}$ or $\dfrac{1}{100}$

63. $-8^0 = -\left(8^0\right) = -1$

64. $-5^{-1} = -\left(5^{-1}\right) = -\left(\dfrac{1}{5^1}\right) = -\dfrac{1}{5}$

65. $\left(-\dfrac{5}{6}\right)^{-2} = \left(-\dfrac{6}{5}\right)^2 = \dfrac{36}{25}$

66. $\left(\dfrac{3}{2}\right)^{-3} = \left(\dfrac{2}{3}\right)^3 = \dfrac{(2)^3}{3^3} = \dfrac{2^3}{3^3}$ or $\dfrac{8}{27}$

67. $4^6 \cdot 4^{-3} = 4^{6+(-3)} = 4^3$

68. $7^{-5} \cdot 7^{-2} = 7^{-5+(-2)} = 7^{-7} = \dfrac{1}{7^7}$

69. $\dfrac{8^{-5}}{8^{-4}} = 8^{-5-(-4)} = 8^{-5+4} = 8^{-1} = \dfrac{1}{8}$

70. $\dfrac{6^{-3}}{6^4} = 6^{-3-4} = 6^{-7} = \dfrac{1}{6^7}$

71. $\dfrac{9^4 \cdot 9^{-5}}{\left(9^{-2}\right)^2} = \dfrac{9^4 \cdot 9^{-5}}{9^{-4}} = 9^{4-5-(-4)} = 9^3$

72. $\dfrac{k^4 \cdot k^{-3}}{\left(k^{-2}\right)^{-3}} = \dfrac{k}{k^6} = \dfrac{1}{k^5}$

73. $5^{-1} + 2^{-1} = \dfrac{1}{5} + \dfrac{1}{2} = \dfrac{7}{10}$

74. $5^{-2} + 5^{-1} = \dfrac{1}{5^2} + \dfrac{1}{5} = \dfrac{1}{25} + \dfrac{1}{5} = \dfrac{6}{25}$

75. $125^{2/3} = \left(125^{1/3}\right)^2 = 5^2 = 25$

76. $128^{3/7} = \left(128^{1/7}\right)^3 = 2^3 = 8$

77. $8^{-5/3} = \dfrac{1}{8^{5/3}} = \dfrac{1}{\left(8^{1/3}\right)^5} = \dfrac{1}{2^5} = \dfrac{1}{32}$

78. $\left(\dfrac{144}{49}\right)^{-1/2} = \left(\dfrac{49}{144}\right)^{1/2} = \dfrac{7}{12}$

79. $\dfrac{5^{1/3} 5^{1/2}}{5^{3/2}} = 5^{1/3+1/2-3/2} = 5^{-2/3} = \dfrac{1}{5^{2/3}}$

80. $\dfrac{2^{3/4} \cdot 2^{-1/2}}{2^{1/4}} = \dfrac{2^{1/4}}{2^{1/4}} = 1$

81. $\left(3a^2\right)^{1/2} \cdot \left(3^2 a\right)^{3/2} = 3^{1/2} a \cdot 3^3 a^{3/2} = 3^{7/2} a^{5/2}$

82. $(4p)^{2/3} \cdot \left(2p^3\right)^{3/2} = 4^{2/3} p^{2/3} \cdot 2^{3/2} \cdot p^{9/2}$

$\qquad = \left(2^2\right)^{2/3} p^{2/3} \cdot 2^{3/2} p^{9/2}$

$\qquad = 2^{4/3} \cdot 2^{3/2} p^{2/3} p^{9/2}$

$\qquad = 2^{17/6} p^{31/6}$

83. $\sqrt[3]{27} = 3$

84. $\sqrt[6]{-64}$ is not a real number.

85. $\sqrt{99} = \sqrt{9 \cdot 11} = \sqrt{9} \cdot \sqrt{11} = 3\sqrt{11}$

86. $\sqrt{63} = \sqrt{9} \cdot \sqrt{7} = 3\sqrt{7}$

87. $\sqrt[3]{54 p^3 q^5} = \sqrt[3]{27 \cdot 2 p^3 q^3 q^2}$

$\qquad = \sqrt[3]{27 p^3 q^3} \cdot \sqrt[3]{2q^2} = 3pq\sqrt[3]{2q^2}$

88. $\sqrt[4]{64 a^5 b^3} = \sqrt[4]{16 a^4} \cdot \sqrt[4]{4ab^3} = 2a\sqrt[4]{4ab^3}$

89. $\sqrt{\dfrac{5n^2}{6m}} = \dfrac{n\sqrt{5}}{\sqrt{6m}} \cdot \dfrac{\sqrt{6m}}{\sqrt{6m}} = \dfrac{n\sqrt{30m}}{6m}$

90. $\sqrt{\dfrac{3x^3}{2z}} = \sqrt{\dfrac{3xx^2}{2z} \cdot \dfrac{\sqrt{2z}}{\sqrt{2z}}} = \dfrac{x\sqrt{3x}\sqrt{2z}}{2z} = \dfrac{x\sqrt{6xz}}{2z}$

91. $3\sqrt{3} - 12\sqrt{12} = 3\sqrt{3} - 12\sqrt{4 \cdot 3} = 3\sqrt{3} - 12 \cdot 2\sqrt{3}$

$\qquad = 3\sqrt{3} - 24\sqrt{3} = -21\sqrt{3}$

92. $8\sqrt{7} + 2\sqrt{63} = 8\sqrt{7} + 2\sqrt{9 \cdot 7}$

$\qquad = 8\sqrt{7} + 6\sqrt{7} = 14\sqrt{7}$

93. $\left(\sqrt{8} - 1\right)\left(\sqrt{8} + 1\right) = \left(\sqrt{8}\right)^2 - 1^2 = 8 - 1 = 7$

94. $\left(\sqrt{7} - \sqrt{3}\right)\left(\sqrt{7} + \sqrt{3}\right) = \left(\sqrt{7}\right)^2 - \left(\sqrt{3}\right)^2 = 7 - 3 = 4$

95. $\dfrac{\sqrt{3}}{1+\sqrt{2}} = \dfrac{\sqrt{3}\left(1-\sqrt{2}\right)}{\left(1+\sqrt{2}\right)\left(1-\sqrt{2}\right)} = \dfrac{\sqrt{3}-\sqrt{6}}{1-2}$

$\qquad = \dfrac{\sqrt{3}-\sqrt{6}}{-1} = \sqrt{6}-\sqrt{3}$

96. $\dfrac{4+\sqrt{2}}{4-\sqrt{5}} = \dfrac{\left(4+\sqrt{2}\right)}{\left(4-\sqrt{5}\right)} \cdot \dfrac{\left(4+\sqrt{5}\right)}{\left(4+\sqrt{5}\right)}$

$= \dfrac{16+4\sqrt{2}+4\sqrt{5}+\sqrt{10}}{16-\left(\sqrt{5}\right)^2}$

$= \dfrac{16+4\sqrt{2}+4\sqrt{5}+\sqrt{10}}{11}$

For Exercises 97–100, use the formula

$\dfrac{\text{Amount for}}{\text{large state}} = \left(\dfrac{E_{\text{large}}}{E_{\text{small}}}\right)^{3/2} \times \dfrac{\text{Amount for}}{\text{small state}}$

and the amount for a small state is \$1,000,000.

97. $E_{\text{large}} = 27,\ E_{\text{small}} = 3$

$\dfrac{\text{Amount for}}{\text{large state}} = \left(\dfrac{27}{3}\right)^{3/2} \times 1,000,000$

$= 9^{3/2} \times 1,000,000$

$= 27 \times 1,000,000$

$= \$27,000,000$

98. $E_{\text{large}} = 31,\ E_{\text{small}} = 6$

$\dfrac{\text{Amount for}}{\text{large state}} = \left(\dfrac{31}{6}\right)^{3/2} \times 1,000,000$

$\approx 5.167^{3/2} \times 1,000,000$

$= 11.74398979 \times 1,000,000$

$= \$11,743,989.79$

99. $E_{\text{large}} = 55,\ E_{\text{small}} = 5$

$\dfrac{\text{Amount for}}{\text{large state}} = \left(\dfrac{55}{5}\right)^{\frac{3}{2}} \times 1,000,000$

$= 36.48287269 \times 1,000,000$

$= \$36,482,872.69$

100. $E_{\text{large}} = 34,\ E_{\text{small}} = 10$

$\dfrac{\text{Amount for}}{\text{large state}} = \left(\dfrac{34}{10}\right)^{3/2} \times 1,000,000$

$= 6.269290231 \times 1,000,000$

$= \$6,269,290.23$

101. $3x - 4(x-2) = 2x+9$

$3x - 4x + 8 = 2x + 9$

$-x + 8 = 2x + 9$

$-1 = 3x \Rightarrow x = -\dfrac{1}{3}$

102. $4y + 9 = -3(1-2y)+5$

$4y + 9 = -3 + 6y + 5$

$4y + 9 = 2 + 6y$

$-2y = -7 \Rightarrow y = \dfrac{7}{2}$

103. $\dfrac{2z}{5} - \dfrac{4z-3}{10} = \dfrac{-z+1}{10}$

$4z - (4z-3) = -z+1$ Multiply by 10.

$3 = -z + 1$

$z = -2$

104. $\dfrac{p}{p+2} - \dfrac{3}{4} = \dfrac{2}{p+2}$

Multiply both sides by the common denominator $(p+2)(4)$.

$4p - 3(p+2) = 8$

$4p - 3p - 6 = 8 \Rightarrow p = 14$

105. $\dfrac{2m}{m-3} = \dfrac{6}{m-3} + 4$

$2m = 6 + 4(m-3)$

$2m = 6 + 4m - 12$

$2m = 4m - 6$

$6 = 2m \Rightarrow 3 = m$

Because $m = 3$ would make the denominators of the fractions equal to 0, making the fractions undefined, the given equation has no solution.

106. $\dfrac{15}{k+5} = 4 - \dfrac{3k}{k+5}$

Multiply both sides of the equation by the common denominator $k+5$.

$15 = 4(k+5) - 3k$

$15 = 4k + 20 - 3k \Rightarrow k = -5$

If $k = -5$, the fractions would be undefined, so the given equation has no solution.

107. $8ax - 3 = 2x$

$8ax - 2x = 3$

$x(8a-2) = 3$

$x = \dfrac{3}{8a-2}$

108. $6x - 5y = 4bx$

$6x - 4bx = 5y$

$(6-4b)x = 5y$

$x = \dfrac{5y}{6-4b}$

109.
$$\frac{2x}{3-c} = ax+1$$

$$(3-c)\left(\frac{2x}{3-c}\right) = (3-c)(ax+1)$$

$$2x = 3ax+3-acx-c$$

$$c-3 = 3ax-acx-2x$$

$$c-3 = x(3a-ac-2)$$

$$x = \frac{c-3}{3a-ac-2}$$

110. $b^2x - 2x = 4b^2$

$$\left(b^2-2\right)x = 4b^2$$

$$x = \frac{4b^2}{b^2-2}$$

111. $|m-4| = 7$

$m-4 = 7$ or $m-4 = -7$

$\quad m = 11 \qquad\qquad m = -3$

The solutions are 11 and –3.

112. $|5-x| = 15$

$5-x = 15$ or $5-x = -15$

$\quad -x = 10$ or $\quad -x = -20$

$\quad x = -10$ or $\quad x = 20$

The solutions are –10 and 20.

113.
$$\left|\frac{2-y}{5}\right| = 8$$

$$\frac{2-y}{5} = 8 \qquad \text{or} \qquad \frac{2-y}{5} = -8$$

$$5\left(\frac{2-y}{5}\right) = 5(8) \quad \text{or} \quad 5\left(\frac{2-y}{5}\right) = -5(-8)$$

$$2-y = 40 \qquad \text{or} \qquad 2-y = -40$$

$$-y = 38 \qquad \text{or} \qquad -y = -42$$

$$y = -38 \quad \text{or} \qquad\quad y = 42$$

The solutions are –38 and 42.

114. $|4k+1| = |6k-3|$

$4k+1 = 6k-3$ or $4k+1 = -(6k-3)$

$4k+1 = 6k-3$ or $4k+1 = -6k+3$

$\quad -2k = -4$ or $\quad 10k = 2$

$\quad\quad k = 2$ or $\quad\quad k = \dfrac{1}{5}$

The solutions are 2 and $\dfrac{1}{5}$.

115. a. Let $D = 259$ (in millions).

$259 = 3.7x+111 \Rightarrow 148 = 3.7x \Rightarrow 40 = x$

Demand will reach 259 million barrels per day in 2010.

b. Let $D = 300$ (in millions).

$300 = 3.7x+111 \Rightarrow 189 = 3.7x \Rightarrow 51.08 \approx x$

Demand will reach 300 million barrels per day in 2021.

116. Let $x =$ the original price.

$x-.15x = 306 \Rightarrow .85x = 306 \Rightarrow x = 360$

The original price of the laser printer was $360.

117. a. Let $N = 1444$.

$1444 = -10.2x+1801 \Rightarrow -357 = -10.2x \Rightarrow$

$x = 35$

There were 1444 daily newspapers 35 years after 1970, in 2005.

b. Answers will vary.

118. Let $x =$ the rate of interest on $750.

Using the interest formula $I = Prt$ with $t = 1$, we have

$$500(.12)+250(.18) = 750x$$

$$60+45 = 750x$$

$$105 = 750x \Rightarrow \frac{105}{750} = .14 = x$$

Borrowing $750 at a 14% annual interest results in the same total amount of annual interest.

119. Let $x =$ the amount invested at 8%.

Then $100,000 - x =$ the amount invested at 5%;

$.08x =$ the interest from 8% investment;

$.05(100,000-x) =$ the interest from 5% investment.

$$.08x+.05(100,000-x) = 6800$$

$$.08x+5000-.05x = 6800$$

$$.03x = 1800 \Rightarrow x = 60,000$$

$$100,000-x = 40,000$$

The firm should invest $60,000 at 8% and $40,000 at 5%.

120. Let $x =$ the number of pounds of chocolate hearts;

$30 - x =$ the number of pounds of candy kisses.

$5x+3.50(30-x) = 4.50(30)$ Total cost

$5x+105-3.5x = 135 \Rightarrow 1.5x = 30 \Rightarrow x = 20$

Use 20 lb of hearts and 10 lb of kisses for the mix.

121. $x^2 - 6x = 4$

$x^2 - 6x - 4 = 0$

$b^2 - 4ac = (-6)^2 - 4(1)(-4) = 36 + 16 = 52$

The discriminant is positive.
There are two real solutions to the equation.

122. $-3x^2 + 5x + 2 = 0$

$b^2 - 4ac = (5)^2 - 4(-3)(2) = 25 + 24 = 49$

The discriminant is positive.
There are two real solutions to the equation.

123. $4x^2 - 12x + 9 = 0$

$b^2 - 4ac = (-12)^2 - 4(4)(9) = 144 - 144 = 0$

The discriminant is 0.
There is one real solution to the equation.

124. $5x^2 + 2x + 1 = 0$

$b^2 - 4ac = 2^2 - 4(5)(1) = 4 - 20 = -16$

The discriminant is negative.
There are no real solutions to the equation.

125. $x^2 + 3x + 5 = 0$

$b^2 - 4ac = 3^2 - 4(1)(5) = 9 - 20 = -11$

The discriminant is negative.
There are no real solutions to the equation.

126. $(b + 7)^2 = 5$

Use the square root property to solve this quadratic equation.

$b + 7 = \sqrt{5}$ or $b + 7 = -\sqrt{5}$

$b = -7 + \sqrt{5}$ or $b = -7 - \sqrt{5}$

The solutions are $-7 + \sqrt{5}$ and $-7 - \sqrt{5}$, which we abbreviate as $-7 \pm \sqrt{5}$.

127. $(2p + 1)^2 = 7$

Solve by the square root property.

$2p + 1 = \sqrt{7}$ or $2p + 1 = -\sqrt{7}$

$2p = -1 + \sqrt{7}$ or $2p = -1 - \sqrt{7}$

$p = \dfrac{-1 + \sqrt{7}}{2}$ or $p = \dfrac{-1 - \sqrt{7}}{2}$

The solutions are $\dfrac{-1 \pm \sqrt{7}}{2}$.

128. $2p^2 + 3p = 2$

Write the equation in standard form and solve by factoring.

$2p^2 + 3p - 2 = 0$

$(2p - 1)(p + 2) = 0$

$2p - 1 = 0$ or $p + 2 = 0$

$p = \dfrac{1}{2}$ or $p = -2$

The solutions are $\dfrac{1}{2}$ and -2.

129. $2y^2 = 15 + y$

Write the equation in standard form and solve by factoring.

$2y^2 - y - 15 = 0$

$(y - 3)(2y + 5) = 0$

$y = 3$ or $y = -\dfrac{5}{2}$

The solutions are 3 and $-\dfrac{5}{2}$.

130. $x^2 - 2x = 2$

Write the equation in standard form.

$x^2 - 2x - 2 = 0$

Use the quadratic formula, with $a = 1$, $b = -2$, and $c = -2$.

$x = \dfrac{2 \pm \sqrt{(-2)^2 - 4(-2)}}{2} = \dfrac{2 \pm \sqrt{12}}{2}$

$= \dfrac{2 \pm 2\sqrt{3}}{2} = 1 \pm \sqrt{3}$

The solutions are $1 \pm \sqrt{3}$.

131. $r^2 + 4r = 1 \Rightarrow r^2 + 4r - 1 = 0$

Using the quadratic formula, we have

$r = \dfrac{-4 \pm \sqrt{16 + 4}}{2} = \dfrac{-4 + \sqrt{20}}{2}$

$= \dfrac{-4 \pm 2\sqrt{5}}{2} = -2 \pm \sqrt{5}$

The solutions are $-2 \pm \sqrt{5}$.

132. $2m^2 - 12m = 11 \Rightarrow 2m^2 - 12m - 11 = 0$
Using the quadratic formula, we have
$$m = \frac{12 \pm \sqrt{144 - 4(2)(-11)}}{2(2)} = \frac{12 \pm \sqrt{232}}{4}$$
$$= \frac{12 \pm 2\sqrt{58}}{4} = \frac{6 \pm \sqrt{58}}{2}$$
The solutions are $\dfrac{6 \pm \sqrt{58}}{2}$.

133. $9k^2 + 6k^2 = 2$
$9k^2 + 6k - 2 = 0$
Using the quadratic formula, we have
$$k = \frac{-6 \pm \sqrt{36 + 72}}{18} = \frac{-6 \pm \sqrt{108}}{18}$$
$$= \frac{-6 \pm \sqrt{36 \cdot 3}}{18} = \frac{-6 \pm 6\sqrt{3}}{18}$$
$$= \frac{6\left(-1 \pm \sqrt{3}\right)}{6 \cdot 3} = \frac{-1 \pm \sqrt{3}}{3}$$
The solutions are $\dfrac{-1 \pm \sqrt{3}}{3}$.

134. $2a^2 + a - 15 = 0$
$(2a - 5)(a + 3) = 0$
$2a - 5 = 0$ or $a + 3 = 0$
$a = \dfrac{5}{2}$ or $a = -3$
The solutions are $\dfrac{5}{2}$ and -3.

135. $12x^2 = 8x - 1$
$12x^2 - 8x + 1 = 0$
$(2x - 1)(6x - 1) = 0$
$2x - 1 = 0$ or $6x - 1 = 0$
$x = \dfrac{1}{2}$ or $x = \dfrac{1}{6}$
The solutions are $\dfrac{1}{2}$ and $\dfrac{1}{6}$.

136. $2q^2 - 11q = 21 \Rightarrow 2q^2 - 11q - 21 = 0 \Rightarrow$
$(2q + 3)(q - 7) = 0$
$2q + 3 = 0$ or $q - 7 = 0$
$q = -\dfrac{3}{2}$ or $q = 7$
The solutions are $-\dfrac{3}{2}$ and 7.

137. $3x^2 + 2x = 16 \Rightarrow 3x^2 + 2x - 16 = 0 \Rightarrow$
$(3x + 8)(x - 2) = 0$
$3x + 8 = 0$ or $x - 2 = 0$
$x = -\dfrac{8}{3}$ or $x = 2$
The solutions are $-\dfrac{8}{3}$ and 2.

138. $6k^4 + k^2 = 1 \Rightarrow 6k^4 + k^2 - 1 = 0$
Let $p = k^2$, so $p^2 = k^4$.
$6p^2 + p - 1 = 0$
$(3p - 1)(2p + 1) = 0$
$3p - 1 = 0$ or $2p + 1 = 0$
$p = \dfrac{1}{3}$ or $p = -\dfrac{1}{2}$
If $p = \dfrac{1}{3}$, $k^2 = \dfrac{1}{3} \Rightarrow k = \pm\sqrt{\dfrac{1}{3}} = \pm\dfrac{\sqrt{3}}{3}$
If $p = -\dfrac{1}{2}$, $k^2 = -\dfrac{1}{2}$ has no real number solution.
The solutions are $\pm\dfrac{\sqrt{3}}{3}$.

139. $21p^4 = 2 + p^2 \Rightarrow 21p^4 - p^2 - 2 = 0$
Let $u = p^2$; then $u^2 = p^4$.
$21u^2 - u - 2 = 0$
$(3u - 1)(7u + 2) = 0$
$3u - 1 = 0$ or $7u + 2 = 0$
$x = \dfrac{1}{3}$ or $x = -\dfrac{2}{7}$
$p^2 = \dfrac{1}{3}$ or $p^2 = -\dfrac{2}{7}$
If $x = -\dfrac{2}{7}$, $p^2 = -\dfrac{2}{7}$ has no real number solution.
$p = \pm\dfrac{1}{\sqrt{3}} = \pm\dfrac{\sqrt{3}}{3}$
The solutions are $\pm\dfrac{\sqrt{3}}{3}$.

140. $2x^4 = 7x^2 + 15 \Rightarrow 2x^4 - 7x^2 - 15 = 0$

Let $p = x^2$, so $p^2 = x^4$.

$2p^2 - 7p - 15 = 0$

$(2p + 3)(p - 5) = 0$

$2p + 3 = 0$ or $p - 5 = 0$

$p = -\dfrac{3}{2}$ or $p = 5$

If $p = -\dfrac{3}{2}$, then $x^2 = -\dfrac{3}{2}$ has no real number solution.

If $p = 5$, then $x^2 = 5 \Rightarrow x = \pm\sqrt{5}$.

The solutions are $\pm\sqrt{5}$.

141. $3m^4 + 20m^2 = 7 \Rightarrow 3m^4 + 20m^2 - 7 = 0$

Let $u = m^2$.

$3u^2 + 20u - 7 = 0$

$(3u - 1)(u + 7) = 0$

$3u - 1 = 0$ or $u + 7 = 0$

$u = \dfrac{1}{3}$ or $u = -7$

$m^2 = \dfrac{1}{3}$ or $m^2 = -7$

For $m^2 = -7$, there is no real number solution.

$m = \pm\dfrac{1}{\sqrt{3}} = \pm\dfrac{\sqrt{3}}{3}$

The solutions are $\pm\dfrac{\sqrt{3}}{3}$.

142.
$$3 = \frac{13}{z} + \frac{10}{z^2}$$
$$3z^2 = 13z + 10 \text{ Multiply by } z^2.$$

$3z^2 - 13z - 10 = 0$

$(3z + 2)(z - 5) = 0$

$3z + 2 = 0$ or $z - 5 = 0$

$z = -\dfrac{2}{3}$ or $z = 5$

The solutions are $-\dfrac{2}{3}$ and 5.

143. $p = \dfrac{E^2 R}{(r + R)^2}$ for r.

$$p(r + R)^2 = E^2 R$$
$$p\left(r^2 + 2rR + R^2\right) = E^2 R$$
$$pr^2 + 2rpR + R^2 p = E^2 R$$
$$pr^2 + 2rpR + R^2 p - E^2 R = 0$$

Use the quadratic formula to solve for r.

$$r = \frac{-2pR \pm \sqrt{4p^2 R^2 - 4p\left(R^2 p - E^2 R\right)}}{2p}$$

$$r = \frac{-2pR \pm \sqrt{4pE^2 R}}{2p} = \frac{-pR \pm E\sqrt{pR}}{p}$$

144. $p = \dfrac{E^2 R}{(r + R)^2}$ for E.

$$p(r + R)^2 = E^2 R \Rightarrow E^2 = \frac{p(r + R)^2}{R} \Rightarrow$$

$$E = \pm\sqrt{\frac{p(r + R)^2}{R}} = \frac{\pm(r + R)\sqrt{pR}}{R}$$

145. $K = s(s - a)$ for s.

$K = s^2 - as \Rightarrow s^2 - as - K = 0$

Use the quadratic formula.

$$s = \frac{a \pm \sqrt{a^2 - 4(-K)}}{2} = \frac{a \pm \sqrt{a^2 + 4K}}{2}$$

146. $kz^2 - hz - t = 0$ for z.

Use the quadratic formula with $a = k$, $b = -h$, and $c = -t$.

$$z = \frac{-(-h) \pm \sqrt{(-h)^2 - 4(k)(-t)}}{2k}$$

$$= \frac{h \pm \sqrt{h^2 + 4kt}}{2k}$$

147. a. $a = .8315(0)^2 - 73.93(0) + 2116.1$

$= 2116.1$

$a = .8315(29.035)^2 - 73.93(29.035)$
$$+ 2116.1$$

≈ 670.5229

At sea level, the atmospheric pressure is 2116.1 pounds per square foot. At the top of Mount Everest, the atmospheric pressure is abou 670.5229 pounds per square foot.

b. $1223.43 = .8315h^2 - 73.93h + 2116.1$

$0 = .8315h^2 - 73.93h + 892.67$

Using the quadratic formula, we have

$$x = \frac{-(-73.93) \pm \sqrt{(-73.93)^2 - 4(.8315)(892.67)}}{2(.8315)}$$

≈ 74.502 or 14.410

Since Mt. Rainier is not as tall as Mt. Everest, the first solution (74.502) is not applicable. Thus, Mt. Rainier is $14.410 \cdot 1000 = 14,410$ feet tall.

148. Let x = the width of the walk.

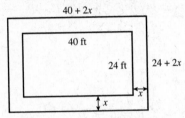

The area $= (24 + 2x)(40 + 2x) - 24(40)$

$= 960 + 48x + 80x + 4x^2 - 960$

$= 4x^2 + 128x$

To use all of the cement, solve

$740 = 4x^2 + 128x$

$4x^2 + 128x - 740 = 0$

$x^2 + 32 - 185 = 0$

$(x + 37)(x - 5) = 0 \Rightarrow x = -37$ or $x = 5$

The width cannot be negative, so the solution is 5 feet.

149. Let x = the dimension of the playground that does not lie along the side of the building.

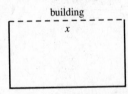

The area is 11,250 square meters so

$x(325 - 2x) = 11,250$

$325x - 2x^2 = 11,250$

$2x^2 - 325x + 11,250 = 0$

$$x = \frac{325 \pm \sqrt{325^2 - 4 \cdot 2 \cdot 11,250}}{2 \cdot 2}$$

$$x = \frac{325 \pm \sqrt{15,625}}{4} = \frac{325 \pm 125}{4}$$

$x = 112.5$ or $x = 50$.

If $x = 112.5$, then $325 - 2x = 100$.
If $x = 50$, then $325 - 2x = 225$.
The width is 100 m and the length is 112.5 m or the width is 50 m and the length is 225 m.

150.

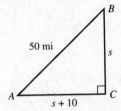

Let s be the speed of the car going north. Then the speed of the car going west is $s + 10$. Use $d = rt$ with $t = 1$ to label the distances shown above. By the Pythagorean theorem,

$$s^2 + (s + 10)^2 = 50^2$$

$$s^2 + s^2 + 20s + 100 = 2500$$

$$2s^2 + 20s - 2400 = 0$$

$$s^2 + 10s - 1200 = 0$$

$$(s - 30)(s + 40) = 0$$

$s = 30$ or $s = -40$.

Reject -40 as a possible solution because speed cannot be negative. The speed of the car headed north is 30 mph and the speed of the car headed west is 40 mph.

151. $v_0 = 200$, $h_0 = 0$, $h = 400$

$400 = -16t^2 + 200t + 0$

$0 = -16t^2 + 200t - 400$

$0 = -8(2t^2 - 25t + 50)$

$0 = -8(x - 10)(2x - 5) \Rightarrow x = 10$ or $x = 2.5$

The rocket reaches 400 feet at 2.5 seconds after liftoff.

152. We use the Pythagorean theorem.

$$(x + 5)^2 + 12^2 = 14.15^2$$

$$x^2 + 10x + 25 + 144 = 200.2225$$

$$x^2 + 10x - 31.2225 = 0$$

Using the quadratic formula, we have

$$x = \frac{-10 \pm \sqrt{10^2 - 4(1)(-31.2225)}}{2(1)}$$

$\approx 2.498, -12.498$

The negative value is not applicable because we are looking for a length. Thus, $x \approx 2.498$.

Case 1 Consumers Often Defy Common Sense

1. The total cost to buy and run this refrigerator for x years is $700 + 85x$.

2. The total cost to buy and run this refrigerator for x years is $1000 + 25x$.

3. Over 10 years, the $700 refrigerator costs
   $700 + 85(10) = 1550$ or $1550,
   and the $1000 refrigerator costs
   $1000 + 25(10) = 1250$ or $1250.
   The $700 refrigerator costs $300 more over 10 years.

4. The total costs for the two refrigerators will be
   equal when $700 + 85x = 1000 + 25x \Rightarrow$
   $60x = 300 \Rightarrow x = 5$.
   The costs will be equal in 5 years.

Chapter 2 Graphs, Lines, and Inequalities

Section 2.1 Graphs, Lines, and Inequalities

1. $(1, -2)$ lies in quadrant IV
$(-2, 1)$ lies in quadrant II
$(3, 4)$ lies in quadrant I
$(-5, -6)$ lies in quadrant III

3. $(1, -3)$ is a solution to $3x - y - 6 = 0$ because
$3(1) - (-3) - 6 = 0$ is a true statement.

5. $(3, 4)$ is not a solution to $(x-2)^2 + (y+2)^2 = 6$
because $(3-2)^2 + (4+2)^2 = 37$, not 6.

7. $4y + 3x = 12$
Find the y-intercept. If $x = 0$,
$4y = -3(0) + 12 \Rightarrow 4y = 12 \Rightarrow y = 3$
The y-intercept is 3.
Next find the x-intercept. If $y = 0$,
$4(0) + 3x = 12 \Rightarrow 3x = 12 \Rightarrow x = 4$
The x-intercept is 4.
Using these intercepts, graph the line.

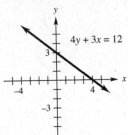

9. $8x + 3y = 12$
Find the y-intercept. If $x = 0$,
$3y = 12 \Rightarrow y = 4$
The y-intercept is 4.
Next, find the x-intercept. If $y = 0$,
$8x = 12 \Rightarrow x = \dfrac{12}{8} = \dfrac{3}{2}$
The x-intercept is $\dfrac{3}{2}$.
Using these intercepts, graph the line.

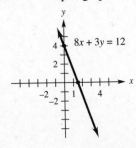

11. $x = 2y + 3$
Find the y-intercept. If $x = 0$,
$$0 = 2y + 3 \Rightarrow 2y = -3 \Rightarrow y = -\frac{3}{2}$$
The y-intercept is $-\dfrac{3}{2}$.
Next, find the x-intercept. If $y = 0$,
$x = 2(0) + 3 \Rightarrow x = 3$
The x-intercept is 3.
Using these intercepts, graph the line.

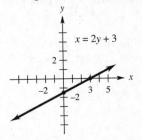

13. The x-intercepts are where the rays cross the x-axis, -2.5 and 3. The y-intercept is where the ray crosses the y-axis, 3.

15. The x-intercepts are -1 and 2. The y-intercept is -2.

17. $3x + 4y = 12$
To find the x-intercept, let $y = 0$:
$3x + 4(0) = 12 \Rightarrow 3x = 12 \Rightarrow x = 4$
The x-intercept is 4.
To find the y-intercept, let $x = 0$:
$3(0) + 4y = 12 \Rightarrow 4y = 12 \Rightarrow y = 3$
The y-intercept is 3.

19. $2x - 3y = 24$
To find the x-intercept, let $y = 0$:
$2x - 3(0) = 24 \Rightarrow 2x = 24 \Rightarrow x = 12$
The x-intercept is 12.
To find the y-intercept, let $x = 0$:
$2(0) - 3y = 24 \Rightarrow -3y = 24 \Rightarrow y = -8$
The y-intercept is -8.

21. $y = x^2 - 9$
To find the x-intercepts, let $y = 0$:
$0 = x^2 - 9 \Rightarrow x^2 = 9 \Rightarrow x = \pm\sqrt{9} = \pm 3$
The x-intercepts are 3 and -3.
To find the y-intercept, let $x = 0$:
$y = 0 - 9 = -9$
The y-intercept is -9.

23. $y = x^2 + x - 20$

To find the x-intercepts, let $y = 0$:

$0 = x^2 + x - 20 \Rightarrow 0 = (x+5)(x-4) \Rightarrow$

$x + 5 = 0 \Rightarrow x = -5$ or $x - 4 = 0 \Rightarrow x = 4$

The x-intercepts are -5 and 4.

To find the y-intercept, let $x = 0$:

$y = 0^2 + 0 - 20 = -20$

The y-intercept is -20.

25. $y = 2x^2 - 5x + 7$

To find the x-intercepts, let $y = 0$:

$0 = 2x^2 - 5x + 7$

This equation does not have real solutions, so there are no x-intercepts.

To find the y-intercept, let $x = 0$:

$y = 2(0)^2 - 5(0) + 7 = 7$

The y-intercept is 7.

27. $y = x^2$

x-intercept: $0 = x^2 \Rightarrow x = 0$

y-intercept: $y = 0$

x	y
-2	4
-1	1
0	0
1	1
2	4

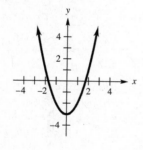

29. $y = x^2 - 3$

x-intercepts: $0 = x^2 - 3 \Rightarrow x^2 = 3 \Rightarrow x = \pm\sqrt{3}$

y-intercepts: $y = 0^2 - 3 = -3$

x	y
-3	6
-1	-2
0	-3
1	-2
3	6

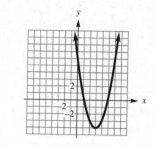

31. $y = x^2 - 6x + 5$

x-intercept:

$0 = x^2 - 6x + 5 \Rightarrow 0 = (x-1)(x-5) \Rightarrow$

$x - 1 = 0 \Rightarrow x = 1$ or $x - 5 = 0 \Rightarrow x = 5$

y-intercept: $y = (0)^2 - 6(0) + 5 = 5$

x	y
-2	21
-1	12
0	5
1	0
-2	-3

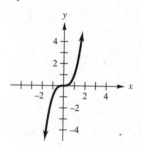

33. $y = x^3$

x-intercept: $0 = x^3 \Rightarrow x = 0$

y-intercept: $y = 0^3 \Rightarrow y = 0$

x	y
-2	-8
-1	-1
0	0
1	1
2	8

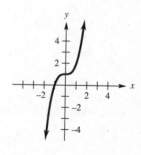

35. $y = x^3 + 1$

x-intercept:

$0 = x^3 + 1 \Rightarrow x^3 = -1 \Rightarrow x = \sqrt[3]{-1} = -1$

y-intercept: $y = 0^3 + 1 = 1$

x	y
-2	-7
-1	0
0	1
1	2
2	9

37. $y = \sqrt{x+4}$

x-intercept: $0 = \sqrt{x+4} \Rightarrow 0 = x+4 \Rightarrow x = -4$

y-intercept: $y = \sqrt{0+4} = \sqrt{4} = 2$

x	y
-2	$\sqrt{2} \approx 1.4$
-1	$\sqrt{3} \approx 1.7$
0	2
2	$\sqrt{6} \approx 2.4$
5	3

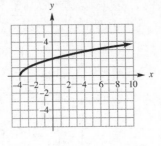

39. $y = \sqrt{4 - x^2}$

x-intercept:

$0 = \sqrt{4 - x^2} \Rightarrow 0 = 4 - x^2 \Rightarrow x^2 = 4 \Rightarrow$

$x = \pm\sqrt{4} = \pm2$

y-intercept: $y = \sqrt{4} = 2$

x	y
-2	0
-1	$\sqrt{3} \approx 1.7$
0	2
2	$\sqrt{3} \approx 1.7$
5	0

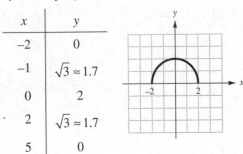

41. San Diego, 9:30AM; Portland, 8:45AM

43. From 10:00AM to 6:15PM

45. (a) about $1,250,000

   (b) $1,750,000

   (c) about $4,250,000

47. (a) about $500,000

   (b) about $1,000,000.

   (c) about $1,500,000.

49. beef, about 63 pounds; chicken, about 61 pounds; pork, about 46 pounds

51. 2002

53. about $21 billion

55. in 2011 and 2012

57. Men, about $740; women, about $600

59. About $620 in 2007

61. No

63. $y = x^2 + x + 1$

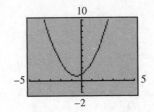

65. $y = (x-3)^3$

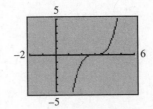

67. $y = x^3 - 3x^2 + x - 1$

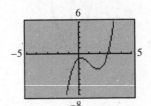

69. $y = x^4 - 2x^3 + 2x$

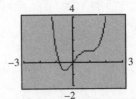

The "flat" part of the graph near $x = 1$ looks like a horizontal line segment, but it is not. The y values increase slightly as you trace along the segment from left to right.

71. $x \approx -1.1038$

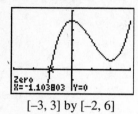

[–3, 3] by [–2, 6]

73. $x \approx 2.1017$

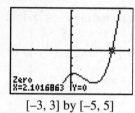

[–3, 3] by [–5, 5]

75. $x \approx -1.7521$

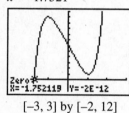

[–3, 3] by [–2, 12]

77. $S = \pi r \sqrt{r^2 + h^2} \Rightarrow 100 = \pi r \sqrt{r^2 + 5^2}$

Plot $Y_1 = \pi X \sqrt{X^2 + 5^2}$ and $Y_2 = 100$ on [0, 10] by [–20, 120], then find the intersection of the two curves.

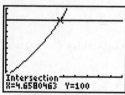

Alternatively, plot $Y_1 = \pi X \sqrt{X^2 + 5^2} - 100$ on [0, 10] by [–20, 20], then find the zero.

The radius should be about 4.6580 inches.

79. Plot $Y_1 = 216.9X^4 - 1202.3X^3 + 3223.9X^2 + 2596.8X - 29,087.2$ and $Y_2 = 150000$ on [0, 10] by [–20,000, 200,000], then find the intersection of the two curves.

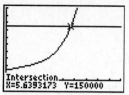

The intersection is approximately $x = 5.6$, which corresponds to the year 2004.

81. Plot $Y_1 = -.000018X^3 + .000488X^2 + .0366X + 1.6$ on [0, 50] by [–1, 5], then find the maximum of the curve.

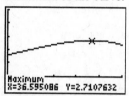

The maximum occurs at $x \approx 36.7$, which corresponds to the year 2006. There are about 2.7 doctors per 1000 residents in mid 2006.

Section 2.2 Equations of Lines

1. Through (2, 5) and (0, 8)

slope $= \dfrac{\Delta y}{\Delta x} = \dfrac{8 - 5}{0 - 2} = \dfrac{3}{-2} = -\dfrac{3}{2}$

3. Through (–4, 14) and (3, 0)

slope $= \dfrac{14 - 0}{-4 - 3} = \dfrac{14}{-7} = -2$

5. Through the origin and (–4, 10); the origin has coordinate (0, 0).

slope $= \dfrac{10 - 0}{-4 - 0} = \dfrac{10}{-4} = -\dfrac{5}{2}$

7. Through (–1, 4) and (–1, 6)

slope $= \dfrac{6 - 4}{-1 - (-1)} = \dfrac{2}{0}$, not defined

The slope is undefined.

9. $b = 5, m = 4$
$y = mx + b$
$y = 4x + 5$

11. $b = 1.5, m = -2.3$
$y = mx + b$
$y = -2.3x + 1.5$

13. $b = 4,\ m = -\dfrac{3}{4}$

$y = mx + b$

$y = -\dfrac{3}{4}x + 4$

15. $2x - y = 9$

Rewrite in slope-intercept form.

$-y = -2x + 9$

$y = 2x - 9$

$m = 2,\ b = -9.$

17. $6x = 2y + 4$

Rewrite in slope-intercept form.

$2y = 6x - 4 \Rightarrow y = 3x - 2$

$m = 3,\ b = -2.$

19. $6x - 9y = 16$

Write in slope-intercept form.

$-9y = -6x + 16$

$9y = 6x - 16$

$y = \dfrac{2}{3}x - \dfrac{16}{9}$

$m = \dfrac{2}{3},\ b = -\dfrac{16}{9}.$

21. $2x - 3y = 0$

Rewrite in slope-intercept form.

$3y = 2x \Rightarrow y = \dfrac{2}{3}x$

$m = \dfrac{2}{3},\ b = 0.$

23. $x = y - 5$

Rewrite in slope-intercept form.

$y = x + 5$

$m = 1,\ b = 5$

25. (a) Largest value of slope is at C.

(b) Smallest value of slope is at B.

(c) Largest absolute value is at B

(d) Closest to 0 is at D

27. $2x - y = -2$

Find the x-intercept by setting $y = 0$ and solving

for x: $2x - 0 = -2 \Rightarrow 2x = -2 \Rightarrow x = -1$

Find the y-intercept by setting $x = 0$ and solving

for y: $2(0) - y = -2 \Rightarrow -y = -2 \Rightarrow y = 2$

Use the points $(-1, 0)$ and $(0, 2)$ to sketch the

graph:

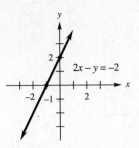

29. $2x + 3y = 4$

Find the x-intercept by setting $y = 0$ and solving

for x: $2x + 3(0) = 4 \Rightarrow 2x = 4 \Rightarrow x = 2$

Find the y-intercept by setting $x = 0$ and solving

for y: $2(0) + 3y = 4 \Rightarrow 3y = 4 \Rightarrow y = \dfrac{4}{3}$

Use the points $(2, 0)$ and $\left(0, \dfrac{4}{3}\right)$ to sketch the

graph:

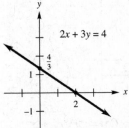

31. $4x - 5y = 2$

Find the x-intercept, by setting $y = 0$ and solving

for x:

$4x - 5(0) = 2 \Rightarrow 4x = 2 \Rightarrow x = \dfrac{1}{2}$

Find the y-intercept by setting $x = 0$ and solving

for y:

$4(0) - 5y = 2 \Rightarrow -5y = 2 \Rightarrow y = -\dfrac{2}{5}$

Use the points $\left(\dfrac{1}{2}, 0\right)$ and $\left(0, -\dfrac{2}{5}\right)$ to sketch

the graph:

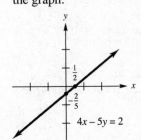

33. For $4x - 3y = 6$, solve for y.

$$y = \frac{4}{3}x - 2$$

For $3x + 4y = 8$, solve for y.

$$y = -\frac{3}{4}x + 2$$

The two slopes are $\frac{4}{3}$ and $-\frac{3}{4}$. Since

$$\left(\frac{4}{3}\right)\left(-\frac{3}{4}\right) = -1,$$

the lines are perpendicular.

35. For $3x + 2y = 8$, solve for y.

$$y = -\frac{3}{2}x + 4$$

For $6y = 5 - 9x$, solve for y.

$$y = -\frac{3}{2}x + \frac{5}{6}$$

Since the slopes are both $-\frac{3}{2}$, the lines are parallel.

37. For $4x = 2y + 3$, solve for y.

$$y = 2x - \frac{3}{2}$$

For $2y = 2x + 3$, solve for y.

$$y = x + \frac{3}{2}$$

Since the two slopes are 2 and 1, the lines are neither parallel nor perpendicular.

39. Triangle with vertices $(9, 6)$, $(-1, 2)$ and $(1, -3)$.

a. Slope of side between vertices $(9, 6)$ and $(-1, 2)$:

$$m = \frac{6-2}{9-(-1)} = \frac{4}{10} = \frac{2}{5}$$

Slope of side between vertices $(-1, 2)$ and $(1, -3)$:

$$m = \frac{2-(-3)}{-1-1} = \frac{5}{-2} = -\frac{5}{2}$$

Slope of side between vertices $(1, -3)$ and $(9, 6)$:

$$m = \frac{-3-6}{1-9} = \frac{-9}{-8} = \frac{9}{8}$$

b. The sides with slopes $\frac{2}{5}$ and $-\frac{5}{2}$ are perpendicular, because $\frac{2}{5}\left(-\frac{5}{2}\right) = -1$. Thus, the triangle is a right triangle.

41. Use point-slope form with

$$(x_1, y_1) = (-3, 2), m = -\frac{2}{3}$$

$$y - y_1 = m(x - x_1)$$

$$y - 2 = -\frac{2}{3}(x - (-3))$$

$$y - 2 = -\frac{2}{3}(x + 3)$$

$$y - 2 = -\frac{2}{3}x - 2$$

$$y = -\frac{2}{3}x \text{ or } 3y = -2x$$

43. $(x_1, y_1) = (2, 3), m = 3$

$$y - y_1 = m(x - x_1)$$

$$y - 3 = 3(x - 2)$$

$$y - 3 = 3x - 6$$

$$y = 3x - 3$$

45. $(x_1, y_1) = (10, 1), m = 0$

$$y - y_1 = m(x - x_1)$$

$$y - 1 = 0(x - 10)$$

$$y - 1 = 0 \Rightarrow y = 1$$

47. Since the slope is undefined, the equation is that of a vertical line through $(-2, 12)$.

$$x = -2$$

49. Through $(-1, 1)$ and $(2, 7)$

Find the slope.

$$m = \frac{7-1}{2-(-1)} = \frac{6}{3} = 2$$

Use the point-slope form with $(2, 7) = (x_1, y_1)$.

$$y - y_1 = m(x - x_1)$$

$$y - 7 = 2(x - 2)$$

$$y - 7 = 2x - 4$$

$$y = 2x + 3$$

51. Through $(1, 2)$ and $(3, 9)$
Find the slope.
$$m = \frac{9-2}{3-1} = \frac{7}{2}$$
Use the point-slope form with $(1, 2) = (x_1, y_1)$.
$$y - y_1 = m(x - x_1)$$
$$y - 2 = \frac{7}{2}(x - 1)$$
$$y - 2 = \frac{7}{2}x - \frac{7}{2}$$
$$y = \frac{7}{2}x - \frac{3}{2}$$

53. Through the origin with slope 5.
$(x_1, y_1) = (0, 0)$; $m = 5$
$$y - y_1 = m(x - x_1)$$
$$y - 0 = 5(x - 0) \Rightarrow y = 5x$$

55. Through $(6, 8)$ and vertical.
A vertical line has undefined slope.
$(x_1, y_1) = (6, 8)$
$x = 6$

57. Through $(3, 4)$ and parallel to $4x - 2y = 5$.
Find the slope of the given line because a line parallel to the line has the same slope.
$(x_1, y_1) = (3, 4)$
$$4x = 2y + 5$$
$$2y = 4x - 5$$
$$y = 2x - \frac{5}{2} \quad m = 2$$
$$y - y_1 = m(x - x_1)$$
$$y - 4 = 2(x - 3)$$
$$y - 4 = 2x - 6$$
$$y = 2x - 2$$

59. x-intercept 6; y-intercept –6
Through the points $(6, 0)$ and $(0, -6)$.
$$m = \frac{0 - (-6)}{6 - 0} = \frac{6}{6} = 1$$
$(x_1, y_1) = (6, 0)$
$$y - y_1 = m(x - x_1)$$
$$y - 0 = 1(x - 6) \Rightarrow y = x - 6$$

61. Through $(-1, 3)$ and perpendicular to the line through $(0, 1)$ and $(2, 3)$.
The slope of the given line is
$$m_1 = \frac{1-3}{0-2} = \frac{-2}{-2} = 1, \text{ so the slope of a line}$$
perpendicular to the line is $m_2 = \frac{-1}{1} = -1$.
$(x_1, y_1) = (-1, 3)$
$$y - y_1 = m(x - x_1)$$
$$y - 3 = -1(x - (-1))$$
$$y - 3 = -x - 1$$
$$y = -x + 2$$

63. Let cost $x = 15,965$ and life: 12 years. Find D.
$$D = \left(\frac{1}{n}\right)x = \frac{1}{12}(15,965) \approx 1330.42$$
The depreciation is $1330.42 per year.

65. Let cost $x = \$201,457$; life: 30 years
$$D = \left(\frac{1}{n}\right)x = \frac{1}{30}(201,457) \approx 6715.23$$
The depreciation is $6715.23 per year.

67. a. $x = 4$
$$y = 83.54(4) + 2920.94 = 3255.1$$
$$3255.1(1,000,000) = 3,255,100,000$$
There were about 3255.1 million or 3,255,100,000 prescriptions in 2004.

b. $x = 9$.
$$y = 83.54(9) + 2920.94 = 3672.8$$
$$3672.8(1,000,000) = 3,672,800,000$$
There were about 3672.8 million or 3,672,8010,000 prescriptions in 2004.

c. $y = 4,750,000,000 = 4750$ million
$$4750 = 83.53x + 2920.94$$
$$1829.06 = 83.53x$$
$$21.9 \approx x$$
This corresponds to the year 2021.

69. a. $x = 6$
$$y = -60.30(60) + 9635.4 = 9273.6$$
$$9273.6(1,000,000) = 9,273,600,000$$
There were approximately $9,273,600,000 in box-office receipts in 2006.

b. $y = 9$ billion $= 9000$ million
$$9000 = -60.30x + 9635.4$$
$$-635.4 = -60.30x \Rightarrow x \approx 10.5$$
This corresponds to the year 2010.
Box-office receipts will reach
$9,000,000,000 in 2010.

71. a. The given data is represented by the points
(0, 119.1) and (2, 132.3).

b. Find the slope.
$$m = \frac{132.3 - 119.1}{2 - 0} = \frac{13.2}{2} = 6.6$$
The y-intercept is 119.1, so the equation is
$y = 6.6x + 119.1$.

c. The year 2004 corresponds to $x = 1$.
$$y = 6.6(1) + 119.1 = 125.7$$
Alcohol sales were about $125.7 billion in
2004.

d. $y = 150$.
$$150 = 6.6x + 119.1$$
$$30.9 = 6.6x \Rightarrow x \approx 4.7$$
This corresponds to the year 2003 + 4 =
2007. Alcohol sales reached $150 billion in
2007.

73. a. $(x_1, y_1) = (0, 41235)$ and
$(x_2, y_2) = (10, 48695)$
Find the slope.
$$m = \frac{48,695 - 41,235}{10 - 0} = \frac{7460}{10} = 746$$
The y-intercept is 41,235, so the equation of
the line is $y = 746x + 41,235$.

b. The year 2012 corresponds to
$x = 2012 - 1995 = 17$.
$$y = 746(17) + 41,235 = 53,917$$
In 2012, there will be 53,917 shopping
centers.

c. $60,000 = 746x + 41,235$
$$18,765 = 746x \Rightarrow x \approx 25.2$$
This corresponds to the year 1995 + 25 =
2020. The number of shopping centers will
reach 60,000 in 2020.

75. a. $(x_1, y_1) = (0, 529000)$ and
$(x_2, y_2) = (16, 487000)$
Find the slope.
$$m = \frac{487,000 - 529,000}{16 - 0} = -2625$$
The y-intercept is 529,000, so the equation
is $y = -2625x + 529,000$.

b. $450,000 = -2625x + 529,000$
$$-79,000 = -2625x \Rightarrow x \approx 30$$
This corresponds to the year 1990 + 30 =
2020. There will be 450,000 employees of
the airline industry in 2020.

77. a. The slope of $-.01723$ indicates that on
average, the 5000-meter run is being run
.01723 seconds faster every year. It is
negative because the times are generally
decreasing as time progresses.

b. $y = -.01723(2012) + 47.61 \approx 12.94$
The model predicts that the time for the
5000-m run will be about 12.94 seconds in
the 2012 Olympics.

Section 2.3 Linear Models

1. a. Let (x_1, y_1) be (32, 0) and (x_2, y_2) be
(68, 20).
Find the slope.
$$m = \frac{20 - 0}{68 - 32} = \frac{20}{36} = \frac{5}{9}$$
Use the point-slope form with (32, 0).
$$y - 0 = \frac{5}{9}(x - 32) \Rightarrow y = \frac{5}{9}(x - 32)$$

b. Let $x = 50$.
$$y = \frac{5}{9}(50 - 32) = \frac{5}{9}(18) = 10°C$$
Let $x = 75$.
$$y = \frac{5}{9}(75 - 32) = \frac{5}{9}(43) \approx 23.89°C$$

3. $F = 867°$
$$C = \frac{5}{9}(867 - 32) = \frac{5}{9}(835) \approx 463.89°C$$

5. Let $(x_1, y_1) = (0, 168.8)$ and

$(x_2, y_2) = (8, 211.1)$. Find the slope.

$$m = \frac{211.1 - 168.8}{8 - 0} = 5.2875$$

The y-intercept is 168.8, so the equation is

$y = 5.2875x + 168.8$.

To estimate the CPI in 2005, let $x = 5$.

$y = 5.2875(5) + 168.8 \approx 195.2$

To estimate the CPI in 2011, let $x = 11$.

$y = 5.2875(11) + 168.8 \approx 227.0$

7. Let $(x_1, y_1) = (0, 132.1)$ and

$(x_2, y_2) = (5, 271.1)$. Find the slope.

$$m = \frac{271.1 - 132.1}{5 - 0} = 27.8$$

The y-intercept is 132.1, so the equation is

$y = 27.8x + 132.1$.

To estimate the amount of online retail sales in 2010, let $x = 2010 - 2006 = 4$.

$y = 27.8(4) + 132.1 = 243.3$

Online retail sales are estimated to be $243.3 billion in 2010.

9. Find the slope of the line.

$(x_1, y_1) = (50, 320)$

$(x_2, y_2) = (80, 440)$

$$m = \frac{y_2 - y_1}{x_2 - x_1} = \frac{440 - 320}{80 - 50} = \frac{120}{30} = 4$$

Each mile per hour increase in the speed of the bat will make the ball travel 4 more feet.

11. a. $y = .12x + 1.1$

Data Point (x, y)	Model Point $(x, \hat{y})$	Residual $y - \hat{y}$	Squared Residual $(y - \hat{y})^2$
(0, 1.1)	(0, 1.1)	0	0
(5, 1.7)	(5, 1.7)	0	0
(7, 2.0)	(7, 1.94)	.06	.0036
(9, 2.2)	(9, 2.18)	.02	.0004
(11, 2.4)	(11, 2.42)	−.02	.0004

$y = .1x + 1.3$

Data Point (x, y)	Model Point $(x, \hat{y})$	Residual $y - \hat{y}$	Squared Residual $(y - \hat{y})^2$
(0, 1.1)	(0, 1.3)	−.2	.04
(5, 1.7)	(5, 1.8)	−.1	.01
(7, 2.0)	(7, 2)	0	0
(9, 2.2)	(9, 2.2)	0	0
(11, 2.4)	(11, 2.4)	0	0

Sum of the residuals for model 1 = .06
Sum of the residuals for model 2 = −.3

b. Sum of the squares of the residuals for model 1 = .0044
Sum of the squares of the residuals for model 2 = .05

c. Model 1 is the better fit.

13. Plot the points.

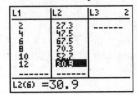 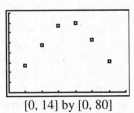

[0, 14] by [0, 80]

Visually, a straight line looks to be a poor model for the data.

```
LinReg
y=ax+b
a=.52
b=45.72666667
r²=.0117596382
r=.1084418654
```

The coefficient of correlation is $r \approx .1$, which indicates that the regression line is a poor fit for the data.

15. a. Let (x_1, y_1) be $(0, 559)$ and (x_2, y_2) be $(44, 217)$. Find the slope.

$$m = \frac{217 - 559}{44 - 0} \approx -7.77$$

The y-intercept is 559, so the equation of the line is $y = -7.77x + 559$.

Using a graphing calculator, the regression-line model is $y = -7.72x + 560.45$.

b. The year 2010 corresponds to $x = 50$. Using the regression-line model generated by using two points, we have

$y = -7.77(50) + 559 = 170.5$, or about 171

deaths per 1000,000 people from heart disease. Using the regression-line model generated by a graphing calculator, we have

$y = -7.72(50) + 560.45 = 174.45$, or about

174 deaths per 1000,000 people from heart disease.

17. a. Let (x_1, y_1) be $(150, 5000)$ and (x_2, y_2) be $(450, 9500)$. Find the slope.

$$m = \frac{9500 - 5000}{450 - 150} = 15$$

Use the point-slope form with $(150, 5000)$.

$y - 5000 = 15(x - 150)$

$\qquad y = 15x + 2750$

Using a graphing calculator, the regression-line model is

$y = 14.9x + 2822$.

b. Using the two-point model:

Let $x = 150 \text{ ft}^2$.

$y = 15(150) + 2750 = 5000$

Let $x = 280 \text{ ft}^2$.

$y = 15(280) + 2750 = 6950$

Let $x = 420 \text{ ft}^2$.

$y = 15(420) + 2750 = 9050$

Using the regression-line model:

Let $x = 150 \text{ ft}^2$.

$y = 14.9(150) + 2822 = 5057$

Let $x = 280 \text{ ft}^2$.

$y = 14.9(280) + 2822 = 6994$

Let $x = 420 \text{ ft}^2$.

$y = 14.9(420) + 2822 = 9080$

Using either model, the predicted values are close to the actual data.

c. Using the two-point model:

Let $x = 235 \text{ ft}^2$.

$y = 15(235) + 2750 = 6275$

Using the regression-line model:

Let $x = 235 \text{ ft}^2$.

$y = 14.9(235) + 2822 = 6323.5$

Adam should choose an air conditioner with a BTU of 6500 if air conditioners are available only with the BTU choices from the table.

19. a. The data points corresponding to the graph are: (9, 18.2), (10, 25.3), (11, 31.9), (12, 31.2), (13, 35.4), (14, 41.4), (15, 49.2), (16, 55.9), (17, 57.4)

b.
```
LinReg
 y=ax+b
 a=4.89
 b=-25.13666667
 r²=.9722869031
 r=.9860460958
■
```

Using a graphing calculator, the regression-line model is $y = 4.89x - 25.14$.

c. The year 2012 corresponds to $x = 22$.

$y = 4.89(22) - 25.14 \approx 82.4$ billion

21. a. Using a graphing calculator, the regression-line model is

$y = .03119x + .5635$.

b. Let $x = 12$, then $y \approx .938$ which is about .938 billion. Now let $x = 16$, then $y \approx 1.06$ which is about 1.06 billion.

23. a. Using a graphing calculator, the regression-line model is $y = .913x + 18.3$.

b. Answers will vary. Sample answer: The slope indicates that the median household income for men living without a spouse increases about $ .913 thousand or $913 per year.

c. The year 2010 corresponds to $x = 30$.

$y = .913(30) + 18.3 \approx 45.7$

The model predicts that the median household income for men living without a spouse is about $45.7 thousand.

25. a. Using a graphing calculator, the regression-line model is $y = 14.7x + 76$.

b. The year 2010 corresponds to $x = 10$.

$y = 14.7(10) + 76 = 223$

The operating revenue is $223 billion in 2010.

c. $180 = 14.7x + 76$

$104 = 14.7x \Rightarrow x \approx 7.1$

This corresponds to the year 2007.

Section 2.4 Linear Inequalities

1. Use brackets if you want to include the endpoint, and parentheses if you want to exclude it.

3. $-8k \leq 32$

   Multiply both sides of the inequality by $-\frac{1}{8}$.

   Since this is a negative number, change the direction of the inequality symbol.

   $$-\frac{1}{8}(-8k) \geq -\frac{1}{8}(32) \Rightarrow k \geq -4$$

   The solution is $[-4, \infty)$.

   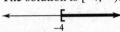

5. $-2b > 0$

   Multiply both sides by $-\frac{1}{2}$.

   $$-2b > 0 \Rightarrow -\frac{1}{2}(-2b) < -\frac{1}{2}(0) \Rightarrow b < 0$$

   The solution is $(-\infty, 0)$. To graph this solution, put a parenthesis at 0 and draw an arrow extending to the left.

   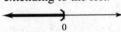

7. $3x + 4 \leq 14$

   Subtract 4 from both sides.
   $$3x + 4 - 4 \leq 14 - 4 \Rightarrow 3x \leq 10$$

   Multiply each side by $\frac{1}{3}$.

   $$\frac{1}{3}(3x) \leq \frac{1}{3}(10) \Rightarrow x \leq \frac{10}{3}$$

   The solution is $\left(-\infty, \frac{10}{3}\right]$.

   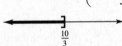

For exercises 9–25, we give the solutions without additional explanation.

9. $-5 - p \geq 3$
   $$-5 + 5 - p \geq 3 + 5$$
   $$-p \geq 8$$
   $$(-1)(-p) \leq (-1)(8)$$
   $$p \leq -8$$
   The solution is $(-\infty, -8]$.

11. $7m - 5 < 2m + 10$
    $$5m - 5 < 10$$
    $$5m < 15$$
    $$\frac{1}{5}(5m) < \frac{1}{5}(15)$$
    $$m < 3$$
    The solution is $(-\infty, 3)$.

13. $m - (4 + 2m) + 3 < 2m + 2$
    $$m - 4 - 2m + 3 < 2m + 2$$
    $$-1 - m < 2m + 2$$
    $$-m - 2m < 2 + 1$$
    $$-3m < 3$$
    $$-\frac{1}{3}(-3m) > -\frac{1}{3}(3)$$
    $$m > -1$$
    The solution is $(-1, \infty)$.

15. $-2(3y - 8) \geq 5(4y - 2)$
    $$-6y + 16 \geq 20y - 10$$
    $$16 + 10 \geq 20y + 6y$$
    $$26 \geq 26y$$
    $$1 \geq y \quad \text{or} \quad y \leq 1$$
    The solution is $(-\infty, 1]$.

17. $3p - 1 < 6p + 2(p - 1)$
    $$3p - 1 < 6p + 2p - 2$$
    $$-1 + 2 < 6p + 2p - 3p$$
    $$1 < 5p$$
    $$\frac{1}{5} < p \quad \text{or} \quad p > \frac{1}{5}$$
    The solution is $\left(\frac{1}{5}, \infty\right)$.

19. $-7 < y - 2 < 5$
    $$-7 + 2 < y - 2 + 2 < 5 + 2$$
    $$-5 < y < 7$$
    The solution is $(-5, 7)$.

21.
$$8 \le 3r + 1 \le 16$$
$$8 - 1 \le 3r \le 16 - 1$$
$$7 \le 3r \le 15$$
$$\frac{7}{3} \le r \le 5$$

The solution is $\left[\frac{7}{3}, 5\right]$.

23.
$$-4 \le \frac{2k-1}{3} \le 2$$
$$-4(3) \le 3\left(\frac{2k-1}{3}\right) \le 2(3)$$
$$-12 \le 2k - 1 \le 6$$
$$-12 + 1 \le 2k \le 6 + 1$$
$$-11 \le 2k \le 7$$
$$-\frac{11}{2} \le k \le \frac{7}{2}$$

The solution is $\left[-\frac{11}{2}, \frac{7}{2}\right]$.

25.
$$\frac{3}{5}(2p+3) \ge \frac{1}{10}(5p+1)$$
$$10 \cdot \frac{3}{5}(2p+3) \ge 10 \cdot \frac{1}{10}(5p+1)$$
$$6(2p+3) \ge 5p + 1$$
$$12p + 18 \ge 5p + 1$$
$$7p \ge -17$$
$$p \ge -\frac{17}{7}$$

The solution is $\left[-\frac{17}{7}, \infty\right)$.

27. $x \ge 2$

29. $-3 < x \le 5$

31. a. Let x represent the number of milligrams per liter of lead in the water.

b. 5% of .040 is .002.
$$.040 - .002 \le x \le .040 + .002$$
$$.038 \le x \le .042$$

c. Since all the samples had a lead content less than or equal to .042 mg per liter, all were less than .050 mg per liter and did meet the federal requirement.

33. a. The six income ranges are:
$$0 < x \le 8025$$
$$8025 < x \le 32{,}550$$
$$32{,}550 < x \le 78{,}850$$
$$78{,}850 < x \le 164{,}550$$
$$164{,}550 < x \le 357{,}700$$
$$x > 357{,}700$$

b. For an income x such that $0 < x \le 8025$, the tax range, T, is
$$0(.10) < T \le 8025(.10) \Rightarrow 0 < T \le 802.50.$$

For an income x such that $8025 < x \le 32{,}550$, the tax range, T, is
$$802.50 + .15(8025 - 8025) < T$$
$$\le 802.50 + .15(32{,}550 - 8025) \Rightarrow$$
$$802.50 < T \le 4481.25.$$
For an income x such that $32{,}550 < x \le 78{,}850$, the tax range, T, is
$$4481.25 + .25(32{,}550 - 32{,}550) < T$$
$$\le 4481.25 + .25(78{,}850 - 32{,}550) \Rightarrow$$
$$4481.25 < T \le 16{,}056.25.$$
For an income x such that $78{,}850 < x \le 164{,}550$, the tax range, T, is
$$16{,}056.25 + .28(78{,}850 - 78{,}850) < T$$
$$\le 16{,}056.25 + .28(164{,}550 - 78{,}850) \Rightarrow$$
$$16{,}056.25 < T \le 40{,}052.25.$$
For an income x such that $164{,}550 < x \le 357{,}700$, the tax range, T, is
$$40{,}052.25 + .33(164{,}550 - 164{,}550) < T$$
$$\le 40{,}052.25 + .33(357{,}700 - 164{,}550) \Rightarrow$$
$$40{,}052.25 < T \le 103{,}791.75.$$
For an income x such that $x > 357{,}700$, the tax range, T, is $T > 103{,}791.75.$

35. $|m| < 2 \Rightarrow -2 < m < 2$
The solution is $(-2, 2)$.

37. $|a| < -2$

Since the absolute value of a number is never negative, the inequality has no solution.

39. $|2x + 5| < 1$

$$-1 < 2x + 5 < 1$$
$$-1 - 5 < 2x < 1 - 5$$
$$-6 < 2x < -4$$
$$-3 < x < -2$$

The solution is $(-3, -2)$.

41. $|3z + 1| \geq 4$

$$3z + 1 \geq 4 \quad \text{or} \quad 3z + 1 \leq -4$$
$$3z \geq 4 - 1 \qquad\qquad 3z \leq -4 - 1$$
$$3z \geq 3 \qquad\qquad 3z \leq -5$$
$$z \geq 1 \qquad\qquad z \leq -\frac{5}{3}$$

The solution is $\left(-\infty, -\dfrac{5}{3}\right]$ or $[1, \infty)$.

43. $\left|5x + \dfrac{1}{2}\right| - 2 < 5$

$$\left|5x + \frac{1}{2}\right| < 7$$
$$-7 < 5x + \frac{1}{2} < 7$$
$$-7 - \frac{1}{2} < 5x < 7 - \frac{1}{2}$$
$$-\frac{15}{2} < 5x < \frac{13}{2}$$
$$-\frac{15}{2} \cdot \frac{1}{5} < x < \frac{13}{2} \cdot \frac{1}{5}$$
$$-\frac{3}{2} < x < \frac{13}{10}$$

The solution is $\left(-\dfrac{3}{2}, \dfrac{13}{10}\right)$.

45. $|T - 83| \leq 7$

$$-7 \leq T - 83 \leq 7$$
$$76 \leq T \leq 90$$

47. $|T - 61| \leq 21$

$$-21 \leq T - 61 \leq 21$$
$$40 \leq T \leq 82$$

49. $|R_L - 26.75| \leq 1.42$

$|R_E - 38.75| \leq 2.17$

a. $|R_L - 26.75| \leq \pm 1.42 \Rightarrow$

$-1.42 \leq R_L - 26.75 \leq 1.42 \Rightarrow$

$25.33 \leq R_L \leq 28.17$

$|R_E - 38.75| \leq 2.17 \Rightarrow$

$-2.17 \leq R_E - 38.75 \leq 2.17 \Rightarrow$

$36.58 \leq R_E \leq 40.92$

b. $225(25.33) \leq T_L \leq 225(28.17)$

$5699.25 \leq T_L \leq 6338.25$

$225(36.58) \leq T_E \leq 225(40.92)$

$8230.5 \leq T_E \leq 9207$

51. $30 \leq 4.89x - 25.13 \leq 40$

$55.13 \leq 4.89x \leq 65.13$

$11.3 \leq x \leq 13.3$

11.3 corresponds to the year 2001, and 13.3 corresponds to the year 2003.

Between the years of 2001 and 2003, revenue was between 30 and 40 billion dollars.

53. $-3.14x + 502 > 470$

$-3.14x > -32 \Rightarrow x < 10.2$

This corresponds to the year 2010.

$-3.14x + 502 < 480$

$-3.14x < -22 \Rightarrow x > 7.0$

This corresponds to the year 2007.

The number of ski resorts was between 480 and 470 between the years 2007 and 2010.

55. a. $2.38x + 12.46 < 40$

$2.38x < 27.54 \Rightarrow x < 11.6$

This corresponds to the year 2001, so the dollars earned per 100 pounds was less than 40 in 2001 and earlier.

b. $2.38x + 12.46 > 50$

$2.38x > 37.54 \Rightarrow x > 15.8$

This corresponds to the year 2005, so the dollars earned per 100 pounds was greater than 50 in 2005 and after.

57. $C = 50x + 6000$; $R = 65x$
To at least break even, $R \geq C$.
$$65x \geq 50x + 6000$$
$$15x \geq 6000 \Rightarrow x \geq 400$$
The number of units of wire must be in the interval $[400, \infty)$.

59. $C = 85x + 1000$; $R = 105x$
$$R \geq C$$
$$105x \geq 85x + 1000$$
$$20x \geq 1000$$
$$x \geq \frac{1000}{20} \Rightarrow x \geq 50$$
x must be in the interval $[50, \infty)$.

61. $C = 1000x + 5000$; $R = 900x$
$$R \geq C$$
$$900x \geq 1000x + 5000$$
$$-100x \geq 5000$$
$$x \leq \frac{5000}{-100} \Rightarrow x \leq -50$$
It is impossible to break even.

Section 2.5 Polynomial and Rational Inequalities

1. $(x + 4)(2x - 3) \leq 0$
Solve the corresponding equation.
$(x + 4)(2x - 3) = 0$
$x + 4 = 0$ or $2x - 3 = 0$

$\qquad x = -4 \qquad\qquad x = \frac{3}{2}$

Note that because the inequality symbol is "$\leq$," -4 and $\frac{3}{2}$ are solutions of the original inequality. These numbers separate the number line into three regions.

In region A, let $x = -6$:
$(-6 + 4)[2(-6)-3] = 30 > 0$.
In region B, let $x = 0$:
$(0 + 4)[2(0) - 3] = -12 < 0$.
In region C, let $x = 2$:
$(2 + 4)[2(2) - 3] = 6 > 0$.
The only region where $(x + 4)(2x - 3)$ is negative is region B, so the solution is $\left[-4, \frac{3}{2}\right]$. To graph this solution, put brackets at -4 and $\frac{3}{2}$ and draw a line segment between these two endpoints.

3. $r^2 + 4r > -3$
Solve the corresponding equation.
$$r^2 + 4r = -3$$
$$r^2 + 4r + 3 = 0$$
$$(r + 1)(r + 3) = 0$$
$$r + 1 = 0 \quad\text{or}\quad r + 3 = 0$$
$$r = -1 \quad\text{or}\qquad r = -3$$
Note that because the inequality symbol is ">," -1 and -3 are not solutions of the original inequality.

In region A, let $r = -4$:
$(-4)^2 + 4(-4) = 0 > -3$.
In region B, let $r = -2$:
$(-2)^2 + 4(-2) = -4 < -3$.
In region C, let $r = 0$:
$0^2 + 4(0) = 0 > -3$.
The solution is $(-\infty, -3)$ or $(-1, \infty)$.
To graph the solution, put a parenthesis at -3 and draw a ray extending to the left, and put a parenthesis at -1 and draw a ray extending to the right.

5. $4m^2 + 7m - 2 \le 0$

Solve the corresponding equation.

$4m^2 + 7m - 2 = 0$

$(4m - 1)(m + 2) = 0$

$4m - 1 = 0 \quad \text{or} \quad m + 2 = 0$

$m = \dfrac{1}{4} \quad \text{or} \quad m = -2$

Because the inequality symbol is "$\le$" $\dfrac{1}{4}$ and -2 are solutions of the original inequality.

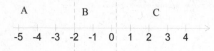

In region A, let $m = -3$:

$4(-3)^2 + 7(-3) - 2 = 13 > 0$.

In region B, let $m = 0$:

$4(0)^2 + 7(0) - 2 = -2 < 0$.

In region C, let $m = 1$:

$4(1)^2 + 7(1) - 2 = 9 > 0$.

The solution is $\left[-2, \dfrac{1}{4}\right]$.

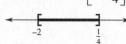

7. $4x^2 + 3x - 1 > 0$

Solve the corresponding equation.

$4x^2 + 3x - 1 = 0$

$(4x - 1)(x + 1) = 0$

$4x - 1 = 0 \quad \text{or} \quad x + 1 = 0$

$x = \dfrac{1}{4} \quad \text{or} \quad x = -1$

Note that $\dfrac{1}{4}$ and -1 are not solutions of the original inequality.

In region A, let $x = -2$:

$4(-2)^2 + 3(-2) - 1 = 9 > 0$.

In region B, let $x = 0$:

$4(0)^2 + 3(0) - 1 = -1 < 0$.

In region C, let $x = 1$:

$4(1)^2 + 3(1) - 1 = 6 > 0$.

The solution is $(-\infty, -1)$ or $\left(\dfrac{1}{4}, \infty\right)$.

9. $x^2 \le 36$

Solve the corresponding equation.

$x^2 = 36 \Rightarrow x = \pm 6$

For region A, let $x = -7$: $(-7)^2 = 49 > 36$.

For region B, let $x = 0$: $0^2 = 0 < 36$.

For region C, let $x = 7$: $7^2 = 49 > 36$.

Both endpoints are included. The solution is $[-6, 6]$.

11. $p^2 - 16p > 0$

Solve the corresponding equation.

$p^2 - 16p = 0 \Rightarrow p(p - 16) = 0 \Rightarrow$

$p = 0$ or $p = 16$

Since the inequality is "$>$", 0 and 16 are not solutions of the original inequality.

For region A, let $p = -1$:

$(-1)^2 - 16(-1) = 17 > 0$.

For region B, let $p = 1$:

$1^2 - 16(1) = -15 < 0$.

For region C, let $p = 17$:

$17^2 - 16(17) = 17 > 0$.

The solution is $(-\infty, 0)$ or $(16, \infty)$.

13. $x^3 - 9x \geq 0$

Solve the corresponding equation.

$x^3 - 9x = 0 \Rightarrow x(x^2 - 9) = 0 \Rightarrow$

$x(x + 3)(x - 3) = 0$

$x = 0$ or $x = -3$ or $x = 3$

Note that 0, –3, and 3 are all solutions of the original inequality.

In region A, let $x = -4$:

$(-4)^3 - 9(-4) = -28 < 0$.

In region B, let $x = -1$:

$(-1)^3 - 9(-1) = 8 > 0$.

In region C, let $x = 1$:

$(1)^3 - 9(1) = -8 < 0$

In region D, let $x = 4$:

$4^3 - 9(4) = 28 > 0$.

The solution is $[-3, 0]$ or $[3, \infty)$.

15. $(x + 7)(x + 2)(x - 2) \geq 0$

Solve the corresponding equation.

$(x + 7)(x + 2)(x - 2) = 0$

$x + 7 = 0$ or $x + 2 = 0$ or $x - 2 = 0$

$x = -7$ or $x = -2$ or $x = 2$

Note that –6, –1 and 4 are all solutions of the original inequality.

In region A, let $x = -8$:

$(-8 + 7)(-8 + 2)(-8 - 2) = -60 < 0$

In region B, let $x = -4$:

$(-4 + 7)(-4 + 2)(-4 - 2) = 36 > 0$

In region C, let $x = 0$:

$(0 + 7)(0 + 2)(0 - 2) = -28 < 0$

In region D, let $x = 3$:

$(3 + 7)(3 + 2)(3 - 2) = 50 > 0$

The solution is $[-7, -2]$ or $[2, \infty)$.

17. $(x + 5)(x^2 - 2x - 3) < 0$

Solve the corresponding equation.

$(x + 5)(x^2 - 2x - 3) = 0$

$(x + 5)(x + 1)(x - 3) = 0$

$x + 5 = 0$ or $x + 1 = 0$ or $x - 3 = 0$

$x = -5$ or $x = -1$ or $x = 3$

Note that –5, –1 and 3 are not solutions of the original inequality.

In region A, let $x = -6$:

$(-6 + 5)\left[(-6)^2 - 2(-6) - 3\right] = (-1)(45)$

$= -45 < 0$

In region B, let $x = -2$:

$(-2 + 5)\left[(-2)^2 - 2(-2) - 3\right] = 3(5) = 15 > 0$

In region C, let $x = 0$:

$(0 + 5)\left[(0)^2 - 2(0) - 3\right] = 5(-3) = -15 < 0$

In region D, let $x = 4$:

$(4 + 5)\left[(4)^2 - 2(4) - 3\right] = 9(5) = 45 > 0$

The solution is $(\infty, -5)$ or $(-1, 3)$.

19. $6k^3 - 5k^2 < 4k \Rightarrow 6k^3 - 5k^2 - 4k < 0$

Solve the corresponding equation.

$6k^3 - 5k^2 - 4k = 0$

$k(6k^2 - 5k - 4) = 0$

$k(3k - 4)(2k + 1) = 0$

$k = 0$ or $k = \dfrac{4}{3}$ or $k = -\dfrac{1}{2}$

Note that 0, $\dfrac{4}{3}$, and $-\dfrac{1}{2}$ are not solutions of the original inequality.

In region A, let $k = -1$:

$6(-1)^3 - 5(-1)^2 - 4(-1) = -7 < 0$

In region B, let $k = -\dfrac{1}{4}$:

$6\left(-\dfrac{1}{4}\right)^3 - 5\left(-\dfrac{1}{4}\right)^2 - 4\left(-\dfrac{1}{4}\right) = \dfrac{19}{32} > 0;$

In region C, let $k = 1$:

$6(1)^3 - 5(1)^2 - 4(1) = -3 < 0$

In region D, let $k = 10$:

$6(10)^3 - 5(10)^2 - 4(10) = 5460.0$

The given inequality is true in regions A and C.

The solution is $\left(-\infty, -\dfrac{1}{2}\right)$ or $\left(0, \dfrac{4}{3}\right)$.

21. The inequality $p^2 < 16$ should be rewritten as

$p^2 - 16 < 0$ and solved by the method shown in this section for solving quadratic inequalities. This method will lead to the correct solution $(-4, 4)$. The student's method and solution are incorrect.

23. To solve $.5x^2 - 1.2x < .2$, write the inequality as

$.5x^2 - 1.2x - .2 < 0$. Graph the equation

$y = .5x^2 - 1.2x - .2$. Enter this equation as y_1 and use $-4 \le x \le 6$ and $-5 \le y \le 5$. On the CALC menu, use "zero" to find the x-values where the graph crosses the x-axis. These values are $x = -.1565$ and $x = 2.5565$. The graph is below the x-axis between these two values. The solution of the inequality is $(-.1565, 2.5565)$.

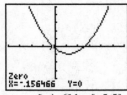 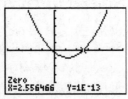

[-4, 6] by [-5,5]

25. To solve $x^3 - 2x^2 - 5x + 7 \ge 2x + 1$, graph

$y_1 = x^3 - 2x^2 - 5x + 7$ and $y_2 = 2x + 1$ in the window [-5, 5] by[-10, 10]. On the CALC menu, use "intersect" to find the x-values where the graphs intersect. These values are $x = -2.2635$, $x = .7556$ and $x = 3.5079$. The graph of y_1 is above the graph of y_2 for $[-2.2635, .7556]$ or $[3.5079, \infty)$.

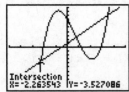 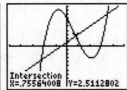

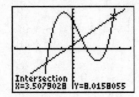

27. To solve $2x^4 + 3x^3 < 2x^2 + 4x - 2$, graph

$y_1 = 2x^4 + 3x^3$ and $y_2 = 2x^2 + 4x - 2$ in the window [-2, 2] by[-5, 5]. On the CALC menu, use "intersect" to find the x-values where the graphs intersect. These values are $x = .5$ and $x = .8393$. The graph of y_1 is below the graph of y_2 to the right of .5 and to the left of .8393. The solution of the inequality is $(.5, .8393)$.

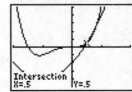 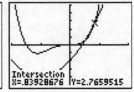

29. $\dfrac{r-4}{r-1} \ge 0$

Solve the corresponding equation.

$\dfrac{r-4}{r-1} = 0$

The quotient can change sign only when the numerator is 0 or the denominator is 0. The numerator is 0 when $r = 4$. The denominator is 0 when $r = 1$. Note that 4 is a solution of the original inequality, but 1 is not.

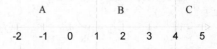

In region A, let $r = 0$:

$\dfrac{0-4}{0-1} = 4 > 0$.

In region B, let $r = 2$:

$\dfrac{2-4}{2-1} = -2 < 0$.

In region C, let $r = 5$:

$\dfrac{5-3}{5-1} = \dfrac{1}{4} > 0$.

The given inequality is true in regions A and C, so the solution is $(-\infty, 1)$ or $[4, \infty)$.

31. $\dfrac{a-2}{a-5} < -1$

Solve the corresponding equation.

$$\frac{a-2}{a-5} = -1$$

$$\frac{a-2}{a-5} + 1 = 0$$

$$\frac{a-2}{a-5} + \frac{a-5}{a-5} = 0$$

$$\frac{2a-7}{a-5} = 0$$

The numerator is 0 when $a = \dfrac{7}{2}$. The denominator is 0 when $a = 5$. Note that $\dfrac{7}{2}$ and 5 are not solutions of the original inequality.

In region A, let $a = 0$:

$\dfrac{0-2}{0-5} = \dfrac{2}{5} > -1$.

In region B, let $a = 4$:

$\dfrac{4-2}{4-5} = \dfrac{2}{-1} = -2 < -1$.

In region C, let $a = 10$:

$\dfrac{10-2}{10-5} = \dfrac{8}{5} > -1$.

The solution is $\left(\dfrac{7}{2}, 5\right)$.

33. $\dfrac{1}{p-2} < \dfrac{1}{3}$

Solve the corresponding equation.

$$\frac{1}{p-2} = \frac{1}{3}$$

$$\frac{1}{p-2} - \frac{1}{3} = 0$$

$$\frac{3 - (p-2)}{3(p-2)} = 0$$

$$\frac{3 - p + 2}{3(p-2)} = 0$$

$$\frac{5 - p}{3(p-2)} = 0$$

The numerator is 0 when $p = 5$. The denominator is 0 when $p = 2$. Note that 2 and 5 are not solutions of the original inequality.

In region A, let $p = 0$: $\dfrac{1}{0-2} = -\dfrac{1}{2} < \dfrac{1}{3}$.

In region B, let $p = 3$: $\dfrac{1}{3-2} = 1 > \dfrac{1}{3}$.

In region C, let $p = 6$: $\dfrac{1}{6-2} = \dfrac{1}{4} < \dfrac{1}{3}$.

The solution is $(-\infty, 2)$ or $(5, \infty)$.

35. $\dfrac{5}{p+1} > \dfrac{12}{p+1}$

Solve the corresponding equation.

$$\frac{5}{p+1} = \frac{12}{p+1}$$

$$\frac{5}{p+1} - \frac{12}{p+1} = 0$$

$$\frac{-7}{p+1} = 0$$

The numerator is never 0. The denominator is 0 when $p = -1$. Therefore, in this case, we separate the number line into only two regions.

In region A, let $p = -2$:

$\dfrac{5}{-2+1} = -5$

$\dfrac{12}{-2+1} = -12$

$-5 > -12$

In region B, let $p = 0$:

$\dfrac{5}{0+1} = 5$

$\dfrac{12}{0+1} = 12$

$12 > 5$

Therefore, the given inequality is true in region A. The only endpoint, -1, is not included because the symbol is ">." Therefore, the solution is $(-\infty, -1)$.

37. $\dfrac{x^2 - x - 6}{x} < 0$

Solve the corresponding equation.

$$\dfrac{x^2 - x - 6}{x} = 0$$

$$x^2 - x - 6 = 0 \quad \text{or} \quad x = 0$$

$$(x - 3)(x + 2) = 0 \quad \text{or} \quad x = 0$$

$$x - 3 = 0 \quad \text{or} \quad x + 2 = 0 \quad \text{or} \quad x = 0$$

$$x = 3 \quad \text{or} \quad x = -2 \quad \text{or} \quad x = 0$$

Note that -2, 0 and 3 are not solutions of the original inequality.

```
     A        B        C        D
  ←──┼────┼────┼────┼────┼────┼────┼────┼────→
    -4   -3   -2   -1    0    1    2    3    4
```

In region A, let $x = -3$:

$$\dfrac{(-3)^2 - (-3) - 6}{-3} = \dfrac{9 + 3 - 6}{-3} = \dfrac{6}{-3} = -2 < 0.$$

In region B, let $x = -1$:

$$\dfrac{(-1)^2 - (-1) - 6}{-1} = \dfrac{1 + 1 - 6}{-1} = \dfrac{-4}{-1} = 4 > 0$$

In region C, let $x = 1$:

$$\dfrac{1^2 - 1 - 6}{1} = -6 < 0.$$

In region D, let $x = 4$:

$$\dfrac{4^2 - 4 - 6}{4} = \dfrac{16 - 10}{4} = \dfrac{6}{4} = \dfrac{3}{2} > 0.$$

The solution is $(-\infty, -2)$ or $(0, 3)$.

39. To solve $\dfrac{2x^2 + x - 1}{x^2 - 4x + 4} \leq 0$, break the inequality

into two inequalities $2x^2 + x - 1 \leq 0$ and

$x^2 - 4x + 4 \leq 0$. Graph the equations

$y = 2x^2 + x - 1$ and $y = x^2 - 4x + 4$. Enter

these equations as y_1 and y_2, and use

$-3 < x < 3$ and $-2 < y < 2$. On the CALC menu,

use "zero" to find the x-values where the graphs

cross the x-axis. These values for y_1 are $x = -1$

and $x = .5$. The graph of y_1 is below the x-axis

to the right of -1 and to the left of $.5$. The graph

of y_2 is never below the x-axis. The solution of

the inequality is $[-1, .5]$.

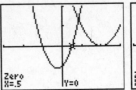
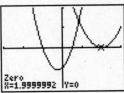

41. $P = 2x^2 - 12x - 32$

The company makes a profit when

$2x^2 - 12x - 32 > 0$.

Solve the corresponding equation.

$$2x^2 - 12x - 32 = 0 \Rightarrow 2\left(x^2 - 6x - 16\right) = 0 \Rightarrow$$

$$(x + 2)(x - 8) = 0 \Rightarrow x = -2 \quad \text{or} \quad x = 8$$

The test regions are $A(-\infty, -2)$, $B(-2, 8)$, and

$C(8, \infty)$. Region A makes no sense in this

context, so we ignore this. Test a number from

regions B and C in the original inequality.

For region B, let $x = 0$.

$$2(0)^2 - 12(0) - 32 = -32 < 0$$

For region C, let $x = 10$.

$$2(10)^2 - 12(10) - 32 = 48 > 0$$

The numbers in region C satisfy the inequality.

The company makes a profit when the amount

spent on advertising in hundreds of thousands of

dollars is in the interval $(8, \infty)$.

43. $P = -x^2 + 280x - 16,000$

The complex makes a profit when

$-x^2 + 280x - 16,000 > 0.$

Solve the corresponding equation.

$$-x^2 + 280x - 16,000 = 0$$

$$-\left(x^2 - 280x + 16,000\right) = 0$$

$$(x - 80)(x - 200) = 0$$

$$x = 80 \quad \text{or} \quad x = 200$$

We only consider positive values of x because

x represents the number of apartments rented.

The test regions are $A(0, 80)$, $B(80, 200)$, and

$C(200, \infty)$. In region A, let $x = 1$:

$$-(1)^2 + 280(1) - 16,000 = -15,721 < 0.$$ In

region B, let $x = 100$:

$$-(100)^2 + 280(100) - 16,000 = 2000 > 0.$$

In region C, let $x = 300$:

$$-(300)^2 + 280(300) - 16,000 = -22,0000 < 0.$$

The complex makes a profit when the number of

units rented is between 80 and 200, exclusive, or

when x is in the interval $(80, 200)$.

45. $1.35x^2 + 3.7x + 183.2 > 250$

Use a graphing calculator to solve

$1.35x^2 + 3.7x + 183.2 = 250 \Rightarrow$

$1.35x^2 + 3.7x - 66.8 = 0$

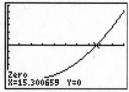

[0, 10] by [–100, 100]

The graph lies above the x-axis for $x > 5.8$, which corresponds to the year 2005. Thus, there will be more than 250,000 employees after 2005.

47. The function representing the cost of x hundred cartons of pens is $C = 114 + 8x - x^2$. The manufacturer can sell the same number of cartons of pens for $.15x$. The manufacturer will break even or make a profit if $.15x \geq C$. So, use a graphing calculator to solve

$.15x \geq 114 + 8x - x^2 \Rightarrow x^2 - 7.85x - 114 \geq 0.$

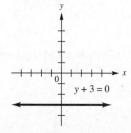

[0, 20] by [–150, 150]

The graph is above the x-axis for $x \geq 15.30$. Therefore, the manufacturer will make a profit if at least $15.30(100) = 1530$ cartons of pens are sold.

Chapter 2 Review Exercises

1. $y = x^2 - 2x - 5$

$(-2, 3): \ (-2)^2 - 2(-2) - 5 = 4 + 4 - 5 = 3$

$(0, -5): \ (0)^2 - 2(0) - 5 = 0 - 0 - 5 = -5$

$(2, -3): \ (2)^2 - 2(2) - 5 = 4 - 4 - 5 = -5 \neq -3$

$(3, -2): \ (3)^2 - 2(3) - 5 = 9 - 6 - 5 = -2$

$(4, 3): \ (4)^2 - 2(4) - 5 = 16 - 8 - 5 = 3$

$(7, 2): \ (7)^2 - 2(7) - 5 = 49 - 14 - 5 = 30 \neq 2$

Solutions are (–2, 3), (0, –5), (3, –2), (4, 3).

2. $x - y = 5$

$(-2, 3): -2 - 3 = -5 \neq 5$

$(0, -5): 0 - (-5) = 0 + 5 = 5$

$(2, -3): 2 - (-3) = 2 + 3 = 5$

$(3, -2): 3 - (-2) = 3 + 2 = 5$

$(4, 3): 4 - 3 = 1 \neq 5$

$(7, 2): 7 - 2 = 5$

Solutions are (0, –5), (2, –3), (3, –2), (7, 2).

3. $5x - 3y = 15$

First, we find the y-intercept. If $x = 0$, $y = -5$, so the y-intercept is –5. Next we find the x-intercept. If $y = 0$, $x = 3$, so the x-intercept is 3. Using these intercepts, we graph the line.

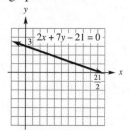

4. $2x + 7y - 21 = 0$

First we find the y-intercept. If $x = 0$, $y = 3$, so the y-intercept is 3. Next we find the x-intercept. If $y = 0$, $x = \dfrac{21}{2}$, so the x-intercept is $\dfrac{21}{2}$. Using these intercepts, we graph the line.

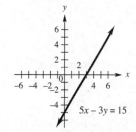

5. $y + 3 = 0$

The equation may be rewritten as $y = -3$. The graph of $y = -3$ is a horizontal line with y-intercept of –3.

6. $y - 2x = 0$

First, we find the y-intercept. If $x = 0$, $y = 0$, so the y-intercept is 0. Since the line passes through the origin, the x-intercept is also 0. We find another point on the line by arbitrarily choosing a value for x. Let $x = 2$. Then $y - 2(2) = 0$, or $y = 4$. The point with coordinates $(2, 4)$ is on the line. Using this point and the origin, we graph the line.

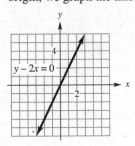

7. $y = .25x^2 + 1$

First we find the y-intercept. If $x = 0$, $y = .25(0)^2 + 1 = 1$, so the y-intercept is 1. Next we find the x-intercepts. If $y = 0$, $0 = .25x^2 + 1 \Rightarrow .25x^2 = -1 \Rightarrow x = \sqrt{-4}$, not a real number. There are no x-intercepts. Make a table of points and plot them.

x	$.25x^2 + 1$
-4	5
-2	2
0	1
2	2
4	5

8. $y = \sqrt{x + 4}$

Make a table of points and plot them.

x	$\sqrt{x + 4}$
-4	0
-3	1
0	2
5	3

9. a. The temperature was over 55° from about 11:30 A.M. to about 7:30 P.M.

b. The temperature was below 40° from midnight until about 5 A.M., and after about 10:30 P.M.

10. At noon in Bratenahl the temperature was about 57°. The temperature in Greenville is 57° when the temperature in Bratenahl is 50°, or at about 10:30 A.M. and 8:30 P.M.

11. Answers vary. A possible answer is "rise over run".

12. Through $(-1, 3)$ and $(2, 6)$

$$\text{slope} = \frac{\Delta y}{\Delta x} = \frac{6 - 3}{2 - (-1)} = \frac{3}{3} = 1$$

13. Through $(4, -5)$ and $(1, 4)$

$$\text{slope} = \frac{-5 - 4}{4 - 1} = \frac{-9}{3} = -3$$

14. Through $(8, -3)$ and the origin

The coordinates of the origin are $(0, 0)$.

$$\text{slope} = \frac{-3 - 0}{8 - 0} = -\frac{3}{8}$$

15. Through $(8, 2)$ and $(0, 4)$

$$\text{slope} = \frac{4 - 2}{0 - 8} = \frac{2}{-8} = -\frac{1}{4}$$

In exercises 16 and 17, we give the solution by rewriting the equation in slope-intercept form. Alternatively, the solution can be obtained by determining two points on the line and then using the definition of slope.

16. $3x + 5y = 25$

First we solve for y.

$$5y = -3x + 25 \Rightarrow y = -\frac{3}{5}x + 5$$

When the equation is written in slope-intercept form, the coefficient of x gives the slope. The slope is $-\frac{3}{5}$.

17. $6x - 2y = 7$

First we solve for y.

$$6x - 2y = 7 \Rightarrow 6x - 7 = 2y \Rightarrow 3x - \frac{7}{2} = y$$

The coefficient of x gives the slope, so the slope is 3.

18. $x - 2 = 0$

The graph of $x - 2 = 0$ is a vertical line. Therefore, the slope is undefined.

19. $y = -4$

The graph of $y = -4$ is a horizontal line.
Therefore, the slope is 0.

20. Parallel to $3x + 8y = 0$

First, find the slope of the given line by solving for y.

$$8y = -3x \Rightarrow y = -\frac{3}{8}x$$

The slope is the coefficient of x, $-\frac{3}{8}$. A line parallel to this line has the same slope, so the slope of the parallel line is also $-\frac{3}{8}$.

21. Perpendicular to $x = 3y$

First, find the slope of the given line by solving for y: $y = \frac{1}{3}x$

The slope of this line is the coefficient of x, $\frac{1}{3}$.

The slope of a line perpendicular to this line is the negative reciprocal of this slope, so the slope of the perpendicular line is –3.

22. Through (0, 5) with $m = -\frac{2}{3}$

Since $m = -\frac{2}{3} = \frac{-2}{3}$, we start at the point with coordinates (0, 5) and move 2 units down and 3 units to the right to obtain a second point on the line. Using these two points, we graph the line.

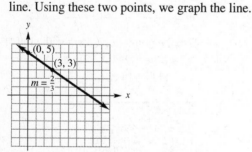

23. Through (–4, 1) with m = 3

Since $m = 3 = \frac{3}{1}$, we start at the point with coordinates (–4, 1) and move 3 units up and 1 unit to the right to obtain a second point on the line. Using these two points, we graph the line.

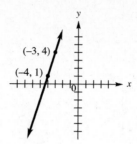

24. Answers vary. One example is:
You need two points; one point and the slope; the y-intercept and the slope.

25. Through (5, –1), slope $\frac{2}{3}$

Use the point slope form with $x_1 = 5$, $y_1 = -1$, and $m = \frac{2}{3}$.

$$y - y_1 = m(x - x_1)$$
$$y - (-1) = \frac{2}{3}(x - 5)$$
$$y + 1 = \frac{2}{3}x - \frac{10}{3}$$

Multiplying by 3 gives
$$3y + 3 = 2x - 10$$
$$3y = 2x - 13$$

26. Through (8, 0), $m = -\frac{1}{4}$

$$y - 0 = -\frac{1}{4}(x - 8)$$
$$4y = -1(x - 8)$$
$$4y = -x + 8$$

27. Through (5, –2) and (1, 3)

$$m = \frac{3 - (-2)}{1 - 5} = \frac{5}{-4} = -\frac{5}{4}$$
$$y - 3 = -\frac{5}{4}(x - 1)$$
$$4(y - 3) = -5(x - 1)$$
$$4y - 12 = -5x + 5$$
$$4y = -5x + 17$$

28. $(2, -3)$ and $(-3, 4)$

$$m = \frac{-3-4}{2-(-3)} = -\frac{7}{5}$$

$$y - (-3) = -\frac{7}{5}(x-2)$$

$$5(y+3) = -7(x-2)$$

$$5y + 15 = -7x + 14$$

$$5y = -7x - 1$$

29. Undefined slope, through $(-1, 4)$
This is a vertical line. Its equation is $x = -1$.

30. Slope 0, $(-2, 5)$
This is a horizontal line. Its equation is $y = 5$.

31. x-intercept -3, y-intercept 5
Use the points $(-3, 0)$ and $(0, 5)$.

$$m = \frac{5-0}{0-(-3)} = \frac{5}{3}$$

$$y = \frac{5}{3}x + 5$$

$$3y = 5(x+3)$$

$$3y = 5x + 15$$

32. x-intercept 3, y-intercept 2.
Use the points $(3, 0)$ and $(0, 2)$.

$$m = \frac{2-0}{0-3} = -\frac{2}{3}$$

$$y = -\frac{2}{3}x + 2$$

$$3y = 3\left(-\frac{2}{3}x + 2\right)$$

$$3y = -2x + 6$$

$$2x + 3y = 6$$

The answer is (d).

33. a. Let (x_1, y_1) be $(0, 30.2)$ and (x_2, y_2) be $(10, 27.2)$. Find the slope.

$$m = \frac{27.2 - 30.2}{10-0} = \frac{-3}{10} = -.3$$

Since the y-intercept is 30.2, the equation is
$y = -.3x + 30.2$.

b. The slope is negative because the amount of wheat exported is decreasing.

c. The year 2011 corresponds to $x = 16$.

$$y = -.3(16) + 30.2 = 25.4$$

If the linear trend continues, there will be 25.4 million metric tons in wheat exports in 2011.

34. a. Let (x_1, y_1) be $(0, 3.0)$ and (x_2, y_2) be $(6, 2.5)$. Find the slope.

$$m = \frac{3.0 - 2.5}{0-6} = \frac{.5}{-6} \approx -.083$$

Since the y-intercept is 3.0, the equation is
$y \approx -.083x + 3.0$.

b.

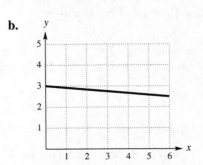

c. The year 2011 corresponds to $x = 11$.

$$y \approx -.083(11) + 3.0 \approx 2.1$$

If the linear trend continues, Canada will produce about 2.1 motor vehicles in 2001.

35. a. Let (x_1, y_1) be $(1, 14.54)$ and (x_2, y_2) be $(6, 16.76)$.
Find the slope.

$$m = \frac{16.76 - 14.54}{6-1} = \frac{2.22}{5} = .444$$

Use the point-slope form with $(1, 14.54)$
$y - 14.54 = .444(x-1) \Rightarrow$
$y = .444x + 14.096$

b.

Using a graphing calculator, the least squares regression line is
$y = .426x + 14.085$.

c. The year 2003 corresponds to $x = 3$. Using the two-point model, we have
$$y = .444(3) + 14.096 \approx 15.43.$$
Using the regression model, we have
$$y = .426(3) + 14.085 \approx 15.36.$$
The two-point model is off by $0.06, while the regression model is off by $0.01.

d. The year 2010 corresponds to $x = 10$. Two-point model:
$$y = .444(10) + 14.096 \approx 18.54$$
Regression model:
$$y = .426(10) + 14.085 \approx 18.35$$
The two-point model predicts that hourly earnings are $18.54 in 2010, while the regression model predicts that hourly earnings are $18.35 in 2010.

36. a. $450 = 11x + 303 \Rightarrow 147 = 11x \Rightarrow 13.4 \approx x$
The weekly median wages for non-union men will earn $450 per week in the year $1996 + 13 = 2009$.

b. $450 = 11.1x + 280 \Rightarrow 170 = 11.1x \Rightarrow$ $15.3 \approx x$
The weekly median wages for non-union women will earn $450 per week in the year $1996 + 15 = 2011$.

37. a.

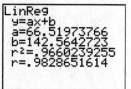

The least-squares regression line is
$$y = 66.52x + 142.6.$$

b. The correlation coefficient is .98. This indicates that the line is a good fit.

c. The year 2012 corresponds to $x = 32$.
$$y = 66.52(32) + 142.6 \approx 2271.24$$
Assuming the trend continues, the total health care expenditures for 2012 will be $2271.2 billion.

38. a.

```
LinReg
y=ax+b
a=84.64166667
b=2132.75
r²=.9874604019
r=.9937104216
```

The least-squares regression line is
$$y = 84.64x + 2132.$$

b. The coefficient of correlation is $r \approx .99$. Since this value is so close to 1, the regression line is a good fit for the data.

c. The year 2011 corresponds to $x = 21$.
$$y = 84.64(21) + 2132 \approx 3909.$$
Assuming the trend continues, the average financial aid award for a student in 2011 will be about $3909.

39.
$$-6x + 3 < 2x$$
$$-6x + 6x + 3 < 2x + 6x$$
$$3 < 8x$$
$$\frac{3}{8} < \frac{8x}{8}$$
$$\frac{3}{8} < x \text{ or } x > \frac{3}{8}$$
The solution is $\left(\frac{3}{8}, \infty\right)$.

40.
$$12z \geq 5z - 7$$
$$12z - 5z \geq 5z - 5z - 7$$
$$7z \geq -7$$
$$\frac{7z}{7} \geq \frac{-7}{7}$$
$$z \geq -1$$
The solution is $[-1, \infty)$.

41.
$$2(3 - 2m) \geq 8m + 3$$
$$6 - 4m \geq 8m + 3$$
$$6 - 4m - 8m \geq 8m - 8m + 3$$
$$6 - 12m \geq 3$$
$$6 - 6 - 12m \geq 3 - 6$$
$$-12m \geq -3$$
$$\frac{-12m}{-12} \leq \frac{-3}{-12} \Rightarrow m \leq \frac{1}{4}$$
The solution is $\left(-\infty, \frac{1}{4}\right]$.

42. $6p - 5 > -(2p + 3)$

$6p - 5 > -2p - 3$

$8p - 5 > -3$

$8p > 2$

$\dfrac{8p}{8} > \dfrac{2}{8} \Rightarrow p > \dfrac{1}{4}$

The solution is $\left(\dfrac{1}{4}, \infty \right)$.

43. $-3 \le 4x - 1 \le 7$

$-2 \le 4x \le 8$

$-\dfrac{1}{2} \le x \le 2$

The solution is $\left[-\dfrac{1}{2}, 2 \right]$.

44. $0 \le 3 - 2a \le 15$

$0 - 3 \le 3 - 3 - 2a \le 15 - 3$

$-3 \le -2a \le 12$

$\dfrac{-3}{-2} \ge \dfrac{-2a}{-2} \ge \dfrac{12}{-2}$

$\dfrac{3}{2} \ge a \ge -6$

The solution is $\left[-6, \dfrac{3}{2} \right]$.

45. $|b| \le 8 \Rightarrow -8 \le b \le 8$

The solution is $[-8, 8]$.

46. $|a| > 7 \Rightarrow a < -7$ or $a > 7$

The solution is $(-\infty, -7)$ or $(7, \infty)$.

47. $|2x - 7| \ge 3$

$2x - 7 \le -3$ or $2x - 7 \ge 3$

$2x \le 4$ or $2x \ge 10$

$x \le 2$ or $x \ge 5$

The solution is $(-\infty, 2]$ or $[5, \infty)$.

48. $|4m + 9| \le 16$

$-16 \le 4m + 9 \le 16$

$-25 \le 4m \le 7$

$-\dfrac{25}{4} \le m \le \dfrac{7}{4}$

The solution is $\left[-\dfrac{25}{4}, \dfrac{7}{4} \right]$.

49. $|5k + 2| - 3 \le 4$

$|5k + 2| \le 7$

$-7 \le 5k + 2 \le 7$

$-9 \le 5k \le 5$

$-\dfrac{9}{5} \le k \le 1$

The solution is $\left[-\dfrac{9}{5}, 1 \right]$.

50. $|3z - 5| + 2 \ge 10$

$|3x - 5| \ge 8$

$3z - 5 \le -8$ or $3z - 5 \ge 8$

$3z \le -3$ or $3z \ge 13$

$z \le -1$ or $z \ge \dfrac{13}{3}$

The solution is $(-\infty, -1]$ or $\left[\dfrac{13}{3}, \infty \right)$.

51. The inequalities that represent the weight of pumpkin that he will not use are $x < 2$ or $x > 10$. This is equivalent to the following inequalities:

$x - 6 < 2 - 6$ or $x - 6 > 10 - 6$

$x - 6 < -4$ or $x - 6 > 4$

$|x - 6| > 4$

Choose answer option (d).

52. Let x = the price of the snow thrower

$|x - 600| \le 55$

53. a. Let (x_1, y_1) be $(0, 492)$ and (x_2, y_2) be $(6, 651)$. Find the slope

$m = \dfrac{651 - 492}{6 - 0} = \dfrac{159}{6} = 26.5$

Since the y-intercept is 492, the equation is $y = 26.5x + 492$.

b. $26.5x + 492 > 550 \Rightarrow 26.5x > 58 \Rightarrow$

$x > 2.19$

Assuming the linear trend continues, the number of barrels exceeds 550 million sometime during 2002 and after.

c. $26.5x + 492 > 700 \Rightarrow 26.5x > 208 \Rightarrow$

$x > 7.85$

Assuming the linear trend continues, the number of barrels exceeds 700 million sometime during 2007 and after.

54. Let m = number of miles driven. The rate for the second rental company is $95 + .2m$. We want to determine when the second company is cheaper than the first.

$125 > 95 + .2m \Rightarrow 30 > .2m \Rightarrow 150 > m$

The second company is cheaper than the first company when the number of miles driven is less than 150.

55. $r^2 + r - 6 < 0$

Solve the corresponding equation.

$r^2 + r - 6 = 0 \Rightarrow (r+3)(r-2) = 0 \Rightarrow$
$r = -3$ or $r = 2$

```
      A           B           C
  ----+----+----+----+----+---->
     -4   -2    0    2    4
```

For region A, test –4:

$(-4)^2 + (-4) - 6 = 6 > 0$.

For region B, test 0:

$0^2 + 0 - 6 = -6 < 0$.

For region C, test 3:

$3^2 + 3 - 6 = 6 > 0$.

The solution is $(-3, 2)$.

56. $y^2 + 4y - 5 \geq 0$

Solve the corresponding equation.

$y^2 + 4y - 5 = 0 \Rightarrow (y+5)(y-1) = 0$
$y = -5$ or $y = 1$

```
      A           B           C
  ----+----+----+----+----+---->
     -6   -4   -2    0    2
```

For region A, test –6:

$(-6)^2 + 4(-6) - 5 = 7 > 0$.

For region B, test 0:

$0^2 + 4(0) - 5 = -5 < 0$.

For region C, test 2:

$2^2 + 4(2) - 5 = 7 > 0$.

Both endpoints are included because the inequality symbol is "$\geq$." The solution is $(-\infty, -5]$ or $[1, \infty)$.

57. $2z^2 + 7z \geq 15$

Solve the corresponding equation.

$$2z^2 + 7z = 15$$
$$2z^2 + 7z - 15 = 0$$
$$(2z - 3)(z + 5) = 0$$
$$z = \frac{3}{2} \text{ or } z = -5$$

These numbers are solutions of the inequality because the inequality symbol is "$\geq$."

```
    A           B           C
 ---+----+----+----+----+---->
   -6   -4   -2    0    2
```

For region A, test –6:

$2(-6)^2 + 7(-6) = 30 > 15$.

For region B, test 0:

$2 \cdot 0^2 + 7 \cdot 0 = 0 < 15$.

For region C, test 2:

$2 \cdot 2^2 + 7 \cdot 2 = 22 > 15$.

The solution is $(-\infty, -5]$ or $\left[\dfrac{3}{2}, \infty\right)$.

58. $3k^2 \leq k + 14$

Solve the corresponding equation.

$$3k^2 = k + 14$$
$$3k^2 - k - 14 = 0$$
$$(3k - 7)(k + 2) = 0$$
$$k = \frac{7}{3} \text{ or } k = -2$$

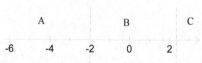

```
        A           B           C
  ----+----+----+----+----+---->
     -6   -4   -2    0    2
```

For region A, test –3:

$3(-3)^2 = 27 \Rightarrow -3 + 14 = 11 \Rightarrow 27 > 11$

For region B, test 0:

$3(0)^2 = 0 \Rightarrow 0 + 14 = 14 \Rightarrow 0 < 14$

For region C, test 3:

$3(3)^2 = 27 \Rightarrow 3 + 14 = 17 \Rightarrow 27 > 17$

The given inequality is true in region B and at both endpoints, so the solution is $\left[-2, \dfrac{7}{3}\right]$.

59. $(x-3)\left(x^2+7x+10\right) \le 0$

Solve the corresponding equation.

$(x-3)\left(x^2+7x+10\right) = 0$

$(x-3)(x+2)(x+5) = 0$

$x-3 = 0$ or $x+2 = 0$ or $x+5 = 0$

   $x = 3$ or $x = -2$ or $x = -5$

Note that -5, -2, and 3 are solutions of the original inequality.

In region A, let $x = -6$:

$(-6-3)\left((-6)^2+7(-6)+10\right)$

$= -9(36-42+10) = -9(4) = -36 < 0$

In region B, let $x = -3$:

$(-3-3)\left((-3)^2+7(-3)+10\right)$

$= -6(9-21+10) = -6(-2) = 12 > 0$.

In region C, let $x = 0$:

$(0-3)\left(0^2+7(0)+10\right) = -3(10) = -30 < 0$.

In region D, let $x = 4$:

$(4-3)\left(4^2+7(4)+10\right)$

$= 1(16+28+10) = 54 > 0$.

The solution is $(-\infty, -5]$ or $[-2, 3]$.

60. $(x+4)\left(x^2-1\right) \ge 0$

Solve the corresponding equation.

$(x+4)\left(x^2-1\right) = 0$

$(x+4)(x+1)(x-1) = 0$

$x+4 = 0$ or $x+1 = 0$ or $x-1 = 0$

   $x = -4$ or $x = -1$ or $x = 1$

Note that -4, -1, and 1 are solutions of the original inequality.

In region A, let $x = -5$:

$(-5+4)\left((-5)^2-1\right) = -1(24) = -24 \le 0$.

In region B, let $x = -2$:

$(-2+4)\left((-2)^2-1\right) = 2(3) = 6 > 0$.

In region C, let $x = 0$:

$(0+4)\left(0^2-1\right) = 4(-1) = -4 < 0$.

In region D, let $x = 2$:

$(2+4)\left(2^2-1\right) = 6(3) = 18 > 0$.

The solution is $[-4, -1]$ or $[1, \infty)$.

61. $\dfrac{m+2}{m} \le 0$

Solve the corresponding equation $\dfrac{m+2}{m} = 0$.

The quotient changes sign when

$m+2 = 0$ or $m = 0$

   $m = -2$ or $m = 0$

-2 is a solution of the inequality, but the inequality is undefined when $m = 0$, so the endpoint 0 must be excluded.

For region A, test -3: $\dfrac{-3+2}{-3} = \dfrac{1}{3} > 0$.

For region B, test -1: $\dfrac{-1+2}{-1} = -1 < 0$.

For region C, test 1: $\dfrac{1+2}{1} = 3 > 0$.

The solution is $[-2, 0)$.

62. $\dfrac{q-4}{q+3} > 0$

Solve the corresponding equation $\dfrac{q-4}{q+3} = 0$.

The numerator is 0 when $q = 4$. The denominator is 0 when $q = -3$.

For region A, test -4: $\dfrac{-4-4}{-4+3} = \dfrac{-8}{-1} = 8 > 0$.

For region B, test 0: $\dfrac{0-4}{0+3} = -\dfrac{4}{3} < 0$.

For region C, test 5: $\dfrac{5-4}{5+3} = \dfrac{1}{8} > 0$.

The inequality is true in regions A and C, and both endpoints are excluded. Therefore, the solution is $(-\infty, -3)$ or $(4, \infty)$.

63. $\dfrac{5}{p+1} > 2$

Solve the corresponding equation.

$$\frac{5}{p+1} = 2 \Rightarrow \frac{5}{p+1} - 2 = 0 \Rightarrow$$

$$\frac{5-2(p+1)}{p+1} = 0 \Rightarrow \frac{3-2p}{p+1} = 0$$

The numerator is 0 when $p = \frac{3}{2}$. The denominator is 0 when $p = -1$.

Neither of these numbers is a solution of the inequality.

```
        A         B         C
  ┼─────┼────┼────┼────┼────►
 -4    -2    0    2    4
```

In region A, test -2: $\dfrac{5}{-2+1} = -5 < 2$.

In region B, test 0: $\dfrac{5}{0+1} = 5 > 2$.

In region C, test 2: $\dfrac{5}{2+1} = \dfrac{5}{3} < 2$.

The solution is $\left(-1, \dfrac{3}{2}\right)$.

64. $\dfrac{6}{a-2} \le -3$

Solve the corresponding equation.

$$\frac{6}{a-2} = -3 \Rightarrow \frac{6}{a-2} + 3 = 0 \Rightarrow$$

$$\frac{6+3(a-2)}{a-2} = 0 \Rightarrow \frac{3a}{a-2} = 0$$

The numerator is 0 when $a = 0$. The denominator is 0 when $a = 2$.

```
        A         B         C
  ┼─────┼────┼────┼────┼────►
 -4    -2    0    2    4
```

For region A, test -1: $\dfrac{6}{-1-2} = -2 \ge -3$.

For region B, test 1: $\dfrac{6}{1-2} = -6 \le -3$.

For region C, test 3: $\dfrac{6}{3-2} = 6 \ge -3$.

The given inequality is true in region B. The endpoint 0 is included because the inequality symbol is "$\le$." However, the endpoint 2 must be excluded because it makes the denominator 0. The solution is [0, 2).

65. $\dfrac{2}{r+5} \le \dfrac{3}{r-2}$

Write the corresponding equation and then set one side equal to zero.

$$\frac{2}{r+5} = \frac{3}{r-2}$$

$$\frac{2}{r+5} - \frac{3}{r-2} = 0$$

$$\frac{2(r-2) - 3(r+5)}{(r+5)(r-2)} = 0$$

$$\frac{2r-4-3r-15}{(r+5)(r-2)} = 0$$

$$\frac{-r-19}{(r+5)(r-2)} = 0$$

The numerator is 0 when $r = 19$. The denominator is 0 when $r = -5$ or $r = 2$.

-19 is a solution of the inequality, but the inequality is undefined when r = -5 or r = 2.

```
    A        B          C     D
  ┼───┼────┼────┼────┼────┼────┼──►
 -20 -16  -12   -8   -4    0    4
```

For region A, test -20:

$$\frac{2}{-20+5} = -\frac{2}{15} \approx -.13 \text{ and}$$

$$\frac{3}{-20-2} = -\frac{3}{22} \approx -.14$$

Since $-.13 > -.14$, -20 is not a solution of the inequality.

For region B, test -6:

$$\frac{2}{-6+5} = -2 \text{ and } \frac{3}{-6-2} = -\frac{3}{8}.$$

Since $-2 < -\dfrac{3}{8}$, -6 is a solution.

For region C, test 0: $\dfrac{2}{0+5} = \dfrac{2}{5}$ and $\dfrac{3}{0-2} = -\dfrac{3}{2}.$

Since $\dfrac{2}{5} > -\dfrac{3}{2}$, 0 is not a solution.

For region D, test 3: $\dfrac{2}{3+5} = \dfrac{1}{4}$ and $\dfrac{3}{3-2} = 3.$

Since $\dfrac{1}{4} < 3$, 3 is a solution. The solution is $[-19, -5)$ or $(2, \infty)$.

66. $\dfrac{1}{z-1} > \dfrac{2}{z+1}$

Write the corresponding equation and then set one side equal to zero.

$$\frac{1}{z-1} = \frac{2}{z+1}$$

$$\frac{1}{z-1} - \frac{2}{z+1} = 0$$

$$\frac{(z+1) - 2(z-1)}{(z-1)(z+1)} = 0$$

$$\frac{3-z}{(z-1)(z+1)} = 0$$

The numerator is 0 when $z = 3$. The denominator is 0 when $z = 1$ and when $z = -1$. These three numbers, -1, 1, and 3, separate the number line into four regions.

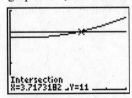

For region A, test -3.

$$\frac{1}{-3-1} > \frac{2}{-3+1} \Rightarrow -\frac{1}{4} > -1, \text{ which is true.}$$

For region B, test 0.

$$\frac{1}{0-1} > \frac{2}{0+1} \Rightarrow -1 > 2, \text{ which is false.}$$

For region C, test 2.

$$\frac{1}{2-1} > \frac{2}{2+1} \Rightarrow 1 > \frac{2}{3}, \text{ which is true.}$$

For region D, test 4.

$$\frac{1}{4-1} > \frac{2}{4+1} \Rightarrow \frac{1}{3} > \frac{2}{5}, \text{ which is false.}$$

Thus, the solution is $(-\infty, -1)$ or $(1, 3)$.

67. $r = 0.145x^2 - .27x + 10.0$

We want to determine when

$0.145x^2 - .27x + 10.0 > 11$ for $0 \le x \le 6$. Using a graphing calculator, plot

$Y_1 = 0.145x^2 - .27x + 10.0$ and $Y_2 = 11$ on $[0, 6]$ by $[0, 15]$. Then determine where the graph of Y_1 lies above the graph of Y_2.

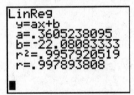

The revenue exceed 11 billion from sometime in 2003 through 2006.

Case 2 Using Extrapolation for Prediction

1. Since $x = 0$ corresponds to the year 1900, enter the following data into a computing device.

x	y
70	3.40
75	4.73
80	6.85
85	8.74
90	10.20
95	11.65
100	14.02
105	16.13

Then determine the least squares regression line.

```
LinReg
 y=ax+b
 a=.3605238095
 b=-22.08083333
 r²=.9957920519
 r=.997893808
■
```

The model is verified.

2. The year 2002 corresponds to $x = 102$.

$y = .361(102) - 22.08 \approx 14.74$.

According to the model, the hourly wage in 2002 was about \$14.74, about 23¢ too low.

3. The year 1960 corresponds to $x = 60$.

$y = .361(60) - 22.08 \approx -0.42$.

The model gives the hourly wage as a negative amount, which is clearly not appropriate.

4.

Year ($x = 0$ is 1900)	Table value	Predicted value	Residual
70	3.40	3.19	.21
75	4.73	4.995	−.265
80	6.85	6.8	.05
85	8.74	8.605	.135
90	10.20	10.41	−.21
95	11.65	12.215	−.565
100	14.02	14.02	0
105	16.13	15.825	.305

(*continued on next page*)

(*continued from page 60*)

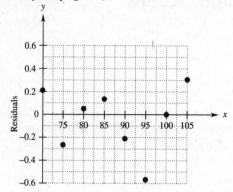

5. You'll get 0 slope and 0 intercept, because the residual represents the vertical distance from the data point to the regression line. Since r is very close to 1, the data points lie very close to the regression line.

```
LinReg
 y=ax+b
 a=-4.761905E-4
 b=-8.333333E-4
 r²=4.126792E-4
 r=-.020314506
```

Chapter 3 Functions and Graphs

Section 3.1 Functions

1.

x	3	2	1	0	–1	–2	–3
y	9	4	1	0	1	4	9

This rule defines y as a function of x because each value of x determines one and only one value of y.

3. The rule $y = x^3$ defines y as a function of x because each value of x determines one and only one value of y.

5. The rule $x = |y + 2|$ does not define y as a function of x because some values of x determine two values for y. For example, if $x = 4$, $4 = |y + 2| \Rightarrow$

$$y + 2 = 4 \quad \text{or} \quad y + 2 = -4$$
$$y = 2 \quad \text{or} \quad y = -6$$

7. The rule $y = \dfrac{-1}{x-1}$ defines y as a function of x because each value of x determines one and only one value of y.

9. $f(x) = 4x - 1$

The domain of f is all real numbers since x may take on any real-number value. Therefore, the domain is $(-\infty, \infty)$.

11. $f(x) = x^4 - 1$

The domain of f is all real numbers since x may take on any real-number value. Therefore, the domain is $(-\infty, \infty)$.

13. $f(x) = \sqrt{-x} + 3$

In order to have $\sqrt{-x}$ be a real number, we must have $-x \geq 0$ or $x \leq 0$. Thus, the domain is all nonpositive real numbers, or $(-\infty, 0]$.

15. $g(x) = \dfrac{1}{x-2}$

Since the denominator cannot be zero, $x \neq 2$. Thus, the domain is all real numbers except 2, which in interval notation is written $(-\infty, 2)$ or $(2, \infty)$.

17. $g(x) = \dfrac{x^2 + 4}{x^2 - 4}$

Solve $x^2 - 4 = 0$ and exclude the solutions from the domain because the solutions make the denominator equal to 0.

$$x^2 - 4 = 0$$
$$(x - 2)(x + 2) = 0$$
$$x - 2 = 0 \quad \text{or} \quad x + 2 = 0$$
$$x = 2 \qquad \text{or} \quad x = -2$$

The domain of gf is all real numbers except $x = \pm 2$ since x cannot take on $x = \pm 2$. Therefore, the domain is $(-\infty, -2)$ or $(-2, 2)$ or $(2, \infty)$.

19. $h(x) = \dfrac{\sqrt{x+4}}{x^2 + x - 12}$

Solve $x^2 + x - 12 = 0$ and exclude the solutions from the domain since x cannot take on these numbers.

$$x^2 + x - 12 = 0$$
$$(x - 3)(x + 4) = 0$$
$$x - 3 = 0 \quad \text{or} \quad x + 4 = 0$$
$$x = 3 \quad \text{or} \quad x = -4$$

For $\sqrt{x+4}$, $x + 4$ must be positive or 0. That is $x + 4 > 0$ or $x > -4$. Therefore, the domain is x such that $x \neq 3$ and $x > -4$.

21. $g(x) = \begin{cases} \dfrac{1}{x} & \text{if } x < 0 \\ \sqrt{x^2 + 1} & \text{if } x \geq 0 \end{cases}$

For $\dfrac{1}{x}$, if $x < 0$, x can take on any real number.

For $\sqrt{x^2 + 1}$, x can take on any real number. The domain is all real numbers or $(-\infty, \infty)$.

23. $f(x) = 8$

For any value of x, the value of $f(x)$ will always be 8. (This is a constant function).

  a. $f(4) = 8$

  b. $f(-3) = 8$

  c. $f(2.7) = 8$

  d. $f(-4.9) = 8$

25. $f(x) = 2x^2 + 4x$

    a. $f(4) = 2(4^2) + 4(4)$
$$= 2(16) + 16$$
$$= 32 + 16 = 48$$

    b. $f(-3) = 2(-3)^2 + 4(-3)$
$$= 2(9) + (-12)$$
$$= 18 - 12 = 6$$

    c. $f(2.7) = 2(2.7)^2 + 4(2.7)$
$$= 2(7.29) + 10.8$$
$$= 14.58 + 10.8 = 25.38$$

    d. $f(-4.9) = 2(-4.9)^2 + 4(-4.9)$
$$= 2(24.01) + (-19.6)$$
$$= 48.02 - 19.6 = 28.42$$

27. $f(x) = \sqrt{x+3}$

    a. $f(4) = \sqrt{4+3} = \sqrt{7} \approx 2.6458$

    b. $f(-3) = \sqrt{-3+3} = \sqrt{0} = 0$

    c. $f(2.7) = \sqrt{2.7+3} = \sqrt{5.7} \approx 2.3875$

    d. $f(-4.9) = \sqrt{-4.9+3} = \sqrt{-1.9}$ not defined

29. $f(x) = \left| x^2 - 6x - 4 \right|$

    a. $f(4) = \left| 4^2 - 6(4) - 4 \right|$
$$f(4) = \left| 16 - 24 - 4 \right| = \left| -12 \right| = 12$$

    b. $f(-3) = \left| (-3)^2 - 6(-3) - 4 \right|$
$$= \left| 9 + 18 - 4 \right| = 23$$

    c. $f(2.7) = \left| (2.7)^2 - 6(2.7) - 4 \right|$
$$= \left| 7.29 - 16.2 - 4 \right| = 12.91$$

    d. $f(-4.9) = \left| (-4.9)^2 - 6(-4.9) - 4 \right|$
$$= \left| 24.01 + 29.4 - 4 \right|$$
$$= \left| 49.41 \right| = 49.41$$

31. $f(x) = \dfrac{\sqrt{x-1}}{x^2 - 1}$

    a. $f(4) = \dfrac{\sqrt{4-1}}{4^2 - 1} = \dfrac{\sqrt{3}}{15} \approx .1155$

    b. $f(-3) = \dfrac{\sqrt{-3-1}}{(-3)^2 - 1} = \dfrac{\sqrt{-4}}{8}$ not defined

    c. $f(2.7) = \dfrac{\sqrt{2.7-1}}{2.7^2 - 1} = \dfrac{\sqrt{1.7}}{6.29} \approx .2073$

    d. $f(-4.9) = \dfrac{\sqrt{-4.9-1}}{(-4.9)^2 - 1}$
$$= \dfrac{\sqrt{-5.9}}{23.01}$$ not defined

33. $f(x) = \begin{cases} x^2 & \text{if } x < 2 \\ 5x - 7 & \text{if } x \geq 2 \end{cases}$

    a. $f(4) = 5(4) - 7 = 13$

    b. $f(-3) = (-3)^2 = 9$

    c. $f(2.7) = 5(2.7) - 7 = 6.5$

    d. $f(-4.9) = (-4.9)^2 = 24.01$

35. $f(x) = 6 - x$

    a. $f(p) = 6 - p$

    b. $f(-r) = 6 - (-r)$
$$= 6 + r$$

    c. $f(m+3) = 6 - (m+3)$
$$= 6 - m - 3$$
$$= 3 - m$$

37. $f(x) = \sqrt{4-x}$

    a. $f(p) = \sqrt{4-p}$ $(p \leq 4)$

    b. $f(-r) = \sqrt{4-(-r)}$
$$= \sqrt{4+r}$$ $(r \geq -4)$

    c. $f(m+3) = \sqrt{4-(m+3)}$
$$= \sqrt{4-m-3}$$
$$= \sqrt{1-m}$$ $(m \leq 1)$

39. $f(x) = x^3 + 1$

   a. $f(p) = p^3 + 1$

   b. $f(-r) = (-r)^3 + 1 = -r^3 + 1$

   c. $f(m+3) = (m+3)^3 + 1$
$$= m^3 + 9m^2 + 27m + 27 + 1$$
$$= m^3 + 9m^2 + 27m + 28$$

41. $f(x) = \dfrac{3}{x-1}$

   a. $f(p) = \dfrac{3}{p-1} \ (p \neq 1)$

   b. $f(-r) = \dfrac{3}{-r-1}$ or $-\dfrac{3}{r+1} \ (r \neq -1)$

   c. $f(m+3) = \dfrac{3}{(m+3)-1} = \dfrac{3}{m+2} \ (m \neq -2)$

43. $f(x) = 2x - 4$
$$\frac{f(x+h) - f(x)}{h} = \frac{[2(x+h) - 4] - (2x-4)}{h}$$
$$= \frac{2x + 2h - 4 - 2x + 4}{h}$$
$$= \frac{2h}{h} = 2$$

45. $f(x) = x^2 + 1$
$$\frac{f(x+h) - f(x)}{h} = \frac{(x+h)^2 + 1 - \left(x^2 + 1\right)}{h}$$
$$= \frac{x^2 + 2hx + h^2 + 1 - x^2 - 1}{h}$$
$$= \frac{2hx + h^2}{h} = 2x + h$$

47. $g(x) = 3x^4 - x^3 + 2x$

X	Y1
3.5	414.31
3.9	642.51
4.3	954.73
4.7	1369.5
5.1	1907.1
5.5	2589.8

Y1■3X^4-X³+2X

49. $T(x) =$
$$\begin{cases} .0535x & \text{if } 0 \leq x \leq 21{,}310 \\ 1140 + .0705(x - 21{,}310) \\ & \text{if } 21{,}310 < x \leq 69{,}990 \\ 4572 + .0785(x - 69{,}990) \\ & \text{if } x > 69{,}990 \end{cases}$$

   a. $.0535(17{,}800) = \$952.30$

   b. $1140 + .0705(58{,}872 - 21{,}310) = \$3788.12$

   c. $4572 + .0785(115{,}412 - 69{,}990)$
$$= \$8137.63$$

51. $R(x) = -.042x^4 + 2.10x^3 - 38.2x^2$
$$+ 302.02x - 846.2$$

   a. The year 2000 corresponds to $x = 10$.
$R(10) =$
$$-.042(10)^4 + 2.10(10)^3 - 38.2(10)^2$$
$$+ 302.02(10) - 846.2 = 34$$
In 2000, the revenue was \$34 billion.

   b. The year 2007 corresponds to $x = 17$.
$R(17) =$
$$-.042(17)^4 + 2.10(17)^3 - 38.2(17)^2$$
$$+ 302.02(17) - 846.2 \approx 57.76$$
In 2007, the revenue was about \$57.76 billion.

53. $g(x) = .011x^2 + 1.05x + 22.8$

   a. The year 2005 corresponds to $x = 5$.
$$g(5) = .011(5)^2 + 1.05(5) + 22.8 \approx 28.3$$
In 2005, about 28.3 million cargo ton miles were transported.

   b. The year 2012 corresponds to $x = 12$.
$$g(12) = .011(12)^2 + 1.05(12) + 22.8 \approx 37.0$$
If the trend continues, in 2012 about 37.0 million cargo ton miles will be transported.

55. $f(t) = 2050 - 500t$

57. $T(x) = .03x$

59. a. The equation for the area of the square is
$$y = x^2$$

b. By the Pythagorean Theorem, $x^2 + x^2 = d^2$.

Solve for x^2: $2x^2 = d^2 \Rightarrow x^2 = \dfrac{d^2}{2}$

Replace x^2 by $\dfrac{d^2}{2}$ in the formula of part a.

$y = \dfrac{d^2}{2}$.

61. $f(26) = .0005456(26)^3 - .014(26)^2$
$$+ .932(26) + 5.703$$
$$\approx 30.060$$
30.060 thousand = 30, 060
There were about 30,060 emergency-room physicians working in 2006.

Section 3.2 Graphs of Functions

1. $f(x) = -.5x + 2$
The graph is a straight line with slope $-.5$ and y-intercept 2.

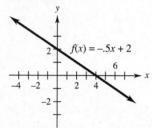

3. $f(x) = \begin{cases} x+3 & \text{if } x \le 1 \\ 4 & \text{if } x > 1 \end{cases}$

Graph the line $y = x + 3$ for $x \le 1$.
Graph the horizontal line $y = 4$ for $x > 1$.

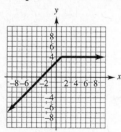

$f(x) = \begin{cases} x+3 & \text{if } x \ 1 \\ 4 & \text{if } x > 1 \end{cases}$

5. $y = \begin{cases} 4-x & \text{if } x \le 0 \\ 3x+4 & \text{if } x > 0 \end{cases}$

Graph the line $y = 4 - x$ for $x \le 0$.
Graph the line $y = 3x + 3$ for $x > 0$.

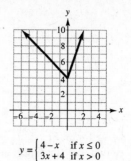

$y = \begin{cases} 4-x & \text{if } x \le 0 \\ 3x+4 & \text{if } x > 0 \end{cases}$

7. $f(x) = \begin{cases} |x| & \text{if } x < 2 \\ -2x & \text{if } x \ge 2 \end{cases}$

Rewrite the function as

$f(x) = \begin{cases} -x & \text{if } x < 0 \\ x & \text{if } 0 \le x < 2 \\ -2x & \text{if } x \ge 2 \end{cases}$

Graph the line $y = -x$ for $x \le 0$.
Graph the line $y = x$ for $0 < x \le 2$.
Graph the line $y = -2x$ for $x \ge 2$.

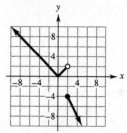

$f(x) = \begin{cases} |x| & \text{if } x < 2 \\ -2x & \text{if } x \ge 2 \end{cases}$

9. $f(x) = |x - 4|$

Using the definition of absolute values gives

$f(x) = \begin{cases} x-4 & \text{if } x-4 \ge 0 \\ -(x-4) & \text{if } x-4 < 0 \end{cases}$ or

$f(x) = \begin{cases} x-4 & \text{if } x \ge 4 \\ -x+4 & \text{if } x < 4 \end{cases}$

We graph the line $y = x - 4$ with slope 1 and y-intercept -4 for $x \ge 4$. We graph the line $y = -x + 4$ with slope -1 and y-intercept 4 for $x < 4$. Note that these partial lines meet at the point $(4, 0)$.

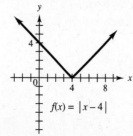

$f(x) = |x - 4|$

11. $f(x) = |3 - 3x|$

$$f(x) = \begin{cases} 3 - 3x & \text{if } 3 - 3x \geq 0 \\ -(3 - 3x) & \text{if } 3 - 3x < 0 \end{cases} \text{ or}$$

$$f(x) = \begin{cases} 3 - 3x & \text{if } x \leq 1 \\ -3 + 3x & \text{if } x > 1 \end{cases}$$

Graph the line $y = 3 - 3x$ for $x \leq 1$.
Graph the line $y = 3x - 3$ for $x > 1$.
The graph consists of two lines that meet at $(1, 0)$.

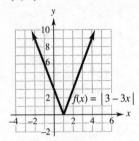

13. $y = -|x - 1|$

$$y = \begin{cases} -(x - 1) & \text{if } x - 1 \geq 0 \\ -[-(x - 1)] & \text{if } x - 1 < 0 \end{cases} \text{ or}$$

$$y = \begin{cases} -x + 1 & \text{if } x \geq 1 \\ x - 1 & \text{if } x < 1 \end{cases}$$

Graph the line $y = -x + 1$ for $x \geq 1$.
Graph the line $y = x - 1$ for $x < 1$.
The graph consists of two lines that meet at $(1, 0)$.

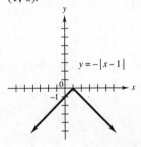

15. $y = |x - 2| + 3$

$$y = \begin{cases} x + 1 & \text{if } x \geq 2 \\ -x + 5 & \text{if } x < 2 \end{cases}$$

Graph the line $y = x + 1$ for $x \geq 2$.
Graph the line $y = -x + 5$ for $x < 2$.
The graph consists of two lines that meet at $(2, 3)$.

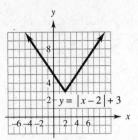

17. $f(x) = [x - 3]$

For x in the interval $[0, 1)$, the value of $[x - 3] = -3$. For x in the interval $[1, 2)$, the value of $[x - 3] = -2$. For x in the interval $[2, 3)$, the value of $[x - 3] = -1$, and so on. The graph consists of a series of line segments. In each case, the left endpoint is included, and the right endpoint is excluded.

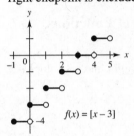

19. $g(x) = [-x]$

For x in the interval $(0, 1]$, the value of $[-x] = -1$. For x in the interval $(1, 2]$, the value of $[-x] = -2$. For x in the interval $(2, 3]$, the value of $[-x] = -3$, and so on. The graph consists of a series of line segments. In each case, the left endpoint is excluded, and the right endpoint is included.

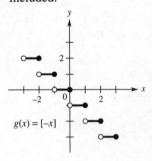

21. $f(x) = [x] + [-x]$

Make a table of intervals and their endpoints:

x	$[x]$	$[-x]$	$f(x)$
-3	-3	3	0
$(-3, -2)$	-3	2	-1
-2	-2	2	0
$(-2, -1)$	-2	1	-1
-1	-1	1	0
$(-1, 0)$	-1	0	-1
0	0	0	0
$(0, 1)$	0	-1	-1
1	1	-1	0
$(1, 2)$	1	-2	-1
2	2	-2	0
$(2, 3)$	2	-3	-1
3	3	-3	0
$(3, 4)$	3	-4	-1

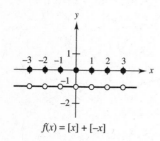

$f(x) = [x] + [-x]$

23. $f(x) = 3 - 2x^2$

Make a table of values and plot the corresponding points.

x	$f(x) = 3 - 2x^2$
-3	-15
-2	-5
-1	1
0	3
1	1
2	-5
3	-15

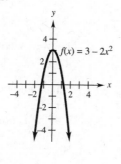

$f(x) = 3 - 2x^2$

25. $h(x) = \dfrac{x^3}{10} + 2$

Make a table of values and plot the corresponding points.

x	$h(x) = \frac{x^3}{10} + 2$
-4	-4.4
-3	-0.7
-2	1.2
-1	1.9
0	2
1	2.1
2	2.8
3	4.7

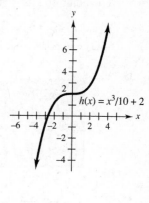

$h(x) = x^3/10 + 2$

27. $g(x) = \sqrt{-x}$

Make a table of values and plot the corresponding points. The function is not defined for $x > 0$.

x	$g(x) = \sqrt{-x}$
-9	3
-4	2
-1	1
0	0

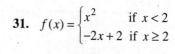

$g(x) = \sqrt{-x}$

29. $f(x) = \sqrt[3]{x}$

Make a table of values and plot the corresponding points.

x	$f(x) = \sqrt[3]{x}$
-8	-2
-1	-1
0	0
1	1
8	2

$f(x) = \sqrt[3]{x}$

31. $f(x) = \begin{cases} x^2 & \text{if } x < 2 \\ -2x + 2 & \text{if } x \geq 2 \end{cases}$

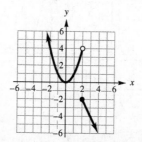

33. Every vertical line intersects this graph in at most one point, so this is the graph of a function.

35. A vertical line intersects the graph in more than one point, so this is not the graph of a function.

37. Every vertical line intersects this graph in at most one point, so this is the graph of a function.

39.

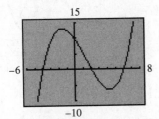

41.

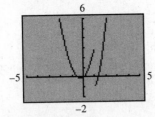

The endpoints are $(1, -1)$, which is on the graph, and $(1, 3)$, which is not on the graph.

43. Draw the graph in the window $[-10, 10]$ by $[-10, 15]$ and locate the x-intercepts.

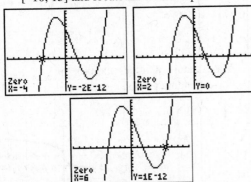

The x-intercepts are $-4, 2, 6$.

45. Draw the graph in the window $[-10, 10]$ by $[-10, 5]$.

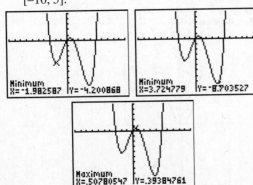

There is a maximum at $(.5078, .3938)$, and there are minima at $(-1.9826, -4.2009)$ and $(3.7248, -8.7035)$.

47. $f(x) = \begin{cases} -50x + 2050 & \text{if } 1 \le x < 22 \\ 84x - 348 & \text{if } 22 \le x < 29 \end{cases}$

a.

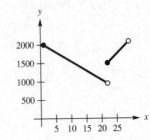

b. When adjusted for inflation, the maximum yearly IRA contribution fell from \$2000 to \$1000 during this period since inflation was increasing during this period.

49. $f(x) = \begin{cases} 18.35 - .18x & \text{if } 0 \le x \le 22 \\ 12.19 + .10x & \text{if } 22 < x \le 44 \end{cases}$

a.

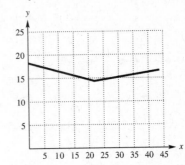

b. The lowest point on the graph occurs when $x = 22$.

$f(22) = 18.35 - .18(22) = 14.39$

The lowest stock price during the period defined by $f(x)$ is \$14.39.

51. a. We must find the equation of the line containing the points $(0, 15.1)$ and $(25, 47.5)$ and the equation of the line containing $(25, 47.5)$ and $(57, 351.05)$.
Line 1:

$m = \dfrac{47.5 - 15.1}{25 - 0} = 1.296 \approx 1.3$

The y-intercept is 15.1, so the equation is $y = 1.3x + 15.1$.

(*continued on next page*)

(continued from page 68)

Line 2:
$$m = \frac{351.05 - 47.5}{57 - 25} \approx 9.5$$
Using the point-slope form, the equation is
$$y - 47.5 = 9.5(x - 25) \Rightarrow$$
$$y - 47.5 = 9.5x - 237.5 \Rightarrow y = 9.5x - 190.$$

$$f(x) = \begin{cases} 1.3x + 15.1 & \text{if } 0 \le x \le 25 \\ 9.5x - 190 & \text{if } x > 25 \end{cases}$$

b.

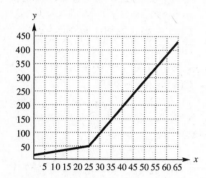

c. The year 2004 corresponds to $x = 54$.
$$f(54) = 9.5(54) - 190 = 323$$
In 2004, the medical-care CPI was 323.

d. The year 2012 corresponds to $x = 62$.
$$f(62) = 9.5(62) - 190 = 399$$
Assuming that this model remains accurate, the medical-care CPI will be 399 in 2012.

53. a. No. The graph of the CPI for all items is always above the x-axis, so the percent of change is always positive; this means that the CPI is always increasing.

b. The CPI for energy was decreasing from early 1997 to mid 1998 and from mid 2001 to mid 2002.

c. The CPI for energy was increasing from 1990 to early 1997; from mid 1998 to mid 2001; and after early 2002. The percentage change was positive during these periods, which means that the CPI was increasing.

55. a.

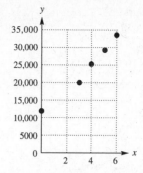

b. Find the slope of the line containing (0, 11999) and (6, 33772).
$$m = \frac{33,772 - 11,999}{6 - 0} \approx 3629.$$
The y-intercept is 11,999, so the equation of the line is $y = 3629x + 11,999$.

c. $f(x) = 3629x + 11,999.$

d. The value $x = 4$ corresponds to the year 2004. According to the table, there were 25,413 loans in 2004. Using the function, $f(4) = 3629(4) + 11,999 = 26,515$, which is close to the actual value.

57. a. According to the graph, $f(2000) = 33$ and $f(2006) = 39$.

b. The figure has vertical line segments, which can't be part of the graph of a function. To make the figure into the graph of f, delete the vertical line segments, then for each horizontal segment of the graph, put a closed dot on the left end and an open-circle dot on the right end.

59. First convert 75 minutes to 1.25 hours, and 15 minutes to .25 hours. Let x represent the number of hours to rent the van. To calculate $C(x)$, subtract 1.25 from x, then divide this number by .25 and round to the next integer. Finally, multiply this result by 5 and add $19.99.

a. 2 hours
$$\frac{2 - 1.25}{.25} = 3$$
$$5(3) + 19.99 = 34.99$$
It costs $34.99 to rent a van for 2 hours.

b. 1.5 hours

$$\frac{1.5 - 1.25}{.25} = 1$$

$$5(1) + 19.99 = 24.99$$

It costs $24.99 to rent a van for 1.5 hours.

c. 3.5 hours

$$\frac{3.5 - 1.25}{.25} = 9$$

$$5(9) + 19.99 = 64.99$$

It costs $64.99 to rent a van for 3.5 hours.

d. 4 hours

$$\frac{4 - 1.25}{.25} = 11$$

$$5(11) + 19.99 = 74.99$$

It costs $74.99 to rent a van for 4 hours.

e.

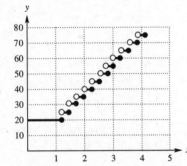

Note that for each step, the left endpoint is not included and the right point is included.

61. Make table of values:

Time	t	$g(t)$	Slope
midnight	0	1,000,000	increasing
noon	12	?	decreasing
4:00 P.M.	16	?	increasing
9:00 P.M.	21	?	vertical

There are many correct answers, including:

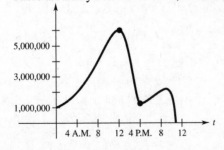

Section 3.3   Applications of Linear Functions

1. The marginal cost is $5, while the fixed cost is $25. Let $C(x)$ be the cost of renting a saw for x hours. Then $C(x) = 5x + 25$.

3. Let x = the number of half-hours and $C(x)$ = the total cost in dollar for x half-hours. Then $C(x) = 2.50x + 8$

5. Fixed cost, $200, 50 items cost $2000 to produce. Since the fixed cost is $200,

$$C(x) = mx + 200 \Rightarrow C(50) = 2000.$$

Therefore,

$$2000 = m(50) + 200 \Rightarrow 50m = 1800 \Rightarrow m = 36$$

$$C(x) = 36x + 200$$

7. Marginal cost, $120; 100 items cost $15,800 to produce.

$$C(x) = mx + b, \ m = 120$$

$$C(100) = 15,800 = 120(100) + b$$

$$b = 15,800 - 12,000 = 3800$$

$$C(x) = 120x + 3800$$

9. $\overline{C}(50) = \dfrac{C(50)}{50} = \dfrac{12(50) + 1800}{50} = 48$

The average cost per item when 50 items are produced is $48.

$\overline{C}(500) = \dfrac{C(500)}{500} = \dfrac{12(500) + 1800}{500} = 15.60$

The average cost per item when 500 items are produced is $15.60.

$\overline{C}(1000) = \dfrac{C(1000)}{1000} = \dfrac{12(1000) + 1800}{1000} = 13.80$

The average cost per item when 1000 items are produced is $13.80.

11. $\overline{C}(200) = \dfrac{C(200)}{200} = \dfrac{6.5(200) + 9800}{200} = 55.50$

The average cost per item when 200 items are produced is $55.50.

$$\overline{C}(2000) = \dfrac{C(2000)}{2000} = \dfrac{6.5(2000) + 9800}{2000}$$

$$= \$11.40$$

The average cost per item when 2000 items are produced is $11.40.

$$\overline{C}(5000) = \dfrac{C(5000)}{5000} = \dfrac{6.5(5000) + 9800}{5000}$$

$$= \$8.46$$

The average cost per item when 5000 items are produced is $8.46.

13. a. Let $(x_1, y_1) = (0, 16615)$ and let
$(x_2, y_2) = (4, 8950)$.

$$m = \frac{16{,}615 - 8950}{0 - 4} \approx -1916$$

The y-intercept is 16,615, so the equation is
$y = -1916x + 16{,}615$.

b. $f(5) = -1916(5) + 16{,}615 = 7035$
The car will be worthe $7035.

c. Because the slope is -1916, the car is
depreciating at a rate of $1916 per year.

15. a. Let $(x_1, y_1) = (0, 120{,}000)$ and
$(x_2, y_2) = (8, 25{,}000)$.

$$m = \frac{25{,}000 - 120{,}000}{8 - 0} = -11{,}875$$
$$y - y_1 = m(x - x_1)$$
$$y - 120{,}000 = -11{,}875(x - 0)$$
$$y - 120{,}000 = -11{,}875x$$
$$y = -11{,}875x + 120{,}000$$
$$f(x) = -11{,}875x + 120{,}000$$

b. The domain ranges from 0 to 8 years, that is
[0, 8].

c. $f(6) = -11{,}875(6) + 120{,}000 = 48{,}750$
The machine will be worth $48,750 in 6
years.

17. a. The fixed costs are
$C(0) = \$42.5(0) + 80{,}000 = \$80{,}000$

b. The slope of $C(x) = 42.5x + 80{,}000$ is 42.5,
so the marginal cost is $42.50.

c. The cost of producing 1000 books is
$C(1000) = 42.5(1000) + 80{,}000 = \$122{,}500$.
The cost of producing 32,000 books is
$C(32{,}000) = 42.5(32{,}000) + 80{,}000$
$= \$1{,}440{,}000$.

d. The average cost per book when 1000 are
produced is:
$$\overline{C}(1000) = \frac{C(1000)}{1000}$$
$$= \frac{42.5(1000) + 80{,}000}{1000} = \$122.50$$

The average cost per book when 32,000 are
produced is,
$$\overline{C}(32{,}000) = \frac{C(32{,}000)}{32{,}000}$$
$$= \frac{42.5(32{,}000) + 80{,}000}{32{,}000} = \$45$$

19. a. Let (x_1, y_1) be $(100, 11.02)$ and (x_2, y_2)
be $(400, 40.12)$.
Find the slope.
$$m = \frac{40.12 - 11.02}{400 - 100} = \frac{29.1}{300} = .097$$
Use the point-slope form with $(100, 11.02)$.
$$y - 11.02 = .097(x - 100)$$
$$y = .097x + 1.32$$
$$C(x) = .097x + 1.32$$

b. The total cost of producing 1000 cups is
$C(1000) = .097(1000) + 1.32 = \$98.32$.

c. The total cost of producing 1001 cups is
$C(1001) = .097(1001) + 1.32 = \$98.42$.

d. The marginal cost of the 1001st cup is
$C(1001) - C(1000) = 98.417 - 98.32 =$
$.097$, which is about 9.7¢.

e. The slope of $C(x) = .097x + 1.32$ is .097, so
the marginal cost is $.097 or 9.7 cents.

21. Each customer pays $7.50 + 1.43x$ per month,
where x is the number of thousands of gallons of
water used. Then, $R(x) = 70{,}000(7.50) + 1.43x$
$= 525{,}000 + 1.43x$.

23. a. $C(x) = 10x + 750$

b. $R(x) = 35x$

c. $P(x) = R(x) - C(x) = 35x - (10x + 750)$
$= 25x - 750$

d. The profit on 100 items is,
$P(100) = 25(100) - 750 = \$1750$

25. a. $C(x) = 18x + 300$

b. $R(x) = 28x$

c. $P(x) = R(x) - C(x) = 28x - (18x + 300)$
$= 10x - 300$

d. The profit on 100 items is,
$P(100) = 10(100) - 300 = \$700$

27. a. $C(x) = 12.5x + 20,000$

b. $R(x) = 30x$

c. $P(x) = R(x) - C(x)$
$= 30x - (12.5x + 20,000)$
$= 17.5x - 20,000$

d. The profit on 100 items is,
$P(100) = 17.5(100) - 20,000 = -\$18,250$ (a loss of \$18,250)

29. $2x - y = 7$ and $y = 8 - 3x$
Solve the first equation for y:
$2x - y = 7 \Rightarrow y = 2x - 7$.
Set the two equations equal and solve for x.
$2x - 7 = 8 - 3x \Rightarrow 5x = 15 \Rightarrow x = 3$
Substitute x into one equation to find y.
$y = 8 - 3(3) = -1$
The lines intersect at $(3, -1)$.

31. $y = 3x - 7$ and $y = 7x + 4$
Set the two equations equal and solve for x.
$$3x - 7 = 7x + 4 \Rightarrow -4x = 11 \Rightarrow x = -\frac{11}{4}$$
Substitute x into one equation to find y.
$$y = 3\left(-\frac{11}{4}\right) - 7 = -\frac{33}{4} - \frac{28}{4} = -\frac{61}{4}$$
The lines intersect at $\left(-\frac{11}{4}, -\frac{61}{4}\right)$.

33. a. The break-even point occurs when $R(x) = C(x)$.
$125x = 100x + 5000 \Rightarrow 25x = 5000 \Rightarrow$
$x = 200$
The break-even point is $x = 200$ or 200,000 policies.

b. To graph the revenue function, graph $y = 125x$. If $x = 0$, $y = 0$. If $x = 20$, $y = 2500$. Use the points $(0, 0)$ and $(20, 2500)$ to graph the line. To graph the cost function, graph $y = 100x + 5000$. If $x = 0$, $y = 5000$. If $x = 30$, $y = 8000$. Use the points $(0, 5000)$ and $(30, 8000)$ to graph the line.

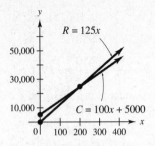

c. From the graph, when $x = 100$, cost is \$15,000 and revenue is \$12,500.

35. Given $R(x) = .21x$ and $P(x) = .084x - 1.5$
a. Cost equals revenue – profit, so
$C(x) = R(x) - P(x) = .21x - (.084x - 1.5)$
$= .21x - .084x + 1.5 = .126x + 1.5$

b. $C(7) = .126(7) + 1.5 = 2.382$
The cost of producing 7 units is \$2.382 million.

c. At the break-even point, profit $P(x) = 0$, so
$C(x) = R(x)$
$.126x + 1.5 = .21x \Rightarrow 1.5 = .084x \Rightarrow$
$x \approx 17.857$
The break-even point occurs at about 17.857 units.

37. $C(x) = 80x + 7000$; $R(x) = 95x$
$95x = 80x + 7000 \Rightarrow 15x = 7000 \Rightarrow x \approx 467$
The break-even point is about 467 units. Do not produce the item since $467 > 400$.

39. $C(x) = 125x + 42,000$; $R(x) = 165.5x$
$125x + 42,000 = 165.5x \Rightarrow 42,000 = 40.5x \Rightarrow$
$x \approx 1037$
The break-even point is about 1037 units. Produce the item since $1037 < 2000$.

41. Workers in both countires had the same compensation, about \$20, in late 2000.

43. a. Domestic production: Let (x_1, y_1) be $(0, 7.4)$ and (x_2, y_2) be $(17, 5.1)$.
$$m = \frac{5.1 - 7.4}{17 - 0} \approx -.14$$
$y = -.14x + 7.4$
Imported oil: Let (x_1, y_1) be $(0, 5.9)$ and (x_2, y_2) be $(17, 10)$.
$$m = \frac{10 - 5.9}{17 - 0} \approx .24$$
$y = .24x + 5.9$

b.

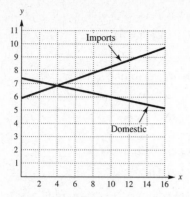

c. The intersection point is at $x = 4$. This means that domestic oil production equaled the amount of imported oil in 1994.

45. On the supply curve, when $q = 20$, $p = 140$. The point $(20, 140)$ is on the graph. When 20 items are supplied, the price is $140.

47. The two curves intersect at the point $(10, 120)$. The equilibrium supply and equilibrium demand are both $q = 10$ or 10 items.

49. $p = 16 - \dfrac{5}{4}q$

  a. If $q = 0$, $p = 16$, so for a demand of 0 units, the price is $16.

  b. If $q = 4$, $p = 16 - \dfrac{5}{4}(4) = 16 - 5 = 11$,

    so for a demand of 4 units, the price is $11.

  c. If $q = 8$, $p = 16 - \dfrac{5}{4}(8) = 16 - 10 = 6$,

    so for a demand of 8 units, the price is $6.

  d. From (c), if $p = 6$, $q = 8$, so at a price of $6, the demand is 8 units.

  e. From (b), if $p = 11$, $q = 4$, so at a price of $11, the demand is 4 units.

  f. From (a), if $p = 16$, $q = 0$, so at a price of $16, the demand is 0 units.

  g.

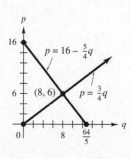

h. $p = \dfrac{3}{4}q$

  If $p = 0$, $q = 0$, so at a price of $0, the supply is 0 units.

i. If $p = 10$,

$$10 = \frac{3}{4}q \Rightarrow \frac{4}{3}(10) = \frac{4}{3}\left(\frac{3}{4}\right)q \Rightarrow \frac{40}{3} = q$$

  When the price is $10, the supply is

  $\dfrac{40}{3}$ units.

j. If $p = 20$,

$$20 = \frac{3}{4}q \Rightarrow \frac{4}{3}(20) = \frac{4}{3}\left(\frac{3}{4}q\right) \Rightarrow \frac{80}{3} = q$$

  When the price is $20, the supply is

  $\dfrac{80}{3}$ units.

k. See(g).

l. The two graphs intersect at the point $(8, 6)$. The equilibrium supply is 8 units.

m. The equilibrium price is $6.

51. a.

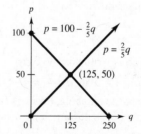

  b. The two graphs intersect at the point $(125, 50)$. The equilibrium demand is 125 units.

  c. The equilibrium price is 50 cents.

  d. $100 - \dfrac{2}{5}q > \dfrac{2}{5}q \Rightarrow 100 > \dfrac{2}{5}q + \dfrac{2}{5}q \Rightarrow$

    $100 > \dfrac{4}{5}q \Rightarrow \dfrac{5}{4}(100) > \dfrac{5}{4}\left(\dfrac{4}{5}q\right) \Rightarrow$

    $125 > q$ or $q < 125$

    Demand exceeds supply when q is in the interval $[0, 125)$.

53. Total cost increases when more items are made (because it includes the cost of all previously made items), so the graph cannot move downward. No; the average cost can decrease as more items are made, so its graph may move downward.

Section 3.4 Quadratic Functions

1. $f(x) = x^2 - 3x - 12$

This function is written in the form

$y = ax^2 + bx + c$ with a = 1, b = −3 and the parabola opens upward since $a = 1 > 0$.

3. $h(x) = -3x^2 + 14x + 1$

The parabola opens downward since $a = -3 < 0$.

5. $g(x) = 2.9x^2 - 12x - 5$

The parabola opens upward since $a = 2.9 > 0$,

7. $f(x) = -2(x - 5)^2 + 7$

This function is written in the form

$y = a(x - h)^2 + k$ with $a = -2$, $h = 5$, and $k = 7$. The vertex of the parabola, (h, k), is $(5, 7)$. The parabola opens downward since $a = -2 < 0$.

9. $h(x) = 4(x + 1)^2 - 9$

This function is written in the form

$y = a(x - h)^2 + k$ with $a = 4$, $h = -1$, and $k = -9$. The vertex of the parabola, (h, k), is $(-1, -9)$. The parabola opens upward since $a = 4 > 0$.

11. $g(x) = 3.2(x - 4.5)^2 + 7.1$

This function is written in the form

$y = a(x - h)^2 + k$ with $a = 3.2$, $h = 4.5$, and $k = 7.1$. The vertex of the parabola, (h, k), is $(4.5, 7.1)$. The parabola opens upward since $a = 3.2 > 0$.

13. I   **15.** K   **17.** J   **19.** F

21. vertex (1, 2); point (5, 6)

$f(x) = a(x - h)^2 + k$

$f(x) = a(x - 1)^2 + 2$

Use (5, 6) to find a.

$6 = a(5 - 1)^2 + 2 \Rightarrow 4 = 16a \Rightarrow a = \dfrac{1}{4}$

$f(x) = \dfrac{1}{4}(x - 1)^2 + 2$

23. vertex (−1, −2); point (1, 2)

$f(x) = a(x - h)^2 + k$

$f(x) = a(x + 1)^2 - 2$

Use (1, 2) to find a.

$2 = a(1 + 1)^2 - 2 \Rightarrow 4 = 4a \Rightarrow a = 1$

$f(x) = (x + 1)^2 - 2$

25. vertex (0, 0); point (2, 12)

$f(x) = a(x - h)^2 + k$

$f(x) = a(x - 0)^2 - 0 \Rightarrow f(x) = ax^2$

Use (2, 12) to find a.

$12 = a(2 - 0)^2 + 0 \Rightarrow 12 = 4a \Rightarrow a = 3$

$f(x) = 3(x - 0)^2 \Rightarrow f(x) = 3x^2$

27. vertex (4, −2); point (6, 6)

$f(x) = a(x - h)^2 + k$

$f(x) = a(x - 4)^2 - 2$

Use (6, 6) to find a.

$6 = a(6 - 4)^2 - 2 \Rightarrow 8 = 4a \Rightarrow a = 2$

$f(x) = 2(x - 4)^2 - 2$

29. $f(x) = -x^2 - 6x + 3$

$a = -1$, $b = -6$

To find the vertex of the function, use the vertex formula $x = -\dfrac{b}{2a}$ and $y = f\left(-\dfrac{b}{2a}\right)$.

$x = -\dfrac{-6}{2(-1)} = -3$

$f(-3) = -(-3)^2 - 6(-3) + 3 = 12$

The vertex is (−3, 12).

31. $f(x) = 3x^2 - 12x + 5$

$a = 3$, $b = -12$

To find the vertex of the function, use the vertex formula $x = -\dfrac{b}{2a}$ and $y = f\left(-\dfrac{b}{2a}\right)$.

$x = -\dfrac{-12}{2(3)} = 2$

$f(2) = 3(2)^2 - 12(2) + 5 = -7$

The vertex is (2, −7).

33. $f(x) = x^2 - 8x + 25$

$a = 1$, $b = -8$

To find the vertex of the function, use the vertex

formula $x = -\dfrac{b}{2a}$ and $y = f\left(-\dfrac{b}{2a}\right)$.

$x = -\dfrac{-8}{2(1)} = 4$

$f(4) = (4)^2 - 8(4) + 25 = 9$

The vertex is (4, 9).

35. $f(x) = 3(x - 2)^2 - 3$

To find the x-intercepts, set $f(x) = 0$ and then solve for x.

$0 = 3(x-2)^2 - 3 \Rightarrow 3(x-2)^2 = 3 \Rightarrow$

$(x-2)^2 = 1 \Rightarrow x - 2 = \pm 1 \Rightarrow x = 1$ or $x = 3$.

The x-intercepts are (1, 0) and (3, 0).

Let $x = 0$ to find the y-intercept.

$f(0) = 3(0-2)^2 - 3 = 9$

The y-intercept is 9.

37. $g(x) = 2x^2 + 8x + 6$

To find the intercepts, set each variable equal to 0. If $x = 0$, $g(0) = 2(0)^2 + 8(0) + 6 = 6$.

The y-intercept is 6.

If $y = 0$, $f(x) = y = 0$, so

$2x^2 + 8x + 6 = 0 \Rightarrow x^2 + 4x + 3 = 0 \Rightarrow$

$(x+3)(x+1) = 0 \Rightarrow x = -3$ or $x = -1$

The x-intercepts are -3 and -1.

39. $f(x) = (x+2)^2 \Rightarrow y = 1(x+2)^2 + 0$

The vertex is $(-2, 0)$. The axis is $x = -2$.

x	-1	0	1
y	1	4	9

Plot these points and use the axis of symmetry to find corresponding points on the other side of the axis. Connect the points with a smooth curve.

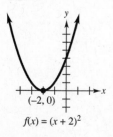

$f(x) = (x + 2)^2$

41. $f(x) = (x-1)^2 - 3$

The vertex is $(1, -3)$. The axis is $x = 1$.

x	2	3	4
y	-2	1	6

Use the axis of symmetry to find corresponding points on the other side of the axis.

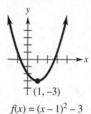

$f(x) = (x - 1)^2 - 3$

In exercises 43–45, we complete the square to write the equation in the form $f(x) = a(x-h)^2 + k$.

Alternatively, we can use the vertex formula,

$h = -\dfrac{b}{2a}$, $k = f(h)$, to find the vertex.

43. $f(x) = x^2 - 4x + 6$

$= \left(x^2 - 4x + 4\right) + (6 - 4)$

$= (x-2)^2 + 2$

The vertex is (2, 2). The axis is $x = 2$.

x	3	4	5
y	3	6	11

Use the axis of symmetry to find corresponding points on the other side of the axis.

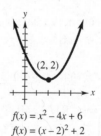

$f(x) = x^2 - 4x + 6$

$f(x) = (x - 2)^2 + 2$

45. $f(x) = 2x^2 - 4x + 5 = \left(2x^2 - 4x\right) + 5$

$= 2\left(x^2 - 2x + 1\right) + (5 - 2)$

$= 2(x-1)^2 + 3$

The vertex is (1, 3). The axis is $x = 1$.

x	2	3	4
y	5	11	21

Use the axis of symmetry to find corresponding points on the other side of the axis.

(continued on next page)

(continued from page 75)

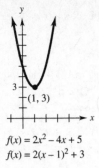

$$f(x) = 2x^2 - 4x + 5$$
$$f(x) = 2(x-1)^2 + 3$$

47. $f(x) = .0328x^2 - 3.55x + 115$

Let x be the age of the driver. Because the graph is a parabola which opens upward, the rate is lowest at the vertex. We must find the x-coordinate of the vertex.

$a = .0328, b = -3.55$

$$x = -\frac{b}{2a} = -\frac{-3.55}{2(.0328)} \approx 54.12$$

The rate is lowest at about age 54.

49. **a.** $y = .057x - .001x^2$

$a = -.001, b = .057$

The number of weeks for splenic artery resistance to reach maximum is when

$$x = -\frac{b}{2a}.$$

$$x = -\frac{.057}{2(-.001)} = 28.5$$

A maximum is reached in 28.5 weeks.

b. The maximum splenic artery resistance is

$$y = .057(28.5) - .001(28.5)^2 = .81225.$$

c. The splenic artery resistance is 0 when $y = 0$. Set $y = 0$ and solve for x.

$$0 = .057x - .001x^2 \Rightarrow x(.057 - .001x) = 0 \Rightarrow$$

$$x = 0 \text{ or } .057 - .001x = 0 \Rightarrow x = \frac{.057}{.001} = 57$$

Explanation varies. Sample answer. No, the answer is not reasonable since babies are born at about 40 weeks of gestation.

51. **a–d.**

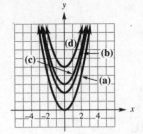

e. The graph of $f(x) = x^2 + c$ is the graph of $k(x) = x^2$ shifted c units upward.

53. **a–d.**

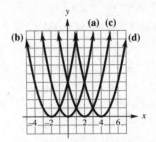

e. The graph of $f(x) = (x+c)^2$ is the graph of $k(x) = x^2$ shifted c units to the left. The graph of $f(x) = (x-c)^2$ is the graph of $g(x) = x^2$ shifted c units to the right.

Section 3.5 Applications of Quadratic Functions

1. **a.** $C(15) = (15)^2 - 40(15) + 405 = 30$

It costs $30 per box to make 15 boxes per day.

$C(18) = (18)^2 - 40(18) + 405 = 9$

It costs $9 per box to make 18 boxes per day.

$C(30) = (30)^2 - 40(30) + 405 = 105$

It cost $105 per box to make 30 boxes per day.

b.

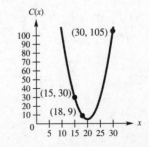

c. The minimum point on the graph corresponds to the number of boxes that will make the cost per box as small as possible.
Complete the square to find the vertex.

$C(x) = x^2 - 40x + 405$

$= (x^2 - 40x + 400 - 400) + 405$

$= (x^2 - 40x + 400) + (-400 + 405)$

$= (x - 20)^2 + 5$

Vertex: (20, 5)

d. She should make 20 boxes per day at a cost of $5 per box.

3. a. $y = -x^2 + 20x - 60 = -\left(x^2 - 20x\right) - 60$

$= -\left(x^2 - 20x + 100\right) - 60 + 100$

$= -(x - 10)^2 + 40$

The graph of this parabola opens downward, so the maximum occurs at the vertex, (10, 40). The maximum firing rate will be reached in 10 milliseconds.

b. When $x = 10$, $y = 40$, so the maximum firing rate is 40 responses/millisec.

5. a. If $P(x) = 0$, then

$0 = -2x^2 + 60x - 120 \Rightarrow 0 = x^2 - 30x + 60$

$x = \dfrac{-(-30) \pm \sqrt{(-30)^2 - 4(1)(60)}}{2(1)}$

$= \dfrac{30 \pm \sqrt{900 - 240}}{2} = \dfrac{30 \pm \sqrt{660}}{2}$

$x = \dfrac{30 + \sqrt{660}}{2} \approx 27.8$ or

$x = \dfrac{30 - \sqrt{660}}{2} \approx 2.2$

Therefore, $P(x) > 0$ if $2.2 < x < 27.8$.
The largest number of cases she can sell and still make a profit is 27.

b. There are many possibilities. For example, she might have to buy additional machinery or pay high overtime wages in order to increase product, thus increasing costs and decreasing profits.

c. Find the vertex to determine the maximum.

$P(x) = -2x^2 + 60x - 120$

$= -2\left(x^2 - 30x\right) - 120$

$= -2\left(x^2 - 30x + 225 - 225\right) - 120$

$= -2\left(x^2 - 30x + 225\right) + 450 - 120$

$= -2(x - 15)^2 + 330$

Since the vertex is at (15, 330), she should sell 15 cases to maximize her profit.

7. Given supply: $p = \dfrac{1}{5}q^2$, and

demand: $p = -\dfrac{1}{5}q^2 + 40$.

a. $10 = -\dfrac{1}{5}q^2 + 40 \Rightarrow -30 = -\dfrac{1}{5}q^2 \Rightarrow$

$150 = q^2 \Rightarrow q \approx 12$

About 12 books are demanded.

b. $20 = -\dfrac{1}{5}q^2 + 40 \Rightarrow -20 = -\dfrac{1}{5}q^2 \Rightarrow$

$100 = q^2 \Rightarrow q = 10$

Ten books are demanded.

c. $30 = -\dfrac{1}{5}q^2 + 40 \Rightarrow -10 = -\dfrac{1}{5}q^2 \Rightarrow$

$50 = q^2 \Rightarrow q \approx 7$

About 7 books are demanded.

d. $40 = -\dfrac{1}{5}q^2 + 40 \Rightarrow 0 = -\dfrac{1}{5}q^2 \Rightarrow q = 0$

No books are demanded.

e. $5 = \dfrac{1}{5}q^2 \Rightarrow 25 = q^2 \Rightarrow 5 = q$

5 books are supplied.

f. $10 = \dfrac{1}{5}q^2 \Rightarrow 50 = q^2 \Rightarrow 7 \approx q$

About 7 books are supplied.

g. $20 = \dfrac{1}{5}q^2 \Rightarrow 100 = q^2 \Rightarrow 10 = q$

10 books are supplied.

h. $30 = \dfrac{1}{5}q^2 \Rightarrow 150 = q^2 \Rightarrow 12 \approx q$

About 12 books are supplied.

i. Using the values from the tables, plot points and graph the functions.

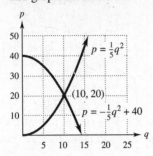

9. a. $p = 640 - 5(0)^2 = 640$

When 0 widgets are demanded, the price is $640.

b. $p = 640 - 5(5)^2 = 515$

When 5 hundred widgets are demanded, the price is $515.

c. $p = 640 - 5(10)^2 = 140$

When 10 hundred widgets are demanded, the price is $140.

d. $p = 640 - 5q^2$

q	0	4	8	10
p	640	560	320	140

$p = 5q^2$

q	0	4	8	10
p	0	80	320	500

Graph these two parabolas on the same axes.

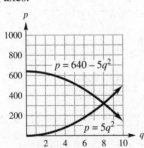

e. supply = demand $\Rightarrow$

$5q^2 = 640 - 5q^2 \Rightarrow 10q^2 = 640 \Rightarrow$

$q^2 = 640 \Rightarrow q = \pm 8$

A negative value of q is not meaningful. The equilibrium supply is 8 hundreds or 800 units.

f. If $q = 8$, $p = 5(8)^2 = 320$.

The equilibrium cost is $320.

11. Set $p = 45q$ and $p = -q^2 + 10,000$ equal.

$45q = -q^2 + 10,000$

Write this quadratic equation in standard form and solve using the quadratic formula.

$q^2 + 45q - 10,000 = 0$

$q = \dfrac{-45 \pm \sqrt{45^2 - 4(1)(-10,000)}}{2(1)}$

$= \dfrac{-45 \pm \sqrt{42025}}{2} = \dfrac{-45 \pm 205}{2}$

$q = \dfrac{-45 + 205}{2} = 80$ or $q = \dfrac{-45 - 205}{2} = -125$

Since q cannot be negative, $q = 80$.
Since $p = 45(80) = 3600$ and

$p = -(80)^2 + 10,000 = 3600$, the equilibrium quantity is $q = 80$ units and the equilibrium price is $p = \$3600$.

13. Set $p = q^2 + 20q$ and $p = -2q^2 + 10q + 3000$ equal.

$q^2 + 20q = -2q^2 + 10q + 3000$

Write this quadratic equation in standard form and solve using the quadratic formula.

$3q^2 + 10q - 3000 = 0$

$q = \dfrac{-10 \pm \sqrt{10^2 - 4(3)(-3000)}}{2(3)}$

$= \dfrac{-10 \pm \sqrt{36,100}}{6} = \dfrac{-10 \pm 190}{6}$

$q = \dfrac{-10 - 190}{6} = \dfrac{-200}{6} = -\dfrac{100}{3}$ or

$q = \dfrac{-10 + 190}{6} = \dfrac{180}{6} = 30$

Since q cannot be negative, $q = 30$.

Since $p = 30^2 + 20(30) = 1500$ and

$p = -2(30)^2 + 10(30) + 3000 = 1500$, the equilibrium quantity is $q = 30$ units and the equilibrium price is $p = \$1500$.

15. Set the revenue function equal to the cost function and solve for x.

$$200x - x^2 = 70x + 2200$$

$$0 = x^2 - 130x + 2200$$

By the quadratic formula,

$$x = \frac{-(-130) \pm \sqrt{(-130)^2 - 4(1)(2200)}}{2(1)}$$

$$= \frac{130 \pm \sqrt{8100}}{2}$$

$x = 20$ or $x = 110$

Since these equations are only valid for $0 \le x \le 100$, 20 is the number of units needed to break even.

17. Set the revenue function equal to the cost function and solve for x.

$$400x - 2x^2 = -x^2 + 200x + 1900$$

$$-x^2 + 200x - 1900 = 0$$

By the quadratic formula,

$$x = \frac{-200 \pm \sqrt{(200)^2 - 4(-1)(-1900)}}{2(-1)}$$

$$= \frac{-200 \pm \sqrt{32400}}{-2}$$

$x = 10$ or $x = 190$

Since these equations are only valid for $0 \le x \le 100$, 10 is the number of units needed to break even.

19. a. Let x be the number of unsold seats. Then the number of people flying is $100 - x$, and the price per ticket is $200 + 4x$. The total revenue is

$$R(x) = (200 + 4x)(100 - x)$$

$$= 20,000 - 200x + 400x - 4x^2$$

$$R(x) = 20,000 + 200x - 4x^2$$

b. $R(x) = -4\left(x^2 - 50x\right) + 20,000$

$$= -4\left(x^2 - 50x + 625 - 625\right) + 20,000$$

$$= -4(x - 25)^2 + 2500 + 20,000$$

$$R(x) = -4(x - 25)^2 + 22,500$$

The vertex is (25, 22,500).

x	5	15	25
$R(x)$	20,900	22,100	22,500

x	35	45
$R(x)$	22,100	20,900

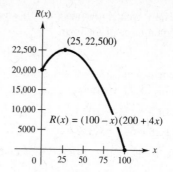

c. The maximum revenue occurs at the vertex, when 25 seats are unsold.

d. $R(25) = 22,500$, so the maximum revenue is $22,500.

21. Let x represent the number of weeks she should wait. Let R represent her revenue in dollars per hog.

$$R(x) = (90 + 5x)(.88 - .02x)$$

$$= 79.2 - 1.8x + 4.4x - .1x^2$$

$$= \left(-1.x^2 + 2.6x\right) + 79.2$$

$$= -1.\left(x^2 - 26x\right) + 79.2$$

$$= -.1\left(x^2 - 26x + 169 - 169\right) + 79.2$$

$$R(x) = -.1\left(x^2 - 26x + 169\right) + 16.9 + 79.2$$

$$= -1.(x - 13)^2 + 96.1$$

The vertex is (13, 96.1). She should wait 13 weeks and will receive $96.10/hog.

23. a. The vertex is given as (20, 0), so the equation is of the form $g(x) = a(x - 20)^2$. The year 2005 corresponds to $x = 35$. Using the point (35, 3), we have

$$3 = a(35 - 20)^2 \Rightarrow 3 = 225a \Rightarrow$$

$$a = \frac{3}{225} \approx .013.$$

The equation is $g(x) = .013(x - 20)^2$.

b. The year 2006 corresponds to $x = 36$.

$$g(36) = .013(36 - 20)^2 \approx 3.$$

According to the model, there were about 3 deaths in 2006.

25. a. The vertex is given as (6, 5.9), so the equation is of the form

$$f(x) = a(x-6)^2 + 5.9.$$ The year 2004 corresponds to $x = 14$. Using the point (14, 28.7), we have

$$28.7 = a(14-6)^2 + 5.9 \Rightarrow$$

$$22.8 = 64a \Rightarrow .356 \approx a$$

The equation is $f(x) = .356(x-6)^2 + 5.9$.

b. The year 2010 corresponds to $x = 20$, and the year 2012 corresponds to $x = 22$.

$$f(20) = .356(20-6)^2 + 5.9 \approx 75.7$$

$$f(22) = .356(22-6)^2 + 5.9 \approx 97.0$$

According to the model, in 2010 expenditures on research and development will be about \$75.7 billion, and in 2012 expenditures will be about \$97.0 billion.

27.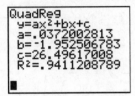

The quadratic regression is

$$g(x) = .037x^2 - 1.95x + 26.5.$$

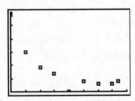

[0, 40] by [0, 30]

$$g(36) = .037(36)^2 - 1.95(36) + 26.5 \approx 4$$

This is 1 less than the model is exercise 23 estimates.

29.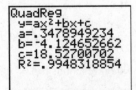

The quadratic regression is

$$f(x) = .348x^2 - 4.12x + 18.5.$$

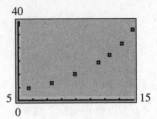

$$f(20) = .348(20)^2 - 4.12(20) + 18.5 = 75.3$$

$$f(22) = .348(22)^2 - 4.12(22) + 18.5 \approx 96.3$$

The quadratic regression estimates are very close to the estimates given by the model in exercise 25.

31. Let x = each of equal sides of the rectangle and y = the side opposite the river.

The perimeter of the rectangle is $2x + y = 320$.

Solve for: $y = 320 - 2x$.

The area of the rectangle is the length multiplied by the width, so we have

$$A = xy = x(320 - 2x) = 320x - 2x^2 \Rightarrow$$

$$A(x) = -2x^2 + 320x$$

The x-value of the vertex leads to the maximum value of the area because the graph opens downward.

$$x = -\frac{b}{2a} = -\frac{320}{2(-2)} = 80$$

$$y = 320 - 2(80) = 160$$

The dimensions of the rectangle are 80 feet by 160 feet.

33. $R(x) = 400x - 2x^2$; $C(x) = 200x + 2000$

a.
$$R(x) = C(x)$$

$$400x - 2x^2 = 200x + 2000$$

$$0 = 2x^2 - 200x + 2000$$

$$0 = x^2 - 100x + 1000$$

$$x = \frac{100 \pm \sqrt{(-100)^2 - 4(1)(1000)}}{2(1)}$$

$$= \frac{100 \pm \sqrt{6000}}{2} = \frac{100 \pm 20\sqrt{15}}{2}$$

$$= 50 \pm 10\sqrt{15}$$

$$x \approx 88.7 \text{ and } 11.3$$

The break-even point occurs when $x \approx 11.3$ and $x \approx 88.7$.

b. $P(x) = R(x) - C(x)$

$$= (400x - 2x^2) - (200x + 2000)$$

$$= -2x^2 + 200x - 2000$$

$$= -2(x^2 - 100x + 2500) - 2000 + 5000$$

$$= -2(x - 50)^2 + 3000$$

The vertex is (50, 3000). Profit is maximum when $x = 50$.

c. From the vertex, the maximum profit is $3000.

d. Since the break-even points are 88.7 and 11.3, a loss will occur if $x < 11.3$ or $x > 88.7$.

e. A profit will occur for $11.3 < x < 88.7$.

Section 3.6 Polynomial Functions

1. $f(x) = x^4$

First we find several ordered pairs.

x	–3	–2	–1	0	1	2	3
y	81	16	1	0	1	16	81

Plot these ordered pairs and draw a smooth curve through them.

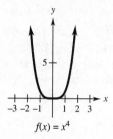

$f(x) = x^4$

3. $h(x) = -.2x^5$

First find several ordered pairs

x	–3	–2	–1	0	1	2	3
y	48.6	6.4	.2	0	–.2	–6.4	–48.6

To graph the function, plot these ordered pairs, and draw a smooth curve through them.

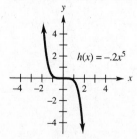

$h(x) = -.2x^5$

5. a. Yes. The graph could possibly be the graph of some polynomial function because it moves sharply away from the x-axis at the far left and far right.

b. No. The total number of peaks and valleys on the graph of a polynomial function of degree n is at most $n - 1$. Since the graph has 2 peaks and 2 valleys, this could not be the graph of a polynomial function of degree 3.

c. No. Since the graph has 2 peaks and 2 valleys, it could not be the graph of a polynomial function of degree 4.

d. Yes. Since the graph has 2 peaks and 2 valleys and the opposite ends of the graph move sharply away from the x-axis in the opposite direction, it could be the graph of a polynomial function of degree 5.

7. a. Yes. The graph could possibly be the graph of some polynomial function because it moves sharply away from the x-axis as $|x|$ gets large.

b. No. Since it has 1 peak and 2 valleys, it could not be the graph of a polynomial function of degree 3 (since $n - 1 = 3 - 1 = 2$).

c. Yes. Since it has 1 peak and 2 valleys, it could be the graph of a polynomial function of degree 4 (since $n - 1 = 4 - 1 = 3$).

d. No. The graph could not be the graph of a polynomial function of degree 5 because the opposite ends of the graph move sharply away from the x-axis in the same direction, indicating an even degree.

9. $f(x) = x^3 - 7x - 9$

Since $f(x)$ is degree 3, it has at most 2 peaks and valleys. When $|x|$ is large, the graph resembles the graph of x^3. Therefore, since the y-intercept is –9, D is the graph of the function.

11. $f(x) = x^4 - 5x^2 + 7$.

Since $f(x)$ is degree 4, it has at most 3 peaks and valleys. When $|x|$ is large, the graph resembles the graph of x^4. Therefore, since the y-intercept is 7, B is the graph of the function.

13. $f(x) = .7x^5 - 2.5x^4 - x^3 + 8x^2 + x + 2$

Since $f(x)$ is degree 5, it has at most 4 peaks and valleys. When $|x|$ is large, the graph resembles the graph of $.7x^5$. Therefore, since the y-intercept is 2, E is the graph of the function.

15. $f(x) = (x+3)(x-4)(x+1)$

First, find x-intercepts by setting $f(x) = 0$ and solving for x.

$f(x) = (x+3)(x-4)(x+1) = 0 \Rightarrow$
$x = -3, 4, -1$

These three numbers divide the x-axis into four regions. Choose an x-value in each region as a test number and compute the function value.

Region	Test No.	Value of $f(x)$	Graph
$x < -3$	-5	-72	below x-axis
$-3 < x < -1$	-2	6	above x-axis
$-1 < x < 4$	2	-30	below x-axis
$4 < x$	5	48	above x-axis

Use this information, the x-intercepts, and the fact that the graph can have a total of at most 2 peaks and valleys to sketch the graph.

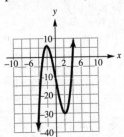

17. $f(x) = x^2(x+3)(x-1)$

First, find x-intercepts by setting $f(x) = 0$ and solving for x.

$f(x) = x^2(x+3)(x-1) = 0 \Rightarrow$
$x = 0, -3, 1$

These three numbers divide the x-axis into four regions. Choose an x-value in each region as a test number and compute the function value.

Region	Test No.	Value of $f(x)$	Graph
$x < -3$	-4	80	above x-axis
$-3 < x < 0$	-1	-4	below x-axis
$0 < x < 1$	0.5	$-.4375$	below x-axis
$1 < x$	2	20	above x-axis

Use this information, the x-intercepts, and the fact that the graph can have a total of at most 3 peaks and valleys to sketch the graph.

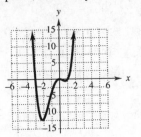

19. $f(x) = x^3 - x^2 - 20x$

First, find x-intercepts by setting $f(x) = 0$ and solving for x.

$f(x) = 0 = x^3 - x^2 - 20x = x(x^2 - x - 20)$
$= x(x-5)(x+4) \Rightarrow x = 0, 5, -4$

These three numbers divide the x-axis into four regions. Choose an x-value in each region as a test number and compute the function value.

Region	Test No.	Value of $f(x)$	Graph
$x < -4$	-5	-50	below x-axis
$-4 < x < 0$	-1	18	above x-axis
$0 < x < 5$	1	-20	below x-axis
$5 < x$	10	700	above x-axis

Use this information, the x-intercepts, and the fact that the graph can have a total of at most 2 peaks and valleys to sketch the graph.

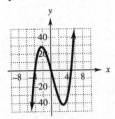

21. $f(x) = x^3 + 4x^2 - 7x$

First, find x-intercepts by setting $f(x) = 0$ and solving for x.

$f(x) = 0 = x^3 + 4x^2 - 7x = x(x^2 + 4x - 7) \Rightarrow$

$x = 0$ or $x^2 + 4x - 7 = 0$.

Use the quadratic formula to solve $x^2 + 4x - 7 = 0$.

$x = \dfrac{-4 \pm \sqrt{4^2 - 4(1)(-7)}}{2(1)} = \dfrac{-4 \pm \sqrt{44}}{2} \Rightarrow$

$x \approx 1.3$ or $x \approx -5.3$

These three numbers divide the x-axis into four regions. Choose an x-value in each region as a test number and compute the function value.

Region	Test No.	Value of $f(x)$	Graph
$x < -5.3$	-6	-30	below x-axis
$-5.3 < x < 0$	-1	10	above x-axis
$0 < x < 1.3$	1	-2	below x-axis
$1.3 < x$	2	10	above x-axis

Use this information, the x-intercepts, and the fact that the graph can have a total of at most 2 peaks and valleys to sketch the graph.

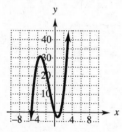

23. To graph $g(x) = x^3 - 3x^2 - 4x - 5$,

enter the function as Y_1. There are many possible viewing windows that will show the complete graph. One such window is $-3 \le x \le 5$ and $-20 \le y \le 5$.

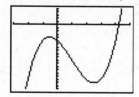

25. To graph

$f(x) = 2x^5 - 3.5x^4 - 10x^3 + 5x^2 + 12x + 6$, enter the function as Y_1. There are many possible viewing windows that will show the complete graph. One such window is $-3 \le x \le 4$ and $-35 \le y \le 20$.

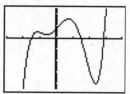

27. a. For a 20% tax rate the revenue is,

$f(20) = \dfrac{20(20 - 100)(20 - 160)}{240}$

$= \dfrac{224,000}{240} \approx \$933.33$ billion

b. For a 40% tax rate the revenue is,

$f(40) = \dfrac{40(40 - 100)(40 - 160)}{240}$

$= \dfrac{288,000}{240} = \$1200$ billion

c. For a 50% tax rate the revenue is,

$f(50) = \dfrac{50(50 - 100)(50 - 160)}{240}$

$= \dfrac{275,000}{240} \approx \$1145.8$ billion

d. For a 70% tax rate the revenue is,

$f(70) = \dfrac{70(70 - 100)(70 - 160)}{240}$

$= \dfrac{189,000}{240} = \$787.5$ billion

e.

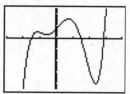

29. a. $P(t) = t^3 - 18t^2 + 81t$

$P(0) = (0)^3 - 18(0)^2 + 81(0) = 0$

$P(3) = (3)^3 - 18(3)^2 + 81(3) = 108$

$P(7) = (7)^3 - 18(7)^2 + 81(7) = 28$

$P(10) = (10)^3 - 18(10)^2 + 81(10) = 10$

b. Notice that $P(9) = 0$. Using this value and the values from (a), plot points and graph the function for $t \geq 0$.

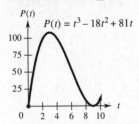

$P(t) = t^3 - 18t^2 + 81t$

c. From the graph, the pressure is increasing for years 0 to 3 and from the ninth year on, and decreasing for years 3 to 9.

31. a. 2008 corresponds to $x = 8$ and 2020 corresponds to $x = 20$.

$$g(x) = -.00096x^3 - .1x^2 + 11.3x + 1274$$

$$g(8) = -.00096(8)^3 - .1(8)^2 + 11.3(8) + 1274$$
$$\approx 1357.500$$

$$g(20) = -.00096(20)^3 - .1(20)^2$$
$$+ 11.3(20) + 1274$$
$$\approx 1452.300$$

In 2008, the population will be about 1,357,500,000, and the population will be about 1,452,300,000 in 2020.

b. When x is large, the graph must resemble the graph of $y = -.00096x^3$, which drops down forever at the far right. This would mean that China's population would become 0 at some point.

33. a.

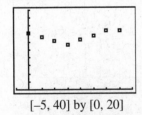

$[-5, 40]$ by $[0, 20]$

b.

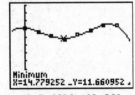

QuarticReg
y=ax⁴+bx³+...+e
a=-3.909091ᴇ-5
b=.0023929293
c=-.0331136364
d=-.084476912
↓e=14.28257576

$$f(x) = -.00003909091x^4 + .0024x^3 - .0331x^2$$
$$-.0845x + 14.2826$$

c.

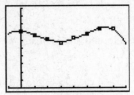

$[-5, 40]$ by $[0, 20]$

The graph fits reasonably well.

d.

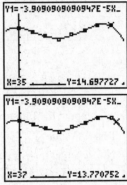

Y1=-3.909090909090947E-5X...

X=35   Y=14.697727

Y1=-3.909090909090947E-5X...

X=37   Y=13.770752

The year 2010 corresponds to $x = 35$, and the year 2012 corresponds to $x = 37$. Using the calculator, we find that $f(35) \approx 14.7$ and $f(37) \approx 13.8$. According to the model, the enrollment will be about 14.7 million in 2010. The enrollment in 2012 will be about 13.8 million.

e.

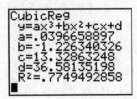

Minimum
X=14.779252 Y=11.660952

$[-5, 40]$ by $[0, 20]$

Using the minimum finder, enrollment was lowest when $x \approx 14.8$, which correspond to late 1989.

35. a.

CubicReg
y=ax³+bx²+cx+d
a=.0396658897
b=-1.226340326
c=13.32863248
d=36.58135198
R²=.7749492858

$$R(x) = .040x^3 - 1.226x^2 + 13.33x + 36.6$$

b.

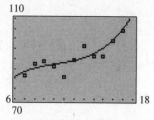

The graph fits the data well.

37. $P(x) = R(x) - C(x)$

$= .040x^3 - 1.226x^2 + 13.33x + 36.6$

$- \left(.035x^3 - 1.137x^2 + 12.98x + 29.2\right)$

$= .005x^3 - .089x^2 + .35x + 7.4$

Section 3.7 Rational Functions

1. $f(x) = \dfrac{1}{x+5}$

Vertical asymptote:
If $x + 5 = 0$, then $x = -5$, so the line $x = -5$ is a vertical asymptote.

Horizontal asymptote: $y = \dfrac{1}{x+5} = \dfrac{0x+1}{1x+5}$

The line $y = \dfrac{a}{c} = \dfrac{0}{1}$ or $y = 0$ is a horizontal asymptote. There are no x-intercepts.

y-intercept: $y = \dfrac{1}{0+5} = \dfrac{1}{5}$

Make a table of values.

x	-8	-7	-6	-5.5	-4.5	-4	-3	-2
y	$-\frac{1}{3}$	$-\frac{1}{2}$	-1	-2	2	1	$\frac{1}{2}$	$\frac{1}{3}$

Using these points and the asymptotes, we graph the function.

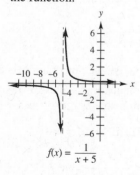

$f(x) = \dfrac{1}{x+5}$

3. $f(x) = \dfrac{-3}{2x+5}$

Vertical asymptote:

If $2x + 5 = 0$, $x = -\dfrac{5}{2}$, so the line $x = -\dfrac{5}{2}$ is a vertical asymptote.

Horizontal asymptote: $y = -\dfrac{3}{2x+5} = \dfrac{0x-3}{2x+5}$

The line $y = \dfrac{0}{2}$ or $y = 0$ is a horizontal asymptote. There are no x-intercepts. If $x = 0$,

then $y = \dfrac{-3}{2(0)+5} = -\dfrac{3}{5}$.

Make a table of values.

x	-5	-4	-3	-2	-1	0
y	$\frac{3}{5}$	1	3	-3	-1	$-\frac{3}{5}$

Using these points and the information above, we graph the function.

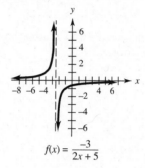

$f(x) = \dfrac{-3}{2x+5}$

5. $f(x) = \dfrac{3x}{x-1}$

Vertical asymptote: If $x - 1 = 0$, $x = 1$, so the line $x = 1$ is a vertical asymptote.

Horizontal asymptote: $y = \dfrac{3x+0}{1x-1}$

The line $y = \dfrac{3}{1}$ or $y = 3$ is a horizontal asymptote.

Find the intercepts. If $x = 0$, then $y = f(x) = 0$.

Next, make a table of values.

x	-3	-2	-1	0	2	3	4	5
y	$\frac{9}{4}$	2	$\frac{3}{2}$	0	6	$\frac{9}{2}$	4	$\frac{15}{4}$

Using these points and the information above, we graph the function.

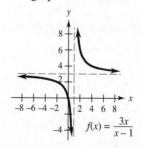

$f(x) = \dfrac{3x}{x-1}$

7. $f(x) = \dfrac{x+1}{x-4}$

Vertical asymptote: If $x - 4 = 0$, $x = 4$, so the line $x = 4$ is a vertical asymptote.

Horizontal asymptote: $y = \dfrac{1x+1}{1x-4}$

The line $y = \dfrac{1}{1}$ or $y = 1$ is a horizontal asymptote.

Find the intercepts. If $x = 0$, then $y = -\frac{1}{4}$. If $y = 0$, then $x = -1$. Now make a table of values.

x	0	1	2	3	5	6	7	8
y	$-\frac{1}{4}$	$-\frac{2}{3}$	$-\frac{3}{2}$	-4	6	$\frac{7}{2}$	$\frac{8}{3}$	$\frac{9}{4}$

Using these points and the information above, we graph the function.

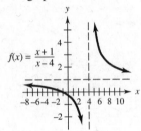

$f(x) = \dfrac{x+1}{x-4}$

9. $f(x) = \dfrac{2-x}{x-3}$

Vertical asymptote:
If $x - 3 = 0$, $x = 3$, so the line $x = 3$ is a vertical asymptote. Horizontal asymptote: $y = \dfrac{-1x+2}{1x-3}$

The line $y = -\dfrac{1}{1}$ or $y = -1$ is a horizontal asymptote. Find the intercepts. If $x = 0$, then

$f(x) = y = -\dfrac{2}{3}$. If $y = 0$, then $x = 2$. Now, make a table of values.

x	-1	0	1	2	4	5	6	7
y	$-\frac{3}{4}$	$-\frac{2}{3}$	$-\frac{1}{2}$	0	-2	$-\frac{3}{2}$	$-\frac{4}{3}$	$-\frac{5}{4}$

Using these points and the information above, we graph the function.

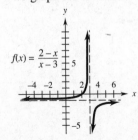

$f(x) = \dfrac{2-x}{x-3}$

11. $f(x) = \dfrac{3x+2}{2x+4}$

Vertical asymptote: Set the denominator equal to 0 and solve for x. $2x + 4 = 0 \Rightarrow x = -2$. The line $x = -2$ is a vertical asymptote. Horizontal asymptote:

$y = \dfrac{3x+2}{2x+4} \Rightarrow y = \dfrac{3}{2}$ is a horizontal asymptote.

Find the intercepts. If $x = 0$, then

$y = \dfrac{3(0)+2}{2(0)+4} = \dfrac{1}{2}$. If $y = 0$, then $0 = \dfrac{3x+2}{2x+4} \Rightarrow$

$3x + 2 = 0 \Rightarrow x = -\dfrac{2}{3}$. Now make a table of values.

x	-10	-6	-4	-3	-1	0	2	8
y	$\frac{7}{4}$	2	$\frac{5}{2}$	$\frac{7}{2}$	$-\frac{1}{2}$	$\frac{1}{2}$	1	$\frac{13}{10}$

Using these points and the information above, we graph the function.

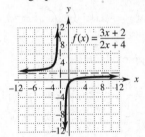

$f(x) = \dfrac{3x+2}{2x+4}$

13. $h(x) = \dfrac{x+1}{x^2+2x-8} = \dfrac{x+1}{(x-4)(x+2)}$

Set the factors in the denominator equal to 0 and solve. $x - 4 = 0 \Rightarrow x = 4$; $x + 2 = 0 \Rightarrow x = -2$.

To find horizontal asymptote, divide the numerator and the denominator by x^2.

$$\dfrac{\frac{x}{x^2} + \frac{1}{x^2}}{\frac{x^2}{x^2} + \frac{2x}{x^2} - \frac{8}{x^2}} = \dfrac{\frac{1}{x} + \frac{1}{x^2}}{1 + \frac{2x}{x^2} - \frac{8}{x^2}}$$

As $|x|$ gets very large, the numerator gets close to 0 and the denominator gets close to 1, so the function has a horizontal asymptote at $y = 0$.

Find the intercepts. If $x = 0$, $y = -\dfrac{1}{8}$.

If $y = 0$, $0 = \dfrac{x+1}{(x-4)(x+2)} \Rightarrow$

$x + 1 = 0 \Rightarrow x = -1$.

(continued on next page)

(*continued from page 86*)

Now, make a table of values.

x	–10	–8	–6	–4
y	$-\frac{1}{8}$	$-\frac{7}{40}$	$-\frac{5}{16}$	undefined

x	–3	–2	–1	0	1
y	$\frac{2}{5}$	$\frac{1}{8}$	0	$-\frac{1}{8}$	$-\frac{2}{5}$

x	2	4	6	8	10
y	undefined	$\frac{5}{16}$	$\frac{7}{40}$	$\frac{1}{8}$	$\frac{11}{112}$

Using these points and the asymptotes, graph the function.

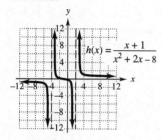

$h(x) = \dfrac{x+1}{x^2 + 2x - 8}$

15. $f(x) = \dfrac{x^2 + 4}{x^2 - 4}$

To find the vertical asymptotes, set the denominator equal to 0 and solve.

$x^2 - 4 = 0 \Rightarrow (x-2)(x+2) = 0 \Rightarrow x = \pm 2$

To find horizontal asymptotes, divide the numerator and the denominator by x^2.

$$\frac{\frac{x^2}{x^2} + \frac{4}{x^2}}{\frac{x^2}{x^2} - \frac{4}{x^2}} = \frac{1 + \frac{4}{x^2}}{1 - \frac{4}{x^2}}$$

As $|x|$ gets very large, the function approaches 1. So the function has a horizontal asymptote at $y = 1$. Find the intercepts.

If $x = 0$, $\dfrac{0+4}{0-4} = -1$, so the y-intercept is –1.

If $y = 0$, $0 = \dfrac{x^2+4}{x^2-4} \Rightarrow x^2 = -4$ is undefined, so there is no x-intercept. Now make a table of values.

x	–6	–4	–3	–1	0	1	3	4
y	$1\frac{1}{4}$	$1\frac{2}{3}$	2.6	$-1\frac{2}{3}$	–1	$-1\frac{2}{3}$	$2\frac{3}{5}$	$1\frac{2}{3}$

Use these points and the information above, graph the function.

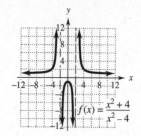

$f(x) = \dfrac{x^2+4}{x^2-4}$

17. $f(x) = \dfrac{x-3}{x^2+x-2} = \dfrac{(x-3)}{(x+2)(x-1)}$

If $(x+2)(x-1) = 0$, then $x = -2$ or $x = 1$. The lines $x = -2$ and $x = 1$ are vertical asymptotes.

19. $g(x) = \dfrac{x^2+2x}{x^2-4x-5} = \dfrac{x(x+2)}{(x-5)(x+1)}$

If $(x-5)(x+1) = 0$, then $x = 5$ or $x = -1$. The lines $x = 5$ and $x = -1$ are vertical asymptotes.

21. $f(x) = \dfrac{4.3x}{100 - x}$

a. $f(50) = \dfrac{4.3(50)}{100 - 50} \Rightarrow = 4.3$ or \$4300

b. $f(70) = \dfrac{4.3(70)}{100 - 70} = 10.03333$ or \$10,033.33

c. $f(80) = \dfrac{4.3(80)}{100 - 80} = 17.2$ or \$17,200

d. $f(90) = \dfrac{4.3(90)}{100 - 90} = 38.7$ or \$38,700

e. $f(95) = \dfrac{4.3(95)}{100 - 95} = 81.7$ or \$81,700

f. $f(98) = \dfrac{4.3(98)}{100 - 98} = 210.7$ or \$210,700

g. $f(99) = \dfrac{4.3(99)}{100 - 99} = 425.7$ or \$425,700

h. Since $f(100)$ is undefined, all the pollutant cannot be removed according to his model.

i.

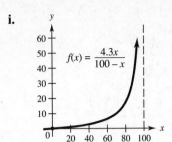

$f(x) = \dfrac{4.3x}{100 - x}$

23. a. Since x represents the size of a generation, x cannot be a negative number. The domain of x is $[0, \infty)$.

b. Let $\lambda = a = b = 1$ and $x \geq 0$.

$$f(x) = \frac{x}{1+x}$$

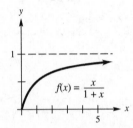

$f(x) = \dfrac{x}{1+x}$

c. Let $\lambda = a = 1$, $b = 2$, and $x \geq 0$

$$f(x) = \frac{x}{1+x^2}$$

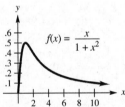

$f(x) = \dfrac{x}{1+x^2}$

d. Increasing b makes the next generation smaller when this generation is larger.

25. $W = \dfrac{S(S - A)}{A} = \dfrac{3(3 - A)}{A}$

a. If $A = 1$, $W = \dfrac{3(3 - 1)}{1} = 6$

The waiting time is 6 min.

b. If $A = 2$, $W = \dfrac{3(3 - 2)}{2} = 1.5$

The waiting time is 1.5 min.

c. If $A = 2.5$, $W = \dfrac{3(3 - 2.5)}{2.5} = .6$

The waiting time is .6 min.

d. The vertical asymptote occurs when the denominator is zero or A = 0.

e. Use the vertical asymptote and the values found in (a), (b), and (c) to graph the function for $0 < A \leq 3$.

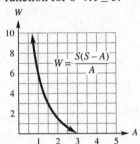

$W = \dfrac{S(S - A)}{A}$

f. When $A < 3$, W is decreasing. The waiting time approaches 0 as A approaches 3. The formula does not apply for $A > 3$ because there will be no waiting if people arrive more than 3 minutes apart

27. $y = \dfrac{900{,}000{,}000 - 30{,}000x}{x + 90{,}000}$

x	0	10,000	20,000	30,000
y	10,000	6000	2727	0

The maximum number of red tranquilizers is 30,000. The maximum number of blue ones is 10,000.

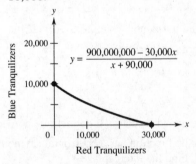

$y = \dfrac{900{,}000{,}000 - 30{,}000x}{x + 90{,}000}$

29. Fixed costs = \$40,000
Marginal cost = \$2.60/unit

a. $C(x) = 2.6x + 40{,}000$

b. Average cost per item:

$$\overline{C}(x) = \frac{C(x)}{x} = \frac{2.6x + 40{,}000}{x}$$

$$= 2.6 + \frac{40{,}000}{x}$$

c. As $|x|$ becomes large, $C(x)$ approaches 2.6. The horizontal asymptote is at $y = 2.6$. This means the average cost per item may get close to \$2.60, but never quite reach it.

31. $C(x) = .2x^3 - 25x^2 + 1531x + 25,000$

$$\bar{C}(x) = \frac{C(x)}{x} = .2x^2 - 25x + 1531 + \frac{25000}{x}$$

Graph the average cost function in the window [0, 150] by [0, 2000], and use the minimum finder to find the point with the smallest y-coordinate.

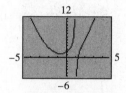

About 73.9 units should be produced to have the lowest possible average cost.

33. a.

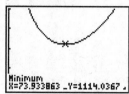

b.

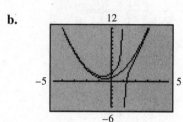

For $|x| \geq 2$, the two graphs appear almost identical because the parabola is an asymptote of the graph.

Chapter 3 Review Exercises

1. This rule does not define y as a function of x because the x-value of 2 corresponds to both 5 and –5.

2. This rule defines y as a function of x because each value of x determines one and only one value of y.

3. $y = \sqrt{x}$

This rule defines y as a function of x because each value of x determines one and only one value of y.

4. $x = |y|$

This rule does not define y as a function of x. A given value of x may define two values for y. For example, if $x = 4$, we have $4 = |y|$, so $y = 4$ and $y = -4$.

5. $x = y^2 + 1$

This rule does not define y as a function of x. A given value of x may define two values for y. For example, if $x = 10$, we have $10 = y^2 + 1$, or $y^2 = 9$, so $y = \pm 3$.

6. $y = 5x - 2$

This rule defines y as a function of x because each value of x determines one and only one value of y.

7. $f(x) = 4x - 1$

   a. $f(6) = 4(6) - 1 = 24 - 1 = 23$

   b. $f(-2) = 4(-2) - 1 = -8 - 1 = -9$

   c. $f(p) = 4p - 1$

   d. $f(r+1) = 4(r+1) - 1 = 4r + 4 - 1 = 4r + 3$

8. $f(x) = 3 - 4x$

   a. $f(6) = 3 - 4(6) = -21$

   b. $f(-2) = 3 - 4(-2) = 3 + 8 = 11$

   c. $f(p) = 3 - 4p$

   d. $f(r+1) = 3 - 4(r+1) = -1 - 4r$

9. $f(x) = -x^2 + 2x - 4$

   a. $f(6) = -6^2 + 2(6) - 4 = -36 + 12 - 4 = -28$

   b. $f(-2) = -(-2)^2 + 2(-2) - 4$
$$= -4 - 4 - 4 = -12$$

   c. $f(p) = -p^2 + 2p - 4$

   d. $f(r+1) = -(r+1)^2 + 2(r+1) - 4$
$$= -\left(r^2 + 2r + 1\right) + 2r + 2 - 4$$
$$= -r^2 - 2r - 1 + 2r + 2 - 4$$
$$= -r^2 - 3$$

10. $f(x) = 8 - x - x^2$

   a. $f(6) = 8 - 6 - 36 = -34$

   b. $f(-2) = 8 - (-2) - (-2)^2 = 8 + 2 - 4 = 6$

   c. $f(p) = 8 - p - p^2$

   d. $f(r+1) = 8 - (r+1) - (r+1)^2$
$$= 8 - r - 1 - \left(r^2 + 2r + 1\right)$$
$$= 8 - r - 1 - r^2 - 2r - 1$$
$$= -r^2 - 3r + 6$$

11. $f(x) = 5x - 3$ and $g(x) = -x^2 + 4x$

   a. $f(-2) = 5(-2) - 3 = -10 - 3 = -13$

   b. $g(3) = -3^2 + 4(3) = -9 + 12 = 3$

   c. $g(-k) = -(-k)^2 + 4(-k) = -k^2 - 4k$

   d. $g(3m) = -(3m)^2 + 4(3m) = -9m^2 + 12m$

   e. $g(k-5) = -(k-5)^2 + 4(k-5)$
$$= -\left(k^2 - 10k + 25\right) + 4k - 20$$
$$= -k^2 + 10k - 25 + 4k - 20$$
$$= -k^2 + 14k - 45$$

   f. $f(3-p) = 5(3-p) - 3 = 15 - 5p - 3$
$$= 12 - 5p$$

12. $f(x) = x^2 + x + 1$

   a. $f(3) = 3^2 + 3 + 1 = 13$

   b. $f(1) = 1^2 + 1 + 1 = 3$

   c. $f(4) = 4^2 + 4 + 1 = 21$

   d. No, $f(3) + f(1) = 13 + 3 = 16$,
      while $f(3 + 1) = f(4) = 21$.

13. $f(x) = |x| - 3$
$$y = \begin{cases} x - 3 & \text{if } x \geq 0 \\ -x - 3 & \text{if } x < 0 \end{cases}$$
We graph the line $y = x - 3$ for $x \geq 0$. We graph the line $y = -x - 3$ for $x < 0$.

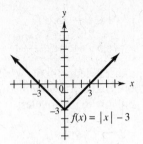

$f(x) = |x| - 3$

14. $f(x) = -|x| - 2$
$$y = \begin{cases} -x - 2 & \text{for } x \geq 0 \\ -(-x) - 2 & \text{for } x < 0 \end{cases}$$
or
$$y = \begin{cases} -x - 2 & \text{for } x \geq 0 \\ x - 2 & \text{for } x < 0 \end{cases}$$
We graph the line $y = -x - 2$ for $x \geq 0$. We graph the line $y = x - 2$ for $x < 0$.

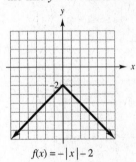

$f(x) = -|x| - 2$

15. $f(x) = -|x+1| + 3$
$$y = \begin{cases} -(x+1) + 3 & \text{if } x + 1 \geq 0 \\ -[-(x+1)] + 3 & \text{if } x + 1 < 0 \end{cases}$$
or
$$y = \begin{cases} -x + 2 & \text{if } x \geq -1 \\ x + 4 & \text{if } x < -1 \end{cases}$$
We graph the line $y = -x + 2$ for $x \geq -1$. We graph the line $y = x + 4$ for $x < -1$.

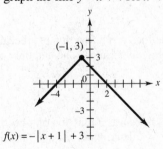

$f(x) = -|x+1| + 3$

16. $f(x) = 2|x-3| - 4$

$y = \begin{cases} 2(x-3) - 4 & \text{if } x-3 \geq 0 \\ -2(x-3) - 4 & \text{if } x-3 < 0 \end{cases}$

or

$y = \begin{cases} 2x - 10 & \text{if } x \geq 3 \\ -2x + 2 & \text{if } x < 3 \end{cases}$

We graph the line $y = 2x - 10$ for $x \geq 3$. We graph the line $y = -2x + 2$ for $x < 3$.

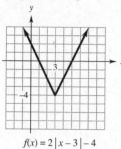

$f(x) = 2|x-3| - 4$

17. $f(x) = [x-3]$

For x in the interval [0, 1), $[x-3] = -3$.
For x in the interval [1, 2), $[x-3] = -2$.
For x in the interval [2, 3), $[x-3] = -1$.
Continue in this pattern. The graph consists of a series of line segments. In each case the left endpoint is included, and the right endpoint is excluded.

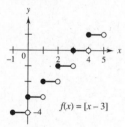

$f(x) = [x-3]$

18. $f(x) = \left[\dfrac{1}{2}x - 2\right]$

$f(x)$ is a step function with breaks in the graph whenever x is an even integer.
If $x \in [-2, 0), f(x) = -3$.
If $x \in [0, 2), f(x) = -2$.
If $x \in [2, 4), f(x) = -1$.
If $x \in [4, 6), f(x) = 0$.
If $x \in [6, 8), f(x) = 1$.

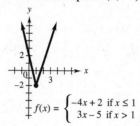

$f(x) = \left[\frac{1}{2}x - 2\right]$

19. $f(x) = \begin{cases} -4x+2 & \text{if } x \leq 1 \\ 3x - 5 & \text{if } x > 1 \end{cases}$

For $x \leq 1$, graph the line $y = -4x + 2$ using the two points (0, 2) and (1, –2). For $x > 1$, graph the line $y = 3x - 5$ using the two points (1, –2) and (4, 7). Note that the two lines meet at their common endpoint (1, –2).

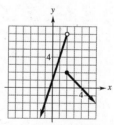

$f(x) = \begin{cases} -4x+2 & \text{if } x \leq 1 \\ 3x-5 & \text{if } x > 1 \end{cases}$

20. $f(x) = \begin{cases} 3x+1 & \text{if } x < 2 \\ -x+4 & \text{if } x \geq 2 \end{cases}$

For $x < 2$, plot the following points to graph the line $y = 3x + 1$.

x	0	1	2
y	1	4	7

For $x \geq 2$, plot the following points to graph the line $y = -x + 4$.

x	2	3	4
y	2	1	0

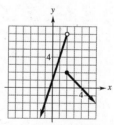

$f(x) = \begin{cases} 3x+1 & \text{if } x < 2 \\ -x+4 & \text{if } x \geq 2 \end{cases}$

21. $f(x) = \begin{cases} |x| & \text{if } x < 3 \\ 6-x & \text{if } x \geq 3 \end{cases}$

For $x < 3$, graph $y = |x|$ using points from the following table.

x	–3	–2	–1	0	1	2	2.9
y	3	2	1	0	1	2	2.9

For $x \geq 3$, graph the line $y = 6 - x$ using the two points (3, 3) and (6, 0).

(continued on next page)

(continued from page 91)

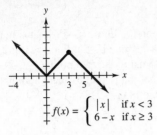

$$f(x) = \begin{cases} |x| & \text{if } x < 3 \\ 6-x & \text{if } x \geq 3 \end{cases}$$

22. $f(x) = \sqrt{x^2}$

Use the table of values to draw the graph.

x	-4	-3	-2	-1	0	1	2	3	4	5
$f(x)$	4	3	2	1	0	1	2	3	4	5

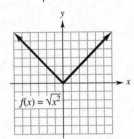

$f(x) = \sqrt{x^2}$

23. $g(x) = \dfrac{x^2}{8} - 3$

Use the table of values to draw the graph.

x	-8	-4	-2	0	2	4	8
$g(x)$	5	-1	$-2\frac{1}{2}$	-3	$-2\frac{1}{2}$	-1	5

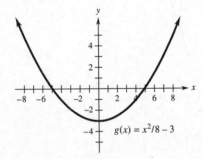

$g(x) = x^2/8 - 3$

24. $h(x) = \sqrt{x} + 2$

Note that x must be greater than or equal to 0.
Use the table of values to draw the graph.

x	0	1	4	9
$h(x)$	2	3	4	5

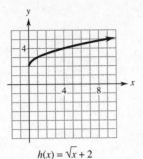

$h(x) = \sqrt{x} + 2$

25. a.

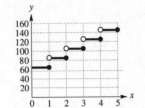

b. The domain is $(0, \infty)$.
The range is $\{65, 85, 105, 125, \ldots\}$.

c. If he can spend no more than $90, he can rent the washer for at most 2 days.

26. a. The charge for 5 hours is
$400 + (80)(5) = 800$.
It costs $800 for 5 hours work, so $750 is not enough.

b.

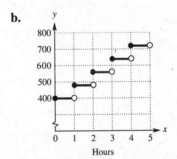

Hours

c. The domain is $(0, \infty)$
The range is $\{400, 480, 560, 640, \ldots\}$.

27.

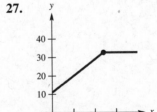

The graph suggests that the number of births to unmarried mothers appear to be leveling off.

28. a. For children less than 5 years old, the general trend is increasing.

b. The year 2007 corresponds to $x = 17$. Using the points $(7, 3.9)$ and $(17, 4.7)$, we have

$$m = \frac{4.7 - 3.9}{17 - 7} = .08.$$

$$y - 3.9 = .08(x - 7) \Rightarrow$$

$$y - 3.9 = .08x - .56 \Rightarrow y = .08x + 3.34$$

c. The year 2014 corresponds to $x = 24$.

$$f(24) = .08(24) + 3.34 = 5.26$$

If the trend continues, about 5.3% of children less than 5 years old will have food allergies in 2014.

29. Eight units cost $300; fixed cost is $60.

a. $C(x) = mx + b$, $b = 60$ and $C(8) = 300$

$$C(x) = mx + 60 \Rightarrow 300 = m(8) + 60 \Rightarrow$$

$$8m = 240 \Rightarrow m = 30$$

$$C(x) = 30x + 60$$

b. The marginal cost is given by the slope, so the marginal cost is $30.

c. Average cost $= \dfrac{C(100)}{100} = \dfrac{30(100) + 60}{100}$

$$= \frac{3060}{100} = 30.6$$

The average cost per unit to produce 100 units is $30.60.

30. Fixed cost is $2000; 36 units cost $8480.

a. $C(x) = mx + 2000$

$$C(36) = 8480 = 36m + 2000$$

$$m = \frac{8480 - 2000}{36} = 180$$

$$C(x) = 180x + 2000$$

b. Since $m = 180$, the marginal cost is $180.

c. Average cost $= \dfrac{C(100)}{100} = \dfrac{180(100) + 2000}{100}$

$$= \frac{20,000}{100} = 200$$

The average cost per unit to produce 100 units is $200.

31. Twelve units cost $445; 50 units cost $1585.

a. $C(x) = mx + b$

Use the two points $(12, 445)$ and $(50, 1585)$.

$$m = \frac{1585 - 445}{50 - 12} = \frac{1140}{38} = 30$$

$$y - 1585 = 30(x - 50) \Rightarrow$$

$$y - 1585 = 30x - 1500 \Rightarrow y = 30x + 85$$

$$C(x) = 30x + 85$$

b. The slope is 30, so the marginal cost is $30.

c. Average cost $= \dfrac{C(100)}{100} = \dfrac{30(100) + 85}{100}$

$$= \frac{3085}{100} = 30.85$$

The average cost per unit to produce 100 units is $30.85.

32. Thirty units cost $1500; 120 units cost $5640.
Use the points $(30, 1500)$ and $(120, 5640)$.

a. $m = \dfrac{5640 - 1500}{120 - 30} = \dfrac{4140}{90} = 46$

$$C(x) = 46x + b \Rightarrow C(30) = 1500 \Rightarrow$$

$$46(30) + b = 1500 \Rightarrow b = 120$$

$$C(x) = 46x + 120$$

b. The slope is 46, so the marginal cost is $46.

c. Average cost $= \dfrac{C(100)}{100} = \dfrac{46(100) + 120}{100}$

$$= \frac{4720}{100} = 47.20$$

The average cost per item to produce 100 units is $47.20.

33. a. The fixed cost is $18,000.

b. Revenue = (Price per item) × (Number of items)

$$R(x) = 28x$$

c. Set the cost function equal to the revenue function and solve for x.

$$24x + 18,000 = 28x$$

$$-4x = -18,000$$

$$x = 4500 \text{ cartridges}$$

d. Let $x = 4500$.

$$R(4500) = 28(4500) = \$126,000$$

34. $68.6x + 3009 > 4300 \Rightarrow 68.6x > 1291 \Rightarrow$

$x > 18.8$

According to the model, energy consumption will exceed 4300 billion kilowatthours sometime during 2009.

35. $-.5q + 30.95 = .3q + 2.15 \Rightarrow -.8q = -28.8 \Rightarrow$

$q = 36$

$-.5(36) + 30.95 = \$12.95$ per month

The equilibrium quantity is 36 million subscribers at an equilibrium price of $12.95 per month.

36. $.0015q + 1 = -.0025q + 64.36 \Rightarrow$

$.004q = 63.36 \Rightarrow q = 15,840$

$p = .0015(15,840) + 1 = 24.76$

The equilibrium quantity is 15,840 prescriptions at an equilibrium price of $24.76.

37. $f(x) = 3(x - 2)^2 + 6$

Here $a = 3$, $h = 2$, and $k = 6$. The vertex is (2, 6). The parabola opens upward since $a = 3 > 0$.

38. $f(x) = 2(x + 3)^2 - 5$

Here $a = 2$, $h = -3$, and $k = -5$. The vertex is $(-3, -5)$. The parabola opens upward since $a = 2 > 0$.

39. $g(x) = -4(x + 1)^2 + 8$

$a = -4$, $h = -1$, $k = 8$. The vertex is $(-1, 8)$. The parabola opens downward since $a = -4 < 0$.

40. $g(x) = -5(x - 4)^2 - 6$

$a = -5$, $h = 4$, $k = -6$. The vertex is $(4, -6)$. The parabola opens downward since $a = -5 < 0$.

41. $f(x) = x^2 - 9$

First, locate the vertex of the parabola and find the axis of the parabola.

$y = 1(x - 0)^2 - 9$

The vertex is at $(0, -9)$. The axis is the line $x = 0$. Make a table of values to find points on one side of the axis and then use the axis of symmetry to find the corresponding points on the other side.

x	1	2	3
y	-8	-5	0

Connect the points with a smooth curve.

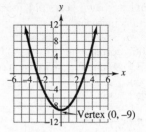

42. $f(x) = 5 - 2x^2 = -2(x - 0)^2 + 5$

The vertex is (0, 5). The axis is the line $x = 0$. Make a table of values to find points on one side of the axis and then use the axis of symmetry to find the corresponding points on the other side.

x	-3	-2	-1	0
y	-13	-3	3	5

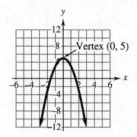

43. $f(x) = x^2 + 2x - 6$

We must complete the square to find the vertex.

$f(x) = \left(x^2 + 2x + 1\right) - 6 - 1 = (x + 1)^2 - 7$

The vertex is $(-1, -7)$. The axis is the line $x = -1$. Make a table of values to find points on one side of the axis and then use the axis of symmetry to find the corresponding points on the other side.

x	-4	-3	-2	-1
y	2	-3	-6	-7

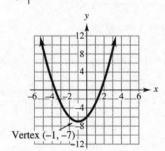

44. $f(x) = -x^2 + 8x - 1$

We must complete the square to find the vertex.

$$f(x) = -\left(x^2 - 8x\right) - 1$$
$$= -\left(x^2 - 8x + 16 - 16\right) - 1$$
$$= -\left(x^2 - 8x + 16\right) - 1 + 16$$
$$= -\left(x - 4\right)^2 + 15$$

The vertex is (4, 15). The axis is the line $x = 4$. Make a table of values to find points on one side of the axis and then use the axis of symmetry to find the corresponding points on the other side.

x	0	1	2	3	4
y	−1	6	11	14	15

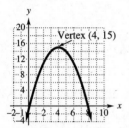

45. $f(x) = -x^2 - 6x + 5$

We must complete the square to find the vertex.

$$f(x) = -\left(x^2 + 6x\right) + 5 = -\left(x^2 + 6x + 9 - 9\right) + 5$$
$$= -\left(x^2 + 6x + 9\right) + 5 + 9 = -\left(x + 3\right)^2 + 14$$

The vertex is (−3, 14). The axis is the line $x = -3$. Make a table of values to find points on one side of the axis and then use the axis of symmetry to find the corresponding points on the other side.

x	−6	−5	−4	−3
y	5	10	13	14

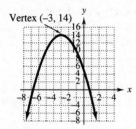

46. $f(x) = 5x^2 + 20x - 2$

We must complete the square to find the vertex.

$$f(x) = 5\left(x^2 + 4x\right) - 2 = 5\left(x^2 + 4x + 4 - 4\right) - 2$$
$$= 5\left(x^2 + 4x + 4\right) - 20 - 2 = 5\left(x + 2\right)^2 - 22$$

The vertex is (−2, −22). The axis is the line $x = -2$. Make a table of values to find points on one side of the axis and then use the axis of symmetry to find the corresponding points on the other side.

x	−5	−4	−3	−2
y	23	−2	−17	−22

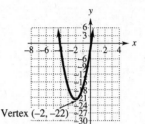

Vertex (−2, −22)

47. $f(x) = 2x^2 - 12x + 10$

We must complete the square to find the vertex.

$$f(x) = 2\left(x^2 - 6x\right) + 10 = 2\left(x^2 - 6x + 9 - 9\right) + 10$$
$$= 2\left(x^2 - 6x + 9\right) - 18 + 10 = 2\left(x - 3\right)^2 - 8$$

The vertex is (3, −8). The axis is the line $x = 3$. Make a table of values to find points on one side of the axis and then use the axis of symmetry to find the corresponding points on the other side.

x	0	1	2	3
y	10	0	−6	−8

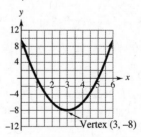

Vertex (3, −8)

48. $f(x) = -3x^2 - 12x - 2$

We must complete the square to find the vertex.

$$f(x) = -3\left(x^2 + 4x\right) - 2$$
$$= -3\left(x^2 + 4x + 4 - 4\right) - 2$$
$$= -3\left(x^2 + 4x + 4\right) + 12 - 2$$
$$= -3\left(x + 2\right)^2 + 10$$

The vertex is (−2, 10). The axis is the line $x = -2$. Make a table of values to find points on one side of the axis and then use the axis of symmetry to find the corresponding points on the other side.

(continued on next page)

(continued from page 95)

x	-5	-4	-3	-2
y	-17	-2	7	10

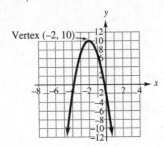

Vertex $(-2, 10)$

49. $f(x) = x^2 + 6x - 2 = \left(x^2 + 6x + 9 - 9\right) - 2$

$= \left(x^2 + 6x + 9\right) - 9 - 2 = (x + 3)^2 - 11$

The vertex is $(-3, -11)$. Because $a = 1 > 0$, the parabola opens upward and the function has a minimum value. This is the y-value of the vertex, which is -11.

50. $f(x) = x^2 + 4x + 5 = \left(x^2 + 4x + 4\right) - 4 + 5$

$== (x + 2)^2 + 1$

The vertex is $(-2, 1)$. Because $a = 1 > 0$, the parabola opens upward and the function has a minimum value, which is 1.

51. $g(x) = -4x^2 + 8x + 3 = \left(-4x^2 + 8x\right) + 3$

$= -4\left(x^2 - 2x\right) + 3 = -4\left(x^2 - 2x + 1\right) + 3 + 4$

$= -4(x - 1)^2 + 7$

The vertex is $(1, 7)$. Because $a = -4 < 0$, the parabola opens downward and the function has a maximum value, which is 7.

52. $g(x) = -3x^2 - 6x + 3 = \left(-3x^2 - 6x\right) + 3$

$= -3\left(x^2 + 2x\right) + 3$

$= -3\left(x^2 + 2x + 1\right) + 3 + 3 = -3(x + 1)^2 + 6$

The vertex is $(-1, 6)$. The parabola opens downward and the function has a maximum value, which is 6.

53. $P(t) = -3t^2 + 18t - 15 = -3\left(t^2 - 6t\right) - 15$

$= -3\left(t^2 - 6t + 9 - 9\right) - 15$

$= -3\left(t^2 - 6t + 9\right) + 27 - 15$

$= -3(t - 3)^2 + 12$

The vertex is $(3, 12)$. The parabola opens downward and the function has a maximum value of $P = 12$ when $t = 3$. So, the time of the largest profit is 3 months after she began.

54. $h = -16t^2 + 800t$

a. Let $h = 3200$ and solve for t.

$$-16t^2 + 800t = 3200$$

$$16t^2 - 800t + 3200 = 0$$

$$16\left(t^2 - 50t + 200\right) = 0 \Rightarrow t^2 - 50t + 200 = 0$$

$$t = \frac{50 \pm \sqrt{(-50)^2 - 4(1)(200)}}{2(1)} = \frac{50 \pm \sqrt{1700}}{2}$$

$$t = \frac{50 + \sqrt{1700}}{2} \quad \text{or} \quad t = \frac{50 - \sqrt{1700}}{2}$$

$$t \approx 45.62 \qquad \text{or} \quad t = 4.38$$

The first time it reaches a height of 3200 ft is about 4.38 sec.

b. Find the vertex of the parabola that is the graph of this function.

$$h = -16t^2 + 800t = -16\left(t^2 - 50t\right)$$

$$= -16\left(t^2 - 50t + 625\right) + 10,000$$

$$= -16(t - 25)^2 + 10,000$$

The vertex is $(25, 10{,}000)$, so the maximum height is 10,000 ft.

55. $P = -x^2 + 250x - 15,000$

$$= -(x^2 - 250x) - 15,000$$

$$= -(x^2 - 250x + 15,625) - 15,000 + 15,625$$

$$= -(x - 125)^2 + 625$$

Vertex: $(125, 625)$
The parabola opens downward and the function has a maximum value of $P = 625$ when $x = 125$. So, the complex produces the largest profit when it rents 125 units.

56. Let x represent the parallel redwood sides. Let y represent the parallel redwood and cement block sides.

$$15x + 15x + 15y + 30y = 900 \Rightarrow$$

$$30x + 45y = 900 \Rightarrow 45y = 900 - 30x \Rightarrow$$

$$y = 20 - \frac{2}{3}x$$

$$\text{Area} = x \cdot y = x\left(20 - \frac{2}{3}x\right)$$

(continued on next page)

(continued from page 96)

$$A = -\frac{2}{3}x^2 + 20x = -\frac{2}{3}\left(x^2 - 30x\right)$$

$$= -\frac{2}{3}\left(x^2 - 30x + 225 - 225\right)$$

$$= -\frac{2}{3}\left(x^2 - 30x + 225\right) + \left(-\frac{2}{3}\right)(-225)$$

$$= -\frac{2}{3}\left(x^2 - 30x + 225\right) + 150$$

$$= -\frac{2}{3}(x - 15)^2 + 150$$

The vertex of the parabola with this equation is (15, 150). The maximum area occurs when

$x = 15$. If $x = 15$, $y = 20 - \frac{2}{3}(15) = 10$. The

dimensions of the enclosure are 10 ft by 15 ft. The maximum possible area is 150 sq ft.

57. a. The year 2007 corresponds to $x = 37$.

$$f(x) = a(x - 7)^2 + 2700 \Rightarrow$$
$$23{,}712 = a(37 - 7)^2 + 2700 \Rightarrow$$
$$23{,}712 = 900a + 2700 \Rightarrow$$
$$900a = 21{,}012 \Rightarrow a \approx 23.35$$

The function is

$$f(x) = 23.35(x - 7)^2 + 2700.$$

b. The year 2012 corresponds to $x = 42$.

$$f(42) = 23.35(42 - 7)^2 + 2700 \approx 31{,}304$$

According to the model, tuition and fees at private colleges will be about \$31,304 in 2012.

58. a.

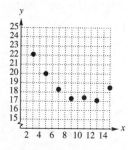

b.

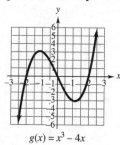

$$g(x) = .079464x^2 - 1.74286x + 26.6491$$

c. The year 2009 corresponds to $x = 19$.

$$g(19) = .079464(19)^2 - 1.74286(19)$$
$$+ 26.6491$$
$$\approx 22.2$$

According to the model about 22.2% of students will be carrying a weapon in 2009.

59. $f(x) = x^4 - 5$

First we find several ordered pairs.

x	-2	-1.5	-1	0	1	1.5	2
y	11	$\frac{1}{16}$	-4	-5	-4	$\frac{1}{16}$	11

Plot these ordered pairs and draw a smooth curve through them.

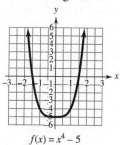

$$f(x) = x^4 - 5$$

60. $g(x) = x^3 - 4x = x\left(x^2 - 4\right) = x(x - 2)(x + 2)$

The x-intercepts are 0, –2, and 2.
These three numbers divide the x-axis into four regions. Choose an x-value in each region as a test number and compute the function value.

Region	Test No.	Value of $f(x)$	Graph
$x < -2$	-3	-15	below x-axis
$-2 < x < 0$	-1	3	above x-axis
$0 < x < 2$	1	-3	below x-axis
$2 < x$	3	15	above x-axis

Use this information, the x-intercepts, and the fact that the graph can have a total of at most 2 peaks and valleys to sketch the graph.

$$g(x) = x^3 - 4x$$

61. $f(x) = x(x-4)(x+1)$

The x-intercepts are 0, 4, and −1. These three numbers divide the x-axis into four regions. Choose an x-value in each region as a test number and compute the function value.

Region	Test No.	Value of $f(x)$	Graph
$x < -1$	−2	−12	below x-axis
$-1 < x < 0$	−.5	1.125	above x-axis
$0 < x < 4$	1	−6	below x-axis
$4 < x$	5	30	above x-axis

Use this information, the x-intercepts, and the fact that the graph can have a total of at most 2 peaks and valleys to sketch the graph.

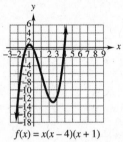

$f(x) = x(x-4)(x+1)$

62. $f(x) = (x-1)(x+2)(x-3)$

The x-intercepts are 1, −2, and 3.
These three numbers divide the x-axis into four regions. Choose an x-value in each region as a test number and compute the function value.

Region	Test No.	Value of $f(x)$	Graph
$x < -2$	−3	−24	below x-axis
$-2 < x < 1$	0	6	above x-axis
$1 < x < 3$	2	−4	below x-axis
$3 < x$	4	18	above x-axis

Use this information, the x-intercepts, and the fact that the graph can have a total of at most 2 peaks and valleys to sketch the graph.

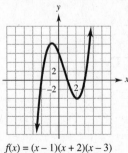

$f(x) = (x-1)(x+2)(x-3)$

63. $f(x) = 3x(3x+2)(x-1)$

The x-intercepts are 0, $-\dfrac{2}{3}$, and 1. These three numbers divide the x-axis into four regions. Choose an x-value in each region as a test number and compute the function value.

Region	Test No.	Value of $f(x)$	Graph
$x < -\frac{2}{3}$	−1	−6	below x-axis
$-\frac{2}{3} < x < 0$	$-\frac{1}{3}$	$\frac{4}{3}$	above x-axis
$0 < x < 1$	$\frac{1}{2}$	$-\frac{21}{8}$	below x-axis
$1 < x$	2	48	above x-axis

Use this information, the x-intercepts, and the fact that the graph can have a total of at most 2 peaks and valleys to sketch the graph.

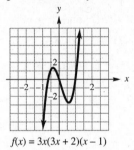

$f(x) = 3x(3x+2)(x-1)$

64. $f(x) = x^3 - 3x^2 - 4x = x(x^2 - 3x - 4)$
    $= x(x-4)(x+1)$

The x-intercepts are 0, 4, and −1. These three numbers divide the x-axis into four regions. Choose an x-value in each region as a test number and compute the function value.

Region	Test No.	Value of $f(x)$	Graph
$x < -1$	−2	−12	below x-axis
$-1 < x < 0$	$-\frac{1}{2}$	$\frac{9}{8}$	above x-axis
$0 < x < 4$	1	−6	below x-axis
$4 < x$	5	30	above x-axis

Use this information, the x-intercepts, and the fact that the graph can have a total of at most 2 peaks and valleys to sketch the graph.

(*continued on next page*)

(continued from page 98)

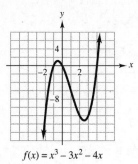

$$f(x) = x^3 - 3x^2 - 4x$$

65. $f(x) = x^4 - 5x^2 - 6 = \left(x^2 - 6\right)\left(x^2 + 1\right)$

First, find x-intercepts by setting $f(x) = 0$ and solving for x. Note that there is no real solution for $x^2 + 1 = 0$.

$$\left(x^2 - 6\right)\left(x^2 + 1\right) = 0 \Rightarrow x^2 = 6 \Rightarrow$$

$$x = \pm\sqrt{6} \approx \pm 2.4$$

These two numbers divide the x-axis into three regions. Choose a point in each region as a test point and compute the function value.

Region	Test No.	Value of $f(x)$	Graph
$x < -\sqrt{6}$	–3	30	above x-axis
$-\sqrt{6} < x < \sqrt{6}$	0	–6	below x-axis
$\sqrt{6} < x$	3	30	above x-axis

Use this information, the x-intercepts, and the fact that the graph can have a total of at most 3 peaks and valleys to sketch the graph.

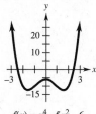

$$f(x) = x^4 - 5x^2 - 6$$

66. $f(x) = x^4 - 7x^2 - 8 = \left(x^2 - 8\right)\left(x^2 + 1\right)$

$$= \left(x - \sqrt{8}\right)\left(x + \sqrt{8}\right)\left(x^2 + 1\right)$$

The x-intercepts are $-\sqrt{8}$ and $\sqrt{8}$.

(Note that $x^2 + 1 \neq 0$ for any x.) These two numbers divide the x-axis into three regions. Choose a point in each region as a test point and compute the function value.

Region	Test No.	Value of $f(x)$	Graph
$x < -\sqrt{8}$	–3	10	above x-axis
$-\sqrt{8} < x < \sqrt{8}$	0	–8	below x-axis
$\sqrt{8} < x$	3	10	above x-axis

Use this information, the x-intercepts, and the fact that the graph can have a total of at most 3 peaks and valleys to sketch the graph.

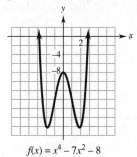

$$f(x) = x^4 - 7x^2 - 8$$

67. Demand equation:

$$p = -.000012q^3 - .00498q^2 + .1264q + 1508$$

Supply equation:

$$p = -.000001q^3 + .00097q^2 + 2q$$

Graph the supply and demand equations in the window [0, 500] by [0, 1500].. Use the intersection finder to determine the equilibrium point.

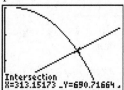

The equilibrium quantity is about 313,152 and the equilibrium price is about $690.72 per thousand.

68. $A(x) = -.000006x^4 + .0017x^3$

$$+ .03x^2 - 24x + 1110$$

Graph the average cost function in the window [0, 150] by [–15, 80] and use the minimum finder to find point where $A(x)$ is smallest.

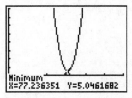

About 77,236 cans should be manufactured for an average cost of about $5.05 per can.

69. $C(x) = -.000006x^3 + .07x^2 + 2x + 1200$

  a. $R(x) = 23x$
$$P(x) = R(x) - C(x)$$
$$= 23x + .000006x^3 - .07x^2$$
$$- 2x - 1200$$
$$= .000006x^3 - .07x^2 + 21x - 1200$$

  b. Graph the profit function in the window [0, 300] by [−100, 450] and use the root finder to find where the graph first crosses the x-axis.

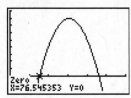

The break-even point is about 76.54, which means that 77 racks must be made to earn a profit.

  c. Use the root finder to find where $P(x)$ next crosses the x-axis.

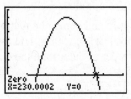

At most about 230 racks can be made without losing money.

  d. Use the maximum finder to find the point where $P(x)$ is the largest.

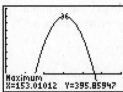

About 153 racks should be made for a maximum profit of about $395.86.

70. a.

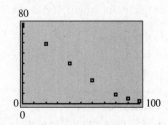

b.

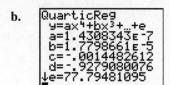

$$f(x) = .000000143x^4 + .0000178x^3$$
$$- .00145x^2 - .9279x + 77.8$$

  c. $f(25) \approx 54.0;\ f(35) \approx 44.5;\ f(50) \approx 30.9$

  d. Answers will vary.

71. $f(x) = \dfrac{1}{x-3}$

Find the vertical asymptote by setting the denominator equal to zero and solving for x.
$$x - 3 = 0 \Rightarrow x = 3$$
The line $x = 3$ is a vertical asymptote. The horizontal asymptote of the function in the form

$y = \dfrac{0x+1}{1x-3}$ is $y = \dfrac{0}{1}$ or $y = 0$.

If $x = 0,\ y = -\dfrac{1}{3}$.

If $y = 0$, there is no solution for x.

x	0	1	2	4	5	6
y	$-\frac{1}{3}$	$-\frac{1}{2}$	-1	1	$\frac{1}{2}$	$\frac{1}{3}$

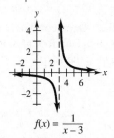

$f(x) = \dfrac{1}{x-3}$

72. $f(x) = \dfrac{-2}{x+4}$

Vertical asymptote: $x = -4$
Horizontal asymptote: $y = 0$

x	−8	−6	−5	$-\frac{9}{2}$	$-\frac{7}{2}$	−3	−2	0
y	$\frac{1}{2}$	1	2	4	−4	−2	−1	$-\frac{1}{2}$

(continued on next page)

(*continued from page 100*)

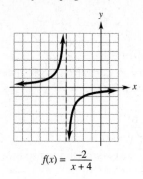

$$f(x) = \frac{-2}{x+4}$$

73. $f(x) = \dfrac{-3}{2x-4}$

Find the vertical asymptote by setting the denominator equal to zero and solving for x.

$2x - 4 = 0 \Rightarrow x = 2$

The line $x = 2$ is a vertical asymptote. The horizontal asymptote of the function in the form

$y = \dfrac{0x - 3}{2x - 4}$ is $y = \dfrac{0}{2}$ or $y = 0$.

If $x = 0$, $y = \dfrac{3}{4}$.

If $y = 0$, there is no solution for x.

x	-1	0	1	3	4	5
y	$\frac{1}{2}$	$\frac{3}{4}$	$\frac{3}{2}$	$-\frac{3}{2}$	$-\frac{3}{4}$	$-\frac{1}{2}$

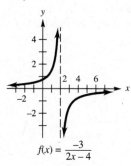

$$f(x) = \frac{-3}{2x-4}$$

74. $f(x) = \dfrac{5}{3x+7}$

Find the vertical asymptote by setting the denominator equal to zero and solving for x.

$3x + 7 = 0 \Rightarrow x = -\dfrac{7}{3}$

Vertical asymptote: $x = -\dfrac{7}{3}$

The horizontal asymptote of the function in the

form $y = \dfrac{0x + 5}{3x + 7}$ is $y = \dfrac{0}{3}$ or $y = 0$.

x	-6	-5	-4	-3	-2	-1	0	1
y	$-\frac{5}{11}$	$-\frac{5}{8}$	-1	$-\frac{5}{2}$	5	$\frac{5}{4}$	$\frac{5}{7}$	$\frac{1}{2}$

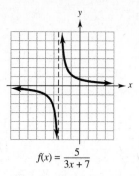

$$f(x) = \frac{5}{3x+7}$$

75. $g(x) = \dfrac{5x - 2}{4x^2 - 4x - 3}$

Find the vertical asymptotes by setting the denominator equal to zero and solving for x.

$4x^2 - 4x - 3 = 0 \Rightarrow (2x - 3)(2x + 1) = 0 \Rightarrow$

$2x - 3 = 0$　or　$2x + 1 = 0$

$2x = 3$　or　$2x = -1$

$x = \dfrac{3}{2}$　or　$x = -\dfrac{1}{2}$

The lines $x = \dfrac{3}{2}$ and $x = -\dfrac{1}{2}$ are vertical asymptotes.

Divide the numerator and denominator by x^2 to find the horizontal asymptotes.

$$\frac{\frac{5x}{x^2} - \frac{2}{x^2}}{\frac{4x^2}{x^2} - \frac{4x}{x^2} - \frac{3}{x^2}} = \frac{\frac{5}{x} - \frac{2}{x^2}}{4 - \frac{4}{x} - \frac{3}{x^2}}$$

As $|x|$ gets very large, the numerator gets close to 0. The function has a horizontal asymptote at $y = 0$. Find the intercepts.

If $x = 0$, $y = \dfrac{-2}{-3} = \dfrac{2}{3}$. The y-intercept is $\dfrac{2}{3}$.

If $y = 0$, $0 = \dfrac{5x - 2}{4x^2 - 4x - 3} \Rightarrow 5x = 2 \Rightarrow x = \dfrac{2}{5}$

The x-intercept is $\dfrac{2}{5}$.

x	-5	-3	-1	$-.3$	0	$\frac{2}{5}$
y	$-.23$	$-.38$	-1.4	2.4	$\frac{2}{3}$	0

x	1	$1\frac{1}{4}$	$1\frac{3}{4}$	2	4
y	-1	-2.4	3	1.6	$.4$

(*continued on next page*)

(continued from page 101)

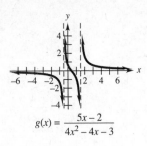

$$g(x) = \frac{5x-2}{4x^2-4x-3}$$

76. $g(x) = \dfrac{x^2}{x^2-1}$

Find the vertical asymptotes by setting the denominator equal to zero and solving for x.

$x^2 - 1 = 0 \Rightarrow x^2 = 1 \Rightarrow x = \pm 1$

The lines $x = 1$ and $x = -1$ are vertical asymptotes.

Divide the numerator and denominator by x^2 to find the horizontal asymptotes.

$$\frac{\frac{x^2}{x^2}}{\frac{x^2}{x^2} - \frac{1}{x^2}} = \frac{1}{1 - \frac{1}{x^2}}$$

As $|x|$ gets very large, the function approaches 1. The function has a horizontal asymptote at $y = 1$. Find the intercepts.

If $x = 0$, $y = 0$. The y-intercept is 0.

If $y = 0$, $0 = \dfrac{x^2}{x^2-1} \Rightarrow x = 0$

The x-intercept is 0.

x	-3	-2	$-\frac{3}{2}$	$-\frac{2}{3}$	$-\frac{1}{2}$	0
y	$\frac{9}{8}$	$\frac{4}{3}$	$\frac{9}{5}$	$-\frac{4}{5}$	$-\frac{1}{3}$	0

x	$\frac{1}{2}$	$\frac{2}{3}$	$\frac{3}{2}$	2	3
y	$-\frac{1}{3}$	$-\frac{4}{5}$	$\frac{9}{5}$	$\frac{4}{3}$	$\frac{9}{8}$

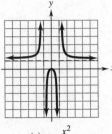

$$g(x) = \frac{x^2}{x^2-1}$$

77. $A(x) = \dfrac{650}{2x+40}$

a. $A(10) = \dfrac{650}{2(10)+40} \approx 10.83$

The average cost per carton to produce 10 cartons is about $10.83.

b. $A(50) = \dfrac{650}{2(50)+40} \approx 4.64$

The average cost per carton to produce 50 cartons is about $4.64.

c. $A(70) = \dfrac{650}{2(70)+40} \approx 3.61$

The average cost per carton to produce 70 cartons is about $3.61.

d. $A(100) = \dfrac{650}{2(100)+40} \approx 2.71$

The average cost per carton to produce 100 cartons is about $2.71.

e. We see that $y = 0$ (the x-axis) is a horizontal asymptote.

x	0	10	30	50	70	100
y	16.25	10.83	6.50	4.64	3.61	2.71

Using these points and the horizontal asymptote, we graph the function.

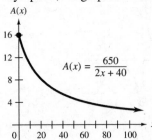

$$A(x) = \frac{650}{2x+40}$$

78. $C(x) = \dfrac{400x+400}{x+4} = \dfrac{400(x+1)}{x+4}$ and

$R(x) = 100x$

a. In graphing the function $C(x)$ we see that $y = 400$ is a horizontal asymptote.

x	0	1	2	3	4	5
y	100	160	200	228.6	250	266.7

Using the asymptote and these points we graph the function $C(x)$. The graph of the function $R(x)$ is a line with a slope of 100 and a y-intercept of 0.

(continued on next page)

(*continued from page 102*)

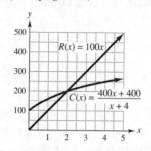

b. To find the break-even point, set $C(x) = R(x)$ and solve for x.

$$\frac{400(x+1)}{x+4} = 100x$$

Divide both sides by 100.

$$\frac{4(x+1)}{x+4} = x \Rightarrow 4x + 4 = x^2 + 4x \Rightarrow \text{Reject}$$

$$x^2 = 4 \Rightarrow x = \pm 2$$

-2 as a solution because the number of hundreds of units cannot be negative. $x = 2$ represents 200 units, so the break-even point is 200 units.

c. $P(1)$ represents a loss. From the graph we see that $C(1) > R(1)$. If the cost is greater than revenue, there is a loss.

d. $P(4)$ represents a profit. From the graph we see that $R(4) > C(4)$. If revenue is greater than cost, there is a profit.

79. Supply: $p = \dfrac{q^2}{4} + 25$

Demand: $p = \dfrac{500}{q}$

a.

q	0	5	10	15
Supply	25	31.25	50	81.25
Demand	No value	100	50	33.3

The equilibrium point is (10, 50).

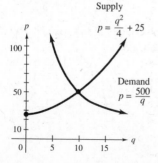

b. Supply exceeds demand if q is in the interval $(10, \infty)$.

c. Demand exceeds supply if q is in the interval $(0, 10)$.

80. $y = \dfrac{9.2x}{106 - x}$

a. If $x = 50$, $y = \dfrac{9.2(50)}{106 - 50} \approx 8.2$ or about $8200.

b. If $x = 98$, $y = \dfrac{9.2(98)}{106 - 98} \approx 112.7$ or about $112,700

c. If $y = 22,000$, then we have

$$22 = \frac{9.2x}{106 - x} \Rightarrow 22(106 - x) = 9.2x \Rightarrow$$

$$2332 - 22x = 9.2x \Rightarrow 2332 = 31.2x \Rightarrow$$

$$x \approx 74.7$$

About 75% of the pollutant can be removed for $22,000.

Case 3 Architectural Arches

1. $f(x) = \dfrac{-k}{c^2} x^2 + k$

If $k = 20$, and $c = 7$, then $f(x) = \dfrac{-20}{49} x^2 + 20$

2. $g(x) = \sqrt{r^2 - x^2} = \sqrt{20^2 - x^2} = \sqrt{400 - x^2}$ The arch is 40 feet wide at its base because the width of the base is twice the radius.

3. The arch is 20 feet tall and 24 feet wide at the base. Thus, $r = 12$ and the vertical sides of the arch are $20 - 12 = 8$ feet tall.

$$h(x) = \sqrt{r^2 - x^2} + 8 = \sqrt{12^2 - x^2} + 8$$

$$= \sqrt{144 - x^2} + 8$$

4. It would fit through the semicircular and Norman arches, but not the parabolic arch. To allow it to fit through the parabolic arch, increase the width of the arch to 15 ft.

Chapter 4 Exponential and Logarithmic Functions

Section 4.1 Exponential Functions

1. $f(x) = 6^x$

This function is exponential, because the variable is in the exponent and the base is a positive constant other than 1.

3. $h(x) = 4x^2 - x + 5$

This function is quadratic because it is a polynomial function of degree 2.

5. $f(x) = 675(1.055^x)$

This function is exponential because the variable is in the exponent and the base is a positive constant other than 1.

7. $f(x) = .6^x$

   a. The equation describing this function has

   the format $f(x) = a^x$ with

   $0 < a < 1$, so the graph lies entirely above the x-axis and falls from left to right. It falls relatively steeply until it reaches the y-intercept 1 and then falls slowly, with the positive x-axis as a horizontal asymptote.

   b. $f(0) = .8^0 = 1; \ (0, 1)$

      $f(1) = .8^1 = .8; \ (1, .8)$

9. $h(x) = 2^{.5x}$

   a. The base of this exponential function is greater than 1, so the graph lies entirely above the x-axis and rises from left to right. The negative x-axis is a horizontal asymptote. The graph rises slowly until it reaches the y-intercept 1 and then rises quite steeply.

   b. $h(0) = 2^{.5(0)} = 2^0 = 1; \ (0, 1)$

      $h(1) = 2^{.5(1)} = 2^{.5}; \ (1, 2^{.5})$

11. $f(x) = e^{-x}$

   a. $e^{-x} = \left(\dfrac{1}{e}\right)^x$, so this exponential function is

   of the form $f(x) = a^x$ with $0 < a < 1$. The graph lies entirely above the x-axis and falls from left to right. It falls very steeply until it reaches the y-intercept 1 and then falls slowly, with the positive x-axis as a horizontal asymptote.

   b. $f(0) = e^{-0} = 1; \ (0,1)$

      $f(1) = e^{-1} = \dfrac{1}{e}; \ \left(1, \dfrac{1}{e}\right)$

13. $f(x) = 3^x$

Construct a table of ordered pairs, plot the points, and connect them to form a smooth curve.

x	-2	-1	0	1	2
y	$\frac{1}{9}$	$\frac{1}{3}$	1	3	9

This graph crosses the y-axis at 1 and it is always above the x-axis.

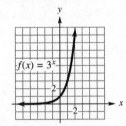

15. $f(x) = 2^{x/2}$

Construct a table of ordered pairs.

x	-4	-2	0	2	4
y	$\frac{1}{4}$	$\frac{1}{2}$	1	2	4

This graph crosses the y-axis at 1 and it is always above the x-axis.

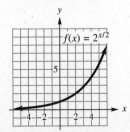

17. $f(x) = \left(\dfrac{1}{5}\right)^x$

x	–2	–1	0	1	2
y	25	5	1	$\frac{1}{5}$	$\frac{1}{25}$

This graph crosses the y-axis at 1 and it is always above the x-axis.

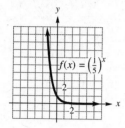

19. a. $f(x) = 2^x$

x	–2	–1	0	1	2
y	$\frac{1}{4}$	$\frac{1}{2}$	1	2	4

b. $g(x) = 2^{x+3}$

x	–2	–1	0	1
y	2	4	8	16

c. $h(x) = 2^{x-4}$

x	–1	0	1	2	3	4
y	$\frac{1}{32}$	$\frac{1}{16}$	$\frac{1}{8}$	$\frac{1}{4}$	$\frac{1}{2}$	1

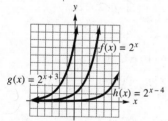

d. The graph of $y = 2^{x+c}$ is the graph of $f(x) = 2^x$ shifted c units to the left. The graph of $y = 2^{x-c}$ is the graph of $f(x) = 2^x$ shifted c units to the right.

21. $y = a^x$

$y = 2.3^x$ is A

23. $y = a^x$

$y = .75^x$ is C

25. $y = a^x$

$y = .31^x$ is E

27. a. $a > 1$

b. Domain: $(-\infty, \infty)$
Range: $(0, \infty)$

c.

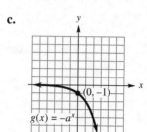

d. Domain: $(-\infty, \infty)$
Range: $(-\infty, 0)$

e.

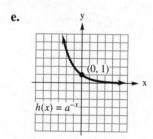

f. Domain: $(-\infty, \infty)$
Range: $(0, \infty)$

29. Since $f(x) = a^x$ and $f(3) = 27$,
$a^3 = 27 \Rightarrow a^3 = 3^3 \Rightarrow a = 3$

a. $f(x) = 3^x$
$f(1) = 3^1 = 3$

b. $f(x) = 3^x$
$f(-1) = 3^{-1} = \dfrac{1}{3}$

c. $f(x) = 3^x$
$f(2) = 3^2 = 9$

d. $f(x) = 3^x$
$f(0) = 3^0 = 1$

31. $f(x) = 2^{-x^2+2}$

x	-2	-1	0	1	2
y	$\frac{1}{4}$	2	4	2	$\frac{1}{4}$

For this graph, the x-axis is an asymptote on the left and on the right.

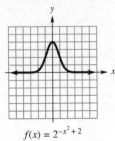

$f(x) = 2^{-x^2+2}$

33. $f(x) = x \cdot 2^x$

x	-4	-2	0	1	2
y	$-\frac{1}{4}$	$-\frac{1}{2}$	0	2	8

This graph passes through the origin, and it has the negative x-axis as a horizontal asymptote towards the left.

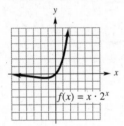

$f(x) = x \cdot 2^x$

35. $y = (1.06)^t$

a.

t	0	1	2	3	4	5
y	1	1.06	1.12	1.19	1.26	1.34

t	6	7	8	9	10
y	1.42	1.50	1.59	1.69	1.79

b. Plot the eleven points in the table and draw a smooth curve connecting them. (Realize that negative values are not acceptable for t, which represents time.)

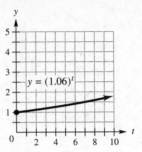

$y = (1.06)^t$

37. $y = (.97)^t$

a. When $t = 10$, $y = (.97)^{10} \approx .74$.

Let x represent the unknown cost.

$.74x = 105,000$

$x = \dfrac{105,000}{.74}$

$x \approx 141,892$

The house will cost about \$141,892.

b. When $t = 8$, $y = (.97)^8 \approx .78$.

Let x represent the unknown cost.

$.78x = 50 \Rightarrow x = \dfrac{50}{.78} \Rightarrow x \approx 64.10$

The book will cost about \$64.10.

39. $f(x) = 110.6e^{0.125x} \quad (0 \le x \le 7)$

$x = 0$ corresponds to 2000.

a. $x = 2002 - 2000 = 2$

$f(2) = 110.6e^{0.125(2)} \approx 142.013$

There were approximately 142,013,000 cell phone accounts in 2002.

b. $x = 2004 - 2000 = 4$

$f(4) = 110.6e^{0.125(4)} \approx 182.349$

There were approximately 182,349,000 cell phone accounts in 2004.

c. $x = 2006 - 2000 = 6$

$f(6) = 110.6e^{0.125(6)} \approx 234.140$

There were approximately 234,000,000 cell phone accounts in 2006.

d. $x = 2010 - 2000 = 10$

$f(10) = 110.6e^{0.125(10)} \approx 386.032$

There are approximately 386,032,000 cell phone accounts in 2010.

e. The answer is unlikely to be accurate since the U.S. population will be less than 386 million in 2010.

41. $W(x) = 2^{-x/24,360}$

    a. $W(1000) = 2^{-1000/24,360} \approx .97$

        After 1000 years, about .97 kg will be left.

    b. $W(10,000) = 2^{-10,000/24,360} \approx .75$

        After 10,000 years, about .75 kg will be left.

    c. $W(15,000) = 2^{-15,000/24,360} \approx .65$

        After 15,000 years, about .65 kg will be left.

    d. $W(24,360) = 2^{-24,360/24,360} \approx .5$

        It will take 24,360 years for the one kilogram to decay to half its original weight.

43. If $C = \$244,000$, $n = 12$, and $r = .15$, then

$$S = C(1-r)^n = 244,000(1-.15)^{12}$$
$$= 244,000(.85)^{12} \approx \$34,706.99$$

The scrap value is about \$34,706.99.

45. $P(t) = 4.834\left(1.01^{(t-1980)}\right)$

    a. $t = 2005$

$$P(2005) = 4.834\left(1.01^{(2005-1980)}\right)$$
$$= 4.834(1.01^{25}) \approx 6.2$$

        In 2005, the population was about 6.2 billion.

    b. $t = 2010$

$$P(2010) = 4.834\left(1.01^{(2010-1980)}\right)$$
$$= 4.834(1.01^{30}) \approx 6.5$$

        In 2010, the population is about 6.5 billion.

    c. $t = 2030$

$$P(2010) = 4.834\left(1.01^{(2030-1980)}\right)$$
$$= 4.834(1.01^{50}) \approx 8.0$$

        In 2030, the population will be about 8.0 billion.

    d. Answers will vary.

47. $f(x) = 1.35\left(1.076^x\right)$, $g(x) = 10.4\left(1.024^x\right)$

$x = 0$ corresponds to 2000.

    a. $x = 2011 - 2000 = 11$

$$f(11) = 1.35\left(1.076^{11}\right) \approx 3.022$$
$$g(11) = 10.4\left(1.024^{11}\right) \approx 13.500$$

According to the model, in 2011, the GDP of China will be about \$3.022 trillion, and the GDP of the U.S. will be about \$13.500 trillion.

    b. $x = 2025 - 2000 = 25$

$$f(25) = 1.35\left(1.076^{25}\right) \approx 8.426$$
$$g(25) = 10.4\left(1.024^{25}\right) \approx 18.816$$

According to the model, in 2025, the GDP of China will be about \$8.426 trillion, and the GDP of the U.S. will be about \$18.816 trillion.

    c. $x = 2047 - 2000 = 47$

$$f(47) = 1.35\left(1.076^{47}\right) \approx 42.219$$
$$g(47) = 10.4\left(1.024^{47}\right) \approx 31.705$$

According to the model, in 2047, the GDP of China will be about \$42.219 trillion, and the GDP of the U.S. will be about \$31.705 trillion.

    d. Graph $Y_1 = 1.35\left(1.076^x\right)$ and $Y_2 = 10.4\left(1.024^x\right)$ in the window [0, 50] by [−5, 35], then find the intersection.

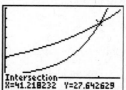

The Chinese GDP will surpass the U.S. GDP sometime during 2041.

49. $N(x) = 100.5\left(1.11^x\right)$

$x = 0$ corresponds to 2000.

    a. $x = 2010 - 2000 = 10$

$$N(10) = 100.5\left(1.11^{10}\right) \approx 285.362$$

According to the model, the projected shortage of full-time equivalent registered nurses will be about 285,362 in 2010.

    b. $x = 2013 - 2000 = 13$

$$N(13) = 100.5\left(1.11^{13}\right) \approx 390.270$$

According to the model, the projected shortage of full-time equivalent registered nurses will be about 390,270 in 2013.

c. $x = 2015 - 2000 = 15$

$N(15) = 100.5(1.11^{15}) \approx 480.851$

According to the model, the projected shortage of full-time equivalent registered nurses will be about 480,851 in 2015.

d. Graph $Y_1 = 100.5(1.11^x)$ and $Y_2 = 600$ in the window [0, 20] by [−100, 700], then find the intersection.

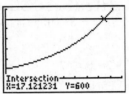

The projected shortage of full-time equivalent registered nurses will exceed 600,000 sometime during 2017.

51. $f(t) = 13.82(.935^t)$

$x = 0$ corresponds to 2000.

a. $x = 2002 - 2000 = 2$

$f(2) = 13.82(.935^2) \approx 12.082$

According to the model, the amount spent on CDs will be about \$12,082,000 in 2002.

b. $x = 2006 - 2000 = 6$

$f(6) = 13.82(.935^6) \approx 9.2337$

According to the model, the amount spent on CDs will be about \$9,233,700 in 2006.

c. $x = 2010 - 2000 = 10$

$f(10) = 13.82(.935^{10}) \approx 7.057$

According to the model, the amount spent on CDs will be about \$7,057,000 in 2010.

d.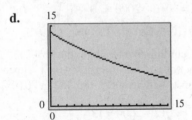

53. $C(t) = 20e^{-.1155t}$

a. $C(4) = 20e^{-.1155(4)} \approx 12.6$

According to the model, there will be about 12.6 mg of aminophylline remaining in the bloodstream after 4 hours.

b. $C(8) = 20e^{-.1155(8)} \approx 7.9$

According to the model, there will be about 7.9 mg of aminophylline remaining in the bloodstream after 8 hours.

c. Graph $Y_1 = 20e^{-.1155x}$ and $Y_2 = 3.5$ in the window [0, 20] by [−5, 20], then find the intersection.

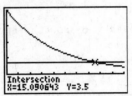

According to the model, there will be 3.59 mg of aminophylline remaining in the bloodstream after about 15.1 hours.

Section 4.2 Applications of Exponential Functions

1. $B(t) = 800(.9898^t)$

a. $B(6) = 800(.9898^6) \approx 752.27$

Your balance after 6 months is \$752.27.

b. $B(12) = 800(.9898^{12}) \approx 707.39$

Your balance after 1 year is \$707.39.

c. $B(60) = 800(.9898^{60}) \approx 432.45$

Your balance after 5 years is \$732.45.

d. $B(96) = 800(.9898^{96}) \approx 298.98$

Your balance after 8 years is \$298.98.

e. Answers vary. Your balance will never reach 0, so never pay the minimum amount.

3. $f(t) = 79.4(1.1^t)$, $t \geq 3$

$t = 3$ corresponds to the year 2003.

a. $t = 2006 - 2003 + 3 = 6$

$f(6) = 79.4(1.1^6) \approx 140.66$

According to the model, about \$140.66 billion will be spent on consumer electronics in 2006.

b. $t = 2009 - 2003 + 3 = 9$

$$f(9) = 79.4\left(1.1^9\right) \approx 187.22$$

According to the model, about \$187.22 billion will be spent on consumer electronics in 2009.

c. $t = 2014 - 2003 + 3 = 14$

$$f(14) = 79.4\left(1.1^{14}\right) \approx 301.52$$

According to the model, about \$301.52 billion will be spent on consumer electronics in 2014. Answers will vary.

5. a.   Let $f(t)$ represent the number of Hispanic-owned businesses in millions in the year t after 2000.
The values of $f(t)$ at $t = 0$ and $t = 2010 - 2000 = 10$ are given, that is, $f(0) = 1.5$ and $f(10) = 3.2$. Solving the first of these equations for y_0 in $f(t) = y_0 b^t$:

$$f(0) = 1.5 \Rightarrow y_0 b^0 = 1.5 \Rightarrow y_0 = 1.5$$

The model has the form $f(t) = 1.5 b^t$.
Solving the second equation, $f(10) = 3.2$, for b:

$$f(10) = 3.2 \Rightarrow 1.5 b^{10} = 3.2 \Rightarrow b^{10} = \frac{3.2}{1.5} \Rightarrow$$

$$b = \left(\frac{3.2}{1.5}\right)^{1/10} \approx 1.0787$$

The model is $f(t) = 1.5\left(1.0787\right)^t$.

b. $t = 2014 - 2000 = 14$

$$f(14) = 1.5\left(1.0787\right)^{14} \approx 4.33$$

According to the model, there will be about 4.33 million Hispanic-owned businesses in 2014.

c.   Graph $Y_1 = 1.5\left(1.0787\right)^x$ and $Y_2 = 6$ in the window [0, 20] by [−2, 10], then find the intersection.

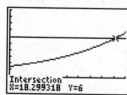

According to the model, there will be 6 million Hispanic-owned businesses sometime during 2018.

7. a.   Let $f(t)$ represent the sales of plasma TVs in billions in year t after 2003. The values of $f(t)$ at $t = 0$ and $t = 2007 - 2003 = 4$ are given, that is, $f(0) = 1.6$ and $f(4) = 6$. Solve the first of these equations for y_0 in

$$f(t) = y_0 b^t. \quad f(0) = 1.6 \Rightarrow y_0 b^0 = 1.6 \Rightarrow$$

$y_0 = 1.6$. Now solve the second equation $f(4) = 6$ for b.

$$f(4) = 6 \Rightarrow 1.6 b^4 = 6 \Rightarrow b^4 = \frac{6}{1.6} \Rightarrow$$

$$b = \left(\frac{6}{1.6}\right)^{1/4} \approx 1.3916$$

The model is $f(t) = 1.6\left(1.3916^t\right)$.

b. $t = 2008 - 2003 = 5$

$$f(5) = 1.6\left(1.3916^5\right) \approx 8.35$$

According to the model, sales will be about \$8.35 billion in 2008.
$t = 2010 - 2003 = 7$

$$f(7) = 1.6\left(1.3916^7\right) \approx 16.17$$

According to the model, sales will be about \$16.17 billion in 2010.

9. a.   Let $t = 0$ correspond to 2000. Using the data for 2000 and 2008 to find a function of the form $f(t) = y_0 b^t$, first solve for y_0:

$$f(0) = 1.00 \Rightarrow y_0 b^0 = 1.00 \Rightarrow y_0 = 1.00$$

The function has the form $f(t) = 1.00 b^t$.
$t = 2008 - 2000 = 8$
Now solve for b.

$$f(8) = 0.79 \Rightarrow 1.00 b^8 = 0.79 \Rightarrow$$

$$b = 0.79^{1/8} \approx 0.9710$$

Thus the function is $f(t) = 0.9710^t$.

Using exponential regression on a graphing calculator, the function produced is

$$g(t) = 1.0103\left(0.9717^t\right).$$

```
ExpReg
 y=a*b^x
 a=1.010279123
 b=.9717427163
 r²=.9819608
 r=-.9909393523
■
```

b. Two-point model:

$t = 2010 - 2000 = 10$

$f(10) = 0.9710^{10} \approx 0.745$

In 2010, a dollar will buy what about 75 cents did in 2000.

$t = 2012 - 2000 = 12$

$f(12) = 0.9710^{12} \approx 0.702$

In 2012, a dollar will buy what about 70 cents did in 2000.

$t = 2014 - 2000 = 14$

$f(14) = 0.9710^{14} \approx 0.662$

In 2014, a dollar will buy what about 66 cents did in 2000.

Exponential regression model:

$g(10) = 1.0103\left(0.9717^{10}\right) \approx 0.758$

In 2010, a dollar will buy what about 76 cents did in 2000.

$g(12) = 1.0103\left(0.9717^{12}\right) \approx 0.715$

In 2012, a dollar will buy what about 72 cents did in 2000.

$g(14) = 1.0103\left(0.9717^{14}\right) \approx 0.676$

In 2014, a dollar will buy what about 38 cents did in 2000.

c. Two point model:

Graph $Y_1 = 0.9710^x$ and $Y_2 = 0.40$ in the window [0, 40] by [–0.1, 1.0], then find the intersection.

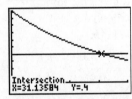

According to the two-point model, the purchasing power of a dollar will drop to 40 cents sometime during 2031.

Exponential regression model:

Graph $Y_1 = 1.0103\left(0.9717^x\right)$ and

$Y_2 = 0.40$ in the window [0, 40] by [–0.1, 1.0], then find the intersection.

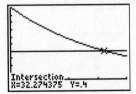

According to the exponential regression model, the purchasing power of a dollar will drop to 40 cents sometime during 2032.

11. a. Let $f(t)$ be the death rate per 100,000 population in the year t after 2000.

Two-point model:

$f(0) = 257.6 \Rightarrow y_0 b^0 = 257.6 \Rightarrow$

$y_0 = 257.6$

$t = 2006 - 2000 = 6$

$f(6) = 210.2 \Rightarrow 257.6 b^6 = 210.2 \Rightarrow$

$b^6 = \dfrac{210.2}{257.6} \Rightarrow b = \left(\dfrac{210.2}{257.6}\right)^{1/6} \approx 0.966676$

$f(t) = 257.6(0.966676)^t$, where $t = 0$ corresponds to 2000.

Exponential regression model:

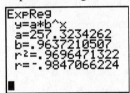

$g(t) = 257.323(0.963721)^t$

b. $t = 2010 - 2000 = 10$

$f(10) = 257.6(0.966676)^{10} \approx 183.55$

$g(10) = 257.323(0.963721)^{10} \approx 177.82$

If the model remains accurate, the death-rate in 2010 will be 183.55 per 100,000 (two-point model) or 177.82 per 100,000 (exponential regression model.

$t = 2015 - 2000 = 15$

$f(15) = 257.6(0.966676)^{15} \approx 154.94$

$g(15) = 257.323(0.963721)^{15} \approx 147.83$

If the model remains accurate, the death-rate in 2015 will be 154.94 per 100,000 (two-point model) or 147.83 per 100,000 (exponential regression model.

c. Two point model:

Graph $Y_1 = 257.6(0.966676)^x$ and

$Y_2 = 100$ in the window [0, 40] by [–50, 300], then find the intersection.

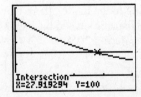

According to the two-point model, the death rate from heart disease will drop to 100 per 100,000 population sometime during 2027. Exponential regression model:

Graph $Y_1 = 257.323(0.963721)^x$ and

$Y_2 = 100$ in the window [0, 40] by [–50, 300], then find the intersection.

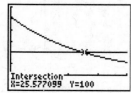

According to the exponential regression model, the death rate from heart disease will drop to 100 per 100,000 population sometime during 2025.

13. $P(t) = 25 - 25e^{-.3t}$

a. $P(1) = 25 - 25e^{-.3(1)} \approx 6$

b. $P(8) = 25 - 25e^{-.3(8)} \approx 23$

c. The maximum number of items that can be produced is 25.

15. $F(t) = T_0 + Cb^t$, where T_0 is the temperature of the constant environment. At $t = 0$, $F(0) = 100$.

So, $100 = T_0 + Cb^0 \Rightarrow 100 = -18 + Cb^0 \Rightarrow$

$100 = -18 + C(1) \Rightarrow C = 118$. The function then becomes $F(t) = -18 + 118b^t$. At $t = 24$,

$F(24) = 50$.

So $50 = -18 + 118b^{24} \Rightarrow$

$68 = 118b^{24} \Rightarrow b^{24} = \dfrac{68}{118} \Rightarrow$

$b = \left(\dfrac{68}{118}\right)^{1/24} \approx 0.9773$

The function now becomes

$F(t) = -18 + 118(0.9773^t)$. At $t = 76$,

$F(76) = -18 + 118(0.9773^{76}) \approx 2.6$. So after 76 minutes, the temperature of the water will be about 26° C.

17. $y(t) = \dfrac{y_0 e^{kt}}{1 - y_0(1 - e^{kt})}$

a. Let $k = 0.1$ and $y_0 = 0.05$. Then

$$y(10) = \frac{0.05e^{.1(10)}}{1 - 0.05(1 - e^{.1(10)})}$$

$$= \frac{0.05e}{1 - 0.05(1 - e)} \approx 0.13.$$

b. Let $k = 0.2$ and $y_0 = 0.1$. Then

$$y(5) = \frac{0.1e^{.2(5)}}{1 - 0.1\left(1 - e^{.2(5)}\right)}$$

$$= \frac{0.1e}{1 - 0.1\left(1 - e\right)} \approx 0.23.$$

c. Plot $Y_1 = \dfrac{0.1e^{.2x}}{1 - 0.1\left(1 - e^{.2x}\right)}$ and $Y_2 = .65$ in

the window [0, 60] by [–.1, 1.1], then find the intersection.

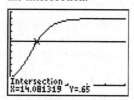

It will take about 14 weeks for 65% of the people to have heard the rumor.

19. $f(x) = \dfrac{38.34}{1 + 17.23e^{-.17243x}}$, $x = 0$ corresponds to 1980.

a. $x = 2005 - 1980 = 25$

$$f(25) = \frac{38.34}{1 + 17.23e^{-.17243(25)}} \approx 31.138$$

The function estimates revenues of about $31.138 billion in 2005.

$x = 2010 - 1970 = 35$

$$f(30) = \frac{38.34}{1 + 17.23e^{-.17243(30)}} \approx 34.929$$

The function estimates revenues of about $34.929 billion in 2005.

b. To graph $f(x) = \dfrac{74.22}{1 + 22.34e^{-.21x}}$ enter this as y_1 and use $0 \le x \le 50$ and $0 \le y \le 50$.

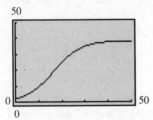

c. To determine the year in which revenues will reach \$37 billion, graph y_1 from part b and $y_2 = 37$ on the same screen. On the CALC menu, select "intersect" and compute the point of intersection.

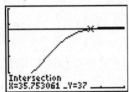

Intersection
X=35.753061 Y=37

The coordinates of this point are (35.75, 37) so the revenues will reach \$37 billion when $t \approx 35.75$ or sometime during 2015.

d. No, the graph of the function appears to level off at about \$38 billion, so it is not likely that revenues will surpass \$40 billion in the foreseeable future.

21. $g(x) = \dfrac{40.35}{1 + 6.39e^{-.0866x}}$, $x = 0$ corresponds to 2000.

a. $x = 2000 - 2000 = 0$

$g(0) = \dfrac{40.35}{1 + 6.39e^{-.0866(0)}} \approx 5.4601$

The function estimates the national debt to be \$5.4601 trillion in 2000.

$x = 2010 - 2000 = 10$

$g(10) = \dfrac{40.35}{1 + 6.39e^{-.0866(10)}} \approx 10.941$

The function estimates the national debt to be \$10.941 trillion in 2010.

$x = 2015 - 2000 = 15$

$g(15) = \dfrac{40.35}{1 + 6.39e^{-.0866(15)}} \approx 14.709$

The function estimates the national debt to be \$14.709 trillion in 2015.

b. To graph $g(x) = \dfrac{40.35}{1 + 6.39e^{-.0866x}}$, enter this as y_1 and use $0 \le x \le 50$ and $0 \le y \le 50$.

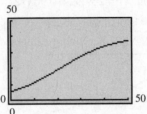

c. To determine the year when the debt will reach \$20 trillion, graph y_1 and $y_2 = 20$ on the same screen. On the CALC menu, select "intersect" and compute the point of intersection.

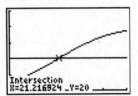

Intersection
X=21.216924 Y=20

The coordinates of this point are (21.2, 20) so the national debt should reach \$20 trillion when $x = 21.2$ or sometime during 2021.

d. Viewing the model and using the graphing calculator's "TRACE" function, the debt does appear to level off after $t = 50$ or around 2050.

Section 4.3 Logarithmic Functions

1. $y = \log_a x$ means $\underline{x = a^y}$.

3. It is missing the value that equals b^y. If that value is x, it should read $y = \log_b x$.

5. $\log 100,000 = 5$ is equivalent to $10^5 = 100,000$. (The base of the logarithm is understood to be 10.)

7. $\log_9 81 = 2$ is equivalent to $9^2 = 81$.

9. $10^{1.9823} = 96$ means $\log 96 = 1.9823$.

11. $3^{-2} = \dfrac{1}{9}$ is equivalent to $\log_3\left(\dfrac{1}{9}\right) = -2$.

13. $\log 1000 = \log_{10} 10^3 = 3$

15. $\log_6 36 = \log_6 6^2 = 2$

17. $\log_4 64 = \log_4 4^3 = 3$

19. $\log_2 \dfrac{1}{4} = \log_2 2^{-2} = -2$

21. $\ln \sqrt{e} = \ln e^{1/2} = \dfrac{1}{2}$

23. $\ln e^{8.77} = 8.77$

25. $\log 53 \approx 1.724$

27. $\ln .0068 \approx -4.991$

29. Answers will vary. Possible answer:
$\log_a 1 = 0$ because $a^0 = 1$ for any valid base a.

31. $\log 6 + \log 8 - \log 2 = \log \dfrac{6(8)}{2} = \log 24$

33. $2 \ln 5 - \dfrac{1}{2} \ln 25 = \ln 5^2 - \ln 25^{1/2} = \ln 25 - \ln 5$
$\qquad = \ln \dfrac{25}{5} = \ln 5$

35. $2 \log u + 3 \log w - 6 \log v$
$\qquad = \log u^2 + \log w^3 - \log v^6 = \log \left(\dfrac{u^2 w^3}{v^6} \right)$

37. $2 \ln(x+2) - \ln(x+3) = \ln(x+2)^2 - \ln(x+3)$
$\qquad = \ln \left(\dfrac{(x+2)^2}{x+3} \right)$

39. $\ln \sqrt{6m^4 n^2} = \ln (6m^4 n^2)^{1/2} = \dfrac{1}{2} \ln 6m^4 n^2$
$\qquad = \dfrac{1}{2}(\ln 6 + \ln m^4 + \ln n^2)$
$\qquad = \dfrac{1}{2}(\ln 6 + 4 \ln m + 2 \ln n)$
$\qquad = \dfrac{1}{2} \ln 6 + 2 \ln m + \ln n$

41. $\log \dfrac{\sqrt{xz}}{z^3} = \log \dfrac{(xz)^{1/2}}{z^3}$
$\qquad = \log(xz)^{1/2} - \log z^3$
$\qquad = \dfrac{1}{2} \log(xz) - 3 \log z$
$\qquad = \dfrac{1}{2} \log x + \dfrac{1}{2} \log z - 3 \log z$
$\qquad = \dfrac{1}{2} \log x - \dfrac{5}{2} \log z$

43. $\ln(x^2 y^5) = \ln x^2 + \ln y^5$
$\qquad = 2 \ln x + 5 \ln y$
$\qquad = 2u + 5v$

45. $\ln \left(\dfrac{x^3}{y^2} \right) = \ln x^3 - \ln y^2 = 3 \ln x - 2 \ln y$
$\qquad = 3u - 2v$

47. $\log_6 384 = \dfrac{\ln 384}{\ln 6} \approx 3.32112$

49. $\log_{35} 5646 = \dfrac{\ln 5646}{\ln 35} \approx 2.429777$

51. $\log (b + c) = \log b + \log c$
Consider the values $b = 1$ and $c = 2$.
Then $\log (b + c)$ becomes
$\log (1 + 2) = \log 3 \approx .4771$,
while $\log b + \log c$ becomes
$\log 1 + \log 2 \approx .3010$.
Many other choices for the b and c values would demonstrate just as clearly that the statement $\log (b + c) = \log b + \log c$ is generally false.

53. $y = \ln (x + 2)$
We must have $x + 2 > 0$, so the domain is $x > -2$.

x	y
-1.99	-4.6
-1.5	$-.7$
-1	0
0	$.7$
2	1.4
4	1.8

Connect these points with a smooth curve.

(*continued on next page*)

(continued from page 113)

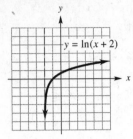

55. $y = \log (x - 3)$

We must have $x - 3 > 0$, so the domain is $x > 3$.

x	y
3.01	−2
3.5	−.30
4	0
6	.48
8	.70

Connect these points with a smooth curve.

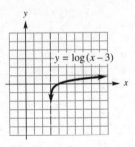

57. Answers will vary. Possible answer:

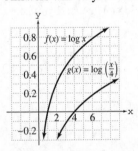

$$\log\left(\frac{x}{4}\right) = \log x - \log 4$$

$g(x)$ equals log 4 subtracted from $f(x)$

59. $\ln 2.75 = 1.0116009$

$e^{1.0116009} = 2.75$

61. $D(r) = \dfrac{\ln 2}{\ln(1 + r)}$

a. $D(4\%) = D(.04) = \dfrac{\ln 2}{\ln(1 + .04)} \approx 17.67$

It takes 17.67 years to double.

b. $D(8\%) = D(.08) = \dfrac{\ln 2}{\ln(1 + .08)} \approx 9.01$

It takes 9.01 years to double.

c. $D(18\%) = D(.18) = \dfrac{\ln 2}{\ln(1 + .18)} \approx 4.19$

It takes 4.19 years to double.

d. $D(36\%) = D(.36) = \dfrac{\ln 2}{\ln(1 + .36)} \approx 2.25$

It takes 2.25 years to double.

e. $17.67 \approx 18 = \dfrac{72}{4}$; $9.01 \approx 9 = \dfrac{72}{8}$;

$4.19 \approx 4 = \dfrac{72}{18}$; $2.25 \approx 2 = \dfrac{72}{36}$

The pattern is that it takes about $72/k$ years for money to double at $k\%$ interest.

63. $g(x) = 10.155 + 3.62 \ln x$, $x = 1$ corresponds to 2001.

a. $x = 2004 - 2001 + 1 = 4$

$g(4) = 10.155 + 3.62 \ln 4 \approx 15.173$

In 2004, sales were about \$15,173 billion.

$x = 2008 - 2001 + 8 = 8$

$g(8) = 10.155 + 3.62 \ln 8 \approx 17.683$

In 2008, sales were about \$17,683 billion.

b.

x	4	8	12	16	20	24
y	15.173	17.683	19.15	20.192	21	21.66

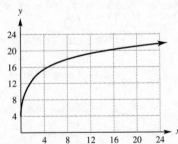

c.   The graph indicates that DVD and videotape sales will increase at a slower rate as time goes on.

65. a. $x = 2007 - 2001 + 1 = 7;$ $g(7) = 15.93 + 2.174 \ln 7 \approx 20.16$

By 2007, approximately 20.2% of the U.S. population will be age 60 or older.

$x = 2015 - 2001 + 1 = 15;$ $g(15) = 15.93 + 2.174 \ln 15 \approx 21.82$

By 2015, approximately 21.8% of the U.S. population will be age 60 or older.

$x = 2030 - 2001 + 1 = 30;$ $g(30) = 15.93 + 2.174 \ln 30 \approx 23.32$

By 2030, approximately 23.3% of the U.S. population will be age 60 or older.

$x = 2050 - 2001 + 1 = 50;$ $g(50) = 15.93 + 2.174 \ln 50 \approx 24.43$

By 2050, approximately 24.4% of the U.S. population will be age 60 or older.

b.

x	7	15	30	50
y	20.16	21.82	23.32	24.43

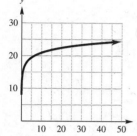

c.   As time goes on, the percentage the U.S. population age 60 or over will level off just below 25%.

67. $n_1 = 2754$, $n_2 = 689$, $n_3 = 4428$, and $n_4 = 629$

$N = n_1 + n_2 + n_3 + n_4 = 8500$

So, the index of diversity is

$$H = \frac{N \log_2 N - \left[n_1 \log_2 n_1 + n_2 \log_2 n_2 + n_3 \log_2 n_3 + n_4 \log_2 n_4 \right]}{N}$$

$$= \frac{8500 \log_2 8500 - \left[2754 \log_2 2754 + 689 \log_2 689 + 4428 \log_2 4428 + 629 \log_2 629 \right]}{8500}$$

$$= \frac{8500 \dfrac{\ln 8500}{\ln 2} - \left[2754 \dfrac{\ln 2754}{\ln 2} + 689 \dfrac{\ln 689}{\ln 2} + 4428 \dfrac{\ln 4428}{\ln 2} + 629 \dfrac{\ln 629}{\ln 2} \right]}{8500} \approx 1.5887$$

69. $f(x) = -2.43 + 1.414 \ln x,$ $x \geq 10,$

$x = 10$ corresponds to 2000.

a. $x = 2006 - 2000 + 10 = 16$

$f(16) = -2.43 + 1.414 \ln 16 \approx 1.4904$

According to the model, in 2006, total assets of bond mutual funds were about $1.4904 trillion.

$x = 2010 - 2000 + 10 = 20$

$f(20) = -2.43 + 1.414 \ln 20 \approx 1.8060$

According to the model, in 2010, total assets of bond mutual funds were about $1.8060 trillion.

b.   Graph $Y_1 = -2.43 + 1.414 \ln x$ and $Y_2 = 2.25$ in the window [10, 35] by [–0.5, 3.5], then find the intersection.

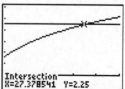

If the model remains accurate, total assets will reach $2.25 trillion sometime during 2017.

71. $g(x) = 16.8 + 2.924\ln(x+1)$,

$x = 0$ corresponds to 1990.

a. $x = 2004 - 1990 = 14$

$g(14) = 16.8 + 2.924\ln(14+1)$

≈ 24.718

According to the model, in 2004, about 24.718 billion pounds of aluminum were shipped.

$x = 2008 - 1990 = 18$

$g(18) = 16.8 + 2.924\ln(18+1)$

≈ 25.410

According to the model, in 2009, about 25.410 billion pounds of aluminum were shipped.

b. Graph $Y_1 = 16.8 + 2.924\ln(x+1)$ and $Y_2 = 27.1$ in the window [0, 45] by [-5, 35], then find the intersection.

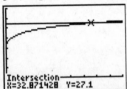

If the model remains accurate, 27.1 billion pounds of aluminum will be shipped sometime during 2022.

Section 4.4 Logarithmic and Exponential Equations

1. $\ln(x+3) = \ln(2x-5) \Rightarrow x+3 = 2x-5$

$x = 8$

3. $\ln(3x+1) - \ln(5+x) = \ln 2$

$\ln\dfrac{3x+1}{5+x} = \ln 2$

$\dfrac{3x+1}{5+x} = 2$

$3x+1 = 10 + 2x$

$x = 9$

5. $2\ln(x-3) = \ln(x+5) + \ln 4$

$\ln(x-3)^2 = \ln[4(x+5)]$

$(x-3)^2 = 4(x+5)$

$x^2 - 6x + 9 = 4x + 20$

$x^2 - 10x - 11 = 0$

$(x-11)(x+1) = 0 \Rightarrow x = 11$ or $x = -1$

Since -1 is not in the domain of $\ln(x-3)$, the only solution is 11.

7. $\log_3(6x-2) = 2 \Rightarrow 6x-2 = 3^2 \Rightarrow$

$6x-2 = 9 \Rightarrow 6x = 11 \Rightarrow x = \dfrac{11}{6}$

9. $\log x - \log(x+4) = -1 \Rightarrow \log\left(\dfrac{x}{x+4}\right) = -1 \Rightarrow$

$\dfrac{x}{x+4} = 10^{-1} \Rightarrow \dfrac{x}{x+4} = \dfrac{1}{10} \Rightarrow 10x = x+4 \Rightarrow$

$9x = 4 \Rightarrow x = \dfrac{4}{9}$

11. $\log_3(y+2) = \log_3(y-7) + \log_3 4$

$\log_3(y+2) = \log_3[4(y-7)]$

$y+2 = 4(y-7)$

$y+2 = 4y - 28$

$30 = 3y$

$y = 10$

13. $\ln(x+9) - \ln x = 1 \Rightarrow \ln\left(\dfrac{x+9}{x}\right) = 1 \Rightarrow$

$\dfrac{x+9}{x} = e^1 \Rightarrow x+9 = ex \Rightarrow x - ex = -9 \Rightarrow$

$(1-e)x = -9 \Rightarrow x = -\dfrac{-9}{e-1} \approx 5.2378$

15. $\log x + \log(x-9) = 1$

$\log[x(x-9)] = 1$

$x(x-9) = 10^1$

$x^2 - 9x = 10$

$x^2 - 9x - 10 = 0$

$(x-10)(x+1) = 0$

$x - 10 = 0$ or $x + 1 = 0$

$x = 10$ or $x = -1$

$x = -1$ is not in the domains of $\log x$ or $\log(x-9)$, so the only solution is $x = 10$.

17. $\log(3+b) = \log(4c-1) \Rightarrow 3+b = 4c-1 \Rightarrow$

$4+b = 4c \Rightarrow \dfrac{4+b}{4} = c \Rightarrow c = \dfrac{4+b}{4}$

19. $2-b = \log(6c+5) \Rightarrow 6c+5 = 10^{2-b} \Rightarrow$

$6c = 10^{2-b} - 5 \Rightarrow c = \dfrac{10^{2-b}-5}{6}$

21. Answers will vary. Sample answer:
Not necessarily. For example, the solution to the example $\log(x+103) = 2$ is $x = -3$. Negative answers must be rejected when they are not in the domain of any of the terms in the equation.

23. $2^{x-1} = 8 \Rightarrow 2^{x-1} = 2^3 \Rightarrow x - 1 = 3 \Rightarrow x = 4$

25. $25^{-3x} = 3125 \Rightarrow (5^2)^{-3x} = 5^5 \Rightarrow$
$5^{-6x} = 5^5 \Rightarrow -6x = 5 \Rightarrow x = -\dfrac{5}{6}$

27. $6^{-x} \Rightarrow 36^{x+6} = 6^{-x} = \left(6^2\right)^{x+6} \Rightarrow$
$6^{-x} = 6^{2(x+6)} \Rightarrow -x = 2(x+6) \Rightarrow$
$-x = 2x + 12 \Rightarrow -3x = 12 \Rightarrow x = -4$

29. $\left(\dfrac{3}{4}\right)^x = \dfrac{16}{9} \Rightarrow \left(\dfrac{3}{4}\right)^x = \left(\dfrac{4}{3}\right)^2 \Rightarrow$
$\left(\dfrac{3}{4}\right)^x = \left(\dfrac{3}{4}\right)^{-2} \Rightarrow x = -2$

31. $2^x = 5$
Take natural logarithms of both sides.
$\ln 2^x = \ln 5$
$x \ln 2 = \ln 5$
$x = \dfrac{\ln 5}{\ln 2} \approx \dfrac{1.6094}{.6931} \approx 2.3219$

33. $2^x = 3^{x-1}$
$\ln 2^x = \ln 3^{x-1}$
$x \ln 2 = (x-1)(\ln 3)$
$x \ln 2 = x \ln 3 - 1 \ln 3$
$x \ln 2 - x \ln 3 = -\ln 3$
$(\ln 2 - \ln 3)x = -\ln 3$
$x = \dfrac{-\ln 3}{\ln 2 - \ln 3} \approx 2.710$

35. $3^{1-2x} = 5^{x+5}$
$\ln 3^{1-2x} = \ln 5^{x+5}$
$(1 - 2x)(\ln 3) = (x + 5)(\ln 5)$
$\ln 3 - 2x \ln 3 = x \ln 5 + 5 \ln 5$
$\ln 3 - 5 \ln 5 = x \ln 5 + 2x \ln 3$
$\ln 3 - 5 \ln 5 = (\ln 5 + 2 \ln 3)x$
$\dfrac{\ln 3 - 5 \ln 5}{\ln 5 + 2 \ln 3} = x$
$x \approx -1.825$

37. $e^{3x} = 6 \Rightarrow \ln e^{3x} = \ln 6 \Rightarrow$
$3x = \ln 6 \Rightarrow x = \dfrac{\ln 6}{3} \approx .597253$

39. $2e^{5a+2} = 8$
$e^{5a+2} = 4$
$\ln e^{5a+2} = \ln 4$
$5a + 2 = \ln 4$
$5a = -2 + \ln 4$
$a = \dfrac{-2 + \ln 4}{5} \approx -.123$

41. $10^{4c-3} = d$
$\log 10^{4c-3} = \log d$
$4c - 3 = \log d$
$4c = \log d + 3$
$c = \dfrac{\log d + 3}{4}$

43. $e^{2c-1} = b \Rightarrow \ln e^{2c-1} = \ln b \Rightarrow$
$2c - 1 = \ln b \Rightarrow 2c = \ln b + 1 \Rightarrow c = \dfrac{\ln b + 1}{2}$

45. $\log_7(r+3) + \log_7(r-3) = 1$
$\log_7\left[(r+3)(r-3)\right] = 1$
$(r+3)(r-3) = 7^1$
$r^2 - 9 = 7$
$r^2 = 16 \Rightarrow r = \pm 4$
Since -4 is not in the domain of $\log_7(r-3)$, 4 is the only solution.

47. $\log_3(a-3) = 1 + \log_3(a+1) \Rightarrow$
$\log_3(a-3) - \log_3(a+1) = 1 \Rightarrow$
$\log_3 \dfrac{a-3}{a+1} = 1 \Rightarrow \dfrac{a-3}{a+1} = 3^1 \Rightarrow$
$a - 3 = 3a + 3 \Rightarrow -6 = 2a \Rightarrow a = -3$
Since -3 is not in the domain of $\log_3(a-3)$ nor in that of $\log_3(a+1)$, there is no solution.

49. $\log_2 \sqrt{2y^2} - 1 = \dfrac{3}{2}$
$\log_2 \sqrt{2y^2} = \dfrac{5}{2}$
$\sqrt{2y^2} = 2^{5/2}$
$2y^2 = 2^5 = 32$
$y^2 = 16 \Rightarrow y = \pm 4$

51. $\log_2\left(\log_3 x\right) = 1 \Rightarrow \log_3 x = 2^1 = 2 \Rightarrow$
$x = 3^2 = 9$

53. $5^{-2x} = \dfrac{1}{25} \Rightarrow 5^{-2x} = 5^{-2} \Rightarrow -2x = -2 \Rightarrow x = 1$

55. $2^{|x|} = 16 \Rightarrow 2^{|x|} = 2^4 \Rightarrow |x| = 4 \Rightarrow x = \pm 4$

57. $2^{x^2-1} = 10 \Rightarrow \ln 2^{x^2-1} = \ln 10 \Rightarrow$
$(x^2-1)\ln 2 = \ln 10 \Rightarrow x^2 - 1 = \dfrac{\ln 10}{\ln 2} \Rightarrow$
$x^2 - 1 \approx \dfrac{2.3026}{.6931} \approx 3.3219 \Rightarrow$
$x^2 = 4.3219 \Rightarrow x = \pm\sqrt{4.3219} \approx \pm 2.0789$

59. $2(e^x + 1) = 10 \Rightarrow e^x + 1 = 5 \Rightarrow e^x = 4 \Rightarrow$
$\ln e^x = \ln 4 \Rightarrow x = \ln 4 \Rightarrow x \approx 1.386$

61. Answers will vary. Possible answer:
Since $x^2 \ge 0$ for every x, $x^2 + 1 \ge 1$ for every x.
Hence $4^{x^2+1} \ge 4^1 = 4$ for every x.

63. $g(t) = .365478(1.073306)^t$,
$t = 0$ corresponds to 1980.

a. $g(t) = 4 \Rightarrow 4 = .365478(1.073306)^t \Rightarrow$
$\dfrac{4}{.365478} = 1.073306^t \Rightarrow$
$\ln\left(\dfrac{4}{.365478}\right) = t \ln 1.073306 \Rightarrow$
$t = \dfrac{\ln\left(\dfrac{4}{.365478}\right)}{\ln 1.073306} \approx 33.82$
Consumer debt will be \$4 trillion sometime during 2013.

b. $g(t) = 5 \Rightarrow 5 = .365478(1.073306)^t \Rightarrow$
$\dfrac{5}{.365478} = 1.073306^t \Rightarrow$
$\ln\left(\dfrac{5}{.365478}\right) = t \ln 1.073306 \Rightarrow$
$t = \dfrac{\ln\left(\dfrac{5}{.365478}\right)}{\ln 1.073306} \approx 36.98$
Consumer debt will be \$5 trillion near the end of 2106.

65. $f(x) = 17.6 + 12.8 \ln x$,
$x = 10$ corresponds to 1910.

a. $f(x) = 75.5 \Rightarrow 75.5 = 17.6 + 12.8 \ln x \Rightarrow$
$57.9 = 12.8 \ln x \Rightarrow \dfrac{57.9}{12.8} = \ln x \Rightarrow$
$x = e^{57.9/12.8} \approx 92.15$
A person whose life expectancy at birth si 75.5 years was born in 1992.

b. $f(x) = 77.5 \Rightarrow 77.5 = 17.6 + 12.8 \ln x \Rightarrow$
$59.9 = 12.8 \ln x \Rightarrow \dfrac{59.9}{12.8} = \ln x \Rightarrow$
$x = e^{59.9/12.8} \approx 107.73$
A person whose life expectancy at birth is 77.5 years was born in 2007.

c. $f(x) = 81 \Rightarrow 81 = 17.6 + 12.8 \ln x \Rightarrow$
$63.4 = 12.8 \ln x \Rightarrow \dfrac{63.4}{12.8} = \ln x \Rightarrow$
$x = e^{63.4/12.8} \approx 141.62$
A person whose life expectancy at birth is 81 years will be born in 2041.

67. $C(x) = 170.63 e^{.03x}$, where $x = 0$ corresponds to 2000.

a. $C(x) = 220 \Rightarrow 170.63 e^{.03x} = 220 \Rightarrow$
$e^{.03x} = \dfrac{220}{170.63} \Rightarrow .03x = \ln\left(\dfrac{220}{170.63}\right) \Rightarrow$
$x = \dfrac{\ln\left(\dfrac{220}{170.63}\right)}{.03} \approx 8.47$
The CPI reached 220 sometime during 2008.

b. $C(x) = 260 \Rightarrow 170.63 e^{.03x} = 260 \Rightarrow$
$e^{.03x} = \dfrac{260}{170.63} \Rightarrow .03x = \ln\left(\dfrac{260}{170.63}\right) \Rightarrow$
$x = \dfrac{\ln\left(\dfrac{260}{170.63}\right)}{.03} \approx 14.039$
If the model remains accurate, the CPI will reach 260 sometime during 2014.

69. $C(t) = 25 e^{-.14t}$

a. $C(0) = 25 e^{-.14(0)} = 25 e^0 = 25$
Initially, 25 g of cobalt was present.

b. We determine the half-life by finding a value of t such that $C(t) = \left(\dfrac{1}{2}\right)(25) = 12.5$.

$$12.5 = 25e^{-.14t} \Rightarrow \frac{1}{2} = e^{-.14t} \Rightarrow$$

$$\ln \frac{1}{2} = \ln e^{-.14t} \Rightarrow \ln \frac{1}{2} = -.14t \Rightarrow$$

$$t = \frac{\ln \frac{1}{2}}{-.14} = \frac{\ln .5}{-.14} \approx 4.95$$

The half-life of cobalt is about 4.95 years.

71. $y = y_0\left(.5^{t/5730}\right)$

36% lost means 64% remains. Since $64\% = .64$, replace y with $.64y_0$ and solve for t.

$$.64y_0 = y_0\left(.5^{t/5730}\right) \Rightarrow .64 = \left(.5^{t/5730}\right) \Rightarrow$$

$$\ln .64 = \ln\left(.5^{t/5730}\right) \Rightarrow \ln .64 = \frac{t}{5730}\ln .5 \Rightarrow$$

$$t = \frac{5730 \ln .64}{\ln .5} \approx 3689.3$$

The ivory is about 3689 years old.

73. Use the formula for the Richter scale given in Example 9 in this section of the textbook.

$$R(i) = \log\left(\frac{i}{i_0}\right)$$

a. $R(i) = 7.9 = \log\left(\dfrac{i}{i_0}\right) \Rightarrow$

$$10^{7.9} = \frac{i}{i_0} \Rightarrow i = 10^{7.9}i_0 \approx 7{,}9432{,}823i_0$$

b. $R(i) = 5.4 = \log\left(\dfrac{i}{i_0}\right) \Rightarrow$

$$10^{5.4} = \frac{i}{i_0} \Rightarrow i = 10^{5.4}i_0 \approx 251{,}189i_0$$

c. The China earthquake was

$$\frac{10^{7.9}}{10^{5.4}} = 10^{7.9-5.4} = 10^{2.5} \approx 316.23$$

times stronger than the Los Angeles earthquake.

75. $D(i) = 10 \cdot \log\left(\dfrac{i}{i_0}\right)$

a. $D(115i_0) = 10\log\left(\dfrac{115i_0}{i_0}\right)$

$$= 10\log 115 \approx 21$$

b. $D(10^{10}i_0) = 10\log\left(\dfrac{10^{10}i_0}{i_0}\right)$

$$= 10 \cdot (10\log 10)$$
$$= 10 \cdot (10 \cdot 1) = 100$$

c. $D(31{,}600{,}000{,}000i_0)$

$$= 10\log\left(\frac{31{,}600{,}000{,}000i_0}{i_0}\right)$$

$$= 10\log 31{,}600{,}000{,}000 \approx 105$$

d. $D(895{,}000{,}000{,}000i_0)$

$$= 10\log\left(\frac{895{,}000{,}000{,}000i_0}{i_0}\right)$$

$$= 10\log(895{,}000{,}000{,}000) \approx 120$$

e. $D(109{,}000{,}000{,}000{,}000i_0)$

$$= 10\log\left(\frac{109{,}000{,}000{,}000{,}000i_0}{i_0}\right)$$

$$= 10\log(109{,}000{,}000{,}000{,}000) \approx 140$$

77. a. $P(T) = 1 - e^{-.0034-.0053T}$

$$P(60) = 1 - e^{-.0034-.0053(60)} \approx .275$$

The reduction will be 27.5% when the tax is $60.

b.
$$P(T) = 1 - e^{-.0034-.0053T}$$
$$.5 = 1 - e^{-.0034-.0053T}$$
$$e^{-.0034-.0053T} = 1 - .5$$
$$-.0034 - .0053T = \ln .5$$
$$T = \frac{\ln(.5) + .0034}{-.0053} \approx 130.14$$

The tax would be $130.14

79. a. To graph
$p = 86.3 \ln h - 680$,
enter this function as y_1 and use
$3000 \le x \le 8500$ and $0 \le y \le 120$.

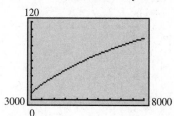

b. Graph $y = 50$, entering this function as y_2. On the CALC menu use "intersect."

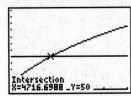

Intersection
X=4716.6988 _Y=50

The point of intersection has approximate coordinates (4716.70, 50). At about 4717 ft, 50% of the moisture is snow. Alternatively, solve

$86.3 \ln h - 680 = 50 \Rightarrow 86.3 \ln h = 730 \Rightarrow$

$\ln h = \dfrac{730}{86.3} \Rightarrow h = e^{730/86.3} \approx 4716.7$.

Chapter 4 Review Exercises

1. $y = a^{x+2}$ is (c).

2. $y = a^x + 2$ is (a).

3. $y = -a^x + 2$ is (d).

4. $y = a^{-x} + 2$ is (b).

5. $0 < a < 1$

6. Domain of f: $(-\infty, \infty)$

7. Range of f: $(0, \infty)$

8. $f(0) = 1$

9. $f(x) = 4^x$

x	-2	-1	0	1	2
y	$\frac{1}{16}$	$\frac{1}{4}$	1	4	16

The negative x-axis is a horizontal asymptote for the graph.

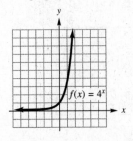

10. $g(x) = 4^{-x}$

x	-2	-1	0	1	2
y	16	4	1	$\frac{1}{4}$	$\frac{1}{16}$

The positive x-axis is a horizontal asymptote for the graph.

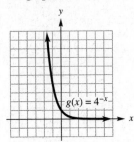

11. $f(x) = \ln x + 5$

x	.01	.1	1	2	4	6	8
y	.4	2.7	5	5.7	6.4	6.8	7.1

The negative y-axis is a vertical asymptote for the graph.

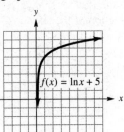

12. $g(x) = \log x - 3$

x	.01	.1	1	2	4	6	8
y	-5	-4	-3	-2.7	-2.4	-2.2	-2.1

The negative y-axis is a vertical asymptote for the graph.

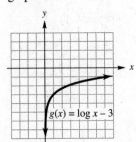

13. $f(x) = 564.13(.9713)^x$,

$x = 0$ corresponds to 2000.

a. $x = 2009 - 2000 = 9$

$f(9) = 564.13(.9713)^9 \approx 434.07$

In 2009, there will be about 434.07 billion cigarettes produced.

$x = 2011 - 2000 = 11$

$f(11) = 564.13(.9713)^{11} \approx 409.51$

In 2011, there will be about 409.51 billion cigarettes produced.

b.   To determine x for $f(x) = 100$, solve:

$f(x) = 100 = 564.13(.9713)^x \Rightarrow$

$\dfrac{100}{564.13} = (.9713)^x \Rightarrow$

$\ln\left(\dfrac{100}{564.13}\right) = x \ln.9713 \Rightarrow$

$x = \dfrac{\ln\left(\dfrac{100}{564.13}\right)}{\ln.9713} \approx 59.41$

Alternatively, graph $Y_1 = 564.13(.9713)^x$ and $Y_2 = 100$ in the window [0, 75] by [−100, 600], then find the intersection.

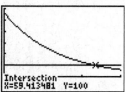

According to the model, cigarette production will fall below 100 billion sometime during 2059. Answers will vary for the second part of the question.

14. $p(t) = 250 - 120(2.8)^{-.5t}$

a. $p(2) = 250 - 120(2.8)^{-.5(2)}$

$p(2) \approx 207$

b. $p(4) = 250 - 120(2.8)^{-.5(4)}$

$p(4) \approx 235$

c. $p(10) = 250 - 120(2.8)^{-.5(10)}$

$p(10) = 249$

d.

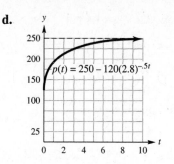

$p(t) = 250 - 120(2.8)^{-.5t}$

15. $10^{2.53148} = 340$ is equivalent to $\log 340 = 2.53148$.

16. $5^4 = 625$ is equivalent to $\log_5 625 = 4$.

17. $e^{3.8067} = 45$ is equivalent to $\ln 45 = 3.8067$.

18. $7^{1/2} = \sqrt{7}$ is equivalent to $\log_7 \sqrt{7} = \dfrac{1}{2}$.

19. $\log 10{,}000 = 4$ is equivalent to $10^4 = 10{,}000$.

20. $\log 26.3 = 1.4200$ is equivalent to $10^{1.4200} = 26.3$.

21. $\ln 81.1 = 4.3957$ is equivalent to $e^{4.3957} = 81.1$.

22. $\log_2 4096 = 12$ is equivalent to $2^{12} = 4096$.

23. $\ln e^5 = 5$ because $\ln e^k = k$ for every real number k.

24. $\log \sqrt[3]{10} = \log_{10} 10^{1/3} = \dfrac{1}{3}$ because $\log_a a^y = y$ for every positive real number y.

25. $10^{\log 8.9} = 8.9$ because $a^{\log_a x} = x$ for every positive real number x.

26. $\ln e^{3t^2} = 3t^2$ because $\ln e^j = j$ for every real number j.

27.   Let $x = \log_8 2$.

$\log_8 2 = x \Rightarrow 8^x = 2 \Rightarrow 2^{3^x} = 2 \Rightarrow 2^{3x} = 2 \Rightarrow$

$3x = 1 \Rightarrow x = \dfrac{1}{3}$

Therefore, $\log_8 2 = \dfrac{1}{3}$.

28. Let $x = \log_8 32$.

$\log_8 32 = x \Rightarrow 8^x = 32 \Rightarrow 2^{3^x} = 2^5 \Rightarrow$

$2^{3x} = 2^5 \Rightarrow 3x = 5 \Rightarrow x = \dfrac{5}{3}$

Therefore, $\log_8 32 = \dfrac{5}{3}$.

29. $\log 4x + \log 5x^5 = \log(4x \cdot 5x^5) = \log\left(20x^6\right)$

30. $4\log u - 5\log u^6 = \log u^4 - \log\left(u^6\right)^5$

$= \log u^4 - \log u^{30}$

$= \log\left(\dfrac{u^4}{u^{30}}\right) = \log\left(\dfrac{1}{u^{26}}\right)$

31. $3\log b - 2\log c = \log b^3 - \log c^2 = \log\left(\dfrac{b^3}{c^2}\right)$

32. $7\ln x - 3(\ln x^3 + 5\ln x) = 7\ln x - 3\ln x^3 - 15\ln x$

$= -3\ln x^3 - 8\ln x$

$= \ln\left(x^3\right)^{-3} + \ln x^{-8}$

$= \ln\left(x^{-9}\right) + \ln x^{-8}$

$= \ln\left(x^{-9} \cdot x^{-8}\right) = \ln x^{-17}$

$= \ln\left(\dfrac{1}{x^{17}}\right)$

33. $\ln(m+8) - \ln m = \ln 3 \Rightarrow$

$\ln\dfrac{m+8}{m} = \ln 3 \Rightarrow \dfrac{m+8}{m} = 3 \Rightarrow 3m = m+8 \Rightarrow$

$2m = 8 \Rightarrow m = 4$

34. $2\ln(y+1) = \ln(y^2-1) + \ln 5$

$\ln(y+1)^2 = \ln\left[(y^2-1)(5)\right]$

$(y+1)^2 = 5(y^2-1)$

$y^2 + 2y + 1 = 5y^2 - 5$

$0 = 4y^2 - 2y - 6$

$0 = 2(2y^2 - y - 3)$

$0 = 2(2y-3)(y+1)$

$y = \dfrac{3}{2}$ or $y = -1$

Since -1 is not in the domain of $\ln(y+1)$ nor in

that of $\ln(y^2-1)$, the only solution is $\dfrac{3}{2}$.

35. $\log(m+3) = 2 \Rightarrow m+3 = 10^2 \Rightarrow$

$m+3 = 100 \Rightarrow m = 97$

36. $\log x^3 = 2 \Rightarrow x^3 = 10^2 \Rightarrow x^3 = 100 \Rightarrow$

$x = \sqrt[3]{100}$

37. $\log_2(3k+1) = 4 \Rightarrow 3k+1 = 2^4 \Rightarrow$

$3k+1 = 16 \Rightarrow 3k = 15 \Rightarrow k = 5$

38. $\log_5\left(\dfrac{5z}{z-2}\right) = 2 \Rightarrow 5^2 = \dfrac{5z}{z-2} \Rightarrow$

$25(z-2) = 5z \Rightarrow 25z - 50 = 5z \Rightarrow$

$20z = 50 \Rightarrow z = \dfrac{50}{20} = 2.5$

39. $\log x + \log(x-3) = 1 \Rightarrow \log\left[x(x-3)\right] = 1 \Rightarrow$

$x(x-3) = 10^1 \Rightarrow x^2 - 3x = 10 \Rightarrow$

$x^2 - 3x - 10 = 0 \Rightarrow (x-5)(x+2) = 0 \Rightarrow$

$x = 5$ or $x = -2$

Since -2 is not in the domain of $\log x$ nor in that of $\log(x-3)$, the only solution is 5.

40. $\log_2 r + \log_2(r-2) = 3$

$\log_2\left[r(r-2)\right] = 3$

$r(r-2) = 2^3$

$r^2 - 2r - 8 = 0$

$(r-4)(r+2) = 0 \Rightarrow r = 4$ or $r = -2$

Since -2 is not in the domain of $\log_2 r$ nor in that of $\log_2(r-2)$, the only solution is 4.

41. $2^{3x} = \dfrac{1}{64} \Rightarrow 2^{3x} = 2^{-6} \Rightarrow 3x = -6 \Rightarrow x = -2$

42. $\left(\dfrac{9}{16}\right)^x = \dfrac{3}{4} \Rightarrow \left[\left(\dfrac{3}{4}\right)^2\right]^x = \left(\dfrac{3}{4}\right)^1 \Rightarrow$

$\left(\dfrac{3}{4}\right)^{2x} = \left(\dfrac{3}{4}\right)^1 \Rightarrow 2x = 1 \Rightarrow x = \dfrac{1}{2}$

43. $9^{2y+1} = 27^y \Rightarrow (3^2)^{2y+1} = (3^3)^y \Rightarrow$

$3^{4y+2} = 3^{3y} \Rightarrow 4y+2 = 3y \Rightarrow y = -2$

44. $\dfrac{1}{2} = \left(\dfrac{b}{4}\right)^{1/4} \Rightarrow \left(\dfrac{1}{2}\right)^4 = \left[\left(\dfrac{b}{4}\right)^{1/4}\right]^4 \Rightarrow$

$\dfrac{1}{16} = \dfrac{b}{4} \Rightarrow 16b = 4 \Rightarrow b = \dfrac{4}{16} = \dfrac{1}{4}$

45. $8^p = 19 \Rightarrow \ln 8^p = \ln 19 \Rightarrow p \ln 8 = \ln 19 \Rightarrow$

$p = \dfrac{\ln 19}{\ln 8} \Rightarrow p \approx \dfrac{2.9444}{2.0794} \Rightarrow p \approx 1.416$

46. $3^z = 11 \Rightarrow \ln 3^z = \ln 11 \Rightarrow z \ln 3 = \ln 11 \Rightarrow$

$z = \dfrac{\ln 11}{\ln 3} \Rightarrow z \approx 2.183$

47. $5 \cdot 2^{-m} = 35 \Rightarrow 2^{-m} = 7 \Rightarrow \ln(2^{-m}) = \ln 7 \Rightarrow$

$-m \ln 2 = \ln 7 \Rightarrow -m = \dfrac{\ln 7}{\ln 2} \Rightarrow$

$-m = \dfrac{1.9459}{.6931} \Rightarrow m \approx -2.807$

48. $2 \cdot 15^{-k} = 18 \Rightarrow 15^{-k} = 9 \Rightarrow \ln 15^{-k} = \ln 9 \Rightarrow$

$-k \ln 15 = \ln 9 \Rightarrow k = -\dfrac{\ln 9}{\ln 15} \Rightarrow k \approx -.811$

49. $e^{-5-2x} = 5 \Rightarrow \ln e^{-5-2x} = \ln 5 \Rightarrow$

$-5 - 2x = \ln 5 \Rightarrow -5 - 2x \approx 1.6094 \Rightarrow$

$-2x = 6.6094 \Rightarrow x \approx -3.305$

50. $e^{3x-1} = 12 \Rightarrow \ln e^{3x-1} = \ln 12 \Rightarrow 3x - 1 = \ln 12 \Rightarrow$

$x = \dfrac{1 + \ln 12}{3} \approx 1.162$

51. $\qquad 6^{2-m} = 2^{3m+1}$

$\ln 6^{2-m} = \ln 2^{3m+1}$

$(2 - m) \ln 6 = (3m + 1) \ln 2$

$2 \ln 6 - m \ln 6 = 3m \ln 2 + \ln 2$

$2 \ln 6 - \ln 2 = 3m \ln 2 + m \ln 6$

$2 \ln 6 - \ln 2 = (3 \ln 2 + \ln 6)m$

$m = \dfrac{2 \ln 6 - \ln 2}{3 \ln 2 + \ln 6} \approx .747$

52. $\qquad 5^{3r-1} = 6^{2r+5}$

$\ln 5^{3r-1} = \ln 6^{2r+5}$

$(3r - 1) \ln 5 = (2r + 5) \ln 6$

$3r \ln 5 - \ln 5 = 2r \ln 6 + 5 \ln 6$

$(3 \ln 5 - 2 \ln 6)r = 5 \ln 6 + \ln 5$

$r = \dfrac{5 \ln 6 + \ln 5}{3 \ln 5 - 2 \ln 6} \approx 8.490$

53. $(1 + .003)^k = 1.089$

$1.003^k = 1.089$

$\ln 1.003^k = \ln 1.089$

$k \ln 1.003 = \ln 1.089$

$k = \dfrac{\ln 1.089}{\ln 1.003} \approx 28.463$

54. $(1 + .094)^z = 2.387$

$1.094^z = 2.387$

$\ln 1.094^z = \ln 2.387$

$z \ln 1.094 = \ln 2.387$

$z = \dfrac{\ln 2.387}{\ln 1.094} \approx 9.684$

55. $g(x) = -41.96 + 9.54 \ln x$,

$x = 90$ corresponds to 1990.

a. $x = 2005 - 1990 + 90 = 105$

$g(105) = -41.96 + 9.54 \ln 105 \approx 2.439$

According to the model, in 2005, National Park expenditures were about \$2.439 billion.

b. $x = 2010 - 1990 + 90 = 110$

$g(110) = -41.96 + 9.54 \ln 110 \approx 2.883$

According to the model, in 2010, National Park expenditures are about \$2.883 billion.

c. $g(x) = 3.5 = -41.96 + 9.54 \ln x \Rightarrow$

$45.46 = 9.54 \ln x \Rightarrow \ln x = \dfrac{45.46}{9.54} \Rightarrow$

$x = e^{45.46/9.54} \approx 117.35$

According to the model, National Park expenditures will reach \$3.5 billion sometime during 2017.

56. $y = 2e^{.02t}$

a. The population will triple, so $y = 3 \cdot 2$. The answer is B.

b. The population will be 3 million, so $y = 3$. The answer is D.

c. $t = 3$

The answer is C.

d. $t = \dfrac{4}{12} = \dfrac{1}{3}$

The answer is A.

57. $A(t) = 10e^{-.00495t}$

a. $A(0) = 10e^{-.00495(0)} = 10e^0 = 10$
The amount of polonium present initially was 10 g.

b. We determine the half-life by finding a value of t such that $A(t) = \left(\dfrac{1}{2}\right)(10) = 5$.

$$5 = 10e^{-.00495(t)}$$

$$\frac{1}{2} = e^{-.00495t}$$

$$\ln\frac{1}{2} = \ln e^{-.00495t}$$

$$\ln\frac{1}{2} = -.00495t$$

$$t = \frac{\ln .5}{-.00495} \approx 140$$

The half-life of polonium is about 140 days.

c. Find t such that $A(t) = 3$.

$$10e^{-.00495(t)} = 3$$

$$e^{-.00495(t)} = .3$$

$$\ln e^{-.00495(t)} = \ln .3$$

$$-.00495t = \ln .3$$

$$t = \frac{\ln .3}{-.00495} \approx 243$$

It will take about 243 days for the polonium to decay to 3 g.

58. $A(x) = 4.23(1.023)^x$, $x = 0$ corresponds to 2000.

a. $A(x) = 8 = 4.23(1.023)^x \Rightarrow$

$$\frac{8}{4.23} = 1.023^x \Rightarrow \ln\left(\frac{8}{4.23}\right) = x\ln 1.023 \Rightarrow$$

$$x = \frac{\ln\left(\dfrac{8}{4.23}\right)}{\ln 1.023} \approx 28.023$$

According to the model, there will be 8 million people with Alzheimer's disease sometime during 2028.

b. $A(x) = 13 = 4.23(1.023)^x \Rightarrow$

$$\frac{13}{4.23} = 1.023^x \Rightarrow \ln\left(\frac{13}{4.23}\right) = x\ln 1.023 \Rightarrow$$

$$x = \frac{\ln\left(\dfrac{13}{4.23}\right)}{\ln 1.023} \approx 49.374$$

According to the model, there will be 13 million people with Alzheimer's disease sometime during 2049.

59. $h(x) = 21.03(.9843)^x$,
$x = 0$ corresponds to 1980.

a. $x = 2002 - 1980 = 22$

$$h(22) = 21.03(.9843)^{22} \approx 14.8472$$

According to the model, there were about 14.8 deaths per 100,000 populations from motor vehicle accidents in 2002.

b. $x = 2008 - 1980 = 28$

$$h(28) = 21.03(.9843)^{28} \approx 13.5023$$

According to the model, there were about 13.5 deaths per 100,000 populations from motor vehicle accidents in 2008.

c. $h(x) = 12 = 21.03(.9843)^x \Rightarrow$

$$\frac{12}{21.03} = .9843^x \Rightarrow \ln\left(\frac{12}{21.03}\right) = x\ln .9843 \Rightarrow$$

$$x = \frac{\ln\left(\dfrac{12}{21.03}\right)}{\ln .9843} \approx 35.4540$$

If the model remains accurate, deaths from motor vehicle accidents will drop below 12 per 100,000 population sometime during 2015.

60. For earthquakes, increasing the ground motion by a factor of 10^k increases the Richter magnitude by k units. In this problem, the second earthquake, with ground motion $1000 = 10^3$ times greater than the first earthquake, will measure $4.6 + 3 = 7.6$ on the Richter scale.

61. $F(t) = T_0 + Ce^{-kt}$

$T_0 = 50, F(0) = 50 + Ce^{-k(0)} = 300, C = 250$

$F(t) = 50 + 250e^{-kt}$

$F(4) = 175 = 50 + 250e^{-k(4)}$

$125 = 250e^{-k(4)}$

$\dfrac{125}{250} = e^{-k(4)} \Rightarrow \ln\left(\dfrac{125}{250}\right) = -k(4) \Rightarrow k \approx .1733$

$F(12) = 50 + 250e^{-.1733(12)} \approx 81.25$

The temperature after 12 minutes is $81.25°$ C.

62. $F(t) = T_0 + Ce^{-kt}, \ T_0 = 18$

$F(0) = 3.4 = 18 + C \Rightarrow C = -14.6$

$F(30) = 18 - 14.6e^{-k(30)} = 7.2$

$-14.6e^{-k(30)} = 7.2 - 18$

$e^{-k(30)} = \dfrac{7.2 - 18}{-14.6} \approx 0.739726$

$k = \dfrac{\ln 0.739726}{-30} \approx 0.01$

$F(t) = 18 - 14.6e^{-(0.01)t} = 10 \Rightarrow$

$-14.6e^{-(0.01)t} = 10 - 18 \Rightarrow$

$e^{-(0.01)t} = \dfrac{-8}{-14.6} \Rightarrow t \approx \dfrac{\ln 0.5479}{-0.01} \approx 60$

It will thaw to $10°C$ in 1 hour.

63. a.

x	0	1	2	3	4	5
y	132.1	157.4	184.4	212.6	241.6	271.1

$f(0) = 132.1, \ f(5) = 271.1$

$f(x) = a(b)^x \Rightarrow f(0) = 132.1 = a(b)^0 \Rightarrow$
$a = 132.1$

$f(5) = a(b)^5 \Rightarrow 271.1 = 132.1(b)^5 \Rightarrow$

$b^5 = \dfrac{271.1}{132.1} \Rightarrow b = \left(\dfrac{271.1}{132.1}\right)^{1/5} \Rightarrow b \approx 1.1546$

The function is $f(x) = 132.1(1.1546)^x$.

b.

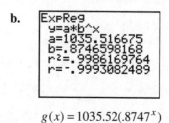

$g(x) = 135.553(1.1543)^x$

c. $x = 2013 - 2006 = 7$

$f(7) = 132.1(1.1546)^7 \approx 361.35$

$g(7) = 135.553(1.1543)^7 \approx 370.12$

Using the two-point model, online retail sales will be about \$361.35 billion in 2013. Using the exponential regression equation, online sales will be about \$370.12 billion in 2013.

d. $f(x) = 500 = 132.1(1.1546)^x \Rightarrow$

$\dfrac{500}{132.1} = (1.1546)^x \Rightarrow$

$\ln\left(\dfrac{500}{132.1}\right) = x\ln 1.1546 \Rightarrow$

$x = \dfrac{\ln\left(\dfrac{500}{132.1}\right)}{\ln 1.1546} \approx 9.3$

$g(x) = 500 = 135.553(1.1543)^x \Rightarrow$

$\dfrac{500}{135.553} = 1.1543^x \Rightarrow$

$\ln\left(\dfrac{500}{135.553}\right) = x\ln 1.1543 \Rightarrow$

$x = \dfrac{\ln\left(\dfrac{500}{135.553}\right)}{\ln 1.1543} \approx 9.1$

Both models predict that online sales will reach \$500 billion per year sometime during 2015.

64. a. At $x = 0, f(0) = 1013$ and at $x = 10$, $f(10) = 265$.

To find $f(x) = a(b^x)$, first solve for a:

$f(0) = 1013 \Rightarrow a(b^0) = 1013 \Rightarrow a = 1013$

Now solve for b:

$f(10) = 265 \Rightarrow 1013b^{10} = 265 \Rightarrow$

$b^{10} = \dfrac{265}{1013} \Rightarrow b = \left(\dfrac{265}{1013}\right)^{1/10} \approx .8745$

The model becomes $f(x) = 1013(.8745^x)$

b.

```
ExpReg
y=a*b^x
a=1035.516675
b=.8746598168
r²=.9986169764
r=-.9993082489
```

$g(x) = 1035.52(.8747^x)$

c. $x = 1.5$

$f(1.5) = 1013(.8745^{1.5}) \approx 828.4$

$g(1.5) = 1035.52(.8747^{1.5}) \approx 847.1$

The models estimate the pressure at 1500m to be 828.4 and 847.1 millibars respectively. The models under and over estimate the actual value of 846 millibars respectively.

$x = 11$

$f(11) = 1013(.8745^{11}) \approx 231.7$

$g(11) = 1035.52(.8747^{11}) \approx 237.5$

The models estimate the pressure at 11,000 m to be 231.7 and 237.5 millibars respectively. Both models overestimate the actual pressure of 227 millibars.

d.   Two-point model:

$1013(.8745^{x}) = 500 \Rightarrow .8745^{x} = \dfrac{500}{1013} \Rightarrow$

$x \ln .8745 = \ln \left(\dfrac{500}{1013} \right) \Rightarrow$

$x = \dfrac{\ln \left(\frac{500}{1013} \right)}{\ln .8745} \approx 5.265$

Regression model:

$1035.52(.8747^{x}) = 500$

$.8747^{x} = \dfrac{500}{1035.52}$

$x \ln (.8747) = \ln \left(\dfrac{500}{1035.52} \right)$

$x = \dfrac{\ln \left(\frac{500}{1035.52} \right)}{\ln .8747} \approx 5.438$

According to the two-point model, the pressure is 500 millibars at 5265 m. According to the regression model, the pressure is 500 millibars at 5438 m.

65. a.

x	1	2	3	4	5	6
y	63	65.7	70.9	77.7	83.8	88.7

$f(1) = 63, \; f(6) = 88.7$

$f(x) = a + b \ln x \Rightarrow$

$f(1) = 63 = a + b \ln 1 \Rightarrow 63 = a$

$f(6) = 63 + b \ln 6 \Rightarrow 88.7 = 63 + b \ln 6 \Rightarrow$

$b \ln 6 = 25.7 \Rightarrow b = \dfrac{25.7}{\ln 6} \approx 14.3434$

The function is $f(x) = 63 + 14.3434 \ln x.$

b.
```
LnReg
y=a+blnx
a=59.12305141
b=14.44870982
r²=.8819097825
r=.9391005178
```

$g(x) = 59.123 + 14.4487 \ln x$

c.   2010 corresponds to $x = 9$

$f(9) = 63 + 14.3434 \ln 9 \approx 94.52$

$g(9) = 59.123 + 14.4487 \ln 9 \approx 90.87$

According to the two-point model, Southwest Airlines will have about 94.52 million passengers in 2010. According to the logarithmic regression, Southwest Airlines will have about 90.87 million passengers in 2010.

d. $f(x) = 100$

$63 + 14.3434 \ln x = 100 \Rightarrow$

$14.3434 \ln x = 37 \Rightarrow \ln x = \dfrac{37}{14.3434} \Rightarrow$

$x = e^{37/14.3434} \approx 13.19$

$g(x) = 100$

$59.123 + 14.4487 \ln x = 100 \Rightarrow$

$14.4487 \ln x = 40.877 \Rightarrow \ln x = \dfrac{40.877}{14.4487} \Rightarrow$

$x = e^{40.877/14.4487} \approx 16.93$

According to the two-point model, Southwest Airlines will have 100 million passengers sometime during 2014. According to the logarithmic regression model, Southwest Airlines will have 100 million passengers sometime during 2017.

66. a.

x	1	5	10	15	16	17
y	9.678	11.083	13.613	16.481	17.091	16.626

$f(1) = 9.678, \; f(17) = 16.626$

$f(x) = a + b \ln x \Rightarrow$

$f(1) = 9.678 = a + b \ln 1 \Rightarrow 9.678 = a$

$f(17) = 9.678 + b \ln 17 \Rightarrow$

$16.626 = 9.678 + b \ln 17 \Rightarrow$

$b \ln 17 = 6.948 \Rightarrow b = \dfrac{6.948}{\ln 17} \approx 2.45234$

The function is

$f(x) = 9.678 + 2.45234 \ln x.$

b.

$$g(x) = 8.69455 + 2.6505 \ln x$$

c. 2010 corresponds to $x = 20$

$$f(20) = 9.678 + 2.45234 \ln 20 \approx 17.025$$

$$g(20) = 8.69455 + 2.6505 \ln 20 \approx 16.635$$

According to the two-point model, there will be about 17,025 kidney transplants in 2010. According to the logarithmic regression, there will be about 16,635 kidney transplants in 2010.

d. $f(x) = 18$ thousand

$$9.678 + 2.45234 \ln x = 18 \Rightarrow$$

$$2.45234 \ln x = 8.322 \Rightarrow \ln x = \frac{8.322}{2.45234} \Rightarrow$$

$$x = e^{8.322/2.45234} \approx 29.8$$

$g(x) = 18$ thousand

$$8.69455 + 2.6505 \ln x = 18$$

$$2.6505 \ln x = 9.30545 \Rightarrow$$

$$\ln x = \frac{9.30545}{2.6505} \Rightarrow$$

$$x = e^{9.30545/2.6505} \approx 33.5$$

According to the two-point model, there will be 18,000 kidney transplants sometime during 2019. According to the logarithmic regression model, there will be 18,000 kidney transplants sometime during 2023.

Case 4 Characteristics of the Monkeyface Prickleback

1. $L_t = L_x(1 - e^{-kt})$

Let $L_x = 71.5$ and $k = .1$; then

$$L_t = 71.5(1 - e^{-.1t}).$$

When $t = 4$,

$$L_t = 71.5(1 - e^{-.1(4)}).$$

$$= 71.5(1 - e^{-.4}) \approx 23.6.$$

When $t = 11$,

$$L_t = 71.5(1 - e^{-.1(11)}).$$

$$= 71.5(1 - e^{-1.1}) \approx 47.7.$$

When $t = 17$,

$$L_t = 71.5(1 - e^{-.1(17)}).$$

$$= 71.5(1 - e^{-1.7}) \approx 58.4.$$

Comparing these answers with the results in Figure 1, the estimates are a bit low.

2. $W = aL^b$

Let $a = .01289$ and $b = 2.9$; then

$$W = .01289L^{2.9}.$$

When $L = 25$, $W = .01289(25)^{2.9} \approx 146.0$.

When $L = 40$, $W = .01289(40)^{2.9} \approx 570.5$.

When $L = 60$, $W = .01289(60)^{2.9} \approx 1848.8$.

Yes, compared to the curve, these answers are reasonable estimates.

Chapter 5 Mathematics of Finance

Section 5.1 Simple Interest and Discount

Note: Exercises in this chapter have been completed with a calculator. To ensure as much accuracy as possible, rounded values have been avoided in intermediate steps. When rounded values have been necessary, several decimal places have been carried throughout the exercise; only the final answer has been rounded to 1 or 2 decimal places. In most cases, answers involving money have been rounded to the nearest cent. Students who use rounded intermediate values should expect their final answers to differ from the answers given here. Depending on the magnitude of the numbers used in the exercise, the difference could be a few pennies or several thousand dollars.

1. The factors are time and interest rate.

3. $2850 at 7% for 8 months

   $P = 2850$, $r = .07$, and $t = \dfrac{8}{12}$.

   $I = Prt$

   $= 2850(.07)\left(\dfrac{8}{12}\right) = 133.00$

   The simple interest is $133.00.

5. $3650 at 6.5% for 11 months

   $P = 3650$, $r = .065$, $t = \dfrac{11}{12}$

   $I = Prt$

   $= 3650(.065)\left(\dfrac{11}{12}\right) = 217.48$

   The simple interest is $217.48.

7. $2830 at 8.9% for 125 days

   $P = 2830$, $r = .089$, $t = \dfrac{125}{365}$

   $I = Prt$

   $= 2830(.089)\left(\dfrac{125}{365}\right) = 86.26$

   The simple interest is $86.26.

9. $5328 at 8%; loan made on August 16 is due December 30.
   The duration of this loan is
   $(31 - 16) + 30 + 31 + 30 + 30 = 136$ days.

   $P = 5328$, $r = .08$, $t = \dfrac{136}{365}$

   $I = Prt$

   $= 5328(.08)\left(\dfrac{136}{365}\right) = 158.82$

   The simple interest is $158.82.

11. $10,000, 15-year bond at 7.5%
    To determine the semiannual interest payment, use the formula $I = Prt$, with $t = \frac{1}{2}$.

    $I = 10,000(0.075)\left(\frac{1}{2}\right) = \$375$.

    Over the 15-year life of the bond, the total interest earned will be $\$375 \cdot 2 \cdot 15 = \$11,250$.

13. $15,000, 3-year bond at 5.15%
    To determine the semiannual interest payment, use the formula $I = Prt$, with $t = \frac{1}{2}$.

    $I = 15,000(0.0515)\left(\frac{1}{2}\right) = \$386.25$.

    Over the 3-year life of the bond, the total interest earned will be $\$386.25 \cdot 2 \cdot 3 = \$2317.50$.

15. $12,000 loan at 3.5% for three months
    Three months is $\frac{1}{4}$ of a year.

    $A = P + I = P + Prt$

    $= 12,000 + 12,000(.035)\left(\frac{1}{4}\right) = \$12,105$

17. $6500 loan at 5.25% for eight months.
    Eight months is $\frac{2}{3}$ of a year.

    $A = P + I = P + Prt$

    $= 6500 + 6500(.0525)\left(\frac{2}{3}\right) \approx \$6727.50$

19. Answers vary. Sample answer: The present value is the amount of money that can be deposited today to yield some larger amount in the future.

21. $15,000 for 9 months; money earns 6%. Use the formula for present value with

$$A = 15,000, t = \frac{9}{12}, r = .06$$

$$P = \frac{A}{1+rt} = \frac{15,000}{1+(.06)\left(\frac{9}{12}\right)}$$

$$= \frac{15,000}{1.045} = \$14,354.07$$

The present value is $14,354.07

23. $15,402 for 120 days; money earns 6.3%.

$$A = 15,402, t = \frac{120}{365} \ r = .063$$

$$P = \frac{A}{1+rt} = \frac{15,402}{1+(.063)\left(\frac{120}{365}\right)}$$

$$= \frac{15,402}{1.0207} = \$15,089.46$$

The present value is $15,089.46

25. Three-month $8000 T-bill with discount rate 1.870%

    a. Three months is $\frac{1}{4}$ of a year.

    Discount $= Prt = 8000 \cdot .01870 \cdot \frac{1}{4} = \$37.40$

    The price of the T-bill is the face value minus the discount. $8000 - 37.40 = \$7962.60$

    b. $I = Prt \Rightarrow 37.40 = 7962.60(r)\left(\frac{1}{4}\right) \Rightarrow$

    $r \approx .01879$
    The actual interest rate is about 1.879%.

27. Six-month $12,000 T-bill with discount rate 2.020%

    a. Six months is $\frac{1}{2}$ of a year.

    Discount $= Prt$

    $= 12,000 \cdot .02020 \cdot \frac{1}{2} = \$121.20$

    The price of the T-bill is the face value minus the discount. $12,000 - 121.20 = \$11,878.80$

    b. $I = Prt \Rightarrow 121.20 = 11,878.80(r)\left(\frac{1}{2}\right) \Rightarrow$

    $r \approx .020406$
    The actual interest rate is about 2.0406%.

29. a. New York City paid the interest from March 1874 until March 2009, 135 years.
$I = Prt = 1000 \cdot .07 \cdot 135 = \$9450$

As of March 2009, New York City paid $9450 in interest on this bond.

    b. The bond that matures in March 2147 will have been earning interest for 2147 – 1874 = 273 years.
$I = Prt = 1000 \cdot .07 \cdot 273 = \$19,110$

This bond will have earned $19,110 in interest when it matures in March 2147.

31. $P = 3000, r = .025, t = \frac{9}{12}$

$$A = P(1+rt)$$

$$= 3000\left[1+(.025)\left(\frac{9}{12}\right)\right] = 3056.25$$

After 9 months, the amount will be $3056.25.

33. Interest $= 67,359.39 - 67,081.20 = 278.19$

$$P = 67,081.20, t = \frac{1}{12}$$

$$I = Prt$$

$$r = \frac{I}{Pt} = \frac{278.19}{(67,081.20)\left(\frac{1}{12}\right)} = .050$$

The interest rate was 5.0%.

35. Interest $= 7675 - 7500 = 175, P = 7500, r = .07$
$I = Prt \Rightarrow 175 = 7500(.07)t \Rightarrow$

$$t = \frac{175}{7500(.07)} = \frac{1}{3}$$

The time period for the loan was $\frac{1}{3}$ of a year, or four months.

37. Want present value of $1769 in 4 months at 6.25% interest.

$$A = 1769, r = .0325, t = \frac{4}{12}$$

$$P = \frac{A}{1+rt} = \frac{1769}{1+(.0325)\left(\frac{4}{12}\right)} = 1750.04$$

The student should deposit $1750.04.

39. $A = 6000, r = .036, t = \frac{10}{12} = \frac{5}{6}$

$$P = \frac{A}{1+rt} = \frac{6000}{1+.036\left(\frac{5}{6}\right)} = 5825.24$$

Yee should deposit $5825.24.

41. $P = 4000, t = \frac{1}{2}$

Discount $= 4000 - 3930 = 70$
Find the discount rate as follows:
$$70 = 4000(r)\left(\frac{1}{2}\right) \Rightarrow r = .035$$
The discount rate is 3.5%

43. Interest $= (24 - 22) + .50 = 2.50$, $P = 22$, $t = 1$

$I = Prt \Rightarrow 2.50 = 22(r)(1) \Rightarrow$

$r = \dfrac{2.50}{22(1)} \approx .114$

The simple interest rate is about 11.4%.

45. You will receive $700 from your tax preparer if you take the offer.

$60 = 760(r)\left(\dfrac{28}{365}\right) \Rightarrow r \approx 1.0291$

The actual interest rate is about 102.91%!

47. First find the maturity value of the loan, the amount that the contractor must repay.

$A = P(1 + rt) \Rightarrow$

$A = 13,500\left(1 + .09\left(\dfrac{9}{12}\right)\right) \approx 14,411.25$

In six months, the bank will receive $14,411.25. Since the bank wants a 10% return, compute the present value of this amount at 10% for six months.

$P = \dfrac{A}{1 + rt} = \dfrac{14,411.25}{1 + .1\left(\dfrac{1}{2}\right)} \approx 13,725$

The bank pays the plumber $13,725 and receives $14,411.25. This is enough for the plumber to pay a bill for $13,650.

49. y_1 is the future value after t years of $100 invested at 8% simple interest. y_2 is the future value after t years of $200 invested at 3%.

a. $y_1 = 100(.08)t + 100 = 8t + 100$

$y_2 = 200(.03)t + 200 = 6t + 200$

b. The graph of y_1 is a line with slope 8 and y-intercept 100. The graph of y_2 is a line with slope 6 and y-intercept 200.

c.

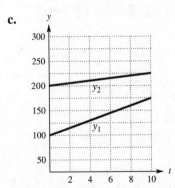

d. The y-intercept of each graph indicates the amount invested. The slope of the graph is the annual amount of interest paid.

Section 5.2 Compound Interest

1. r is the interest rate per year, while i is the interest rate per compounding period. t is the number of years, while n is the number of compounding periods.

3. The interest rate and number of compounding periods determine the amount of interest earned on a fixed principal.

5. The present value is the amount of money to be invested to earn a certain amount of return.

7. $1000 at 4% compounded annually for 6 years

$P = 1000$, $i = \dfrac{4\%}{1} = .04$, and $n = 6(1) = 6$

$A = P(1 + i)^n = 1000(1.04)^6 \approx 1265.32$

The compound amount is $1265.32.

9. $470 at 8% compounded semiannually for 12 years

$P = 470$, $i = \dfrac{8\%}{2} = .04$, $n = 12(2) = 24$

$A = P(1 + i)^n = 470(1.04)^{24} \approx 1204.75$

The compound amount is $1204.75.

11. $6500 at 4.5% compounded quarterly for 8 years

$P = 6500$, $i = \dfrac{4.5\%}{4} = .01125$, $n = 8(4) = 32$

$A = P(1 + i)^n = 6500(1.01125)^{32} \approx 9297.93$

The compound amount is $9297.93.

13. $26,000 at 6% compounded annually for 5 yr

$P = 26,000$, $i = .06$, $n = 5$

Find the compound amount and then amount of interest.

$A = P(1 + i)^n = 26,000(1.06)^5 \approx 34,793.87$

The compound amount is $34,793.87. The amount of interest earned is $34,793.87 - $26,000 = $8793.87.

15. $6000 at 4% compounded semiannually for 6.4 years

Find the compound amount and then amount of interest.

$P = 6000$, $i = \dfrac{.04}{2} = .02$, $n = 6.4(2) = 12.8$

$A = P(1 + i)^n = 6000(1.02)^{12.8} \approx 7730.96$

The amount of interest earned is $7730.96 - $6000 = $1730.96.

17. $5124.98 at 6.3% compounded quarterly for 5.2 years

$P = 5124.98,\ i = \dfrac{.063}{4} = .01575,$

$n = 5.2(4) = 20.8$

$A = P(1+i)^n = 5124.98(1.01575)^{20.8}$

≈ 7093.46

The amount of interest earned is
$7093.46 - $5124.98 = $1968.48.

19. $P = 3000,\ A = 3606,\ n = 5$. Solve for i.

$P(1+i)^n = A \Rightarrow 3000(1+i)^5 = 3606 \Rightarrow$

$(1+i)^5 = \dfrac{3606}{3000} \Rightarrow 1+i = \sqrt[5]{\dfrac{3606}{3000}} \Rightarrow$

$i = \sqrt[5]{\dfrac{3606}{3000}} - 1 \approx .0375$

The interest rate is about 3.75%

21. $P = 8500,\ A = 12{,}161,\ n = 7$. Solve for i.

$P(1+i)^n = A \Rightarrow 8500(1+i)^7 = 12{,}161 \Rightarrow$

$(1+i)^7 = \dfrac{12{,}161}{8500} \Rightarrow \sqrt[7]{(1+i)^7} = \sqrt[7]{\dfrac{12{,}161}{8500}} \Rightarrow$

$1+i = \sqrt[7]{\dfrac{12{,}161}{8500}} \Rightarrow i = \sqrt[7]{\dfrac{12{,}161}{8500}} - 1 \approx 0.0525$

The interest rate is about 5.25%.

For exercises 23–27, we use the compound interest formula. Interest is paid twice per year.

23. $P = 4630,\ r = .052,\ n = 2(15) = 30$

$A = P\left(1 + \dfrac{i}{n}\right)^{nt} = 4630\left(1 + \dfrac{.052}{2}\right)^{30} \approx 10{,}000$

The face value is $10,000.

25. $P = 9992,\ r = .035,\ t = 2(20) = 40$

$A = P\left(1 + \dfrac{i}{n}\right)^{n} = 9992\left(1 + \dfrac{.035}{2}\right)^{40} \approx 20{,}000$

The face value is $20,000.

27. $P = 3888.50,\ r = .063,\ t = 2(30) = 60$

$A = P\left(1 + \dfrac{i}{n}\right)^{n} = 3888.50\left(1 + \dfrac{.063}{2}\right)^{60} \approx 25{,}000$

The face value is $25,000.

29. 4% compounded semiannually
Use the formula for effective rate with $r = .04$ and $m = 2$.

$r_E = \left(1 + \dfrac{r}{m}\right)^m - 1 = \left(1 + \dfrac{.04}{2}\right)^2 - 1$

$= (1.02)^2 - 1 = 1.0404 - 1 = .0404$

The effective rate is 4.04%.

31. 5% compounded quarterly
$r = .05,\ m = 4$

$r_E = \left(1 + \dfrac{r}{m}\right)^m - 1 = \left(1 + \dfrac{.05}{4}\right)^4 - 1 \approx .0509$

The effective rate is 5.09%.

33. 5.2% compounded monthly
$r = .052,\ m = 12$

$r_E = \left(1 + \dfrac{r}{m}\right)^m - 1 = \left(1 + \dfrac{.052}{12}\right)^{12} - 1 \approx .05326$

The effective rate is 5.326%.

35. $12,000 at 5% compounded annually for 6 yr
Use the present value formula for compound interest with
$A = 12{,}000,\ i = .05,$ and $n = 6$.

$P = A(1+i)^{-n} = 12{,}000(1.05)^{-6} \approx 8954.58$

The present value is $8954.58.

37. $17,230 at 4% compounded quarterly for 10 yr

$A = 17{,}230,\ i = \dfrac{.04}{4} = .01,\ n = 10(4) = 40$

$P = A(1+i)^{-n} = 17{,}230(1.01)^{-40} \approx 11{,}572.58$

The present value is $11,572,58.

39. $A = 5000,\ i = \dfrac{.035}{2},\ n = 5(2) = 10$

$P = \dfrac{A}{(1+i)^n} = \dfrac{5000}{\left(1 + \dfrac{.035}{2}\right)^{10}} \approx 4203.64$

A fair price would be $4203.64.

41. $A = 20{,}000,\ i = \dfrac{.047}{2},\ n = 15(2) = 30$

$P = \dfrac{A}{(1+i)^n} = \dfrac{20{,}000}{\left(1 + \dfrac{.047}{2}\right)^{30}} \approx 9963.10$

A fair price would be $9963.10.

43. $A = 1210,\ i = \dfrac{.08}{4} = .02,\ n = 5(4) = 20$

$P = A(1+i)^{-n} = 1210(1.02)^{-20} \approx 814.30$

Since this amount is less than $1000, "$1000 now" is greater.

45. $P = 50,000$, $i = \dfrac{.09}{12} = .0075$, $n = 4(12) = 48$

Find the compound amount and then the amount of interest.

$A = P(1+i)^n = 50,000(1.0075)^{48} \approx \$71,570.27$

The business will pay interest in the amount of $71,570.27 – $50,000 = $21,570.27.

47. $P = 10,000$

Money market fund:

$i = \dfrac{.058}{365}$, $n = 365(2) = 730$

$A = P(1+i)^n = 10,000\left(1+\dfrac{.058}{365}\right)^{730}$

$\approx 11,229.86$

Interest $= 11,229.86 - 10,000 = 1229.86$

Treasury note (recall that treasury notes pay simple interest–see section 5.1):

$A = P(1+rt)$

$A = 10,000(1+.06(2)) \approx 11,200$

Interest $= 11,200 - 10,000 = 1200$

The treasury note pays the most interest.

49. $P = 10,000$, $r = 0.05$, $t = 10$

a. $A = P(1+i)^n$

$= 10,000\left(1+\dfrac{.05}{1}\right)^{10(1)}$

$\approx 16,288.95$

The future value is $16,288.95.

b. $A = 10,000\left(1+\dfrac{.05}{4}\right)^{10(4)} \approx 16,436.19$

The future value is $16,436.19.

c. $A = 10,000\left(1+\dfrac{.05}{12}\right)^{10(12)} \approx 16,470.09$

The future value is $16,470.09.

d. $A = 10,000\left(1+\dfrac{.05}{365}\right)^{10(365)} = 16,486.65$

The future value is $16,486.65.

51. $P = 1000$, $i = .06$, $n = 5$

$A = P(1+i)^n = 1000(1.06)^5 \approx 1338.23$

Since this amount is greater than $1210, "$1000 now" is larger.

53. $A = P(1+i)^n$

$A = 10,000\left(1+\dfrac{.08}{4}\right)^{20(4)} \approx 48,754$

$A = 149,000\left(1+\dfrac{.08}{4}\right)^{20(4)} \approx 726,440$

$A = 150,000\left(1+\dfrac{.08}{4}\right)^{20(4)} \approx 731,316$

$A = 1,000,000\left(1+\dfrac{.08}{4}\right)^{20(4)} \approx 4,875,439$

The range of amounts saved are $0–$48,754; $48,754–$726,440; $731,316–$4,875,439; more than $4,875,439

55. $A = P(1+i)^n$

$420,000,000 + 100 = 100\left(1+\dfrac{r}{1}\right)^{160}$

$\dfrac{420,000,100}{100} = (1+r)^{160}$

$\sqrt[160]{\dfrac{420,000,100}{100}} - 1 = r$

$0.1000 = r$

The interest rate was 10%.

57. $A = 3901$, $P = 10,000$, $n = 8$

$A = P(1+i)^n \Rightarrow 3901 = 10,000(1+i)^8 \Rightarrow$

$.3901 = (1+i)^8 \Rightarrow i = \sqrt[8]{.3901} - 1 \approx -.111$

The interest rate is about –11.1%.

59. For Flagstar Bank, $r = 4.38\%$, $m = 4$.

$r_E = \left(1+\dfrac{r}{m}\right)^m - 1 = \left(1+\dfrac{.0438}{4}\right)^4 - 1 \approx .0445$

The effective rate for Flagstar Bank is about 4.45%.

For Principal Bank, $r = 4.37\%$, $m = 12$.

$r_E = \left(1+\dfrac{r}{m}\right)^m - 1 = \left(1+\dfrac{.0437}{12}\right)^{12} - 1 \approx .04459$

The effective rate for Principal Bank is about 4.46%.

Principal Bank pays a higher APY.

61. $A = 2.9$, $i = \dfrac{.05}{12}$, $n = 5(12) = 60$

Use the formula for present value with compound interest.

$P = A(1+i)^{-n} \approx 2.9\left(1+\dfrac{.05}{12}\right)^{-60} \approx 2.259696$

The company should invest about $2,259,696 now.

63. $A = 23,500$, $i = .0375$, $n = 3$

Use the formula for present value with compound interest.

$$P = A(1+i)^{-n}$$

$$= 23,500(1+.0375)^{-3} \approx 21,042.80$$

The car would have cost about $21,043 three years ago.

65. To find the number of years it will take $1 to inflate to $2, use the formula for compound amount with $A = 2$, $P = 1$, and $i = .03$.

$$A = P(1+i)^n \Rightarrow 2 = 1(1.03)^n \Rightarrow 2 = (1.03)^n \Rightarrow$$

$$\ln 2 = n \ln 1.03 \Rightarrow n = \frac{\ln 2}{\ln 1.03} \approx 23.45$$

Prices will double in about 23.45 yr.

67. Find the number of years, it will take for $1 to inflate to $2 using the formula for compound amount with $A = 2$, $P = 1$, $i = .05$

$$A = P(1+i)^n \Rightarrow 2 = 1(1.05)^n \Rightarrow 2 = (1.05)^n \Rightarrow$$

$$\ln 2 = n \ln 1.05 \Rightarrow n = \frac{\ln 2}{\ln 1.05} \approx 14.21$$

Prices will double in about 14.2 yr.

69. Find the number of years it will take for a demand of 1 unit of electricity to increase to a demand of 2 units using the formula for compound amount with $A = 2$, $P = 1$, $i = .02$

$$A = P(1+i)^n \Rightarrow 2 = 1(1.02)^n \Rightarrow 2 = (1.02)^n \Rightarrow$$

$$\ln 2 = n \ln 1.02 \Rightarrow n = \frac{\ln 2}{\ln 1.02} \approx 35.0$$

The electric utilities will need to double their generating capacity in about 35 yr.

71. First consider the case of earning interest at a rate of k per annum compounded quarterly for all eight years and earning $2203.76 interest on the $1000 investment.

$$2203.76 = 1000\left(1+\frac{k}{4}\right)^{8(4)} \Rightarrow$$

$$2.20376 = \left(1+\frac{k}{4}\right)^{32}$$

Use a calculator to raise both sides to the power of $\frac{1}{32}$.

$$1.025 = 1+\frac{k}{4} \Rightarrow .025 = \frac{k}{4} \Rightarrow .1 = k$$

Next consider the actual investments. The $1000 was invested for the first five years at a rate of j per annum compounded semiannually.

$$A = 1000\left(1+\frac{j}{2}\right)^{5(2)} = 1000\left(1+\frac{j}{2}\right)^{10}$$

This amount was then invested for the remaining three years at $k = .1$ per annum compounded quarterly for a final compound amount of $1990.76.

$$1990.76 = A\left(1+\frac{.1}{4}\right)^{3(4)}$$

$$1990.76 = A(1.025)^{12}$$

$$1480.24 \approx A$$

Recall that $A = 1000\left(1+\dfrac{j}{2}\right)^{10}$ and substitute this value into the above equation.

$$1480.24 = 1000\left(1+\frac{j}{2}\right)^{10}$$

$$1.48024 = \left(1+\frac{j}{2}\right)^{10}$$

Use a calculator to raise both sides to the power of $\frac{1}{10}$.

$$1.04 \approx 1+\frac{j}{2} \Rightarrow .04 = \frac{j}{2} \Rightarrow .08 = j$$

The ratio of k to j is $\dfrac{k}{j} = \dfrac{.1}{.08} = 1.25$, which is choice (a).

Section 5.3 Annuities, Future Value, and Sinking Funds

1. $1 + 1.05 + 1.05^2 + 1.05^3 + \cdots + 1.05^{14}$

$$= \frac{1.05^{15} - 1}{1.05 - 1} \approx 21.5786$$

In Exercises 3–7, use the formula

$$S = R \cdot s_{\overline{n}|i} \text{ or } S = R\left[\frac{(1+i)^n - 1}{i}\right].$$

3. $R = 12,000$, $i = .062$, $n = 8$

$$S = 12,000\left[\frac{(1.062)^8 - 1}{.062}\right] \approx 119,625.61$$

The future value is $119,625.61.

5. $R = 865$, $i = \dfrac{.06}{2} = .03$, $n = 10(2) = 20$

$$S = 865\left[\frac{(1.03)^{20} - 1}{.03}\right] \approx 23,242.87$$

The future value is $23,242.87.

7. $R = 1200$, $i = \dfrac{.08}{4} = .02$, $n = 10(4) = 40$

$$S = 1200\left[\dfrac{(1.02)^{40} - 1}{.02}\right] \approx 72,482.38$$

The future value is $72,482.38.

9. $R = 400$, $i = \dfrac{.04}{12}$, $n = 12(10) = 120$

$$S = 400\left[\dfrac{\left(1 + \frac{.04}{12}\right)^{120} - 1}{\frac{.04}{12}}\right] \approx 58,899.92$$

At the end of 10 years, there will be about $58,899.92 in the account. Using the compound amount formula, the future value of this money is

$58,899.92\left(1 + \frac{.06}{12}\right)^{120} \approx 107,162.32$, for $i = \dfrac{.06}{12}$,

and $n = 12(10) = 120$. Now compute the amount in the account for $R = 600$, $i = \dfrac{.06}{12}$, and

$n = 12(10) = 120$.

$$S = 600\left[\dfrac{\left(1 + \frac{.06}{12}\right)^{120} - 1}{\frac{.06}{12}}\right] \approx 98,327.61.$$

$98,327.61 + 107,162.32 = 205,489.93$
There will be about $205,490 in the account at the end of 20 years.

11. $R = 1000$, $i = \dfrac{.042}{4}$, $n = 4(10) = 40$

$$S = 1000\left[\dfrac{\left(1 + \frac{.042}{4}\right)^{40} - 1}{\frac{.042}{4}}\right] \approx 49,393.58$$

At the end of 10 years, there will be about $49,393.58 in the account. Using the compound amount formula, the future value of this money is

$49,393.58\left(1 + \frac{.074}{4}\right)^{60} \approx 148,365.36$, for

$i = \dfrac{.074}{4}$, and $n = 4(15) = 60$. Now compute the

amount in the account for $R = 1500$, $i = \dfrac{.074}{4}$,

and $n = 4(15) = 60$.

$$S = 1500\left[\dfrac{\left(1 + \frac{.074}{4}\right)^{60} - 1}{\frac{.074}{4}}\right] \approx 162,465.22.$$

$148,365.36 + 162,465.22 = 310,830.58$
There will be about $310,831 in the account at the end of 25 years.

13. $S = 11,000$, $i = \dfrac{.05}{2} = .025$, $n = 6(2) = 12$

$$11,000 = R\left[\dfrac{(1.025)^{12} - 1}{.025}\right]$$

$11,000 \approx R(13.79555)$
$R \approx 797.36$
The periodic payment is $797.36.

15. $S = 50,000$, $i = \dfrac{.08}{4} = .02$, $n = \left(2\dfrac{1}{2}\right)(4) = 10$

$$50,000 = R\left[\dfrac{(1.02)^{10} - 1}{.02}\right]$$

$50,000 \approx R(10.94972)$
$R \approx 4566.33$
The periodic payment is $4566.33.

17. $S = 6000$, $i = \dfrac{.06}{12}$, $= .005$, $n = 3(12) = 36$

$$6000 = R\left[\dfrac{(1.005)^{36} - 1}{.005}\right]$$

$6000 \approx R(39.33610)$
$R \approx 152.53$
The periodic payment is $152.53.

For exercises 19–21, we use the future value formula,

$$S = R\left[\dfrac{(1+i)^n - 1}{i}\right].$$

19. $R = 3940$, $n = 10$, $S = 50,000$, $i = r$

$$50,000 = 3940\left[\dfrac{(1+i)^{10} - 1}{i}\right]$$

Graphing each side of the equation, we find that the intersection is $x \approx .0519$

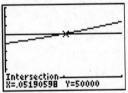

[0, .1] by [−10,000, 50,000]
An interest rate of about 5.19% is needed.

21. $R = 1675$, $n = 5(4) = 20$, $S = 38,000$, $i = \frac{r}{4}$

$$38,000 = 1675 \left[\frac{\left(1 + \frac{r}{4}\right)^{20} - 1}{\frac{r}{4}} \right]$$

Graphing each side of the equation, we find that the intersection is $x \approx .05223$

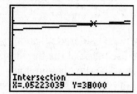

[0, .075] by [−10,000, 50,000]

An interest rate of about 5.223% is needed.

23. A sinking fund is a fund that is set up to receive periodic payments. Businesses and corporations use sinking funds to repay bond issues or to provide for replacement of fixed assets.

25. $R = 500$, $i = .05$, $n = 10$

$$S = 500 \left[\frac{(1.05)^{11} - 1}{.05} \right] - 500 \approx 6603.39$$

The future value is $6603.39.

27. $R = 16,000$, $i = .047$, $n = 11$

$$S = 16,000 \left[\frac{(1.047)^{12} - 1}{.047} \right] - 16,000$$
$$\approx 234,295.32$$

The future value is $234,295.32.

29. $R = 1000$, $i = \frac{.08}{2} = .04$, $n = 9(2) = 18$

$$S = 1000 \left[\frac{(1.04)^{19} - 1}{.04} \right] - 1000$$
$$\approx 26,671.23$$

The future value is $26,671.23.

31. $R = 100$, $i = \frac{.09}{4} = .0225$, $n = 7(4) = 28$

$$S = 100 \left[\frac{(1.0225)^{29} - 1}{.0225} \right] - 100$$
$$\approx 3928.88$$

The future value is $3928.88.

For exercises 33–35, we use the future value of an annuity due formula, $S = R \left[\dfrac{(1+i)^{n+1} - 1}{i} \right] - R$.

33. $S = 8000$, $n = 3(4) = 12$, $i = \frac{.044}{4}$

$$8000 = R \left[\frac{\left(1 + \frac{.044}{4}\right)^{13} - 1}{\frac{.044}{4}} \right] - R \Rightarrow$$

$$8000 = R \left(\frac{\left(1 + \frac{.044}{4}\right)^{13} - 1}{\frac{.044}{4}} - 1 \right) \Rightarrow$$

$$R = \frac{8000}{\dfrac{\left(1 + \frac{.044}{4}\right)^{13} - 1}{\frac{.044}{4}} - 1} \approx 620.46$$

The payment should be $620.46.

35. $S = 55,000$, $n = 12(12) = 144$, $i = \frac{.057}{12}$

$$55,000 = R \left[\frac{\left(1 + \frac{.057}{12}\right)^{145} - 1}{\frac{.057}{12}} \right] - R \Rightarrow$$

$$55,000 = R \left(\frac{\left(1 + \frac{.057}{12}\right)^{145} - 1}{\frac{.057}{12}} - 1 \right) \Rightarrow$$

$$R = \frac{55,000}{\dfrac{\left(1 + \frac{.057}{12}\right)^{145} - 1}{\frac{.057}{12}} - 1} \approx 265.71$$

The payment should be $265.71.

37. $R = 150$, $i = \frac{.048}{12}$, $n = 40(12) = 480$

Use the formula for future value of an ordinary annuity.

$$S = R \cdot s_{\overline{n}|i}$$

$$= 150 \left[\frac{\left(1 + \frac{.048}{12}\right)^{480} - 1}{\frac{.048}{12}} \right] \approx 217,308.21$$

After 40 years, the amount in the account will be about $217,308.21.

39. $R = 12,000$, $i = \dfrac{.06}{1}$, $n = 9$

   a. Use the formula for future value of an ordinary annuity.

$$S = R \cdot s_{\overline{n}|i}$$

$$= 12,000\left[\frac{(1+.06)^9 - 1}{.06}\right] \approx 137,895.79$$

     She will have \$137,895.79 on deposit after 9 years.

   b. $i = \dfrac{.05}{1}$

$$S = 12,000\left[\frac{(1+.05)^9 - 1}{.05}\right] \approx 132,318.77$$

     She will have \$132,318.77 on deposit after 9 years.

   c. $137,895.79 - 132,318.77 = 5577.02$
     She would lose \$5577.02.

41. From ages 50 to 60, we have an ordinary annuity with $R = 1200$, $i = \dfrac{.07}{4} = .0175$, $n = 10(4) = 40$. Use the formula for the future value of an ordinary annuity.

$$S = \left[\frac{(1+i)^n - 1}{i}\right] = 1200\left[\frac{(1.0175)^{40} - 1}{.0175}\right]$$
$$= 68,680.96$$

At age 60, the value of the retirement account is \$68,680.96. This amount now earns 9% interest compounded monthly for 5 years.
Use the formula for compound amount with $P = 68,680.96$, $i = \dfrac{.09}{12} = .0075$, and $n = 5(12) = 60$ to find the value of this amount after 5 years.

$$A = P(1+i)^n$$
$$= 68,680.96(1.0075)^{60} = 107,532.48$$

The value of the amount she withdraws from the retirement account will be \$107,532.48 when she reaches 65. The deposits of \$300 at the end of each month into the mutual fund form another ordinary annuity. Use the formula for the future value of an ordinary annuity with $R = 300$, $i = \dfrac{.09}{12} = .0075$, and $n = 5(12) = 60$.

$$S = R\left[\frac{(1+i)^n - 1}{i}\right] = 300\left[\frac{(1.0075)^{60} - 1}{.0075}\right]$$
$$= 22,627.24$$

The value of this annuity after 5 years is \$22,627.24.
The total amount in the mutual fund account when the woman reaches age 65 will be \$107,532.48 + \$22,627.24 = \$130,159.72.

43. This may be considered an annuity due, since payments are made at the beginning of each year, starting with the day the daughter is born. However, a payment should not be subtracted at the end, since a twenty-second payment is made on her twenty-first birthday. Thus, the future value is given by

$$S = R\left[\frac{(1+i)^{n+1} - 1}{i}\right],$$

where $R = 1000$, $i = .065$, and $n = 21$. Therefore,

$$S = 1000\left[\frac{(1.065)^{22} - 1}{.065}\right] = 46,101.64$$

There will be \$46,101.64 in the account at the end of the day on the daughter's twenty-first birthday.

45. For the first 12 years, we have an annuity due. To find the amount in this account after 12 years, use the formula for the future value of an annuity due with $R = 10,000$, $i = .05$, and $n = 12$.

$$S = R\left[\frac{(1+i)^{n+1} - 1}{i}\right] - R$$
$$= 10,000\left[\frac{(1.05)^{13} - 1}{.05}\right] - 10,000$$
$$\approx 167,129.83$$

This amount, \$167,129.83, now earns 6% interest compounded semiannually for another 9 yr, but no new deposits are made. Use the formula for compound amount with $P = 167,129.83$, $i = \dfrac{.06}{2} = .03$, and $n = 9(2) = 18$.

$$A = P(1+i)^n = 167,129.83(1.03)^{18}$$
$$= 284,527.35.$$

The final amount on deposit after 21 yr is \$284,527.35.

47. $S = 10,000$, $n = 8(4) = 32$

  a. $i = \dfrac{.05}{4} = .0125$

$$S = R \cdot s_{\overline{n}|i}$$

$$10,000 = R\left[\frac{(1.0125)^{32} - 1}{.0125}\right] \Rightarrow$$

$$10,000 \approx R(39.05044) \Rightarrow R = \$256.08$$

He should deposit $256.08 at the end of each quarter.

  b. $i = \dfrac{.058}{4} = .0145$

$$S = R \cdot s_{\overline{n}|i}$$

$$10,000 = R\left[\frac{(1.0145)^{32} - 1}{.0145}\right]$$

$$10,000 \approx R(40.35398) \Rightarrow R = \$247.81$$

He should deposit $247.81 quarterly.

49. $S = 24,000$, $i = \dfrac{.05}{4} = .0125$, $n = 6(4) = 24$

$$S = R \cdot s_{\overline{n}|i}$$

$$24,000 = R\left[\frac{(1.0125)^{24} - 1}{.0125}\right] \Rightarrow$$

$$R = \frac{24,000}{\dfrac{(1.0125)^{24} - 1}{.0125}} \approx 863.68$$

She should deposit $863.68 at the end of each quarter.

51. **a.** $P = 60,000$, $r = .08$, $t = \dfrac{1}{4}$

$$I = Prt = 60,000(.08)\left(\frac{1}{4}\right) = \$1200$$

Each quarterly interest payment is $1200.

  b. $S = 60,000$, $i = \dfrac{.06}{2} = .03$,

$n = 7(2) = 14$

$$S = R \cdot s_{\overline{n}|i}$$

$$60,000 = R\left[\frac{(1.03)^{14} - 1}{.03}\right]$$

$$60,000 \approx R(17.08632)$$

$$R = \$3511.58$$

The amount of each payment is $3511.58.

53. $S = 147,126$, $R = 300$, $n = 12(20)$, $i = \dfrac{r}{12}$

$$S = R\left[\frac{(1+i)^n - 1}{i}\right]$$

$$147,126 = 300\left[\frac{\left(1 + \frac{r}{12}\right)^{12(20)} - 1}{\frac{r}{12}}\right]$$

Graph each side of this equation. The intersection is (.065, 147,126).

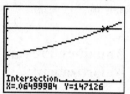

Intersection
X=.06499984   Y=147126

[0, .075] by [−20,000, 200,000]

She received an interest rate of 6.5%.

55. **a.**   Answers will vary.

  b.   Joe contributed $(600)(40) = \$24,000$ to his retirement.

  c.   For the first ten years, Joe has an ordinary annuity with $R = 600$, $i = \dfrac{.081}{4}$, and $n = 10(4) = 40$. Use the formula for the future value of an ordinary annuity.

$$S = \left[\frac{(1+i)^n - 1}{i}\right] = 600\left[\frac{\left(1 + \frac{.081}{4}\right)^{40} - 1}{\frac{.081}{4}}\right]$$

$$= 36,438.25$$

He makes no further deposits, but leaves this amount in the account at the same interest rate. Use the formula for compound amount with $P = 36,438.25$, $i = \dfrac{.081}{4}$, and $n = 35(4) = 140$ to find the value of this amount after 35 years.

$$A = P(1+i)^n$$

$$= 36,438.25\left(1 + \frac{.081}{4}\right)^{140} \approx 603,229$$

After 45 years, there will be about $603,229 in Joe's account.

  d.   Sarah contributed $(600)(4)(35) = \$84,000$ to her retirement.

e. Sarah has an ordinary annuity for 35 years with $R = 600$, $i = \dfrac{.081}{4}$, and $n = 35(4) = 140$.

$$S = \left[\frac{(1+i)^n - 1}{i}\right] = 600\left[\frac{\left(1 + \frac{.081}{4}\right)^{140} - 1}{\frac{.081}{4}}\right]$$

$$\approx 460,884$$

Sarah will have about \$460,884 in her account after 45 years.

Section 5.4 Annuities, Present Value, and Amortization

1. The present value of an annuity is the amount that must be deposited today to provide all the payments for the term of the annuity. The future value of an annuity is the final amount in the account before any payouts.

In Exercises 3–5, use the formula

$$P = R\left[\frac{1 - (1+i)^{-n}}{i}\right].$$

3. $R = 1400$, $i = .06$, $n = 8$

$$P = 1400\left[\frac{1 - (1.06)^{-8}}{.06}\right] \approx 8693.71$$

The present value is \$8693.71.

5. $R = 50,000$, $i = \dfrac{.05}{4} = .0125$, $n = 10(4) = 40$

$$P = 50,000\left[\frac{1 - (1.0125)^{-40}}{.0125}\right] \approx 1,566,346.66$$

The present value is \$1,566,346.66.

7. $R = 650$, $i = \dfrac{.049}{4}$, $n = 5(4) = 20$

$$P = 650\left[\frac{1 - \left(1 + \frac{.049}{4}\right)^{-20}}{\frac{.049}{4}}\right] \approx 11,468.10$$

About \$11,468.10 is needed to fund the withdrawals.

9. $R = 425$, $i = \dfrac{.061}{12}$, $n = 10(12) = 120$

$$P = 425\left[\frac{1 - \left(1 + \frac{.061}{12}\right)^{-120}}{\frac{.061}{12}}\right] \approx 38,108.61$$

About \$38,108.61 is needed to fund the withdrawals.

11. $P = 90,000$, $i = \dfrac{.049}{12}$, $n = 22(12) = 264$

$$90,000 = R\left[\frac{1 - \left(1 + \frac{.049}{12}\right)^{-264}}{\frac{.049}{12}}\right] \Rightarrow$$

$$R = \frac{90,000}{\frac{1 - \left(1 + \frac{.049}{12}\right)^{-264}}{\frac{.049}{12}}} \approx 557.68$$

The monthly payment will be \$557.68.

13. $P = 275,000$, $i = \dfrac{.06}{4} = .015$, $n = 18(4) = 72$

$$275,000 = R\left[\frac{1 - (1+.015)^{-72}}{.015}\right] \Rightarrow$$

$$R = \frac{275,000}{\frac{1 - (1+.015)^{-72}}{.015}} \approx 6272.14$$

The monthly payment will be \$6272.14.

In Exercises 15–17, use the formula

$$P = R \cdot a_{\overline{n}|i} \text{ or } P = R\left[\frac{1 - (1+i)^{-n}}{i}\right].$$

15. $R = 10,000$, $i = .03$, $n = 15$

The lump sum is the same as the present value of the annuity.

$$P = 10,000\left[\frac{1 - (1.03)^{-15}}{.03}\right] \approx 119,379.35$$

The required lump sum is \$119,379.35.

17. $R = 10,000$, $i = .06$, $n = 15$

$$P = 10,000\left[\frac{1 - (1.06)^{-15}}{.06}\right] \approx 97,122.49$$

The required lump sum is \$97,122.49.

19. Find the future value of an annuity with

$R = 4000$, $i = \dfrac{.06}{2} = .03$, and $n = 10(2) = 20$.

$S = R\left[\dfrac{(1+i)^n - 1}{i}\right] = 4000\left[\dfrac{(1.03)^{20} - 1}{.03}\right]$

$\approx 107,481.50$

Now find the present value of $107,481.50 at 8% compounded quarterly for 10 years.

$A = 107,481.50$, $i = \dfrac{.08}{4} = .02$, $n = 10(4) = 40$

$A = P(1+i)^n \Rightarrow 107,481.50 = P(1.02)^{40} \Rightarrow$

$P = \dfrac{107,481.50}{(1.02)^{40}} \approx 48,677.34$

The required lump sum is $48,677.34.

21. The interest that is paid each half-year is

$I = Prt = 15,000 \cdot .06 \cdot \dfrac{1}{2} = \$450$. So, we have a

two-part investment, an annuity that pays $450 every six months for 4 years and the $15,000 face value of the bond which will be paid when the bond matures 4 years from now. The purchaser should be willing to pay the present value of each part of the investment, assuming 5% interest compounded semiannually.
Present value of annuity

$R = 450$, $i = \dfrac{.05}{2} = .025$, $n = 4(2) = 8$

$P = 450\left[\dfrac{1 - (1 + .025)^{-8}}{.025}\right] \approx 3226.56$

Present value of $15,000 in 4 years

$P = A(1+i)^{-n}$

$P = 15,000(1 + .025)^{-8} \approx 12,311.20$

The purchaser should be willing to pay
$3226.56 + $12,311.20 = $15,537.76.

23. The interest that is paid each half-year is

$I = Prt = 10,000 \cdot .054 \cdot \frac{1}{2} = \$270$. So, we have

a two-part investment, an annuity that pays $270 every six months for 12 years and the $10,000 face value of the bond which will be paid when the bond matures 12 years from now. The purchaser should be willing to pay the present value of each part of the investment, assuming 6.5% interest compounded semiannually.

Present value of annuity

$R = 270$, $i = \dfrac{.065}{2}$, $n = 12(2) = 24$

$P = 270\left[\dfrac{1 - \left(1 + \frac{.065}{2}\right)^{-24}}{\frac{.065}{2}}\right] \approx 4451.85$

Present value of $10,000 in 12 years

$P = A(1+i)^{-n} = 10,000\left(1 + \frac{.065}{2}\right)^{-24} \approx 4641.29$

The purchaser should be willing to pay
$4451.85 + $4641.29 = $9093.14.

25. $P = 2500$, $i = \dfrac{.08}{4} = .02$, $n = 6$

$R = 2500\left[\dfrac{.02}{1 - (1.02)^{-6}}\right] \approx 446.31$

Quarterly payments of $446.31 are required to amortize this loan.

27. $P = 90,000$, $i = \dfrac{.07}{1} = .07$, $n = 12$

$R = 90,000\left[\dfrac{.07}{1 - (1.07)^{-12}}\right] \approx 11,331.18$

Annual payments of $11,331.18 are required to amortize this loan.

29. $P = 7400$, $i = \dfrac{.082}{2} = .041$, $n = 18$

$R = 7400\left[\dfrac{.041}{1 - (1.041)^{-18}}\right] \approx 589.31$

Semiannual payments of $589.31 are required to amortize this loan.

For exercises 31–37, we use the amortization payment

formula $R = \dfrac{Pi}{1 - (1+i)^{-n}}$.

31. $P = 149,560$, $i = \dfrac{.0675}{12}$, $n = 25(12) = 300$

$R = 149,560\left[\dfrac{\frac{.0675}{12}}{1 - \left(1 + \frac{.0675}{12}\right)^{-300}}\right] \approx 1033.33$

Monthly payments of $1033.33 are required to amortize this loan.

33. $P = 153{,}762,\ i = \dfrac{.0545}{12},\ n = 30(12) = 360$

$$R = 153{,}762\left[\frac{\frac{.0545}{12}}{1-\left(1+\frac{.0545}{12}\right)^{-360}}\right] \approx 868.23$$

Monthly payments of $868.23 are required to amortize this loan.

35. $P = 8500,\ i = \dfrac{.066}{12},\ n = 3(12) = 36$

$$R = 8500\left[\frac{\frac{.066}{12}}{1-\left(1+\frac{.066}{12}\right)^{-36}}\right] \approx 260.90$$

After 2 years (24 payments), the remaining 12 payments can be thought of as annuity. The present value of this annuity is the remaining balance.
So, we use the present value formula with

$R = 260.90,\ i = \dfrac{.066}{12},\ n = 12$

$$P = 260.90\left[\frac{1-\left(1+\frac{.066}{12}\right)^{-12}}{\frac{.066}{12}}\right] \approx 3021.69$$

The monthly payment is $260.90 and the balance remaining after two years is about $3022.

37. $P = 130{,}000,\ i = \dfrac{.068}{12},\ n = 30(12) = 360$

$$R = 130{,}000\left[\frac{\frac{.068}{12}}{1-\left(1+\frac{.068}{12}\right)^{-360}}\right] \approx 847.50$$

After 12 years (144 payments), the remaining 216 payments can be thought of as annuity. The present value of this annuity is the remaining balance. So, we use the present value formula

with $R = 847.50,\ i = \dfrac{.068}{12},\ n = 216$

$$P = 847.50\left[\frac{1-\left(1+\frac{.068}{12}\right)^{-216}}{\frac{.068}{12}}\right] \approx 105{,}428.61$$

The monthly payment is $847.50 and the balance remaining after two years is about $105,429.

39. Locate the table entry that is in the row labeled Payment Number 5 and in the column labeled Interest for Period. Note that $6.80 of the fifth payment is interest.

41. To find how much interest is paid in the first 5 months of the loan, add the first five nonzero entries in the column labeled Interest for Period.

$10.00
 9.21
 8.42
 7.61
+ 6.80
―――――
$42.04

In the first 5 months of the loan, $42.04 is paid in interest.

For exercises 43–45, we use the formula for the present

value of an annuity due, $S = R + R\left[\dfrac{1-(1+i)^{-(n-1)}}{i}\right]$

43. The yearly payment for $57.6 million is

$\dfrac{57{,}600{,}000}{30} = 1{,}920{,}000.$

$R = 1{,}920{,}000,\ i = .051,\ n = 30$

$$S = 1{,}920{,}000 + 1{,}920{,}000\left[\frac{1-(1.051)^{-(30-1)}}{.051}\right]$$

$\approx 30{,}669{,}881$

The cash value is about $30,669,881.
The figure shows the solution using the TVM on a TI-84 Plus calculator.

```
N=30
I%=5.1
▪PV=30669880.65
PMT=-1920000
FV=0▪
P/Y=1
C/Y=1
PMT:END BEGIN
```

45. The yearly payment for $41.6 million is

$\dfrac{41{,}600{,}000}{26} = 1{,}600{,}000.$

$R = 1{,}600{,}000,\ i = .04735,\ n = 26$

$$S = 1{,}600{,}000 + 1{,}600{,}000\left[\frac{1-(1.04735)^{-25}}{.04735}\right]$$

$\approx 24{,}761{,}633$

The cash value is about $24,761,633.

The figure shows the solution using the TVM on a TI-84 Plus calculator.

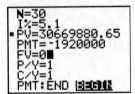

```
N=26
I%=4.735
▪PV=24761633.17
PMT=-1600000
FV=0
P/Y=1
C/Y=1
PMT:END BEGIN
```

47. The monthly payments form an annuity with $R = 30$, $i = .0125$, and $n = 36$.

    a. The price of the stereo system is $600 plus the present value of the annuity.

$$P = 600 + 30\left[\frac{1-(1.0125)^{-36}}{.0125}\right]$$
$$= 600 + 865.42$$
$$= 1465.42$$

    The stereo costs $1465.42.

    b. The total amount paid is
$$600 + 30(36) = 600 + 1080$$
$$= \$1680$$
$$\text{Interest} = 1680.00 - 1465.42 = 214.58$$
    The total amount of interest paid is $214.58.

49. $P = 15,000$, $i = \dfrac{.10}{2} = .05$, $n = 4(2) = 8$

$$R = \frac{P}{a_{\overline{n}|i}} = \frac{15,000}{\left[\frac{1-(1.05)^{-8}}{.05}\right]} \approx \frac{15,000}{6.46321} = \$2320.83$$

Each payment is $2320.83.

51. $R = \dfrac{35,000}{\left[\frac{1-(1+\frac{.0743}{12})^{-240}}{\frac{.0743}{12}}\right]} \approx 280.46$

Total payments $= 20(12)(280.46) = 67,310.40$
Interest $= 67,310.40 - 35,000 = 32,310.40$
The monthly payments are $280.46 and the total interest is $32,310.40.

53. **a.** $P = 14,000 + 7200 - 1200 = 20,000$

$$i = \frac{.12}{2} = .06,\ n = 5(2) = 10$$

$$R = \frac{20,000}{\left[\frac{1-(1.06)^{-10}}{.06}\right]} \approx 2717.36$$

Each payment is $2717.36.

    b. There were 2 payments left at the time she decided to pay off the loan.

55. **a.** $P = 212,000 - .2(212,000) = 169,600$

$$i = \frac{.072}{12} = .006, n = 30(12) = 360$$

$$R = \frac{169,600}{\left[\frac{1-(1.006)^{-360}}{.006}\right]} \approx 1151.22$$

    Their monthly payments are $1151.22.

b. There are $360 - 96 = 264$ payments left to be made. This can be thought of as an annuity consisting of 264 payments of $1151.22 at 7.2% interest

$$\frac{.072}{12} = .006 \text{ per period.}$$

$$1151.22\left[\frac{1-(1.006)^{-264}}{.006}\right] = 152,320.58$$

Their loan balance is $152,320.58.

c. They will have made $6(12) = 72$ payments. There will be $360 - 72 = 288$ payments remaining. The loan balance at the beginning of the seventh year will be:

$$1151.22\left[\frac{1-(1.006)^{-288}}{.006}\right] = 157,609.90$$

The loan balance at the end of the seventh year will be:

$$1151.22\left[\frac{1-(1.006)^{-276}}{.006}\right] = 155,060.12$$

So they pay
$157,609.90 - \$155,060.12 = \$2549.78$
interest during the 7th year.

d. $1151.22 + 150 = \dfrac{169,600}{\left[\frac{1-(1.006)^{-12x}}{.006}\right]}$

Solve by graphing $y_1 = 1301.22$ and

$y_2 = \dfrac{169,600}{\left[\frac{1-(1.006)^{-12x}}{.006}\right]}$. It will take 21.22 years

to pay off the loan.

57. We can think of these as two ordinary annuities. First, we determine the future value of her retirement account with $R = 400$, $i = \dfrac{.07}{12}$, and $n = 45(12) = 540$.

$$S = R\left[\frac{(1+i)^n - 1}{i}\right] = 400\left[\frac{\left(1+\frac{.07}{12}\right)^{540} - 1}{\frac{.07}{12}}\right]$$
$$\approx 1,517,038$$

Now we compute the amount of each withdrawal, using $P = 1,517,038$, $i = \dfrac{.05}{12}$, and $n = 30(12) = 360$.

(continued on next page)

(continued from page 141)

$$P = R\left[\frac{1-(1+i)^{-n}}{i}\right] \Rightarrow 1,517,038 = R\left[\frac{1-\left(1+\frac{.05}{12}\right)^{-360}}{\frac{.05}{12}}\right] \Rightarrow R = \frac{1,517,038}{\frac{1-\left(1+\frac{.05}{12}\right)^{-360}}{\frac{.05}{12}}} \approx 8143.79$$

Her maximum monthly withdrawal is $8143.79.

59. Amount needed for retirement:

$$3500 = \frac{P}{\left[\frac{1-(1+\frac{.105}{12})^{-120}}{\frac{.105}{12}}\right]} \Rightarrow P = 259,384.15$$

Monthly contribution to acquire $259,384.15: $259,384.15 = R\left[\frac{(1.00875)^{240}-1}{.00875}\right] \Rightarrow R = 320.03$

She must make monthly payments of $320.03.

61. Balance of loan after 12 years of payments:

Monthly payment: $R = \dfrac{160,000}{\left[\frac{1-(1+\frac{.098}{12})^{-30(12)}}{\frac{.098}{12}}\right]} \approx 1380.53$

since 216 payments need to be made after they have paid 144 payments, they can be thought of as an annuity consisting of 216 payments of $1380.53 at .81667% interest $\left(\frac{.098}{12} = .0081667\right)$.

$$1380.53\left[\frac{1-(1.0081667)^{-216}}{.0081667}\right] = 139,868.73$$

Monthly payments on refinancing $139,868.73: $R = \dfrac{139,868.73}{\left[\frac{1-(1+\frac{.072}{12})^{-25(12)}}{\frac{.072}{12}}\right]} \approx 1006.48$

Balance 5 years after the refinance:
They have made 60 payments and have 240 remaining.

$$1006.48\left[\frac{1-(1.006)^{-240}}{.006}\right] = 127,831.45$$

Their balance is $127,831.45.

In Exercises 63–65, the computer program "Explorations in Finite Mathematics" has been used to prepare the amortization schedules. Answers found using other computer programs or calculators may differ slightly.

63. $P = 4000$, $i = .08$, $n = 4$
Each annual payment is

$$R = \frac{P}{a_{\overline{n}|i}} = \frac{4000}{\left[\frac{1-(1.08)^{-4}}{.08}\right]} = \$1207.68$$

(although the last payment may differ slightly).
The interest for each period is 8% of the principal at the end of the previous period.
The portion to principal for each period is the difference between the amount of the payment and the interest for the period.
The principal at the end of each new period is obtained by subtracting the new portion to principal from the principal at the end of the previous period.
Repeat these steps four times to obtain the following amortization schedule.

(continued on next page)

(continued from page 142)

Payment Number	Amount of Payment	Interest for Period	Portion to Principal	Principal at End of Period
0	------	------	------	$4000.00
1	$1207.68	$320.00	$887.68	3112.32
2	1207.68	248.99	958.69	2153.63
3	1207.68	172.29	1035.39	1118.24
4	1207.70	89.46	1118.24	0

65. $P = 8(1048) - 1200 = 7184, \ i = \dfrac{.12}{12} = .01, \ n = 4(12) = 48$

Each monthly payment is

$$R = \frac{P}{a_{\overline{n}|i}} = \frac{7184}{\left[\dfrac{1-(1.01)^{-48}}{.01}\right]} = \$189.18.$$

The interest for the first period is $.01(7184) = \$71.84$.
The portion to principal for the first period is $\$189.18 - 71.84 = \$117.34$.
The principal at the end of the first period is $\$7184 - 117.34 = \$7066.66$.
Repeat these steps to construct three more rows of the table, which will look as follows.

Payment Number	Amount of Payment	Interest for Period	Portion to Principal	Principal at End of Period
0	------	------	------	$7184.00
1	$189.18	$71.84	$117.34	7066.66
2	189.18	70.67	118.51	6948.15
3	189.18	69.48	119.70	6828.45
4	189.18	68.28	120.90	6707.55

Chapter 5 Review Exercises

1. $P = 4902, \ r = .065, \ t = \dfrac{11}{12}$

$$I = Prt = 4902(.065)\left(\frac{11}{12}\right) \approx 292.08$$

The simple interest is $292.08.

2. $P = 42,368, \ r = .0922, \ t = \dfrac{5}{12}$

$$I = Prt = (42,368)(.0922)\left(\frac{5}{12}\right) \approx 1627.64$$

The simple interest is $1627.64.

3. $P = 3478, \ r = .074, \ t = \dfrac{88}{365}$

$$I = Prt = 3478(.074)\left(\frac{88}{365}\right) \approx 62.05$$

The simple interest is $62.05.

4. $P = 2390, \ r = .087, \ t = \dfrac{[(31-3)+30+28]}{365} = \dfrac{86}{365}$

$$I = Prt = (2390)(.087)\left(\frac{86}{365}\right) \approx 48.99$$

The simple interest is $48.99.

5. $I = Prt = 12,000 \cdot .0475 \cdot \dfrac{1}{2} = 285$

The semiannual simple interest payment is $285.
The total interest earned over the life of the loan is
($285)(12) = $3420.

6. $I = Prt = 20,000 \cdot .0525 \cdot \dfrac{1}{2} = 525$

The semiannual simple interest payment is $525.
The total interest earned over the life of the loan is
($525)(18) = $9450.

7. $A = P(1 + rt) = 7750\left(1 + .068 \cdot \frac{4}{12}\right) = \$7925.67$
The maturity value is $7925.67.

8. $A = P(1 + rt) = 15,600\left(1 + .082 \cdot \frac{9}{12}\right)$
$= \$16,559.40$
The maturity value is $16,559.40.

9. The present value of an amount A is the amount that must be invested now to obtain a final balance of A.

10. $A = 459.57$, $r = .055$, $t = \dfrac{7}{12}$

$$P = \frac{A}{1+rt} = \frac{459.57}{1+.055\left(\frac{7}{12}\right)} \approx 445.28$$

The present value is $445.28.

11. $A = 80,612$, $r = .0677$, $t = \dfrac{128}{365}$

$$P = \frac{A}{1+rt} = \frac{80,612}{1+(.0677)\left(\frac{128}{365}\right)} \approx 78,742.54$$

The present value is $78,742.54.

12. Discount $= Prt = 7000 \cdot .035 \cdot \frac{9}{12} = \$183.75$

The price of the T-bill is the face value – the discount. $7000 - 183.75 = \$6816.25$

13. Discount $= Prt = 10,000 \cdot .04 \cdot \frac{6}{12} = \$200$

The price of the T-bill is the face value – the discount. $10,000 - 200 = \$9800$

Now compute the actual interest rate:

$$I = Prt \Rightarrow 200 = 980(r)\left(\tfrac{1}{2}\right) \Rightarrow$$

$r \approx .040816$

The actual interest rate is about 4.082%.

14. Answers vary. Possible answer: Compound interest produces more interest because interest is paid on the previously earned interest as well as on the original principal.

15. $P = 2800$, $i = .06$, $n = 12$

$A = P(1+i)^n = 2800(1.06)^{12} \approx 5634.15$

The compound amount is $5634.15. The amount of interest earned is
$5634.15 - $2800 = $2834.15.

16. $P = 57,809.34$, $i = \dfrac{.04}{4} = .01$, $n = 6(4) = 24$

$A = P(1+i)^n = 57,809.34(1.01)^{24} \approx 73,402.52$

The compound amount is $73,402.52.
The amount of interest earned is
$73,402.52 - $57,809.34 = $15,593.18.

17. $P = 12,903.45$, $i = \dfrac{.0637}{4} = .015925$, $n = 29$

$A = P(1+i)^n$

$= 12,903.45(1.015925)^{29} \approx \$20,402.98$

The compound amount is $20,402.98. The amount of interest earned is
$20,402.98 - $12,903.45 = $7499.53.

18. $P = 4677.23$, $i = \dfrac{.0457}{12}$, $n = 32$

$A = P(1+i)^n$

$$= 4677.23\left(1+\frac{.0457}{12}\right)^{32} \approx 5282.19$$

The compound amount is $5282.19. The amount of interest earned is
$5282.19 - $4677.23 = $604.96.

19. $P = 22,000$, $i = \dfrac{.055}{4}$, $n = 6(4) = 24$

$A = P(1+i)^n$

$$= 22,000\left(1+\frac{.055}{4}\right)^{24} \approx 30,532.58$$

The compound amount is $30,532.52. The amount of interest earned is
$30,532.52 - $22,000 = $8532.58.

20. $P = 2975$, $i = \dfrac{.047}{12}$, $n = 4(12) = 48$

$A = P(1+i)^n$

$$= 2975\left(1+\frac{.047}{12}\right)^{48} \approx 3589.01$$

The compound amount is $3580.01. The amount of interest earned is
$3589.01 - $2975 = $614.01.

21. $P = 12,366$, $i = \dfrac{.039}{2}$, $n = 5(2) = 10$

$$A = 12,366\left(1+\frac{.039}{2}\right)^{10} \approx 15,000$$

The face value is $15,000.

22. $P = 11,575$, $i = \dfrac{.052}{2}$, $n = 15(2) = 30$

$$A = 11,575\left(1+\frac{.052}{2}\right)^{30} \approx 25,000$$

The face value is $25,000.

23. Use the formula for effective rate with $r = .05$ and $m = 2$.

$$r_E = \left(1 + \frac{r}{m}\right)^m - 1 = \left(1 + \frac{.05}{2}\right)^2 - 1$$
$$= .050625$$

The effective rate is 5.0625%.

24. Use the formula for effective rate with $r = .065$ and $m = 365$.

$$r_E = \left(1 + \frac{r}{m}\right)^m - 1 = \left(1 + \frac{.065}{365}\right)^{365} - 1$$
$$= .067153$$

The effective rate is 6.7153%.

25. $A = 42,000$, $i = \frac{.12}{12} = .01$, $n = 7(12) = 84$

$$P = \frac{A}{(1+i)^n} = \frac{42,000}{(1.01)^{84}} \approx 18,207.65$$

The present value is $18,207.65.

26. $A = 17,650$, $i = \frac{.08}{4} = .02$, $n = 4(4) = 16$

$$P = \frac{A}{(1+i)^n} = \frac{17,650}{(1.02)^{16}} \approx 12,857.07$$

The present value is $12,857.07.

27. $A = 1347.89$, $i = \frac{.062}{2} = .031$,
$n = (3.5)(2) = 7$

$$P = \frac{A}{(1+i)^n} = \frac{1347.89}{(1.031)^7} \approx 1088.54$$

The present value is $1088.54.

28. $A = 2388.90$, $i = \frac{.0575}{12}$, $n = 44$

$$P = \frac{A}{(1+i)^n} = \frac{2388.90}{\left(1 + \frac{.0575}{12}\right)^{44}} \approx 1935.77$$

The present value is $1935.77.

29. $P = \frac{A}{(1+i)^n} = \frac{15,000}{\left(1 + \frac{.044}{2}\right)^{2(10)}} \approx 9706.74$

A purchaser should be willing to pay about $9706.74.

30. $P = \frac{A}{(1+i)^n} = \frac{30,000}{\left(1 + \frac{.062}{2}\right)^{2(25)}} \approx 6519.11$

A purchaser should be willing to pay about $6519.11.

31. The future value of an annuity is the balance in the account at the end of the term.

32. $R = 1288$, $i = .07$, $n = 14$

$$S = R \cdot s_{\overline{n}|i} = 1288\left[\frac{(1.07)^{14} - 1}{.07}\right] \approx 29,045.03$$

The future value of this ordinary annuity is $29,045.03.

33. $R = 4000$, $i = \frac{.06}{4} = .015$, $n = 8(4) = 32$

$$S = R \cdot s_{\overline{n}|i} = 4000\left[\frac{(1.015)^{32} - 1}{.015}\right]$$
$$\approx 162,753.15$$

The future value of this ordinary annuity is $162,753.15.

34. $R = 233$, $i = \frac{.06}{12} = .005$, $n = 4(12) = 48$

$$S = R \cdot s_{\overline{n}|i} = 233\left[\frac{(1.005)^{48} - 1}{.005}\right] \approx 12,604.79$$

The future value of this ordinary annuity is $12,604.79.

35. $R = 672$, $i = \frac{.05}{4} = .0125$, $n = 7(4) = 28$

Because deposits are made at the beginning of each time period, this is an annuity due.

$$S = R \cdot s_{\overline{n+1}|i} - R = 672\left[\frac{(1.0125)^{29} - 1}{.0125}\right] - 672$$
$$\approx 22,643.29$$

The future value of this annuity due is $22,643.29.

36. $R = 11,900$, $i = \frac{.07}{12} = .005833$, $n = 13$

$$S = R \cdot s_{\overline{n+1}|i} - R$$
$$= 11,900\left[\frac{1.0058333^{14} - 1}{.0058333}\right] - 11,900$$
$$\approx 161,166.70$$

The future value of this annuity due is $161,166.70.

37. Answers vary. Possible answer: The purpose of a sinking fund is to receive funds deposited regularly for some goal.

38. $S = 6500$, $i = .05$, $n = 6$

$$S = R \cdot s_{\overline{n}|i} \Rightarrow R = \frac{S}{s_{\overline{n}|i}} = \frac{6500}{\left[\frac{(1.05)^6 - 1}{.05}\right]} \approx 955.61$$

The amount of each payment into this sinking fund is $955.61.

39. $S = 57{,}000$, $i = \frac{.06}{2} = .03$, $n = \left(8\frac{1}{2}\right)(2) = 17$

$$R = \frac{S}{s_{\overline{n}|i}} = \frac{57{,}000}{\left[\frac{(1.03)^{17} - 1}{.03}\right]} \approx 2619.29$$

The amount of each payment is $2619.29.

40. $S = 233{,}188$, $i = \frac{.057}{4} = .01425$,

$n = \left(7\frac{3}{4}\right)(4) = 31$

$$R = \frac{S}{s_{\overline{n}|i}} = \frac{233{,}188}{\left[\frac{(1.01425)^{31} - 1}{.01425}\right]} \approx 6035.27$$

The amount of each payment is $6035.27.

41. $S = 56{,}788$, $i = \frac{.0612}{12} = .0051$,

$n = \left(4\frac{1}{2}\right)(12) = 54$

$$R = \frac{S}{s_{\overline{n}|i}} = \frac{56{,}788}{\left[\frac{(1.0051)^{54} - 1}{.0051}\right]} \approx 916.12$$

The amount of each payment is $916.12.

42. $R = 850$, $i = .05$, $n = 4$

$$P = R \cdot a_{\overline{n}|i} = 850\left[\frac{1 - (1.05)^{-4}}{.05}\right] \approx 3014.06$$

The present value of this ordinary annuity is $3014.06.

43. $R = 1500$, $i = \frac{.08}{4} = .02$, $n = 7(4) = 28$

$$P = R \cdot a_{\overline{n}|i} = 1500\left[\frac{1 - (1.02)^{-28}}{.02}\right] \approx 31{,}921.91$$

The present value is $31,921.91.

44. $R = 4210$, $i = \frac{.056}{2} = .028$, $n = 8(2) = 16$

$$P = R \cdot a_{\overline{n}|i} = 4210\left[\frac{1 - (1.028)^{-16}}{.028}\right]$$

$\approx \$53{,}699.94$

The present value is $53,699.94.

45. $R = 877.34$, $i = \frac{.064}{12} \approx .0053$, $n = 17$

$$P = R \cdot a_{\overline{n}|i} = 877.34\left[\frac{1 - \left(1 + \frac{.064}{12}\right)^{-17}}{\frac{.064}{12}}\right]$$

$\approx 14{,}226.42$

The present value is $14,226.42.

For exercises 46–48, recall that an annuity in which the payments are made at the end of each period is an ordinary annuity. We are seeking the present value.

46. $R = 800$, $i = \frac{.046}{4}$, $n = 4(4) = 16$

$$P = 800\left[\frac{1 - \left(1 + \frac{.046}{4}\right)^{-16}}{\frac{.046}{4}}\right] \approx 11{,}630.63$$

About $11,630.63 is needed to fund the withdrawals.

47. $R = 1500$, $i = \frac{.058}{12}$, $n = 10(12) = 120$

$$P = 1500\left[\frac{1 - \left(1 + \frac{.058}{12}\right)^{-120}}{\frac{.058}{12}}\right] \approx 136{,}340.32$$

About $136,340.32 is needed to fund the withdrawals.

48. $R = 3000$, $i = .062$, $n = 15$

$$P = 3000\left[\frac{1 - (1.062)^{-15}}{.062}\right] \approx 28{,}759.74$$

About $28,759.74 is needed to fund the withdrawals.

49. $P = 150,000, \ i = \dfrac{.051}{12}, \ n = 15(12) = 180$

$$150,000 = R\left[\dfrac{1 - \left(1 + \frac{.051}{12}\right)^{-180}}{\frac{.051}{12}}\right] \Rightarrow$$

$$R = \dfrac{150,000}{\dfrac{1 - \left(1 + \frac{.051}{12}\right)^{-180}}{\frac{.051}{12}}} \approx 1194.02$$

The monthly payment will be $1194.02.

50. $P = 25,000, \ i = \dfrac{.049}{4}, \ n = 8(4) = 32$

$$25,000 = R\left[\dfrac{1 - \left(1 + \frac{.049}{4}\right)^{-32}}{\frac{.049}{4}}\right] \Rightarrow$$

$$R = \dfrac{25,000}{\dfrac{1 - \left(1 + \frac{.049}{4}\right)^{-32}}{\frac{.049}{4}}} \approx 949.07$$

The monthly payment will be $949.07.

51. Find the future value of an annuity with $R = 4200, \ i = .045,$ and $n = 12.$

$$S = R\left[\dfrac{(1 + i)^n - 1}{i}\right] = 4200\left[\dfrac{(1.045)^{12} - 1}{.045}\right]$$
$$\approx 64,948.93$$

Now find the present value of $64,948.93 at 4.5% compounded annually for 12 years.
$A = 64,948.93, \ i = .045, \ n = 12$

$$A = P(1 + i)^n \Rightarrow 64,948.93 = P(1.045)^{12} \Rightarrow$$
$$P = \dfrac{64,948.93}{(1.045)^{12}} \approx 38,298.04$$

The required lump sum is $38,298.04.

52. The interest that is paid each half-year is

$$I = Prt = 24,000 \cdot .05 \cdot \dfrac{1}{2} = \$600. \ \text{So, we have a}$$

two-part investment, an annuity that pays $600 every six months for 6 years and the $24,000 face value of the bond which will be paid when the bond matures 6 years from now. The purchaser should be willing to pay the present value of each part of the investment, assuming 6.5% interest compounded semiannually.

Present value of annuity

$$R = 600, \ i = \dfrac{.065}{2} = .0325, \ n = 6(2) = 12$$

$$P = 600\left[\dfrac{1 - (1.0325)^{-12}}{.0325}\right] \approx 5884.25$$

Present value of $24,000 in 6 years

$$P = A(1 + i)^{-n}$$
$$P = 24,000(1.0325)^{-12} \approx 16,350.48$$

The purchaser should be willing to pay
$5884.25 + $16,350.48 = $22,234.73, or about $22,235.

53. $P = 32,000, \ i = \dfrac{.084}{4} = .021, \ n = 10$

$$P = R \cdot a_{\overline{n}|i} \Rightarrow R = \dfrac{P}{a_{\overline{n}|i}} = \dfrac{32,000}{\left[\dfrac{1 - (1.021)^{-10}}{.021}\right]}$$

$$\approx 3581.11$$

Quarterly payments of $3581.11 are necessary to amortize this loan.

54. $P = 5607, \ i = \dfrac{.076}{12}, \ n = 32$

$$R = \dfrac{P}{a_{\overline{n}|i}} = \dfrac{5607}{\left[\dfrac{1 - \left(1 + \frac{.076}{12}\right)^{-32}}{\frac{.076}{12}}\right]} \approx 194.13$$

Monthly payments of $194.13 are needed to amortize this loan.

55. $P = 56,890, \ i = \dfrac{.0674}{12}, \ n = 25(12) = 300$

$$R = \dfrac{P}{a_{\overline{n}|i}} = \dfrac{56,890}{\left[\dfrac{1 - \left(1 + \frac{.0674}{12}\right)^{-300}}{\frac{.0674}{12}}\right]} \approx 392.70$$

The monthly payment for this mortgage is $392.70.

56. $P = 77,110, \ i = \dfrac{.0845}{12}, \ n = 30(12) = 360$

$$R = \dfrac{P}{a_{\overline{n}|i}} = \dfrac{77,110}{\left[\dfrac{1 - \left(1 + \frac{.0845}{12}\right)^{-300}}{\frac{.0845}{12}}\right]} \approx 590.18$$

The monthly payment for this mortgage is $590.18.

57. After 5 payments, the remaining 5 payments can be thought of as annuity. The present value of this annuity is the remaining balance. So, we use the present value formula with $R = 3581.11$,

$$i = \frac{.084}{4} = .021, \; n = 5$$

$$P = 3581.11 \left[\frac{1 - (1 + .021)^{-5}}{.021} \right] \approx 16,830.54$$

The balance remaining after five payments is about \$16,830.54.

58. After 15 year (180 payments), the remaining 120 payments can be thought of as annuity. The present value of this annuity is the remaining balance. So, we use the present value formula with $R = 392.70$, $i = \frac{.0674}{12}$, $n = 120$

$$P = 392.70 \left[\frac{1 - \left(1 + \frac{.0674}{12}\right)^{-120}}{\frac{.0674}{12}} \right] \approx 34,215.39$$

The balance remaining after five payments is about \$34,215.39.

59. Locate the entry of the table that is in the row labeled "Payment Number 5" and in the column labeled "Interest for Period," to observe that \$896.06 of the fifth payment is interest.

60. Locate the entry of the table that is in the row labeled "Payment Number 6" and in the column labeled "Portion to Principal," to observe that \$127.48 of the sixth payment is used to reduce the debt.

61. To find out how much interest is paid in the first 3 months of the loan, add the first three nonzero entries in the column labeled "Interest for Period."

$899.58
 898.71
+897.83
——————
$2696.12

62. To find out how much the debt has been reduced by the end of the first 6 months, add the first six nonzero entries in the column labeled "Portion to Principal."

$123.06
 123.93
 124.81
 125.69
 126.58
 127.48
——————
$751.55

63. a. $\dfrac{86,280,000}{30} = 2,876,000$

Each yearly payment would have been \$2,876,000

b. We use the formula for the present value of an annuity due, $S = R + R \left[\dfrac{1 - (1 + i)^{-(n-1)}}{i} \right]$.

$R = 2,876,000$, $i = .0587$, $n = 30$

$$S = 2,876,000 + 2,876,000 \left[\frac{1 - (1.0587)^{-29}}{.0587} \right]$$

$$\approx 42,500,960$$

The cash value is about \$42,500,960.

64. $P = 15,000$, $i = \dfrac{.06}{2} = .03$, n $= 7.5(2) = 15$

$$I = P(1 + i)^n - P$$
$$= 15,000(1 + .03)^{15} - 15,000 \approx 8369.51$$

65. Starting at age 23:

$P = 500$, $i = \dfrac{.05}{4} = .0125$,

$n = (65 - 23)4 = 42(4) = 168$

$A = P(1 + i)^n = 500(1.0125)^{168} \approx 4030.28$

Starting at age 40:

$t = (65 - 40)4 = 25(4) = 100$

$A = 500(1.0125)^{100} \approx 1731.70$

$4030.28 - 1731.70 = 2298.58$

He will have \$2298.58 more if he invests now.

66. $r_F = \left(1 + \dfrac{r}{m}\right)^m - 1$; $r_E = \left(1 + \dfrac{r}{m}\right)^m - 1$

Frontenac Bank:

$$r_F = \left(1 + \frac{.0394}{4}\right)^4 - 1 \approx .04$$

The effective rate is about 4.00%.
E*TRADE Bank:

$$r_E = \left(1 + \frac{.0393}{365}\right)^{365} - 1 \approx .0401$$

The effective rate is about 4.01%.
E*TRADE Bank paid a higher effective rate.

67. $R = 3200$, $i = \dfrac{.068}{2} = .034$, $n = (3.5)(2) = 7$

Use the formula for the future value of an ordinary annuity.

$$S = R \cdot s_{\overline{n}|i} = 3200 \left[\frac{(1.034)^7 - 1}{.034} \right] \approx 24{,}818.76$$

The final amount in the account will be $24,818.76.
The interest earned will be
$24,818.76 – 7($3200) = $2418.76.

68. $S = 52{,}000$, $i = \dfrac{.075}{12} = .00625$, $n = 20(12) = 240$

Use the formula for future value of an ordinary annuity to find the periodic payment.

$$S = R \cdot s_{\overline{n}|i} \Rightarrow$$

$$R = \frac{S}{s_{\overline{n}|i}} = \frac{52{,}000}{\left[\frac{(1.00625)^{240} - 1}{.00625} \right]} \approx 93.91$$

The firm should invest $93.91 monthly.

69. $S = 150{,}000$, $i = \dfrac{.0525}{4} = .013125$

$n = 79(4) = 316$

Use the formula for future value of an annuity.

$$R = S \left[\frac{i}{(1+i)^n - 1} \right]$$

$$\approx 150{,}000 \left[\frac{.013125}{(1.013125)^{316} - 1} \right] \approx 32.49$$

She would have to put $32.49 into her savings account every three months.

70. Age 55: $P = 6000$, $i = \dfrac{.08}{1} = .08$, $n = 20(1) = 20$

$$S = 6000 \left[\frac{(1.08)^{20} - 1}{.08} \right] \approx 274{,}571.79$$

Age 65:

$$P = 12{,}000, \ i = \frac{.08}{1} = .08, \ n = 10(1) = 10$$

$$S = 12{,}000 \left[\frac{(1.08)^{10} - 1}{.08} \right] \approx 173{,}838.75$$

If the pension starts at age 55, the total is $274,571.79, whereas, if the pension starts at age 65, the total is $173,838.75. Thus, taking the option at age 55 produces the larger amount.

71. $A = 7500$, $i = \dfrac{.10}{2} = .05$, $n = 3(2) = 6$

$$A = P(1 + i)^n$$

$$P = \frac{A}{(1 + i)^n} = \frac{7500}{(1.05)^6} \approx 5596.62$$

The required lump sum is $5596.62.

72. $P = 15{,}000$, $i = \dfrac{.072}{12} = .006$, $n = 36$

This is an ordinary annuity.

$$P = R \cdot a_{\overline{n}|i} \Rightarrow$$

$$R = \frac{P}{a_{\overline{n}|i}} = \frac{15{,}000}{\left[\frac{1 - (1.006)^{-36}}{.006} \right]} \approx 464.53$$

The amount of each payment will be $464.53.

73. $P = 40{,}000$, $i = \dfrac{.09}{2} = .045$, $n = 8(2) = 16$

This is an ordinary annuity.

$$R = \frac{P}{a_{\overline{n}|i}} = \frac{40{,}000}{\left[\frac{1 - (1.045)^{-16}}{.045} \right]} \approx \$3560.61$$

The amount of each payment will be $3560.61.

74. Amount of loan = $91,000 – $20,000 = $71,000

a. Use the formula for amortization payments with $P = 71{,}000$, $i = \dfrac{.09}{12} = .0075$, and $n = 30(12) = 360$.

$$R = \frac{Pi}{1 - (1 + i)^{-n}} = \frac{71{,}000(.0075)}{1 - (1.0075)^{-360}} \approx 571.28$$

The monthly payment for this mortgage is $571.28.

b. To find the amount of the first payment that goes to interest, use $I = Prt$ with $P = 71{,}000$, $i = .0075$, and $t = 1$.
$I = (71{,}000)(.0075)(1) = 532.50$
Of the first payment, $532.50 is interest.

c. Using method 1, since 180 payments were made, there are 180 remaining payments. The present value is

$$571.28\left[\frac{1-(1.0075)^{-180}}{.0075}\right] \approx 56,324.44,$$

so the remaining balance is $56,324.44. Using method 2, since 180 payments were already made, we have

$$571.28\left[\frac{1-(1.0075)^{-180}}{.0075}\right] \approx 56,324.44.$$

They still owe
$71,000 - $56,324.44 = $14,675.56.
Furthermore, they owe the interest on this amount for 180 months, for a total remaining balance of
$(14,675.56)(1.0075)^{180} = $56,325.43.$

d. Closing costs
= $3700 + (.025)($136,000)
= $3700 + $3400 = $7100

e. Amount of money received
   = Selling price − Closing costs −
       Current mortgage balance
Using method 1, the amount received is
$136,000 − $7100 − $56,324.44
= $72,575.56.
Using method 2, the amount received is
$136,000 − $7100 − $56,325.43
= $72,574.57.

75. This is an ordinary annuity. We are seeking the future value.

$$R = 250, \ i = \frac{.112}{12}, \ n = 20(12) = 240$$

$$S = 250\left[\frac{\left(1+\frac{.112}{12}\right)^{240}-1}{\frac{.112}{12}}\right] \approx 222,221.02$$

There will be about $222,221.02 in the account at the end of 20 years.

76. Total amount after 16 deposits:

$$R = 500, \ i = \frac{.09}{4} = .0225, \ n = 16$$

$$S = 500\left[\frac{(1.0225)^{16}-1}{.0225}\right] = 9502.70$$

Total amount after next 52 quarters:
$A = 9502.70(1+.0225)^{52} = 30.223.13$

Total of 32 $750 deposits:

$$R = 750, \ i = \frac{.09}{4} = .0225, \ n = 32$$

$$S = 750\left[\frac{(1.0225)^{32}-1}{.0225}\right] \approx 34.603.43$$

Gene's account balance is
$30,223.13 + $34,603.43 = $64,826.56.

77. $P = 10,000, \ i = .05, \ n = 7$

$A = P(1+i)^n = 10,000(1.05)^7 \approx \$14,071.00$

At the end of 7 yr, the value of the death benefit will be $14,071.00. This balance of $14,071.00 is to be paid out in 120 equal monthly payments, with $i = \frac{.03}{12} = .0025$. The full balance will be paid out, so we use the amortization formula. Let X be the amount of each payment.

$$X = \frac{14,071.00}{a_{\overline{120}|.0025}} = \frac{14,071.00}{\left[\frac{1-(1.0025)^{-120}}{.0025}\right]} \approx \$135.87$$

This corresponds to choice (d).

78. a. $r_E = \left(1+\frac{r}{m}\right)^m - 1 \Rightarrow r_E + 1 = \left(1+\frac{r}{m}\right)^m \Rightarrow$

$$\log(r_E+1) = m\log\left(1+\frac{r}{m}\right) \Rightarrow$$

$$\frac{\log(r_E+1)}{m} = \log\left(1+\frac{r}{m}\right) \Rightarrow$$

$$10^{\frac{\log(r_E+1)}{m}} = 1+\frac{r}{m} \Rightarrow r = m\left[10^{\frac{\log(r_E+1)}{m}}-1\right]$$

$$r = 12\left(10^{(\log 1.1)/12}-1\right) \approx .09569$$

The annual interest rate is about 9.569%.

b. $P = 140,000, \ i = \frac{.06625}{12}, \ n = 30(12) = 360$

Use the formula for amortization payments.

$$R = P\left[\frac{i}{1-(1+i)^{-n}}\right]$$

$$= 140,000\left[\frac{\frac{.06625}{12}}{1-\left(1+\frac{.06625}{12}\right)^{-360}}\right] \approx 896.44$$

Her payment is $896.44.

c. $R = 1200 - 896.44 = 303.56, i = \dfrac{.09569}{12},$

$n = 30(12) = 360$

Use the formula for the future value of an annuity.

$$S = R\left[\dfrac{(1+i)^n - 1}{i}\right]$$

$$= 303.56\left[\dfrac{\left(1 + \frac{.09569}{12}\right)^{360} - 1}{\frac{.09569}{12}}\right]$$

$$\approx 626,200.88$$

She will have $626,200.88 in the fund.

d. $R = 140,000\left[\dfrac{\frac{.0625}{12}}{1 - \left(1 + \frac{.0625}{12}\right)^{-12(15)}}\right]$

≈ 1200.39

His payment is $1200.39.

e. $S = 1200\left[\dfrac{\left(1 + \frac{.09569}{12}\right)^{15(12)} - 1}{\frac{.09569}{12}}\right]$

$\approx 478,134.14$

He will have $478,134.14 in the fund.

f. Sue is ahead by
$626,200.88 - 478,134.14 = 148,066.74$

g. Answer varies.

79. Amount needed for retirement:

$$55,000 = \dfrac{x}{\left[\dfrac{1 - (1.09)^{-20}}{.09}\right]}$$

$x = 502,070.01$

Annual contribution to acquire $502,070.01:

$$502,070.01 = R\left[\dfrac{(1.09)^{25} - 1}{.09}\right]$$

$$R \approx 5927.56$$

She will have to make annual deposits of $5927.56.

Case 5 Continuous Compounding

1. a. $A = 20,000e^{.06 \cdot 2} \approx 22,549.94$

b. $A = 20,000e^{.06 \cdot 10} \approx 36,442.38$

c. $A = 20,000e^{.06 \cdot 20} \approx 66,402.34$

2. a. $A = 2500\left(1 + \dfrac{.055}{12}\right)^{12 \cdot 2} \approx 2789.99$

b. $A = 2500e^{.055 \cdot 2} \approx 2790.70$

3. $25,000 at 5% compounded continuously for 5 years: $A = 25,000e^{.05 \cdot 5} \approx 32,100.64$
$25,000 at 5% compounded daily for 5 years:

$$A = 25,000\left(1 + \dfrac{.05}{365}\right)^{365 \cdot 5} \approx 32,100.09$$

$32,100.64 - 32,100.09 = .55$
The deposit earns about 55¢ more when compounded continuously than when compounded daily.

4. a. $r_E = e^{.045} - 1 \approx .04603$
The effective rate is about 4.603%.

b. $r_E = e^{.057} - 1 \approx .05866$
The effective rate is about 5.866%.

c. $r_E = e^{.074} - 1 \approx .07681$
The effective rate is about 7.681%.

5. a. $5000 = Pe^{.0375 \cdot 8} \Rightarrow P = \dfrac{5000}{e^{.0375 \cdot 8}} \approx 3704.09$
The present value of the $5000 is about $3704.09.

b. Yes, because $4000 is more than the present value.

Chapter 6 Systems of Linear Equations and Matrices

Section 6.1 Systems of Two Linear Equations in Two Variables

1. Is $(-1, 3)$ a solution of
$$2x + y = 1 \quad (1)$$
$$-3x + 2y = 9 \quad (2)$$
Check $(-1, 3)$ in equation (1).
$$2(-1) + 3 = 1 \,?$$
$$-2 + 3 = 1 \,? \quad \text{True}$$
Check $(-1, 3)$ in equation (2).
$$-3(-1) + 2(3) = 9 \,?$$
$$3 + 6 = 9 \,? \quad \text{True}$$
The ordered pair $(-1, 3)$ is a solution of this system of equations.

3. $3x - y = 1 \quad (1)$
$x + 2y = -9 \quad (2)$
Solve for y from equation (1) to get $y = 3x - 1$.
Then substitute $y = 3x - 1$
into (2) $x + 2(3x - 1) = -9$. Solve for x.
$$x + 2(3x - 1) = -9 \Rightarrow x + 6x - 2 = -9 \Rightarrow$$
$$7x - 2 = -9 \Rightarrow 7x = -7 \Rightarrow x = -1$$
Substitute $x = -1$ into $y = 3x - 1$ to solve for y.
$$y = 3(-1) - 1 \Rightarrow y = -4.$$
The solution set is $(-1, -4)$.

5. $3x - 2y = 4 \quad (1)$
$2x + y = -1 \quad (2)$
Solve for y from equation 2 to get
$y = -1 - 2x$.
Substitute $y = -1 - 2x$ into equation (1)
$$3x - 2(-1 - 2x) = 4 \Rightarrow 3x + 2 + 4x = 4 \Rightarrow$$
$$7x + 2 = 4 \Rightarrow 7x = 2 \Rightarrow x = \frac{2}{7}$$
Substitute $x = \frac{2}{7}$ into $y = -1 - 2x$ to solve for y.
$$y = -1 - 2\left(\frac{2}{7}\right) = -\frac{11}{7}$$
The solution set is $\left(\frac{2}{7}, -\frac{11}{7}\right)$.

7. $x - 2y = 5 \quad (1)$
$2x + y = 3 \quad (2)$
Multiply equation (2) by 2 and add the result to equation (1).
$$x - 2y = 5 \quad (1)$$
$$\underline{4x + 2y = 6}$$
$$5x \qquad = 11 \Rightarrow x = \frac{11}{5}$$
Substitute $\frac{11}{5}$ for x in equation (1) to get
$$\frac{11}{5} - 2y = 5 \Rightarrow -2y = \frac{14}{5} \Rightarrow y = -\frac{7}{5}.$$
The solution of the system is $\left(\frac{11}{5}, -\frac{7}{5}\right)$.

9. $2x - 2y = 12 \quad (1)$
$-2x + 3y = 10 \quad (2)$
Add the two equations to get $y = 22$.
Substitute 22 for y in equation (1) to get
$$2x - 2(22) = 12 \Rightarrow 2x - 44 = 12 \Rightarrow$$
$$2x = 56 \Rightarrow x = 28$$
The solution of the system is $(28, 22)$.

11. $x + 3y = -1 \quad (1)$
$2x - y = 5 \quad (2)$
Multiply equation (2) by 3, then add the result to equation (1)
$$x + 3y = -1 \quad (1)$$
$$\underline{6x - 3y = 15 \quad (2)}$$
$$7x = 14 \Rightarrow x = 2$$
Substitute 2 for x in equation (1) and solve for y.
$$2 + 3y = -1 \Rightarrow 3y = -3 \Rightarrow y = -1$$
The solution of the system is $(2, -1)$.

13. $2x + 3y = 15 \quad (1)$
$8x + 12y = 40 \quad (2)$
Multiply equation (1) by -4 and add the result to equation (1).
$$-8x - 12y = -60$$
$$\underline{8x + 12y = 40 \quad (2)}$$
$$0 = -20 \quad \text{False}$$
The system is inconsistent. There is no solution.

15. $2x - 8y = 2$ (1)
$3x - 12y = 3$ (2)

Multiply equation (1) by 3 and equation (2) by –2 to get

$6x - 24y = 6$ (3)
$-6x + 24y = -6.$ (4)

Add the two equations.

$6x - 24y = 6$ (3)
$\underline{-6x + 24y = -6}$ (4)
$\qquad\qquad 0 = 0$

The system is dependent, so there is an infinite number of solutions. To solve the system in terms of one of the parameters, say y, solve equation (1) for x in terms of y.

$2x - 8y = 2 \Rightarrow 2x = 8y + 2 \Rightarrow x = 4y + 1$

The solution to the system is the infinite set of pairs of the form $(4y + 1, y)$ for any real number y.

Note: the system could also be solved in terms of x. The ordered pairs would all have the form

$\left(x, \dfrac{1}{4}x - \dfrac{1}{4}\right).$

17. $4x - 3y = -1 \Rightarrow -3y = -4x - 1 \Rightarrow y = \dfrac{4}{3}x + \dfrac{1}{3}$

The slope is $\dfrac{4}{3}$ and the y-intercept is $\dfrac{1}{3}$.

$x + 2y = 19 \Rightarrow 2y = -x + 19 \Rightarrow$

$y = -\dfrac{1}{2}x + \dfrac{19}{2}$

The slope is $-\dfrac{1}{2}$ and the y-intercept is $\dfrac{19}{2}$.

The slopes and y-intercepts of these two equations match graphical solution (a).

19. $\dfrac{x}{5} + 3y = 31$ (1)

$2x - \dfrac{y}{5} = 8$ (2)

Multiply both equations (1) and (2) by 5 to clear the fractions.

$x + 15y = 155$ (3)
$10x - y = 40$ (4)

Multiply equation (4) by 15, then add the result to equation (3) and solve for x.

$150x - 15y = 600$ (5)
$\underline{\ \ \ x + 15y = 155}$ (3)
$151x \qquad\ \ = 755 \Rightarrow x = 5$

Substitute 5 for x in equation (1) to find y.

$\dfrac{x}{5} + 3y = 31 \Rightarrow \dfrac{5}{5} + 3y = 31 \Rightarrow$
$3y = 30 \Rightarrow y = 10$

The solution is $(5, 10)$.

21. a. Multiply the second equation by –1 and then add the equations:

$200{,}000 - .5r - .3b = 0$
$\underline{-350{,}000 + .5r + .7b = 0}$
$-150{,}000 \qquad\ \ + .4b = 0 \Rightarrow .4b = 150{,}000 \Rightarrow$
$b = 375{,}000$

Substitute $b = 375{,}000$ into the first equation and solve for r.

$200{,}000 - .5r - .3(375{,}000) = 0 \Rightarrow$
$-.5r = -87{,}500 \Rightarrow r = 175{,}000$

b. Answers vary.

23. $-247x + 2y = 61{,}552$ (1)
$-854x + 3y = 45{,}765$ (2)

We are asked to find when the median income of men and women will be the same, i.e., the year. This is represented by the x-coordinate of the solution. Multiply equation (1) by –3 and equation (2) by 2, then add the resulting equations and solve for x.

$741x - 6y = -184{,}656$
$\underline{-1708x + 6y = \ \ \ 91{,}530}$
$-967x \qquad = -93{,}126 \Rightarrow x \approx 96.3$

$1990 + 96.3 = 2086.3$

According to the model, the median income of men and women will be the same in 2086.

25. There are 5 soundtrack discs, so $33 - 5 = 28$ discs contain episodes. Six of the discs contained four episodes each, so there are $86 - 4(6) = 62$ episodes on the remaining $28 - 6 = 22$ discs.

Let $x =$ the number of discs with 2 episodes and let $y =$ the number of discs with 3 episodes. Then we have the system

$2x + 3y = 62$ (1)
$\ x + \ y = 22$ (2)

Solve equation (2) for y, then substitute that expression into equation (1) and solve for x.

$y = 22 - x$
$2x + 3(22 - x) = 62 \Rightarrow 2x + 66 - 3x = 62 \Rightarrow$
$-x = -4 \Rightarrow x = 4$

Substitute $x = 4$ into equation (2) and solve for y.

$4 + y = 22 \Rightarrow y = 18$

There are 4 discs with two episodes and 18 discs with three episodes.

27. We use the formula rate × time = distance. Let x = the speed of the plane and let y = the speed of the wind.

$5(x+y) = 3000 \Rightarrow x + y = 600$

$6(x-y) = 3000 \Rightarrow x - y = 500$

Add the equations and solve for x.

$2x = 1100 \Rightarrow x = 550$

Substitute $x = 550$ into the first equation and solve for y.

$550 + y = 600 \Rightarrow y = 50$

The plane's speed is 550 mph and the wind speed is 50 mph.

29. Let x = the number of shares of Boeing stock and y = the number of shares of GE stock.

$$30x + 70y = 16,000 \quad (1)$$
$$1.50(30x) + 3(70y) = 345,000 \quad (2)$$

Simplify equation (2)

$30x + 70y = 16,000 \quad (1)$

$45x + 210y = 34,500 \quad (2)$

Multiply equation (1) by -3, add the resulting equation to equation (2) and solve for x.

$-90x - 210y = -48,000$

$\underline{45x + 210y = 34,500}$

$ -45x = -13,500 \Rightarrow x = 300$

Substitute $x = 300$ into (1) and solve for y.

$30(300) + 70y = 16000$

$9000 + 70y = 16000 \Rightarrow 70y = 7000 \Rightarrow y = 100$

Shirley owns 300 shares of Boeing stock and 100 of GE stock.

31. a. To find the equation of the line through (1, 2) and (3, 4), first find the slope.

$m = \dfrac{4-2}{3-1} = \dfrac{2}{2} = 1$

Use the point-slope form with $(x_1, y_1) = (1, 2)$.

$y - y_1 = m(x - x_1) \Rightarrow y - 2 = 1(x - 1) \Rightarrow$
$y - 2 = x - 1 \Rightarrow y = x + 1$

An alternate solution can be written using a system of equations. Using the form $y = mx + b$, the line through (1, 2) and (3, 4) would yield the system of equations.

$2 = m(1) + b \quad (1)$

$4 = m(3) + b. \quad (2)$

This simplifies to

$m + b = 2 \quad (1)$

$3m + b = 4. \quad (2)$

This system can be solved by multiplying equation (1) by -1 and adding the result to equation (2).

$-m - b = -2$

$\underline{3m + b = 4}$

$2m = 2 \Rightarrow m = 1$

Substitute 1 for m in equation (1).

$m + b = 2 \Rightarrow 1 + b = 2 \Rightarrow b = 1$

So, the equation is $y = x + 1$.

b. To find the equation of the line with slope $m = 3$ through (–1, 1), use the form $y = mx + b$. Since $m = 3$, the equation becomes $y = 3x + b$.

Substitute –1 for x and 1 for y.

$1 = 3(-1) + b \Rightarrow 1 = -3 + b \Rightarrow 4 = b$

The equation is $y = 3x + 4$.

c. To find a point on both lines, solve the system

$y = x + 1 \quad (1)$

$y = 3x + 4 \quad (2)$

Equate the expressions for y, then solve.

$x + 1 = 3x + 4 \Rightarrow -2x = 3 \Rightarrow x = -\dfrac{3}{2}$

Substitute $-\dfrac{3}{2}$ for x in equation (1).

$y = x + 1 \Rightarrow y = -\dfrac{3}{2} + 1 \Rightarrow y = -\dfrac{1}{2}$

The point $\left(-\dfrac{3}{2}, -\dfrac{1}{2}\right)$ is on both lines.

Section 6.2 Larger Systems of Linear Equations

1. $\begin{aligned} x - 3z &= 2 \\ 2x - 4y + 5z &= 1 \\ 5x - 8y + 7z &= 6 \\ 3x - 4y + 2z &= 3 \end{aligned}$

3. $\begin{aligned} 3x + z + 2w + 18v &= 0 \\ -4x + y - w - 24v &= 0 \\ 7x - y + z + 3w + 42v &= 0 \\ 4x + z + 2w + 24v &= 0 \end{aligned}$

5. $\begin{aligned} x + y + 2z + 3w &= 1 \\ -y - z - 2w &= -1 \\ 3x + y + 4z + 5w &= 2 \end{aligned}$

7. $x + 12y - 3z + 4w = 10$
$2y + 3z \quad + w = 4$
$-z \quad\quad = -7$
$6y - 2z - 3w = 0$

9. $x + 3y - 4z + 2w = 1$ (1)
$y + z - w = 4$ (2)
$2z + 2w = -6$ (3)
$3w = 9$ (4)

Divide (4) by 3 to obtain $w = 3$. Substitute
$w = 3$ into (3) and solve for z.
$2z + 2(3) = -6 \Rightarrow 2z = -12 \Rightarrow z = -6$
Substitute $w = 3$ and $z = -6$ into (2) to solve for y.
$y + (-6) - 3 = 4 \Rightarrow y = 13$
Substitute $y = 13$, $w = 3$, and $z = -6$ into (1) to
solve for x.
$x + 3(13) - 4(-6) + 2(3) = 1 \Rightarrow x = -68$
The solution set is $(-68, 13, -6, 3)$.

11. $2x + 2y - 4z + w = -5$ (1)
$3y + 4z - w = 0$ (2)
$2z - 7w = -6$ (3)
$5w = 15$ (4)

Divide (4) by 5 to obtain $w = 3$. Substitute
$w = 3$ into (3) to solve for z.

$2z - 7(3) = -6 \Rightarrow 2z - 21 = -6 \Rightarrow z = \dfrac{15}{2}$

Substitute $w = 3$ and $z = \frac{15}{2}$ into (2) to solve for
y.

$3y + 4\left(\dfrac{15}{2}\right) - 3 = 0 \Rightarrow 3y + 30 = 3 \Rightarrow$
$3y = -27 \Rightarrow y = -9$

Substitute $y = -9$, $z = \frac{15}{2}$, and $w = 3$ into (1) to
solve for x.

$2x + 2(-9) - 4\left(\dfrac{15}{2}\right) + 3 = -5 \Rightarrow$

$2x - 18 - 30 + 3 = -5 \Rightarrow 2x - 45 = -5 \Rightarrow$
$2x = 40 \Rightarrow x = 20$

The solution set is $\left(20, -9, \dfrac{15}{2}, 3\right)$.

13. The augmented matrix for the system
$2x + y + z = 3$
$3x - 4y + 2z = -5$
$x + y + z = 2$

is $\begin{bmatrix} 2 & 1 & 1 & 3 \\ 3 & -4 & 2 & -5 \\ 1 & 1 & 1 & 2 \end{bmatrix}$.

15. The system of equations for the augmented
matrix

$\begin{bmatrix} 2 & 3 & 8 & 20 \\ 1 & 4 & 6 & 12 \\ 0 & 3 & 5 & 10 \end{bmatrix}$ is

$2x + 3y + 8z = 20$
$x + 4y + 6z = 12$
$3y + 5z = 10.$

17. $\begin{bmatrix} 1 & 2 & 3 & -1 \\ 6 & 5 & 4 & 6 \\ 2 & 0 & 7 & -4 \end{bmatrix}$

Interchange R_2 and R_3 to obtain

$\begin{bmatrix} 1 & 2 & 3 & -1 \\ 2 & 0 & 7 & -4 \\ 6 & 5 & 4 & 6 \end{bmatrix}$.

19. $\begin{bmatrix} -4 & -3 & 1 & -1 & 2 \\ 8 & 2 & 5 & 0 & 6 \\ 0 & -2 & 9 & 4 & 5 \end{bmatrix}$

Replace R_2 with $2R_1 + R_2$ to obtain

$\begin{bmatrix} -4 & -3 & 1 & -1 & 2 \\ 0 & -4 & 7 & -2 & 10 \\ 0 & -2 & 9 & 4 & 5 \end{bmatrix}$.

21. $\begin{bmatrix} 1 & 0 & 0 & 0 & \frac{3}{2} \\ 0 & 1 & 0 & 0 & 17 \\ 0 & 0 & 1 & 0 & -5 \\ 0 & 0 & 0 & 1 & 0 \end{bmatrix}$

The solution of the system is $\left(\frac{3}{2}, 17, -5, 0\right)$.

23. $\begin{bmatrix} 1 & 0 & 0 & 1 & 2 \\ 0 & 1 & 0 & 2 & -3 \\ 0 & 0 & 1 & 0 & 5 \\ 0 & 0 & 0 & 0 & 0 \end{bmatrix}$

The last row, which represents $0 = 0$, indicates
that the system has no unique solution. The
linear system associated with this matrix is
$x + w = 2$ (1)
$y + 2w = -3$ (2)
$z = 5$ (3)
To put the solution in terms of w, solve for x in
equation (1) and y in equation (2).
The solution is $(2 - w, -3 - 2w, 5, w)$.

25.
$$x + 2y \quad = 0 \quad (1)$$
$$y - z = 2 \quad (2)$$
$$x + \ y + z = -2 \ (3)$$

Multiply equation (3) by -1, then add the result to equation (1).
$$x + 2y \quad = 0$$
$$\underline{-x - \ y - z = 2}$$
$$y - z = 2$$

The system now becomes
$$y - z = 2 \quad (4)$$
$$y - z = 2 \quad (2)$$
$$x + y + z = -2 \quad (3)$$

Multiply equation (4) by -1, then add the result to equation (2).
$$y - z = 2$$
$$\underline{-y + z = -2}$$
$$0 = 0 \ (5)$$

Since equation (5) is true, the system is dependent.

27.
$$x + 2y + 4z = 6 \quad (1)$$
$$y + z = 1 \quad (2)$$
$$x + 3y + 5z = 10 \ (3)$$

Multiply equation (3) by -1 and add the result to equation (1).
$$x + 2y + 4z = 6$$
$$\underline{-x - 3y - 5z = -10}$$
$$-y - z = -4$$

The system now becomes
$$-y - \ z = -4 \quad (1)$$
$$y + \ z = 1 \quad (2)$$
$$x + 3y + 5z = 10 \quad (4)$$

Add equation (2) to equation (4).
$$y + z = 1 \quad (2)$$
$$\underline{-y - z = -4 \ (4)}$$
$$0 = -3 \ (5)$$

Since equation (5) is false, the system is inconsistent.

29.
$$a - 3b - 2c = -3 \quad (1)$$
$$3a + 2b - c = 12 \quad (2)$$
$$-a - b + 4c = 3 \quad (3)$$

Multiply equation (1) by -3 and add the result to equation (2). Also, add equation (1) to equation (3).
The new system is
$$a - 3b - 2c = -3 \quad (1)$$
$$11b + 5c = 21 \quad (4)$$
$$-4b + 2c = 0. \quad (5)$$

Multiply equation (4) by $\frac{1}{11}$. This gives
$$a - 3b - 2c = -3 \ (1)$$
$$b + \frac{5}{11}c = \frac{21}{11} \ (6)$$
$$-4b + 2c = 0. \quad (5)$$

Multiply equation (6) by 4 and add the result to equation (5). This gives
$$a - 3b - 2c = -3 \ (1)$$
$$b + \frac{5}{11}c = \frac{21}{11} \ (6)$$
$$\frac{42}{11}c = \frac{84}{11}. \ (7)$$

Multiply equation (7) by $\frac{11}{42}$ to get
$$a - 3b - 2c = -3 \ (1)$$
$$b + \frac{5}{11}c = \frac{21}{11} \ (6)$$
$$c = 2. \quad (8)$$

Substitute 2 for c in equation (6) to get
$$b + \frac{10}{11} = \frac{21}{11}, \text{ or } b = \frac{11}{11} = 1.$$

Substitute 2 for c and 1 for b in equation (1) to get $a = 4$. The system is independent.

31. The augmented matrix is
$$\begin{bmatrix} -1 & 3 & 2 & | & 0 \\ 2 & -1 & -1 & | & 3 \\ 1 & 2 & 3 & | & 0 \end{bmatrix}.$$

Now use the matrix method to solve the system.
$$\begin{bmatrix} 1 & -3 & -2 & | & 0 \\ 2 & -1 & -1 & | & 3 \\ 1 & 2 & 3 & | & 0 \end{bmatrix} \begin{matrix} -R_1 \\ \\ \\ \end{matrix}$$

$$\begin{bmatrix} 1 & -3 & -2 & | & 0 \\ 0 & 5 & 3 & | & 3 \\ 1 & 2 & 3 & | & 0 \end{bmatrix} \begin{matrix} \\ -2R_1 + R_2 \\ \\ \end{matrix}$$

$$\begin{bmatrix} 1 & -3 & -2 & | & 0 \\ 0 & 5 & 3 & | & 3 \\ 0 & -5 & -5 & | & 0 \end{bmatrix} \begin{matrix} \\ \\ -R_3 + R_1 \end{matrix}$$

$$\begin{bmatrix} 1 & -3 & -2 & | & 0 \\ 0 & 5 & 3 & | & 3 \\ 0 & 0 & -2 & | & 3 \end{bmatrix} \begin{matrix} \\ \\ R_2 + R_3 \end{matrix}$$

(*continued on next page*)

(continued from page 156)

$$\left[\begin{array}{ccc|c} 1 & -3 & -2 & 0 \\ 0 & 1 & \dfrac{3}{5} & \dfrac{3}{5} \\ 0 & 0 & 1 & -\dfrac{3}{2} \end{array}\right] \begin{array}{l} \\ \frac{1}{5}R_2 \\ \\ -\frac{1}{2}R_3 \end{array}$$

The system is now in row echelon form, and

$$z = -\frac{3}{2}.$$

Using back substitution, we find that

$$y + \frac{3}{5}\left(-\frac{3}{2}\right) = \frac{3}{5} \Rightarrow y = \frac{3}{2}$$

Now solve for x.

$$-x - 3\left(-\frac{3}{2}\right) - 2\left(-\frac{3}{2}\right) = 0 \Rightarrow x = \frac{3}{2}$$

The solution set is $\left(\dfrac{3}{2}, \dfrac{3}{2}, -\dfrac{3}{2}\right)$.

33. The augmented matrix is

$$\left[\begin{array}{ccc|c} 1 & -2 & 4 & 6 \\ 1 & 2 & 13 & 6 \\ -2 & 6 & -1 & -10 \end{array}\right].$$

Now use the matrix method to solve the system.

$$\left[\begin{array}{ccc|c} 1 & -2 & 4 & 6 \\ 0 & -4 & -9 & 0 \\ 0 & 2 & 7 & 2 \end{array}\right] \begin{array}{l} \\ R_1 - R_2 \\ 2R_1 + R_3 \end{array}$$

$$\left[\begin{array}{ccc|c} 1 & -2 & 4 & 6 \\ 0 & -4 & -9 & 0 \\ 0 & 0 & 5 & 4 \end{array}\right] \begin{array}{l} \\ \\ R_2 + 2R_3 \end{array}$$

$$\left[\begin{array}{ccc|c} 1 & -2 & 4 & 6 \\ 0 & 1 & \dfrac{9}{4} & 0 \\ 0 & 0 & 1 & \dfrac{4}{5} \end{array}\right] \begin{array}{l} \\ -\frac{1}{4}R_2 \\ \frac{1}{5}R_3 \end{array}$$

The system is now in row echelon form, and $z = \frac{4}{5}$. Using back substitution, we find that

$$y + \frac{9}{4}\left(\frac{4}{5}\right) = 0 \Rightarrow y = -\frac{9}{5}$$

$$x - 2\left(-\frac{9}{5}\right) + 4\left(\frac{4}{5}\right) = 6 \Rightarrow x = -\frac{4}{5}$$

The solution set for the system is

$$\left(-\frac{4}{5}, -\frac{9}{5}, \frac{4}{5}\right).$$

35. Note that we have four equations in three unknowns. The augmented matrix is

$$\left[\begin{array}{ccc|c} 1 & 1 & 1 & 200 \\ 1 & -2 & 0 & 0 \\ 2 & 3 & 5 & 600 \\ 2 & -1 & 1 & 200 \end{array}\right]$$

Use the matrix method to solve the system.

$$\left[\begin{array}{ccc|c} 1 & 1 & 1 & 200 \\ 0 & 3 & 1 & 200 \\ 0 & 4 & 4 & 400 \\ 2 & -1 & 1 & 200 \end{array}\right] \begin{array}{l} \\ R_1 - R_2 \\ R_3 - R_4 \\ \end{array}$$

$$\left[\begin{array}{ccc|c} 1 & 1 & 1 & 200 \\ 0 & 3 & 1 & 200 \\ 0 & 4 & 4 & 400 \\ 0 & 3 & 1 & 200 \end{array}\right] \begin{array}{l} \\ \\ \\ 2R_1 - R_4 \end{array}$$

$$\left[\begin{array}{ccc|c} 1 & 1 & 1 & 200 \\ 0 & 3 & 1 & 200 \\ 0 & 1 & 1 & 100 \\ 0 & 0 & 0 & 0 \end{array}\right] \begin{array}{l} \\ \\ \frac{1}{4}R_3 \\ R_2 - R_4 \end{array}$$

$$\left[\begin{array}{ccc|c} 1 & 1 & 1 & 200 \\ 0 & 3 & 1 & 200 \\ 0 & 0 & 2 & 100 \\ 0 & 0 & 0 & 0 \end{array}\right] \begin{array}{l} \\ \\ 3R_3 - R_2 \\ \end{array}$$

$$\left[\begin{array}{ccc|c} 1 & 1 & 1 & 200 \\ 0 & 1 & \dfrac{1}{3} & \dfrac{200}{3} \\ 0 & 0 & 1 & 50 \\ 0 & 0 & 0 & 0 \end{array}\right] \begin{array}{l} \\ \frac{1}{3}R_2 \\ \frac{1}{2}R_3 \\ \end{array}$$

The system is now in row echelon form, and $z = 50$. Use back-substitution to solve for x and y.

$$y + \frac{1}{3}(50) = \frac{200}{3} \Rightarrow y = 50$$
$$x + 1(50) + 1(50) = 200 \Rightarrow x = 100$$

The solution to the system is $(100, 50, 50)$.

37.
$$x + y + z = 5$$
$$2x + y - z = 2$$
$$x - y + z = -2$$

The augmented matrix is

$$\begin{bmatrix} 1 & 1 & 1 & 5 \\ 2 & 1 & -1 & 2 \\ 1 & -1 & 1 & -2 \end{bmatrix}$$

Now use the matrix method to solve the system:

$$\begin{bmatrix} 1 & 1 & 1 & 5 \\ 0 & -1 & -3 & -8 \\ 0 & -2 & 0 & -7 \end{bmatrix} \begin{matrix} \\ -2R_1 + R_2 \\ -R_1 + R_3 \end{matrix}$$

$$\begin{bmatrix} 1 & 1 & 1 & 5 \\ 0 & -1 & -3 & -8 \\ 0 & 0 & 6 & 9 \end{bmatrix} \begin{matrix} \\ \\ -2R_2 + R_3 \end{matrix}$$

$$\begin{bmatrix} 1 & 1 & 1 & 5 \\ 0 & 1 & 3 & 8 \\ 0 & 0 & 1 & \frac{3}{2} \end{bmatrix} \begin{matrix} \\ -R_2 \\ \frac{1}{6}R_3 \end{matrix}$$

The system is now in row echelon form and $z = \frac{3}{2}$. Use back-substitution to solve for x and y.

$$y + 3\left(\frac{3}{2}\right) = 8 \Rightarrow y = \frac{7}{2}$$

$$x + \frac{7}{2} + \frac{3}{2} = 5 \Rightarrow x = 0$$

The solution set is $\left(0, \frac{7}{2}, \frac{3}{2}\right)$.

39.
$$x + 2y + z = 5$$
$$2x + y - 3z = -2$$
$$3x + y + 4z = -5$$

The matrix is

$$\begin{bmatrix} 1 & 2 & 1 & 5 \\ 2 & 1 & -3 & -2 \\ 3 & 1 & 4 & -5 \end{bmatrix}$$

$$\begin{bmatrix} 1 & 2 & 1 & 5 \\ 0 & -3 & -5 & -12 \\ 0 & -5 & 1 & -20 \end{bmatrix} \begin{matrix} \\ -2R_1 + R_2 \\ -3R_1 + R_3 \end{matrix}$$

$$\begin{bmatrix} 1 & 2 & 1 & 5 \\ 0 & 1 & \frac{5}{3} & 4 \\ 0 & -5 & 1 & -20 \end{bmatrix} \begin{matrix} \\ -\frac{1}{3}R_2 \\ \end{matrix}$$

$$\begin{bmatrix} 1 & 0 & -\frac{7}{3} & -3 \\ 0 & 1 & \frac{5}{3} & 4 \\ 0 & 0 & \frac{28}{3} & 0 \end{bmatrix} \begin{matrix} -2R_2 + R_1 \\ \\ 5R_2 + R_3 \end{matrix}$$

$$\begin{bmatrix} 1 & 0 & -\frac{7}{3} & -3 \\ 0 & 1 & \frac{5}{3} & 4 \\ 0 & 0 & 1 & 0 \end{bmatrix} \begin{matrix} \\ \\ \frac{3}{28}R_3 \end{matrix}$$

$$\begin{bmatrix} 1 & 0 & 0 & -3 \\ 0 & 1 & 0 & 4 \\ 0 & 0 & 1 & 0 \end{bmatrix} \begin{matrix} \frac{7}{3}R_3 + R_1 \\ -\frac{5}{3}R_3 + R_2 \\ \end{matrix}$$

The solution is $(-3, 4, 0)$.

41.
$$x + 3y - 6z = 7$$
$$2x - y + 2z = 0$$
$$x + y + 2z = -1$$

The matrix is

$$\begin{bmatrix} 1 & 3 & -6 & 7 \\ 2 & -1 & 2 & 0 \\ 1 & 1 & 2 & -1 \end{bmatrix}$$

$$\begin{bmatrix} 1 & 3 & -6 & 7 \\ 0 & -7 & 14 & -14 \\ 0 & -2 & 8 & -8 \end{bmatrix} \begin{matrix} \\ -2R_1 + R_2 \\ -1R_1 + R_3 \end{matrix}$$

$$\begin{bmatrix} 1 & 3 & -6 & 7 \\ 0 & 1 & -2 & 2 \\ 0 & -2 & 8 & -8 \end{bmatrix} \begin{matrix} \\ -\frac{1}{7}R_2 \\ \end{matrix}$$

$$\begin{bmatrix} 1 & 0 & 0 & 1 \\ 0 & 1 & -2 & 2 \\ 0 & 0 & 4 & -4 \end{bmatrix} \begin{matrix} -3R_2 + R_1 \\ \\ 2R_2 + R_3 \end{matrix}$$

$$\begin{bmatrix} 1 & 0 & 0 & 1 \\ 0 & 1 & -2 & 2 \\ 0 & 0 & 1 & -1 \end{bmatrix} \begin{matrix} \\ \\ \frac{1}{4}R_3 \end{matrix}$$

$$\begin{bmatrix} 1 & 0 & 0 & 1 \\ 0 & 1 & 0 & 0 \\ 0 & 0 & 1 & -1 \end{bmatrix} \begin{matrix} \\ 2R_3 + R_2 \\ \end{matrix}$$

The solution is $(1, 0, -1)$.

43.
$$x - 2y + 4z = 9$$
$$x + y + 13z = 6$$
$$-2x + 6y - z = -10$$

The augmented matrix is

$$\begin{bmatrix} 1 & -2 & 4 & 9 \\ 1 & 1 & 13 & 6 \\ -2 & 6 & -1 & -10 \end{bmatrix}$$

$$\begin{bmatrix} 1 & -2 & 4 & 9 \\ 0 & 3 & 9 & -3 \\ 0 & 2 & 7 & 8 \end{bmatrix} \begin{matrix} \\ R_2 - R_1 \\ 2R_1 + R_3 \end{matrix}$$

(continued on next page)

(*continued on from page 158*)

$$\begin{bmatrix} 1 & -2 & 4 & | & 9 \\ 0 & 1 & 3 & | & -1 \\ 0 & 2 & 7 & | & 8 \end{bmatrix} \tfrac{1}{3}R_2$$

$$\begin{bmatrix} 1 & -2 & 4 & | & 9 \\ 0 & 1 & 3 & | & -1 \\ 0 & 0 & 1 & | & 10 \end{bmatrix} R_3 - 2R_2$$

$$\begin{bmatrix} 1 & -2 & 0 & | & -31 \\ 0 & 1 & 0 & | & -31 \\ 0 & 0 & 1 & | & 10 \end{bmatrix} \begin{matrix} R_1 - 4R_3 \\ R_2 - 3R_3 \end{matrix}$$

$$\begin{bmatrix} 1 & 0 & 0 & | & -93 \\ 0 & 1 & 0 & | & -31 \\ 0 & 0 & 1 & | & 10 \end{bmatrix} R_1 + 2R_2$$

The solution is $(-93, -31, 10)$.

45. $x + 3y + 4z = 14$
$2x - 3y + 2z = 10$
$3x - y + z = 9$
$4x + 2y + 5z = 9$

The augmented matrix is

$$\begin{bmatrix} 1 & 3 & 4 & | & 14 \\ 2 & -3 & 2 & | & 10 \\ 3 & -1 & 1 & | & 9 \\ 4 & 2 & 5 & | & 9 \end{bmatrix}$$

Now use the matrix method to solve the system.

$$\begin{bmatrix} 1 & 3 & 4 & | & 14 \\ 0 & -9 & -6 & | & -18 \\ 0 & -10 & -11 & | & -33 \\ 0 & -10 & -11 & | & -47 \end{bmatrix} \begin{matrix} -2R_1 + R_2 \\ -3R_1 + R_3 \\ -4R_1 + R_4 \end{matrix}$$

$$\begin{bmatrix} 1 & 3 & 4 & | & 14 \\ 0 & -9 & -6 & | & -18 \\ 0 & -10 & -11 & | & -33 \\ 0 & 0 & 0 & | & -14 \end{bmatrix} -R_3 + R_4$$

The last row is equivalent to the equation $0 = -14$, which is false. Therefore the system is inconsistent and has no solution.

47. $x + 8y + 8z = 8$
$3x - y + 3z = 5$
$-2x - 4y - 6z = 5$

The augmented matrix is

$$\begin{bmatrix} 1 & 8 & 8 & | & 8 \\ 3 & -1 & 3 & | & 5 \\ -2 & -4 & -6 & | & 5 \end{bmatrix}$$

Now use the matrix method to solve the system.

$$\begin{bmatrix} 1 & 8 & 8 & | & 8 \\ 0 & 25 & 21 & | & 19 \\ 0 & 12 & 10 & | & 21 \end{bmatrix} \begin{matrix} 3R_1 - R_2 \\ 2R_1 + R_3 \end{matrix}$$

$$\begin{bmatrix} 1 & 8 & 8 & | & 8 \\ 0 & 1 & 0.84 & | & 0.76 \\ 0 & 12 & 10 & | & 21 \end{bmatrix} 0.04R_2$$

$$\begin{bmatrix} 1 & 0 & 1.28 & | & 1.92 \\ 0 & 1 & 0.84 & | & 0.76 \\ 0 & 0 & 0.08 & | & -11.88 \end{bmatrix} \begin{matrix} R_1 - 8R_2 \\ \\ 12R_2 - R_3 \end{matrix}$$

$$\begin{bmatrix} 1 & 0 & 1.28 & | & 1.92 \\ 0 & 1 & 0.84 & | & 0.76 \\ 0 & 0 & 1 & | & -148.5 \end{bmatrix} \tfrac{R_3}{.08}$$

$$\begin{bmatrix} 1 & 0 & 0 & | & 192 \\ 0 & 1 & 0 & | & 125.5 \\ 0 & 0 & 1 & | & -148.5 \end{bmatrix} \begin{matrix} R_1 - 1.28R_3 \\ R_2 - 0.84R_3 \end{matrix}$$

The solution is $(192, 125.5, -148.5)$.

49. $5x + 3y + 4z = 19$
$3x - y + z = -4$

The augmented matrix is

$$\begin{bmatrix} 5 & 3 & 4 & | & 19 \\ 3 & -1 & 1 & | & -4 \end{bmatrix}$$

Now use the matrix method to solve the system:

$$\begin{bmatrix} 14 & 0 & 7 & | & 7 \\ 3 & -1 & 1 & | & -4 \end{bmatrix} 3R_2 + R_1$$

$$\begin{bmatrix} 1 & 0 & \tfrac{1}{2} & | & \tfrac{1}{2} \\ 3 & -1 & 1 & | & -4 \end{bmatrix} \tfrac{1}{14}R_1$$

Since there are only two equations, the system has an infinite number of solutions. Solve the first equation for x in terms of the parameter z.

$$x + \frac{1}{2}z = \frac{1}{2} \Rightarrow x = \frac{1}{2} - \frac{1}{2}z = \frac{1-z}{2}$$

Substitute $\dfrac{1-z}{2}$ for x in the second equation and solve for y in terms of the parameter z.

$$3\left(\frac{1-z}{2}\right) - y + z = -4$$

$$3\left(\frac{1-z}{2}\right) + z + 4 = y$$

$$\frac{3(1-z) + 2z + 8}{2} = y$$

$$\frac{3 - 3z + 2z + 8}{2} = y \Rightarrow \frac{11-z}{2} = y$$

The solution is $\left(\dfrac{1-z}{2}, \dfrac{11-z}{2}, z\right)$ for any real number z.

51.
$$x - 2y + z = 5$$
$$2x + y - z = 2$$
$$-2x + 4y - 2z = 2$$

The matrix is

$$\begin{bmatrix} 1 & -2 & 1 & | & 5 \\ 2 & 1 & -1 & | & 2 \\ -2 & 4 & -2 & | & 2 \end{bmatrix}$$

$$\begin{bmatrix} 1 & -2 & 1 & | & 5 \\ 0 & 5 & -3 & | & -8 \\ 0 & 0 & 0 & | & 12 \end{bmatrix} \begin{matrix} \\ -2R_1 + R_2 \\ 2R_1 + R_3 \end{matrix}$$

The third row represents the equation
$0 = 12$, which is false. The system is inconsistent and has no solution.

53.
$$-8x - 9y = 11$$
$$24x + 34y = 2$$
$$16x + 11y = -57$$

The matrix is

$$\begin{bmatrix} -8 & -9 & | & 11 \\ 24 & 34 & | & 2 \\ 16 & 11 & | & -57 \end{bmatrix}$$

$$\begin{bmatrix} -8 & -9 & | & 11 \\ 0 & 7 & | & 35 \\ 0 & -7 & | & -35 \end{bmatrix} \begin{matrix} \\ 3R_1 + R_2 \\ 2R_1 + R_3 \end{matrix}$$

$$\begin{bmatrix} 1 & \frac{9}{8} & | & -\frac{11}{8} \\ 0 & 1 & | & 5 \\ 0 & 1 & | & 5 \end{bmatrix} \begin{matrix} -\frac{1}{8}R_1 \\ \frac{1}{7}R_2 \\ -\frac{1}{7}R_3 \end{matrix}$$

$$\begin{bmatrix} 1 & 0 & | & -7 \\ 0 & 1 & | & 5 \\ 0 & 0 & | & 0 \end{bmatrix} \begin{matrix} -\frac{9}{8}R_2 + R_1 \\ \\ -1R_2 + R_3 \end{matrix}$$

The solution is $(-7, 5)$.

55.
$$x + 2y = 3$$
$$2x + 3y = 4$$
$$3x + 4y = 5$$
$$4x + 5y = 6$$

The matrix is

$$\begin{bmatrix} 1 & 2 & | & 3 \\ 2 & 3 & | & 4 \\ 3 & 4 & | & 5 \\ 4 & 5 & | & 6 \end{bmatrix}$$

$$\begin{bmatrix} 1 & 2 & | & 3 \\ 0 & -1 & | & -2 \\ 0 & -2 & | & -4 \\ 0 & -3 & | & -6 \end{bmatrix} \begin{matrix} \\ -2R_1 + R_2 \\ -3R_1 + R_3 \\ -4R_1 + R_4 \end{matrix}$$

$$\begin{bmatrix} 1 & 2 & | & 3 \\ 0 & 1 & | & 2 \\ 0 & -2 & | & -4 \\ 0 & -3 & | & -6 \end{bmatrix} -R_2$$

$$\begin{bmatrix} 1 & 0 & | & -1 \\ 0 & 1 & | & 2 \\ 0 & 0 & | & 0 \\ 0 & 0 & | & 0 \end{bmatrix} \begin{matrix} R_1 - 2R_2 \\ \\ 2R_2 + R_3 \\ 3R_2 + R_4 \end{matrix}$$

The solution is $(-1, 2)$.

57.
$$x + y - z = -20$$
$$2x - y + z = 11$$

The matrix is

$$\begin{bmatrix} 1 & 1 & -1 & | & -20 \\ 2 & -1 & 1 & | & 11 \end{bmatrix}$$

$$\begin{bmatrix} 1 & 1 & -1 & | & -20 \\ 0 & -3 & 3 & | & 51 \end{bmatrix} -2R_1 + R_2$$

$$\begin{bmatrix} 1 & 1 & -1 & | & -20 \\ 0 & 1 & -1 & | & -17 \end{bmatrix} -\frac{1}{3}R_2$$

$$\begin{bmatrix} 1 & 0 & 0 & | & -3 \\ 0 & 1 & -1 & | & -17 \end{bmatrix} -R_2 + R_1$$

This last matrix is the augmented matrix for the system

$$x = -3 \quad (1)$$
$$y - z = -17 \quad (2)$$

Solve equation (2) for y in terms of z.
$y = z - 17$. The solution is all ordered triples of the form $(-3, z - 17, z)$ for any real number z.

59.
$$2x + y + 3z - 2w = -6$$
$$4x + 3y + z - w = -2$$
$$x + y + z + w = -5$$
$$-2x - 2y - 2z + 2w = -10$$

The matrix is

$$\begin{bmatrix} 2 & 1 & 3 & -2 & | & -6 \\ 4 & 3 & 1 & -1 & | & -2 \\ 1 & 1 & 1 & 1 & | & -5 \\ -2 & -2 & -2 & 2 & | & -10 \end{bmatrix}$$

Interchange R_1 and R_3.

$$\begin{bmatrix} 1 & 1 & 1 & 1 & | & -5 \\ 4 & 3 & 1 & -1 & | & -2 \\ 2 & 1 & 3 & -2 & | & -6 \\ -2 & -2 & -2 & 2 & | & -10 \end{bmatrix}$$

(*continued on next page*)

(continued from page 160)

$$\left[\begin{array}{cccc|c} 1 & 1 & 1 & 1 & -5 \\ 0 & -1 & -3 & -5 & 18 \\ 0 & -1 & 1 & -4 & 4 \\ 0 & 0 & 0 & 4 & -20 \end{array}\right] \begin{array}{l} \\ -4R_1 + R_2 \\ -2R_1 + R_3 \\ 2R_1 + R_4 \end{array}$$

$$\left[\begin{array}{cccc|c} 1 & 1 & 1 & 1 & -5 \\ 0 & 1 & 3 & 5 & -18 \\ 0 & -1 & 1 & -4 & 4 \\ 0 & 0 & 0 & 1 & -5 \end{array}\right] \begin{array}{l} \\ -R_2 \\ \\ \frac{1}{4}R_4 \end{array}$$

$$\left[\begin{array}{cccc|c} 1 & 0 & -2 & -4 & 13 \\ 0 & 1 & 3 & 5 & -18 \\ 0 & 0 & 4 & 1 & -14 \\ 0 & 0 & 0 & 1 & -5 \end{array}\right] \begin{array}{l} -R_2 + R_1 \\ \\ R_2 + R_3 \\ \\ \end{array}$$

$$\left[\begin{array}{cccc|c} 1 & 0 & -2 & -4 & 13 \\ 0 & 1 & 3 & 5 & -18 \\ 0 & 0 & 1 & \frac{1}{4} & -\frac{7}{2} \\ 0 & 0 & 0 & 1 & -5 \end{array}\right] \begin{array}{l} \\ \\ \frac{1}{4}R_3 \\ \\ \end{array}$$

$$\left[\begin{array}{cccc|c} 1 & 0 & -2 & -4 & 13 \\ 0 & 1 & 3 & 5 & -18 \\ 0 & 0 & 1 & 0 & -\frac{9}{4} \\ 0 & 0 & 0 & 1 & -5 \end{array}\right] \begin{array}{l} \\ \\ -\frac{1}{4}R_4 + R_3 \\ \\ \end{array}$$

$$\left[\begin{array}{cccc|c} 1 & 0 & -2 & 0 & -7 \\ 0 & 1 & 3 & 0 & 7 \\ 0 & 0 & 1 & 0 & -\frac{9}{4} \\ 0 & 0 & 0 & 1 & -5 \end{array}\right] \begin{array}{l} 4R_4 + R_1 \\ -5R_4 + R_2 \\ \\ \end{array}$$

$$\left[\begin{array}{cccc|c} 1 & 0 & 0 & 0 & -\frac{23}{2} \\ 0 & 1 & 0 & 0 & \frac{55}{4} \\ 0 & 0 & 1 & 0 & -\frac{9}{4} \\ 0 & 0 & 0 & 1 & -5 \end{array}\right] \begin{array}{l} 2R_3 + R_1 \\ -3R_3 + R_2 \\ \\ \end{array}$$

The solution is $\left(-\dfrac{23}{2}, \dfrac{55}{4}, -\dfrac{9}{4}, -5\right)$.

61. $\begin{aligned} x + 2y - z &= 3 \\ 3x + y + w &= 4 \\ 2x - y + z + w &= 2 \end{aligned}$

The matrix is

$$\left[\begin{array}{cccc|c} 1 & 2 & -1 & 0 & 3 \\ 3 & 1 & 0 & 1 & 4 \\ 2 & -1 & 1 & 1 & 2 \end{array}\right]$$

$$\left[\begin{array}{cccc|c} 1 & 2 & -1 & 0 & 3 \\ 0 & -5 & 3 & 1 & -5 \\ 0 & -5 & 3 & 1 & -4 \end{array}\right] \begin{array}{l} \\ -3R_1 + R_2 \\ -2R_1 + R_3 \end{array}$$

$$\left[\begin{array}{cccc|c} 1 & 2 & -1 & 0 & 3 \\ 0 & -5 & 3 & 1 & -5 \\ 0 & 0 & 0 & 0 & 1 \end{array}\right] \begin{array}{l} \\ \\ -1R_2 + R_3 \end{array}$$

Since the last row yields a false statement, there is no solution.

63. $\begin{aligned} \dfrac{3}{x} - \dfrac{1}{y} + \dfrac{4}{z} &= -13 \\ \dfrac{1}{x} + \dfrac{2}{y} - \dfrac{1}{z} &= 12 \\ \dfrac{4}{x} - \dfrac{1}{y} + \dfrac{3}{z} &= -7 \end{aligned}$

Let $u = \dfrac{1}{x}, v = \dfrac{1}{y}, w = \dfrac{1}{z}$. Then the system becomes

$$\begin{aligned} 3u - v + 4w &= -13 \\ u + 2v - w &= 12 \\ 4u - v + 3w &= -7 \end{aligned}$$

The matrix is

$$\left[\begin{array}{ccc|c} 3 & -1 & 4 & -13 \\ 1 & 2 & -1 & 12 \\ 4 & -1 & 3 & -7 \end{array}\right]$$

Interchange rows 1 and 2:

$$\left[\begin{array}{ccc|c} 1 & 2 & -1 & 12 \\ 3 & -1 & 4 & -13 \\ 4 & -1 & 3 & -7 \end{array}\right]$$

$$\left[\begin{array}{ccc|c} 1 & 2 & -1 & 12 \\ 0 & -7 & 7 & -49 \\ 0 & -9 & 7 & -55 \end{array}\right] \begin{array}{l} \\ -3R_1 + R_2 \\ -4R_1 + R_3 \end{array}$$

$$\left[\begin{array}{ccc|c} 1 & 2 & -1 & 12 \\ 0 & 1 & -1 & 7 \\ 0 & -9 & 7 & -55 \end{array}\right] \begin{array}{l} \\ -\frac{1}{7}R_2 \\ \\ \end{array}$$

$$\left[\begin{array}{ccc|c} 1 & 2 & -1 & 12 \\ 0 & 1 & -1 & 7 \\ 0 & 0 & -2 & 8 \end{array}\right] \begin{array}{l} \\ \\ 9R_2 + R_3 \end{array}$$

$$\left[\begin{array}{ccc|c} 1 & 2 & -1 & 12 \\ 0 & 1 & -1 & 7 \\ 0 & 0 & 1 & -4 \end{array}\right] \begin{array}{l} \\ \\ -\frac{1}{2}R_3 \end{array}$$

$$\left[\begin{array}{ccc|c} 1 & 2 & 0 & 8 \\ 0 & 1 & 0 & 3 \\ 0 & 0 & 1 & -4 \end{array}\right] \begin{array}{l} R_1 + R_3 \\ R_2 + R_3 \\ \\ \end{array}$$

$$\left[\begin{array}{ccc|c} 1 & 0 & 0 & 2 \\ 0 & 1 & 0 & 3 \\ 0 & 0 & 1 & -4 \end{array}\right] \begin{array}{l} -2R_2 + R_1 \\ \\ \\ \end{array}$$

(continued on next page)

(continued from page 161)

Thus, $u = 2$, $v = 3$, and $w = -4$. Substituting, we have $2 = \dfrac{1}{x} \Rightarrow x = \dfrac{1}{2}$, $3 = \dfrac{1}{y} \Rightarrow y = \dfrac{1}{3}$,

$-4 = \dfrac{1}{z} \Rightarrow z = -\dfrac{1}{4}$. The solution is

$\left(\dfrac{1}{2}, \dfrac{1}{3}, -\dfrac{1}{4}\right)$.

65. We are given the system
$$-5.7x + y = 242.9$$
$$5.4x + y = 498.2 .$$
$$-1.6x + y = 337.2$$
The associated matrix is
$$\begin{bmatrix} -5.7 & 1 & 242.9 \\ 5.4 & 1 & 498.2 \\ -1.6 & 1 & 337.2 \end{bmatrix}.$$

$$\begin{bmatrix} -5.7 & 1 & 242.9 \\ 11.1 & 0 & 255.3 \\ 4.1 & 0 & 94.3 \end{bmatrix} \quad \begin{matrix} -R_1 + R_2 \\ -R_1 + R_3 \end{matrix}$$

$$\begin{bmatrix} -5.7 & 1 & 242.9 \\ 1 & 0 & 23 \\ 1 & 0 & 23 \end{bmatrix} \quad \begin{matrix} \frac{1}{11.1} R_2 \\ \frac{1}{4.1} R_3 \end{matrix}$$

$$\begin{bmatrix} 1 & 0 & 23 \\ -5.7 & 1 & 242.9 \\ 1 & 0 & 23 \end{bmatrix} \quad \text{Interchange } R_1 \text{ and } R_2$$

$$\begin{bmatrix} 1 & 0 & 23 \\ 0 & 1 & 374 \\ 1 & 0 & 23 \end{bmatrix} \quad 5.7R_1 + R_2$$

Thus, $x = 23$ and $y = 374$.
The TI-84 Plus calculator solution is shown below.

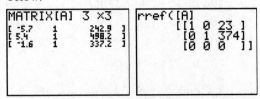

All three cities had the same population in $1970 + 23 = 1993$. The population was 374,000.

67.
$$x + 1.25y + .25z = 457.5$$
$$x + .6y + .4z = 390$$
$$3.16x + 3.48y + .4z = 1297.2$$
The associated matrix is
$$\begin{bmatrix} 1 & 1.25 & .25 & 457.5 \\ 1 & .6 & .4 & 390 \\ 3.16 & 3.48 & .4 & 1297.2 \end{bmatrix}.$$

$$\begin{bmatrix} 1 & 1.25 & .25 & 457.5 \\ 0 & .65 & -.15 & 67.5 \\ 0 & -.47 & -.39 & -148.5 \end{bmatrix} \quad \begin{matrix} -R_2 + R_1 \\ -3.16R_1 + R_3 \end{matrix}$$

$$\begin{bmatrix} 1 & 1.25 & .25 & 457.5 \\ 0 & 1 & -\frac{3}{13} & \frac{1350}{13} \\ 0 & -.47 & -.39 & -148.5 \end{bmatrix} \quad \frac{1}{.65} R_2$$

At this point, it is probably easier to convert to fractions.

$$\begin{bmatrix} 1 & \frac{5}{4} & \frac{1}{4} & \frac{4575}{10} \\ 0 & 1 & -\frac{3}{13} & \frac{1350}{13} \\ 0 & -\frac{47}{100} & -\frac{39}{100} & -\frac{1485}{10} \end{bmatrix}$$

$$\begin{bmatrix} 1 & \frac{5}{4} & \frac{1}{4} & \frac{4575}{10} \\ 0 & 1 & -\frac{3}{13} & \frac{1350}{13} \\ 0 & 0 & -\frac{162}{325} & -\frac{1296}{13} \end{bmatrix} \quad \frac{47}{100} R_2 + R_3$$

$$\begin{bmatrix} 1 & 0 & \frac{7}{13} & \frac{4260}{13} \\ 0 & 1 & -\frac{3}{13} & \frac{1350}{13} \\ 0 & 0 & 1 & 200 \end{bmatrix} \quad \begin{matrix} -\frac{5}{4}R_2 + R_1 \\[4pt] -\frac{325}{162} R_3 \end{matrix}$$

$$\begin{bmatrix} 1 & 0 & 0 & 220 \\ 0 & 1 & 0 & 150 \\ 0 & 0 & 1 & 200 \end{bmatrix} \quad \begin{matrix} -\frac{7}{13} R_3 + R_1 \\[4pt] \frac{3}{13} R_3 + R_2 \end{matrix}$$

(continued on next page)

(continued from page 162)

The TI-84 Plus calculator solution is shown below.

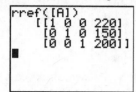

Left three columns of A right three columns of A

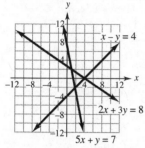

There were 220 adults, 150 teenagers, and 200 preteen children present.

69.

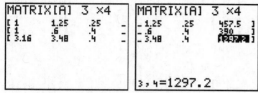

Answers vary. Sample answer: The three lines do not have one common intersection point.

Section 6.3 Applications of Systems of Linear Equations

For these exercises, we show just one of the various solution methods that could be used.

1. From example 1, let x = the number of vans, let y = the number of small trucks, and let z = the number of large trucks. We have the system

$$
\begin{aligned}
x + \quad y + \quad z &= 182 &&(1)\\
13{,}000x + 18{,}000y + 25{,}000z &= 3{,}000{,}000 &&(2)\\
x - \quad 2y \quad\;\; &= 0 &&(3)
\end{aligned}
$$

We divide equation (2) by 1000, so the system becomes

$$
\begin{aligned}
x + \quad y + \quad z &= 182\\
13x + 18y + 25z &= 3000\\
x - \quad 2y \quad\;\; &= 0
\end{aligned}
$$

Form the augmented matrix and transform it into reduced row echelon form.

$$
\left[\begin{array}{ccc|c}
1 & 1 & 1 & 182\\
13 & 18 & 25 & 3000\\
1 & -2 & 0 & 0
\end{array}\right]
$$

$$
\left[\begin{array}{ccc|c}
1 & 1 & 1 & 182\\
0 & 5 & 12 & 634\\
0 & 3 & 1 & 182
\end{array}\right]
\begin{array}{l}
\\ -13R_1 + R_2 \\ -R_3 + R_1
\end{array}
$$

$$
\left[\begin{array}{ccc|c}
1 & 1 & 1 & 182\\
0 & 5 & 12 & 634\\
0 & 0 & -\dfrac{31}{5} & -\dfrac{992}{5}
\end{array}\right]
\begin{array}{l}
\\ \\ -\frac{3}{5}R_2 + R_3
\end{array}
$$

$$
\left[\begin{array}{ccc|c}
1 & 1 & 1 & 182\\
0 & 5 & 12 & 634\\
0 & 0 & 1 & 32
\end{array}\right]
\begin{array}{l}
\\ \\ -\frac{5}{31}R_3
\end{array}
$$

$$
\left[\begin{array}{ccc|c}
1 & 1 & 0 & 150\\
0 & 5 & 0 & 250\\
0 & 0 & 1 & 32
\end{array}\right]
\begin{array}{l}
-R_3 + R_1 \\ -12R_3 + R_2 \\
\end{array}
$$

$$
\left[\begin{array}{ccc|c}
1 & 1 & 0 & 150\\
0 & 1 & 0 & 50\\
0 & 0 & 1 & 32
\end{array}\right]
\begin{array}{l}
\\ \frac{1}{5}R_2 \\
\end{array}
$$

$$
\left[\begin{array}{ccc|c}
1 & 0 & 0 & 100\\
0 & 1 & 0 & 50\\
0 & 0 & 1 & 32
\end{array}\right]
\begin{array}{l}
-R_2 + R_1 \\ \\
\end{array}
$$

The company should buy 100 vans, 50 small trucks, and 32 large trucks.

3. From example 3, let x = the number of units of corn, let y = the number of units of soybeans, and let z = the number of units of cottonseed. We have the system

$$
\begin{aligned}
10x + 20y + 30z &= 1800\\
30x + 20y + 40z &= 2400\\
20x + 40y + 25z &= 2200
\end{aligned}
$$

Form the augmented matrix and transform it into reduced row echelon form.

$$
\left[\begin{array}{ccc|c}
10 & 20 & 30 & 1800\\
30 & 20 & 40 & 2400\\
20 & 40 & 25 & 2200
\end{array}\right]
$$

$$
\left[\begin{array}{ccc|c}
10 & 20 & 30 & 1800\\
0 & 40 & 50 & 3000\\
0 & 0 & 35 & 1400
\end{array}\right]
\begin{array}{l}
\\ 3R_1 - R_2 \\ 2R_1 - R_3
\end{array}
$$

(continued on next page)

(continued from page 163)

$$\begin{bmatrix} 1 & 2 & 3 & | & 180 \\ 0 & 40 & 50 & | & 3000 \\ 0 & 0 & 1 & | & 40 \end{bmatrix} \begin{matrix} \frac{1}{10}R_1 \\ \\ \frac{1}{35}R_3 \end{matrix}$$

$$\begin{bmatrix} 1 & 2 & 0 & | & 60 \\ 0 & 40 & 0 & | & 1000 \\ 0 & 0 & 1 & | & 40 \end{bmatrix} \begin{matrix} -3R_3 + R_1 \\ -50R_3 + R_2 \\ \end{matrix}$$

$$\begin{bmatrix} 1 & 2 & 0 & | & 60 \\ 0 & 1 & 0 & | & 25 \\ 0 & 0 & 1 & | & 40 \end{bmatrix} \begin{matrix} \\ \frac{1}{40}R_2 \\ \end{matrix}$$

$$\begin{bmatrix} 1 & 0 & 0 & | & 10 \\ 0 & 1 & 0 & | & 25 \\ 0 & 0 & 1 & | & 40 \end{bmatrix} -2R_2 + R_1$$

The feed should contain 10 units of corn, 25 units of soybeans, and 40 units of cottonseed.

5. Let x = the amount charged for a shirt, let y = the amount charged for a pair of slack, and let z = the amount charged for a sport coat. We have the system

$$3x + y \quad\quad = 10.96$$
$$7x + 2y + z = 30.40$$
$$4x \quad\quad + z = 14.45$$

which has the augmented matrix

$$\begin{bmatrix} 3 & 1 & 0 & | & 10.96 \\ 7 & 2 & 1 & | & 30.40 \\ 4 & 0 & 1 & | & 14.45 \end{bmatrix}.$$

$$\begin{bmatrix} 4 & 0 & 1 & | & 14.45 \\ 7 & 2 & 1 & | & 30.40 \\ 3 & 1 & 0 & | & 10.96 \end{bmatrix} \text{interchange } R_1 \text{ and } R_3$$

$$\begin{bmatrix} 1 & 0 & \frac{1}{4} & | & 3.6125 \\ 7 & 2 & 1 & | & 30.40 \\ 3 & 1 & 0 & | & 10.96 \end{bmatrix} \frac{1}{4}R_1$$

$$\begin{bmatrix} 1 & 0 & \frac{1}{4} & | & 3.6125 \\ 0 & 2 & -\frac{3}{4} & | & 5.1125 \\ 0 & 1 & -\frac{3}{4} & | & .1225 \end{bmatrix} \begin{matrix} \\ -7R_1 + R_2 \\ -3R_1 + R_3 \end{matrix}$$

$$\begin{bmatrix} 1 & 0 & \frac{1}{4} & | & 3.6125 \\ 0 & 1 & 0 & | & 4.99 \\ 0 & 1 & -\frac{3}{4} & | & .1225 \end{bmatrix} R_2 - R_3$$

$$\begin{bmatrix} 1 & 0 & \frac{1}{4} & | & 3.6125 \\ 0 & 1 & 0 & | & 4.99 \\ 0 & 0 & 1 & | & 6.49 \end{bmatrix} \frac{4}{3}(R_2 - R_3)$$

$$\begin{bmatrix} 1 & 0 & 0 & | & 1.99 \\ 0 & 1 & 0 & | & 4.99 \\ 0 & 0 & 1 & | & 6.49 \end{bmatrix} R_1 - \frac{1}{4}R_2$$

Shirts cost \$1.99 each, slacks cost \$4.99 each, and sports coats cost \$6.49 each.

7. Let x = the number of adult tickets, let y = the number of teenager tickets, let z = the number of preteen tickets. Then we have

$$x + y + z = 570 \quad (1)$$
$$5x + 3y + 2z = 1950 \quad (2).$$
$$\frac{3}{4}z = y \quad (3)$$

Using substitution, we have

$$x + \frac{3}{4}z + z = 570 \quad (4)$$
$$5x + 3\left(\frac{3}{4}z\right) + 2z = 1950 \quad (5)$$

Clear the fractions and combine like terms.

$$4x + 7z = 2280 \quad (6)$$
$$20x + 17z = 7800 \quad (7)$$

Multiply (6) by -5, then add the result to (7) and solve for z.

$$-20x - 35z = -11400$$
$$\underline{20x + 17z = \quad 7800}$$
$$-18z = -3600 \Rightarrow z = 200$$

Substitute $z = 200$ into equation (6) and solve for x, and into equation (3) to solve for y.

$$4x + 7(200) = 2280 \Rightarrow 4x = 880 \Rightarrow x = 220$$
$$y = \frac{3}{4}(200) = 150$$

There were 220 adult tickets, 150 teenager tickets, and 200 preteen tickets sold.

9. Let x = investment in B bonds, and
let $2x$ = investment in AAA bonds. Then,
$30,000 - 3x$ = the investment in A bonds.
The sum of the interests of the three is \$2000.
Thus, we have

$0.10(x) + 0.06(30,000 - 3x) + 0.05(2x) = 2000.00$

$0.10x + 1800 - 0.18x + 0.10x = 2000.00$

$0.02x + 1800 = 2000 \Rightarrow 0.02x = 200 \Rightarrow$

$x = 10,000; \quad 2x = 20,000; \quad 30,000 - 3x = 0$

\$10,000 was invested in B bonds, \$20,000 was invested in the AAA bonds, and no money was invested in A bonds.

11. Let x = the number of pounds of pretzels,
y = the number of pounds of dried fruit, and
z = the number of pounds of nuts.
The system of equations is

$$x + y + z = 140$$
$$3x + 4y + 8z = 6(140)$$
$$x = 2y.$$

The system simplifies to

$$x + y + z = 140$$
$$3x + 4y + 8z = 840$$
$$x - 2y = 0.$$

The matrix is

$$\begin{bmatrix} 1 & 1 & 1 & | & 140 \\ 3 & 4 & 8 & | & 840 \\ 1 & -2 & 0 & | & 0 \end{bmatrix}$$

$$\begin{bmatrix} 1 & 1 & 1 & | & 140 \\ 0 & 1 & 5 & | & 420 \\ 0 & -3 & -1 & | & -140 \end{bmatrix} \begin{matrix} \\ -3R_1 + R_2 \\ -R_1 + R_3 \end{matrix}$$

$$\begin{bmatrix} 1 & 0 & -4 & | & -280 \\ 0 & 1 & 5 & | & 420 \\ 0 & 0 & 14 & | & 1120 \end{bmatrix} \begin{matrix} -R_2 + R_1 \\ \\ 3R_2 + R_3 \end{matrix}$$

$$\begin{bmatrix} 1 & 0 & -4 & | & -280 \\ 0 & 1 & 5 & | & 420 \\ 0 & 0 & 1 & | & 80 \end{bmatrix} \begin{matrix} \\ \\ \frac{1}{14}R_3 \end{matrix}$$

$$\begin{bmatrix} 1 & 0 & 0 & | & 40 \\ 0 & 1 & 0 & | & 20 \\ 0 & 0 & 1 & | & 80 \end{bmatrix} \begin{matrix} 4R_3 + R_1 \\ -5R_3 + R_2 \\ \end{matrix}$$

Use 40 lb of pretzels, 20 lb of dried fruit, and 80 lb of nuts.

13. a. The two equations are

$$a(25)^2 + b(25) = 625a + 25b = 61.7$$
$$a(35)^2 + b(35) = 1225a + 35b = 106$$

The matrix is

$$\begin{bmatrix} 625 & 25 & | & 61.7 \\ 1225 & 35 & | & 106 \end{bmatrix}$$

$$\begin{bmatrix} 1 & \frac{1}{25} & | & \frac{617}{6250} \\ 1225 & 35 & | & 106 \end{bmatrix} \frac{1}{625}R_1$$

$$\begin{bmatrix} 1 & \frac{1}{25} & | & \frac{617}{6250} \\ 0 & -14 & | & \frac{-3733}{250} \end{bmatrix} -1225R_1 + R_2$$

$$\begin{bmatrix} 1 & \frac{1}{25} & | & \frac{617}{6250} \\ 0 & 1 & | & \frac{3733}{3500} \end{bmatrix} -\frac{1}{14}R_2$$

$$\begin{bmatrix} 1 & 0 & | & \frac{981}{17,500} \\ 0 & 1 & | & \frac{3733}{3500} \end{bmatrix} -\frac{1}{25}R_2 + R_1$$

Therefore, $a = \dfrac{981}{17,500} \approx .056057$ and

$b = \dfrac{3733}{3500} \approx 1.06657.$

b. Let $x = 55$

$y = .056057(55)^2 + 1.06657(55) \approx 228$ ft

The stopping distance for a car traveling at 55 mph is about 228 ft.

15. Let x = amount in mutual fund
let y = amount in corporate bonds
let z = amount in fast food franchise

$x + y + z = x + 2x + z = 3x + z = 70,000$

$.02x + .1(2x) + .06z = .22x + .06z = 4800$

The matrix is

$$\begin{bmatrix} 3 & 1 & | & 70,000 \\ .22 & .06 & | & 4800 \end{bmatrix}$$

$$\begin{bmatrix} 3 & 1 & | & 70,000 \\ 22 & 6 & | & 480,000 \end{bmatrix}$$

$$\begin{bmatrix} 1 & \frac{1}{3} & | & \frac{70,000}{3} \\ 22 & 6 & | & 480,000 \end{bmatrix} \frac{1}{3}R_1$$

$$\begin{bmatrix} 1 & \frac{1}{3} & | & \frac{70,000}{3} \\ 0 & -\frac{4}{3} & | & -\frac{100,000}{3} \end{bmatrix} -22R_1 + R_2$$

$$\begin{bmatrix} 1 & \frac{1}{3} & | & \frac{70,000}{3} \\ 0 & 1 & | & 25,000 \end{bmatrix} -\frac{3}{4}R_2$$

$$\begin{bmatrix} 1 & 0 & | & 15,000 \\ 0 & 1 & | & 25,000 \end{bmatrix} -\frac{1}{3}R_2 + R_1$$

Therefore, she should invest \$15,000 in the mutual fund, \$30,000 in bonds, and \$25,000 in the food franchise.

17. a. The attenuation value for beam 3 is $b + c$.

b. The system of equations is
$$a + b = .8$$
$$a + c = .55$$
$$b + c = .65$$
The augmented matrix is
$$\begin{bmatrix} 1 & 1 & 0 & | & .8 \\ 1 & 0 & 1 & | & .55 \\ 0 & 1 & 1 & | & .65 \end{bmatrix}$$
$$\begin{bmatrix} 1 & 1 & 0 & | & .8 \\ 0 & 1 & -1 & | & .25 \\ 0 & 1 & 1 & | & .65 \end{bmatrix} \begin{matrix} \\ -R_2 + R_1 \\ \\ \end{matrix}$$
$$\begin{bmatrix} 1 & 1 & 0 & | & .8 \\ 0 & 1 & 0 & | & .45 \\ 0 & 1 & 1 & | & .65 \end{bmatrix} \begin{matrix} \\ \frac{1}{2}(R_2 + R_3) \\ \\ \end{matrix}$$
$$\begin{bmatrix} 1 & 0 & 0 & | & .35 \\ 0 & 1 & 0 & | & .45 \\ 0 & 0 & 1 & | & .2 \end{bmatrix} \begin{matrix} R_1 - R_2 \\ \\ R_3 - R_2 \end{matrix}$$
So $a = .35$, $b = .45$, and $c = .2$. Therefore, A is tumorous, B is bone, and C is healthy tissue.

19. a. For intersection C, x_2 cars leave on 11th street and x_3 leave on N street. The number of cars entering C must equal the number leaving, so that
$$x_2 + x_3 = 300 + 400$$
$$x_2 + x_3 = 700$$

For intersection D, x_3 cars enter on N street and x_4 cars enter on 10th street. The figure shows that 200 cars leave D on N street and 400 on 10th street.
$$x_3 + x_4 = 200 + 400$$
$$x_3 + x_4 = 600$$

b. The system of equations is
$$x_1 + x_4 = 1000$$
$$x_1 + x_2 = 1100$$
$$x_2 + x_3 = 700$$
$$x_3 + x_4 = 600$$
Solve each equation in terms of x_4.
$$x_1 = 1000 - x_4$$
$$x_3 = 600 - x_4$$
$$x_2 = 1100 - (1000 - x_4) = 100 + x_4$$
The solution is
$$(1000 - x_4, 100 + x_4, 600 - x_4, x_4).$$

c. The largest possible value is 600 and the smallest possible value is 0 for x_4, the number of cars leaving intersection A on 10th street.

d. For x_1: the largest possible value: 1000
        smallest possible value: 400

    x_2: largest possible value: 700
        smallest possible value: 100

    x_3: largest possible value: 600
        smallest possible value: 0

e. If you know the number of cars entering or leaving three of the intersections, then the number entering or leaving the fourth is automatically determined, because the number leaving must equal the number entering.

21. Let x = number of cups of Hearty Chicken Rotini let y = number of cups of Hearty Chicken, and let z = number of cups of Chunky Chicken Noodle
The system of equations is
$$100x + 130y + 110z = 1710$$
$$960x + 480y + 890z = 11,580$$
$$7x + 9y + 8z = 121$$
After simplifying the equations, the matrix is
$$\begin{bmatrix} 10 & 13 & 11 & | & 171 \\ 96 & 48 & 89 & | & 1158 \\ 7 & 9 & 8 & | & 121 \end{bmatrix}$$
$$\begin{bmatrix} 1 & \frac{13}{10} & \frac{11}{10} & | & \frac{171}{10} \\ 96 & 48 & 89 & | & 1158 \\ 7 & 9 & 8 & | & 121 \end{bmatrix} \frac{1}{10}R_1$$
$$\begin{bmatrix} 1 & \frac{13}{10} & \frac{11}{10} & | & \frac{171}{10} \\ 0 & -\frac{384}{5} & -\frac{83}{5} & | & -\frac{2418}{5} \\ 0 & -\frac{1}{10} & \frac{3}{10} & | & \frac{13}{10} \end{bmatrix} \begin{matrix} \\ -96R_1 + R_3 \\ -7R_1 + R_3 \end{matrix}$$
$$\begin{bmatrix} 1 & \frac{13}{10} & \frac{11}{10} & | & \frac{171}{10} \\ 0 & -\frac{384}{5} & -\frac{83}{5} & | & -\frac{2418}{5} \\ 0 & 1 & -3 & | & -13 \end{bmatrix} \begin{matrix} \\ \\ -10R_3 \end{matrix}$$
Interchange R_3 and R_2.
$$\begin{bmatrix} 1 & \frac{13}{10} & \frac{11}{10} & | & \frac{171}{10} \\ 0 & 1 & -3 & | & -13 \\ 0 & -\frac{384}{5} & -\frac{83}{5} & | & -\frac{2418}{5} \end{bmatrix}$$

(continued on next page)

(continued from page 166)

$$\begin{bmatrix} 1 & 0 & 5 & | & 34 \\ 0 & 1 & -3 & | & -13 \\ 0 & 0 & -247 & | & -1482 \end{bmatrix} \begin{matrix} -\frac{13}{10}R_2 + R_1 \\ \\ \frac{384}{5}R_2 + R_3 \end{matrix}$$

$$\begin{bmatrix} 1 & 0 & 5 & | & 34 \\ 0 & 1 & -3 & | & -13 \\ 0 & 0 & 1 & | & 6 \end{bmatrix} -\frac{1}{247}R_3$$

$$\begin{bmatrix} 1 & 0 & 0 & | & 4 \\ 0 & 1 & 0 & | & 5 \\ 0 & 0 & 1 & | & 6 \end{bmatrix} \begin{matrix} R_1 - 5R_3 \\ R_2 + 3R_3 \end{matrix}$$

Therefore, use 4 cups of Hearty Chicken Rotini, 5 cups of Hearty Chicken, and 6 cups of Chunky Chicken Noodle.

$$\text{Serving size} = \frac{4+5+6}{10} = 1.5 \text{ cups}$$

23. Let x = the amount invested in AAA bonds; let y = the amount invested in A bonds; and let z = the amount invested in B bonds.

a. Solve the system
$$x + y + z = 25,000 \quad (1)$$
$$.06x + .07y + .1z = 1810 \quad (2)$$
$$x = 2z. \quad (3)$$

Equation (3) should be rewritten so that the system becomes
$$x + y + z = 25,000 \quad (1)$$
$$.06x + .07y + .1z = 1810 \quad (2)$$
$$x - 2z = 0. \quad (3)$$

The matrix is
$$\begin{bmatrix} 1 & 1 & 1 & | & 25,000 \\ .06 & .07 & .01 & | & 1810 \\ 1 & 0 & -2 & | & 0 \end{bmatrix}$$

$$\begin{bmatrix} 1 & 1 & 1 & | & 25,000 \\ 6 & 7 & 10 & | & 181,000 \\ 1 & 0 & -2 & | & 0 \end{bmatrix} 100R_2$$

$$\begin{bmatrix} 1 & 1 & 1 & | & 25,000 \\ 0 & 1 & 4 & | & 31,000 \\ 0 & -1 & -3 & | & -25,000 \end{bmatrix} \begin{matrix} -6R_1 + R_2 \\ -R_1 + R_3 \end{matrix}$$

$$\begin{bmatrix} 1 & 0 & -3 & | & -6000 \\ 0 & 1 & 4 & | & 31,000 \\ 0 & 0 & 1 & | & 6000 \end{bmatrix} \begin{matrix} -R_2 + R_1 \\ \\ R_2 + R_3 \end{matrix}$$

$$\begin{bmatrix} 1 & 0 & 0 & | & 12,000 \\ 0 & 1 & 0 & | & 7000 \\ 0 & 0 & 1 & | & 6000 \end{bmatrix} \begin{matrix} 3R_3 + R_1 \\ -4R_3 + R_2 \end{matrix}$$

The client should invest $12,000 in AAA bonds at 6%, $7000 in A bonds at 7%, and $6000 in B bonds at 10%.

b. The new system is
$$x + y + z = 30,000$$
$$.06x + .07y + .1z = 2150$$
$$x - 2z = 0.$$

The matrix is
$$\begin{bmatrix} 1 & 1 & 1 & | & 30,000 \\ .06 & .07 & .1 & | & 2150 \\ 1 & 0 & -2 & | & 0 \end{bmatrix}$$

$$\begin{bmatrix} 1 & 1 & 1 & | & 30,000 \\ 6 & 7 & 10 & | & 215,000 \\ 1 & 0 & -2 & | & 0 \end{bmatrix} 100R_2$$

$$\begin{bmatrix} 1 & 1 & 1 & | & 30,000 \\ 0 & 1 & 4 & | & 35,000 \\ 0 & -1 & -3 & | & -30,000 \end{bmatrix} \begin{matrix} -6R_1 + R_2 \\ -R_1 + R_3 \end{matrix}$$

$$\begin{bmatrix} 1 & 0 & -3 & | & -5000 \\ 0 & 1 & 4 & | & 35,000 \\ 0 & 0 & 1 & | & 5000 \end{bmatrix} \begin{matrix} -R_2 + R_1 \\ \\ R_2 + R_3 \end{matrix}$$

$$\begin{bmatrix} 1 & 0 & 0 & | & 10,000 \\ 0 & 1 & 0 & | & 15,000 \\ 0 & 0 & 1 & | & 5000 \end{bmatrix} \begin{matrix} 3R_3 + R_1 \\ -4R_3 + R_2 \end{matrix}$$

The client should invest $10,000 in AAA bonds at 6%, $15,000 in A bonds at 7%, and $5000 in B bonds at 10%.

c. The new system is
$$x + y + z = 40,000$$
$$.06x + .07y + .1z = 2900$$
$$x - 2z = 0.$$

The matrix is
$$\begin{bmatrix} 1 & 1 & 1 & | & 40,000 \\ .06 & .07 & .1 & | & 2900 \\ 1 & 0 & -2 & | & 0 \end{bmatrix}$$

$$\begin{bmatrix} 1 & 1 & 1 & | & 40,000 \\ 6 & 7 & 10 & | & 290,000 \\ 1 & 0 & -2 & | & 0 \end{bmatrix} 100R_2$$

(continued on next page)

(continued from page 167)

$$\begin{bmatrix} 1 & 1 & 1 & | & 40,000 \\ 0 & 1 & 4 & | & 50,000 \\ 0 & -1 & -3 & | & -40,000 \end{bmatrix} \begin{matrix} \\ -6R_1 + R_2 \\ -R_1 + R_3 \end{matrix}$$

$$\begin{bmatrix} 1 & 0 & -3 & | & -10,000 \\ 0 & 1 & 4 & | & 50,000 \\ 0 & 0 & 1 & | & 10,000 \end{bmatrix} \begin{matrix} -R_2 + R_1 \\ \\ R_2 + R_3 \end{matrix}$$

$$\begin{bmatrix} 1 & 0 & 0 & | & 20,000 \\ 0 & 1 & 0 & | & 10,000 \\ 0 & 0 & 1 & | & 10,000 \end{bmatrix} \begin{matrix} 3R_3 + R_1 \\ -4R_3 + R_2 \\ \\ \end{matrix}$$

The client should invest $20,000 in AAA bonds at 6%, $10,000 in A bonds at 7%, and $10,000 in B bonds at 10%.

25. a. An equation $y = ax^2 + bx + c$ is sought.
When $x = 6$, $y = 2.80$. Therefore
$36a + 6b + c = 2.80$.
When $x = 8$, $y = 2.48$. Therefore,
$64a + 8b + c = 2.48$.
When $x = 10$, $y = 2.24$. Therefore,
$100a + 10b + c = 2.24$.
Thus, the system of equations is
$$\begin{aligned} 36a + 6b + c &= 2.80 \quad (1) \\ 64a + 8b + c &= 2.48 \quad (2) \\ 100a + 10b + c &= 2.24. \quad (3) \end{aligned}$$
Solve the system by elimination. Subtract equation (1) from equation (2) to obtain
$28a + 2b = -.32$,
and subtract equation (2) from equation (3) to obtain
$36a + 2b = -.24$.
Now the system is
$$\begin{aligned} 28a + 2b &= -.32 \quad (4) \\ 36a + 2b &= -.24 \quad (5) \end{aligned}$$
Subtract equation (5) from equation (4).
$-8a = -.08$
Thus, $a = .01$. Substituting into
$28a + 2b = -.32$ gives
$$\begin{aligned} 28(.01) + 2b &= -.32 \\ 2b &= -.60 \\ b &= -.30. \end{aligned}$$
Substituting into equation (1) gives
$$\begin{aligned} 36(.01) + 6(-.30) + c &= 2.80 \\ .36 - 1.80 + c &= 2.80 \\ -1.44 + c &= 2.80 \\ c &= 4.24. \end{aligned}$$
Thus, the equation is
$y = .01x^2 - .3x + 4.24.$

b. Write $y = .01x^2 - .3x + 4.24$ in the form
$y = a(x - h)^2 + k.$
$y = .01(x^2 - 30x) + 4.24$
$y = .01(x^2 - 30x + 225) + 4.24 - 2.25$
$y = .01(x - 15)^2 + 1.99$
The minimum value of y is 1.99, occurring when $x = 15$. Thus, 15 platters should be fired at one time to minimize the fuel cost. The minimum fuel cost is $1.99.

27. a. $f(x) = ax^2 + bx + c$
$11 = a(3)^2 + b(3) + c$
$20 = a(28)^2 + b(28) + c$
$30 = a(44)^2 + b(44) + c$
The three equations reduce to:
$$\begin{aligned} 9a + 3b + c &= 11 \quad (1) \\ 784a + 28b + c &= 20 \quad (2) \\ 1936a + 44b + c &= 30 \quad (3) \end{aligned}$$
Subtract (1) from (2) and (3) to get
$$\begin{aligned} 775a + 25b &= 9 \quad (4) \\ 1927a + 41b &= 19 \quad (5) \end{aligned}$$
Multiply equation (4) by -41, and equation (5) by 25 and then add to obtain
$$\begin{aligned} -31775a - 1025b &= -369 \\ \underline{48175a + 1025b} &= \underline{475} \\ 16400a &= 106 \end{aligned}$$
$$a = \frac{106}{16400} \approx .006463$$
Substitute .006463 for a in equation (4) and solve for b.
$775(.006463) + 25b = 9$
$b = .159634$
Substitute .006463 for a and .159634 for b in equation (1) and solve for c.
$9(.006463) + 3(.159634) + c = 11$
$c = 10.4629$
The quadratic function is
$f(x) = .006463x^2 + .159634x + 10.4629$

b. Let $x = 9$. Then the GDP estimate in 2009 is
$$f(9) = .006463(9)^2 + .15963(9) + 10.4629$$
$$= 12.4231$$
This is about $12.4 trillion.
Let $x = 15$. Then the GDP estimate in 2015 is
$$f(15) = .006463(15)^2 + .15963(15) + 10.4629$$
$$= 14.3115$$
This is about $14.3 trillion.

c. Use the quadratic formula to solve
$.006463x^2 + .159634x + 10.4629 = 25$
where $a = .006463$, $b = .159634$, and
$c = -14.5371$ to get $x = 36.6505$. The GDP
will reach \$25 trillion in 2036.

29. a. $C = aS^2 + bS + c$

$33 = a(320)^2 + b(320) + c$

$40 = a(600)^2 + b(600) + c$

$50 = a(1283)^2 + b(1283) + c$

Simplify the equations

$c + 320b + 102,400a = 33$

$c + 600b + 360,000a = 40$

$c + 1283b + 1,646,089a = 50$

The matrix is

$$\begin{bmatrix} 1 & 320 & 102,400 & | & 33 \\ 1 & 600 & 360,000 & | & 40 \\ 1 & 1283 & 1,646,089 & | & 50 \end{bmatrix}$$

$$\begin{bmatrix} 1 & 320 & 102,400 & | & 33 \\ 0 & 280 & 257,600 & | & 7 \\ 0 & 963 & 1,543,689 & | & 17 \end{bmatrix} \begin{matrix} \\ -R_1 + R_2 \\ -R_1 + R_3 \end{matrix}$$

$$\begin{bmatrix} 1 & 320 & 102,400 & | & 33 \\ 0 & 280 & 257,600 & | & 7 \\ 0 & 0 & 657,729 & | & -\frac{283}{40} \end{bmatrix} \begin{matrix} \\ \\ -\frac{963}{280}R_2 + R_3 \end{matrix}$$

$$\begin{bmatrix} 1 & 320 & 102,400 & | & 33 \\ 0 & 1 & 920 & | & \frac{1}{40} \\ 0 & 0 & 1 & | & -\frac{283}{26,309,160} \end{bmatrix} \begin{matrix} \\ \frac{1}{280}R_2 \\ \frac{1}{657,729}R_3 \end{matrix}$$

$$\begin{bmatrix} 1 & 0 & -192,000 & | & 25 \\ 0 & 1 & 0 & | & \frac{918,089}{26,309,160} \\ 0 & 0 & 1 & | & -\frac{283}{26,309,160} \end{bmatrix} \begin{matrix} -320R_2 + R_1 \\ -920R_3 + R_2 \\ \\ \end{matrix}$$

$$\begin{bmatrix} 1 & 0 & 0 & | & \frac{5,028,275}{219,243} \\ 0 & 1 & 0 & | & \frac{918,089}{26,309,160} \\ 0 & 0 & 1 & | & -\frac{283}{26,309,160} \end{bmatrix} 192,000R_3 + R_1$$

The relationship is expressed as
$C = aS^2 + bS + c$
$C = -.0000108S^2 + .034896S + 22.9$

b. $45 = -.0000108S^2 + .034896S + 22.9$
Plot $y_1 = 45$ and

$y_2 = -.0000108x^2 + .034896x + 22.9$

They intersect at $(864.7, 45)$. The top speed
is approximately 865 knots.

Section 6.4 Basic Matrix Operations

1. $\begin{bmatrix} 7 & -8 & 4 \\ 0 & 13 & 9 \end{bmatrix}$ is a 2×3 matrix.

Its additive inverse is $\begin{bmatrix} -7 & 8 & -4 \\ 0 & -13 & -9 \end{bmatrix}$.

3. $\begin{bmatrix} -3 & 0 & 11 \\ 1 & \frac{1}{4} & -7 \\ 5 & -3 & 9 \end{bmatrix}$ is a 3×3 square matrix.

Its additive inverse is $\begin{bmatrix} 3 & 0 & -11 \\ -1 & -\frac{1}{4} & 7 \\ -5 & 3 & -9 \end{bmatrix}$.

5. $\begin{bmatrix} 7 \\ 11 \end{bmatrix}$ is a 2×1 column matrix.

Its additive inverse is $\begin{bmatrix} -7 \\ -11 \end{bmatrix}$.

7. If $A + B = A$, then B must be a zero matrix.
Because A is a 5×3 matrix and only matrices of
the same size can be added, B must also be 5×3.
Therefore, B is a 5×3 zero matrix.

9. $\begin{bmatrix} 1 & 2 & 7 & -1 \\ 8 & 0 & 2 & -4 \end{bmatrix} + \begin{bmatrix} -8 & 12 & -5 & 5 \\ -2 & -3 & 0 & 0 \end{bmatrix}$

$= \begin{bmatrix} 1+(-8) & 2+12 & 7+(-5) & -1+5 \\ 8+(-2) & 0+(-3) & 2+0 & -4+0 \end{bmatrix}$

$= \begin{bmatrix} -7 & 14 & 2 & 4 \\ 6 & -3 & 2 & -4 \end{bmatrix}$

11. $\begin{bmatrix} -1 & -5 & 9 \\ 2 & 2 & 3 \end{bmatrix} + \begin{bmatrix} 4 & 4 & -7 \\ 1 & -1 & 2 \end{bmatrix}$

$= \begin{bmatrix} -1+4 & -5+4 & 9+(-7) \\ 2+1 & 2+(-1) & 3+2 \end{bmatrix}$

$= \begin{bmatrix} 3 & -1 & 2 \\ 3 & 1 & 5 \end{bmatrix}$

13. $\begin{bmatrix} -3 & -2 & 5 \\ 3 & 9 & 0 \end{bmatrix} - \begin{bmatrix} 1 & 5 & -2 \\ -3 & 6 & 8 \end{bmatrix}$

$= \begin{bmatrix} -3-1 & -2-5 & 5-(-2) \\ 3-(-3) & 9-6 & 0-8 \end{bmatrix}$

$= \begin{bmatrix} -4 & -7 & 7 \\ 6 & 3 & -8 \end{bmatrix}$

15. $2A = 2\begin{bmatrix} -2 & 0 \\ 5 & 3 \end{bmatrix} = \begin{bmatrix} -4 & 0 \\ 10 & 6 \end{bmatrix}$

17. $-4B = -4\begin{bmatrix} 0 & 2 \\ 4 & -6 \end{bmatrix} = \begin{bmatrix} 0 & -8 \\ -16 & 24 \end{bmatrix}$

19. $-4A + 5B = -4\begin{bmatrix} -2 & 0 \\ 5 & 3 \end{bmatrix} + 5\begin{bmatrix} 0 & 2 \\ 4 & -6 \end{bmatrix} = \begin{bmatrix} 8 & 0 \\ -20 & -12 \end{bmatrix} + \begin{bmatrix} 0 & 10 \\ 20 & -30 \end{bmatrix} = \begin{bmatrix} 8 & 10 \\ 0 & -42 \end{bmatrix}$

21. $A = \begin{bmatrix} 1 & -2 \\ 4 & 3 \end{bmatrix}, B = \begin{bmatrix} 2 & -1 \\ 0 & 5 \end{bmatrix}$

$2A + 3B = 2\begin{bmatrix} 1 & -2 \\ 4 & 3 \end{bmatrix} + 3\begin{bmatrix} 2 & -1 \\ 0 & 5 \end{bmatrix} = \begin{bmatrix} 2 & -4 \\ 8 & 6 \end{bmatrix} + \begin{bmatrix} 6 & -3 \\ 0 & 15 \end{bmatrix} = \begin{bmatrix} 8 & -7 \\ 8 & 21 \end{bmatrix}$

If $2X = 2A + 3B$, then

$2X = \begin{bmatrix} 8 & -7 \\ 8 & 21 \end{bmatrix}$ and $X = \begin{bmatrix} 4 & -\frac{7}{2} \\ 4 & \frac{21}{2} \end{bmatrix}$.

23. $X + T = \begin{bmatrix} x & y \\ z & w \end{bmatrix} + \begin{bmatrix} r & s \\ t & u \end{bmatrix} = \begin{bmatrix} x+r & y+s \\ z+t & w+u \end{bmatrix}$, which is another 2×2 matrix.

25. Show that $X + (T + P) = (X + T) + P$. On the left-hand side, the sum of $T + P$ is obtained first, and then $X + (T + P)$. This gives the matrix

$\begin{bmatrix} x+(r+m) & y+(s+n) \\ z+(t+p) & w+(u+q) \end{bmatrix}$.

For the right-hand side, first the sum $X + T$ is obtained, and then $(X + T) + P$. This gives the matrix

$\begin{bmatrix} (x+r)+m & (y+s)+n \\ (z+t)+p & (w+u)+q \end{bmatrix}$.

Comparing corresponding elements shows that they are equal by the associative property of addition of real numbers. Thus, $X + (T + P) = (X + T) + P$.

27. Show that $P + O = P$.

$P + O = \begin{bmatrix} m & n \\ p & q \end{bmatrix} + \begin{bmatrix} 0 & 0 \\ 0 & 0 \end{bmatrix} = \begin{bmatrix} m+0 & n+0 \\ p+0 & q+0 \end{bmatrix} = \begin{bmatrix} m & n \\ p & q \end{bmatrix} = P$

29. Several possible answers, including:

	basketball	hockey	football	baseball
percent of no shows	16	16	20	18
lost revenue per fan ($)	18.20	18.25	19	15.40
lost annual revenue (millions $)	22.7	35.8	51.9	96.3

31.

	2000	2003	2006
heart	3929	3444	2814
lung	3514	3800	2857
liver	16,095	16,927	16,861
kidney	44,589	53,563	66,961

33. Possible answer.

	1995	2000	2003	2005
Ages 15-24	2.8	2.6	2.7	2.7
Ages 45-54	109.6	94.2	92.5	89.7
Ages 65-74	795.4	665.6	585.0	518.9

35. a. A matrix for the death rate of male drivers is

$$A = \begin{bmatrix} 2.61 & 4.39 & 6.29 & 9.08 \\ 1.63 & 2.77 & 4.61 & 6.92 \\ .92 & .75 & .62 & .54 \end{bmatrix}$$

b. A matrix for the death rate of female drivers is

$$B = \begin{bmatrix} 1.38 & 1.72 & 1.94 & 3.31 \\ 1.26 & 1.48 & 2.82 & 2.28 \\ .41 & .33 & .27 & .40 \end{bmatrix}$$

c. Subtract matrix B from matrix A to see the difference between the death rate of males and females.

$$\begin{bmatrix} 2.61 & 4.39 & 6.29 & 9.08 \\ 1.63 & 2.77 & 4.61 & 6.92 \\ .92 & .75 & .62 & .54 \end{bmatrix} - \begin{bmatrix} 1.38 & 1.72 & 1.94 & 3.31 \\ 1.26 & 1.48 & 2.82 & 2.28 \\ .41 & .33 & .27 & .40 \end{bmatrix} = \begin{bmatrix} 1.23 & 2.67 & 4.35 & 5.77 \\ .37 & 1.29 & 1.79 & 4.64 \\ .51 & .42 & .35 & .14 \end{bmatrix}$$

Section 6.5 Matrix Products and Inverses

1. AB is 2×2. A has 2 rows, B has 2 columns.
BA is 2×2. B has 2 rows, A has 2 columns.

3. AB is 3×3. A has 3 rows, B has 3 columns.
BA is 5×5. B has 3 columns, A has 3 rows.

5. AB does not exist because the number of columns of A is not the same as the number of rows of B.
BA is 3×2. B has 3 rows, A has 2 columns.

7. Columns; rows

9. $\begin{bmatrix} 1 & 2 \\ 3 & 4 \end{bmatrix}\begin{bmatrix} -1 \\ 3 \end{bmatrix} = \begin{bmatrix} 1(-1) + 2(3) \\ 3(-1) + 4(3) \end{bmatrix} = \begin{bmatrix} 5 \\ 9 \end{bmatrix}$

11. $\begin{bmatrix} 2 & 2 & -1 \\ 5 & 0 & 1 \end{bmatrix}\begin{bmatrix} 0 & -2 \\ -1 & 5 \\ 0 & 2 \end{bmatrix} = \begin{bmatrix} (2)(0) + (2)(-1) + (-1)(0) & (2)(-2) + (2)(5) + (-1)(2) \\ (5)(0) + (0)(-1) + (1)(0) & (5)(-2) + (0)(5) + (1)(2) \end{bmatrix} = \begin{bmatrix} -2 & 4 \\ 0 & -8 \end{bmatrix}$

13. $\begin{bmatrix} -4 & 1 \\ 2 & -3 \end{bmatrix}\begin{bmatrix} 1 & 0 \\ 0 & 1 \end{bmatrix} = \begin{bmatrix} (-4)(1) + (1)(0) & (-4)(0) + (1)(1) \\ (2)(1) + (-3)(0) & (2)(0) + (-3)(1) \end{bmatrix} = \begin{bmatrix} -4 & 1 \\ 2 & -3 \end{bmatrix}$

15.
$$\begin{bmatrix} 1 & 0 & 0 \\ 0 & 1 & 0 \\ 0 & 0 & 1 \end{bmatrix}\begin{bmatrix} 3 & -5 & 7 \\ -2 & 1 & 6 \\ 0 & -3 & 4 \end{bmatrix} = \begin{bmatrix} (1)(3)+(0)(-2)+(0)(0) & (1)(-5)+(0)(1)+(0)(-3) & (1)(7)+(0)(6)+(0)(4) \\ (0)(3)+(1)(-2)+(0)(0) & (0)(-5)+(1)(1)+(0)(-3) & (0)(7)+(1)(6)+(0)(4) \\ (0)(3)+(0)(-2)+(1)(0) & (0)(-5)+(0)(1)+(1)(-3) & (0)(7)+(0)(6)+(1)(4) \end{bmatrix}$$

$$= \begin{bmatrix} 3 & -5 & 7 \\ -2 & 1 & 6 \\ 0 & -3 & 4 \end{bmatrix}$$

17.
$$\begin{bmatrix} 1 & 2 & 3 \\ 4 & 0 & 6 \\ 7 & 8 & 9 \end{bmatrix}\begin{bmatrix} -1 & 4 \\ 7 & 0 \\ 1 & 2 \end{bmatrix} = \begin{bmatrix} (1)(-1)+(2)(7)+(3)(1) & (1)(4)+(2)(0)+(3)(2) \\ (4)(-1)+(0)(7)+(6)(1) & (4)(4)+(0)(0)+(6)(2) \\ (7)(-1)+(8)(7)+(9)(1) & (7)(4)+(8)(0)+(9)(2) \end{bmatrix} = \begin{bmatrix} 16 & 10 \\ 2 & 28 \\ 58 & 46 \end{bmatrix}$$

19. $AB = \begin{bmatrix} -3 & -9 \\ 2 & 6 \end{bmatrix}\begin{bmatrix} 4 & 6 \\ 2 & 3 \end{bmatrix} = \begin{bmatrix} -30 & -45 \\ 20 & 30 \end{bmatrix}, \quad BA = \begin{bmatrix} 4 & 6 \\ 2 & 3 \end{bmatrix}\begin{bmatrix} -3 & -9 \\ 2 & 6 \end{bmatrix} = \begin{bmatrix} 0 & 0 \\ 0 & 0 \end{bmatrix}$

Since $\begin{bmatrix} -30 & -45 \\ 20 & 30 \end{bmatrix} \neq \begin{bmatrix} 0 & 0 \\ 0 & 0 \end{bmatrix}$, $AB \neq BA$.

Therefore, matrix multiplication is not commutative.

21.

$$A + B = \begin{bmatrix} -3 & -9 \\ 2 & 6 \end{bmatrix} + \begin{bmatrix} 4 & 6 \\ 2 & 3 \end{bmatrix} = \begin{bmatrix} 1 & -3 \\ 4 & 9 \end{bmatrix}$$

$$A - B = \begin{bmatrix} -3 & -9 \\ 2 & 6 \end{bmatrix} - \begin{bmatrix} 4 & 6 \\ 2 & 3 \end{bmatrix} = \begin{bmatrix} -7 & -15 \\ 0 & 3 \end{bmatrix}$$

$$(A + B)(A - B) = \begin{bmatrix} 1 & -3 \\ 4 & 9 \end{bmatrix}\begin{bmatrix} -7 & -15 \\ 0 & 3 \end{bmatrix} = \begin{bmatrix} -7 & -24 \\ -28 & -33 \end{bmatrix}$$

$$A^2 = \begin{bmatrix} -3 & -9 \\ 2 & 6 \end{bmatrix}\begin{bmatrix} -3 & -9 \\ 2 & 6 \end{bmatrix} = \begin{bmatrix} -9 & -27 \\ 6 & 18 \end{bmatrix}$$

$$B^2 = \begin{bmatrix} 4 & 6 \\ 2 & 3 \end{bmatrix}\begin{bmatrix} 4 & 6 \\ 2 & 3 \end{bmatrix} = \begin{bmatrix} 28 & 42 \\ 14 & 21 \end{bmatrix}$$

$$A^2 - B^2 = \begin{bmatrix} -9 & -27 \\ 6 & 18 \end{bmatrix} - \begin{bmatrix} 28 & 42 \\ 14 & 21 \end{bmatrix} = \begin{bmatrix} -37 & -69 \\ -8 & -3 \end{bmatrix}$$

Since $\begin{bmatrix} -7 & -24 \\ -28 & -33 \end{bmatrix} \neq \begin{bmatrix} -37 & -69 \\ -8 & -3 \end{bmatrix}$, $(A + B)(A - B) \neq A^2 - B^2$.

23. Verify that $(PX)T = P(XT)$.

$$(PX)T = \left(\begin{bmatrix} m & n \\ p & q \end{bmatrix}\begin{bmatrix} x & y \\ z & w \end{bmatrix}\right)\begin{bmatrix} r & s \\ t & u \end{bmatrix} = \begin{bmatrix} mx+nz & my+nw \\ px+qz & py+qw \end{bmatrix}\begin{bmatrix} r & s \\ t & u \end{bmatrix}$$

$$= \begin{bmatrix} (mx+nz)r+(my+nw)t & (mx+nz)s+(my+nw)u \\ (px+qz)r+(py+qw)t & (px+qz)s+(py+qw)u \end{bmatrix}$$

$$= \begin{bmatrix} mxr+nzr+myt+nwt & mxs+nzs+myu+nwu \\ pxr+qzr+pyt+qwt & pxs+qzs+pyu+qwu \end{bmatrix}\begin{matrix} \text{Distributive property} \\ \text{for real numbers} \end{matrix}$$

(continued on next page)

(continued from page 172)

$$P(XT) = \begin{bmatrix} m & n \\ p & q \end{bmatrix} \left(\begin{bmatrix} x & y \\ z & w \end{bmatrix} \begin{bmatrix} r & s \\ t & u \end{bmatrix} \right) = \begin{bmatrix} m & n \\ p & q \end{bmatrix} \begin{bmatrix} xr + yt & xs + yu \\ zr + wt & zs + wu \end{bmatrix}$$

$$= \begin{bmatrix} m(xr + yt) + n(zr + wt) & m(xs + yu) + n(zs + wu) \\ p(xr + yt) + q(zr + wt) & p(xs + yu) + q(zs + wu) \end{bmatrix}$$

$$= \begin{bmatrix} mxr + myt + nzr + nwt & mxs + myu + nzs + nwu \\ pxr + pyt + qzr + qwt & pxs + pyu + qzs + qwu \end{bmatrix} \begin{array}{l}\text{Distributive property} \\ \text{for real numbers}\end{array}$$

$$= \begin{bmatrix} mxr + nzr + myt + nwt & mxs + nzs + myu + nwu \\ pxr + qzr + pyt + qwt & pxs + qzs + pyu + qwu \end{bmatrix} \begin{array}{l}\text{Commutative property} \\ \text{for real numbers}\end{array}$$

Thus, $(PX)T = P(XT)$.

25. Verify $k(X + T) = kX + kT$ for any real number k.

$$k(X + T) = k \begin{bmatrix} x & y \\ z & w \end{bmatrix} + \begin{bmatrix} r & s \\ t & u \end{bmatrix}$$

$$= k \begin{bmatrix} x + r & y + s \\ z + t & w + u \end{bmatrix}$$

$$= \begin{bmatrix} k(x + r) & k(y + s) \\ k(z + t) & k(w + u) \end{bmatrix}$$

$$= \begin{bmatrix} kx + kr & ky + ks \\ kz + kt & kw + ku \end{bmatrix}$$

$$= \begin{bmatrix} kx & ky \\ kz & kw \end{bmatrix} + \begin{bmatrix} kr & ks \\ kt & ku \end{bmatrix}$$

$$= k \begin{bmatrix} x & y \\ z & w \end{bmatrix} + k \begin{bmatrix} r & s \\ t & u \end{bmatrix} = kX + kT$$

27. $\begin{bmatrix} 5 & 2 \\ 3 & -1 \end{bmatrix} \begin{bmatrix} -1 & 2 \\ 3 & -4 \end{bmatrix} = \begin{bmatrix} 1 & 2 \\ -6 & 10 \end{bmatrix} \neq I$,

so the given matrices are not inverses of each other.

29. $\begin{bmatrix} 3 & -1 \\ -4 & 2 \end{bmatrix} \begin{bmatrix} 1 & \frac{1}{2} \\ 2 & \frac{3}{2} \end{bmatrix} = \begin{bmatrix} 1 & 0 \\ 0 & 1 \end{bmatrix} = I$

$\begin{bmatrix} 1 & \frac{1}{2} \\ 2 & \frac{3}{2} \end{bmatrix} \begin{bmatrix} 3 & -1 \\ -4 & 2 \end{bmatrix} = \begin{bmatrix} 1 & 0 \\ 0 & 1 \end{bmatrix} = I$

Therefore, the given matrices are inverses of each other.

31. $\begin{bmatrix} 1 & 1 & 1 \\ 2 & 3 & 0 \\ 1 & 2 & 1 \end{bmatrix} \begin{bmatrix} 1.5 & .5 & -1.5 \\ -1 & 0 & 1 \\ .5 & -.5 & .5 \end{bmatrix} = \begin{bmatrix} 1 & 0 & 0 \\ 0 & 1 & 0 \\ 0 & 0 & 1 \end{bmatrix} = I$

The given matrices are inverses of each other.

33. To find the inverse of $\begin{bmatrix} 2 & -3 \\ -1 & 2 \end{bmatrix}$, write the augmented matrix $[A \mid I]$.

$\begin{bmatrix} 2 & -3 & | & 1 & 0 \\ -1 & 2 & | & 0 & 1 \end{bmatrix}$.

$\begin{bmatrix} 1 & -1 & | & 1 & 1 \\ -1 & 2 & | & 0 & 1 \end{bmatrix} R_1 + R_2$

$\begin{bmatrix} 1 & -1 & | & 1 & 1 \\ 0 & 1 & | & 1 & 2 \end{bmatrix} R_1 + R_2$

$\begin{bmatrix} 1 & 0 & | & 2 & 3 \\ 0 & 1 & | & 1 & 2 \end{bmatrix} R_1 + R_2$

The inverse is $\begin{bmatrix} 2 & 3 \\ 1 & 2 \end{bmatrix}$.

35. To find the inverse of $\begin{bmatrix} -1 & 2 \\ 1 & -1 \end{bmatrix}$, write the augmented matrix $[A \mid I]$.

$\begin{bmatrix} -1 & 2 & | & 1 & 0 \\ 1 & -1 & | & 0 & 1 \end{bmatrix}$

$\begin{bmatrix} -1 & 2 & | & 1 & 0 \\ 0 & 1 & | & 1 & 1 \end{bmatrix} R_1 + R_2$

$\begin{bmatrix} -1 & 0 & | & -1 & -2 \\ 0 & 1 & | & 1 & 1 \end{bmatrix} -2R_2 + R_1$

$\begin{bmatrix} 1 & 0 & | & 1 & 2 \\ 0 & 1 & | & 1 & 1 \end{bmatrix} -R_1$

The inverse is $\begin{bmatrix} 1 & 2 \\ 1 & 1 \end{bmatrix}$.

37. To find the inverse of $\begin{bmatrix} 1 & 2 \\ 3 & 6 \end{bmatrix}$, write the

augmented matrix $\begin{bmatrix} A \mid I \end{bmatrix}$.

$\begin{bmatrix} 1 & 2 & 1 & 0 \\ 3 & 6 & 0 & 1 \end{bmatrix}$

$\begin{bmatrix} 1 & 2 & 1 & 0 \\ 0 & 0 & -3 & 1 \end{bmatrix} -3R_1 + R_2$

Since there is no way to continue the
transformation, the given matrix has no inverse.

39. To find the inverse of $\begin{bmatrix} 1 & -1 & 0 \\ -1 & 2 & 3 \\ 1 & 0 & 2 \end{bmatrix}$,

write the augmented matrix $\begin{bmatrix} A \mid I \end{bmatrix}$.

$\begin{bmatrix} 1 & -1 & 0 & 1 & 0 & 0 \\ -1 & 2 & 3 & 0 & 1 & 0 \\ 1 & 0 & 2 & 0 & 0 & 1 \end{bmatrix}$

$\begin{bmatrix} 1 & -1 & 0 & 1 & 0 & 0 \\ 0 & 1 & 3 & 1 & 1 & 0 \\ 0 & -1 & -2 & 1 & 0 & -1 \end{bmatrix} \begin{matrix} R_1 + R_2 \\ R_1 - R_3 \end{matrix}$

$\begin{bmatrix} 1 & 0 & 3 & 2 & 1 & 0 \\ 0 & 1 & 3 & 1 & 1 & 0 \\ 0 & 0 & 1 & 2 & 1 & -1 \end{bmatrix} \begin{matrix} R_1 + R_2 \\ \\ R_2 + R_3 \end{matrix}$

$\begin{bmatrix} 1 & 0 & 0 & -4 & -2 & 3 \\ 0 & 1 & 0 & -5 & -2 & 3 \\ 0 & 0 & 1 & 2 & 1 & -1 \end{bmatrix} \begin{matrix} R_1 - 3R_3 \\ R_2 - 3R_3 \end{matrix}$

The inverse of the given matrix is

$\begin{bmatrix} -4 & -2 & 3 \\ -5 & -2 & 3 \\ 2 & 1 & -1 \end{bmatrix}$.

41. To find the inverse of $\begin{bmatrix} 1 & 4 & 3 \\ 1 & -3 & -2 \\ 2 & 5 & 4 \end{bmatrix}$,

write the augmented matrix $\begin{bmatrix} A \mid I \end{bmatrix}$.

$\begin{bmatrix} 1 & 4 & 3 & 1 & 0 & 0 \\ 1 & -3 & -2 & 0 & 1 & 0 \\ 2 & 5 & 4 & 0 & 0 & 1 \end{bmatrix}$

$\begin{bmatrix} 1 & 4 & 3 & 1 & 0 & 0 \\ 0 & -7 & -5 & -1 & 1 & 0 \\ 0 & -3 & -2 & -2 & 0 & 1 \end{bmatrix} \begin{matrix} -1R_1 + R_2 \\ -2R_1 + R_3 \end{matrix}$

$\begin{bmatrix} 1 & 4 & 3 & 1 & 0 & 0 \\ 0 & 1 & \frac{5}{7} & \frac{1}{7} & -\frac{1}{7} & 0 \\ 0 & -3 & -2 & -2 & 0 & 1 \end{bmatrix} -\frac{1}{7}R_2$

$\begin{bmatrix} 1 & 0 & \frac{1}{7} & \frac{3}{7} & \frac{4}{7} & 0 \\ 0 & 1 & \frac{5}{7} & \frac{1}{7} & -\frac{1}{7} & 0 \\ 0 & 0 & \frac{1}{7} & -\frac{11}{7} & -\frac{3}{7} & 1 \end{bmatrix} \begin{matrix} -4R_2 + R_1 \\ \\ 3R_2 + R_3 \end{matrix}$

$\begin{bmatrix} 1 & 0 & \frac{1}{7} & \frac{3}{7} & \frac{4}{7} & 0 \\ 0 & 1 & \frac{5}{7} & \frac{1}{7} & -\frac{1}{7} & 0 \\ 0 & 0 & 1 & -11 & -3 & 7 \end{bmatrix} 7R_3$

$\begin{bmatrix} 1 & 0 & 0 & 2 & 1 & -1 \\ 0 & 1 & 0 & 8 & 2 & -5 \\ 0 & 0 & 1 & -11 & -3 & 7 \end{bmatrix} \begin{matrix} -\frac{1}{7}R_3 + R_1 \\ -\frac{5}{7}R_3 + R_2 \\ \end{matrix}$

The inverse of the given matrix is

$\begin{bmatrix} 2 & 1 & -1 \\ 8 & 2 & -5 \\ -11 & -3 & 7 \end{bmatrix}$.

43. To find the inverse of $\begin{bmatrix} 1 & -1 & 4 \\ 0 & 1 & 3 \\ 2 & -3 & 4 \end{bmatrix}$,

write the augmented matrix $\begin{bmatrix} A \mid I \end{bmatrix}$.

$\begin{bmatrix} 1 & -1 & 4 & 1 & 0 & 0 \\ 0 & 1 & 3 & 0 & 1 & 0 \\ 2 & -3 & 4 & 0 & 0 & 1 \end{bmatrix}$

$\begin{bmatrix} 1 & 0 & 7 & 1 & 1 & 0 \\ 0 & 1 & 3 & 0 & 1 & 0 \\ 0 & -1 & -4 & -2 & 0 & 1 \end{bmatrix} \begin{matrix} R_1 + R_2 \\ \\ -2R_1 + R_3 \end{matrix}$

$\begin{bmatrix} 1 & 0 & 7 & 1 & 1 & 0 \\ 0 & 1 & 3 & 0 & 1 & 0 \\ 0 & 0 & 1 & 2 & -1 & -1 \end{bmatrix} -(R_2 + R_3)$

$\begin{bmatrix} 1 & 0 & 0 & -13 & 8 & 7 \\ 0 & 1 & 0 & -6 & 4 & 3 \\ 0 & 0 & 1 & 2 & -1 & -1 \end{bmatrix} \begin{matrix} -7R_3 + R_1 \\ -3R_3 + R_2 \end{matrix}$

The inverse of the given matrix is

$\begin{bmatrix} -13 & 8 & 7 \\ -6 & 4 & 3 \\ 2 & -1 & -1 \end{bmatrix}$.

45.

$$A = \begin{bmatrix} 1 & 2 & 3 \\ 1 & 4 & 2 \\ 0 & 1 & -1 \end{bmatrix}$$

```
[A]⁻¹
   [[6   -5 8 ]
    [-1  1  -1]
    [-1  1  -2]]
```

47. Use a graphing calculator to find the inverse of

$$\begin{bmatrix} 1 & 0 & -2 & 0 \\ -2 & 1 & 2 & 2 \\ 3 & -1 & -2 & -3 \\ 0 & 1 & 4 & 1 \end{bmatrix}.$$

Using a graphing calculator, we find that the inverse of the given matrix is

$$\begin{bmatrix} \frac{1}{2} & -1 & -\frac{1}{2} & \frac{1}{2} \\ -\frac{1}{2} & 4 & \frac{5}{2} & -\frac{1}{2} \\ -\frac{1}{4} & -\frac{1}{2} & -\frac{1}{4} & \frac{1}{4} \\ \frac{1}{2} & -2 & -\frac{3}{2} & \frac{1}{2} \end{bmatrix}.$$

49. a. $R = \begin{bmatrix} .024 & .008 \\ .025 & .007 \\ .015 & .009 \\ .011 & .011 \end{bmatrix}$

b. $P = \begin{bmatrix} 1996 & 286 & 226 & 460 \\ 2440 & 365 & 252 & 484 \\ 2906 & 455 & 277 & 499 \\ 3683 & 519 & 310 & 729 \\ 4723 & 697 & 364 & 702 \end{bmatrix}$

c. $PR = \begin{bmatrix} 1996 & 286 & 226 & 460 \\ 2440 & 365 & 252 & 484 \\ 2906 & 455 & 277 & 499 \\ 3683 & 519 & 310 & 729 \\ 4723 & 697 & 364 & 702 \end{bmatrix} \begin{bmatrix} .024 & .008 \\ .025 & .007 \\ .015 & .009 \\ .011 & .011 \end{bmatrix}$

$$= \begin{bmatrix} 1996(.024)+286(.025)+226(.015)+460(.011) & 1996(.008)+286(.007)+226(.009)+460(.011) \\ 2440(.024)+365(.025)+252(.015)+484(.011) & 2440(.008)+365(.007)+252(.009)+484(.011) \\ 2906(.024)+455(.025)+277(.015)+499(.011) & 2906(.008)+455(.007)+277(.009)+499(.011) \\ 3683(.024)+519(.025)+310(.015)+729(.011) & 3683(.008)+519(.007)+310(.009)+729(.011) \\ 4723(.024)+697(.025)+364(.015)+702(.011) & 4723(.008)+697(.007)+364(.009)+702(.011) \end{bmatrix}$$

$$= \begin{bmatrix} 63.504 & 25.064 \\ 76.789 & 29.667 \\ 90.763 & 34.415 \\ 114.036 & 43.906 \\ 143.959 & 53.661 \end{bmatrix}$$

d. The rows represent the years 1970, 1980, 1990, 2000, 2025. Column 1 gives the total births in those years, column 2 the total deaths.

e. The total number of births in 2000 was 114,036,000.
The total number of deaths projected for 2025 is 53,661,000.

51. a. $A = \begin{bmatrix} 278.1 & 31.6 & 37.4 & 126.8 \\ 300.1 & 34.3 & 41.1 & 127.3 \end{bmatrix}$

b. $B = \begin{bmatrix} .01425 & .00865 \\ .01145 & .00775 \\ .0175 & .00755 \\ .00945 & .00925 \end{bmatrix}$

c. $AB = \begin{bmatrix} 278.1 & 31.6 & 37.4 & 126.8 \\ 300.1 & 34.3 & 41.1 & 127.3 \end{bmatrix} \begin{bmatrix} .01425 & .00865 \\ .01145 & .00775 \\ .0175 & .00755 \\ .00945 & .00925 \end{bmatrix} = \begin{bmatrix} 6.177505 & 4.105735 \\ 6.591395 & 4.349520 \end{bmatrix}$

d. The rows correspond to years. The entries in each column give the total number of births and deaths, respectively, in the four countries, taken together.

e. The total number of people born in these four countries combined in 2001 is 6.177505 million or 6,177,505.

53. a. $A = \begin{bmatrix} 1259 & 600 \\ 1339 & 642 \\ 1424 & 673 \\ 1493 & 723 \end{bmatrix}$

b. $B = \begin{bmatrix} 82.6 & 82.7 & 82.7 & 82.6 \\ 65.4 & 66.1 & 66.2 & 66.6 \end{bmatrix}$

c. $AB = \begin{bmatrix} 1259 & 600 \\ 1339 & 642 \\ 1424 & 673 \\ 1493 & 723 \end{bmatrix} \begin{bmatrix} 82.6 & 82.7 & 82.7 & 82.6 \\ 65.4 & 66.1 & 66.2 & 66.6 \end{bmatrix} = \begin{bmatrix} 143233.4 & 143779.3 & 143839.3 & 143953.4 \\ 152588.2 & 153171.5 & 153235.7 & 153358.6 \\ 161636.6 & 162250.1 & 162317.4 & 162444.2 \\ 170606 & 171261.4 & 171333.7 & 171473.6 \end{bmatrix}$

d. $BA = \begin{bmatrix} 82.6 & 82.7 & 82.7 & 82.6 \\ 65.4 & 66.1 & 66.2 & 66.6 \end{bmatrix} \begin{bmatrix} 1259 & 600 \\ 1339 & 642 \\ 1424 & 673 \\ 1493 & 723 \end{bmatrix} = \begin{bmatrix} 455815.3 & 218030.3 \\ 364549.1 & 174380.6 \end{bmatrix}$

e. In AB, row 1, column 1 is the total per-capita dollar amount paid by private insurance for physician services and prescription drugs in 2003; row 2, column 2 is the same total for 2004; row 3, column 3 is the same total for 2005; row 4, column 4 is the same total for 2006. All other entries in AB are not meaningful here. In BA, row 1, column 1 is the per capita total dollar amount paid by private insurance for physician services over the four-year period; row 2, column 2 is the per-capita total dollar amount paid by private insurance for prescription drugs over the four-year period. The other two entries in BA are not meaningful here.

55. a. Let matrix P contain the amount of products needed.

$$P = \begin{array}{c} \text{Dept. 1} \\ \text{Dept. 2} \\ \text{Dept. 3} \\ \text{Dept. 4} \end{array} \begin{bmatrix} 10 & 4 & 3 & 5 & 6 \\ 7 & 2 & 2 & 3 & 8 \\ 4 & 5 & 1 & 0 & 10 \\ 0 & 3 & 4 & 5 & 5 \end{bmatrix}$$

with columns Paper, Tape, Print Rib., Memo Pads, Pens.

Let matrix C contain the cost from each supplier.

$$C = \begin{array}{c} \text{Paper} \\ \text{Tape} \\ \text{Ink Cartridges} \\ \text{Memo Pads} \\ \text{Pens} \end{array} \begin{bmatrix} 9 & 12 \\ 6 & 6 \\ 24 & 18 \\ 4 & 4 \\ 8 & 12 \end{bmatrix}$$

with columns A B.

To find the total departmental cost from each supplier, multiply P times C.

$$PC = \begin{bmatrix} 10 & 4 & 3 & 5 & 6 \\ 7 & 2 & 2 & 3 & 8 \\ 4 & 5 & 1 & 0 & 10 \\ 0 & 3 & 4 & 5 & 5 \end{bmatrix} \begin{bmatrix} 9 & 12 \\ 6 & 6 \\ 24 & 18 \\ 4 & 4 \\ 8 & 12 \end{bmatrix} = \begin{array}{c} \text{Dept. 1} \\ \text{Dept. 2} \\ \text{Dept. 3} \\ \text{Dept. 4} \end{array} \begin{bmatrix} 254 & 290 \\ 199 & 240 \\ 170 & 216 \\ 174 & 170 \end{bmatrix}$$

with columns A B.

b. The total cost from each supplier would be found by adding each column. The total cost from Supplier A is \$797; the total cost from Supplier B is \$916. The company should buy from Supplier A.

Section 6.6 Applications of Matrices

1. The solution to the matrix equation $AX = B$ is
$X = A^{-1}B$.
$$A = \begin{bmatrix} 1 & -1 \\ 5 & 6 \end{bmatrix}, B = \begin{bmatrix} 4 \\ -2 \end{bmatrix}$$
Use a graphing calculator or row operations on
$[A \mid I]$ to find A^{-1}.
$$\begin{bmatrix} 1 & -1 & | & 1 & 0 \\ 5 & 6 & | & 0 & 1 \end{bmatrix}$$
$$\begin{bmatrix} 1 & -1 & | & 1 & 0 \\ 0 & 11 & | & -5 & 1 \end{bmatrix} -5R_1 + R_2$$
$$\begin{bmatrix} 1 & -1 & | & 1 & 0 \\ 0 & 1 & | & -\frac{5}{11} & \frac{1}{11} \end{bmatrix} \frac{1}{11}R_2$$
$$\begin{bmatrix} 1 & 0 & | & \frac{6}{11} & \frac{1}{11} \\ 0 & 1 & | & -\frac{5}{11} & \frac{1}{11} \end{bmatrix} R_1 + R_2$$
$$A^{-1} = \begin{bmatrix} \frac{6}{11} & \frac{1}{11} \\ -\frac{5}{11} & \frac{1}{11} \end{bmatrix}$$
$$X = A^{-1}B = \begin{bmatrix} \frac{6}{11} & \frac{1}{11} \\ -\frac{5}{11} & \frac{1}{11} \end{bmatrix}\begin{bmatrix} 4 \\ -2 \end{bmatrix} = \begin{bmatrix} 2 \\ -2 \end{bmatrix}.$$

3. The solution to the matrix equation $AX = B$ is
$X = A^{-1}B$.
$$A = \begin{bmatrix} 3 & 1 \\ 4 & 2 \end{bmatrix}, B = \begin{bmatrix} 3 & 4 \\ 5 & 6 \end{bmatrix}.$$
Use a graphing calculator or row operations on
$[A \mid I]$ to find A^{-1}.
$$\begin{bmatrix} 3 & 1 & | & 1 & 0 \\ 4 & 2 & | & 0 & 1 \end{bmatrix}$$
$$\begin{bmatrix} 1 & \frac{1}{3} & | & \frac{1}{3} & 0 \\ 4 & 2 & | & 0 & 1 \end{bmatrix} \frac{1}{3}R_1$$
$$\begin{bmatrix} 1 & \frac{1}{3} & | & \frac{1}{3} & 0 \\ 0 & \frac{2}{3} & | & -\frac{4}{3} & 1 \end{bmatrix} R_2 - 4R_1$$
$$\begin{bmatrix} 1 & \frac{1}{3} & | & \frac{1}{3} & 0 \\ 0 & 1 & | & -2 & \frac{3}{2} \end{bmatrix} \frac{3}{2}R_2$$
$$\begin{bmatrix} 1 & 0 & | & 1 & -\frac{1}{2} \\ 0 & 1 & | & -2 & \frac{3}{2} \end{bmatrix} R_1 - \frac{1}{3}R_2$$
$$A^{-1} = \begin{bmatrix} 1 & -\frac{1}{2} \\ -2 & \frac{3}{2} \end{bmatrix}$$
$$X = A^{-1}B = \begin{bmatrix} 1 & -\frac{1}{2} \\ -2 & \frac{3}{2} \end{bmatrix}\begin{bmatrix} 3 & 4 \\ 5 & 6 \end{bmatrix} = \begin{bmatrix} \frac{1}{2} & 1 \\ \frac{3}{2} & 1 \end{bmatrix}.$$

5. The solution to the matrix equation $AX = B$ is
$X = A^{-1}B$.
$$A = \begin{bmatrix} 2 & 1 & 0 \\ -4 & -1 & 3 \\ 3 & 1 & -2 \end{bmatrix}, B = \begin{bmatrix} 1 \\ 4 \\ 0 \end{bmatrix}.$$
Use a graphing calculator or row operations on
$[A \mid I]$ to find that
$$A^{-1} = \begin{bmatrix} 1 & -2 & -3 \\ -1 & 4 & 6 \\ 1 & -1 & -2 \end{bmatrix}.$$
$$X = A^{-1}B = \begin{bmatrix} 1 & -2 & -3 \\ -1 & 4 & 6 \\ 1 & -1 & -2 \end{bmatrix}\begin{bmatrix} 1 \\ 4 \\ 0 \end{bmatrix} = \begin{bmatrix} -7 \\ 15 \\ -3 \end{bmatrix}.$$

7. $x + 2y + 3z = 10$
$2x + 3y + 2z = 6$
$-x - 2y - 4z = -1$
has coefficient matrix
$$A = \begin{bmatrix} 1 & 2 & 3 \\ 2 & 3 & 2 \\ -1 & -2 & -4 \end{bmatrix}.$$
Find A^{-1}:
$$[A \mid I] = \begin{bmatrix} 1 & 2 & 3 & | & 1 & 0 & 0 \\ 2 & 3 & 2 & | & 0 & 1 & 0 \\ -1 & -2 & -4 & | & 0 & 0 & 1 \end{bmatrix}$$
$$\begin{bmatrix} 1 & 2 & 3 & | & 1 & 0 & 0 \\ 0 & -1 & -4 & | & -2 & 1 & 0 \\ 0 & 0 & -1 & | & 1 & 0 & 1 \end{bmatrix} \begin{matrix} -2R_1 + R_2 \\ R_1 + R_3 \end{matrix}$$
$$\begin{bmatrix} 1 & 0 & -5 & | & -3 & 2 & 0 \\ 0 & 1 & 4 & | & 2 & -1 & 0 \\ 0 & 0 & 1 & | & -1 & 0 & -1 \end{bmatrix} \begin{matrix} 2R_2 + R_1 \\ -1R_2 \\ -1R_3 \end{matrix}$$
$$\begin{bmatrix} 1 & 0 & 0 & | & -8 & 2 & -5 \\ 0 & 1 & 0 & | & 6 & -1 & 4 \\ 0 & 0 & 1 & | & -1 & 0 & -1 \end{bmatrix} \begin{matrix} 5R_3 + R_1 \\ -4R_3 + R_2 \end{matrix}$$
$$A^{-1} = \begin{bmatrix} -8 & 2 & -5 \\ 6 & -1 & 4 \\ -1 & 0 & -1 \end{bmatrix}$$
$$X = A^{-1}B = \begin{bmatrix} -8 & 2 & -5 \\ 6 & -1 & 4 \\ -1 & 0 & -1 \end{bmatrix}\begin{bmatrix} 10 \\ 6 \\ -1 \end{bmatrix} = \begin{bmatrix} -63 \\ 50 \\ -9 \end{bmatrix}$$
The solution is $(-63, 50, -9)$.

9. $x + 4y + 3z = -12$

$x - 3y - 2z = 0$

$2x + 5y + 4z = 7$

has the coefficient matrix

$$A = \begin{bmatrix} 1 & 4 & 3 \\ 1 & -3 & -2 \\ 2 & 5 & 4 \end{bmatrix}.$$

From Exercise 41 of Section 6.5,

$$A^{-1} = \begin{bmatrix} 2 & 1 & -1 \\ 8 & 2 & -5 \\ -11 & -3 & 7 \end{bmatrix}.$$

$$X = A^{-1}B = \begin{bmatrix} 2 & 1 & -1 \\ 8 & 2 & -5 \\ -11 & -3 & 7 \end{bmatrix} \begin{bmatrix} -12 \\ 0 \\ 7 \end{bmatrix} = \begin{bmatrix} -31 \\ -131 \\ 181 \end{bmatrix}.$$

The solution is $(-31, -131, 181)$.

11. $x + 2y + 3z = 4$

$x + 4y + 2z = 8$

$y - z = -4$

has coefficient matrix

$$A = \begin{bmatrix} 1 & 2 & 3 \\ 1 & 4 & 2 \\ 0 & 1 & -1 \end{bmatrix}.$$

From Exercise 45 of Section 6.5,

$$A^{-1} = \begin{bmatrix} 6 & -5 & 8 \\ -1 & 1 & -1 \\ -1 & 1 & -2 \end{bmatrix}.$$

$$X = A^{-1}B = \begin{bmatrix} 6 & -5 & 8 \\ -1 & 1 & -1 \\ -1 & 1 & -2 \end{bmatrix} \begin{bmatrix} 4 \\ 8 \\ -4 \end{bmatrix} = \begin{bmatrix} -48 \\ 8 \\ 12 \end{bmatrix}.$$

The solution is $(-48, 8, 12)$.

13. $x - 2z = 4$

$-2x + y + 2z + 2w = -8$

$3x - y - 2z - 3w = 12$

$y + 4z + w = -4$

The coefficient matrix is

$$A = \begin{bmatrix} 1 & 0 & -2 & 0 \\ -2 & 1 & 2 & 2 \\ 3 & -1 & -2 & -3 \\ 0 & 1 & 4 & 1 \end{bmatrix}.$$

From Exercise 47 in Section 6.5,

$$A^{-1} = \begin{bmatrix} \frac{1}{2} & -1 & -\frac{1}{2} & \frac{1}{2} \\ \frac{1}{2} & 4 & \frac{5}{2} & -\frac{1}{2} \\ -\frac{1}{4} & -\frac{1}{2} & -\frac{1}{4} & \frac{1}{4} \\ \frac{1}{2} & -2 & -\frac{3}{2} & \frac{1}{2} \end{bmatrix}$$

$$X = \begin{bmatrix} \frac{1}{2} & -1 & -\frac{1}{2} & \frac{1}{2} \\ \frac{1}{2} & 4 & \frac{5}{2} & -\frac{1}{2} \\ -\frac{1}{4} & -\frac{1}{2} & -\frac{1}{4} & \frac{1}{4} \\ \frac{1}{2} & -2 & -\frac{3}{2} & \frac{1}{2} \end{bmatrix} \begin{bmatrix} 4 \\ -8 \\ 12 \\ -4 \end{bmatrix} = \begin{bmatrix} 2 \\ 2 \\ -1 \\ -2 \end{bmatrix}.$$

The solution is $(2, 2, -1, -2)$.

15. Since $N = X - MX$, $N = IX - MX \Rightarrow$

$N = (I - M)X \Rightarrow$

$(I - M)^{-1}N = (I - M)^{-1}(I - M)X \Rightarrow$

$(I - M)^{-1}N = IX.$

Thus, $X = (I - M)^{-1}N$.

$$I - M = \begin{bmatrix} 1 & 0 \\ 0 & 1 \end{bmatrix} - \begin{bmatrix} 0 & 1 \\ -2 & 1 \end{bmatrix} = \begin{bmatrix} 1 & -1 \\ 2 & 0 \end{bmatrix}$$

If $I - M = \begin{bmatrix} 1 & -1 \\ 2 & 0 \end{bmatrix}$,

$$(I - M)^{-1} = \begin{bmatrix} 0 & \frac{1}{2} \\ -1 & \frac{1}{2} \end{bmatrix}.$$

Since $X = (I - M)^{-1}N$,

$$X = \begin{bmatrix} 0 & \frac{1}{2} \\ -1 & \frac{1}{2} \end{bmatrix} \begin{bmatrix} 8 \\ -12 \end{bmatrix} = \begin{bmatrix} -6 \\ -14 \end{bmatrix}.$$

17. Let $x =$ the number of buffets, let $y =$ the number of chairs, and let $z =$ the number of tables. The information can be summarized in a table.

	Cutting hours	Assembly hours	Finishing hours
Buffets	$15x$	$20x$	$5x$
Chairs	$5y$	$8y$	$5y$
Tables	$10z$	$6z$	$6z$
Total	4900	6600	3900

The system of equations is

$15x + 5y + 10z = 4900$

$20x + 8y + 6z = 6600$

$5x + 5y + 6z = 3900$

(continued on next page)

(*continued from page 179*)

Then, we have

$$A = \begin{bmatrix} 15 & 5 & 10 \\ 20 & 8 & 6 \\ 5 & 5 & 6 \end{bmatrix} \text{ and } B = \begin{bmatrix} 4900 \\ 6600 \\ 3900 \end{bmatrix}.$$

Using a calculator, we find that the inverse is

$$\begin{bmatrix} \frac{3}{70} & \frac{1}{21} & -\frac{5}{42} \\ -\frac{3}{14} & \frac{2}{21} & \frac{11}{42} \\ \frac{1}{7} & -\frac{5}{42} & \frac{1}{21} \end{bmatrix}.$$

$$X = A^{-1}B = \begin{bmatrix} \frac{3}{70} & \frac{1}{21} & -\frac{5}{42} \\ -\frac{3}{14} & \frac{2}{21} & \frac{11}{42} \\ \frac{1}{7} & -\frac{5}{42} & \frac{1}{21} \end{bmatrix} \begin{bmatrix} 4900 \\ 6600 \\ 3900 \end{bmatrix} = \begin{bmatrix} 60 \\ 600 \\ 100 \end{bmatrix}$$

Therefore, 60 buffets, 600 chairs, and 100 tables should be produced each week.

19. The following solution presupposes the use of a TI-82 graphing calculator. Similar results can be obtained from other graphing calculators.
Let x = number of bacterium 1, let y = number of bacterium 2, and let z = number of bacterium 3.

	Food		
	I	II	III
Bacterium 1	1.3	1.3	2.3
Bacterium 2	1.1	2.4	3.7
Bacterium 3	8.1	2.9	5.1
Totals	16,000	28,000	44,000

The data in the table produce the system of equations

$$1.3x + 1.1y + 8.1z = 16,000$$
$$1.3x + 2.4y + 2.9z = 28,000$$
$$2.3x + 3.7y + 5.1z = 44,000.$$

We store the following matrix as matrix A:

$$\begin{bmatrix} 1.3 & 1.1 & 8.1 & | & 16,000 \\ 1.3 & 2.4 & 2.9 & | & 28,000 \\ 2.3 & 3.7 & 5.1 & | & 44,000 \end{bmatrix}.$$

Using row operations, we transform this to obtain

$$\begin{bmatrix} 1 & 0 & 0 & | & 2339.74359 \\ 0 & 1 & 0 & | & 10,128.20513 \\ 0 & 0 & 1 & | & 224.3589744 \end{bmatrix}.$$

Thus, 2340 of the first species, 10,128 of the second species, and 224 of the third species can be maintained.

21. Let x = the wholesale price of jeans, let y = the wholesale price of jackets, let z = the wholesale price of sweaters, and let w = the wholesale price of shirts. The system is

$$3000x + 3000y + 2200z + 4200w = 507,650$$
$$2700x + 2500y + 2100z + 4300w = 459,075$$
$$5000x + 2000y + 1400z + 7500w = 541,225$$
$$7000x + 1800y + 600z + 8000w = 571,500$$

Store the following matrix as matrix A.

$$\begin{bmatrix} 3000 & 3000 & 2200 & 4200 & | & 507,650 \\ 2700 & 2500 & 2100 & 4300 & | & 459,075 \\ 5000 & 2000 & 1400 & 7500 & | & 541,225 \\ 7000 & 1800 & 600 & 8000 & | & 571,500 \end{bmatrix}$$

Using row operations, transform this matrix to obtain

$$\begin{bmatrix} 1 & 0 & 0 & 0 & | & 34.5 \\ 0 & 1 & 0 & 0 & | & 72 \\ 0 & 0 & 1 & 0 & | & 44 \\ 0 & 0 & 0 & 1 & | & 21.75 \end{bmatrix}$$

Therefore, the wholesale price for jeans is $34.50, jacket is $72, sweater is $44, and a shirt is $21.75.

23. $A = \begin{bmatrix} \frac{1}{2} & \frac{2}{5} \\ \frac{1}{4} & \frac{1}{5} \end{bmatrix}$, $D = \begin{bmatrix} 2 \\ 4 \end{bmatrix}$

To find the production matrix, first calculate $I - A$.

$$I - A = \begin{bmatrix} 1 & 0 \\ 0 & 1 \end{bmatrix} - \begin{bmatrix} \frac{1}{2} & \frac{2}{5} \\ \frac{1}{4} & \frac{1}{5} \end{bmatrix} = \begin{bmatrix} \frac{1}{2} & -\frac{2}{5} \\ -\frac{1}{4} & \frac{4}{5} \end{bmatrix}$$

Using row operations, find the inverse of $I - A$.

$$(I - A)^{-1} = \begin{bmatrix} \frac{8}{3} & \frac{4}{3} \\ \frac{5}{6} & \frac{5}{3} \end{bmatrix}$$

Since $X = (I - A)^{-1}D$,

$$X = \begin{bmatrix} \frac{8}{3} & \frac{4}{3} \\ \frac{5}{6} & \frac{5}{3} \end{bmatrix} \begin{bmatrix} 2 \\ 4 \end{bmatrix} = \begin{bmatrix} \frac{32}{3} \\ \frac{25}{3} \end{bmatrix}.$$

25. $A = \begin{bmatrix} .25 & .08 \\ .33 & .11 \end{bmatrix}$, $D = \begin{bmatrix} 690 \\ 920 \end{bmatrix}$

$(I - A)^{-1} = \begin{bmatrix} 1.388 & 0.1248 \\ 0.5147 & 1.1698 \end{bmatrix}$

$X = (I - A)^{-1} D = \begin{bmatrix} 1.388 & 0.1248 \\ 0.5147 & 1.1699 \end{bmatrix} \begin{bmatrix} 690 \\ 920 \end{bmatrix}$

$= \begin{bmatrix} 1073 \\ 1431 \end{bmatrix}$

Produce 1073 metric tons of wheat, and 1431 metric tons of oil.

27. First, write the input-output matrix A.

$A = \begin{bmatrix} .4 & .6 \\ .5 & .25 \end{bmatrix} = \begin{bmatrix} \frac{2}{5} & \frac{3}{5} \\ \frac{1}{2} & \frac{1}{4} \end{bmatrix}$

$I - A = \begin{bmatrix} 1 & 0 \\ 0 & 1 \end{bmatrix} - \begin{bmatrix} \frac{2}{5} & \frac{3}{5} \\ \frac{1}{2} & \frac{1}{4} \end{bmatrix} = \begin{bmatrix} \frac{3}{5} & -\frac{3}{5} \\ -\frac{1}{2} & \frac{3}{4} \end{bmatrix}$

Form $[A - I \mid I]$.

$\begin{bmatrix} \frac{3}{5} & -\frac{3}{5} & 1 & 0 \\ -\frac{1}{2} & \frac{3}{4} & 0 & 1 \end{bmatrix}$

$\begin{bmatrix} 1 & -1 & \frac{5}{3} & 0 \\ -\frac{1}{2} & \frac{3}{4} & 0 & 1 \end{bmatrix} \frac{5}{3} R_1$

$\begin{bmatrix} 1 & -1 & \frac{5}{3} & 0 \\ 0 & \frac{1}{4} & \frac{5}{6} & 1 \end{bmatrix} \frac{1}{2} R_1 + R_2$

$\begin{bmatrix} 1 & -1 & \frac{5}{3} & 0 \\ 0 & 1 & \frac{10}{3} & 4 \end{bmatrix} 4R_2$

$\begin{bmatrix} 1 & 0 & 5 & 4 \\ 0 & 1 & \frac{10}{3} & 4 \end{bmatrix} R_2 + R_1$

Thus, $(I - A)^{-1} = \begin{bmatrix} 5 & 4 \\ \frac{10}{3} & 4 \end{bmatrix}$

Since $D = \begin{bmatrix} 15 \text{ million} \\ 12 \text{ million} \end{bmatrix}$ and $X = (I - A)^{-1} D$,

$X = \begin{bmatrix} 5 & 4 \\ \frac{10}{3} & 4 \end{bmatrix} \begin{bmatrix} 15 \text{ million} \\ 12 \text{ million} \end{bmatrix} = \begin{bmatrix} 123 \text{ million} \\ 98 \text{ million} \end{bmatrix}$.

The output should be $123 million of electricity and $98 million of gas.

29. $A = \begin{bmatrix} \frac{1}{4} & \frac{1}{6} \\ \frac{1}{2} & 0 \end{bmatrix}$, $I - A = \begin{bmatrix} \frac{3}{4} & -\frac{1}{6} \\ -\frac{1}{2} & 1 \end{bmatrix}$

$(I - A)^{-1} = \begin{bmatrix} \frac{3}{2} & \frac{1}{4} \\ \frac{3}{4} & \frac{9}{8} \end{bmatrix}$

a. $X = \begin{bmatrix} \frac{3}{2} & \frac{1}{4} \\ \frac{3}{4} & \frac{9}{8} \end{bmatrix} \begin{bmatrix} 1 \\ 1 \end{bmatrix} = \begin{bmatrix} \frac{7}{4} \\ \frac{15}{8} \end{bmatrix}$

Thus, $\frac{7}{4}$ bushes of yams and $\frac{15}{8} \approx 2$ pigs should be produced.

b. $X = \begin{bmatrix} \frac{3}{2} & \frac{1}{4} \\ \frac{3}{4} & \frac{9}{8} \end{bmatrix} \begin{bmatrix} 100 \\ 70 \end{bmatrix} = \begin{bmatrix} 167.5 \\ 153.75 \end{bmatrix}$

Thus, 167.5 bushels of yams and $153.75 \approx 154$ pigs should be produced.

31. a. From the input-output matrix, we see that .40 unit of agriculture, .12 unit of manufacturing, and 3.60 units of households are required for the manufacturing sector to produce 1 unit.

b. $A = \begin{bmatrix} .25 & .40 & .133 \\ .14 & .12 & .100 \\ .80 & 3.60 & .133 \end{bmatrix}$

$I - A = \begin{bmatrix} .75 & -.40 & -.133 \\ -.14 & .88 & -.100 \\ -.80 & -3.60 & .867 \end{bmatrix}$

Next, calculate $(I - A)^{-1}$.

$(I - A)^{-1} \approx \begin{bmatrix} 6.61 & 13.53 & 2.57 \\ 3.3 & 8.91 & 1.53 \\ 19.8 & 49.5 & 9.9 \end{bmatrix}$

$X = (I - A)^{-1} D$

$\approx \begin{bmatrix} 6.61 & 13.53 & 2.57 \\ 3.3 & 8.91 & 1.53 \\ 19.8 & 49.5 & 9.9 \end{bmatrix} \begin{bmatrix} 35 \\ 38 \\ 40 \end{bmatrix} \approx \begin{bmatrix} 848 \\ 516 \\ 2970 \end{bmatrix}$

Therefore, 848 units of agriculture, 516 units of manufacturing, and 2970 units of households need to be produced.

c. Since .25 units of agriculture are used in producing each unit of agriculture, $(848)(.25) = 212$ units are used. Since .4 units of agriculture are used in producing each unit of manufacturing, $(.4)(516) = 206.4$ units are used. Since .133 units of agriculture are used in producing each unit of households, $(.133)(2970) = 395.01$. Thus, $212 + 206 + 395 = 813$ units of agriculture are used in the production process.

33. a. The energy sector requires .017 unit of manufacturing and .216 unit of energy to produce one unit.

b. We are given the input-output matrix A and the production matrix X.

$$A = \begin{bmatrix} .293 & 0 & 0 \\ .014 & .207 & .017 \\ .044 & .010 & .216 \end{bmatrix}, \quad X = \begin{bmatrix} 175,000 \\ 22,000 \\ 12,000 \end{bmatrix}$$

$$D = (I - A)X$$

$$= \begin{bmatrix} .707 & 0 & 0 \\ -.014 & .793 & -.017 \\ .044 & -.010 & .784 \end{bmatrix}\begin{bmatrix} 175,000 \\ 22,000 \\ 12,000 \end{bmatrix}$$

$$= \begin{bmatrix} 123,725 \\ 14,792 \\ 1488 \end{bmatrix}$$

Therefore, about 123,725,000 pounds of agriculture, 14,792,000 pounds of manufacturing, and 1,488,000 pounds of energy should be produced.

c. $X = (I - A)^{-1} D$

$$\approx \begin{bmatrix} 1.414 & 0 & 0 \\ .027 & 1.261 & .027 \\ .080 & .016 & 1.276 \end{bmatrix}\begin{bmatrix} 138,213 \\ 17,597 \\ 1786 \end{bmatrix}$$

$$\approx \begin{bmatrix} 195,492 \\ 25,933 \\ 13,580 \end{bmatrix}$$

Therefore, about 195,492,000 pounds of agriculture, 25,933,000 pounds of manufacturing, and 13,580,000 pounds of energy should be produced.

35. $A = \begin{bmatrix} .1045 & .0428 & .0029 & .0031 \\ .0826 & .1087 & .0584 & .0321 \\ .0867 & .1019 & .2032 & .3555 \\ .6253 & .3448 & .6106 & .0798 \end{bmatrix}, \quad D = \begin{bmatrix} 450 \\ 300 \\ 125 \\ 100 \end{bmatrix}$

$$(I - A)^{-1} \approx \begin{bmatrix} 1.133 & .062 & .019 & .013 \\ .202 & 1.19 & .171 & .108 \\ .748 & .536 & 1.87 & .742 \\ 1.343 & .845 & 1.315 & 1.63 \end{bmatrix}$$

$$X = (I - A)^{-1} D$$

$$X \approx \begin{bmatrix} 1.133 & .062 & .019 & .013 \\ .202 & 1.19 & .171 & .108 \\ .748 & .536 & 1.87 & .742 \\ 1.343 & .845 & 1.315 & 1.63 \end{bmatrix}\begin{bmatrix} 450 \\ 300 \\ 125 \\ 100 \end{bmatrix} \approx \begin{bmatrix} 532 \\ 481 \\ 805 \\ 1185 \end{bmatrix}$$

This means \$532 million of natural resources, \$481 million of manufacturing, \$805 million of trade and services, and \$1185 million of personal consumption.

37. a. The input-output matrix A, and the matrix $I - A$, are

$$A = \begin{bmatrix} .2 & .1 & .1 \\ .1 & .1 & 0 \\ .5 & .6 & .7 \end{bmatrix} \text{ and }$$

$$I - A = \begin{bmatrix} .8 & -.1 & -.1 \\ -.1 & .9 & 0 \\ -.5 & -.6 & .3 \end{bmatrix}.$$

Next, calculate $(I - A)^{-1}$.

$$(I - A)^{-1} \approx \begin{bmatrix} 1.67 & .56 & .56 \\ .19 & 1.17 & .06 \\ 3.15 & 3.27 & 4.38 \end{bmatrix}$$

b. These multipliers imply that if the demand for one community's output increases by \$1 then the output in the other community will increase by the amount in the row and column of that matrix. For example, if the demand for Hermitage's output increases by \$1, then output from Sharon will increase by \$.56, Farrell by \$.06, and Hermitage by \$4.38.

39. The message *Head for the hills* broken into groups of 2 letters would have the following matrices.

$$\begin{bmatrix} 8 \\ 5 \end{bmatrix}, \begin{bmatrix} 1 \\ 4 \end{bmatrix}, \begin{bmatrix} 27 \\ 6 \end{bmatrix}, \begin{bmatrix} 15 \\ 18 \end{bmatrix}, \begin{bmatrix} 27 \\ 20 \end{bmatrix},$$

$$\begin{bmatrix} 8 \\ 5 \end{bmatrix}, \begin{bmatrix} 27 \\ 8 \end{bmatrix}, \begin{bmatrix} 9 \\ 12 \end{bmatrix}, \begin{bmatrix} 12 \\ 19 \end{bmatrix}.$$

Multiply $M = \begin{bmatrix} 1 & 3 \\ 2 & 7 \end{bmatrix}$ by each matrix in the code to obtain

$$\begin{bmatrix} 23 \\ 51 \end{bmatrix}, \begin{bmatrix} 13 \\ 30 \end{bmatrix}, \begin{bmatrix} 45 \\ 96 \end{bmatrix}, \begin{bmatrix} 69 \\ 156 \end{bmatrix}, \begin{bmatrix} 87 \\ 194 \end{bmatrix},$$

$$\begin{bmatrix} 23 \\ 51 \end{bmatrix}, \begin{bmatrix} 51 \\ 110 \end{bmatrix}, \begin{bmatrix} 45 \\ 102 \end{bmatrix}, \begin{bmatrix} 69 \\ 157 \end{bmatrix}.$$

41. $A = \begin{bmatrix} 0 & 1 & 2 & 2 \\ 1 & 0 & 1 & 0 \\ 2 & 1 & 0 & 1 \\ 2 & 0 & 1 & 0 \end{bmatrix}; \quad A^2 = \begin{bmatrix} 9 & 2 & 3 & 2 \\ 2 & 2 & 2 & 3 \\ 3 & 2 & 6 & 4 \\ 2 & 3 & 4 & 5 \end{bmatrix}$

a. The number of ways to travel from city 1 to city 3 by passing through exactly one city is the entry in row 1, column 3 of A^2, which is 3.

b. The number of ways to travel from city 2 to city 4 by passing through exactly one city is the entry in row 2, column 4 of A^2, which is 3.

c. The number of ways to travel from city 1 to city 3 by passing through at most one city is the sum of the entries in row 1, column 3 of A and A^2, which is 2 + 3 or 5.

d. The number of ways to travel from city 2 to city 4 by passing through at most one city is the sum of the entries in row 2, column 4 of A and A^2, which is 0 + 3 or 3.

43. a.
$$B = \begin{array}{c} \\ 1 \\ 2 \\ 3 \end{array} \begin{array}{ccc} 1 & 2 & 3 \\ \left[\begin{array}{ccc} 0 & 2 & 3 \\ 2 & 0 & 4 \\ 3 & 4 & 0 \end{array}\right] \end{array}$$

b.
$$B^2 = \begin{bmatrix} 0 & 2 & 3 \\ 2 & 0 & 4 \\ 3 & 4 & 0 \end{bmatrix}\begin{bmatrix} 0 & 2 & 3 \\ 2 & 0 & 4 \\ 3 & 4 & 0 \end{bmatrix}$$
$$= \begin{bmatrix} 13 & 12 & 8 \\ 12 & 20 & 6 \\ 8 & 6 & 25 \end{bmatrix}$$

c. Use the entry in row 1, column 2 of B^2, which is 12.

d. Use the sum of the entries in row 1, column 2 of B and B^2, which is 2 + 12 or 14.

45. a.
$$C = \begin{array}{c} \\ d \\ r \\ c \\ m \end{array} \begin{array}{cccc} d & r & c & m \\ \left[\begin{array}{cccc} 0 & 1 & 1 & 1 \\ 0 & 0 & 0 & 1 \\ 0 & 1 & 0 & 1 \\ 0 & 0 & 0 & 0 \end{array}\right] \end{array}$$

b.
$$C^2 = \begin{bmatrix} 0 & 1 & 1 & 1 \\ 0 & 0 & 0 & 1 \\ 0 & 1 & 0 & 1 \\ 0 & 0 & 0 & 0 \end{bmatrix}\begin{bmatrix} 0 & 1 & 1 & 1 \\ 0 & 0 & 0 & 1 \\ 0 & 1 & 0 & 1 \\ 0 & 0 & 0 & 0 \end{bmatrix}$$
$$= \begin{bmatrix} 0 & 1 & 0 & 2 \\ 0 & 0 & 0 & 0 \\ 0 & 0 & 0 & 1 \\ 0 & 0 & 0 & 0 \end{bmatrix}$$

C^2 gives the number of food sources once removed from the feeder. Thus, since dogs eat rats and rats eat mice, mice are an indirect as well as a direct food source of dogs

Chapter 6 Review Exercises

1. $-5x - 3y = -3$ (1)
$2x + y = 4$ (2)

Multiply equation (2) by 3 and add to equation (1).
$$-5x - 3y = -3$$
$$\underline{6x + 3y = 12}$$
$$x = 9$$

Substitute 9 for x in equation (2) to solve for y.
$$2(9) + y = 4 \Rightarrow y = 14$$
The solution is $(9, -14)$.

2. $3x - y = 8$ (1)
$2x + 3y = 6$ (2)

Multiply equation (1) by 3 and add to equation (2).
$$9x - 3y = 24$$
$$\underline{2x + 3y = 6}$$
$$11x = 30 \Rightarrow x = \frac{30}{11}$$

Substitute $\dfrac{30}{11}$ for x in equation (1) to solve for y.
$$3\left(\frac{30}{11}\right) - y = 8 \Rightarrow y = \frac{2}{11}$$

The solution is $\left(\dfrac{30}{11}, \dfrac{2}{11}\right)$.

3. $3x - 5y = 16$ (1)
$2x + 3y = -2$ (2)

Multiply equation (1) by 2 and equation (2) by -3 and add the results.
$$6x - 10y = 32$$
$$\underline{-6x - 9y = 6}$$
$${-19y = 38} \Rightarrow y = -2$$

Substitute -2 for y in equation (1) to solve for x.
$$3x - 5(-2) = 16 \Rightarrow 3x = 6 \Rightarrow x = 2$$
The solution is $(2, -2)$.

4. $\frac{1}{4}x - \frac{1}{3}y = -\frac{1}{4}$ (1)

$\frac{1}{10}x + \frac{2}{5}y = \frac{2}{5}$ (2)

Multiply equation (1) by 12 and equation (2) by 10 and add the results.

$3x - 4y = -3$ (3)

$\underline{x + 4y = \ \ 4}$ (4)

$4x \qquad = 1 \Rightarrow x = \frac{1}{4}$

Substitute $\frac{1}{4}$ for x in equation (4) to solve for y.

$\frac{1}{4} + 4y = 4 \Rightarrow 4y = \frac{15}{4} \Rightarrow y = \frac{15}{16}$

The solution is $\left(\frac{1}{4}, \frac{15}{16}\right)$.

5. Let x = the number of standard paper clips (in 1000s), and let y = the number of extra large paper clips (in 1000s).
We will solve the following system.

$\frac{1}{4}x + \frac{1}{3}y = 4$ (1)

$\frac{1}{2}x + \frac{1}{3}y = 6$ (2)

Multiply equation (1) by 12 and equation (2) by –6 and add the results.

$3x + 4y = 48$

$\underline{-3x - 2y = -36}$

$\qquad 2y = \ 12 \Rightarrow y = \ 6$

Substitute 6 for y in equation (1) and solve for x.

$\frac{1}{4}x + \frac{1}{3}(6) = 4 \Rightarrow \frac{1}{4}x = 2 \Rightarrow x = 8$

The manufacturer can make 8000 standard and 6000 extra large paper clips.

6. Let x = the number of shares of the first stock, and let y = the number of shares of the second stock.
We will solve the following system.

$32x + \ 23y = 10,100$ (1)

$1.2x + 1.4y = \ \ 540$ (2)

Multiply equation (1) by 1.2 and equation (2) by –32 and add the results.

$38.4x + 27.6y = \ \ 12,120$

$\underline{-38.4x - 44.8y = -17,280}$

$\qquad -17.2y = \ -5160 \Rightarrow y = 300$

Substitute 300 for y in equation (2) and solve for x.

$1.2x + 1.4(300) = 540 \Rightarrow 1.2x = 120 \Rightarrow x = 100$

She should buy 100 shares of the first stock and 300 shares of the second stock.

7. Let x = the amount invested in the first fund, and let y = the amount invested in the second fund. We will solve the system

$.08x + .02y = 780$ (1)

$.1x + .01y = 810$ (2)

Multiply equation (2) by –2 and add to equation (1).

$.08x + \ .02y = \ \ \ 780$

$\underline{-.2x - \ .02y = -1620}$

$-.12x \qquad = \ -840 \Rightarrow x = 7000$

Substitute 7000 for x in equation (1) and solve for y.

$.08(7000) + .02y = 780 \Rightarrow .02y = 220 \Rightarrow y = 11,000$

Joyce has $7000 invested in the first fund and $11,000 in the second.

8. $x - 2y = 1$ (1)

$4x + 4y = 2$ (2)

$10x + 8y = 4$ (3)

Multiply equation (1) by 2 and add to equation (2).

$2x - 4y = 2$

$\underline{4x + 4y = 2}$

$6x \qquad = 4 \Rightarrow x = \frac{2}{3}$

Substitute $\frac{2}{3}$ for x in equation (1) to solve for y.

$\frac{2}{3} - 2y = 1 \Rightarrow -2y = \frac{1}{3} \Rightarrow y = -\frac{1}{6}$

The ordered pair $\left(\frac{2}{3}, -\frac{1}{6}\right)$ satisfies equations (1) and (2), but not equation (3). The system is inconsistent; there is no solution.

9. $x + y - 4z = 0$ (1)

$2x + y - 3z = 2$ (2)

Multiply equation (1) by –1 and add to equation (2).

$-x - y + 4z = 0$ (3)

$\underline{2x + y - 3z = 2}$ (2)

$x \quad + z = 2$ (4)

(*continued on next page*)

(continued from page 184)

Solve equation (4) for x in terms of z.
$x + z = 2 \Rightarrow x = 2 - z$
Substitute $2 - z$ for x in equation (1) and solve for y in terms of z.
$(2 - z) + y - 4z = 0$
$\quad 2 + y - 5z = 0$
$\qquad\qquad y = 5z - 2$

The system is dependent, and the solution is all ordered triples of the form $(2 - z, 5z - 2, z)$ for any real number z.

10. $3x + y - z = 3 \quad (1)$
$\quad x \quad + 2z = 6 \quad (2)$
$-3x - y + 2z = 9 \quad (3)$

Add equations (1) and (3).
$3x + y - z = 3$
$\underline{-3x - y + 2z = 9}$
$\qquad\qquad z = 12$

Substitute 12 for z in equation (2) and solve for x.
$x + 2(12) = 6 \Rightarrow x = -18$

Substitute -18 for x and 12 for z in equation (1) and solve for y.
$3(-18) + y - 12 = 3 \Rightarrow y = 69$
The solution is $(-18, 69, 12)$.

11. $4x - y - 2z = 4 \quad (1)$
$\quad x - y - \dfrac{1}{2}z = 1 \quad (2)$
$\quad 2x - y - \quad z = 8 \quad (3)$

Multiply equation (2) by -4 and add to equation (1).
$4x - y - 2z = \quad 4 \quad (1)$
$\underline{-4x + 4y + 2z = -4} \quad (4)$
$\qquad 3y = \quad 0 \Rightarrow y = 0$

Substitute 0 for y in equations (1) and (3).
$4x - 2z = 4 \quad (5)$
$2x - z = 8 \quad (6)$

Multiply equation (6) by -2 and add to equation (5).
$4x - 2z = \quad 4 \quad (5)$
$\underline{-4x + 2z = -16} \quad (7)$
$\qquad 0 = -12 \quad (8)$

Since equation (8) is false, this system is inconsistent; there is no solution.

We solve exercises 12–16 using matrix methods.

12. $x + z = -3$
$\quad\quad y - z = \ 6$
$2x + 3z = 5$

The augmented matrix is
$$\begin{bmatrix} 1 & 0 & 1 & | & -3 \\ 0 & 1 & -1 & | & 6 \\ 2 & 0 & 3 & | & 5 \end{bmatrix}$$

$$\begin{bmatrix} 1 & 0 & 1 & | & -3 \\ 0 & 1 & -1 & | & 6 \\ 0 & 0 & 1 & | & 11 \end{bmatrix} \begin{matrix} \\ \\ -2R_1 + R_3 \end{matrix}$$

$$\begin{bmatrix} 1 & 0 & 0 & | & -14 \\ 0 & 1 & 0 & | & 17 \\ 0 & 0 & 1 & | & 11 \end{bmatrix} \begin{matrix} -1R_3 + R_1 \\ R_3 + R_2 \\ \\ \end{matrix}$$

The solution is $(-14, 17, 11)$.

13. $2x + 3y + 4z = 8$
$-x + \ y - 2z = -9$
$2x + 2y + 6z = 16$

The augmented matrix is
$$\begin{bmatrix} 2 & 3 & 4 & | & 8 \\ -1 & 1 & -2 & | & -9 \\ 2 & 2 & 6 & | & 16 \end{bmatrix}$$

$$\begin{bmatrix} 2 & 3 & 4 & | & 8 \\ -1 & 1 & -2 & | & -9 \\ 1 & 1 & 3 & | & 8 \end{bmatrix} \begin{matrix} \\ \\ \frac{1}{2}R_3 \end{matrix}$$

$$\begin{bmatrix} 1 & 1 & 3 & | & 8 \\ -1 & 1 & -2 & | & -9 \\ 2 & 3 & 4 & | & 8 \end{bmatrix} \text{Interchange } R_1 \text{ and } R_3$$

$$\begin{bmatrix} 1 & 1 & 3 & | & 8 \\ 0 & 2 & 1 & | & -1 \\ 0 & 1 & -2 & | & -8 \end{bmatrix} \begin{matrix} \\ R_1 + R_2 \\ -2R_1 + R_3 \end{matrix}$$

$$\begin{bmatrix} 1 & 0 & 5 & | & 16 \\ 0 & 5 & 0 & | & -10 \\ 0 & 1 & -2 & | & -8 \end{bmatrix} \begin{matrix} R_1 - R_3 \\ 2R_2 + R_3 \\ \\ \end{matrix}$$

$$\begin{bmatrix} 1 & 0 & 5 & | & 16 \\ 0 & 1 & 0 & | & -2 \\ 0 & 1 & -2 & | & -8 \end{bmatrix} \begin{matrix} \\ \frac{1}{5}R_2 \\ \\ \end{matrix}$$

$$\begin{bmatrix} 1 & 0 & 5 & | & 16 \\ 0 & 1 & 0 & | & -2 \\ 0 & 0 & -2 & | & -6 \end{bmatrix} \begin{matrix} \\ \\ -R_2 + R_3 \end{matrix}$$

(continued on next page)

(continued from page 185)

$$\begin{bmatrix} 1 & 0 & 5 & | & 16 \\ 0 & 1 & 0 & | & -2 \\ 0 & 0 & 1 & | & 3 \end{bmatrix} -\tfrac{1}{2}R_3$$

$$\begin{bmatrix} 1 & 0 & 0 & | & 1 \\ 0 & 1 & 0 & | & -2 \\ 0 & 0 & 1 & | & 3 \end{bmatrix} -5R_3 + R_1$$

The solution is $(1, -2, 3)$.

14.
$$\begin{aligned} 5x - 8y + z &= 1 \\ 3x - 2y + 4z &= 3 \\ 10x - 16y + 2z &= 3 \end{aligned}$$

The augmented matrix is

$$\begin{bmatrix} 5 & -8 & 1 & | & 1 \\ 3 & -2 & 4 & | & 3 \\ 10 & -16 & 2 & | & 3 \end{bmatrix}$$

$$\begin{bmatrix} 1 & -\tfrac{8}{5} & \tfrac{1}{5} & | & \tfrac{1}{5} \\ 3 & -2 & 4 & | & 3 \\ 10 & -16 & 2 & | & 3 \end{bmatrix} \tfrac{1}{5}R_1$$

$$\begin{bmatrix} 1 & -\tfrac{8}{5} & \tfrac{1}{5} & | & \tfrac{1}{5} \\ 0 & \tfrac{14}{5} & \tfrac{17}{5} & | & \tfrac{12}{5} \\ 0 & 0 & 0 & | & 1 \end{bmatrix} \begin{matrix} \\ -3R_1 + R_2 \\ -10R_1 + R_3 \end{matrix}$$

Since row 3 has all zeros except for the last entry, there is no solution.

15.
$$\begin{aligned} x - 2y + 3z &= 4 \\ 2x + y - 4z &= 3 \\ -3z + 4y - z &= -2 \end{aligned}$$

The augmented matrix is

$$\begin{bmatrix} 1 & -2 & 3 & | & 4 \\ 2 & 1 & -4 & | & 3 \\ -3 & 4 & -1 & | & -2 \end{bmatrix}$$

$$\begin{bmatrix} 1 & -2 & 3 & | & 4 \\ 0 & 5 & -10 & | & -5 \\ 0 & -2 & 8 & | & 10 \end{bmatrix} \begin{matrix} \\ -2R_1 + R_2 \\ 3R_1 + R_3 \end{matrix}$$

$$\begin{bmatrix} 1 & -2 & 3 & | & 4 \\ 0 & 1 & -2 & | & -1 \\ 0 & -2 & 8 & | & 10 \end{bmatrix} \tfrac{1}{5}R_2$$

$$\begin{bmatrix} 1 & 0 & -1 & | & 2 \\ 0 & 1 & -2 & | & -1 \\ 0 & 0 & 4 & | & 8 \end{bmatrix} \begin{matrix} 2R_2 + R_1 \\ \\ 2R_2 + R_3 \end{matrix}$$

$$\begin{bmatrix} 1 & 0 & -1 & | & 2 \\ 0 & 1 & -2 & | & -1 \\ 0 & 0 & 1 & | & 2 \end{bmatrix} \tfrac{1}{4}R_3$$

$$\begin{bmatrix} 1 & 0 & 0 & | & 4 \\ 0 & 1 & 0 & | & 3 \\ 0 & 0 & 1 & | & 2 \end{bmatrix} \begin{matrix} R_3 + R_1 \\ 2R_3 + R_2 \\ \end{matrix}$$

The solution is $(4, 3, 2)$.

16.
$$\begin{aligned} 3x + 2y - 6z &= 3 \\ x + y + 2z &= 2 \\ 2x + 2y + 5z &= 0 \end{aligned}$$

The augmented matrix is

$$\begin{bmatrix} 3 & 2 & -6 & | & 3 \\ 1 & 1 & 2 & | & 2 \\ 2 & 2 & 5 & | & 0 \end{bmatrix}$$

Interchange R_1 and R_2.

$$\begin{bmatrix} 1 & 1 & 2 & | & 2 \\ 3 & 2 & -6 & | & 3 \\ 2 & 2 & 5 & | & 0 \end{bmatrix}$$

$$\begin{bmatrix} 1 & 1 & 2 & | & 2 \\ 0 & -1 & -12 & | & -3 \\ 0 & 0 & 1 & | & -4 \end{bmatrix} \begin{matrix} \\ -3R_1 + R_2 \\ -2R_1 + R_3 \end{matrix}$$

$$\begin{bmatrix} 1 & 1 & 2 & | & 2 \\ 0 & 1 & 12 & | & 3 \\ 0 & 0 & 1 & | & -4 \end{bmatrix} -1R_2$$

$$\begin{bmatrix} 1 & 0 & -10 & | & -1 \\ 0 & 1 & 12 & | & 3 \\ 0 & 0 & 1 & | & -4 \end{bmatrix} -1R_2 + R_1$$

$$\begin{bmatrix} 1 & 0 & 0 & | & -41 \\ 0 & 1 & 0 & | & 51 \\ 0 & 0 & 1 & | & -4 \end{bmatrix} \begin{matrix} 10R_3 + R_1 \\ -12R_3 + R_2 \\ \end{matrix}$$

The solution is $(-41, 51, -4)$.

Exercises 17–20 can be solved using either the elimination method or matrix methods.

17. Let x = the number of one dollar bills, let y = the number of five dollar bills, and let z = the number of ten dollar bills.
From the given information, we have the system

$$\begin{aligned} x + y + z &= 35 \quad (1) \\ x + 5y + 10z &= 144 \quad (2) \\ z &= y + 2 \quad (3) \end{aligned}$$

(continued on next page)

(continued from page 186

Substitute $y + 2$ for z in equation (1) to get
$$x + y + y + 2 = 35$$
$$x + 2y + 2 = 35$$
$$x + 2y = 33 \quad (4)$$

Substitute $y + 2$ for z in equation (2) to get
$$x + 5y + 10(y + 2) = 144$$
$$x + 5y + 10y + 20 = 144$$
$$x + 15y = 124 \quad (5)$$

Multiply equation (4) by -1 and add to equation (5).
$$-x - 2y = -33$$
$$\underline{x + 15y = 124}$$
$$13y = 91 \Rightarrow y = 7$$

Substitute 7 for y in equation (4) and solve for x.
$$x + 2(7) = 33 \Rightarrow x = 19$$

Substitute 7 for y in equation (3) and solve for z.
$$z = 7 + 2 = 9$$

The solution is 19 ones, 7 fives, and 9 tens.

18. Organize the information into a table.

	I	II	III
Food	100	200	150
Shelter	250	0	200
Counseling	0	100	100

Let x = the number of clients form source I, let y = the number of clients from source II, and let z = the number of clients from source III.
The system is
$$100x + 200y + 150z = 50,000$$
$$250x + 200z = 32,500$$
$$ 100y + 100z = 25,000$$

Using matrix methods, we have

$$\begin{bmatrix} 100 & 200 & 150 & | & 50,000 \\ 250 & 0 & 200 & | & 32,500 \\ 0 & 100 & 100 & | & 25,000 \end{bmatrix}$$

$$\begin{bmatrix} 1 & 2 & \frac{3}{2} & | & 500 \\ 1 & 0 & \frac{4}{5} & | & 130 \\ 0 & 1 & 1 & | & 250 \end{bmatrix} \begin{matrix} \frac{1}{100}R_1 \\ \frac{1}{250}R_2 \\ \frac{1}{100}R_3 \end{matrix}$$

$$\begin{bmatrix} 1 & 0 & -\frac{1}{2} & | & 0 \\ 0 & 2 & \frac{7}{10} & | & 370 \\ 0 & 1 & 1 & | & 250 \end{bmatrix} \begin{matrix} R_1 - 2R_3 \\ R_1 - R_2 \\ \end{matrix}$$

$$\begin{bmatrix} 1 & 0 & -\frac{1}{2} & | & 0 \\ 0 & 2 & \frac{7}{10} & | & 370 \\ 0 & 0 & \frac{13}{10} & | & 130 \end{bmatrix} 2R_3 - R_2$$

$$\begin{bmatrix} 1 & 0 & -\frac{1}{2} & | & 0 \\ 0 & 2 & \frac{7}{10} & | & 370 \\ 0 & 0 & 1 & | & 100 \end{bmatrix} \frac{10}{13}R_3$$

$$\begin{bmatrix} 1 & 0 & 0 & | & 50 \\ 0 & 2 & 0 & | & 300 \\ 0 & 0 & 1 & | & 100 \end{bmatrix} \begin{matrix} R_1 + \frac{1}{2}R_3 \\ R_2 - \frac{7}{10}R_3 \\ \end{matrix}$$

$$\begin{bmatrix} 1 & 0 & 0 & | & 50 \\ 0 & 1 & 0 & | & 150 \\ 0 & 0 & 1 & | & 100 \end{bmatrix} \frac{1}{2}R_2$$

Thus, the agency can serve 50 clients from source I, 150 from source II, and 100 from source III.

19. Let x = the number of blankets, let y = the number of rugs, and let z = the number of skirts. Solve the following system.
$$24x + 30y + 12z = 306 \quad (1)$$
$$4x + 5y + 3z = 59 \quad (2)$$
$$15x + 18y + 9z = 201 \quad (3)$$

Simplify the system by dividing equation (1) by 6 and equation (3) by 3.
$$4x + 5y + 2z = 51 \quad (4)$$
$$4x + 5y + 3z = 59 \quad (2)$$
$$5x + 6y + 3z = 67 \quad (5)$$

Multiply equation (4) by -1 and add to equation (2).
$$-4x - 5y - 2z = -51$$
$$\underline{4x + 5y + 3z = 59}$$
$$z = 8$$

Substitute 8 for z in equation (4).
$$4x + 5y + 2(8) = 51$$
$$4x + 5y = 35 \quad (5)$$

Substitute 8 for z in equation (5).
$$5x + 6y + 3(8) = 67$$
$$5x + 6y = 43 \quad (6)$$

Multiply equation (5) by 5 and equation (6) by -4 and add the results.
$$20x + 25y = 175$$
$$\underline{-20x - 24y = -172}$$
$$y = 3$$

(continued on next page)

(continued from page 187)

Substitute 3 for y in equation (5) and solve for x.
$4x + 5(3) = 35 \Rightarrow 4x = 20 \Rightarrow x = 5$

They can make 5 blankets, 3 rugs, and 8 skirts.

20. Let x = the number of chairs, and let y = number tables, and let z = number of chests.

	Construction	Painting	Packing
Chair	2	1	2
Table	4	3	3
Chest	8	6	4
Totals	2000	1400	1300

$2x + 4y + 8z = 2000$

$x + 3y + 6z = 1400$

$2x + 3y + 4z = 1300$

Divide the first equation by 2 and write the augmented matrix.

$$\begin{bmatrix} 1 & 2 & 4 & | & 1000 \\ 1 & 3 & 6 & | & 1400 \\ 2 & 3 & 4 & | & 1300 \end{bmatrix}$$

$$\begin{bmatrix} 1 & 2 & 4 & | & 1000 \\ 0 & 1 & 2 & | & 400 \\ 0 & -1 & -4 & | & -700 \end{bmatrix} \begin{matrix} \\ -1R_1 + R_2 \\ -2R_1 + R_3 \end{matrix}$$

$$\begin{bmatrix} 1 & 0 & 0 & | & 200 \\ 0 & 1 & 2 & | & 400 \\ 0 & 0 & -2 & | & -300 \end{bmatrix} \begin{matrix} -2R_2 + R_1 \\ \\ R_2 + R_3 \end{matrix}$$

$$\begin{bmatrix} 1 & 0 & 0 & | & 200 \\ 0 & 1 & 0 & | & 100 \\ 0 & 0 & -2 & | & -300 \end{bmatrix} \begin{matrix} \\ R_3 + R_2 \\ \\ \end{matrix}$$

$$\begin{bmatrix} 1 & 0 & 0 & | & 200 \\ 0 & 1 & 0 & | & 100 \\ 0 & 0 & 1 & | & 150 \end{bmatrix} -\tfrac{1}{2}R_3$$

The factory can produce 200 chairs, 100 tables, and 150 chests.

21. $\begin{bmatrix} 2 & 3 \\ 5 & 9 \end{bmatrix}$

The matrix is 2×2. Since the matrix is 2×2, it is square.

22. $\begin{bmatrix} 2 & -1 \\ 4 & 6 \\ 5 & 7 \end{bmatrix}$

The size of this matrix is 3×2.

23. $\begin{bmatrix} 12 & 4 & -8 & -1 \end{bmatrix}$

The matrix is 1×4. The matrix is a row matrix.

24. $\begin{bmatrix} -7 & 5 & 6 & 4 \\ 3 & 2 & -1 & 2 \\ -1 & 12 & 8 & -1 \end{bmatrix}$

This is a 3×4 matrix.

25. $\begin{bmatrix} 6 & 8 & 10 \\ 5 & 3 & -2 \end{bmatrix}$

This matrix is 2×3.

26. $\begin{bmatrix} -9 \\ 15 \\ 4 \end{bmatrix}$

This matrix is 3×1. It is a column matrix.

27. $\begin{bmatrix} 8 & 8 & 8 \\ 10 & 5 & 9 \\ 7 & 10 & 7 \\ 8 & 9 & 7 \end{bmatrix}$

28. $\begin{bmatrix} 5 & 7 & 2532 & 52\frac{3}{8} & -\frac{1}{4} \\ 3 & 9 & 1464 & 56 & \frac{1}{8} \\ 2.50 & 5 & 4974 & 41 & -1\frac{1}{2} \\ 1.36 & 10 & 1754 & 18 & \frac{1}{2} \end{bmatrix}$

29. $B = \begin{bmatrix} 1 & 2 & -3 \\ 2 & 3 & 0 \\ 0 & 1 & 4 \end{bmatrix}$; $-B = \begin{bmatrix} -1 & -2 & 3 \\ -2 & -3 & 0 \\ 0 & -1 & -4 \end{bmatrix}$

30. $D = \begin{bmatrix} 6 \\ 1 \\ 0 \end{bmatrix}$; $-D = \begin{bmatrix} -6 \\ -1 \\ 0 \end{bmatrix}$

31. $A = \begin{bmatrix} 4 & 6 \\ -2 & -2 \\ 5 & 9 \end{bmatrix}$, $C = \begin{bmatrix} 5 & 0 \\ -1 & 3 \\ 4 & 7 \end{bmatrix}$

$3A - 2C$

$3A - 2C = \begin{bmatrix} 12 & 18 \\ -6 & -6 \\ 15 & 27 \end{bmatrix} - \begin{bmatrix} 10 & 0 \\ -2 & 6 \\ 8 & 14 \end{bmatrix}$

$= \begin{bmatrix} 2 & 18 \\ -4 & -12 \\ 7 & 13 \end{bmatrix}$

32. $F = \begin{bmatrix} -1 & 2 \\ 6 & 7 \end{bmatrix}$, $G = \begin{bmatrix} 2 & 5 \\ 1 & 6 \end{bmatrix}$

$F + 3G = \begin{bmatrix} -1 & 2 \\ 6 & 7 \end{bmatrix} + \begin{bmatrix} 6 & 15 \\ 3 & 18 \end{bmatrix} = \begin{bmatrix} 5 & 17 \\ 9 & 25 \end{bmatrix}$

33. $B = \begin{bmatrix} 1 & 2 & -3 \\ 2 & 3 & 0 \\ 0 & 1 & 4 \end{bmatrix}$, $C = \begin{bmatrix} 5 & 0 \\ -1 & 3 \\ 4 & 7 \end{bmatrix}$

$2B - 5C$ is not defined, since the matrices have different sizes.

34. $G = \begin{bmatrix} 2 & 5 \\ 1 & 6 \end{bmatrix}$, $F = \begin{bmatrix} -1 & 2 \\ 6 & 7 \end{bmatrix}$

$G - 2F = \begin{bmatrix} 2 & 5 \\ 1 & 6 \end{bmatrix} - \begin{bmatrix} -2 & 4 \\ 12 & 14 \end{bmatrix} = \begin{bmatrix} 4 & 1 \\ -11 & -8 \end{bmatrix}$

35. Using the data from Exercise 28, the first day matrix is the following.

$\begin{array}{c} \\ \text{ATT} \\ \text{GE} \\ \text{GO} \\ \text{S} \end{array} \begin{array}{c} \text{Sales} \quad \text{Price} \\ \text{change} \\ \begin{bmatrix} 2532 & -\frac{1}{4} \\ 1464 & \frac{1}{8} \\ 4974 & -\frac{3}{2} \\ 1754 & \frac{1}{2} \end{bmatrix} \end{array}$

The next day matrix is the following.

$\begin{bmatrix} 2310 & -\frac{1}{4} \\ 1258 & -\frac{1}{4} \\ 5061 & \frac{1}{2} \\ 1812 & \frac{1}{2} \end{bmatrix}$

The total sales and price changes for the two days are given by the sum

$\begin{bmatrix} 2532 & -\frac{1}{4} \\ 1464 & \frac{1}{8} \\ 4974 & -\frac{3}{2} \\ 1754 & \frac{1}{2} \end{bmatrix} + \begin{bmatrix} 2310 & -\frac{1}{4} \\ 1258 & -\frac{1}{4} \\ 5061 & \frac{1}{2} \\ 1812 & \frac{1}{2} \end{bmatrix} = \begin{bmatrix} 4842 & -\frac{1}{2} \\ 2722 & -\frac{1}{8} \\ 10,035 & -1 \\ 3566 & 1 \end{bmatrix}$

36. a. First shipment:

$\begin{array}{c} \\ \text{Chicago} \\ \text{Dallas} \\ \text{Atlanta} \end{array} \begin{array}{c} \text{Tulsa} \quad \text{New Orleans} \\ \begin{bmatrix} 110,000 & 85,000 \\ 73,000 & 108,000 \\ 95,000 & 69,000 \end{bmatrix} \end{array}$

Second shipment:

$\begin{bmatrix} 58,000 & 40,000 \\ 33,000 & 52,000 \\ 80,000 & 30,000 \end{bmatrix}$

b. Add the 2 matrices in part (a).

$\begin{bmatrix} 168,000 & 125,000 \\ 106,000 & 160,000 \\ 175,000 & 99,000 \end{bmatrix}$

37. $A = \begin{bmatrix} 4 & 6 \\ -2 & -2 \\ 5 & 9 \end{bmatrix}$, $G = \begin{bmatrix} 2 & 5 \\ 1 & 6 \end{bmatrix}$

$AG = \begin{bmatrix} (4)(2)+(6)(1) & (4)(5)+(6)(6) \\ (-2)(2)+(-2)(1) & (-2)(5)+(-2)(6) \\ (5)(2)+(9)(1) & (5)(5)+(9)(6) \end{bmatrix}$

$= \begin{bmatrix} 14 & 56 \\ -6 & -22 \\ 19 & 79 \end{bmatrix}$

38. $E = \begin{bmatrix} 1 & 3 & -4 \end{bmatrix}$, $B = \begin{bmatrix} 1 & 2 & -3 \\ 2 & 3 & 0 \\ 0 & 1 & 4 \end{bmatrix}$

$EB = \begin{bmatrix} 7 & 7 & -19 \end{bmatrix}$

39. $G = \begin{bmatrix} 2 & 5 \\ 1 & 6 \end{bmatrix}$, $F = \begin{bmatrix} -1 & 2 \\ 6 & 7 \end{bmatrix}$

$GF = \begin{bmatrix} (2)(-1)+(5)(6) & (2)(2)+(5)(7) \\ (1)(-1)+(6)(6) & (1)(2)+(6)(7) \end{bmatrix}$

$= \begin{bmatrix} 28 & 39 \\ 35 & 44 \end{bmatrix}$

40. $C = \begin{bmatrix} 5 & 0 \\ -1 & 3 \\ 4 & 7 \end{bmatrix}$, $A = \begin{bmatrix} 4 & 6 \\ -2 & -2 \\ 5 & 9 \end{bmatrix}$

CA is not defined since the number of columns in C is not equal to the number of rows in A.

41. $A = \begin{bmatrix} 4 & 6 \\ -2 & -2 \\ 5 & 9 \end{bmatrix}$, $G = \begin{bmatrix} 2 & 5 \\ 1 & 6 \end{bmatrix}$, $F = \begin{bmatrix} -1 & 2 \\ 6 & 7 \end{bmatrix}$

From Exercise 39, $GF = \begin{bmatrix} 28 & 39 \\ 35 & 44 \end{bmatrix}$.

Then, $AGF = A(GF)$

$AGF = A(GF) = \begin{bmatrix} 4 & 6 \\ -2 & -2 \\ 5 & 9 \end{bmatrix} \begin{bmatrix} 28 & 39 \\ 35 & 44 \end{bmatrix}$

$= \begin{bmatrix} 322 & 420 \\ -126 & -166 \\ 455 & 591 \end{bmatrix}$

42. The matrix is

$\begin{bmatrix} 3.54 & 1.41 \\ 1.53 & 1.57 \\ .34 & .29 \\ 7.53 & 6.21 \end{bmatrix} \cdot \begin{bmatrix} 8 \\ 8 \end{bmatrix} = \begin{bmatrix} 3.54(8) + 1.41(8) \\ 1.53(8) + 1.57(8) \\ .34(8) + .29(8) \\ 7.53(8) + 6.21(8) \end{bmatrix}$

$= \begin{bmatrix} 39.6 \\ 24.8 \\ 5.04 \\ 109.92 \end{bmatrix}$

Therefore, there are about 40 head and face injuries, about 25 concussions, about 5 neck injuries, and about 110 other injuries.

43. a.

	Cutting	Shaping
Standard	$\frac{1}{4}$	$\frac{1}{2}$
Extra Large	$\frac{1}{3}$	$\frac{1}{3}$

b. $\begin{bmatrix} 48 & 66 \end{bmatrix} \begin{bmatrix} \frac{1}{4} & \frac{1}{2} \\ \frac{1}{3} & \frac{1}{3} \end{bmatrix} = \begin{bmatrix} 34 & 46 \end{bmatrix}$

The cutting machine will operate for 34 hr and the shaping machine for 46 hr.

44. a.

	Cost Per Share	Earnings Per Share
Stock 1	32	1.20
Stock 2	23	1.49
Stock 3	54	2.10

b.

	Stock		
	1	2	3
Number of shares	$\begin{bmatrix} 50$	20	$15 \end{bmatrix}$

c. $\begin{bmatrix} 50 & 20 & 15 \end{bmatrix} \begin{bmatrix} 32 & 1.20 \\ 23 & 1.49 \\ 54 & 2.10 \end{bmatrix} = \begin{bmatrix} 2870 & 121.30 \end{bmatrix}$

Total cost = \$2870.
Total dividend = \$121.30

45. There are many correct answers. Here is one example.

$A = \begin{bmatrix} 3 & 0 \\ 2 & 1 \end{bmatrix}$; let $B = \begin{bmatrix} 1 & 2 \\ 3 & 4 \end{bmatrix}$.

$AB = \begin{bmatrix} 3 & 6 \\ 5 & 8 \end{bmatrix}$; $BA = \begin{bmatrix} 7 & 2 \\ 17 & 4 \end{bmatrix}$

Thus, AB and BA are both defined, and $AB \neq BA$.

46. No. $A = 4I$, so $AB = BA = 4B$.

47. Let $A = \begin{bmatrix} -2 & 2 \\ 0 & 5 \end{bmatrix}$.

$[A \mid I] = \begin{bmatrix} -2 & 2 & | & 1 & 0 \\ 0 & 5 & | & 0 & 1 \end{bmatrix}$

$= \begin{bmatrix} 1 & -1 & | & -\frac{1}{2} & 0 \\ 0 & 1 & | & 0 & \frac{1}{5} \end{bmatrix} \begin{matrix} -\frac{1}{2}R_1 \\ \frac{1}{5}R_2 \end{matrix}$

$= \begin{bmatrix} 1 & 0 & | & -\frac{1}{2} & \frac{1}{5} \\ 0 & 1 & | & 0 & \frac{1}{5} \end{bmatrix} \begin{matrix} R_2 + R_1 \\ \end{matrix}$

$A^{-1} = \begin{bmatrix} -\frac{1}{2} & \frac{1}{5} \\ 0 & \frac{1}{5} \end{bmatrix}$

48. Let $A = \begin{bmatrix} 3 & -1 \\ -5 & 2 \end{bmatrix}$.

$[A \mid I] = \begin{bmatrix} 3 & -1 & | & 1 & 0 \\ -5 & 2 & | & 0 & 1 \end{bmatrix}$

$\begin{bmatrix} 1 & -\frac{1}{3} & | & \frac{1}{3} & 0 \\ -5 & 2 & | & 0 & 1 \end{bmatrix} \frac{1}{3}R_1$

$\begin{bmatrix} 1 & -\frac{1}{3} & | & \frac{1}{3} & 0 \\ 0 & \frac{1}{3} & | & \frac{5}{3} & 1 \end{bmatrix} 5R_1 + R_2$

$\begin{bmatrix} 1 & 0 & | & 2 & 1 \\ 0 & 1 & | & 5 & 3 \end{bmatrix} \begin{matrix} R_1 + R_2 \\ 3R_2 \end{matrix}$

$A^{-1} = \begin{bmatrix} 2 & 1 \\ 5 & 3 \end{bmatrix}$

49. Let $A = \begin{bmatrix} 6 & 4 \\ 3 & 2 \end{bmatrix}$.

$[A \mid I] = \begin{bmatrix} 6 & 4 & 1 & 0 \\ 3 & 2 & 0 & 1 \end{bmatrix}$

$\begin{bmatrix} 1 & \frac{2}{3} & \frac{1}{6} & 0 \\ 3 & 2 & 0 & 1 \end{bmatrix} \frac{1}{6}R_1$

$\begin{bmatrix} 1 & \frac{2}{3} & \frac{1}{6} & 0 \\ 0 & 0 & -\frac{1}{2} & 1 \end{bmatrix} -3R_1 + R_2$

The second row can never become $\begin{bmatrix} 0 & 1 \end{bmatrix}$, so A has no inverse.

50. Let $A = \begin{bmatrix} 3 & 0 \\ -1 & 4 \end{bmatrix}$.

$\begin{bmatrix} 3 & 0 & 1 & 0 \\ -1 & 4 & 0 & 1 \end{bmatrix}$

$\begin{bmatrix} 1 & 0 & \frac{1}{3} & 0 \\ -1 & 4 & 0 & 1 \end{bmatrix} \frac{1}{3}R_1$

$\begin{bmatrix} 1 & 0 & \frac{1}{3} & 0 \\ 0 & 4 & \frac{1}{3} & 1 \end{bmatrix} R_1 + R_2$

$\begin{bmatrix} 1 & 0 & \frac{1}{3} & 0 \\ 0 & 1 & \frac{1}{12} & \frac{1}{4} \end{bmatrix} \frac{1}{4}R_2$

$A^{-1} = \begin{bmatrix} \frac{1}{3} & 0 \\ \frac{1}{12} & \frac{1}{4} \end{bmatrix}$

51. Let $A = \begin{bmatrix} 2 & 0 & 6 \\ 1 & -1 & 0 \\ 0 & 1 & -3 \end{bmatrix}$.

$[A \mid I] = \begin{bmatrix} 2 & 0 & 6 & 1 & 0 & 0 \\ 1 & -1 & 0 & 0 & 1 & 0 \\ 0 & 1 & -3 & 0 & 0 & 1 \end{bmatrix}$

Interchange rows.

$\begin{bmatrix} 1 & -1 & 0 & 0 & 1 & 0 \\ 0 & 1 & -3 & 0 & 0 & 1 \\ 2 & 0 & 6 & 1 & 0 & 0 \end{bmatrix}$

$\begin{bmatrix} 1 & -1 & 0 & 0 & 1 & 0 \\ 0 & 1 & -3 & 0 & 0 & 1 \\ 0 & 2 & 6 & 1 & -2 & 0 \end{bmatrix} -2R_1 + R_3$

$\begin{bmatrix} 1 & 0 & -3 & 0 & 1 & 1 \\ 0 & 1 & -3 & 0 & 0 & 1 \\ 0 & 0 & 12 & 1 & -2 & -2 \end{bmatrix} \begin{matrix} R_2 + R_1 \\ \\ -2R_2 + R_3 \end{matrix}$

$\begin{bmatrix} 1 & 0 & -3 & 0 & 1 & 1 \\ 0 & 1 & -3 & 0 & 0 & 1 \\ 0 & 0 & 1 & \frac{1}{12} & -\frac{1}{6} & -\frac{1}{6} \end{bmatrix} \frac{1}{12}R_3$

$[I \mid B] = \begin{bmatrix} 1 & 0 & 0 & \frac{1}{4} & \frac{1}{2} & \frac{1}{2} \\ 0 & 1 & 0 & \frac{1}{4} & -\frac{1}{2} & \frac{1}{2} \\ 0 & 0 & 1 & \frac{1}{12} & -\frac{1}{6} & -\frac{1}{6} \end{bmatrix} \begin{matrix} 3R_3 + R_1 \\ 3R_3 + R_2 \\ \\ \end{matrix}$

$A^{-1} = \begin{bmatrix} \frac{1}{4} & \frac{1}{2} & \frac{1}{2} \\ \frac{1}{4} & -\frac{1}{2} & \frac{1}{2} \\ \frac{1}{12} & -\frac{1}{6} & -\frac{1}{6} \end{bmatrix}$

52. Let $A = \begin{bmatrix} 2 & -1 & 0 \\ 1 & 0 & 2 \\ 1 & -4 & 0 \end{bmatrix}$.

$\begin{bmatrix} 2 & -1 & 0 & 1 & 0 & 0 \\ 1 & 0 & 2 & 0 & 1 & 0 \\ 1 & -4 & 0 & 0 & 0 & 1 \end{bmatrix}$

Interchange rows 1 and 2.

$\begin{bmatrix} 1 & 0 & 2 & 0 & 1 & 0 \\ 2 & -1 & 0 & 1 & 0 & 0 \\ 1 & -4 & 0 & 0 & 0 & 1 \end{bmatrix}$

$\begin{bmatrix} 1 & 0 & 2 & 0 & 1 & 0 \\ 0 & -1 & -4 & 1 & -2 & 0 \\ 0 & -4 & -2 & 0 & -1 & 1 \end{bmatrix} \begin{matrix} \\ -2R_1 + R_2 \\ -1R_1 + R_3 \end{matrix}$

Multiply row 2 by −1 and continue.

$\begin{bmatrix} 1 & 0 & 2 & 0 & 1 & 0 \\ 0 & 1 & 4 & -1 & 2 & 0 \\ 0 & 0 & 14 & -4 & 7 & 1 \end{bmatrix} -4R_2 + R_3$

$\begin{bmatrix} 1 & 0 & 2 & 0 & 1 & 0 \\ 0 & 1 & 4 & -1 & 2 & 0 \\ 0 & 0 & 1 & -\frac{2}{7} & \frac{1}{2} & \frac{1}{14} \end{bmatrix} \frac{1}{14}R_3$

$\begin{bmatrix} 1 & 0 & 0 & \frac{4}{7} & 0 & -\frac{1}{7} \\ 0 & 1 & 0 & \frac{1}{7} & 0 & -\frac{2}{7} \\ 0 & 0 & 1 & -\frac{2}{7} & \frac{1}{2} & \frac{1}{14} \end{bmatrix} \begin{matrix} -2R_3 + R_1 \\ -4R_3 + R_2 \\ \\ \end{matrix}$

$A^{-1} = \begin{bmatrix} \frac{4}{7} & 0 & -\frac{1}{7} \\ \frac{1}{7} & 0 & -\frac{2}{7} \\ -\frac{2}{7} & \frac{1}{2} & \frac{1}{14} \end{bmatrix}$

53. Let $A = \begin{bmatrix} 2 & 3 & 5 \\ -2 & -3 & -5 \\ 1 & 4 & 2 \end{bmatrix}$.

$[A \mid I] = \begin{bmatrix} 2 & 3 & 5 & | & 1 & 0 & 0 \\ -2 & -3 & -5 & | & 0 & 1 & 0 \\ 1 & 4 & 2 & | & 0 & 0 & 1 \end{bmatrix}$

$\begin{bmatrix} 1 & 4 & 2 & | & 0 & 0 & 1 \\ -2 & -3 & -5 & | & 0 & 1 & 0 \\ 2 & 3 & 5 & | & 1 & 0 & 0 \end{bmatrix}$ Interchange R_1 and R_3

$\begin{bmatrix} 1 & 4 & 2 & | & 0 & 0 & 1 \\ -2 & -3 & -5 & | & 0 & 1 & 0 \\ 0 & 0 & 0 & | & 1 & 1 & 0 \end{bmatrix} R_2 + R_3$

The third row can never become $\begin{bmatrix} 0 & 0 & 1 \end{bmatrix}$, so A does not have an inverse.

54. Let $A = \begin{bmatrix} 1 & 3 & 6 \\ 4 & 0 & 9 \\ 5 & 15 & 30 \end{bmatrix}$

$\begin{bmatrix} 1 & 3 & 6 & | & 1 & 0 & 0 \\ 4 & 0 & 9 & | & 0 & 1 & 0 \\ 5 & 15 & 30 & | & 0 & 0 & 1 \end{bmatrix}$

$\begin{bmatrix} 1 & 3 & 6 & | & 1 & 0 & 0 \\ 0 & -12 & -15 & | & -4 & 1 & 0 \\ 0 & 0 & 0 & | & -5 & 0 & 1 \end{bmatrix} \begin{matrix} \\ -4R_1 + R_2 \\ -5R_1 + R_3 \end{matrix}$

The last row is all zeros, so no inverse exists.

55. Let $A = \begin{bmatrix} 1 & 3 & -2 & -1 \\ 0 & 1 & 1 & 2 \\ -1 & -1 & 1 & -1 \\ 1 & -1 & -3 & -2 \end{bmatrix}$.

Use a graphing calculator to find A^{-1}.

$A^{-1} = \begin{bmatrix} -\frac{2}{3} & -\frac{17}{3} & -\frac{14}{3} & -3 \\ \frac{1}{3} & \frac{1}{3} & \frac{1}{3} & 0 \\ -\frac{1}{3} & -\frac{10}{3} & -\frac{7}{3} & -2 \\ 0 & 2 & 1 & 1 \end{bmatrix}$

56. Let $A = \begin{bmatrix} 3 & 2 & 0 & -1 \\ 2 & 0 & 1 & 2 \\ 1 & 2 & -1 & 0 \\ 2 & -1 & 1 & 1 \end{bmatrix}$.

Use a graphing calculator to find A^{-1}.

$A^{-1} = \begin{bmatrix} 0 & -\frac{1}{3} & \frac{1}{3} & \frac{2}{3} \\ \frac{1}{3} & \frac{2}{3} & -\frac{1}{3} & -1 \\ \frac{2}{3} & 1 & -\frac{4}{3} & -\frac{4}{3} \\ -\frac{1}{3} & \frac{1}{3} & \frac{1}{3} & 0 \end{bmatrix}$

57. $F = \begin{bmatrix} -1 & 2 \\ 6 & 7 \end{bmatrix}$

Form the augmented matrix $\begin{bmatrix} F \mid I \end{bmatrix}$.

$\begin{bmatrix} -1 & 2 & | & 1 & 0 \\ 6 & 7 & | & 0 & 1 \end{bmatrix}$

$\begin{bmatrix} 1 & -2 & | & -1 & 0 \\ 6 & 7 & | & 0 & 1 \end{bmatrix} -1R_1$

$\begin{bmatrix} 1 & -2 & | & -1 & 0 \\ 0 & 19 & | & 6 & 1 \end{bmatrix} -6R_1 + R_2$

$\begin{bmatrix} 1 & -2 & | & -1 & 0 \\ 0 & 1 & | & \frac{6}{19} & \frac{1}{19} \end{bmatrix} \frac{1}{19}R_2$

$\begin{bmatrix} 1 & 0 & | & -\frac{7}{19} & \frac{2}{19} \\ 0 & 1 & | & \frac{6}{19} & \frac{1}{19} \end{bmatrix} R_1 + 2R_2$

$F^{-1} = \begin{bmatrix} -\frac{7}{19} & \frac{2}{19} \\ \frac{6}{19} & \frac{1}{19} \end{bmatrix}$

58. $G = \begin{bmatrix} 2 & 5 \\ 1 & 6 \end{bmatrix}$

Form the augmented matrix $\begin{bmatrix} G \mid I \end{bmatrix}$.

$\begin{bmatrix} 2 & 5 & | & 1 & 0 \\ 1 & 6 & | & 0 & 1 \end{bmatrix}$

Interchange R_1 and R_2.

$\begin{bmatrix} 1 & 6 & | & 0 & 1 \\ 2 & 5 & | & 1 & 0 \end{bmatrix}$

$\begin{bmatrix} 1 & 6 & | & 0 & 1 \\ 0 & -7 & | & 1 & -2 \end{bmatrix} -2R_1 + R_2$

$\begin{bmatrix} 1 & 6 & | & 0 & 1 \\ 0 & 1 & | & -\frac{1}{7} & \frac{2}{7} \end{bmatrix} -\frac{1}{7}R_2$

(continued on next page)

(*continued from page 192*)

$$\begin{bmatrix} 1 & 0 & \frac{6}{7} & -\frac{5}{7} \\ 0 & 1 & -\frac{1}{7} & \frac{2}{7} \end{bmatrix} -6R_2 + R_1$$

$$G^{-1} = \begin{bmatrix} \frac{6}{7} & -\frac{5}{7} \\ -\frac{1}{7} & \frac{2}{7} \end{bmatrix}$$

59. $G = \begin{bmatrix} 2 & 5 \\ 1 & 6 \end{bmatrix}$, $F = \begin{bmatrix} -1 & 2 \\ 6 & 7 \end{bmatrix}$

$$G - F = \begin{bmatrix} 3 & 3 \\ -5 & -1 \end{bmatrix}$$

Form the augmented matrix $[G - F \mid I]$.

$$\begin{bmatrix} 3 & 3 & 1 & 0 \\ -5 & -1 & 0 & 1 \end{bmatrix}$$

$$\begin{bmatrix} 1 & 1 & \frac{1}{3} & 0 \\ -5 & -1 & 0 & 1 \end{bmatrix} \frac{1}{3}R_1$$

$$\begin{bmatrix} 1 & 1 & \frac{1}{3} & 0 \\ 0 & 4 & \frac{5}{3} & 1 \end{bmatrix} 5R_1 + R_2$$

$$\begin{bmatrix} 1 & 1 & \frac{1}{3} & 0 \\ 0 & 1 & \frac{5}{12} & \frac{1}{4} \end{bmatrix} \frac{1}{4}R_2$$

$$\begin{bmatrix} 1 & 0 & -\frac{1}{12} & -\frac{1}{4} \\ 0 & 1 & \frac{5}{12} & \frac{1}{4} \end{bmatrix} -R_2 + R_1$$

$$(G - F)^{-1} = \begin{bmatrix} -\frac{1}{12} & -\frac{1}{4} \\ \frac{5}{12} & \frac{1}{4} \end{bmatrix}$$

60. $F = \begin{bmatrix} -1 & 2 \\ 6 & 7 \end{bmatrix}$, $G = \begin{bmatrix} 2 & 5 \\ 1 & 6 \end{bmatrix}$

$$F + G = \begin{bmatrix} 1 & 7 \\ 7 & 13 \end{bmatrix}$$

Form the augmented matrix $[F + G \mid I]$.

$$\begin{bmatrix} 1 & 7 & 1 & 0 \\ 7 & 13 & 0 & 1 \end{bmatrix}$$

$$\begin{bmatrix} 1 & 7 & 1 & 0 \\ 0 & -36 & -7 & 1 \end{bmatrix} -7R_1 + R_2$$

$$\begin{bmatrix} 1 & 7 & 1 & 0 \\ 0 & 1 & \frac{7}{36} & -\frac{1}{36} \end{bmatrix} -\frac{1}{36}R_2$$

$$\begin{bmatrix} 1 & 0 & -\frac{13}{36} & \frac{7}{36} \\ 0 & 1 & \frac{7}{36} & -\frac{1}{36} \end{bmatrix} -7R_2 + R_1$$

$$(F + G)^{-1} = \begin{bmatrix} -\frac{13}{36} & \frac{7}{36} \\ \frac{7}{36} & -\frac{1}{36} \end{bmatrix}$$

61. $B = \begin{bmatrix} 1 & 2 & -3 \\ 2 & 3 & 0 \\ 0 & 1 & 4 \end{bmatrix}$

Form the augmented matrix $[B \mid I]$.

$$\begin{bmatrix} 1 & 2 & -3 & 1 & 0 & 0 \\ 2 & 3 & 0 & 0 & 1 & 0 \\ 0 & 1 & 4 & 0 & 0 & 1 \end{bmatrix}$$

$$\begin{bmatrix} 1 & 0 & -11 & 1 & 0 & -2 \\ 0 & 1 & -6 & 2 & -1 & 0 \\ 0 & 1 & 4 & 0 & 0 & 1 \end{bmatrix} \begin{matrix} -2R_3 + R_1 \\ -R_2 + 2R_1 \end{matrix}$$

$$\begin{bmatrix} 1 & 0 & -11 & 1 & 0 & -2 \\ 0 & 1 & -6 & 2 & -1 & 0 \\ 0 & 0 & 1 & -\frac{1}{5} & \frac{1}{10} & \frac{1}{10} \end{bmatrix} \frac{1}{10}(-R_2 + R_3)$$

$$\begin{bmatrix} 1 & 0 & 0 & -\frac{6}{5} & \frac{11}{10} & -\frac{9}{10} \\ 0 & 1 & 0 & \frac{4}{5} & -\frac{2}{5} & \frac{3}{5} \\ 0 & 0 & 1 & -\frac{1}{5} & \frac{1}{10} & \frac{1}{10} \end{bmatrix} \begin{matrix} R_1 + 11R_3 \\ R_2 + 6R_3 \end{matrix}$$

$$B^{-1} = \begin{bmatrix} -\frac{6}{5} & \frac{11}{10} & -\frac{9}{10} \\ \frac{4}{5} & -\frac{2}{5} & \frac{3}{5} \\ -\frac{1}{5} & \frac{1}{10} & \frac{1}{10} \end{bmatrix}$$

62. Answers vary. Sample answer: There is no way to transform the given matrix into the identity matrix using row operations

63. $A = \begin{bmatrix} -3 & 4 \\ -1 & 2 \end{bmatrix}$, $B = \begin{bmatrix} 3 \\ -1 \end{bmatrix}$

If $AX = B$, then $X = A^{-1}B$.

If $A = \begin{bmatrix} -3 & 4 \\ -1 & 2 \end{bmatrix}$, then $A^{-1} = \begin{bmatrix} -1 & 2 \\ -\frac{1}{2} & \frac{3}{2} \end{bmatrix}$.

Thus, $X = \begin{bmatrix} -1 & 2 \\ -\frac{1}{2} & \frac{3}{2} \end{bmatrix} \begin{bmatrix} 3 \\ -1 \end{bmatrix} = \begin{bmatrix} -5 \\ -3 \end{bmatrix}$.

64. $A = \begin{bmatrix} 1 & 3 \\ -2 & 4 \end{bmatrix}$, $B = \begin{bmatrix} 9 \\ 6 \end{bmatrix}$

$$A^{-1} = \begin{bmatrix} \frac{2}{5} & -\frac{3}{10} \\ \frac{1}{5} & \frac{1}{10} \end{bmatrix}$$

Then $X = A^{-1}B$

$$\begin{bmatrix} \frac{2}{5} & -\frac{3}{10} \\ \frac{1}{5} & \frac{1}{10} \end{bmatrix} \begin{bmatrix} 9 \\ 6 \end{bmatrix} = \begin{bmatrix} \frac{9}{5} \\ \frac{12}{5} \end{bmatrix}.$$

65. $A = \begin{bmatrix} 1 & 0 & 2 \\ -1 & 1 & 0 \\ 3 & 0 & 4 \end{bmatrix}$, $B = \begin{bmatrix} 8 \\ 4 \\ -6 \end{bmatrix}$

If $AX = B$, then $X = A^{-1}B$.

If $A = \begin{bmatrix} 1 & 0 & 2 \\ -1 & 1 & 0 \\ 3 & 0 & 4 \end{bmatrix}$,

then $A^{-1} = \begin{bmatrix} -2 & 0 & 1 \\ -2 & 1 & 1 \\ \frac{3}{2} & 0 & -\frac{1}{2} \end{bmatrix}$.

$X = \begin{bmatrix} -2 & 0 & 1 \\ -2 & 1 & 1 \\ \frac{3}{2} & 0 & -\frac{1}{2} \end{bmatrix} \begin{bmatrix} 8 \\ 4 \\ -6 \end{bmatrix} = \begin{bmatrix} -22 \\ -18 \\ 15 \end{bmatrix}$.

66. $A = \begin{bmatrix} 2 & 4 & 0 \\ 1 & -2 & 0 \\ 0 & 0 & 3 \end{bmatrix}$, $B = \begin{bmatrix} 72 \\ -24 \\ 48 \end{bmatrix}$

$A^{-1} = \begin{bmatrix} \frac{1}{4} & \frac{1}{2} & 0 \\ \frac{1}{8} & -\frac{1}{4} & 0 \\ 0 & 0 & \frac{1}{3} \end{bmatrix}$

$X = A^{-1}B = \begin{bmatrix} \frac{1}{4} & \frac{1}{2} & 0 \\ \frac{1}{8} & -\frac{1}{4} & 0 \\ 0 & 0 & \frac{1}{3} \end{bmatrix} \begin{bmatrix} 72 \\ -24 \\ 48 \end{bmatrix} = \begin{bmatrix} 6 \\ 15 \\ 16 \end{bmatrix}$.

67. $x + y = -2$

$2x + 5y = 2$

The system as a matrix equation is

$\begin{bmatrix} 1 & 1 \\ 2 & 5 \end{bmatrix} \begin{bmatrix} x \\ y \end{bmatrix} = \begin{bmatrix} -2 \\ 2 \end{bmatrix}$.

Let $A = \begin{bmatrix} 1 & 1 \\ 2 & 5 \end{bmatrix}$ and $B = \begin{bmatrix} -2 \\ 2 \end{bmatrix}$.

$A^{-1} = \begin{bmatrix} \frac{5}{3} & -\frac{1}{3} \\ -\frac{2}{3} & \frac{1}{3} \end{bmatrix}$

$\begin{bmatrix} x \\ y \end{bmatrix} = \begin{bmatrix} \frac{5}{3} & -\frac{1}{3} \\ -\frac{2}{3} & \frac{1}{3} \end{bmatrix} \begin{bmatrix} -2 \\ 2 \end{bmatrix} = \begin{bmatrix} -4 \\ 2 \end{bmatrix}$.

The solution is $(-4, 2)$.

68. $5x - 3y = -2$

$2x + 7y = -9$

The system as a matrix equation is

$\begin{bmatrix} 5 & -3 \\ 2 & 7 \end{bmatrix} \begin{bmatrix} x \\ y \end{bmatrix} = \begin{bmatrix} -2 \\ -9 \end{bmatrix}$.

Let $A = \begin{bmatrix} 5 & -3 \\ 2 & 7 \end{bmatrix}$ and $B = \begin{bmatrix} -2 \\ -9 \end{bmatrix}$.

$A^{-1} = \begin{bmatrix} \frac{7}{41} & \frac{3}{41} \\ -\frac{2}{41} & \frac{5}{41} \end{bmatrix}$

$X = A^{-1}B = \begin{bmatrix} \frac{7}{41} & \frac{3}{41} \\ -\frac{2}{41} & \frac{5}{41} \end{bmatrix} \begin{bmatrix} -2 \\ -9 \end{bmatrix} = \begin{bmatrix} -1 \\ -1 \end{bmatrix}$

The solution is $(-1, -1)$.

69. $2x + y = 10$

$3x - 2y = 8$

The system as a matrix equation is

$\begin{bmatrix} 2 & 1 \\ 3 & -2 \end{bmatrix} \begin{bmatrix} x \\ y \end{bmatrix} = \begin{bmatrix} 10 \\ 8 \end{bmatrix}$.

Let $A = \begin{bmatrix} 2 & 1 \\ 3 & -2 \end{bmatrix}$ and $B = \begin{bmatrix} 10 \\ 8 \end{bmatrix}$.

$A^{-1} = \begin{bmatrix} \frac{2}{7} & \frac{1}{7} \\ \frac{3}{7} & -\frac{2}{7} \end{bmatrix}$

$X = \begin{bmatrix} \frac{2}{7} & \frac{1}{7} \\ \frac{3}{7} & -\frac{2}{7} \end{bmatrix} \begin{bmatrix} 10 \\ 8 \end{bmatrix} = \begin{bmatrix} 4 \\ 2 \end{bmatrix}$.

The solution is $(4, 2)$.

70. $x - 2y = 7$

$3x + y = 7$

The system as a matrix equation is

$\begin{bmatrix} 1 & -2 \\ 3 & 1 \end{bmatrix} \begin{bmatrix} x \\ y \end{bmatrix} = \begin{bmatrix} 7 \\ 7 \end{bmatrix}$.

$A = \begin{bmatrix} 1 & -2 \\ 3 & 1 \end{bmatrix}$, $B = \begin{bmatrix} 7 \\ 7 \end{bmatrix}$. $A^{-1} = \begin{bmatrix} \frac{1}{7} & \frac{2}{7} \\ -\frac{3}{7} & \frac{1}{7} \end{bmatrix}$

$X = A^{-1}B = \begin{bmatrix} 3 \\ -2 \end{bmatrix}$

The solution is $(3, -2)$.

71. $x + y + z = 1$
$2x - y \quad = -2$
$\quad 3y + z = 2$

The system as a matrix equation is

$$\begin{bmatrix} 1 & 1 & 1 \\ 2 & -1 & 0 \\ 0 & 3 & 1 \end{bmatrix} \begin{bmatrix} x \\ y \\ z \end{bmatrix} = \begin{bmatrix} 1 \\ -2 \\ 2 \end{bmatrix}.$$

Let $A = \begin{bmatrix} 1 & 1 & 1 \\ 2 & -1 & 0 \\ 0 & 3 & 1 \end{bmatrix}$ and $B = \begin{bmatrix} 1 \\ -2 \\ 2 \end{bmatrix}.$

$$A^{-1} = \begin{bmatrix} -\frac{1}{3} & \frac{2}{3} & \frac{1}{3} \\ -\frac{2}{3} & \frac{1}{3} & \frac{2}{3} \\ 2 & -1 & -1 \end{bmatrix}$$

$$X = \begin{bmatrix} -\frac{1}{3} & \frac{2}{3} & \frac{1}{3} \\ -\frac{2}{3} & \frac{1}{3} & \frac{2}{3} \\ 2 & -1 & -1 \end{bmatrix} \begin{bmatrix} 1 \\ -2 \\ 2 \end{bmatrix} = \begin{bmatrix} -1 \\ 0 \\ 2 \end{bmatrix}.$$

The solution is $(-1, 0, 2)$.

72. $x \quad = -3$
$\quad y + z = 6$
$2x - 3z = -9$

The system as a matrix equation is

$$\begin{bmatrix} 1 & 0 & 0 \\ 0 & 1 & 1 \\ 2 & 0 & -3 \end{bmatrix} \begin{bmatrix} x \\ y \\ z \end{bmatrix} = \begin{bmatrix} -3 \\ 6 \\ -9 \end{bmatrix}.$$

$$A = \begin{bmatrix} 1 & 0 & 0 \\ 0 & 1 & 1 \\ 2 & 0 & -3 \end{bmatrix}, B = \begin{bmatrix} -3 \\ 6 \\ -9 \end{bmatrix}$$

$$A^{-1} = \begin{bmatrix} 1 & 0 & 0 \\ -\frac{2}{3} & 1 & \frac{1}{3} \\ \frac{2}{3} & 0 & -\frac{1}{3} \end{bmatrix}$$

$$X = A^{-1}B = \begin{bmatrix} -3 \\ 5 \\ 1 \end{bmatrix}$$

The solution is $(-3, 5, 1)$.

73. $3x - 2y + 4z = 4$
$4x + y - 5z = 2$
$-6x + 4y - 8z = -2$

The system as a matrix equation is

$$\begin{bmatrix} 3 & -2 & 4 \\ 4 & 1 & -5 \\ -6 & 4 & -8 \end{bmatrix} \begin{bmatrix} x \\ y \\ z \end{bmatrix} = \begin{bmatrix} 4 \\ 2 \\ -2 \end{bmatrix}.$$

Let $A = \begin{bmatrix} 3 & -2 & 4 \\ 4 & 1 & -5 \\ -6 & 4 & -8 \end{bmatrix}.$

Since row 3 is -2 times row 1, the matrix will have no inverse, and the system cannot be solved by this method. Another method should be used to complete the solution. Use the elimination method. Multiply equation (1) by 2 and add the result to equation (3).

$6x - 4y + 8z = 8$
$\underline{-6x + 4y - 8z = -2}$
$\quad\quad\quad\quad 0 = 6$

This false result indicates that the system has no solution.

74. $x + 2y \quad = -1$
$\quad 3y - z = -5$
$x + 2y - z = -3$

$$A = \begin{bmatrix} 1 & 2 & 0 \\ 0 & 3 & -1 \\ 1 & 2 & -1 \end{bmatrix}, B = \begin{bmatrix} -1 \\ -5 \\ -3 \end{bmatrix}$$

$$A^{-1} = \begin{bmatrix} \frac{1}{3} & -\frac{2}{3} & \frac{2}{3} \\ \frac{1}{3} & \frac{1}{3} & -\frac{1}{3} \\ 1 & 0 & -1 \end{bmatrix}$$

$$X = A^{-1}B = \begin{bmatrix} 1 \\ -1 \\ 2 \end{bmatrix}$$

The solution is $(1, -1, 2)$.

Exercises 75–82 can be solved using either the elimination method or matrix methods.

75. Let x = the amount of the 9% wine and let y = the amount of the 14% wine. We have the system

$x + y = 40$
$.09x + .14y = .12(40).$

This system can be simplified to

$x + y = 40 \quad (1)$
$.09x + .14y = 4.8. \quad (2)$

Multiply equation (1) by $-.09$ and add to equation (2).

$-.09x - .09y = -3.6$
$\underline{.09x + .14y = 4.8}$
$\quad\quad .05y = 1.2 \Rightarrow y = 24$

Substitute 24 for y in equation (1) to get $x = 16$. The wine maker should mix 16 liters of the 9% wine and 24 liters of the 14% wine.

76. Let x = number of grams of 12 carat gold and let y = number of grams of 22 carat gold.
The system is

$$x + y = 25 \qquad (1)$$
$$\frac{12}{24}x + \frac{22}{24}y = \frac{15}{24}(25) \quad (2)$$

Multiply equation (2) by 24.

$$x + y = 25 \quad (1)$$
$$12x + 22y = 375 \quad (3)$$

Multiply equation (1) by -12 and add the result to equation (3).

$$-12x - 12y = -300$$
$$\underline{12x + 22y = 375}$$
$$10y = 75 \Rightarrow y = 7.5$$

Substitute 7.5 for y in equation (1)
$$x + 7.5 = 25 \Rightarrow x = 17.5$$
The merchant should mix 17.5 grams of 12 carat gold and 7.5 grams of 22 carat gold.

77. Let x = the number of liters of 40% solution and let y = the number of liters of 60% solution.

$$x + y = 40$$
$$.4x + .6y = .45(40) = 18$$

Multiply the second equation by 5.

$$x + y = 40$$
$$2x + 3y = 90$$

Use the Gauss-Jordan method.

$$\begin{bmatrix} 1 & 1 & | & 40 \\ 2 & 3 & | & 90 \end{bmatrix}$$

$$\begin{bmatrix} 1 & 1 & | & 40 \\ 0 & 1 & | & 10 \end{bmatrix} -2R_1 + R_2$$

$$\begin{bmatrix} 1 & 0 & | & 30 \\ 0 & 1 & | & 10 \end{bmatrix} -R_2 + R_1$$

The chemist should use 30 liters of the 40% solution and 10 liters of the 60% solution.

78. Let x = the number of pounds of tea worth $4.60 a pound and let y = the number of pounds of tea worth $6.50 a pound.
The system is

$$x + y = 10 \qquad (1)$$
$$4.60x + 6.50y = 5.74(10). \quad (2)$$

Multiply equation (1) by -4.60 and add the result to equation (2).

$$-4.60x - 4.60y = -46.0$$
$$\underline{4.60x + 6.50y = 57.4}$$
$$1.90y = 11.4 \Rightarrow y = 6$$

Substitute 6 for y in equation (1) and solve for x.
$$x + 6 = 10 \Rightarrow x = 4$$

Thus, 4 pounds of the tea worth $4.60 a pound should be used.

79. Let x = the number of bowls and let y = the number of plates.
We will solve the system

$$3x + 2y = 480 \qquad (1)$$
$$.25x + .2y = 44 \qquad (2)$$

Multiply equation (2) by -10 and add to equation (1).

$$3x + 2y = 480$$
$$\underline{-2.5x - 2y = -440}$$
$$.5x = 40 \Rightarrow x = 80$$

Substitute 80 for x in equation (1) and solve for y.
$$3(80) + 2y = 480 \Rightarrow y = 120$$

The factory can produce 80 bowls and 120 plates.

80. Let x = the speed of the boat and let y = the speed of the current. The system is

$$3(x + y) = 57 \quad (1)$$
$$5(x - y) = 55. \quad (2)$$

Dividing equation (1) by 3 and equation (2) by 5 gives

$$x + y = 19 \quad (3)$$
$$x - y = 11. \quad (4)$$

Adding the above two equations gives
$$2x = 30 \Rightarrow x = 15.$$
Substituting into (3) gives
$$15 + y = 19 \Rightarrow y = 4.$$

The speed of the boat is 15 km/hr, and the speed of the current is 4 km/hr.

81. Let x = the amount invested at 8%, let y = the amount invested at $8\frac{1}{2}$%, and let z = the amount invested at 11%. The system is

$$x + y + z = 50,000$$
$$.08x + .085y + .11z = 4436.25$$
$$.11z = .08x + 80.$$

Write the system in the proper form.

$$x + y + z = 50,000$$
$$.08x + .085y + .11z = 4436.25$$
$$-.08x + .11z = 80$$

(continued on next page)

(continued from page 196)

Write the augmented matrix of the system.

$$\begin{bmatrix} 1 & 1 & 1 & \big| & 50,000 \\ .08 & .085 & .11 & \big| & 4436.25 \\ -.08 & 0 & .11 & \big| & 80 \end{bmatrix}$$

$$\begin{bmatrix} 1 & 1 & 1 & \big| & 50,000 \\ 0 & .005 & .03 & \big| & 436.25 \\ 0 & .08 & .19 & \big| & 4080 \end{bmatrix} \begin{matrix} \\ -.08R_1 + R_2 \\ .08R_1 + R_3 \end{matrix}$$

$$\begin{bmatrix} 1 & 1 & 1 & \big| & 50,000 \\ 0 & 1 & 6 & \big| & 87,250 \\ 0 & .08 & .19 & \big| & 4080 \end{bmatrix} \begin{matrix} \\ \frac{1}{.005}R_2 \\ \\ \end{matrix}$$

$$\begin{bmatrix} 1 & 0 & -5 & \big| & -37,250 \\ 0 & 1 & 6 & \big| & 87,250 \\ 0 & 0 & -.29 & \big| & -2900 \end{bmatrix} \begin{matrix} -1R_2 + R_1 \\ \\ -.08R_2 + R_3 \end{matrix}$$

$$\begin{bmatrix} 1 & 0 & -5 & \big| & -37,250 \\ 0 & 1 & 6 & \big| & 87,250 \\ 0 & 0 & 1 & \big| & 10,000 \end{bmatrix} \begin{matrix} \\ \\ -\frac{1}{.29}R_3 \end{matrix}$$

$$\begin{bmatrix} 1 & 0 & 0 & \big| & 12,750 \\ 0 & 1 & 0 & \big| & 27,250 \\ 0 & 0 & 1 & \big| & 10,000 \end{bmatrix} \begin{matrix} 5R_3 + R_1 \\ -6R_3 + R_2 \\ \\ \end{matrix}$$

Thus, $x = 12{,}750$, $y = 27{,}250$, and $z = 10{,}000$. Ms. Tham invested \$12,750 at 8%, \$27,250 at $8\frac{1}{2}$%, and \$10,000 at 11%.

82. Let $x =$ the number of student tickets, let $y =$ the number of alumni tickets, let $z =$ the number of other adult tickets, and let $w =$ the number of children's tickets.

$$\begin{aligned} x + y + z + w &= 3750 & (1) \\ 4x + 10y + 12z + 6w &= 29{,}100 & (2) \\ x &= 6w & (3) \\ y &= \tfrac{4}{5}x & (4) \end{aligned}$$

Solve equation (3) for w in terms of x.

$$x = 6w \Rightarrow w = \tfrac{1}{6}x. \quad (5)$$

Now substitute the values for y and w into equations (1) and (2)

$$x + \tfrac{4}{5}x + z + \tfrac{1}{6}x = 3750 \quad (6)$$

$$4x + 10\left(\tfrac{4}{5}x\right) + 12z + 6\left(\tfrac{1}{6}x\right) = 29{,}100 \quad (7)$$

Simplify by multiplying equation (6) by 30 and combining like terms, and combining like terms in (7).

$$\begin{aligned} 59x + 30z &= 112{,}500 & (8) \\ 13x + 12z &= 29{,}100 & (9) \end{aligned}$$

Multiply (8) by 2 and (9) by –5, then add the resulting equations and solve for x.

$$\begin{array}{r} 118x + 60z = 225{,}000 \\ -65x - 60z = -145{,}500 \\ \hline 53x \quad\quad = 79{,}500 \Rightarrow x = 1500 \end{array}$$

Substitute $x = 1500$ into (3), (4), and (8) to solve for the remaining variables.

$$1500 = 6w \Rightarrow w = 250$$

$$y = \tfrac{4}{5}(1500) = 1200$$

$$59(1500) + 30z = 112{,}500 \Rightarrow 30z = 24{,}000 \Rightarrow z = 800$$

There were 1500 students, 1200 alumni, 800 other adults, and 250 children at the game.

83. $A = \begin{bmatrix} 0 & \frac{1}{4} \\ \frac{1}{2} & 0 \end{bmatrix}$, $D = \begin{bmatrix} 2100 \\ 1400 \end{bmatrix}$

a. $I - A = \begin{bmatrix} 1 & 0 \\ 0 & 1 \end{bmatrix} - \begin{bmatrix} 0 & \frac{1}{4} \\ \frac{1}{2} & 0 \end{bmatrix} = \begin{bmatrix} 1 & -\frac{1}{4} \\ -\frac{1}{2} & 1 \end{bmatrix}$

b. $\begin{bmatrix} 1 & -\frac{1}{4} & \big| & 1 & 0 \\ -\frac{1}{2} & 1 & \big| & 0 & 1 \end{bmatrix}$

$\begin{bmatrix} 1 & -\frac{1}{4} & \big| & 1 & 0 \\ 0 & \frac{7}{8} & \big| & \frac{1}{2} & 1 \end{bmatrix} \frac{1}{2}R_1 + R_2$

$\begin{bmatrix} 1 & -\frac{1}{4} & \big| & 1 & 0 \\ 0 & 1 & \big| & \frac{4}{7} & \frac{8}{7} \end{bmatrix} \frac{8}{7}R_2$

$\begin{bmatrix} 1 & 0 & \big| & \frac{8}{7} & \frac{2}{7} \\ 0 & 1 & \big| & \frac{4}{7} & \frac{8}{7} \end{bmatrix} \frac{1}{4}R_2 + R_1$

$(I - A)^{-1} = \begin{bmatrix} \frac{8}{7} & \frac{2}{7} \\ \frac{4}{7} & \frac{8}{7} \end{bmatrix}.$

c. $X = (I - A)^{-1}D = \begin{bmatrix} \frac{8}{7} & \frac{2}{7} \\ \frac{4}{7} & \frac{8}{7} \end{bmatrix} \begin{bmatrix} 2100 \\ 1400 \end{bmatrix} = \begin{bmatrix} 2800 \\ 2800 \end{bmatrix}$

84. a. The input-output matrix is

$$\begin{array}{cc} & c \quad g \end{array}$$
$$\begin{array}{c} c \\ g \end{array} \begin{bmatrix} 0 & \frac{1}{2} \\ \frac{2}{3} & 0 \end{bmatrix} = A.$$

b. $I - A = \begin{bmatrix} 1 & -\frac{1}{2} \\ -\frac{2}{3} & 1 \end{bmatrix}$, $D = \begin{bmatrix} 400 \\ 800 \end{bmatrix}$

$(I - A)^{-1} = \begin{bmatrix} \frac{3}{2} & \frac{3}{4} \\ 1 & \frac{3}{2} \end{bmatrix}$

$X = (I - A)^{-1} D = \begin{bmatrix} 1200 \\ 1600 \end{bmatrix}$

The production required is 1200 units of cheese and 1600 units of goats.

85. Write the input-output matrix.

$$\begin{matrix} & A & M \\ A = \begin{matrix} A \\ M \end{matrix} & \begin{bmatrix} .10 & .70 \\ .40 & .20 \end{bmatrix} \end{matrix} = \begin{bmatrix} \frac{1}{10} & \frac{7}{10} \\ \frac{4}{10} & \frac{2}{10} \end{bmatrix}$$

$I - A = \begin{bmatrix} 1 & 0 \\ 0 & 1 \end{bmatrix} - \begin{bmatrix} \frac{1}{10} & \frac{7}{10} \\ \frac{4}{10} & \frac{2}{10} \end{bmatrix} = \begin{bmatrix} \frac{9}{10} & -\frac{7}{10} \\ -\frac{4}{10} & \frac{8}{10} \end{bmatrix}$

Find $(I - A)^{-1}$.

$\begin{bmatrix} \frac{9}{10} & -\frac{7}{10} & 1 & 0 \\ -\frac{4}{10} & \frac{8}{10} & 0 & 1 \end{bmatrix}$

$\begin{bmatrix} 1 & -\frac{7}{9} & \frac{10}{9} & 0 \\ -\frac{4}{10} & \frac{8}{10} & 0 & 1 \end{bmatrix} \frac{10}{9} R_1$

$\begin{bmatrix} 1 & -\frac{7}{9} & \frac{10}{9} & 0 \\ 0 & \frac{44}{90} & \frac{4}{9} & 1 \end{bmatrix} \frac{4}{10} R_1 + R_2$

$\begin{bmatrix} 1 & -\frac{7}{9} & \frac{10}{9} & 0 \\ 0 & 1 & \frac{10}{11} & \frac{90}{44} \end{bmatrix} \frac{90}{44} R_2$

$\begin{bmatrix} 1 & 0 & \frac{180}{99} & \frac{70}{44} \\ 0 & 1 & \frac{10}{11} & \frac{90}{44} \end{bmatrix} \frac{7}{9} R_2 + R_1$

$(I - A)^{-1} = \begin{bmatrix} \frac{180}{99} & \frac{70}{44} \\ \frac{10}{11} & \frac{90}{44} \end{bmatrix}$

$X = (I - A)^{-1} D$

$X = \begin{bmatrix} \frac{180}{99} & \frac{70}{44} \\ \frac{10}{11} & \frac{90}{44} \end{bmatrix} \begin{bmatrix} 60,000 \\ 20,000 \end{bmatrix}$

$X = \begin{bmatrix} 140,909 \\ 95,455 \end{bmatrix}$ Rounded

The agriculture industry should produce $140,909, while the manufacturing industry should produce $95,455.

86. a. From the input-output matrix, we know that .9 unit agriculture; .4 unit services; .02 unit mining; .9 unit manufacturing are required to produce 1 unit.

b. The input-output matrix, A, and the matrix, $I - A$, are

$$A = \begin{bmatrix} .02 & .9 & 0 & .001 \\ 0 & .4 & 0 & .06 \\ .01 & .02 & .06 & .07 \\ .25 & .9 & .9 & .4 \end{bmatrix}$$

$$I - A = \begin{bmatrix} .980 & -.900 & 0 & -.001 \\ 0 & .600 & 0 & -.060 \\ -.010 & -.020 & .940 & -.070 \\ -.250 & -.900 & -.900 & .600 \end{bmatrix}$$

Next, calculate $(I - A)^{-1}$.

$$(I - A)^{-1} \approx \begin{bmatrix} 1.079 & 1.960 & .2132 & .223 \\ .064 & 2.129 & .230 & .240 \\ .060 & .4107 & 1.242 & .186 \\ .635 & 4.627 & 2.296 & 2.398 \end{bmatrix}$$

$$D = \begin{bmatrix} 760 \\ 1600 \\ 1000 \\ 2000 \end{bmatrix}$$

$$X = (I - A)^{-1} D \approx \begin{bmatrix} 4615 \\ 4165 \\ 2317 \\ 14,979 \end{bmatrix}$$

This is about 4615 units of agriculture, 4165 units of services, 2317 units of mining, and 14,979 units of manufacturing.

c. Since .25 units of manufacturing are used in producing each unit of agriculture, $(4615)(.25) = 1153.75$ units are used. Since .9 units of manufacturing are used in producing each unit of service, $(.9)(4165) = 3748.5$ units are used. Since .9 units of manufacturing are used in producing each unit of mining, $(.9)(2317) = 2085.3$. Since .4 units of manufacturing are used in producing each unit of manufacturing, $(.4)(14,979) = 5991.6$.
Thus, $1153.75 + 3748.5 + 2085.3 + 5991.6 \approx 12,979$ units of manufacturing are used in the production process.

87. a. From the input-output matrix, we know that .4 unit agriculture; .09 unit construction, .4 unit energy, .1 unit manufacturing, .9 unit transportation are required to produce 1 unit.

b. The input-output matrix, A and the matrix, $I - A$, are

$$A = \begin{bmatrix} .18 & .017 & .4 & .005 & 0 \\ .14 & .018 & .09 & .001 & 0 \\ .9 & 0 & .4 & .06 & .002 \\ .19 & .16 & .1 & .008 & .5 \\ .14 & .25 & .9 & .4 & .12 \end{bmatrix}$$

$(I - A)$

$$= \begin{bmatrix} .82 & -.017 & -.4 & -.005 & 0 \\ -.14 & .982 & -.09 & -.001 & 0 \\ -.9 & 0 & .6 & -.06 & -.002 \\ -.19 & -.16 & -.1 & .992 & -.5 \\ -.14 & -.25 & -.9 & -.4 & .88 \end{bmatrix}$$

$$X = \begin{bmatrix} 28,067 \\ 9383 \\ 51,372 \\ 61,364 \\ 90,403 \end{bmatrix}$$

$$D = (I - A)X = \begin{bmatrix} 1999.809 \\ 599.882 \\ 1700.254 \\ 3700.378 \\ 2499.110 \end{bmatrix}$$

This represents about 2000 units of agriculture, 600 units of construction, 1700 units of energy, 3700 units of manufacturing, and 2500 units of transportation.

c. Multiply matrix $(I - A)^{-1}$ by matrix D where

$$D = \begin{bmatrix} 2400 \\ 850 \\ 1400 \\ 3200 \\ 1800 \end{bmatrix}$$

$(I - A)^{-1}$

$$\approx \begin{bmatrix} 7.744 & .294 & 5.745 & .509 & .302 \\ 2.319 & 1.111 & 1.890 & .167 & .099 \\ 13.100 & .543 & 11.546 & 1.006 & .598 \\ 14.119 & .976 & 12.001 & 2.357 & 1.367 \\ 21.707 & 1.361 & 18.714 & 2.229 & 2.445 \end{bmatrix}$$

$$X = (I - A)^{-1}D \approx \begin{bmatrix} 29,049 \\ 9869 \\ 52,362 \\ 61,520 \\ 90,987 \end{bmatrix}$$

This represents 29,049 units of agriculture, 9869 units of construction, 52,362 units of energy, 61,520 units of manufacturing, and 90,987 units of transportation.

88.

$$M = \begin{array}{c} \\ A \\ B \\ C \\ D \end{array} \begin{array}{cccc} A & B & C & D \\ \begin{bmatrix} 0 & 1 & 0 & 1 \\ 1 & 0 & 0 & 1 \\ 0 & 0 & 0 & 1 \\ 1 & 1 & 1 & 0 \end{bmatrix} \end{array}$$

a. M^2 gives the number of one-step flights between cities.

$$M^2 = \begin{bmatrix} 0 & 1 & 0 & 1 \\ 1 & 0 & 0 & 1 \\ 0 & 0 & 0 & 1 \\ 1 & 1 & 1 & 0 \end{bmatrix}\begin{bmatrix} 0 & 1 & 0 & 1 \\ 1 & 0 & 0 & 1 \\ 0 & 0 & 0 & 1 \\ 1 & 1 & 1 & 0 \end{bmatrix}$$

$$= \begin{bmatrix} 2 & 1 & 1 & 1 \\ 1 & 2 & 1 & 1 \\ 1 & 1 & 1 & 0 \\ 1 & 1 & 0 & 3 \end{bmatrix}$$

$m_{13}^2 = 1$ so there is 1 one-stop flight between cities A and C.

b. The number of direct or one-stop flights from B to C is $m_{23}^2 + m_{23} = 1 + 0 = 1$.

c. The number of two-stop flights is given by

$$M^3 = M^2 \cdot M = \begin{bmatrix} 2 & 3 & 1 & 4 \\ 3 & 2 & 1 & 4 \\ 1 & 1 & 0 & 3 \\ 4 & 4 & 3 & 2 \end{bmatrix}$$

89. a. The message "leave now" broken into groups of 2 letters would have the following matrices:

$$\begin{bmatrix} 12 \\ 5 \end{bmatrix}, \begin{bmatrix} 1 \\ 22 \end{bmatrix}, \begin{bmatrix} 5 \\ 27 \end{bmatrix}, \begin{bmatrix} 14 \\ 15 \end{bmatrix}, \begin{bmatrix} 23 \\ 27 \end{bmatrix}$$

Multiply $M = \begin{bmatrix} 2 & 6 \\ 1 & 4 \end{bmatrix}$ by each matrix in the code to obtain $\begin{bmatrix} 54 \\ 32 \end{bmatrix}, \begin{bmatrix} 134 \\ 89 \end{bmatrix}, \begin{bmatrix} 172 \\ 113 \end{bmatrix}, \begin{bmatrix} 118 \\ 74 \end{bmatrix}, \begin{bmatrix} 208 \\ 131 \end{bmatrix}$.

b. To decode the message, use $M^{-1} = \begin{bmatrix} 2 & -3 \\ -\frac{1}{2} & 1 \end{bmatrix}$.

Case 6 Matrix Operations and Airline Route Maps

1. As the matrix A^2 indicates, the only city *not* reachable from New Bedford in a two-flight sequence is Provincetown, since this is the only city whose column has a zero in the row for New Bedford. Thus, the cities reachable by a two-flight sequence from New Bedford are Boston, Hyannis, Martha's Vineyard, Nantucket, New Bedford, and Providence.

In the matrix A^3, there are no zero entries in the row for New Bedford, so all Cape Air cities may be reached by a three-flight sequence.

2. Hyannis—Boston—Nantucket—New Bedford
Hyannis—Boston—Martha's Vineyard—New Bedford
Hyannis—Nantucket—Martha's Vineyard—New Bedford
Hyannis—Martha's Vineyard—Nantucket—New Bedford

3. The trips between two different cities for which the matrix entries remain zero until A^3 are trips between Provincetown and Providence, and Provincetown and New Bedford. These trips each take three flights.

4. There are 14 vertices, so the adjacency matrix is a 14×14 matrix. If the vertices in the Big Sky Airlines graph respectively correspond to Spokane (S), Kalispell (K), Missoula (M), Helena (H), Great Falls (GF), Billings (B), Lewistown (L), Havre (HV), Glasgow (G), Miles City (MC), Sidney (SY), Wolf (W), Glendine (GD), and Bismarck (BK), then the adjacency matrix for Big Sky Airlines is

	S	K	M	H	GF	B	L	HV	G	MC	SY	W	GD	BK	
	0	1	0	0	1	0	0	0	0	0	0	0	0	0	S
	1	0	1	1	0	0	0	0	0	0	0	0	0	0	K
	0	1	0	1	0	1	0	0	0	0	0	0	0	0	M
	0	1	1	0	0	1	0	0	0	0	0	0	0	0	H
	1	0	0	0	0	1	0	0	0	0	0	0	0	0	GF
	0	0	1	1	1	0	1	0	1	1	1	1	0	0	B
$A =$	0	0	0	0	0	1	0	1	0	0	0	0	0	0	L
	0	0	0	0	0	0	1	0	0	0	0	0	0	0	HV
	0	0	0	0	0	1	0	0	0	0	0	1	0	0	G
	0	0	0	0	0	1	0	0	0	0	0	0	1	0	MC
	0	0	0	0	0	1	0	0	0	0	0	0	1	1	SY
	0	0	0	0	0	1	0	0	1	0	0	0	0	0	W
	0	0	0	0	0	0	0	0	0	1	1	0	0	0	GD
	0	0	0	0	0	0	0	0	0	0	1	0	0	0	BK

5. To find the cities that may be reached by a three-flight sequence from Helena, find all the columns which have a nonzero element in row 4 of matrix $S = A + A^2 + A^3$.

Since row 4 of $S_3 = A + A^2 + A^3$ is
$$\begin{array}{cccccccccccccc} S & K & M & H & GF & B & L & HV & G & MC & SY & W & GD & BK \\ [3 & 8 & 8 & 7 & 3 & 13 & 2 & 1 & 3 & 2 & 2 & 3 & 2 & 1] \end{array}H,$$

all Big Sky cities may be reached by a three-flight sequence from Helena. The cities that need at least three flights to get from them to Helena are the cities represented by columns with zeros in row 4 of $S_2 = A + A^2$.

Since row 4 of $S_2 = A + A^2$ is
$$\begin{array}{cccccccccccccc} S & K & M & H & GF & B & L & HV & G & MC & SY & W & GD & BK \\ [1 & 2 & 3 & 3 & 1 & 2 & 1 & 0 & 1 & 1 & 1 & 1 & 0 & 0] \end{array}H,$$

at least three flights must be used to get to Helena from Havre, Glendine, and Bismarck.

6. From Exercise 5 we know that some trips take 3 flights. By looking for zeros in S_3, we can determine if any trips need more than 3 flights.

$S_3 =$

	S	K	M	H	GF	B	L	HV	G	MC	SY	W	GD	BK	
	2	5	3	3	4	3	1	0	1	1	1	1	0	0	S
	5	5	8	8	3	5	2	0	2	2	2	2	0	0	K
	3	8	7	8	3	13	2	1	3	2	2	3	2	1	M
	3	8	8	7	3	13	2	1	3	2	2	3	2	1	H
	4	3	3	3	2	10	1	1	2	1	1	2	2	1	GF
	3	5	13	13	10	12	10	1	11	11	12	11	2	1	B
	1	2	2	2	1	10	2	3	2	1	1	2	2	1	L
	0	0	1	1	1	1	3	1	1	1	1	1	0	0	HV
	1	2	3	3	2	11	2	1	4	2	2	5	2	1	G
	1	2	2	2	1	11	1	1	2	2	2	2	5	2	MC
	1	2	2	2	1	12	1	1	2	2	3	2	6	4	SY
	1	2	3	3	2	11	2	1	5	2	2	4	2	1	W
	0	0	2	2	2	2	2	0	2	5	6	2	2	1	GD
	0	0	1	1	1	1	1	0	1	2	4	1	1	1	BK

From S_3 we see that trips between Spokane and Havre, Spokane and Glendine, Spokane and Bismarck, Kalispell and Havre, Kalispell and Glendine, Kalispell and Bismarck, Havre and Glendine, and Havre and Bismarck require more than 3 flights. By looking at $S_4 = A + A^2 + A^3 + A^4$, we find that all entries have positive values; thus, these trips can be made in 4 flights.

Chapter 7 Linear Programming

Section 7.1 Graphing Linear Inequalities in Two Variables

1. $y \geq -x - 2$

Graph of $y = -x - 2$ is a solid line. The intercepts are $(-2, 0)$ and $(0, -2)$. Use the origin as a test point. Since $0 \geq -0 - 2$ is true, the origin will be included in the region, so shade the half-plane above the line. This matches choice F.

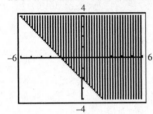

3. $y \leq x + 2$

Graph $y = x + 2$ as a solid line. The intercepts are $(-2, 0)$ and $(0, 2)$. Use the origin as a test point. Since $0 \leq 0 + 2$ is true, the origin will be included in the region, so shade the half-plane below the line. This matches choice A.

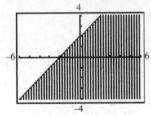

5. $6x + 4y \geq -12$

Graph $6x + 4y = -12$ is as a solid line. The intercepts are $(-2, 0)$ and $(0, -3)$. Use the origin as a test point. Since $6(0) + 4(0) \geq -12$ is true, the origin will be included in the region, so shade the half-plane above the line. This matches choice E.

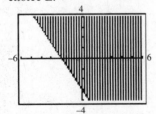

7. $y < 5 - 2x$

Graph $y = 5 - 2x$ as a dashed line. The intercepts are $(0, 5)$ and $\left(\frac{5}{2}, 0\right)$. Test $(0, 0)$ in the original inequality. The result, $0 < 5$, is true, so shade the region that contains $(0, 0)$.

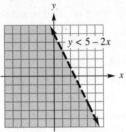

9. $3x - 2y \geq 18$

Graph $3x - 2y = 18$ as a solid line. The intercepts are $(0, -9)$ and $(6, 0)$. Use the origin as a test point. Since $3(0) - 2(0) \geq 18$ is false, the origin will not be included in the region, so shade the half-plane below the line.

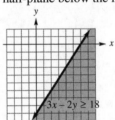

11. $2x - y \leq 4$

Graph $2x - y = 4$ as a solid line. The intercepts are $(0, -4)$ and $(2, 0)$. Use the origin as a test point. Since $2(0) - 0 \leq 4$ is true, the origin will be included in the region, so shade the half-plane above the line.

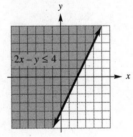

13. $y \leq -4$

Graph $y = -4$ as a solid line. $y = -4$ is a horizontal line crossing the y-axis at $(0, -4)$. Since the inequality symbol is $\leq$, shade the half-plane below the line.

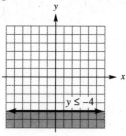

15. $3x - 2y \geq 18$

Graph the line $3x - 2y = 18$ as a solid line. The intercepts are $(0, -9)$ and $(6, 0)$. Use the origin as a test point. Since $3(0) - 2(0) \geq 18$ is false, the origin will not be included in the region, so shade the half-plane below the line.

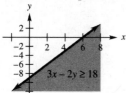

17. $3x + 4y \geq 12$

Graph $3x + 4y = 12$ as a solid line. The intercepts are $(0, 3)$ and $(4, 0)$. Use the origin as a test point. Since $3(0) + 4(0) > 12$ is false, the origin will not be included in the region, so shade the half-plane above the line.

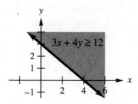

19. $2x - 4y \leq 3$

Graph $2x - 4y = 3$ as a solid line. The intercepts are $\left(\dfrac{3}{2}, 0\right)$ and $\left(0, -\dfrac{3}{4}\right)$. Use the origin as a test point. $2(0) - 4(0) < 3$ is true, so the region above the line, which includes the origin, is the correct region to shade.

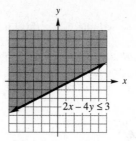

21. $x \leq 5y$

Graph $x = 5y$ as a solid line. Since this line contains the origin, some point other than $(0, 0)$ must be used as a test point. The point $(1, 2)$ gives $1 \leq 5(2)$ or $1 \leq 10$, a true sentence. Shade the side of the line containing $(1, 2)$, that is, the side above the line.

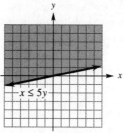

23. $-3x \leq y$

Graph $y = -3x$ as a solid line. Since this line contains the origin, use some point other than $(0, 0)$ as a test point. $(1, 1)$ used as a test point gives $-3 < 1$, a true sentence. Shade the region containing $(1, 1)$, which is the region above the line.

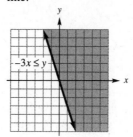

25. $y \leq x$

Graph $y = x$ as a dashed line. Since this line contains the origin, choose a point other than $(0, 0)$ as a test point. $(2, 3)$ gives $3 < 2$, which is false. Shade the region that does not contain the test point, that is, the region below the line.

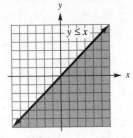

27. Answers may vary.
Possible answer: When the inequality is $<$ or $>$, the line is dashed. When the inequality is $\leq$ or $\geq$, the line is solid.

29. $y \geq 3x - 6$
$y \geq -x + 1$

Graph $y \geq 3x - 6$ as the region on or above the solid line $y = 3x - 6$. Graph $y \geq -x + 1$ as the region on or above the solid line $y = -x + 1$. The feasible region is the overlap of the two half-planes.

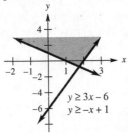

31. $2x + y \leq 5$
$x + 2y \leq 5$

Graph $2x + y \leq 5$ as the region on or below the solid line $2x + y = 5$, which has intercepts $(0, 5)$ and $\left(\frac{5}{2}, 0\right)$. Graph $x + 2y \leq 5$ as the region on or below the solid line $x + 2y = 5$, which has intercepts $\left(0, \frac{5}{2}\right)$ and $(5, 0)$. The feasible region is the overlap of the two half-planes.

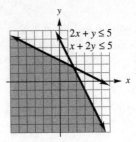

33. $2x + y \geq 8$
$4x - y \leq 3$

Graph $2x + y \geq 8$ as the region on or above the dashed line $2x + y = 8$, which has intercepts $(0, 8)$ and $(4, 0)$. Graph $4x - y \leq 3$ as the region above the solid line $4x - y = 3$, which has intercepts $(0, -3)$ and $\left(\frac{3}{4}, 0\right)$.

The overlap of these two regions is the feasible region.

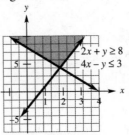

35. $2x - y \leq 1$
$3x + y \leq 6$

Graph $2x - y \leq 1$ as the region on or above the dashed line $2x - y = 1$. Graph $3x + y \leq 6$ as the region on or below the solid line $3x + y = 6$. Shade the overlapping part of these two regions to show the feasible region.

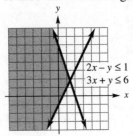

37. $-x - y \le 5$
$2x - y \le 4$

Graph $-x - y \le 5$ as the region on or above the solid line $-x - y = 5$. Graph $2x - y \le 4$ as the region on or above the solid line $2x - y = 4$. Shade the overlapping part of these two regions to show the feasible region.

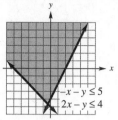

39. $3x + y \ge 6$
$x + 2y \ge 7$
$x \ge 0$
$y \ge 0$

The inequalities $x \ge 0$ and $y \ge 0$ restrict the feasible region to the first quadrant. Graph $3x + y \ge 6$ as the region on or above the solid line $3x + y = 6$. Graph $x + 2y \ge 7$ as the region on or above the solid line $x + 2y = 7$. The feasible region is the overlap of these half-planes.

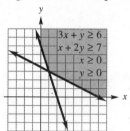

41. $-2 \le x \le 3$
$-1 \le y \le 5$
$2x + y \le 6$

The graph of $-2 \le x \le 3$ is the region between the vertical line $x = -2$ and $x = 3$, including the lines. The graph of $-1 \le y \le 5$ is the region between the horizontal lines $y = -1$ and $y = 5$, including the lines. The graph of $2x + y \le 6$ is the region on or below the solid line $2x + y = 6$. Shade the region common to all three graphs to show the feasible region.

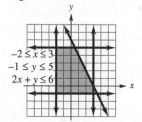

43. $2y - x \ge -5$
$y \le 3 + x$
$x \ge 0, \; y \ge 0$

The graph of $2y - x \ge -5$ consists of the boundary line $2y - x = 5$ and the region above it. The graph of $y \le 3 + x$ consists of the boundary line $y = 3 + x$ and the region below it. The inequalities $x \ge 0$ and $y \ge 0$ restrict the feasible region to the first quadrant. Shade the region common to all of these graphs to show the feasible region.

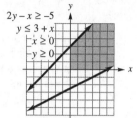

45. $3x + 4y \ge 12$
$2x - 3y \le 6$
$0 \le y \le 2$
$x \ge 0$

$3x + 4y \ge 12$ is the set of points on or above the solid line $3x + 4y = 12$. $2x - 3y \le 6$ is the set of points on or above the solid line $2x - 3y = 6$. $0 \le y \le 2$ is the rectangular strip of points lying on or between the horizontal lines $y = 0$ and $y = 2$. $x \ge 0$ consists of all the points on and to the right of the y-axis. The feasible region is the triangular region satisfying all the inequalities.

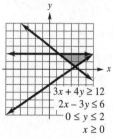

47. The shaded region lies between the horizontal lines $y = 0$ and $y = 4$, so this inequality is $0 \le y \le 4$. The shaded region also lies to the right of the vertical line $x = 0$, so this inequality is $x \ge 0$. The shaded region also lies below the line containing the points (3, 4) and (6, 0) which has the equation $4x + 3y = 24$. Using the origin as a test point, we determine that $4(0) + 3(0) \le 24$ is true, so the inequality must be $\le$. So the third inequality is $4x + 3y \le 24$. So, the inequalities are
$x \ge 0$
$0 \le y \le 4$
$4x + 3y \le 24$

49. The feasible region is the interior of the rectangle with vertices (2, 3), (2, –1), (7, 3), and (7, –1). The *x*-values range from 2 to 7, not including 2 or 7. The *y*-values range from –1 to 3, not including –1 or 3. Thus, the system of inequalities is

$$2 < x < 7$$
$$-1 < y < 3.$$

51. a.

	Number	Hours Spinning	Hours Dyeing	Hours Weaving
Shawls	x	1	1	1
Afghans	y	2	1	4
Maximum number of hours available		8	6	14

b.

$$x + 2y \leq 8 \qquad \text{Spinning inequality}$$
$$x + y \leq 6 \qquad \text{Dyeing inequality}$$
$$x + 4y \leq 14 \qquad \text{Weaving inequality}$$
$$x \geq 0 \qquad \text{Ensures a non-negative number of each}$$
$$y \geq 0$$

c. Graph the solid lines $x + 2y = 8$, $x + y = 6$, $x + 4y = 14$, $x = 0$, and $y = 0$, and shade the appropriate half-planes to get the feasible region.

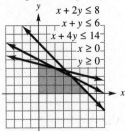

53. The system is

$$x \geq 3000$$
$$y \geq 5000$$
$$x + y \leq 10,000.$$

The first inequality gives the region to the right of the vertical line $x = 3000$, including the points on the line. The second inequality gives the region above the horizontal line $y = 5000$, including the points on the line. The third inequality gives the region below the line $x + y = 10,000$, including the points on the line.

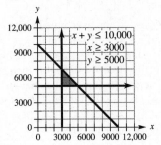

55.

	Number	Emissions	Cost
Type 1	x	.5 lb	$.16
Type 2	y	.3 lb	$.20
Maximum		1.8 lb	$.8

The manufacturer produces at least 3.2 million barrels annually, so $x + y \geq 3.2$. The cost is not to exceed $.8 million, so $.16x + .20y \leq .8$. Total emissions must not exceed 1.8 million lb, so $.5x + .3y \leq 1.8$. We obtain

$$x + y \geq 3.2$$
$$.16x + .20y \leq .8$$
$$.5x + .3y \leq 1.8$$
$$x \geq 0, \ y \geq 0.$$

Using the above system, graph solid lines and shade appropriate half-planes to get the feasible region bounded by the *y* axis on the left, lines $.16x + .20y = .8$ and $.5x + .3y = 1.8$ above, and the line $x + y = 3.2$ below.

(continued on next page)

(*continued from page 206*)

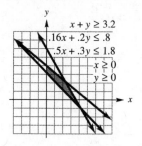

$x + y \geq 3.2$
$.16x + .2y \leq .8$
$.5x + .3y \leq 1.8$
$x \geq 0$
$y \geq 0$

Section 7.2 Linear Programming: The Graphical Method

1. Make a table indicating the value of the objective function $z = 6x + y$ at each corner point.

Corner Point	Value of $z = 6x + y$
(1, 1)	$6(1) + 1 = 7$ Minimum
(2, 7)	$6(2) + 1(7) = 19$
(5, 10)	$6(5) + 10 = 40$ Maximum
(6, 3)	$6(6) + 3 = 39$

The maximum value of 40 occurs at (5, 10). The minimum value of 7 occurs as (1, 1).

3.

Corner Point	Value of $z = .3x + .5y$	
(0, 0)	0	Minimum
(0, 12)	6	Maximum
(4, 8)	5.2	
(7, 3)	3.6	
(9, 0)	2.7	

The maximum is 6 at (0, 12); the minimum is 0 at (0, 0).

5. a.

Corner Point	Value of $z = x + 5y$	
(0, 8)	40	
(3, 4)	23	
$\left(\frac{13}{2}, 2\right)$	16.5	
(12, 0)	12	Minimum

The minimum is 12 at (12, 0). There is no maximum because the feasible region is unbounded.

b.

Corner Point	Value of $z = 2x + 3y$	
(0, 8)	24	
(3, 4)	18	Minimum
$\left(\frac{13}{2}, 2\right)$	19	
(12, 0)	24	

The minimum is 18 at (3, 4). There is no maximum because the feasible region is unbounded.

c.

Corner Point	Value of $z = 2x + 4y$	
(0, 8)	32	
(3, 4)	22	
$\left(\frac{13}{2}, 2\right)$	21	Minimum
(12, 0)	24	

The minimum is 21 at $\left(\frac{13}{2}, 2\right)$. There is no maximum.

d.

Corner Point	Value of $z = 4x + y$	
(0, 8)	8	Minimum
(3, 4)	16	
$\left(\frac{13}{2}, 2\right)$	28	
(12, 0)	48	

The minimum is 8 at (0, 8). There is no maximum.

7. Maximize $z = 4x + 3y$
subject to: $2x + 3y \leq 6$
$4x + y \leq 6$
$x \geq 0, \; y \geq 0.$

Graph the feasible region, and identify the corner points.

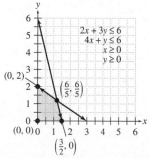

$2x + 3y \leq 6$
$4x + y \leq 6$
$x \geq 0$
$y \geq 0$

$(0, 2)$
$\left(\frac{6}{5}, \frac{6}{5}\right)$
$(0, 0)$
$\left(\frac{3}{2}, 0\right)$

The graph shows the feasible region is bounded.
(*continued on next page*)

(continued from page 207)

The corner points are

$(0, 0), (0, 2), \left(\frac{3}{2}, 0\right), \left(\frac{6}{5}, \frac{6}{5}\right)$, which is the

intersection of $2x + 3y = 6$ and
$4x + y = 6$. Use the corner points to find the
maximum value of the objective function.

Corner Point	Value of $z = 4x + 3y$
$(0, 0)$	0
$(0, 2)$	6
$\left(\frac{6}{5}, \frac{6}{5}\right)$	$\frac{42}{5} = 8.4$ Maximum
$\left(\frac{3}{2}, 0\right)$	6

The maximum value of $z = 4x + 3y$ is 8.4 at
$x = 1.2, y = 1.2$.

9. Minimize $z = 2x + y$
   subject to: $3x - y \geq 12$
   $\qquad\qquad x + y \leq 15$
   $\qquad\qquad x \geq 2, \ y \geq 3.$

Graph the feasible region, and identify the corner
points

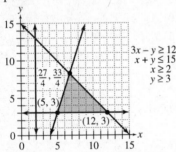

The feasible region is bounded with corner points

$(5, 3), \left(\frac{27}{4}, \frac{33}{4}\right), (12, 3)$.

Corner Point	Value of $z = 2x + y$
$(5, 3)$	13 Minimum
$\left(\frac{27}{4}, \frac{33}{4}\right)$	$\frac{87}{4}$
$(12, 3)$	27

The minimum is 13 at $(5, 3)$.

11. Maximize $z = 5x + y$
    subject to: $x - y \leq 10$
    $\qquad\qquad 5x + 3y \leq 75$
    $\qquad\qquad x \geq 0, \ y \geq 0.$

Graph the feasible region, and identify the corner
points.

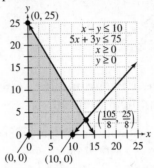

This region is bounded, with corner

points $(0, 0), (0, 25), \left(\frac{105}{8}, \frac{25}{8}\right), (10, 0)$.

Corner Point	Value of $z = 5x + y$
$(0, 0)$	0
$(0, 25)$	25
$\left(\frac{105}{8}, \frac{25}{8}\right)$	$\frac{275}{4} = 68.75$ Maximum
$(10, 0)$	50

The maximum is 68.75 at $\left(\frac{105}{8}, \frac{25}{8}\right)$.

13. $3x + 2y \geq 6$
    $x + 2y \geq 4$
    $x \geq 0, \ y \geq 0$

Graph the feasible region and identify the corner
points.

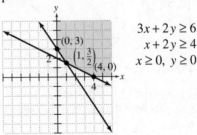

$3x + 2y \geq 6$
$x + 2y \geq 4$
$x \geq 0, \ y \geq 0$

(continued on next page)

(*continued from page 208*)

Corner Point	Value of $z = 3x + 4y$
$(0, 3)$	12
$\left(1, \dfrac{3}{2}\right)$	9 Minimum
$(4, 0)$	12

The minimum value of 9 occurs at $\left(1, \dfrac{3}{2}\right)$.

There is no maximum value because the feasible region is unbounded.

15. $x + y \le 6$
$-x + y \le 2$
$2x - y \le 8$

Graph the feasible region, and identify the corner points.

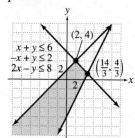

Corner Point	Value of $z = 3x + 4y$
$(2, 4)$	22 Maximum
$\left(\dfrac{14}{3}, \dfrac{4}{3}\right)$	$\dfrac{58}{3}$

The maximum value of z is 22 at $(2, 4)$. There is no minimum because the feasible region is unbounded.

17. a. $x + y \le 20$
$x + 3y \le 24$

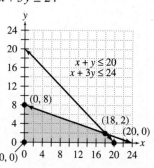

Corner Point	Value of $z = 10x + 12y$
$(0, 0)$	0
$(0, 8)$	96
$(18, 2)$	204 Maximum
$(20, 0)$	200

The maximum value of 204 occurs when $x = 18$ and $y = 2$, or at $(18, 2)$.

b. $3x + y \le 15$
$x + 2y \le 18$

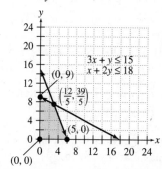

Corner Point	Value of $z = 10x + 12y$
$(0, 0)$	0
$(0, 9)$	108
$\left(\dfrac{12}{5}, \dfrac{39}{5}\right)$	$\dfrac{588}{5}$
$(5, 0)$	50

The maximum value of $\dfrac{588}{5}$ or $117\dfrac{3}{5}$

occurs when $x = \dfrac{12}{5}$ and $y = \dfrac{39}{5}$, or at

$\left(\dfrac{12}{5}, \dfrac{39}{5}\right)$.

c. $x + 2y \geq 10$
$2x + y \geq 12$
$x - y \leq 8$

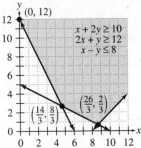

The feasible region is unbounded, so there is no maximum.

19. Answer varies. Sample response: The constraints do not describe a feasible region, i.e. there does not exist a point that satisfies all five constraints.

Section 7.3 Applications of Linear Programming

1. Let x = the number of canoes and let y = the number of rowboats.
   The constraints are $8x + 5y \leq 110$, $x \geq 0$, $y \geq 0$.

3. Let x = the number of radio spots and let y = the number of TV ads.
   The constraints are $250x + 750y \leq 9500$, $x \geq 0$ $y \geq 0$.

5. Let x = the number of chain saws and let y = the number of wood chippers.
   Assembling x chain saws at 4 hours each takes $4x$ hours while assembling y wood chippers at 6 hours each takes $6y$ hours. There are only 48 available hours, the first constraint is $4x + 6y \leq 48$.
   The number of chain saws and wood chippers assembled cannot be negative, so $x \geq 0$, $y \geq 0$.
   Each chain saw produces a profit of \$150 and each wood chipper, \$220. If z represents total profit, then $z = 150x + 220y$ is the objective function to be maximized.
   Maximize $z = 150x + 220y$
   subject to: $4x + 6y \leq 48$
   $\qquad\qquad x \geq 0, \; y \geq 0$

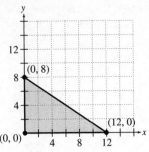

Corner Point	$z = 150x + 220y$
(0, 0)	$150(0) + 220(0) = 0$
(12, 0)	$150(12) + 220(0) = 1800$ (maximum)
(0, 8)	$150(0) + 220(8) = 1760$

12 chain saws and no wood chippers should be assembled for maximum profit.

7. Let x = the number of pounds of deluxe coffee and let y = the number of pounds of regular coffee. The mixture of deluxe and regular coffee needs to be at least 50 pounds, so $x + y \geq 50$ The mixture must have at least 10 pounds of deluxe coffee, so $x \geq 10$.
   The pounds of coffee cannot be negative, so $x \geq 0$, $y \geq 0$.
   At \$6 per pound of deluxe coffee and \$5 per pound of regular coffee, the total cost, z, is $z = 6x + 5y$, which is the objective function to be minimized.
   Minimize: $z = 6x + 5y$
   subject to: $x + y \geq 50$
   $\qquad\qquad x \geq 10$
   $\qquad\qquad y \geq 0$

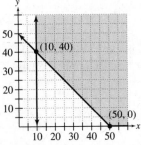

Corner Point	$z = 6x + 5y$
(50, 0)	$6(50) + 5(0) = 300$
(10, 40)	$6(10) + 5(40) = 260$ (minimum)

The mixture should contain 10 pounds of deluxe and 40 pounds of regular coffee to minimize cost.

9. From exercise 1, we let x = the number of canoes and y = the number of rowboats. The constraint imposed by the number of canoes sold is $x \le 10$. The company always sells at least 6 rowboats, so $y \ge 6$. A negative number of boats cannot be sold, so $x \ge 0$. The constraint imposed by the number of hours of labor available each week is $8x + 5y \le 110$. Thus, the constraints are

$$x \le 10$$
$$8x + 5y \le 110$$
$$y \ge 6$$
$$x \ge 0$$

The profit function, z, is $z = 400x + 225y$, which is the objective function to be maximized.

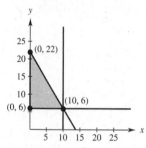

Corner Point	$z = 400x + 225y$
(0, 6)	$400(0) + 225(6) = 1350$
(10, 6)	$400(10) + 225(6) = 5350$ maximum
(0, 22)	$400(0) + 225(22) = 4950$

The company should sell 10 canoes and 6 rowboats to maximize profits.

11. From exercise 3, we let x = the number of radio spots and let y = the number of TV ads. Since each radio spot costs \$250 and each TV ad cost \$750, the constraint imposed by the cost is $250x + 750y \le 9500$. The constraints imposed by the number of ads are $x \ge 8$ and $y \ge 3$. Thus, the constraints are

$250x + 750y \le 9500$

$x \ge 8$, $y \ge 3$

The function that describes the number of people reached by the ads is $z = 600x + 2000y$. This is the function to be maximized.

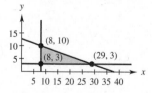

Corner Point	$z = 600x + 2000y$
(8, 3)	$600(8) + 2000(3) = 10{,}800$
(8, 10)	$600(8) + 2000(10) = 24{,}800$ maximum
(29, 3)	$600(29) + 2000(3) = 23{,}400$

The candidate should use 8 radio spots and 10 tv ads to maximize the number of people reached.

13. Let x = the number of units of Policy A to purchase and let y = the number of units of Policy B to purchase.

a. Minimize: $z = 50x + 40y$

subject to: $10x + 15y \ge 300$
$180x + 120y \ge 3000$
$x \ge 0$
$y \ge 0.$

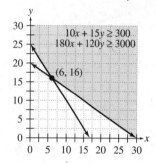

Corner Point	$z = 50x + 40y$
(0, 25)	$50(0) + 40(25) = 1000$
(6, 16)	$50(6) + 40(16) = 940$ minimum
(30, 0)	$50(30) + 40(0) = 1500$

The minimum premium cost is from purchasing 6 units of Policy A and 16 units of Policy B and is \$940.

b. Minimize: $z = 25x + 40y$

Corner Point	$z = 25x + 40y$
(0, 25)	$25(0) + 40(25) = 1000$
(6, 16)	$25(6) + 40(16) = 790$ minimum
(30, 0)	$25(30) + 40(0) = 750$

The constraints and graph remain unchanged. Now the minimum premium cost is from purchasing 30 units of Policy A and no units of Policy B and is \$750.

15. Let x = the number of brand X pills, let y = the number of brand Z pills. The constraint imposed by the amount of vitamin A is $8x + 2y \geq 16$. The constraint imposed by the amount of vitamin B-1 is $x + y \geq 5$. The constraint imposed by the amount of vitamin C is $2x + 7y \geq 20$. The cost function z is $z = 15x + 30y$. This is the function to be minimized subject to the constraints

$$8x + 2y \geq 16$$
$$x + y \geq 5$$
$$2x + 7y \geq 20$$
$$x \geq 0, \ y \geq 0$$

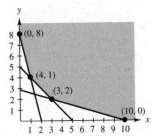

Corner Point	$z = .15x + .30y$
(0, 8)	$.15(0) + .30(8) = 2.40$
(4, 1)	$.15(4) + .30(1) = .90$ (minimum)
(3, 2)	$.15(3) + .30(2) = 1.05$
(10, 0)	$.15(10) + .30(0) = 1.50$

He can satisfy the requirements with 4 brand X pills and 1 brand Z pill for a minimum cost of $0.90.

17. Let x = the number of Type 1 bolts and let y = the number of Type 2 bolts.

Maximize $z = .10x + .12y$

subject to: $.1x + .1y \leq 240$
$.1x + .4y \leq 720$
$.1x + .02y \leq 160$
$x \geq 0$
$y \geq 0$

Graph the feasible region, and label the corner points. Use the corner points to find the maximum value of the objective function.

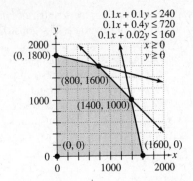

Corner Point	Value of $z = .10x + .12y$
(0, 0)	$.10(0) + .12(0) = 0$
(0, 1800)	$.10(0) + .12(1800) = 216$
(800, 1600)	$.10(800) + .12(1600) = 272$ Maximum
(1400, 1600)	$.10(1400) + .12(1600) = 260$
(1600, 0)	$.1(1600) + .12(0) = 160$

Manufacture 800 Type 1 bolts and 1600 Type 2 bolts for a maximum revenue of $272/day.

19. Summarizing the data provided,

	Energy	Protein	Hides	Availability	Cost(hrs)
plant	30	10	0	25	30
animal	20	25	1	25	15
Requirements	360	300	8		

Let x = the number of plants to gather and let y = the number of animals to gather.

Minimize $z = 30x + 15y$ subject to:

$$30x + 20y \geq 360$$
$$10x + 25y \geq 300$$
$$y \geq 8$$
$$x \leq 25$$
$$y \leq 25$$
$$x \geq 0, \ y \geq 0$$

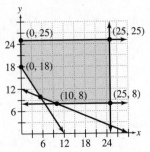

(continued on next page)

(continued from page 212)

Corner Point	$z = 30x + 15y$
(0, 25)	$30(0) + 15(25) = 375$
(0, 18)	$30(0) + 15(18) = 270$
	minimum
$\left(5\frac{5}{11}, 9\frac{9}{11}\right)$	$30\left(5\frac{5}{11}\right) + 15\left(9\frac{9}{11}\right) = 310\frac{10}{11}$
(10, 8)	$30(10) + 15(8) = 420$
(25, 8)	$30(25) + 15(8) = 870$
(25, 25)	$30(25) + 15(25) = 1125$

The population should gather no plants and 18 animals for a minimum 270 hours.

21. Let x = the number of cards from warehouse I to San Jose. Then $350 - x$ is the number of cards from warehouse II to San Jose. Let y = the number of cards from warehouse I to Memphis. Then $250 - y$ is the number of cards from warehouse II to Memphis. The constraints are represented by the inequalities
$$x \geq 0, \ y \geq 0$$
$$x \leq 350$$
$$y \leq 250$$
$$x + y \leq 500$$
$(350 - x) + (250 - y) \leq 290 \Rightarrow x + y \geq 310.$
Minimize
$.25x + .23(350 - x) + .22y + .21(250 - y)$
$= 133 + .02x + .01y.$
The graph and corner points are shown below.

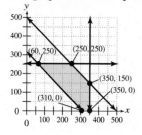

Corner Point	Value of $z = 133 + .02x + .01y$
(60, 250)	136.7 Minimum
(250, 250)	140.5
(350, 150)	141.5
(350, 0)	140
(310, 0)	139.2

The minimum cost is $136.70. From warehouse I ship 60 boxes to San Jose and 250 boxes to Memphis. From warehouse II ship 290 boxes to San Jose and none to Memphis.

23. Let x = the amount invested in bonds and let y = the amount invested in mutual funds. (Both are in millions of dollars.)
The amount of annual interest is $.04x + .06y$.
Maximize $z = .04x + .06y$

subject to: $x \geq 20$
$y \geq 6$
$300x + 100y \leq 8400$
$x + y \leq 50$
$x \geq 0, \ y \geq 0$

Graph the feasible region, and label the corner points.

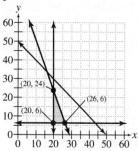

Corner Point	Value of $z = .04x + .06y$
(20, 24)	2.24 Maximum
(26, 6)	14.
(20, 6)	1.16

He should invest $20 million in bonds and $24 million in mutual funds for maximum annual interest of $2.24 million.

25. Let x = the number of humanities courses and let y = the number of science courses.

Maximize $z = 3(5y) + 2\left(\frac{1}{2}\right)(4x)$

$+3\left(\frac{1}{4}\right)(4x) + 4\left(\frac{1}{4}\right)(4x)$

subject to: $x \geq 4$
$4 \leq y \leq 12$
$4x + 5y \leq 92.$

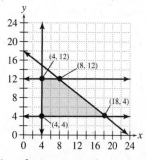

(continued on next page)

(continued from page 213)

The objective function simplifies to $z = 11x + 15y$.

Corner Point	Value of $z = 11x + 15y$
$(4, 4)$	104
$(4, 12)$	224
$(8, 12)$	268 Maximum
$(18, 4)$	258

She should take 8 humanities and 12 science courses.

27. 1 Zeta + 2 Beta must not exceed 1000; thus (b) is the correct answer.

29. $4 Zeta + $5.25 Beta equals the total contribution margin; (c) is the correct answer.

Section 7.4 The Simplex Method: Maximization

1. Maximize $z = 32x_1 + 9x_2$

subject to: $4x_1 + 2x_2 \le 20$
$5x_1 + x_2 \le 50$
$2x_1 + 3x_2 \le 25$
$x_1 \ge 0,\ x_2 \ge 0$.

a. There are 3 constraints, so 3 slack variables are needed.

b. Use s_1, s_2, and s_3 for the slack variables.

c. $4x_1 + 2x_2 + s_1 = 20$
$5x_1 + x_2 + s_2 = 50$
$2x_1 + 3x_2 + s_3 = 25$

3. Maximize $z = 8x_1 + 3x_2 + x_3$

subject to: $3x_1 - x_2 + 4x_3 \le 95$
$7x_1 + 6x_2 + 8x_3 \le 118$
$4x_1 + 5x_2 + 10x_3 \le 220$
$x_1 \ge 0,\ x_2 \ge 0,\ x_3 \ge 0$.

a. There are 3 constraints, so 3 slack variables are needed.

b. Use s_1, s_2, and s_3 for the slack variables.

c. $3x_1 - x_2 + 4x_3 + s_1 = 95$
$7x_1 + 6x_2 + 8x_3 + s_2 = 118$
$4x_1 + 5x_2 + 10x_3 + s_3 = 220$

5. Maximize $z = 5x_1 + x_2$

subject to: $2x_1 + 5x_2 \le 6$
$4x_1 + x_2 \le 6$
$5x_1 + 3x_2 \le 15$
$x_1 \ge 0,\ x_2 \ge 0$.

Since there are 3 constraints, 3 slack variables are needed: s_1, s_2, and s_3.

The constraints are now
$2x_1 + 5x_2 + s_1 = 6$
$4x_1 + x_2 + s_2 = 6$
$5x_1 + 3x_2 + s_3 = 15$.

The initial simplex tableau is

$$
\begin{array}{cccccc}
x_1 & x_2 & s_1 & s_2 & s_3 & z \\
\end{array}
$$
$$
\left[\begin{array}{cccccc|c}
2 & 5 & 1 & 0 & 0 & 0 & 6 \\
4 & 1 & 0 & 1 & 0 & 0 & 6 \\
5 & 3 & 0 & 0 & 1 & 0 & 15 \\
\hline
-5 & -1 & 0 & 0 & 0 & 1 & 0
\end{array}\right].
$$

7. Maximize $z = x_1 + 5x_2 + 10x_3$

subject to: $x_1 + 2x_2 + 3x_3 \le 10$
$2x_1 + x_2 + x_3 \le 8$
$3x_1 + 4x_3 \le 6$
$x_1 \ge 0,\ x_2 \ge 0,\ x_3 \ge 0$.

Since there are 3 constraints, 3 slack variables are needed.: s_1, s_2, s_3.

The constraints are now
$x_1 + 2x_2 + 3x_3 + s_1 = 10$
$2x_1 + x_2 + x_3 + s_2 = 8$
$3x_1 + 4x_3 + s_3 = 6$

The initial tableau is

$$
\begin{array}{ccccccc}
x_1 & x_2 & x_3 & s_1 & s_2 & s_3 & z \\
\end{array}
$$
$$
\left[\begin{array}{ccccccc|c}
1 & 2 & 3 & 1 & 0 & 0 & 0 & 10 \\
2 & 1 & 1 & 0 & 1 & 0 & 0 & 8 \\
3 & 0 & 4 & 0 & 0 & 1 & 0 & 6 \\
\hline
-1 & -5 & -10 & 0 & 0 & 0 & 1 & 0
\end{array}\right].
$$

9.
$$
\begin{array}{cccccc}
x_1 & x_2 & x_3 & s_1 & s_2 & z \\
\end{array}
$$
$$
\left[\begin{array}{cccccc|c}
2 & 2 & 0 & 3 & 1 & 0 & 15 \\
3 & 4 & 1 & 6 & 0 & 0 & 20 \\
\hline
-2 & -3 & 0 & 1 & 0 & 1 & 10
\end{array}\right]
$$

The most negative indicator is -3, so the pivot column is column 2. Since $\frac{20}{4} = 5$ is the smaller quotient, the pivot row is row two. The pivot is the 4 in row two, column two.

11.

$$\begin{array}{cccccccc} x_1 & x_2 & x_3 & s_1 & s_2 & s_3 & z \end{array}$$

$$\left[\begin{array}{ccccccc|c} 6 & 2 & 1 & 3 & 0 & 0 & 0 & 8 \\ 0 & 2 & 0 & 1 & 0 & 1 & 0 & 7 \\ 6 & 1 & 0 & 3 & 1 & 0 & 0 & 6 \\ \hline -3 & -2 & 0 & 2 & 0 & 0 & 1 & 12 \end{array}\right]$$

The most negative indicator is –3, so the pivot column is column one. Since $\frac{6}{6} = 1$ is the smallest quotient, the pivot row is row three. Thus, the pivot is the 6 in row three, column one.

13.

$$\begin{array}{cccccc} x_1 & x_2 & x_3 & s_1 & s_2 & z \end{array}$$

$$\left[\begin{array}{cccccc|c} 1 & 2 & 4 & 1 & 0 & 0 & 56 \\ 2 & \boxed{2} & 1 & 0 & 1 & 0 & 40 \\ \hline -1 & -3 & -2 & 0 & 0 & 1 & 0 \end{array}\right]$$

Start by multiplying each entry of row 2 by $\frac{1}{2}$ in order to change the pivot to 1.

$$\begin{array}{cccccc} x_1 & x_2 & x_3 & s_1 & s_2 & z \end{array}$$

$$\left[\begin{array}{cccccc|cl} 1 & 2 & 4 & 1 & 0 & 0 & 56 & \\ 1 & 1 & \frac{1}{2} & 0 & \frac{1}{2} & 0 & 20 & \frac{1}{2}R_2 \\ \hline -1 & -3 & -2 & 0 & 0 & 1 & 0 & \end{array}\right]$$

Now use row operations to change the entry in row one column two and the indicator –3 to 0.

$$\begin{array}{cccccc} x_1 & x_2 & x_3 & s_1 & s_2 & z \end{array}$$

$$\left[\begin{array}{cccccc|cl} -1 & 0 & 3 & 1 & -1 & 0 & 16 & -2R_2 + R_1 \\ 1 & 1 & \frac{1}{2} & 0 & \frac{1}{2} & 0 & 20 & \\ \hline 2 & 0 & -\frac{1}{2} & 0 & \frac{3}{2} & 1 & 60 & 3R_2 + R_3 \end{array}\right]$$

15.

$$\begin{array}{ccccccc} x_1 & x_2 & x_3 & s_1 & s_2 & s_3 & z \end{array}$$

$$\left[\begin{array}{ccccccc|c} 1 & 1 & 1 & 1 & 0 & 0 & 0 & 60 \\ 3 & 1 & \boxed{2} & 0 & 1 & 0 & 0 & 100 \\ 1 & 2 & 3 & 0 & 0 & 1 & 0 & 200 \\ \hline -1 & -1 & -2 & 0 & 0 & 0 & 1 & 0 \end{array}\right]$$

$$\begin{array}{ccccccc} x_1 & x_2 & x_3 & x_4 & x_5 & x_6 & z \end{array}$$

$$\left[\begin{array}{ccccccc|cl} 1 & 1 & 1 & 1 & 0 & 0 & 0 & 60 & \\ \frac{3}{2} & \frac{1}{2} & 1 & 0 & \frac{1}{2} & 0 & 0 & 50 & \frac{1}{2}R_2 \\ 1 & 2 & 3 & 0 & 0 & 1 & 0 & 200 & \\ \hline -1 & -1 & -2 & 0 & 0 & 0 & 1 & 0 & \end{array}\right]$$

$$\begin{array}{ccccccc} x_1 & x_2 & x_3 & s_1 & s_2 & s_3 & z \end{array}$$

$$\left[\begin{array}{ccccccc|cl} -\frac{1}{2} & \frac{1}{2} & 0 & 1 & -\frac{1}{2} & 0 & 0 & 10 & -1R_2 + R_1 \\ \frac{3}{2} & \frac{1}{2} & 1 & 0 & \frac{1}{2} & 0 & 0 & 50 & \\ -\frac{7}{2} & \frac{1}{2} & 0 & 0 & -\frac{3}{2} & 1 & 0 & 50 & -3R_2 + R_3 \\ \hline 2 & 0 & 0 & 0 & 1 & 0 & 1 & 100 & 2R_2 + R_4 \end{array}\right]$$

17.

$$\begin{array}{cccccc} x_1 & x_2 & x_3 & s_1 & s_2 & z \end{array}$$

$$\left[\begin{array}{cccccc|c} 3 & 2 & 0 & -3 & 1 & 0 & 29 \\ 4 & 0 & 1 & -2 & 0 & 0 & 16 \\ \hline -5 & 0 & 0 & -1 & 0 & 1 & 11 \end{array}\right]$$

a. The basic variables are x_3, s_2, and z. x_1, x_2, and s_1 are nonbasic.

b. The basic feasible solution is $x_1 = 0$, $x_2 = 0$, $x_3 = 16$, $s_1 = 0$, $s_2 = 29$, $z = 11$

c. Because there are still negative indicators, this solution is not the maximum.

19.

$$\begin{array}{ccccccc} x_1 & x_2 & x_3 & s_1 & s_2 & s_3 & z \end{array}$$

$$\left[\begin{array}{ccccccc|c} 1 & 0 & 2 & \frac{1}{2} & 0 & \frac{1}{3} & 0 & 6 \\ 0 & 1 & -1 & 5 & 0 & -1 & 0 & 13 \\ 0 & 0 & 1 & \frac{3}{2} & 1 & -\frac{1}{3} & 0 & 21 \\ \hline 0 & 0 & 2 & \frac{1}{2} & 0 & 3 & 1 & 18 \end{array}\right]$$

a. The basic variables are x_1, x_2, s_2, and z. x_3, s_1, and s_3 are nonbasic.

b. The basic feasible solution is $x_1 = 6$, $x_2 = 13$, $x_3 = 0$, $s_1 = 0$, $s_2 = 21$, $s_3 = 0$, $z = 18$.

c. Because there are no negative indicators, this solution is the maximum.

21. Maximize $z = x_1 + 3x_2$

subject to: $\quad x_1 + x_2 \le 10$
$\qquad\qquad 5x_1 + 2x_2 \le 20$
$\qquad\qquad x_1 + 2x_2 \le 36$
$\qquad\qquad x_1 \ge 0, \ x_2 \ge 0.$

Using slack variables, s_1, s_2, s_3. the constraints become:

$$\begin{array}{l} x_1 + x_2 + s_1 \qquad\qquad = 10 \\ 5x_1 + 2x_2 \quad + s_2 \qquad = 20 \\ x_1 + 2x_2 \qquad\quad + s_3 = 36. \end{array}$$

The initial simplex tableau is

$$\begin{array}{cccccc} x_1 & x_2 & s_1 & s_2 & s_3 & z \end{array}$$

$$\left[\begin{array}{cccccc|c} 1 & \boxed{1} & 1 & 0 & 0 & 0 & 10 \\ 5 & 2 & 0 & 1 & 0 & 0 & 20 \\ 1 & 2 & 0 & 0 & 1 & 0 & 36 \\ \hline -1 & -3 & 0 & 0 & 0 & 1 & 0 \end{array}\right].$$

Pivot on the 1 in row one, column two.

(*continued on next page*)

(continued from page 215)

$$\begin{array}{cccccc} x_1 & x_2 & s_1 & s_2 & s_3 & z \end{array}$$

$$\left[\begin{array}{cccccc|c} 1 & 1 & 1 & 0 & 0 & 0 & 10 \\ 3 & 0 & -2 & 1 & 0 & 0 & 0 \\ -1 & 0 & -2 & 0 & 1 & 0 & 16 \\ \hline 2 & 0 & 3 & 0 & 0 & 1 & 30 \end{array}\right] \begin{array}{l} \\ -2R_1 + R_2 \\ -2R_1 + R_3 \\ 3R_1 + R_4 \end{array}$$

Since there are no negative indicators, this matrix is the final tableau. The maximum is 30 when $x_1 = 0$, $x_2 = 10$, $s_1 = 0$, $s_2 = 0$, and $s_3 = 16$.

23. Maximize $z = 2x_1 + x_2$
    subject to: $x_1 + 3x_2 \le 12$
    $2x_1 + x_2 \le 10$
    $x_1 + x_2 \le 4$
    $x_1 \ge 0$, $x_2 \ge 0$.

    Using slack variables s_1, s_2, s_3. the constraints become:

    $$\begin{array}{rcl} x_1 + 3x_2 + s_1 & = & 12 \\ 2x_1 + x_2 + s_2 & = & 10 \\ x_1 + x_2 + s_3 & = & 4. \end{array}$$

    The initial simplex tableau is

    $$\begin{array}{cccccc} x_1 & x_2 & s_1 & s_2 & s_3 & z \end{array}$$

    $$\left[\begin{array}{cccccc|c} 1 & 3 & 1 & 0 & 0 & 0 & 12 \\ 2 & 1 & 0 & 1 & 0 & 0 & 10 \\ \boxed{1} & 1 & 0 & 0 & 1 & 0 & 4 \\ \hline -2 & -1 & 0 & 0 & 0 & 1 & 0 \end{array}\right].$$

    $\uparrow$

    Pivot on the 1 in row three, column one.

    $$\begin{array}{cccccc} x_1 & x_2 & s_1 & s_2 & s_3 & z \end{array}$$

    $$\left[\begin{array}{cccccc|c} 0 & 2 & 1 & 0 & -1 & 0 & 8 \\ 0 & -1 & 0 & 1 & -2 & 0 & 2 \\ 1 & 1 & 0 & 0 & 1 & 0 & 4 \\ \hline 0 & 1 & 0 & 0 & 2 & 1 & 8 \end{array}\right] \begin{array}{l} -1R_3 + R_1 \\ -2R_3 + R_2 \\ \\ 2R_3 + R_4 \end{array}$$

    Since there are no negative indicators, this matrix is the final tableau.
    The maximum is 8 when $x_1 = 4$, $x_2 = 0$, $s_1 = 8$, $s_2 = 2$, $s_3 = 0$.

25. Maximize $z = 5x_1 + 4x_2 + x_3$
    subject to: $-2x_1 + x_2 + 2x_3 \le 3$
    $x_1 - x_2 + x_3 \le 1$
    $x_1 \ge 0$, $x_2 \ge 0$, $x_3 \ge 0$.

    Using the slack variables s_1 and s_2, the constraints become:

    $$\begin{array}{rcl} -2x_1 + x_2 + 2x_3 + s_1 & = & 3 \\ x_1 - x_2 + x_3 + s_2 & = & 1. \end{array}$$

    The initial simplex tableau is

    $$\begin{array}{cccccc} x_1 & x_2 & x_3 & s_1 & s_2 & z \end{array}$$

    $$\left[\begin{array}{cccccc|c} -2 & 1 & 2 & 1 & 0 & 0 & 3 \\ \boxed{1} & -1 & 1 & 0 & 1 & 0 & 1 \\ \hline -5 & -4 & -1 & 0 & 0 & 1 & 0 \end{array}\right].$$

    $\uparrow$

    Pivot on the 1 in row two, column one.

    $$\begin{array}{cccccc} x_1 & x_2 & x_3 & s_1 & s_2 & z \end{array}$$

    $$\left[\begin{array}{cccccc|c} 0 & -1 & 4 & 1 & 2 & 0 & 5 \\ 1 & -1 & 1 & 0 & 1 & 0 & 1 \\ \hline 0 & -9 & 4 & 0 & 5 & 1 & 5 \end{array}\right] \begin{array}{l} 2R_2 + R_1 \\ \\ 5R_2 + R_3 \end{array}$$

    There is a negative indicator in column two, but all entries in that column are negative, so there is no place to continue pivoting. Therefore, there is no maximum.

27. Maximize $z = 2x_1 + x_2 + x_3$
    subject to: $x_1 - 3x_2 + x_3 \le 3$
    $x_1 - 2x_2 + 2x_3 \le 12$
    $x_1 \ge 0$, $x_2 \ge 0$, $x_3 \ge 0$.

    Using slack variables s_1 and s_2, the constraints become:

    $$\begin{array}{rcl} x_1 - 3x_2 + x_3 + s_1 & = & 3 \\ x_1 - 2x_2 + 2x_3 + s_2 & = & 12. \end{array}$$

    The initial simplex tableau is

    $$\begin{array}{cccccc} x_1 & x_2 & x_3 & s_1 & s_2 & z \end{array}$$

    $$\left[\begin{array}{cccccc|c} \boxed{1} & -3 & 1 & 1 & 0 & 0 & 3 \\ 1 & -2 & 2 & 0 & 1 & 0 & 12 \\ \hline -2 & -1 & -1 & 0 & 0 & 1 & 0 \end{array}\right].$$

    Pivot on the 1 in row one, column one.

    $$\begin{array}{cccccc} x_1 & x_2 & x_3 & s_1 & s_2 & z \end{array}$$

    $$\left[\begin{array}{cccccc|c} 1 & -3 & 1 & 1 & 0 & 0 & 3 \\ 0 & \boxed{1} & 1 & -1 & 1 & 0 & 9 \\ \hline 0 & -7 & 1 & 2 & 0 & 1 & 6 \end{array}\right] \begin{array}{l} \\ -R_1 + R_2 \\ 2R_1 + R_3 \end{array}$$

    $\uparrow$

(continued on next page)

(continued from page 216)

Pivot on the 1 in row two, column two.

$$\begin{array}{cccccc} x_1 & x_2 & x_3 & s_1 & s_2 & z \end{array}$$

$$\left[\begin{array}{cccccc|c} 1 & 0 & 4 & -2 & 3 & 0 & 30 \\ 0 & 1 & 1 & -1 & 1 & 0 & 9 \\ \hline 0 & 0 & 8 & -5 & 7 & 1 & 69 \end{array}\right] \begin{array}{l} 3R_2 + R_1 \\ \\ 7R_2 + R_3 \end{array}$$

The only negative indicator is in column four, which has all negative entries, so there is no place to continue pivoting. Therefore, there is no maximum.

29. Maximize $z = 2x_1 + 2x_2 - 4x_3$

Subject to: $3x_1 + 3x_2 - 6x_3 \le 51$
$5x_1 + 5x_2 + 10x_3 \le 99$
$x_1 \ge 0,\ x_2 \ge 0,\ x_3 \ge 0.$

Using slack variables s_1 and s_2, the constraints become:

$3x_1 + 3x_2 - 6x_3 + s_1 \qquad = 51$
$5x_1 + 5x_2 + 10x_3 \qquad + s_2 = 99.$

The initial simplex tableau is

$$\begin{array}{cccccc} x_1 & x_2 & x_3 & s_1 & s_2 & z \end{array}$$

$$\left[\begin{array}{cccccc|c} \boxed{3} & 3 & -6 & 1 & 0 & 0 & 51 \\ 5 & 5 & 10 & 0 & 1 & 0 & 99 \\ \hline -2 & -2 & 4 & 0 & 0 & 1 & 0 \end{array}\right].$$
$\uparrow$

Pivot on the 3 in row one, column one.

$$\begin{array}{cccccc} x_1 & x_2 & x_3 & s_1 & s_2 & z \end{array}$$

$$\left[\begin{array}{cccccc|c} 1 & 1 & -2 & \frac{1}{3} & 0 & 0 & 17 \\ 5 & 5 & 10 & 0 & 1 & 0 & 99 \\ \hline -2 & -2 & 4 & 0 & 0 & 1 & 0 \end{array}\right] \frac{1}{3}R_1$$

$$\begin{array}{cccccc} x_1 & x_2 & x_3 & s_1 & s_2 & z \end{array}$$

$$\left[\begin{array}{cccccc|c} 1 & 1 & -2 & \frac{1}{3} & 0 & 0 & 17 \\ 0 & 0 & 20 & -\frac{5}{3} & 1 & 0 & 14 \\ \hline 0 & 0 & 0 & \frac{2}{3} & 0 & 1 & 34 \end{array}\right] \begin{array}{l} \\ -5R_1 + R_2 \\ 2R_1 + R_3 \end{array}$$

The maximum is 34 when
$x_1 = 17,\ x_2 = 0,\ x_3 = 0,\ s_1 = 0,\ s_2 = 14$
or when $x_1 = 0,\ x_2 = 17,\ x_3 = 0,\ s_1 = 0,$
and $s_2 = 14.$

31. Maximize $z = 300x_1 + 200x_2 + 100x_3$

subject to: $x_1 + x_2 + x_3 \le 100$
$2x_1 + 3x_2 + 4x_3 \le 320$
$2x_1 + x_2 + x_3 \le 160$
$x_1 \ge 0,\ x_2 \ge 0,\ x_3 \ge 0.$

Using slack variables s_1, s_2, s_3. the constraints become:

$x_1 + x_2 + x_3 + s_1 \qquad\qquad = 100$
$2x_1 + 3x_2 + 4x_3 + \qquad s_2 \qquad = 320$
$2x_1 + x_2 + x_3 + \qquad\qquad s_3 = 160.$

The initial simplex tableau is

$$\begin{array}{ccccccc} x_1 & x_2 & x_3 & s_1 & s_2 & s_3 & z \end{array}$$

$$\left[\begin{array}{ccccccc|c} 1 & 1 & 1 & 1 & 0 & 0 & 0 & 100 \\ 2 & 3 & 4 & 0 & 1 & 0 & 0 & 320 \\ \boxed{2} & 1 & 1 & 0 & 0 & 1 & 0 & 160 \\ \hline -300 & -200 & -100 & 0 & 0 & 0 & 1 & 0 \end{array}\right].$$

Pivot on the 2 in row three, column one.

$$\begin{array}{ccccccc} x_1 & x_2 & x_3 & s_1 & s_2 & s_3 & z \end{array}$$

$$\left[\begin{array}{ccccccc|c} 1 & 1 & 1 & 1 & 0 & 0 & 0 & 100 \\ 2 & 3 & 4 & 0 & 1 & 0 & 0 & 320 \\ \boxed{1} & \frac{1}{2} & \frac{1}{2} & 0 & 0 & \frac{1}{2} & 0 & 80 \\ \hline -300 & -200 & -100 & 0 & 0 & 0 & 1 & 0 \end{array}\right] \frac{1}{2}R_3$$

$$\begin{array}{ccccccc} x_1 & x_2 & x_3 & s_1 & s_2 & s_3 & z \end{array}$$

$$\left[\begin{array}{ccccccc|c} 0 & \frac{1}{2} & \frac{1}{2} & 1 & 0 & -\frac{1}{2} & 0 & 20 \\ 0 & 2 & 3 & 0 & 1 & -1 & 0 & 160 \\ 1 & \frac{1}{2} & \frac{1}{2} & 0 & 0 & \frac{1}{2} & 0 & 80 \\ \hline 0 & -50 & 50 & 0 & 0 & 150 & 1 & 24{,}000 \end{array}\right] \begin{array}{l} -1R_3 + R_1 \\ -2R_3 + R_2 \\ \\ 300R_3 + R_4 \end{array}$$

Pivot on the $\frac{1}{2}$ in row one, column two.

$$\begin{array}{ccccccc} x_1 & x_2 & x_3 & s_1 & s_2 & s_3 & z \end{array}$$

$$\left[\begin{array}{ccccccc|c} 0 & 1 & 1 & 2 & 0 & -1 & 0 & 40 \\ 0 & 2 & 3 & 0 & 1 & -1 & 0 & 160 \\ 1 & \frac{1}{2} & \frac{1}{2} & 0 & 0 & \frac{1}{2} & 0 & 80 \\ \hline 0 & -50 & 50 & 0 & 0 & 150 & 1 & 24{,}000 \end{array}\right] 2R_1$$

$$\begin{array}{ccccccc} x_1 & x_2 & x_3 & s_1 & s_2 & s_3 & z \end{array}$$

$$\left[\begin{array}{ccccccc|c} 0 & 1 & 1 & 2 & 0 & -1 & 0 & 40 \\ 0 & 0 & 1 & -4 & 1 & 1 & 0 & 80 \\ 1 & 0 & 0 & -1 & 0 & 1 & 0 & 60 \\ \hline 0 & 0 & 100 & 100 & 0 & 100 & 1 & 26{,}000 \end{array}\right] \begin{array}{l} \\ -2R_1 + R_2 \\ -\frac{1}{2}R_1 + R_3 \\ 50R_1 + R_4 \end{array}$$

The maximum value is 26,000 when $x_1 = 60,\ x_2 = 40,$
$x_3 = 0,\ s_1 = 0,\ s_2 = 80,$ and $s_3 = 0.$

33. Maximize $z = 4x_1 - 3x_2 + 2x_3$

subject to:
$$2x_1 - x_2 + 8x_3 \le 40$$
$$4x_1 - 5x_2 + 6x_3 \le 60$$
$$2x_1 - 2x_2 + 6x_3 \le 24$$
$$x_1 \ge 0, \ x_2 \ge 0, \ x_3 \ge 0.$$

Note: The third constraint simplifies to
$x_1 - x_2 + 3x_3 \le 12.$

Using slack variables s_1, s_2, s_3. the constraints become:

$$2x_1 - x_2 + 8x_3 + s_1 \qquad\qquad = 40$$
$$4x_1 - 5x_2 + 6x_3 + \qquad s_2 \qquad = 60$$
$$x_1 - x_2 + 3x_3 + \qquad\qquad s_3 = 12.$$

The initial simplex tableau is

$$\begin{array}{ccccccc|c}
x_1 & x_2 & x_3 & s_1 & s_2 & s_3 & z & \\
2 & -1 & 8 & 1 & 0 & 0 & 0 & 40 \\
4 & -5 & 6 & 0 & 1 & 0 & 0 & 60 \\
\boxed{1} & -1 & 3 & 0 & 0 & 1 & 0 & 12 \\
\hline
-4 & 3 & -2 & 0 & 0 & 0 & 1 & 0
\end{array}$$

Pivot on the 1 in row three, column one.

$$\begin{array}{ccccccc|cl}
x_1 & x_2 & x_3 & s_1 & s_2 & s_3 & z & \\
0 & \boxed{1} & 2 & 1 & 0 & -2 & 0 & 16 & -2R_3 + R_1 \\
0 & -1 & -6 & 0 & 1 & -4 & 0 & 12 & -4R_3 + R_2 \\
1 & -1 & 3 & 0 & 0 & 1 & 0 & 12 & \\
\hline
0 & -1 & 10 & 0 & 0 & 4 & 1 & 48 & 4R_3 + R_4
\end{array}$$

Pivot on the 1 in row one, column two.

$$\begin{array}{ccccccc|cl}
x_1 & x_2 & x_3 & s_1 & s_2 & s_3 & z & \\
0 & 1 & 2 & 1 & 0 & -2 & 0 & 16 & \\
0 & 0 & -4 & 1 & 1 & -6 & 0 & 28 & R_1 + R_2 \\
1 & 0 & 5 & 1 & 0 & -1 & 0 & 28 & R_1 + R_3 \\
\hline
0 & 0 & 12 & 1 & 0 & 2 & 1 & 64 & R_1 + R_4
\end{array}$$

The maximum is 64 when
$x_1 = 28, \ x_2 = 16, \ x_3 = 0, \ s_1 = 0, \ s_2 = 28,$ and
$s_3 = 0.$

35. Maximize $z = x_1 + 2x_2 + x_3 + 5x_4$

subject to:
$$x_1 + 2x_2 + x_3 + x_4 \le 50$$
$$3x_1 + x_2 + 2x_3 + x_4 \le 100$$
$$x_1 \ge 0, \ x_2 \ge 0, \ x_3 \ge 0, x_4 \ge 0.$$

Using slack variables s_1 and s_2 the constraints become:

$$x_1 + 2x_2 + x_3 + x_4 + s_1 \qquad = 50$$
$$3x_1 + x_2 + 2x_3 + x_4 \qquad + s_2 = 100$$

The initial simplex tableau is

$$\begin{array}{ccccccc|c}
x_1 & x_2 & x_3 & x_4 & s_1 & s_2 & z & \\
1 & 2 & 1 & \boxed{1} & 1 & 0 & 0 & 50 \\
3 & 1 & 2 & 1 & 0 & 1 & 0 & 100 \\
\hline
-1 & -2 & -1 & -5 & 0 & 0 & 1 & 0
\end{array}$$

The pivot is the 1 in row one, column four.

$$\begin{array}{ccccccc|cl}
x_1 & x_2 & x_3 & x_4 & s_1 & s_2 & z & \\
1 & 2 & 1 & 1 & 1 & 0 & 0 & 50 & \\
2 & -1 & 1 & 0 & -1 & 1 & 0 & 50 & -1R_1 + R_2 \\
\hline
4 & 8 & 4 & 0 & 5 & 0 & 1 & 250 & 5R_1 + R_3
\end{array}$$

The maximum is 250 when
$x_1 = 0, x_2 = 0, \ x_3 = 0, \ x_4 = 50, \ s_1 = 0,$ and
$s_2 = 50.$

37.
$$\begin{array}{ccccc|c}
x_1 & x_2 & x_3 & s_1 & s_2 & z & \\
1 & 1 & 1 & 1 & 0 & 0 & 12 \\
2 & 1 & 2 & 0 & 1 & 0 & 30 \\
\hline
-2 & -2 & -1 & 0 & 0 & 1 & 0
\end{array}$$

a. Pivot on the 1 in row one, column one.

$$\begin{array}{cccccc|cl}
x_1 & x_2 & x_3 & s_1 & s_2 & z & \\
1 & 1 & 1 & 1 & 0 & 0 & 12 & \\
0 & -1 & 0 & -2 & 1 & 0 & 6 & -2R_1 + R_2 \\
\hline
0 & 0 & 1 & 2 & 0 & 1 & 24 & 2R_1 + R_3
\end{array}$$

The maximum is 24 when
$x_1 = 12, \ x_2 = 0, \ x_3 = 0, \ s_1 = 0, \ s_2 = 6.$

b. Pivot on the 1 in row one, column two.

$$\begin{array}{cccccc|cl}
x_1 & x_2 & x_3 & s_1 & s_2 & z & \\
1 & 1 & 1 & 1 & 0 & 0 & 12 & \\
1 & 0 & 1 & -1 & 1 & 0 & 18 & -1R_1 + R_2 \\
\hline
0 & 0 & 1 & 2 & 0 & 1 & 24 & 2R_1 + R_3
\end{array}$$

The maximum is 24 when
$x_1 = 0, \ x_2 = 12, \ x_3 = 0, \ s_1 = 0, \ s_2 = 18.$

c. This problem has a unique maximum value of z, which is 24, but it occurs at two different basic feasible solutions.

Section 7.5 Maximization Applications

1. Let x_1 = the number of Siamese cats and
let x_2 = the number of Persian cats.

The problem is to maximize $z = 12x_1 + 10x_2$
subject to:
$$2x_1 + x_2 \le 90$$
$$x_1 + 2x_2 \le 80$$
$$x_1 + x_2 \le 50$$
$$x_1 \ge 0,\ x_2 \ge 0.$$

There are three constraints to be changed into equalities, so introduce three slack variables, s_1, s_2, and s_3. The problem can now be restated as:

Find $x_1 \ge 0$, $x_2 \ge 0$, $s_1 \ge 0$, $s_2 \ge 0$, $s_3 \ge 0$, such that
$$2x_1 + x_2 + s_1 \qquad\qquad = 90$$
$$x_1 + 2x_2 \qquad + s_2 \qquad = 80$$
$$x_1 + x_2 \qquad\qquad + s_3 = 50$$
and $z = 12x_1 + 10x_2$ is maximized.

The initial simplex tableau is

$$
\begin{array}{ccccc}
x_1 & x_2 & s_1 & s_2 & s_3 \\
\end{array}
$$
$$
\left[\begin{array}{ccccc|c}
2 & 1 & 1 & 0 & 0 & 90 \\
1 & 2 & 0 & 1 & 0 & 80 \\
1 & 1 & 0 & 0 & 1 & 50 \\
\hline
-12 & -10 & 0 & 0 & 0 & 0 \\
\end{array}\right].
$$

3. Let x_1 = the number of kg of P,
x_2 = the number of kg of Q,
x_3 = the number of kg of R, and
x_4 = the number of kg of S

The constraints are
$$.375x_3 + .625x_4 \le 500$$
$$.75x_2 + .50x_3 + .375x_4 \le 600$$
$$x_1 + .25x_2 + .125x_3 \qquad \le 300$$
$$x_1 \ge 0,\ x_2 \ge 0,\ x_3 \ge 0,\ x_4 \ge 0.$$

(Notice the food contents are given in *percent* of nutrient per kilogram.) The objective function to maximize is $z = 90x_1 + 70x_2 + 60x_3 + 50x_4$.

The initial tableau is

$$
\begin{array}{ccccccc}
x_1 & x_2 & x_3 & x_4 & s_1 & s_2 & s_3 \\
\end{array}
$$
$$
\left[\begin{array}{ccccccc|c}
0 & 0 & .375 & .625 & 1 & 0 & 0 & 500 \\
0 & .75 & .5 & .375 & 0 & 1 & 0 & 600 \\
1 & .25 & .125 & 0 & 0 & 0 & 1 & 300 \\
\hline
-90 & -70 & -60 & -50 & 0 & 0 & 0 & 0 \\
\end{array}\right]
$$

5. a. The information is contained in the table.

	Aluminum	Steel	Profit
1 Speed	12	20	$8
3 Speed	21	30	$12
10 Speed	16	40	$24
Amount Available	42,000	91,800	

Let x_1 = number of 1-speed bikes,
let x_2 = number of 3-speed bikes, and
let x_3 = number of 10-speed bikes.

The problem is to maximize
$z = 8x_1 + 12x_2 + 24x_3$
subject to:
$$12x_1 + 21x_2 + 16x_3 \le 42,000$$
$$20x_1 + 30x_2 + 40x_3 \le 91,800$$
$$x_1 \ge 0,\ x_2 \ge 0,\ x_3 \ge 0.$$

The initial simplex tableau is

$$
\begin{array}{ccccc}
x_1 & x_2 & x_3 & s_1 & s_2 \\
\end{array}
$$
$$
\left[\begin{array}{ccccc|c}
12 & 21 & 16 & 1 & 0 & 42,000 \\
20 & 30 & 40 & 0 & 1 & 91,800 \\
\hline
-8 & -12 & -24 & 0 & 0 & 0 \\
\end{array}\right].
$$

Pivot on the 40 in row two, column three.

$$
\begin{array}{ccccc}
x_1 & x_2 & x_3 & s_1 & s_2 \\
\end{array}
$$
$$
\left[\begin{array}{ccccc|c}
12 & 21 & 16 & 1 & 0 & 42,000 \\
\frac{1}{2} & \frac{3}{4} & 1 & 0 & \frac{1}{40} & 2295 \\
\hline
-8 & -12 & -24 & 0 & 0 & 0 \\
\end{array}\right] \begin{array}{c} \\ \frac{1}{40}R_2 \\ \\ \end{array}
$$

$$
\begin{array}{ccccc}
x_1 & x_2 & x_3 & s_1 & s_2 \\
\end{array}
$$
$$
\left[\begin{array}{ccccc|c}
4 & 9 & 0 & 1 & -\frac{2}{5} & 5280 \\
\frac{1}{2} & \frac{3}{4} & 1 & 0 & \frac{1}{40} & 2295 \\
\hline
4 & 6 & 0 & 0 & \frac{3}{5} & 55,080 \\
\end{array}\right]
\begin{array}{l}
-16R_2 + R_1 \\
\frac{1}{40}R_2 \\
24R_2 + R_3 \\
\end{array}
$$

From the final tableau, the maximum is 55,080 when $x_3 = 2295$, $s_1 = 5280$, and $x_1 = x_2 = s_2 = 0$. Thus, the manufacturer should make no 1-speed or 3-speed bicycles, and should make 2295 10-speed bicycles for a maximum profit of $55,080.

b. In the optimal solution, $s_1 = 5280$ and $s_2 = 0$. $s_1 = 5280$ means 5280 units of aluminum should be left unused. $s_2 = 0$ means all of the steel is used.

7. a. Let x_1 = the number of minutes allotted to the senator, let x_2 = the number of minutes allotted to the congresswoman, and let x_3 = the number of minutes allotted to the governor.

Maximize $z = 40x_1 + 60x_2 + 50x_3$

Subject to: $x_1 + x_2 + x_3 \le 27$

$$x_1 \ge 2x_3$$
$$x_1 + x_3 \ge 2x_2$$
$$x_1 \ge 0,\ x_2 \ge 0,\ x_3 \ge 0$$

Rewrite problem in standard maximum form.

Maximize $z = 40x_1 + 60x_2 + 50x_3$

$$x_1 + x_2 + x_3 \le 27$$
$$-x_1 \qquad + 2x_3 \le 0$$
$$-x_1 + 2x_2 - x_3 \le 0$$
$$x_1 \ge 0,\ x_2 \ge 0,\ x_3 \ge 0$$

The initial simplex tableau is

$$
\begin{array}{cccccc}
x_1 & x_2 & x_3 & s_1 & s_2 & s_3 \\
\end{array}
$$
$$
\left[
\begin{array}{cccccc|c}
1 & 1 & 1 & 1 & 0 & 0 & 27 \\
-1 & 0 & 2 & 0 & 1 & 0 & 0 \\
-1 & \boxed{2} & -1 & 0 & 0 & 1 & 0 \\
\hline
-40 & -60 & -50 & 0 & 0 & 0 & 0 \\
\end{array}
\right]
$$

Pivot on the 2 in row three, column two.

Multiply row three by $\frac{1}{2}$ and complete the pivot to get

$$
\begin{array}{cccccc}
x_1 & x_2 & x_3 & s_1 & s_2 & s_3 \\
\end{array}
$$
$$
\left[
\begin{array}{cccccc|c}
\frac{3}{2} & 0 & \frac{3}{2} & 1 & 0 & -\frac{1}{2} & 27 \\
-1 & 0 & \boxed{2} & 0 & 1 & 0 & 0 \\
-\frac{1}{2} & 1 & -\frac{1}{2} & 0 & 0 & \frac{1}{2} & 0 \\
\hline
-70 & 0 & -80 & 0 & 0 & 30 & 0 \\
\end{array}
\right]
\begin{array}{l}
-R_3 + R_1 \\
\\
\frac{1}{2}R_3 \\
60R_3 + R_4 \\
\end{array}
$$

Pivot on the 2 in row two, column three.

Multiply row 2 by $\frac{1}{2}$ and complete the pivot to get

$$
\begin{array}{cccccc}
x_1 & x_2 & x_3 & s_1 & s_2 & s_3 \\
\end{array}
$$
$$
\left[
\begin{array}{cccccc|c}
\boxed{\frac{9}{4}} & 0 & 0 & 1 & -\frac{3}{4} & -\frac{1}{2} & 27 \\
-\frac{1}{2} & 0 & 1 & 0 & \frac{1}{2} & 0 & 0 \\
-\frac{3}{4} & 1 & 0 & 0 & \frac{1}{4} & \frac{1}{2} & 0 \\
\hline
-110 & 0 & 0 & 0 & 40 & 30 & 0 \\
\end{array}
\right]
\begin{array}{l}
-\frac{3}{2}R_2 + R_1 \\
\frac{1}{2}R_2 \\
\frac{1}{2}R_2 + R_3 \\
80R_2 + R_4 \\
\end{array}
$$

Pivot on the $\frac{9}{4}$ in row one, column one.

$$
\begin{array}{cccccc}
x_1 & x_2 & x_3 & s_1 & s_2 & s_3 \\
\end{array}
$$
$$
\left[
\begin{array}{cccccc|c}
1 & 0 & 0 & \frac{4}{9} & -\frac{1}{3} & -\frac{2}{9} & 12 \\
0 & 0 & 1 & \frac{2}{9} & \frac{1}{3} & -\frac{1}{9} & 6 \\
0 & 1 & 0 & \frac{1}{3} & 0 & \frac{1}{3} & 9 \\
\hline
0 & 0 & 0 & \frac{440}{9} & \frac{10}{3} & \frac{50}{9} & 1320 \\
\end{array}
\right]
\begin{array}{l}
\frac{4}{9}R_1 \\
\frac{1}{2}R_1 + R_2 \\
\frac{3}{4}R_1 + R_3 \\
110R_1 + R_4 \\
\end{array}
$$

The senator should get 12 minutes of airtime, the congresswoman 9 minutes and the governor 6 minutes for a maximum of 1,320,000 viewers.

b. In the optimal solution, $s_1 = 0$, $s_2 = 0$ and $s_3 = 0$. $s_1 = 0$ means that all of the 27 minutes will be used by the politicians, i.e. there is no slack. $s_2 = 0$ means that the senator has exactly twice as much time as the governor. $s_3 = 0$ means that the senator and governor have exactly twice as much time as the congresswoman.

9. a. Let x_1 = the number of Flexscan tvs and let x_2 = the number of Panoramic I tvs. We want to maximize the profit function, $z = 350x_1 + 500x_2$ subject to the constraints

$$5x_1 + 7x_2 \le 3600$$
$$x_1 + 2x_2 \le 900$$
$$4x_1 + 4x_2 \le 2600.$$

The third constraint simplifies to

$$x_1 + x_2 \le 650$$

The initial simplex tableau is

$$
\begin{array}{ccccc}
x_1 & x_2 & s_1 & s_2 & s_3 \\
\end{array}
$$
$$
\left[
\begin{array}{ccccc|c}
5 & 7 & 1 & 0 & 0 & 3600 \\
1 & \boxed{2} & 0 & 1 & 0 & 900 \\
1 & 1 & 0 & 0 & 1 & 650 \\
\hline
-350 & -500 & 0 & 0 & 0 & 0 \\
\end{array}
\right]
$$

Pivot on the 2 in row two, column two. First divide each entry in row 2 by 2.

$$
\begin{array}{ccccc}
x_1 & x_2 & s_1 & s_2 & s_3 \\
\end{array}
$$
$$
\left[
\begin{array}{ccccc|c}
5 & 7 & 1 & 0 & 0 & 3600 \\
\frac{1}{2} & 1 & 0 & \frac{1}{2} & 0 & 450 \\
1 & 1 & 0 & 0 & 1 & 650 \\
\hline
-350 & -500 & 0 & 0 & 0 & 0 \\
\end{array}
\right]
$$

(continued on next page)

(continued from page 220)

$$\begin{bmatrix} \begin{array}{ccccc|c} x_1 & x_2 & s_1 & s_2 & s_3 & \\ \boxed{\frac{3}{2}} & 0 & 1 & -\frac{7}{2} & 0 & 450 \\ \frac{1}{2} & 1 & 0 & \frac{1}{2} & 0 & 450 \\ \frac{1}{2} & 0 & 0 & -\frac{1}{2} & 1 & 200 \\ \hline -100 & 0 & 0 & 250 & 0 & 225,000 \end{array} \end{bmatrix} \begin{array}{l} -7R_2 + R_1 \\ \\ -R_2 + R_3 \\ 500R_2 + R_4 \end{array}$$

Now pivot on $\frac{3}{2}$ in row 1 column 1. First multiply each entry in row 1 by $\frac{1}{3}$.

$$\begin{bmatrix} \begin{array}{ccccc|c} x_1 & x_2 & s_1 & s_2 & s_3 & \\ \frac{1}{2} & 0 & \frac{1}{3} & -\frac{7}{6} & 0 & 150 \\ \frac{1}{2} & 1 & 0 & \frac{1}{2} & 0 & 450 \\ \frac{1}{2} & 0 & 0 & -\frac{1}{2} & 1 & 200 \\ \hline -100 & 0 & 0 & 250 & 0 & 225,000 \end{array} \end{bmatrix}$$

$$\begin{bmatrix} \begin{array}{ccccc|c} x_1 & x_2 & s_1 & s_2 & s_3 & \\ \frac{1}{2} & 0 & \frac{1}{3} & -\frac{7}{6} & 0 & 150 \\ 0 & 1 & -\frac{1}{3} & \frac{5}{3} & 0 & 300 \\ 0 & 0 & -\frac{1}{3} & \frac{2}{3} & 1 & 50 \\ \hline 0 & 0 & \frac{200}{3} & \frac{50}{3} & 0 & 255,000 \end{array} \end{bmatrix} \begin{array}{l} \\ -R_1 + R_2 \\ -R_1 + R_3 \\ 200R_1 + R_4 \end{array}$$

$$\begin{bmatrix} \begin{array}{ccccc|c} x_1 & x_2 & s_1 & s_2 & s_3 & \\ 1 & 0 & \frac{2}{3} & -\frac{7}{3} & 0 & 300 \\ 0 & 1 & -\frac{1}{3} & \frac{5}{3} & 0 & 300 \\ 0 & 0 & -\frac{1}{3} & \frac{2}{3} & 1 & 50 \\ \hline 0 & 0 & \frac{200}{3} & \frac{50}{3} & 0 & 255,000 \end{array} \end{bmatrix} 2R_1$$

From the final tableau, we have $x_1 = 300$ and $x_2 = 300$. Thus, the company should produce 300 Flexscan tvs and 300 Panoramic I tvs, for a maximum profit of $255,000

b. s_1 and s_2 both are zero. $4(300) + 4(300) + s_3 = 2600 \Rightarrow s_3 = 200$

Thus, there are 200 unused hours in the testing and packaging department.

11. Let x_1, x_2, and x_3 = the number of newspapers, radio, and TV ads respectively.

Maximize $z = 2000x_1 + 1200x_2 + 10,000x_3$

subject to:

$400x_1 + 200x_2 + 1200x_3 \le 8000$

$x_1 \le 20$, $x_2 \le 30$, $x_3 \le 6$

$x_1 \ge 0$, $x_2 \ge 0$, $x_3 \ge 0$.

The initial simplex tableau is

$$\begin{bmatrix} \begin{array}{cccccccc|c} x_1 & x_2 & x_3 & s_1 & s_2 & s_3 & s_4 & \\ 400 & 200 & 1200 & 1 & 0 & 0 & 0 & 8000 \\ 1 & 0 & 0 & 0 & 1 & 0 & 0 & 20 \\ 0 & 1 & 0 & 0 & 0 & 1 & 0 & 30 \\ 0 & 0 & \boxed{1} & 0 & 0 & 0 & 1 & 6 \\ \hline -2000 & -1200 & -10,000 & 0 & 0 & 0 & 0 & 0 \end{array} \end{bmatrix}$$

(continued on next page)

(continued from page 221)

Pivot on the 1 in row four, column three.

x_1	x_2	x_3	s_1	s_2	s_3	s_4		
400	200	0	1	0	0	-1200	800	$-1200R_4 + R_1$
1	0	0	0	1	0	0	20	
0	1	0	0	0	1	0	30	
0	0	1	0	0	0	1	6	
-2000	-1200	0	0	0	0	10,000	60,000	$10{,}000R_4 + R_5$

Pivot on the 400 in row one, column one.

x_1	x_2	x_3	s_1	s_2	s_3	s_4		
1	$\frac{1}{2}$	0	$\frac{1}{400}$	0	0	-3	2	$\frac{1}{400}R_1$
1	0	0	0	1	0	0	20	
0	1	0	0	0	1	0	30	
0	0	1	0	0	0	1	6	
-2000	-1200	0	0	0	0	10,000	60,000	

x_1	x_2	x_3	s_1	s_2	s_3	s_4		
1	$\frac{1}{2}$	0	$\frac{1}{400}$	0	0	-3	2	
0	$-\frac{1}{2}$	0	$-\frac{1}{400}$	1	0	3	18	$-1R_1 + R_2$
0	1	0	0	0	1	0	30	
0	0	1	0	0	0	1	6	
0	-200	0	5	0	0	4000	64,000	$2000R_1 + R_5$

Pivot on the $\frac{1}{2}$ in row one, column two.

x_1	x_2	x_3	s_1	s_2	s_3	s_4		
2	1	0	$\frac{1}{200}$	0	0	-6	4	$2R_1$
0	$-\frac{1}{2}$	0	$-\frac{1}{400}$	1	0	3	18	
0	1	0	0	0	1	0	30	
0	0	1	0	0	0	1	6	
0	-200	0	5	0	0	4000	64,000	

x_1	x_2	x_3	s_1	s_2	s_3	s_4		
2	1	0	$\frac{1}{200}$	0	0	-6	4	
1	0	0	0	1	0	0	20	$\frac{1}{2}R_1 + R_2$
-2	0	0	$-\frac{1}{200}$	0	1	6	26	$-1R_1 + R_3$
0	0	1	0	0	0	1	6	
400	0	0	6	0	0	2800	64,800	$200R_1 + R_5$

From the final tableau, we have $x_1 = 0$, $x_2 = 4$, and $x_3 = 6$.

No newspaper ads, 4 radio ads, and 6 TV ads will give a maximum exposure of 64,800 people.

13. a. Let x_1 = the number of fund-raising parties, let x_2 = the number of mailings, and let x_3 = the number of dinner parties.

Maximize $z = 200{,}000x_1 + 100{,}000x_2 + 600{,}000x_3$

Subject to:
$$x_1 + x_2 + x_3 \le 25$$
$$3000x_1 + 1000x_2 + 12{,}000x_3 \le 102{,}000$$
$$x_1 \ge 0,\ x_2 \ge 0,\ x_3 \ge 0$$

The initial simplex tableau is

$$
\begin{array}{c}
\begin{array}{ccccc}
x_1 & x_2 & x_3 & s_1 & s_2
\end{array} \\
\left[
\begin{array}{ccccc|c}
1 & 1 & 1 & 1 & 0 & 25 \\
3000 & 1000 & \boxed{12{,}000} & 0 & 1 & 102{,}000 \\
\hline
-200{,}000 & -100{,}000 & -600{,}000 & 0 & 0 & 0
\end{array}
\right]
\end{array}
$$

Pivot on the 12,000, row two, column three.

$$
\begin{array}{c}
\begin{array}{ccccc}
x_1 & x_2 & x_3 & s_1 & s_2
\end{array} \\
\left[
\begin{array}{ccccc|c}
\frac{3}{4} & \boxed{\frac{11}{12}} & 0 & 1 & -\frac{1}{12{,}000} & \frac{33}{2} \\
\frac{1}{4} & \frac{1}{12} & 1 & 0 & \frac{1}{12{,}000} & \frac{17}{2} \\
\hline
-50{,}000 & -50{,}000 & 0 & 0 & 50 & 5{,}100{,}000
\end{array}
\right]
\end{array}
\begin{array}{l}
-R_2 + R_1 \\
\frac{1}{12{,}000}R_2 \\
600{,}000R_2 + R_3
\end{array}
$$

Pivot on $\frac{11}{12}$ in row one, column two.

$$
\begin{array}{c}
\begin{array}{ccccc}
x_1 & x_2 & x_3 & s_1 & s_2
\end{array} \\
\left[
\begin{array}{ccccc|c}
\boxed{\frac{9}{11}} & 1 & 0 & \frac{12}{11} & -\frac{1}{11{,}000} & 18 \\
\frac{2}{11} & 0 & 1 & -\frac{1}{11} & \frac{1}{11{,}000} & 7 \\
\hline
-\frac{100{,}000}{11} & 0 & 0 & \frac{600{,}000}{11} & \frac{500}{11} & 6{,}000{,}000
\end{array}
\right]
\end{array}
\begin{array}{l}
\frac{12}{11}R_1 \\
-\frac{1}{12}R_1 + R_2 \\
50{,}000R_1 + R_3
\end{array}
$$

Pivot on $\frac{9}{11}$ in row one, column one.

$$
\begin{array}{c}
\begin{array}{ccccc}
x_1 & x_2 & x_3 & s_1 & s_2
\end{array} \\
\left[
\begin{array}{ccccc|c}
1 & \frac{11}{9} & 0 & \frac{4}{3} & -\frac{1}{9000} & 22 \\
0 & -\frac{2}{9} & 1 & -\frac{1}{3} & \frac{1}{9000} & 3 \\
\hline
0 & \frac{100{,}000}{9} & 0 & \frac{200{,}000}{3} & \frac{400}{9} & 6{,}200{,}000
\end{array}
\right]
\end{array}
\begin{array}{l}
\frac{11}{9}R_1 \\
-\frac{2}{11}R_1 + R_2 \\
\frac{100{,}000}{11}R_1 + R_3
\end{array}
$$

The party should plan 22 fund raising parties and 3 dinner parties and no mailings to raise a maximum of $6,200,000.

b. Answers will vary.

15. Let x_1 = the number of hours running, x_2 = the number of hours biking, and x_3 = the number of hours walking. We are looking to maximize $z = 531x_1 + 472x_2 + 354x_3$ subject to the constraints

$$x_1 + x_2 + x_3 \le 15$$
$$x_1 \qquad\qquad \le 3$$
$$2x_2 - x_3 \le 0$$

$x_1 \ge 0, x_2 \ge 0, x_3 \ge 0.$

The initial simplex tableau is

$$
\begin{array}{ccccccc|c}
x_1 & x_2 & x_3 & s_1 & s_2 & s_3 & \\
1 & 1 & 1 & 1 & 0 & 0 & 15 \\
\boxed{1} & 0 & 0 & 0 & 1 & 0 & 3 \\
0 & 2 & -1 & 0 & 0 & 1 & 0 \\
\hline
-531 & -472 & -354 & 0 & 0 & 0 & 0
\end{array}
$$

Pivot on the 1 in row 2, column 1.

$$
\begin{array}{ccccccc|cl}
x_1 & x_2 & x_3 & s_1 & s_2 & s_3 & \\
0 & 1 & 1 & 1 & -1 & 0 & 12 & -R_2 + R_1 \\
1 & 0 & 0 & 0 & 1 & 0 & 3 & \\
0 & \boxed{2} & -1 & 0 & 0 & 1 & 0 & \\
\hline
0 & -472 & -354 & 0 & 531 & 0 & 1593 & 531R_2 + R_4
\end{array}
$$

Pivot on the 2 in row 3, column 2.

$$
\begin{array}{ccccccc|cl}
x_1 & x_2 & x_3 & s_1 & s_2 & s_3 & \\
0 & 0 & \boxed{3} & 2 & -2 & -1 & 24 & -R_3 + 2R_1 \\
1 & 0 & 0 & 0 & 1 & 0 & 3 & \\
0 & 2 & -1 & 0 & 0 & 1 & 0 & \\
\hline
0 & 0 & -590 & 0 & 531 & 236 & 1593 & 236R_2 + R_4
\end{array}
$$

Finally, pivot on the 3 in row 1, column 3.

$$
\begin{array}{ccccccc|cl}
x_1 & x_2 & x_3 & s_1 & s_2 & s_3 & \\
0 & 0 & 3 & 2 & -2 & -1 & 24 & \\
1 & 0 & 0 & 0 & 1 & 0 & 3 & \\
0 & 6 & 0 & 2 & -2 & 2 & 24 & R_1 + 3R_3 \\
\hline
0 & 0 & 0 & 1180 & 413 & 118 & 18,939 & 590R_1 + 3R_4
\end{array}
$$

$$
\begin{array}{ccccccc|cl}
x_1 & x_2 & x_3 & s_1 & s_2 & s_3 & \\
0 & 0 & 1 & \frac{2}{3} & -\frac{2}{3} & -\frac{1}{3} & 8 & \frac{1}{3}R_1 \\
1 & 0 & 0 & 0 & 1 & 0 & 3 & \\
0 & 1 & 0 & \frac{1}{3} & -\frac{1}{3} & \frac{1}{3} & 4 & \frac{1}{6}R_3 \\
\hline
0 & 0 & 0 & 1180 & 413 & 118 & 18,939 &
\end{array}
$$

Thus, $x_1 = 3$, $x_2 = 4$, and $x_3 = 8$, so

$z = 531(3) + 472(4) + 354(8) = 6313.$

Rachel should run 3 hours, bike 4 hours, and walk 8 hours for a maximum calorie expenditure of 6313 calories.

17. Using the data given, the problem should be stated as follows: Maximize $5x_1 + 4x_2 + 3x_3$ subject to:

$$2x_1 + 3x_2 + x_3 \le 400$$
$$4x_1 + 2x_2 + 3x_3 \le 600$$
$$x_1 \ge 0, \ x_2 \ge 0, \ x_3 \ge 0.$$

when x_1 = the number of type A lamps, x_2 = the number of type B lamps, and x_3 = the number of type C lamps.

a. The coefficients of the objective function would be choice (3): 5, 4, 3

b. The constraints in the model would be choice (4): 400, 600

c. The constraint imposed by the available number of person-hours in department I could be expressed as choice (3):

$$2x_1 + 3x_2 + 1x_3 \le 400$$

19. From Exercise 1, the initial simplex tableau is

$$
\begin{array}{ccccc|c}
x_1 & x_2 & s_1 & s_2 & s_3 & \\
\boxed{2} & 1 & 1 & 0 & 0 & 90 \\
1 & 2 & 0 & 1 & 0 & 80 \\
1 & 1 & 0 & 0 & 1 & 50 \\
\hline
-12 & -10 & 0 & 0 & 0 & 0
\end{array}
$$

Pivot on the 2 in row one, column one.

$$
\begin{array}{ccccc|cl}
x_1 & x_2 & s_1 & s_2 & s_3 & \\
1 & \frac{1}{2} & \frac{1}{2} & 0 & 0 & 45 & \frac{1}{2}R_1 \\
1 & 2 & 0 & 1 & 0 & 80 & \\
1 & 1 & 0 & 0 & 1 & 50 & \\
\hline
-12 & -10 & 0 & 0 & 0 & 0
\end{array}
$$

$$
\begin{array}{ccccc|cl}
x_1 & x_2 & s_1 & s_2 & s_3 & \\
1 & \frac{1}{2} & \frac{1}{2} & 0 & 0 & 45 & \\
0 & \frac{3}{2} & -\frac{1}{2} & 1 & 0 & 35 & -1R_1 + R_2 \\
0 & \boxed{\frac{1}{2}} & -\frac{1}{2} & 0 & 1 & 5 & -1R_1 + R_3 \\
\hline
0 & -4 & 6 & 0 & 0 & 540 & 12R_1 + R_4
\end{array}
$$

Pivot on the $\frac{1}{2}$ in row three, column two.

$$
\begin{array}{ccccc|cl}
x_1 & x_2 & s_1 & s_2 & s_3 & \\
1 & \frac{1}{2} & \frac{1}{2} & 0 & 0 & 45 & \\
0 & \frac{3}{2} & -\frac{1}{2} & 1 & 0 & 35 & \\
0 & 1 & -1 & 0 & 2 & 10 & 2R_3 \\
\hline
0 & -4 & 6 & 0 & 0 & 540
\end{array}
$$

(continued on next page)

(continued from page 224)

$$
\begin{array}{ccccc}
x_1 & x_2 & s_1 & s_2 & s_3 \\
\end{array}
$$

$$
\left[\begin{array}{ccccc|c}
1 & 0 & 1 & 0 & -1 & 40 \\
0 & 0 & 1 & 1 & -3 & 20 \\
0 & 1 & -1 & 0 & 2 & 10 \\
\hline
0 & 0 & 2 & 0 & 8 & 580
\end{array}\right]
\begin{array}{l}
-\frac{1}{2}R_3 + R_1 \\
-\frac{3}{2}R_3 + R_2 \\
\\
4R_3 + R_4
\end{array}
$$

The breeder should raise 40 Siamese and 10 Persian cats for a maximum gross income of $580.

21. The initial simplex tableau is

$$
\begin{array}{ccccccc}
x_1 & x_2 & x_3 & x_4 & s_1 & s_2 & s_3 \\
\end{array}
$$

$$
\left[\begin{array}{ccccccc|c}
0 & 0 & .375 & .625 & 1 & 0 & 0 & 500 \\
0 & .75 & .5 & .375 & 0 & 1 & 0 & 600 \\
1 & .25 & .125 & 0 & 0 & 0 & 1 & 300 \\
\hline
-90 & -70 & -60 & -50 & 0 & 0 & 0 & 0
\end{array}\right]
$$

Using the SIMPLEX program on a TI-84 Plus, we get the maximum total growth value is 87,454.55 when 163.6 kg of food P, none of food Q, 1090.9 kg of food R, and 145.5 kg of food S are used.

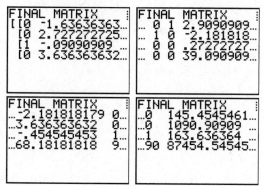

Section 7.6   The Simplex Method: Duality and Minimization

1. The transpose of

$$
\begin{bmatrix}
3 & -4 & 5 \\
1 & 10 & 7 \\
0 & 3 & 6
\end{bmatrix}
\text{ is }
\begin{bmatrix}
3 & 1 & 0 \\
-4 & 10 & 3 \\
5 & 7 & 6
\end{bmatrix}.
$$

3. The transpose of

$$
\begin{bmatrix}
3 & 0 & 14 & -5 & 3 \\
4 & 17 & 8 & -6 & 1
\end{bmatrix}
\text{ is }
\begin{bmatrix}
3 & 4 \\
0 & 17 \\
14 & 8 \\
-5 & -6 \\
3 & 1
\end{bmatrix}.
$$

5. Minimize $w = 3y_1 + 5y_2$

subject to: $3y_1 + y_2 \geq 4$
$\qquad\qquad -y_1 + 2y_2 \geq 6$
$\qquad\qquad y_1 \geq 0, \; y_2 \geq 0.$

The augmented matrix is

$$
\begin{bmatrix}
3 & 1 & 4 \\
-1 & 2 & 6 \\
\hline
3 & 5 & 0
\end{bmatrix}.
$$

The transpose of this matrix is

$$
\begin{bmatrix}
3 & -1 & 3 \\
1 & 2 & 5 \\
\hline
4 & 6 & 0
\end{bmatrix}.
$$

The entries in this second matrix can be used to write the following dual maximization problem:

Maximize $z = 4x_1 + 6x_2$

subject to: $3x_1 - x_2 \leq 3$
$\qquad\qquad x_1 + 2x_2 \leq 5$
$\qquad\qquad x_1 \geq 0, \; x_2 \geq 0.$

7. Minimize $w = 2y_1 + 8y_2$

subject to: $y_1 + 7y_2 \geq 18$
$\qquad\qquad 4y_1 + y_2 \geq 15$
$\qquad\qquad 5y_1 + 3y_2 \geq 20$
$\qquad\qquad y_1 \geq 0, \; y_2 \geq 0.$

The augmented matrix is

$$
\begin{bmatrix}
1 & 7 & 18 \\
4 & 1 & 15 \\
5 & 3 & 20 \\
2 & 8 & 0
\end{bmatrix}.
$$

The transpose of this matrix is

$$
\begin{bmatrix}
1 & 4 & 5 & 2 \\
7 & 1 & 3 & 8 \\
\hline
18 & 15 & 20 & 0
\end{bmatrix}.
$$

The entries of this second matrix can be used to write the following dual maximization problem:

Maximize $z = 18x_1 + 15x_2 + 20x_3$

subject to: $x_1 + 4x_2 + 5x_3 \leq 2$
$\qquad\qquad 7x_1 + x_2 + 3x_3 \leq 8$
$\qquad\qquad x_1 \geq 0, \; x_2 \geq 0, \; x_3 \geq 0.$

9. Minimize $w = 5y_1 + y_2 + 3y_3$

subject to: $7y_1 + 6y_2 + 8y_3 \geq 18$
$4y_1 + 5y_2 + 10y_3 \geq 20$
$y_1 \geq 0, \ y_2 \geq 0, \ y_3 \geq 0.$

The augmented matrix is

$$\left[\begin{array}{ccc|c} 7 & 6 & 8 & 18 \\ 4 & 5 & 10 & 20 \\ \hline 5 & 1 & 3 & 0 \end{array}\right].$$

The transpose of this matrix is

$$\left[\begin{array}{cc|c} 7 & 4 & 5 \\ 6 & 5 & 1 \\ 8 & 10 & 3 \\ \hline 18 & 20 & 0 \end{array}\right].$$

The entries of this second matrix can be used to write the following dual maximization problem.

Maximize $z = 18x_1 + 20x_2$

subject to: $7x_1 + 4x_2 \leq 5$
$6x_1 + 5x_2 \leq 1$
$8x_1 + 10x_2 \leq 3$
$x_1 \geq 0, \ x_2 \geq 0.$

11. Minimize $w = 8y_1 + 9y_2 + 3y_3$

subject to: $y_1 + y_2 + y_3 \geq 5$
$y_1 + y_2 \qquad \geq 4$
$2y_1 + y_2 + 3y_3 \geq 15$
$y_1 \geq 0, \ y_2 \geq 0, \ y_3 \geq 0.$

The augmented matrix is

$$\left[\begin{array}{ccc|c} 1 & 1 & 1 & 5 \\ 1 & 1 & 0 & 4 \\ 2 & 1 & 3 & 15 \\ \hline 8 & 9 & 3 & 0 \end{array}\right].$$

The transpose of this matrix is

$$\left[\begin{array}{ccc|c} 1 & 1 & 2 & 8 \\ 1 & 1 & 1 & 9 \\ 1 & 0 & 3 & 3 \\ \hline 5 & 4 & 15 & 0 \end{array}\right].$$

The entries of this second matrix can be used to write the following dual maximization problem:

Maximize $z = 5x_1 + 4x_2 + 15x_3$

subject to: $x_1 + x_2 + 2x_3 \leq 8$
$x_1 + x_2 + x_3 \leq 9$
$x_1 \qquad + 3x_3 \leq 3$
$x_1 \geq 0, \ x_2 \geq 0, \ x_3 \geq 0.$

13. From exercise 9, we are to minimize

$w = 5y_1 + y_2 + 3y_3$ subject to the constraints

$7y_1 + 6y_2 + 8y_3 \geq 18$
$4y_1 + 5y_2 + 10y_3 \geq 20$
$y_1 \geq 0, \ y_2 \geq 0, \ y_3 \geq 0$

We write the dual maximization problem, maximize $z = 18x_1 + 20x_2$

subject to: $7x_1 + 4x_2 \leq 5$
$6x_1 + 5x_2 \leq 1$
$8x_1 + 10x_2 \leq 3$
$x_1 \geq 0, \ x_2 \geq 0.$

The initial simplex tableau is

$$\begin{array}{cccccc} x_1 & x_2 & s_1 & s_2 & s_3 & \\ \left[\begin{array}{ccccc|c} 7 & 4 & 1 & 0 & 0 & 5 \\ 6 & \boxed{5} & 0 & 1 & 0 & 1 \\ 8 & 10 & 0 & 0 & 1 & 3 \\ \hline -18 & -20 & 0 & 0 & 0 & 0 \end{array}\right] \end{array}.$$

Pivot on the 5 in row 2, column 2.

$$\begin{array}{ccccc} x_1 & x_2 & s_1 & s_2 & s_3 & \\ \left[\begin{array}{ccccc|c} 11 & 0 & 5 & -4 & 0 & 21 \\ 6 & 5 & 0 & 1 & 0 & 1 \\ -4 & 0 & 0 & -2 & 1 & 1 \\ \hline 6 & 0 & 0 & 4 & 0 & 4 \end{array}\right] \begin{array}{l} -4R_2 + 5R_1 \\ \\ -2R_2 + R_3 \\ 4R_2 + R_4 \end{array} \end{array}$$

From the final tableau, the minimum is 4 for $y_1 = 0, \ y_2 = 4,$ and $y_3 = 0.$

15. From exercise 11, we are to minimize
$$w = 8y_1 + 9y_2 + 3y_3$$
subject to:
$$y_1 + y_2 + y_3 \geq 5$$
$$y_1 + y_2 \geq 4$$
$$2y_1 + y_2 + 3y_3 \geq 15$$
$$y_1 \geq 0,\ y_2 \geq 0,\ y_3 \geq 0.$$

We write the dual maximization problem,
maximize $z = 5x_1 + 4x_2 + 15x_3$
subject to:
$$x_1 + x_2 + 2x_3 \leq 8$$
$$x_1 + x_2 + x_3 \leq 9$$
$$x_1 + 3x_3 \leq 3$$
$$x_1 \geq 0,\ x_2 \geq 0,\ x_3 \geq 0.$$

The initial simplex tableau is

$$
\begin{array}{cccccc}
x_1 & x_2 & x_3 & s_1 & s_2 & s_3 \\
\end{array}
$$

$$
\left[\begin{array}{cccccc|c}
1 & 1 & 2 & 1 & 0 & 0 & 8 \\
1 & 1 & 1 & 0 & 1 & 0 & 9 \\
1 & 0 & \boxed{3} & 0 & 0 & 1 & 3 \\
\hline
-5 & -4 & -15 & 0 & 0 & 0 & 0 \\
\end{array}\right].
$$

Pivot on the 3 in row 3, column 3.

$$
\begin{array}{cccccc}
x_1 & x_2 & x_3 & s_1 & s_2 & s_3 \\
\end{array}
$$

$$
\left[\begin{array}{cccccc|c}
1 & 1 & 2 & 1 & 0 & 0 & 8 \\
1 & 1 & 1 & 0 & 1 & 0 & 9 \\
\frac{1}{3} & 0 & 1 & 0 & 0 & \frac{1}{3} & 1 \\
\hline
-5 & -4 & -15 & 0 & 0 & 0 & 0 \\
\end{array}\right]
\begin{array}{l}
\\ \\ \frac{1}{3}R_3 \\ \\
\end{array}
$$

$$
\begin{array}{cccccc}
x_1 & x_2 & x_3 & s_1 & s_2 & s_3 \\
\end{array}
$$

$$
\left[\begin{array}{cccccc|c}
\frac{1}{3} & \boxed{1} & 0 & 1 & 0 & -\frac{2}{3} & 6 \\
\frac{2}{3} & 1 & 0 & 0 & 1 & -\frac{1}{3} & 8 \\
\frac{1}{3} & 0 & 1 & 0 & 0 & \frac{1}{3} & 1 \\
\hline
0 & -4 & 0 & 0 & 0 & 5 & 15 \\
\end{array}\right]
\begin{array}{l}
-2R_3 + R_1 \\ -R_3 + R_2 \\ \\ 15R_3 + R_4 \\
\end{array}
$$

Now pivot on the 1 in row 1, column 2.

$$
\begin{array}{cccccc}
x_1 & x_2 & x_3 & s_1 & s_2 & s_3 \\
\end{array}
$$

$$
\left[\begin{array}{cccccc|c}
\frac{1}{3} & \boxed{1} & 0 & 1 & 0 & -\frac{2}{3} & 6 \\
\frac{2}{3} & 1 & 0 & 0 & 1 & -\frac{1}{3} & 8 \\
\frac{1}{3} & 0 & 1 & 0 & 0 & \frac{1}{3} & 1 \\
\hline
\frac{4}{3} & 0 & 0 & 4 & 0 & \frac{7}{3} & 39 \\
\end{array}\right]
\begin{array}{l}
\\ \\ \\ 4R_1 + R_4 \\
\end{array}
$$

From the final tableau, the minimum is 39 for
$y_1 = 4,\ y_2 = 0,$ and $y_3 = \frac{7}{3}.$

17. Minimize $w = 2y_1 + y_2 + 3y_3$
subject to:
$$y_1 + y_2 + y_3 \geq 100$$
$$2y_1 + y_2 \geq 50$$
$$y_1 \geq 0,\ y_2 \geq 0,\ y_3 \geq 0.$$

The augmented matrix is

$$
\left[\begin{array}{ccc|c}
1 & 1 & 1 & 100 \\
2 & 1 & 0 & 50 \\
\hline
2 & 1 & 3 & 0 \\
\end{array}\right].
$$

The transpose is

$$
\left[\begin{array}{cc|c}
1 & 2 & 2 \\
1 & 1 & 1 \\
1 & 0 & 3 \\
\hline
100 & 50 & 0 \\
\end{array}\right].
$$

The dual maximization problem is:
Maximize $z = 100x_1 + 50x_2$
subject to:
$$x_1 + 2x_2 \leq 2$$
$$x_1 + x_2 \leq 1$$
$$x_1 \leq 3$$
$$x_1 \geq 0,\ x_2 \geq 0.$$

The initial simplex tableau is

$$
\begin{array}{ccccc}
x_1 & x_2 & s_1 & s_2 & s_3 \\
\end{array}
$$

$$
\left[\begin{array}{ccccc|c}
1 & 2 & 1 & 0 & 0 & 2 \\
\boxed{1} & 1 & 0 & 1 & 0 & 1 \\
1 & 0 & 0 & 0 & 1 & 3 \\
\hline
-100 & -50 & 0 & 0 & 0 & 0 \\
\end{array}\right].
$$

Pivot on the 1 in row two, column one.

$$
\begin{array}{ccccc}
x_1 & x_2 & s_1 & s_2 & s_3 \\
\end{array}
$$

$$
\left[\begin{array}{ccccc|c}
0 & 1 & 1 & -1 & 0 & 1 \\
1 & 1 & 0 & 1 & 0 & 1 \\
0 & -1 & 0 & -1 & 1 & 2 \\
\hline
0 & 50 & 0 & 100 & 0 & 100 \\
\end{array}\right]
\begin{array}{l}
-1R_2 + R_1 \\ \\ -R_2 + R_3 \\ 100R_2 + R_4 \\
\end{array}
$$

The solution to the minimization problem is found in the bottom row of the final matrix in the entries corresponding to the slack variables.
The minimum is 100 when
$y_1 = 0,\ y_2 = 100,$ and $y_3 = 0.$

19. Minimize $w = 3y_1 + y_2 + 4y_3$

subject to:
$$2y_1 + y_2 + y_3 \geq 6$$
$$y_1 + 2y_2 + y_3 \geq 8$$
$$2y_1 + y_2 + 2y_3 \geq 12$$
$$y_1 \geq 0,\ y_2 \geq 0,\ y_3 \geq 0$$

The augmented matrix is

$$\begin{bmatrix} 2 & 1 & 1 & 6 \\ 1 & 2 & 1 & 8 \\ 2 & 1 & 2 & 12 \\ \hline 3 & 1 & 4 & 0 \end{bmatrix}.$$

The transpose is

$$\begin{bmatrix} 2 & 1 & 2 & 3 \\ 1 & 2 & 1 & 1 \\ 1 & 1 & 2 & 4 \\ \hline 6 & 8 & 12 & 0 \end{bmatrix}.$$

The dual maximization problem is

Maximize $z = 6x_1 + 8x_2 + 12x_3$

subject to:
$$2x_1 + x_2 + 2x_3 \leq 3$$
$$x_1 + 2x_2 + x_3 \leq 1$$
$$x_1 + x_2 + 2x_3 \leq 4$$
$$x_1 \geq 0,\ x_2 \geq 0,\ x_3 \geq 0.$$

The initial tableau is

$$\begin{array}{cccccc} x_1 & x_2 & x_3 & s_1 & s_2 & s_3 \\ \end{array}$$
$$\begin{bmatrix} 2 & 1 & 2 & 1 & 0 & 0 & 3 \\ 1 & 2 & \boxed{1} & 0 & 1 & 0 & 1 \\ 1 & 1 & 2 & 0 & 0 & 1 & 4 \\ \hline -6 & -8 & -12 & 0 & 0 & 0 & 0 \end{bmatrix}.$$

Pivot on the 1 in row two, column three.

$$\begin{array}{cccccc} x_1 & x_2 & x_3 & s_1 & s_2 & s_3 \\ \end{array}$$
$$\begin{bmatrix} 0 & -3 & 0 & 1 & -2 & 0 & 1 \\ 1 & 2 & 1 & 0 & 1 & 0 & 1 \\ -1 & -3 & 0 & 0 & -2 & 1 & 2 \\ \hline 6 & 16 & 0 & 0 & 12 & 0 & 12 \end{bmatrix} \begin{array}{l} -2R_2 + R_1 \\ \\ -2R_2 + R_3 \\ 12R_2 + R_4 \end{array}$$

The minimum is 12 when
$y_1 = 0,\ y_2 = 12,\ y_3 = 0.$

21. Minimize $w = 6y_1 + 4y_2 + 2y_3$

subject to:
$$2y_1 + 2y_2 + y_3 \geq 2$$
$$y_1 + 3y_2 + 2y_3 \geq 3$$
$$y_1 + y_2 + 2y_3 \geq 4$$
$$y_1 \geq 0,\ y_2 \geq 0,\ y_3 \geq 0.$$

The augmented matrix is

$$\begin{bmatrix} 2 & 2 & 1 & 2 \\ 1 & 3 & 2 & 3 \\ 1 & 1 & 2 & 4 \\ \hline 6 & 4 & 2 & 0 \end{bmatrix}.$$

The transpose is

$$\begin{bmatrix} 2 & 1 & 1 & 6 \\ 2 & 3 & 1 & 4 \\ 1 & 2 & 2 & 2 \\ \hline 2 & 3 & 4 & 0 \end{bmatrix}.$$

The dual maximization problem is

Maximize $z = 2x_1 + 3x_2 + 4x_3$

subject to:
$$2x_1 + x_2 + x_3 \leq 6$$
$$2x_1 + 3x_2 + x_3 \leq 4$$
$$x_1 + 2x_2 + 2x_3 \leq 2$$
$$x_1 \geq 0,\ x_2 \geq 0,\ x_3 \geq 0.$$

The initial simplex tableau is

$$\begin{array}{cccccc} x_1 & x_2 & x_3 & s_1 & s_2 & s_3 \\ \end{array}$$
$$\begin{bmatrix} 2 & 1 & 1 & 1 & 0 & 0 & 6 \\ 2 & 3 & 1 & 0 & 1 & 0 & 4 \\ 1 & 2 & \boxed{2} & 0 & 0 & 1 & 2 \\ \hline -2 & -3 & -4 & 0 & 0 & 0 & 0 \end{bmatrix}.$$

Pivot on the 2 in row three, column three.

$$\begin{array}{cccccc} x_1 & x_2 & x_3 & s_1 & s_2 & s_3 \\ \end{array}$$
$$\begin{bmatrix} 2 & 1 & 1 & 1 & 0 & 0 & 6 \\ 2 & 3 & 1 & 0 & 1 & 0 & 4 \\ \frac{1}{2} & 1 & 1 & 0 & 0 & \frac{1}{2} & 1 \\ \hline -2 & -3 & -4 & 0 & 0 & 0 & 0 \end{bmatrix} \begin{array}{l} \\ \\ \frac{1}{2}R_3 \\ \\ \end{array}$$

$$\begin{array}{cccccc} x_1 & x_2 & x_3 & s_1 & s_2 & s_3 \\ \end{array}$$
$$\begin{bmatrix} \frac{3}{2} & 0 & 0 & 1 & 0 & -\frac{1}{2} & 5 \\ \frac{3}{2} & 2 & 0 & 0 & 1 & -\frac{1}{2} & 3 \\ \frac{1}{2} & 1 & 1 & 0 & 0 & \frac{1}{2} & 1 \\ \hline 0 & 1 & 0 & 0 & 0 & 2 & 4 \end{bmatrix} \begin{array}{l} -R_3 + R_1 \\ -R_3 + R_2 \\ \\ 4R_3 + R_4 \end{array}$$

The minimum is 4 when $y_1 = 0$, $y_2 = 0$, and
$y_3 = 2.$

23. Minimize $w = 20y_1 + 12y_2 + 40y_3$

subject to: $y_1 + y_2 + 5y_3 \geq 20$
$2y_1 + y_2 + y_3 \geq 30$
$y_1 \geq 0,\ y_2 \geq 0,\ y_3 \geq 0.$

The augmented matrix is

$$\begin{bmatrix} 1 & 1 & 5 & | & 20 \\ 2 & 1 & 1 & | & 30 \\ \hline 20 & 12 & 40 & | & 0 \end{bmatrix}.$$

The transpose is

$$\begin{bmatrix} 1 & 2 & | & 20 \\ 1 & 1 & | & 12 \\ 5 & 1 & | & 40 \\ \hline 20 & 30 & | & 0 \end{bmatrix}.$$

The dual maximization problem is
Maximize $z = 20x_1 + 30x_2$

subject to: $x_1 + 2x_2 \leq 20$
$x_1 + x_2 \leq 12$
$5x_1 + x_2 \leq 40$
$x_1 \geq 0,\ x_2 \geq 0.$

The initial simplex tableau is

$$\begin{array}{ccccc} x_1 & x_2 & s_1 & s_2 & s_3 \\ \end{array}$$
$$\begin{bmatrix} 1 & \boxed{2} & 1 & 0 & 0 & | & 20 \\ 1 & 1 & 0 & 1 & 0 & | & 12 \\ 5 & 1 & 0 & 0 & 1 & | & 40 \\ \hline -20 & -30 & 0 & 0 & 0 & | & 0 \end{bmatrix}.$$

Pivot on the 2 in row one, column two.

$$\begin{array}{ccccc} x_1 & x_2 & s_1 & s_2 & s_3 \\ \end{array}$$
$$\begin{bmatrix} \frac{1}{2} & 1 & \frac{1}{2} & 0 & 0 & | & 10 \\ 1 & 1 & 0 & 1 & 0 & | & 12 \\ 5 & 1 & 0 & 0 & 1 & | & 40 \\ \hline -20 & -30 & 0 & 0 & 0 & | & 0 \end{bmatrix} \frac{1}{2}R_1$$

$$\begin{array}{ccccc} x_1 & x_2 & s_1 & s_2 & s_3 \\ \end{array}$$
$$\begin{bmatrix} \frac{1}{2} & 1 & \frac{1}{2} & 0 & 0 & | & 10 \\ \boxed{\frac{1}{2}} & 0 & -\frac{1}{2} & 1 & 0 & | & 2 \\ \frac{9}{2} & 0 & -\frac{1}{2} & 0 & 1 & | & 30 \\ \hline -5 & 0 & 15 & 0 & 0 & | & 300 \end{bmatrix} \begin{array}{l} \\ -R_1 + R_2 \\ -R_1 + R_3 \\ 30R_1 + R_4 \end{array}$$

Pivot on the $\frac{1}{2}$ in row two, column one.

$$\begin{array}{ccccc} x_1 & x_2 & s_1 & s_2 & s_3 \\ \end{array}$$
$$\begin{bmatrix} \frac{1}{2} & 1 & \frac{1}{2} & 0 & 0 & | & 10 \\ 1 & 0 & -1 & 2 & 0 & | & 4 \\ \frac{9}{2} & 0 & -\frac{1}{2} & 0 & 1 & | & 30 \\ \hline -5 & 0 & 15 & 0 & 0 & | & 300 \end{bmatrix} 2R_2$$

$$\begin{array}{ccccc} x_1 & x_2 & s_1 & s_2 & s_3 \\ \end{array}$$
$$\begin{bmatrix} 0 & 1 & 1 & -1 & 0 & | & 8 \\ 1 & 0 & -1 & 2 & 0 & | & 4 \\ 0 & 0 & 4 & -9 & 1 & | & 12 \\ \hline 0 & 0 & 10 & 10 & 0 & | & 320 \end{bmatrix} \begin{array}{l} -\frac{1}{2}R_2 + R_1 \\ \\ -\frac{9}{2}R_2 + R_3 \\ 5R_2 + R_4 \end{array}$$

The minimum is 320 when $y_1 = 10$, $y_2 = 10$, and $y_3 = 0$.

25. Minimize $w = 4y_1 + 2y_2 + y_3$

subject to: $y_1 + y_2 + y_3 \geq 4$
$3y_1 + y_2 + 3y_3 \geq 6$
$y_1 + y_2 + 3y_3 \geq 5$
$y_1 \geq 0,\ y_2 \geq 0,\ y_3 \geq 0.$

The augmented matrix is

$$\begin{bmatrix} 1 & 1 & 1 & | & 4 \\ 3 & 1 & 3 & | & 6 \\ 1 & 1 & 3 & | & 5 \\ \hline 4 & 2 & 1 & | & 0 \end{bmatrix}.$$

The transpose is

$$\begin{bmatrix} 1 & 3 & 1 & | & 4 \\ 1 & 1 & 1 & | & 2 \\ 1 & 3 & 3 & | & 1 \\ \hline 4 & 6 & 5 & | & 0 \end{bmatrix}.$$

The dual maximization problem is
Maximize $z = 4x_1 + 6x_2 + 5x_3$

subject to: $x_1 + 3x_2 + x_3 \leq 4$
$x_1 + x_2 + x_3 \leq 2$
$x_1 + 3x_2 + 3x_3 \leq 1$
$x_1 \geq 0,\ x_2 > 0,\ x_3 \geq 0.$

The initial simplex tableau is

$$\begin{array}{cccccc} x_1 & x_2 & x_3 & s_1 & s_2 & s_3 \\ \end{array}$$
$$\begin{bmatrix} 1 & 3 & 1 & 1 & 0 & 0 & | & 4 \\ 1 & 1 & 1 & 0 & 1 & 0 & | & 2 \\ 1 & \boxed{3} & 3 & 0 & 0 & 1 & | & 1 \\ \hline -4 & -6 & -5 & 0 & 0 & 0 & | & 0 \end{bmatrix}.$$

Pivot on the 3 in row three, column two.

$$\begin{array}{cccccc} x_1 & x_2 & x_3 & s_1 & s_2 & s_3 \\ \end{array}$$
$$\begin{bmatrix} 1 & 3 & 1 & 1 & 0 & 0 & | & 4 \\ 1 & 1 & 1 & 0 & 1 & 0 & | & 2 \\ \frac{1}{3} & 1 & 1 & 0 & 0 & \frac{1}{3} & | & \frac{1}{3} \\ \hline -4 & -6 & -5 & 0 & 0 & 0 & | & 0 \end{bmatrix} \frac{1}{3}R_3$$

(continued on next page)

(*continued from page 229*)

$$
\begin{array}{cccccc|c}
x_1 & x_2 & x_3 & s_1 & s_2 & s_3 & \\
0 & 0 & -2 & 1 & 0 & -1 & 3 \\
\frac{2}{3} & 0 & 0 & 0 & 1 & -\frac{1}{3} & \frac{5}{3} \\
\boxed{\frac{1}{3}} & 1 & 1 & 0 & 0 & \frac{1}{3} & \frac{1}{3} \\
\hline
-2 & 0 & 1 & 0 & 0 & 2 & 2
\end{array}
\begin{array}{l}
-3R_3 + R_1 \\
-1R_3 + R_2 \\
\\
6R_3 + R_4
\end{array}
$$

Pivot on the $\frac{1}{3}$ in row three, column one.

$$
\begin{array}{cccccc|c}
x_1 & x_2 & x_3 & s_1 & s_2 & s_3 & \\
0 & 0 & -2 & 1 & 0 & -1 & 3 \\
\frac{2}{3} & 0 & 0 & 0 & 1 & -\frac{1}{3} & \frac{5}{3} \\
1 & 3 & 3 & 0 & 0 & 1 & 1 \\
\hline
-2 & 0 & 1 & 0 & 0 & 2 & 2
\end{array}
\;3R_3
$$

$$
\begin{array}{cccccc|c}
x_1 & x_2 & x_3 & s_1 & s_2 & s_3 & \\
0 & 0 & -2 & 1 & 0 & -1 & 3 \\
0 & -2 & -2 & 0 & 1 & -1 & 1 \\
1 & 3 & 3 & 0 & 0 & 1 & 1 \\
\hline
0 & 6 & 7 & 0 & 0 & 4 & 4
\end{array}
\begin{array}{l}
\\
-\frac{2}{3}R_3 + R_2 \\
\\
2R_3 + R_4
\end{array}
$$

The minimum is 4 when $y_1 = 0$, $y_2 = 0$, and

$y_3 = 4$.

27. Let y_1 = the amount of product A and let

y_2 = the amount of product B.

The given information can be expressed as the following maximization problem:

Minimize $w = 24y_1 + 40y_2$

subject to: $4y_1 + 2y_2 \geq 20$

$\qquad\qquad 2y_1 + 5y_2 \geq 18$

$\qquad\qquad y_1 \geq 0,\ y_2 \geq 0.$

The augmented matrix is

$$
\begin{bmatrix}
4 & 2 & 20 \\
2 & 5 & 18 \\
\hline
24 & 40 & 0
\end{bmatrix}.
$$

The transpose is

$$
\begin{bmatrix}
4 & 2 & 24 \\
2 & 5 & 40 \\
\hline
20 & 18 & 0
\end{bmatrix}.
$$

The dual maximization problem is

Maximize $z = 20x_1 + 18x_2$

subject to: $4x_1 + 2x_2 \leq 24$

$\qquad\qquad 2x_1 + 5x_2 \leq 40$

$\qquad\qquad x_1 \geq 0,\ x_2 \geq 0.$

The initial simplex tableau is

$$
\begin{array}{cccc|c}
x_1 & x_2 & s_1 & s_2 & \\
\boxed{4} & 2 & 1 & 0 & 24 \\
2 & 5 & 0 & 1 & 40 \\
\hline
-20 & -18 & 0 & 0 & 0
\end{array}.
$$

Pivot on the 4 in row one, column one.

$$
\begin{array}{ccccc|c}
x_1 & x_2 & x_3 & x_4 & x_5 & \\
1 & \frac{1}{2} & \frac{1}{4} & 0 & & 6 \\
2 & 5 & 0 & 1 & & 40 \\
\hline
-20 & -18 & 0 & 0 & & 0
\end{array}
\;\tfrac{1}{4}R_1
$$

$$
\begin{array}{cccc|c}
x_1 & x_2 & s_1 & s_2 & \\
1 & \frac{1}{2} & \frac{1}{4} & 0 & 6 \\
0 & \boxed{4} & -\frac{1}{2} & 1 & 28 \\
\hline
0 & -8 & 5 & 0 & 120
\end{array}
\begin{array}{l}
\\
-2R_1 + R_2 \\
20R_1 + R_3
\end{array}
$$

Pivot on the 4 in row two, column two.

$$
\begin{array}{ccccc|c}
x_1 & x_2 & x_3 & x_4 & x_5 & \\
1 & \frac{1}{2} & \frac{1}{4} & 0 & & 6 \\
0 & 1 & -\frac{1}{8} & \frac{1}{4} & & 7 \\
\hline
0 & -8 & 5 & 0 & & 120
\end{array}
\;\tfrac{1}{4}R_2
$$

$$
\begin{array}{cccc|c}
x_1 & x_2 & x_3 & x_4 & \\
1 & 0 & \frac{5}{16} & -\frac{1}{8} & \frac{5}{2} \\
0 & 1 & -\frac{1}{8} & \frac{1}{4} & 7 \\
\hline
0 & 0 & 4 & 2 & 176
\end{array}
\begin{array}{l}
-\frac{1}{2}R_2 + R_1 \\
\\
8R_2 + R_3
\end{array}
$$

Glenn should use 4 servings of product A and 2 servings of product B for a minimum cost of $1.76.

29. Let y_1 = the number of additional units of

regular beer to produce, and let y_2 = the number of additional units of light beer to produce

The sale of 12 units of regular beer and 10 units of light beer already generates

$12 \cdot 100{,}000 + 10 \cdot 300{,}000 = \$4{,}200{,}000$ in revenue. Additional units need only to generate $7{,}000{,}000 - 4{,}200{,}000 = \$2{,}800{,}000$ in revenue.

Minimize $w = 36{,}000y_1 + 48{,}000y_2$

subject to: $100{,}000y_1 + 300{,}000y_2 \geq 2{,}800{,}000$

$\qquad\qquad y_1 \qquad\quad + y_2 \geq 20$

$\qquad\qquad y_1 \geq 0,\ y_2 \geq 0$

The augmented matrix is

$$
\begin{bmatrix}
100{,}000 & 300{,}000 & 2{,}800{,}000 \\
1 & 1 & 20 \\
\hline
36{,}000 & 48{,}000 & 0
\end{bmatrix}
$$

(*continued on next page*)

(continued from page 230)

The transpose is

$$\begin{bmatrix} 100{,}000 & 1 & | & 36{,}000 \\ 300{,}000 & 1 & | & 48{,}000 \\ \hline 2{,}800{,}000 & 20 & | & 0 \end{bmatrix}$$

The dual maximization problem is

Maximize $z = 2{,}800{,}000x_1 + 20x_2$

subject to: $100{,}000x_1 + x_2 \le 36{,}000$
$\qquad\qquad 300{,}000x_1 + x_2 \le 48{,}000$
$\qquad\qquad x_1 \ge 0,\ x_2 \ge 0$

The initial simplex tableau is

$$\begin{array}{cccc}
x_1 & x_2 & s_1 & s_2 \\
\end{array}$$
$$\begin{bmatrix} 100{,}000 & 1 & 1 & 0 & | & 36{,}000 \\ \boxed{300{,}000} & 1 & 0 & 1 & | & 48{,}000 \\ \hline -2{,}800{,}000 & -20 & 0 & 0 & | & 0 \end{bmatrix}$$

Pivot on the 300,000 in row two, column one.

$$\begin{array}{cccc}
x_1 & x_2 & s_1 & s_2 \\
\end{array}$$
$$\begin{bmatrix} 0 & \boxed{\tfrac{2}{3}} & 1 & -\tfrac{1}{3} & | & 20{,}000 \\ 1 & \tfrac{1}{300{,}000} & 0 & \tfrac{1}{300{,}000} & | & \tfrac{16}{100} \\ \hline 0 & -\tfrac{32}{3} & 0 & \tfrac{28}{3} & | & 448{,}000 \end{bmatrix}$$
$$\begin{array}{l} -100{,}000R_2 + R_1 \\ \tfrac{1}{300{,}000}R_2 \\ 2{,}800{,}000R_2 + R_3 \end{array}$$

Pivot on the $\tfrac{2}{3}$ in row one, column two.

$$\begin{bmatrix} 0 & 1 & \tfrac{3}{2} & -\tfrac{1}{2} & | & 30{,}000 \\ 1 & 0 & -\tfrac{1}{200{,}000} & \tfrac{1}{200{,}000} & | & \tfrac{3}{50} \\ \hline 0 & 0 & 16 & 4 & | & 768{,}000 \end{bmatrix}$$
$$\begin{array}{l} \tfrac{3}{2}R_1 \\ -\tfrac{1}{300{,}000}R_1 + R_2 \\ \tfrac{32}{3}R_1 + R_3 \end{array}$$

The total cost is the cost of the original 12 and 10 units of beer plus the cost of the additional units of beer.

$768{,}000 + 12 \cdot 36{,}000 + 10 \cdot 48{,}000 = \$1{,}680{,}000$

The brewery should make 16 additional units of regular beer and 4 additional units of light beer for a total of 28 units of regular beer and 14 units of light beer at a minimum cost of $1,680,000.

31. The linear program (P)

Minimize $z = x_1 + 2x_2$

$\qquad\qquad -2x_1 + x_2 \ge 1$
$\qquad\qquad x_1 - 2x_2 \ge 1$
$\qquad\qquad x_1 \ge 0,\ x_2 \ge 0$

has no feasible solution. Graphing the constraints of (P) shows that the constraints do not form a feasible space.

To determine the dual of (P), (D), we have the

augmented matrix $\begin{bmatrix} -2 & 1 & | & 1 \\ 1 & -2 & | & 1 \\ \hline 1 & 2 & | & 0 \end{bmatrix}$, the transpose

$\begin{bmatrix} -2 & 1 & | & 1 \\ 1 & -2 & | & 2 \\ \hline 1 & 1 & | & 0 \end{bmatrix}$, and dual maximization problem

Maximize $w = y_1 + y_2$

subject to: $-2y_1 + y_2 \le 1$
$\qquad\qquad y_1 - 2y_2 \le 2$
$\qquad\qquad y_1 \ge 0,\ y_2 \ge 0$

Graphing the constraints of (D) does show a feasible space but the objective function is unbounded.

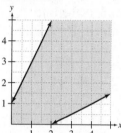

The answer is (a).

33. Let x_1 = the number of toy bears and let x_2 = the number of monkeys.

Maximize $z = x_1 + 1.5x_2$

subject to: $x_1 + 2x_2 \le 200$
$\qquad\qquad 4x_1 + 3x_2 \le 600$
$\qquad\qquad x_2 \le 90$
$\qquad\qquad x_1 \ge 0,\ x_2 \ge 0.$

a. The augmented matrix is

$$\begin{bmatrix} 1 & 2 & | & 200 \\ 4 & 3 & | & 600 \\ 0 & 1 & | & 90 \\ \hline 1 & 1.5 & | & 0. \end{bmatrix}$$

The transpose is

$$\begin{bmatrix} 1 & 4 & 0 & | & 1 \\ 2 & 3 & 1 & | & 1.5 \\ \hline 200 & 600 & 90 & | & 0 \end{bmatrix}.$$

The dual minimization problem is

Minimize $w = 200y_1 + 600y_2 + 90y_3$

subject to: $\qquad y_1 + 4y_2 \qquad \ge 1$
$\qquad\qquad 2y_1 + 3y_2 + y_3 \ge 1.5$
$\qquad\qquad y_1 \ge 0,\ y_2 \ge 0,\ y_3 \ge 0.$

b. The solution to the dual minimization problem is found in the last row of the final simplex. tableau of the original problem. The minimum is 180 when $y_1 = .6$, $y_2 = .1$, and $y_3 = 0$.

c. The shadow cost for felt is in row four, column three of the final matrix, $.60/unit. So, an increase of 10 squares of felt would increase the profit by $6. The profit would be $186.

d. The shadow cost for stuffing is $.10/unit; the shadow cost for trim is $0/unit. So, a decrease of 10 oz of stuffing and 10 ft of trim would decrease the profit by $1. The profit would be $179.

Section 7.7   The Simplex Method: Nonstandard Problems

1. Maximize $z = -5x_1 + 4x_2 - 2x_3$

subject to
$$-2x_2 + 5x_3 \geq 8$$
$$4x_1 - x_2 + 3x_3 \leq 12$$
$$x_1 \geq 0,\ x_2 \geq 0,\ x_3 \geq 0.$$

a. Insert a surplus variable in the first constraint and slack variables in the last constrains. The problem then becomes:

Maximize $z = -5x_1 + 4x_2 - 2x_3$
subject to:
$$-2x_2 + 5x_3 - s_1 \qquad = 8$$
$$4x_1 - x_2 + 3x_3 \qquad + s_2 = 12$$
$$x_1 \geq 0,\ x_2 \geq 0,\ x_3 \geq 0,\ s_1 \geq 0,\ s_2 \geq 0.$$

b. The initial simplex tableau is

x_1	x_2	x_3	s_1	s_2	
0	-2	5	-1	0	8
4	-1	3	0	1	12
5	-4	2	0	0	0

3. Maximize $z = 2x_1 - 3x_2 + 4x_3$

subject to:
$$x_1 + x_2 + x_3 \leq 100$$
$$x_1 + x_2 + x_3 \geq 75$$
$$x_1 + x_2 \geq 27$$
$$x_1 \geq 0,\ x_2 \geq 0,\ x_3 \geq 0.$$

a. Insert a slack variable in the first constraints and surplus variables in the last two constraints. The problem then becomes:

Maximize $z = 2x_1 - 3x_2 + 4x_3$
subject to:
$$x_1 + x_2 + x_3 + s_1 \qquad = 100$$
$$x_1 + x_2 + x_3 \qquad - s_2 \qquad = 75$$
$$x_1 + x_2 \qquad\qquad - s_3 = 27$$
$$x_1 \geq 0,\ x_2 \geq 0,\ x_3 \geq 0,\ s_1 \geq 0,\ s_2 \geq 0,\ s_3 \geq 0.$$

b. The initial simplex tableau is

x_1	x_2	x_3	s_1	s_2	s_3	
1	1	1	1	0	0	100
1	1	1	0	-1	0	75
1	1	0	0	0	-1	27
-2	3	-4	0	0	0	0

5. Minimize $w = 2y_1 + 5y_2 - 3y_3$

subject to:
$$y_1 + 2y_2 + 3y_3 \geq 115$$
$$2y_1 + y_2 + y_3 \leq 200$$
$$y_1 + y_3 \geq 50$$
$$y_1 \geq 0,\ y_2 \geq 0,\ y_3 \geq 0.$$

Rewrite the objective function to get a maximization problem.

Maximize $z = -2y_1 - 5y_2 + 3y_3$
subject to:
$$y_1 + 2y_2 + 3y_3 \geq 115$$
$$2y_1 + y_2 + y_3 \leq 200$$
$$y_1 + y_3 \geq 50$$
$$y_1 \geq 0,\ y_2 \geq 0,\ y_3 \geq 0.$$

Insert surplus variables in the first and third constraints and a slack variable in the second to get the initial simplex tableau.

y_1	y_2	y_3	s_1	s_2	s_3	
1	2	3	-1	0	0	115
2	1	1	0	1	0	200
1	0	1	0	0	-1	50
2	5	-3	0	0	0	0

7. Minimize $w = 10y_1 + 8y_2 + 15y_3$

subject to:
$$y_1 + y_2 + y_3 \geq 12$$
$$5y_1 + 4y_2 + 9y_3 \geq 48$$
$$y_1 \geq 0, \ y_2 \geq 0, \ y_3 \geq 0.$$

Rewrite the objective function to get a maximization problem.

Maximize $z = -10y_1 - 8y_2 - 15y_3$

subject to:
$$y_1 + y_2 + y_3 \geq 12$$
$$5y_1 + 4y_2 + 9y_3 \geq 48$$
$$y_1 \geq 0, \ y_2 \geq 0, \ y_3 \geq 0.$$

Insert surplus variables in both constraints to get the initial simplex tableau.

$$
\begin{array}{ccccc}
y_1 & y_2 & y_3 & s_1 & s_2
\end{array}
$$
$$
\left[
\begin{array}{ccccc|c}
1 & 1 & 1 & -1 & 0 & 12 \\
5 & 4 & 9 & 0 & -1 & 48 \\
\hline
10 & 8 & 15 & 0 & 0 & 0
\end{array}
\right]
$$

9. Maximize $z = 12x_1 + 10x_2$

subject to:
$$x_1 + 2x_2 \geq 24$$
$$x_1 + x_2 \leq 40$$
$$x_1 \geq 0, \ x_2 \geq 0.$$

The initial simplex tableau is

$$
\begin{array}{cccc}
x_1 & x_2 & s_1 & s_2
\end{array}
$$
$$
\left[
\begin{array}{cccc|c}
\boxed{1} & 2 & -1 & 0 & 24 \\
1 & 1 & 0 & 1 & 40 \\
\hline
-12 & -10 & 0 & 0 & 0
\end{array}
\right]
$$

For Stage I pivoting, pivot on the 1 in row one, column one.

$$
\begin{array}{cccc}
x_1 & x_2 & s_1 & s_2
\end{array}
$$
$$
\left[
\begin{array}{cccc|c}
1 & 2 & -1 & 0 & 24 \\
0 & -1 & \boxed{1} & 1 & 16 \\
\hline
0 & 14 & -12 & 0 & 288
\end{array}
\right]
\begin{array}{l}
\\ -R_1 + R_2 \\ 12R_2 + R_3
\end{array}
$$

For Stage II pivoting, pivot on the 1 in row two, column three.

$$
\begin{array}{cccc}
x_1 & x_2 & x_3 & x_4
\end{array}
$$
$$
\left[
\begin{array}{cccc|c}
1 & 1 & 0 & 1 & 40 \\
0 & -1 & 1 & 1 & 16 \\
\hline
0 & 2 & 0 & 12 & 480
\end{array}
\right]
\begin{array}{l}
R_2 + R_1 \\ \\ 12R_1 + R_3
\end{array}
$$

The maximum is 480 when $x_1 = 40$ and $x_2 = 0$.

11. Find $x_1 \geq 0$, $x_2 \geq 0$, and $x_3 \geq 0$ such that
$$x_1 + x_2 + 2x_3 \leq 38$$
$$2x_1 + x_2 + x_3 \geq 24$$
and $z = 3x_1 + 2x_2 + 2x_3$ is maximized.

The initial simplex tableau is

$$
\begin{array}{ccccc}
x_1 & x_2 & x_3 & s_1 & s_2
\end{array}
$$
$$
\left[
\begin{array}{ccccc|c}
1 & 1 & 2 & 1 & 0 & 38 \\
\boxed{2} & 1 & 1 & 0 & -1 & 24 \\
\hline
-3 & -2 & -2 & 0 & 0 & 0
\end{array}
\right]
$$

For Stage I pivoting, pivot on the 2 in row two, column one.

$$
\begin{array}{ccccc}
x_1 & x_2 & x_3 & s_1 & s_2
\end{array}
$$
$$
\left[
\begin{array}{ccccc|c}
1 & 1 & 2 & 1 & 0 & 38 \\
1 & \frac{1}{2} & \frac{1}{2} & 0 & -\frac{1}{2} & 12 \\
\hline
-3 & -2 & -2 & 0 & 0 & 0
\end{array}
\right]
\begin{array}{l}
\\ \frac{1}{2}R_2 \\ \\
\end{array}
$$

$$
\begin{array}{ccccc}
x_1 & x_2 & x_3 & s_1 & s_2
\end{array}
$$
$$
\left[
\begin{array}{ccccc|c}
0 & \frac{1}{2} & \frac{3}{2} & 1 & \boxed{\frac{1}{2}} & 26 \\
1 & \frac{1}{2} & \frac{1}{2} & 0 & -\frac{1}{2} & 12 \\
\hline
0 & -\frac{1}{2} & -\frac{1}{2} & 0 & -\frac{3}{2} & 36
\end{array}
\right]
\begin{array}{l}
-1R_2 + R_1 \\ \\ 3R_2 + R_3
\end{array}
$$

For Stage II pivoting, pivot on the $\frac{1}{2}$ in row one, column five.

$$
\begin{array}{ccccc}
x_1 & x_2 & x_3 & s_1 & s_2
\end{array}
$$
$$
\left[
\begin{array}{ccccc|c}
0 & 1 & 3 & 2 & 1 & 52 \\
1 & \frac{1}{2} & \frac{1}{2} & 0 & -\frac{1}{2} & 12 \\
\hline
0 & -\frac{1}{2} & -\frac{1}{2} & 0 & -\frac{3}{2} & 36
\end{array}
\right]
\begin{array}{l}
2R_1 \\ \\ \\
\end{array}
$$

$$
\begin{array}{ccccc}
x_1 & x_2 & x_3 & s_1 & s_2
\end{array}
$$
$$
\left[
\begin{array}{ccccc|c}
0 & 1 & 3 & 2 & 1 & 52 \\
1 & 1 & 2 & 1 & 0 & 38 \\
\hline
0 & 1 & 4 & 3 & 0 & 114
\end{array}
\right]
\begin{array}{l}
\\ \frac{1}{2}R_1 + R_2 \\ \frac{3}{2}R_1 + R_3
\end{array}
$$

The maximum is 114 when $x_1 = 38$, $x_2 = 0$, and $x_3 = 0$.

13. Find $x_1 \geq 0$ and $x_2 \geq 0$ such that

$$x_1 + 2x_2 \leq 18$$
$$x_1 + 3x_2 \geq 12$$
$$2x_1 + 2x_2 \leq 30$$

and $z = 5x_1 + 10x_2$ is maximized.

The initial simplex tableau is

$$\begin{array}{ccccc} x_1 & x_2 & s_1 & s_2 & s_3 \\ \end{array}$$
$$\left[\begin{array}{ccccc|c} 1 & 2 & 1 & 0 & 0 & 18 \\ \boxed{1} & 3 & 0 & -1 & 0 & 12 \\ 2 & 2 & 0 & 0 & 1 & 30 \\ \hline -5 & -10 & 0 & 0 & 0 & 0 \end{array} \right].$$

For Stage I pivoting, we can pivot on the 1 in row two, column one or the 3 in row two, column two. Choosing the 1 in row two, column one as the pivot, we proceed as follows.

$$\begin{array}{ccccc} x_1 & x_2 & s_1 & s_2 & s_3 \\ \end{array}$$
$$\left[\begin{array}{ccccc|c} 0 & -1 & 1 & 1 & 0 & 6 \\ 1 & 3 & 0 & -1 & 0 & 12 \\ 0 & -4 & 0 & \boxed{2} & 1 & 6 \\ \hline 0 & 5 & 0 & -5 & 0 & 60 \end{array} \right] \begin{array}{l} -R_2 + R_1 \\ \\ -2R_2 + R_3 \\ 5R_2 + R_4 \end{array}$$

For Stage II, pivot on the 2 in row three, column four.

$$\begin{array}{ccccc} x_1 & x_2 & s_1 & s_2 & s_3 \\ \end{array}$$
$$\left[\begin{array}{ccccc|c} 0 & -1 & 1 & 1 & 0 & 6 \\ 1 & 3 & 0 & -1 & 0 & 12 \\ 0 & -2 & 0 & 1 & \frac{1}{2} & 3 \\ \hline 0 & 5 & 0 & -5 & 0 & 60 \end{array} \right] \begin{array}{l} \\ \\ \frac{1}{2}R_3 \\ \\ \end{array}$$

$$\begin{array}{ccccc} x_1 & x_2 & s_1 & s_2 & s_3 \\ \end{array}$$
$$\left[\begin{array}{ccccc|c} 0 & \boxed{1} & 1 & 0 & -\frac{1}{2} & 3 \\ 1 & 1 & 0 & 0 & \frac{1}{2} & 15 \\ 0 & -2 & 0 & 1 & \frac{1}{2} & 3 \\ \hline 0 & -5 & 0 & 0 & \frac{5}{2} & 75 \end{array} \right] \begin{array}{l} -R_3 + R_1 \\ R_3 + R_2 \\ \\ 5R_3 + R_4 \end{array}$$

Now pivot on the 1 in row one, column two.

$$\begin{array}{ccccc} x_1 & x_2 & s_1 & s_2 & s_3 \\ \end{array}$$
$$\left[\begin{array}{ccccc|c} 0 & 1 & 1 & 0 & -\frac{1}{2} & 3 \\ 1 & 0 & -1 & 0 & 1 & 12 \\ 0 & 0 & 2 & 1 & -\frac{1}{2} & 9 \\ \hline 0 & 0 & 5 & 0 & 0 & 90 \end{array} \right] \begin{array}{l} \\ -R_1 + R_2 \\ 2R_1 + R_3 \\ 5R_1 + R_4 \end{array}$$

The above tableau gives the solution:
The maximum is 90 when $x_1 = 12$ and $x_2 = 3$.

However, in Stage I, we could also choose the 3 in row two, column two as the pivot and proceed as follows.

$$\begin{array}{ccccc} x_1 & x_2 & s_1 & s_2 & s_3 \\ \end{array}$$
$$\left[\begin{array}{ccccc|c} 1 & 2 & 1 & 0 & 0 & 18 \\ \frac{1}{3} & \boxed{1} & 0 & -\frac{1}{3} & 0 & 4 \\ 2 & 2 & 0 & 0 & 1 & 30 \\ \hline -5 & -10 & 0 & 0 & 0 & 0 \end{array} \right] \begin{array}{l} \\ \frac{1}{3}R_2 \\ \\ \end{array}$$

$$\begin{array}{ccccc} x_1 & x_2 & s_1 & s_2 & s_3 \\ \end{array}$$
$$\left[\begin{array}{ccccc|c} \frac{1}{3} & 0 & 1 & \boxed{\frac{2}{3}} & 0 & 10 \\ \frac{1}{3} & 1 & 0 & -\frac{1}{3} & 0 & 4 \\ \frac{4}{3} & 0 & 0 & \frac{2}{3} & 1 & 22 \\ \hline -\frac{5}{3} & 0 & 0 & -\frac{10}{3} & 0 & 40 \end{array} \right] \begin{array}{l} -2R_2 + R_1 \\ \\ -2R_2 + R_3 \\ 10R_2 + R_4 \end{array}$$

For Stage II, pivot on the $\frac{2}{3}$ in row one, column four.

$$\begin{array}{ccccc} x_1 & x_2 & s_1 & s_2 & s_3 \\ \end{array}$$
$$\left[\begin{array}{ccccc|c} \frac{1}{2} & 0 & \frac{3}{2} & 1 & 0 & 15 \\ \frac{1}{3} & 1 & 0 & -\frac{1}{3} & 0 & 4 \\ \frac{4}{3} & 0 & 0 & \frac{2}{3} & 1 & 22 \\ \hline -\frac{5}{3} & 0 & 0 & -\frac{10}{3} & 0 & 40 \end{array} \right] \begin{array}{l} \frac{3}{2}R_1 \\ \\ \\ \end{array}$$

$$\begin{array}{ccccc} x_1 & x_2 & s_1 & s_2 & s_3 \\ \end{array}$$
$$\left[\begin{array}{ccccc|c} \frac{1}{2} & 0 & \frac{3}{2} & 1 & 0 & 15 \\ \frac{1}{2} & 1 & \frac{1}{2} & 0 & 0 & 9 \\ 1 & 0 & -1 & 0 & 1 & 12 \\ \hline 0 & 0 & 5 & 0 & 0 & 90 \end{array} \right] \begin{array}{l} \\ \frac{1}{3}R_1 + R_2 \\ -\frac{2}{3}R_1 + R_3 \\ \frac{10}{3}R_1 + R_4 \end{array}$$

This tableau gives the solution:
The maximum is 90 when $x_1 = 0$ and $x_2 = 9$.

Thus, the maximum is 90 when $x_1 = 12$ and $x_2 = 3$ or when $x_1 = 0$ and $x_2 = 9$.

15. Minimize $w = 3y_1 + 2y_2$

subject to: $2y_1 + 3y_2 \geq 60$
$y_1 + 4y_2 \geq 40$
$y_1 \geq 0, \ y_2 \geq 0.$

The initial simplex tableau is

$$\begin{array}{cccc} y_1 & y_2 & s_1 & s_2 \\ \end{array}$$
$$\left[\begin{array}{cccc|c} \boxed{2} & 3 & -1 & 0 & 60 \\ 1 & 4 & 0 & -1 & 40 \\ \hline 3 & 2 & 0 & 0 & 0 \end{array}\right].$$

For Stage I, pivot on the 2 in row one, column one.

$$\begin{array}{cccc} y_1 & y_2 & s_1 & s_2 \\ \end{array}$$
$$\left[\begin{array}{cccc|c} 1 & \frac{3}{2} & -\frac{1}{2} & 0 & 30 \\ 0 & \boxed{\frac{5}{2}} & \frac{1}{2} & -1 & 10 \\ \hline 0 & -\frac{5}{2} & \frac{3}{2} & 0 & -90 \end{array}\right]\begin{array}{l} \frac{1}{2}R_1 \\ -R_1 + R_2 \\ -3R_1 + R_3 \end{array}$$

To continue Stage I, pivot on the $\frac{5}{2}$ in row two, column two.

$$\begin{array}{cccc} y_1 & y_2 & s_1 & s_2 \\ \end{array}$$
$$\left[\begin{array}{cccc|c} 1 & 0 & -\frac{4}{5} & \boxed{\frac{3}{5}} & 24 \\ 0 & 1 & \frac{1}{5} & -\frac{2}{5} & 4 \\ \hline 0 & 0 & 2 & -1 & -80 \end{array}\right]\begin{array}{l} -\frac{3}{5}R_2 + R_1 \\ \frac{2}{5}R_2 \\ \frac{5}{2}R_2 + R_3 \end{array}$$

For Stage II, pivot on the $\frac{3}{5}$ in row one, column four.

$$\begin{array}{cccc} y_1 & y_2 & s_1 & s_2 \\ \end{array}$$
$$\left[\begin{array}{cccc|c} \frac{5}{3} & 0 & -\frac{4}{3} & 1 & 40 \\ \frac{2}{3} & 1 & -\frac{1}{3} & 0 & 20 \\ \hline \frac{5}{3} & 0 & \frac{2}{3} & 0 & -40 \end{array}\right]\begin{array}{l} \frac{5}{3}R_1 \\ \frac{2}{5}R_1 + R_2 \\ R_1 + R_3 \end{array}$$

The minimum is 40 when $y_1 = 0$ and $y_2 = 20$.

17. Maximize $z = 3x_1 + 2x_2$

subject to: $x_1 + x_2 = 50$
$4x_1 + 2x_2 \geq 120$
$5x_1 + 2x_2 \leq 200$
$x_1 \geq 0, \ x_2 \geq 0$

The initial simplex tableau is

$$\begin{array}{cccccc} x_1 & x_2 & s_1 & s_2 & s_3 & s_4 \\ \end{array}$$
$$\left[\begin{array}{cccccc|c} 1 & 1 & -1 & 0 & 0 & 0 & 50 \\ 1 & 1 & 0 & 1 & 0 & 0 & 50 \\ \boxed{4} & 2 & 0 & 0 & -1 & 0 & 120 \\ 5 & 2 & 0 & 0 & 0 & 1 & 200 \\ \hline -3 & -2 & 0 & 0 & 0 & 0 & 0 \end{array}\right]$$

For Stage I, pivot on the 4 in row three, column one.

$$\begin{array}{cccccc} x_1 & x_2 & s_1 & s_2 & s_3 & s_4 \\ \end{array}$$
$$\left[\begin{array}{cccccc|c} 0 & \boxed{\frac{1}{2}} & -1 & 0 & \frac{1}{4} & 0 & 20 \\ 0 & \frac{1}{2} & 0 & 1 & \frac{1}{4} & 0 & 20 \\ 1 & \frac{1}{2} & 0 & 0 & -\frac{1}{4} & 0 & 30 \\ 0 & -\frac{1}{2} & 0 & 0 & \frac{5}{4} & 1 & 50 \\ \hline 0 & -\frac{1}{2} & 0 & 0 & -\frac{3}{4} & 0 & 90 \end{array}\right]\begin{array}{l} -R_3 + R_1 \\ -R_3 + R_2 \\ \frac{1}{4}R_3 \\ -5R_3 + R_4 \\ 3R_3 + R_5 \end{array}$$

To continue Stage I, pivot on the $\frac{1}{2}$ in row one column two.

$$\begin{array}{cccccc} x_1 & x_2 & s_1 & s_2 & s_3 & s_4 \\ \end{array}$$
$$\left[\begin{array}{cccccc|c} 0 & 1 & -2 & 0 & \frac{1}{2} & 0 & 40 \\ 0 & 0 & \boxed{1} & 1 & 0 & 0 & 0 \\ 1 & 0 & 1 & 0 & -\frac{1}{2} & 0 & 10 \\ 0 & 0 & -1 & 0 & \frac{3}{2} & 1 & 70 \\ \hline 0 & 0 & -1 & 0 & -\frac{1}{2} & 0 & 110 \end{array}\right]\begin{array}{l} 2R_1 \\ -\frac{1}{2}R_1 + R_2 \\ -\frac{1}{2}R_1 + R_3 \\ \frac{1}{2}R_1 + R_4 \\ \frac{1}{2}R_1 + R_5 \end{array}$$

For Stage II, pivot on the 1 in row two, column three.

$$\begin{array}{cccccc} x_1 & x_2 & s_1 & s_2 & s_3 & s_4 \\ \end{array}$$
$$\left[\begin{array}{cccccc|c} 0 & 1 & 0 & 2 & \frac{1}{2} & 0 & 40 \\ 0 & 0 & 1 & 1 & 0 & 0 & 0 \\ 1 & 0 & 0 & -1 & -\frac{1}{2} & 0 & 10 \\ 0 & 0 & 0 & 1 & \boxed{\frac{3}{2}} & 1 & 70 \\ \hline 0 & 0 & 0 & 1 & -\frac{1}{2} & 0 & 110 \end{array}\right]\begin{array}{l} 2R_2 + R_1 \\ \\ -R_2 + R_3 \\ R_2 + R_4 \\ R_2 + R_5 \end{array}$$

Pivot on the $\frac{3}{2}$ in row four, column five.

$$\begin{array}{cccccc} x_1 & x_2 & s_1 & s_2 & s_3 & s_4 \\ \end{array}$$
$$\left[\begin{array}{cccccc|c} 0 & 1 & 0 & \frac{5}{3} & 0 & -\frac{1}{3} & \frac{50}{3} \\ 0 & 0 & 1 & 1 & 0 & 0 & 0 \\ 1 & 0 & 0 & -\frac{2}{3} & 0 & \frac{1}{3} & \frac{100}{3} \\ 0 & 0 & 0 & \frac{2}{3} & 1 & \frac{2}{3} & \frac{140}{3} \\ \hline 0 & 0 & 0 & \frac{4}{3} & 0 & \frac{1}{3} & \frac{400}{3} \end{array}\right]\begin{array}{l} -\frac{1}{2}R_4 + R_1 \\ \\ \frac{1}{2}R_4 + R_3 \\ \frac{2}{3}R_4 \\ \frac{1}{2}R_4 + R_5 \end{array}$$

The maximum is $133\frac{1}{3}$ when $x_1 = 33\frac{1}{3}$ and

$x_2 = 16\frac{2}{3}.$

19. Minimize $w = 32y_1 + 40y_2$

Maximize $z = -w = -32y_1 - 40y_2$.

subject to: $20y_1 + 10y_2 = 200$
$25y_1 + 40y_2 \le 500$
$18y_1 + 24y_2 \ge 300$

$y_1 \ge 0, \ y_2 \ge 0$

The initial simplex tableau is

$$
\begin{array}{cccccc}
y_1 & y_2 & s_1 & s_2 & s_3 & s_4 \\
\end{array}
$$
$$
\left[\begin{array}{cccccc|c}
\boxed{20} & 10 & -1 & 0 & 0 & 0 & 200 \\
20 & 10 & 0 & 1 & 0 & 0 & 200 \\
25 & 40 & 0 & 0 & 1 & 0 & 500 \\
18 & 24 & 0 & 0 & 0 & -1 & 300 \\
\hline
32 & 40 & 0 & 0 & 0 & 0 & 0
\end{array}\right]
$$

For Stage I, pivot on the 20 in row one, column one.

$$
\begin{array}{cccccc}
y_1 & y_2 & s_1 & s_2 & s_3 & s_4 \\
\end{array}
$$
$$
\left[\begin{array}{cccccc|c}
1 & \frac{1}{2} & -\frac{1}{20} & 0 & 0 & 0 & 10 \\
0 & 0 & 1 & 1 & 0 & 0 & 0 \\
0 & \frac{55}{2} & \frac{5}{4} & 0 & 1 & 0 & 250 \\
0 & \boxed{15} & \frac{9}{10} & 0 & 0 & -1 & 120 \\
\hline
0 & 24 & \frac{8}{5} & 0 & 0 & 0 & -320
\end{array}\right]
\begin{array}{l}
\frac{1}{20}R_1 \\
-20R_1 + R_2 \\
-25R_1 + R_3 \\
-18R_1 + R_4 \\
-32R_1 + R_5
\end{array}
$$

Pivot on the 15 in row four, column two.

$$
\begin{array}{cccccc}
y_1 & y_2 & s_1 & s_2 & s_3 & s_4 \\
\end{array}
$$
$$
\left[\begin{array}{cccccc|c}
1 & 0 & -\frac{4}{50} & 0 & 0 & \frac{1}{30} & 6 \\
0 & 0 & 1 & 1 & 0 & 0 & 0 \\
0 & 0 & -\frac{2}{5} & 0 & 1 & \frac{11}{6} & 30 \\
0 & 1 & \frac{3}{50} & 0 & 0 & -\frac{1}{15} & 8 \\
\hline
0 & 0 & \frac{4}{25} & 0 & 0 & \frac{8}{5} & -512
\end{array}\right]
\begin{array}{l}
-\frac{1}{2}R_4 + R_1 \\
\\
-\frac{55}{2}R_4 + R_3 \\
\frac{1}{15}R_4 \\
-24R_4 + R_5
\end{array}
$$

The program is optimal after Stage I.

The minimum is 512 when $y_1 = 6$ and $y_2 = 8$.

21. Maximize $z = -5x_1 + 4x_2 - 2x_3$

subject to $-2x_2 + 5x_3 \ge 8$
$4x_1 - x_2 + 3x_3 \le 12$
$x_1 \ge 0, \ x_2 \ge 0, \ x_3 \ge 0.$

Insert a surplus variable in the first constraint and slack variables in the last two constraints. The problem then becomes:

Maximize $z = -5x_1 + 4x_2 - 2x_3$

subject to: $-2x_2 + 5x_3 - s_1 = 8$
$4x_1 - x_2 + 3x_3 + s_2 = 12$
$x_1 \ge 0, \ x_2 \ge 0, \ x_3 \ge 0, \ s_1 \ge 0, \ s_2 \ge 0.$

The initial simplex tableau is

$$
\begin{array}{ccccc}
x_1 & x_2 & x_3 & s_1 & s_2 \\
\end{array}
$$
$$
\left[\begin{array}{ccccc|c}
0 & -2 & \boxed{5} & -1 & 0 & 8 \\
4 & -1 & 3 & 0 & 1 & 12 \\
\hline
5 & -4 & 2 & 0 & 0 & 0
\end{array}\right].
$$

For Stage I, pivot on the 5 in row 1, column 3.

$$
\begin{array}{ccccc}
x_1 & x_2 & x_3 & s_1 & s_2 \\
\end{array}
$$
$$
\left[\begin{array}{ccccc|c}
0 & -\frac{2}{5} & 1 & -\frac{1}{5} & 0 & \frac{8}{5} \\
4 & -1 & 3 & 0 & 1 & 12 \\
\hline
5 & -4 & 2 & 0 & 0 & 0
\end{array}\right]
\begin{array}{l}
\frac{1}{5}R_1 \\
\\
\\
\end{array}
$$

$$
\begin{array}{ccccc}
x_1 & x_2 & x_3 & s_1 & s_2 \\
\end{array}
$$
$$
\left[\begin{array}{ccccc|c}
0 & -\frac{2}{5} & 1 & -\frac{1}{5} & 0 & \frac{8}{5} \\
4 & \boxed{\frac{1}{5}} & 0 & \frac{3}{5} & 1 & \frac{36}{5} \\
\hline
5 & -\frac{16}{5} & 0 & \frac{2}{5} & 0 & -\frac{16}{5}
\end{array}\right]
\begin{array}{l}
\\
-3R_1 + R_2 \\
-2R_1 + R_3
\end{array}
$$

This completes Stage I because. the solution given in the usual way from the matrix has nonnegative values for all variables. Since there are negative indicators in the objective row, we continue with Stage II. Now pivot on the $\frac{1}{5}$ in row 1, column 2.

$$
\begin{array}{ccccc}
x_1 & x_2 & x_3 & s_1 & s_2 \\
\end{array}
$$
$$
\left[\begin{array}{ccccc|c}
8 & 0 & 1 & 1 & 2 & 16 \\
4 & \frac{1}{5} & 0 & \frac{3}{5} & 1 & \frac{36}{5} \\
\hline
69 & 0 & 0 & 10 & 16 & 112
\end{array}\right]
\begin{array}{l}
R_1 + 2R_2 \\
\\
R_3 + 16R_2
\end{array}
$$

$$
\begin{array}{ccccc}
x_1 & x_2 & x_3 & s_1 & s_2 \\
\end{array}
$$
$$
\left[\begin{array}{ccccc|c}
8 & 0 & 1 & 1 & 2 & 16 \\
20 & 1 & 0 & 3 & 5 & 36 \\
\hline
69 & 0 & 0 & 10 & 16 & 112
\end{array}\right]
\begin{array}{l}
\\
5R_2 \\
\end{array}
$$

The maximum is 112 when $x_1 = 0, \ x_2 = 36, x_3 = 16.$

23. Maximize $z = 2x_1 - 3x_2 + 4x_3$

subject to: $x_1 + x_2 + x_3 \le 100$
$x_1 + x_2 + x_3 \ge 75$
$x_1 + x_2 \ge 27$
$x_1 \ge 0, \ x_2 \ge 0, \ x_3 \ge 0.$

The initial simplex tableau is

$$
\begin{array}{cccccc}
x_1 & x_2 & x_3 & s_1 & s_2 & s_3 \\
\end{array}
$$
$$
\left[\begin{array}{cccccc|c}
1 & 1 & 1 & 1 & 0 & 0 & 100 \\
1 & 1 & \boxed{1} & 0 & -1 & 0 & 75 \\
1 & 1 & 0 & 0 & 0 & -1 & 27 \\
\hline
-2 & 3 & -4 & 0 & 0 & 0 & 0
\end{array}\right].
$$

(continued on next page)

(continued from page 236)

For Stage I, pivot on the 1 in row 2, column 3.

$$\begin{array}{cccccc} x_1 & x_2 & x_3 & s_1 & s_2 & s_3 \end{array}$$
$$\begin{bmatrix} 0 & 0 & 0 & 1 & 1 & 0 & 25 \\ 1 & 1 & 1 & 0 & -1 & 0 & 75 \\ \boxed{1} & 1 & 0 & 0 & 0 & -1 & 27 \\ \hline 2 & 7 & 0 & 0 & -4 & 0 & 300 \end{bmatrix} \begin{array}{l} -R_2 + R_1 \\ \\ \\ 4R_2 + R_4 \end{array}$$

Now pivot on the 1 in row 3 column 1.

$$\begin{array}{cccccc} x_1 & x_2 & x_3 & s_1 & s_2 & s_3 \end{array}$$
$$\begin{bmatrix} 0 & 0 & 0 & 1 & 1 & 0 & 25 \\ 0 & 0 & 1 & 0 & -1 & 1 & 48 \\ 1 & 1 & 0 & 0 & 0 & -1 & 27 \\ \hline 0 & 5 & 0 & 0 & -4 & 2 & 246 \end{bmatrix} \begin{array}{l} \\ -R_3 + R_2 \\ \\ -2R_3 + R_4 \end{array}$$

This completes Stage I because the solution given in the usual way from the matrix has nonnegative values for all variables. Since there is a negative indicator in the objective row, we continue with Stage II.
Pivot on the 1 in row 1 column 5.

$$\begin{array}{cccccc} x_1 & x_2 & x_3 & s_1 & s_2 & s_3 \end{array}$$
$$\begin{bmatrix} 0 & 0 & 0 & 1 & \boxed{1} & 0 & 25 \\ 0 & 0 & 1 & 1 & 0 & 1 & 73 \\ 1 & 1 & 0 & 0 & 0 & -1 & 27 \\ \hline 0 & 5 & 0 & 4 & 0 & 2 & 346 \end{bmatrix} \begin{array}{l} \\ R_1 + R_2 \\ \\ 4R_1 + R_4 \end{array}$$

The maximum is 346 when
$x_1 = 27$, $x_2 = 0$, $x_3 = 73$.

In exercises 25–27, the two-stage program in Appendix A was used to produce the matrix giving a feasible solution. (We used the LINPROG program on a TI84 Plus.) A different program might produce a different feasible solution, but will produce the same final solution.

25. Minimize $w = 2y_1 + 5y_2 - 3y_3$

subject to: $y_1 + 2y_2 + 3y_3 \ge 115$
$2y_1 + y_2 + y_3 \le 200$
$y_1 + y_3 \ge 50$
$y_1 \ge 0,\ y_2 \ge 0,\ y_3 \ge 0.$

The initial simplex tableau is

$$\begin{array}{cccccc} y_1 & y_2 & y_3 & s_1 & s_2 & s_3 \end{array}$$
$$\begin{bmatrix} 1 & 2 & 3 & -1 & 0 & 0 & 115 \\ 2 & 1 & 1 & 0 & 1 & 0 & 200 \\ \boxed{1} & 0 & 1 & 0 & 0 & -1 & 50 \\ \hline 2 & 5 & -3 & 0 & 0 & 0 & 0 \end{bmatrix}$$

Stage I of the two-stage program produces the matrix

$$\begin{array}{cccccc} y_1 & y_2 & y_3 & s_1 & s_2 & s_3 \end{array}$$
$$\begin{bmatrix} 0 & 1 & 1 & -\frac{1}{2} & 0 & \frac{1}{2} & \frac{65}{2} \\ 0 & 0 & -2 & \frac{1}{2} & 1 & \frac{3}{2} & \frac{135}{2} \\ 1 & 0 & 1 & 0 & 0 & -1 & 50 \\ \hline 0 & 0 & -10 & \frac{5}{2} & 0 & -\frac{1}{2} & -\frac{525}{2} \end{bmatrix}.$$

This gives the feasible solution $y_1 = 50$,

$y_2 = \frac{65}{2}$, $y_3 = 0$, $s_1 = 0$, $s_2 = \frac{135}{2}$, and $s_3 = 0$.
Stage II produces the final matrix

$$\begin{array}{cccccc} y_1 & y_2 & y_3 & s_1 & s_2 & s_3 \end{array}$$
$$\begin{bmatrix} 2 & 1 & 1 & 0 & 1 & 0 & 200 \\ 1 & 1 & 0 & 0 & 1 & 1 & 150 \\ 5 & 1 & 0 & 1 & 3 & 0 & 485 \\ \hline 8 & 8 & 0 & 0 & 3 & 0 & 600 \end{bmatrix}.$$

The minimum is –600 when $y_1 = 0$, $y_2 = 0$, and $y_3 = 200$.

27. Minimize $w = 10y_1 + 8y_2 + 15y_3$

subject to: $y_1 + y_2 + y_3 \ge 12$
$5y_1 + 4y_2 + 9y_3 \ge 48$
$y_1 \ge 0,\ y_2 \ge 0,\ y_3 \ge 0.$

Insert surplus variables in both constraints to get the initial simplex tableau.

$$\begin{array}{ccccc} y_1 & y_2 & y_3 & s_1 & s_2 \end{array}$$
$$\begin{bmatrix} 1 & 1 & 1 & -1 & 0 & 12 \\ 5 & 4 & 9 & 0 & -1 & 48 \\ \hline 10 & 8 & 15 & 0 & 0 & 0 \end{bmatrix}$$

Stage I of the two-stage program produces the matrix

$$\begin{array}{ccccc} y_1 & y_2 & y_3 & s_1 & s_2 \end{array}$$
$$\begin{bmatrix} 0 & 1 & -4 & -5 & 1 & 12 \\ 1 & 0 & 5 & 4 & -1 & 0 \\ \hline 0 & 0 & -3 & 0 & 2 & -96 \end{bmatrix}.$$

This gives the feasible solution
$y_1 = 0$, $y_2 = 12$, $y_3 = 0$, $s_1 = 0$, and $s_2 = 0$.
Stage II produces the final matrix

$$\begin{array}{ccccc} y_1 & y_2 & y_3 & s_1 & s_2 \end{array}$$
$$\begin{bmatrix} \frac{4}{5} & 1 & 0 & -\frac{9}{5} & \frac{1}{5} & 12 \\ \frac{1}{5} & 0 & 1 & \frac{4}{5} & -\frac{1}{5} & 0 \\ \hline \frac{3}{5} & 0 & 0 & \frac{12}{5} & \frac{7}{5} & -96 \end{bmatrix}.$$

The minimum is 96 when $y_1 = 0$, $y_2 = 12$, and $y_3 = 0$.

29. Let y_1 = the amount of ingredient I per barrel of gasoline, let y_2 = the amount of ingredient II per barrel of gasoline, let y_3 = the amount of ingredient III per barrel of gasoline.

Minimize $w = .30y_1 + .09y_2 + .27y_3$

Maximize $z = -w = -.30y_1 - .09y_2 - .27y_3$

subject to:
$$y_1 + y_2 + y_3 \geq 10$$
$$y_1 + y_2 + y_3 \leq 15$$
$$y_1 - \tfrac{1}{4}y_2 \qquad \geq 0 \quad \left(\text{Since } y_1 \geq \tfrac{1}{4}y_2\right)$$
$$-y_1 \qquad + y_3 \geq 0 \quad (\text{Since } y_3 \geq y_1)$$
$$y_1 \geq 0, \ y_2 \geq 0, \ y_3 \geq 0$$

The initial simplex tableau is

$$
\begin{array}{ccccccc}
y_1 & y_2 & y_3 & s_1 & s_2 & s_3 & s_4 \\
\end{array}
$$

$$
\left[
\begin{array}{ccccccc|c}
1 & 1 & 1 & -1 & 0 & 0 & 0 & 10 \\
1 & 1 & 1 & 0 & 1 & 0 & 0 & 15 \\
1 & -\tfrac{1}{4} & 0 & 0 & 0 & -1 & 0 & 0 \\
-1 & 0 & 1 & 0 & 0 & 0 & -1 & 0 \\
\hline
.30 & .09 & .27 & 0 & 0 & 0 & 0 & 0
\end{array}
\right].
$$

31. Let y_1 = the number of computers from W_1 to D_1; let y_2 = the number of computers from W_2 to D_1; let y_3 = the number of computers from W_1 to D_2; let y_4 = the number of computers from W_2 to D_2.

Minimize $w = 14y_1 + 12y_2 + 22y_3 + 10y_4$

Maximize $z = -w = -14y_1 - 12y_2 - 22y_3 - 10y_4$

subject to:
$$y_1 + y_2 \qquad\qquad = 32$$
$$y_3 + y_4 = 20$$
$$y_1 + \qquad y_3 \qquad \leq 25$$
$$y_2 + \qquad y_4 \leq 30$$
$$y_1 \geq 0, \ y_2 \geq 0, y_3 \geq 0, \ y_4 \geq 0.$$

The initial simplex tableau is

$$
\begin{array}{cccccccccc}
y_1 & y_2 & y_3 & y_4 & s_1 & s_2 & s_3 & s_4 & s_5 & s_6 \\
\end{array}
$$

$$
\left[
\begin{array}{cccccccccc|c}
1 & 1 & 0 & 0 & -1 & 0 & 0 & 0 & 0 & 0 & 32 \\
1 & 1 & 0 & 0 & 0 & 1 & 0 & 0 & 0 & 0 & 32 \\
0 & 0 & 1 & 1 & 0 & 0 & -1 & 0 & 0 & 0 & 20 \\
0 & 0 & 1 & 1 & 0 & 0 & 0 & 1 & 0 & 0 & 20 \\
1 & 0 & 1 & 0 & 0 & 0 & 0 & 0 & 1 & 0 & 25 \\
0 & 1 & 0 & 1 & 0 & 0 & 0 & 0 & 0 & 1 & 30 \\
\hline
14 & 12 & 22 & 10 & 0 & 0 & 0 & 0 & 0 & 0 & 0
\end{array}
\right].
$$

33. Let x_1 = the number of barrels of oil supplied by S_1 to D_1.

Let x_2 = the number of barrels of oil supplied by S_2 to D_1.

Let x_3 = the number of barrels of oil supplied by S_1 to D_2.

Let x_4 = the number of barrels of oil supplied by S_2 to D_2.

Minimize $w = 30x_1 + 25x_2 + 20x_3 + 22x_4$

subject to:
$$x_1 + x_2 \qquad\qquad \geq 3000$$
$$x_3 + x_4 \geq 5000$$
$$x_1 \quad + x_3 \qquad \leq 5000$$
$$x_2 \qquad + x_4 \leq 5000$$
$$2x_1 + 5x_2 + 6x_3 + 4x_4 \leq 40,000$$

The initial simplex tableau is

x_1	x_2	x_3	x_4	s_1	s_2	s_3	s_4	s_5	
[1]	1	0	0	−1	0	0	0	0	3000
0	0	1	1	0	−1	0	0	0	5000
1	0	1	0	0	0	1	0	0	5000
0	1	0	1	0	0	0	1	0	5000
2	5	6	4	0	0	0	0	1	40,000
30	25	20	22	0	0	0	0	0	0

For Stage I, pivot on the 1 in row one, column one.

x_1	x_2	x_3	x_4	s_1	s_2	s_3	s_4	s_5		
1	1	0	0	−1	0	0	0	0	3000	
0	0	1	1	0	−1	0	0	0	5000	
0	−1	[1]	0	1	0	1	0	0	2000	$-R_1 + R_3$
0	1	0	1	0	0	0	1	0	5000	
0	3	6	4	2	0	0	0	1	34,000	$-2R_1 + R_5$
0	−5	20	22	30	0	0	0	0	−90,000	$-30R_1 + R_6$

To continue Stage I, pivot on the 1 in row three, column three.

x_1	x_2	x_3	x_4	s_1	s_2	s_3	s_4	s_5		
1	1	0	0	−1	0	0	0	0	3000	
0	1	0	[1]	−1	−1	−1	0	0	3000	$-R_3 + R_2$
0	−1	1	0	1	0	1	0	0	2000	
0	1	0	1	0	0	0	1	0	5000	
0	9	0	4	−4	0	−6	0	1	22,000	$-6R_3 + R_5$
0	15	0	22	10	0	−20	0	0	−130,000	$-20R_3 + R_6$

To continue Stage I, pivot on the 1 in row two, column four.

x_1	x_2	x_3	x_4	s_1	s_2	s_3	s_4	s_5		
1	1	0	0	−1	0	0	0	0	3000	
0	1	0	1	−1	−1	−1	0	0	3000	
0	−1	1	0	1	0	1	0	0	2000	
0	0	0	0	1	1	1	1	0	2000	$-R_2 + R_4$
0	[5]	0	0	0	4	−2	0	1	10,000	$-4R_2 + R_5$
0	−7	0	0	32	22	2	0	0	−196,000	$-22R_2 + R_6$

(continued on next page)

(*continued from page 239*)

We have completed Stage I. For Stage II, pivot on the 5 in row five, column two.

$$
\begin{array}{ccccccccc}
x_1 & x_2 & x_3 & x_4 & s_1 & s_2 & s_3 & s_4 & s_5 \\
\end{array}
$$

$$
\left[\begin{array}{ccccccccc|c}
1 & 1 & 0 & 0 & -1 & 0 & 0 & 0 & 0 & 3000 \\
0 & 1 & 0 & 1 & -1 & -1 & -1 & 0 & 0 & 3000 \\
0 & -1 & 1 & 0 & 1 & 0 & 1 & 0 & 0 & 2000 \\
0 & 0 & 0 & 0 & 1 & 1 & 1 & 1 & 0 & 2000 \\
0 & 1 & 0 & 0 & 0 & \frac{4}{5} & -\frac{2}{5} & 0 & \frac{1}{5} & 2000 \\
\hline
0 & -7 & 0 & 0 & 32 & 22 & 2 & 0 & 0 & -196{,}000
\end{array}\right]\begin{array}{l} \\ \\ \\ \\ \frac{1}{5}R_5 \\ \\ \end{array}
$$

$$
\begin{array}{ccccccccc}
x_1 & x_2 & x_3 & x_4 & s_1 & s_2 & s_3 & s_4 & s_5 \\
\end{array}
$$

$$
\left[\begin{array}{ccccccccc|c}
1 & 0 & 0 & 0 & -1 & -\frac{4}{5} & \frac{2}{5} & 0 & -\frac{1}{5} & 1000 \\
0 & 0 & 0 & 1 & -1 & -\frac{9}{5} & -\frac{3}{5} & 0 & -\frac{1}{5} & 1000 \\
0 & 0 & 1 & 0 & 1 & \frac{4}{5} & \frac{3}{5} & 0 & \frac{1}{5} & 4000 \\
0 & 0 & 0 & 0 & 1 & 1 & \boxed{1} & 1 & 0 & 2000 \\
0 & 1 & 0 & 0 & 0 & \frac{4}{5} & -\frac{2}{5} & 0 & \frac{1}{5} & 2000 \\
\hline
0 & 0 & 0 & 0 & 32 & \frac{138}{5} & -\frac{4}{5} & 0 & \frac{7}{5} & -182{,}000
\end{array}\right]\begin{array}{l} -R_5+R_1 \\ -R_5+R_2 \\ R_5+R_3 \\ \\ \\ 7R_5+R_6 \end{array}
$$

Finally, pivot on the 1 in row four, column seven.

$$
\begin{array}{ccccccccc}
x_1 & x_2 & x_3 & x_4 & s_1 & s_2 & s_3 & s_4 & s_5 \\
\end{array}
$$

$$
\left[\begin{array}{ccccccccc|c}
1 & 0 & 0 & 0 & -\frac{7}{5} & -\frac{6}{5} & 0 & -\frac{2}{5} & -\frac{1}{5} & 200 \\
0 & 0 & 0 & 1 & -\frac{2}{5} & -\frac{6}{5} & 0 & \frac{3}{5} & -\frac{1}{5} & 2200 \\
0 & 0 & 1 & 0 & \frac{2}{5} & \frac{1}{5} & 0 & -\frac{3}{5} & \frac{1}{5} & 2800 \\
0 & 0 & 0 & 0 & 1 & 1 & 1 & 1 & 0 & 2000 \\
0 & 1 & 0 & 0 & \frac{2}{5} & \frac{6}{5} & 0 & \frac{2}{5} & \frac{1}{5} & 2800 \\
\hline
0 & 0 & 0 & 0 & \frac{164}{5} & \frac{142}{5} & 0 & \frac{4}{5} & \frac{7}{5} & -180{,}400
\end{array}\right]\begin{array}{l} -\frac{2}{5}R_4+R_1 \\ \frac{3}{5}R_4+R_2 \\ -\frac{3}{5}R_4+R_3 \\ \\ \frac{2}{5}R_4+R_5 \\ \frac{4}{5}R_4+R_6 \end{array}
$$

This gives $x_1 = 200$, $x_2 = 2800$, $x_3 = 2800$, and $x_4 = 2200$ for a minimum cost of \$180,400.

35. Let y_1 = the amount of bluegrass seed, y_2 = the amount of rye seed, and y_3 = the amount of Bermuda seed. The problem is to minimize $w = .12y_1 + .15y_2 + .05y_3$ subject to

$$y_1 \geq .2(y_1 + y_2 + y_3)$$
$$y_3 \leq \frac{2}{3}y_2$$
$$y_1 + y_2 + y_3 \geq 5000$$

Or, maximize: $z = -w = -.12y_1 - .15y_2 - .05y_3$ subject to

$$.8y_1 - .2y_2 - .2y_3 \geq 0$$
$$2y_2 - 3y_3 \geq 0$$
$$y_1 + y_2 + y_3 \geq 5000$$

Adding surplus variables s_1, s_2, and s_3, the initial tableau is

$$
\begin{array}{cccccc}
y_1 & y_2 & y_3 & s_1 & s_2 & s_3 \\
\end{array}
$$

$$
\left[
\begin{array}{cccccc|c}
\boxed{.8} & -.2 & -.2 & -1 & 0 & 0 & 0 \\
0 & 2 & -3 & 0 & -1 & 0 & 0 \\
1 & 1 & 1 & 0 & 0 & -1 & 5000 \\
\hline
.12 & .15 & .05 & 0 & 0 & 0 & 0 \\
\end{array}
\right]
$$

Since y_4 through y_6 are negative, this does not have a feasible solution. For Stage I, use the .8 in row one, column one.

$$
\begin{array}{cccccc}
y_1 & y_2 & y_3 & s_1 & s_2 & s_3 \\
\end{array}
$$

$$
\left[
\begin{array}{cccccc|c}
1 & -.25 & -.25 & -1.25 & 0 & 0 & 0 \\
0 & \boxed{2} & -3 & 0 & -1 & 0 & 0 \\
0 & 1.25 & 1.25 & 1.25 & 0 & -1 & 5000 \\
\hline
0 & .18 & .08 & .15 & 0 & 0 & 0 \\
\end{array}
\right]
\begin{array}{l}
1.25R_1 \\
\\
-R_1 + R_3 \\
-.12R_1 + R_4 \\
\end{array}
$$

Continue Stage 1 by pivoting on the 2 in row two, column two.

$$
\begin{array}{cccccc}
y_1 & y_2 & y_3 & s_1 & s_2 & s_3 \\
\end{array}
$$

$$
\left[
\begin{array}{cccccc|c}
1 & 0 & -.625 & -1.25 & -.125 & 0 & 0 \\
0 & 1 & -1.5 & 0 & -.5 & 0 & 0 \\
0 & 0 & \boxed{3.125} & 1.25 & .625 & -1 & 5000 \\
\hline
0 & 0 & .35 & .15 & .09 & 0 & 0 \\
\end{array}
\right]
\begin{array}{l}
.25R_2 + R_1 \\
.5R_2 \\
-1.25R_1 + R_3 \\
-.18R_1 + R_4 \\
\end{array}
$$

To finish Stage 1, pivot on the 3.125 in row three, column three.

$$
\begin{array}{cccccc}
y_1 & y_2 & y_3 & s_1 & s_2 & s_3 \\
\end{array}
$$

$$
\left[
\begin{array}{cccccc|c}
1 & 0 & 0 & -1 & 0 & -.2 & 1000 \\
0 & 1 & 0 & .6 & -.2 & -.48 & 2400 \\
0 & 0 & 1 & .4 & .2 & -.32 & 1600 \\
\hline
0 & 0 & 0 & .01 & .02 & .112 & -560 \\
\end{array}
\right]
\begin{array}{l}
.625R_3 + R_1 \\
1.5R_3 + R_2 \\
.32R_3 \\
-.35R_3 + R_4 \\
\end{array}
$$

Stage II pivoting is not necessary. The solution is $y_1 = 1000$, $y_2 = 2400$, $y_3 = 1600$ and $w = 560$. In other words, Topgrade Turf should use 1000 lbs. of bluegrass seed, 2400 lbs. of rye seed, and 1600 lbs. of Bermuda seed for a minimum cost of $560.

37. Let x_1 = amount allotted to commercial loans (in millions) and x_2 = amount allotted to home loans (in millions)

Maximize $z = .10x_1 + .12x_2$

subject to: $x_1 + x_2 \le 25$
$2x_1 + 3x_2 \le 72$
$-4x_1 + x_2 \ge 0$
$x_1 + x_2 \ge 10$
$x_1 \ge 0, x_2 \ge 0$

The initial simplex tableau is

$$
\begin{array}{cccccc}
x_1 & x_2 & s_1 & s_2 & s_3 & s_4 \\
\end{array}
$$
$$
\left[\begin{array}{cccccc|c}
1 & 1 & 1 & 0 & 0 & 0 & 25 \\
2 & 3 & 0 & 1 & 0 & 0 & 72 \\
-4 & \boxed{1} & 0 & 0 & -1 & 0 & 0 \\
1 & 1 & 0 & 0 & 0 & -1 & 10 \\
\hline
-.10 & -.12 & 0 & 0 & 0 & 0 & 0
\end{array}\right]
$$

For Stage I, pivot on the 1 in row three, column two.

$$
\begin{array}{cccccc}
x_1 & x_2 & s_1 & s_2 & s_3 & s_4 \\
\end{array}
$$
$$
\left[\begin{array}{cccccc|c}
5 & 0 & 1 & 0 & 1 & 0 & 25 \\
14 & 0 & 0 & 1 & 3 & 0 & 72 \\
-4 & 1 & 0 & 0 & -1 & 0 & 0 \\
\boxed{5} & 0 & 0 & 0 & 1 & -1 & 10 \\
\hline
-\frac{29}{50} & 0 & 0 & 0 & -\frac{3}{25} & 0 & 0
\end{array}\right]
\begin{array}{l}
-R_3 + R_1 \\
-3R_3 + R_2 \\
\\
-R_3 + R_4 \\
\frac{3}{25}R_3 + R_5
\end{array}
$$

To continue Stage I, pivot on the 5 in row four, column one.

$$
\begin{array}{cccccc}
x_1 & x_2 & s_1 & s_2 & s_3 & s_4 \\
\end{array}
$$
$$
\left[\begin{array}{cccccc|c}
0 & 0 & 1 & 0 & 0 & \boxed{1} & 15 \\
0 & 0 & 0 & 1 & \frac{1}{5} & \frac{14}{5} & 44 \\
0 & 1 & 0 & 0 & -\frac{1}{5} & -\frac{4}{5} & 8 \\
1 & 0 & 0 & 0 & \frac{1}{5} & -\frac{1}{5} & 2 \\
\hline
0 & 0 & 0 & 0 & -\frac{1}{250} & -\frac{29}{250} & \frac{29}{25}
\end{array}\right]
\begin{array}{l}
-5R_4 + R_1 \\
-14R_4 + R_2 \\
4R_4 + R_3 \\
\frac{1}{5}R_4 \\
\frac{29}{50}R_4 + R_5
\end{array}
$$

For Stage II, pivot on the 1 in row one, column six.

$$
\begin{array}{cccccc}
x_1 & x_2 & s_1 & s_2 & s_3 & s_4 \\
\end{array}
$$
$$
\left[\begin{array}{cccccc|c}
0 & 0 & 1 & 0 & 0 & 1 & 15 \\
0 & 0 & -\frac{14}{5} & 1 & \boxed{\frac{1}{5}} & 0 & 2 \\
0 & 1 & \frac{4}{5} & 0 & -\frac{1}{5} & 0 & 20 \\
1 & 0 & \frac{1}{5} & 0 & \frac{1}{5} & 0 & 5 \\
\hline
0 & 0 & \frac{29}{250} & 0 & -\frac{1}{250} & 0 & \frac{29}{10}
\end{array}\right]
\begin{array}{l}
\\
-\frac{14}{5}R_1 + R_2 \\
\frac{4}{5}R_1 + R_3 \\
\frac{1}{5}R_1 + R_4 \\
\frac{29}{250}R_1 + R_5
\end{array}
$$

To continue stage II, pivot on the $\frac{1}{5}$ in row two, column five.

$$
\begin{array}{cccccc}
x_1 & x_2 & s_1 & s_2 & s_3 & s_4 \\
\end{array}
$$
$$
\left[\begin{array}{cccccc|c}
0 & 0 & 1 & 0 & 0 & 1 & 15 \\
0 & 0 & -14 & 5 & 1 & 0 & 10 \\
0 & 1 & -2 & 1 & 0 & 0 & 22 \\
1 & 0 & 3 & -1 & 0 & 0 & 3 \\
\hline
0 & 0 & \frac{3}{50} & \frac{1}{50} & 0 & 0 & \frac{147}{50}
\end{array}\right]
\begin{array}{l}
\\
5R_2 \\
\frac{1}{5}R_2 + R_3 \\
-\frac{1}{5}R_2 + R_4 \\
\frac{1}{250}R_2 + R_5
\end{array}
$$

This gives $x_1 = 3$, $x_2 = 22$, and the maximum is

$\frac{147}{50} = 2.94$. Allot \$3,000,000 for commercial loans and \$22,000,000 for home loans for a maximum interest income of \$2,940,000.

39. Let y_1 = the amount of regular beer and y_2 = the amount of light beer.

Minimize $36,000y_1 + 48,000y_2$

subject to: $y_1 \ge 12$
$y_2 \ge 10$
$-2y_1 + y_2 \le 0$ (Since $y_2 \le 2y_1$)
$y_1 + y_2 \ge 42.$

(Since the company already produces at least $12 + 10 = 22$ units and can produce at least 20 additional units: $y_1 + y_2 \ge 22 + 20 = 42.$)

The initial simplex tableau is

$$
\begin{array}{cccccc}
y_1 & y_2 & s_1 & s_2 & s_3 & s_4 \\
\end{array}
$$
$$
\left[\begin{array}{cccccc|c}
\boxed{1} & 0 & -1 & 0 & 0 & 0 & 12 \\
0 & 1 & 0 & -1 & 0 & 0 & 10 \\
-2 & 1 & 0 & 0 & 1 & 0 & 0 \\
1 & 1 & 0 & 0 & 0 & -1 & 42 \\
\hline
36 & 48 & 0 & 0 & 0 & 0 & 0
\end{array}\right].
$$

For Stage I, pivot on the 1 in row one, column one.

$$
\begin{array}{cccccc}
y_1 & y_2 & s_1 & s_2 & s_3 & s_4 \\
\end{array}
$$
$$
\left[\begin{array}{cccccc|c}
1 & 0 & -1 & 0 & 0 & 0 & 12 \\
0 & \boxed{1} & 0 & -1 & 0 & 0 & 10 \\
0 & 1 & -2 & 0 & 1 & 0 & 24 \\
0 & 1 & 1 & 0 & 0 & -1 & 30 \\
\hline
0 & 48 & 36 & 0 & 0 & 0 & -432
\end{array}\right]
\begin{array}{l}
\\
\\
2R_1 + R_3 \\
-R_1 + R_4 \\
-36R_1 + R_5
\end{array}
$$

To continue Stage I, pivot on the 1 in row two, column two.

(continued on next page)

(continued from page 242)

$$
\begin{array}{cccccc}
y_1 & y_2 & s_1 & s_2 & s_3 & s_4 \\
\end{array}
$$

$$
\left[\begin{array}{cccccc|c}
1 & 0 & -1 & 0 & 0 & 0 & 12 \\
0 & 1 & 0 & -1 & 0 & 0 & 10 \\
0 & 0 & -2 & 1 & 1 & 0 & 14 \\
0 & 0 & 1 & \boxed{1} & 0 & -1 & 20 \\
\hline
0 & 0 & 36 & 48 & 0 & 0 & -912
\end{array}\right]
\begin{array}{l}
\\
\\
-R_2+R_3 \\
-R_2+R_4 \\
-48R_2+R_5
\end{array}
$$

Now pivot on the 1 in row four, column four.

$$
\begin{array}{cccccc}
y_1 & y_2 & s_1 & s_2 & s_3 & s_4 \\
\end{array}
$$

$$
\left[\begin{array}{cccccc|c}
1 & 0 & -1 & 0 & 0 & 0 & 12 \\
0 & 1 & 1 & 0 & 0 & -1 & 30 \\
0 & 0 & \boxed{-3} & 0 & 1 & 1 & -6 \\
0 & 0 & 1 & 1 & 0 & -1 & 20 \\
\hline
0 & 0 & -12 & 0 & 0 & 48 & -1872
\end{array}\right]
\begin{array}{l}
\\
R_4+R_2 \\
-R_4+R_3 \\
\\
-48R_4+R_5
\end{array}
$$

For Stage II, pivot on the –3 in row three, column three.

$$
\begin{array}{cccccc}
y_1 & y_2 & s_1 & s_2 & s_3 & s_4 \\
\end{array}
$$

$$
\left[\begin{array}{cccccc|c}
1 & 0 & 0 & 0 & -\frac{1}{3} & -\frac{1}{3} & 14 \\
0 & 1 & 0 & 0 & \frac{1}{3} & -\frac{2}{3} & 28 \\
0 & 0 & 1 & 0 & -\frac{1}{3} & -\frac{1}{3} & 2 \\
0 & 0 & 0 & 1 & \boxed{\frac{1}{3}} & -\frac{2}{3} & 18 \\
\hline
0 & 0 & 0 & 0 & -4 & 44 & -1848
\end{array}\right]
\begin{array}{l}
R_3+R_1 \\
-R_3+R_2 \\
-\frac{1}{3}R_3 \\
-R_3+R_4 \\
12R_3+R_5
\end{array}
$$

Finally, pivot on the $\frac{1}{3}$ in row four, column five.

$$
\begin{array}{cccccc}
y_1 & y_2 & s_1 & s_2 & s_3 & s_4 \\
\end{array}
$$

$$
\left[\begin{array}{cccccc|c}
1 & 0 & 0 & 1 & 0 & -1 & 32 \\
0 & 1 & 0 & -1 & 0 & 0 & 10 \\
0 & 0 & 1 & 1 & 0 & -1 & 20 \\
0 & 0 & 0 & 3 & 1 & -2 & 54 \\
\hline
0 & 0 & 0 & 12 & 0 & 36 & -1632
\end{array}\right]
\begin{array}{l}
\frac{1}{3}R_4+R_1 \\
-\frac{1}{3}R_4+R_2 \\
\frac{1}{3}R_4+R_3 \\
3R_4 \\
4R_4+R_5
\end{array}
$$

Make 32 units of regular beer and 10 units of light beer for a minimum cost of $1,632,000.

41. From Exercise 29, we know the initial simplex tableau is

$$
\begin{array}{ccccccc}
y_1 & y_2 & y_3 & s_1 & s_2 & s_3 & s_4 \\
\end{array}
$$

$$
\left[\begin{array}{ccccccc|c}
1 & 1 & 1 & -1 & 0 & 0 & 0 & 10 \\
1 & 1 & 1 & 0 & 1 & 0 & 0 & 15 \\
1 & -\frac{1}{4} & 0 & 0 & 0 & -1 & 0 & 0 \\
-1 & 0 & 1 & 0 & 0 & 0 & -1 & 0 \\
\hline
.30 & .09 & .27 & 0 & 0 & 0 & 0 & 0
\end{array}\right].
$$

Using the LINPROG program on a TI-84 Plus, we find the optimal solution is

$$y_1 = 1\frac{2}{3}, \quad y_2 = 6\frac{2}{3}, \quad \text{and} \quad y_3 = 1\frac{2}{3}.$$

For each barrel of gasoline, the mixture should contain $1\frac{2}{3}$ ounces of ingredient I, $6\frac{2}{3}$ ounces of ingredient II, and $1\frac{2}{3}$ ounces of ingredient III at a minimum cost of $1.55 per barrel.

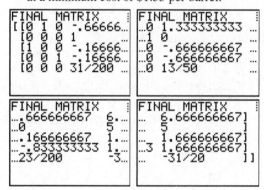

43. From Exercise 31, we know the initial simplex tableau is

y_1	y_2	y_3	y_4	s_1	s_2	s_3	s_4	s_5	s_6	
1	1	0	0	-1	0	0	0	0	0	32
1	1	0	0	0	1	0	0	0	0	32
0	0	1	1	0	0	-1	0	0	0	20
0	0	1	1	0	0	0	1	0	0	20
1	0	1	0	0	0	0	0	1	0	25
0	1	0	1	0	0	0	0	0	1	30
14	12	22	10	0	0	0	0	0	0	0

Using the LINPROG program on a TI-84 Plus, we find the optimal solution is $y_1 = 22$, $y_2 = 0$, $y_3 = 10$, and $y_4 = 20$. The manufacturer should send 22 computers from W_1 to D_1, 0 from W_1 to D_2, 10 from W_2 to D_1, and 20 computers from W_2 to D_2 for a minimum cost of $628.

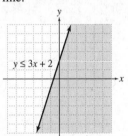

Chapter 7 Review Exercises

1. $y \le 3x + 2$

Graph the solid boundary line $y = 3x + 2$ which goes through the points (0, 2) and (2, 8). The inequality is $\le$, so shade the half-plane below the line.

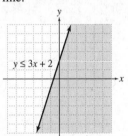

2. $2x - y \ge 6$

Graph the solid boundary line $2x - y = 6$, which goes through the points (0, –6) and (3, 0). Testing the origin gives the statement $2(0) - 0 \ge 6$, which is false, so shade the half-plane not containing the origin, which is the region below the line.

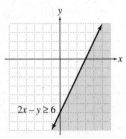

3. $3x + 4y \ge 12$

Graph the solid boundary line $3x + 4y = 12$, which goes through (0, 3) and (4, 0). Testing the origin gives $3(0) + 4(0) \ge 12$, which is false, so shade the half-plane not containing the origin, which is the region above the line.

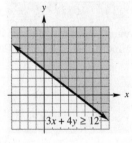

4. $y \leq 4$

Graph the solid boundary line $y = 4$, which is the horizontal line crossing the y-axis at (0, 4). Shade the half-plane below the line.

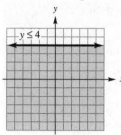

5. $x + y \leq 6$
    $2x - y \geq 3$

$x + y \leq 6$ is the region on or below the line $x + y = 6$; $2x - y \geq 3$ is the region on or below the line $2x - y = 3$. The system of inequalities must meet both conditions so we shade the overlap of the two half-planes.

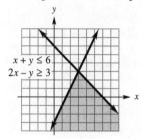

6. $4x + y \geq 8$
    $2x - 3y \leq 6$

Graph $4x + y = 8$ as a solid line using (2, 0) and (0, 8), and $2x - 3y = 6$ as a solid line using (3, 0) and (0, –2). The test point (0, 0) gives $0 + 0 \geq 8$, which is false and $0 - 0 \leq 6$, which is true. Shade all points above $4x + y = 8$ and above $2x - 3y = 6$. The solution region is to the right of $4x + y = 8$ and above $2x - 3y = 6$.

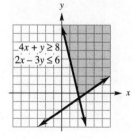

7. $2 \leq x \leq 5$
    $1 \leq y \leq 6$
    $x - y \leq 3$

The graph of $2 \leq x \leq 5$ is the region lying on or between the two vertical lines $x = 2$ and $x = 5$. The graph of $1 \leq y \leq 6$ is the region lying on or between the two horizontal lines $y = 1$ and $y = 6$. The graph of $x - y \leq 3$ is the region lying on or below the line $x - y = 3$. Shade the region that is common to all three graphs.

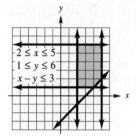

8. $x + 2y \leq 4$
    $2x - 3y \leq 6$
        $x \geq 0$
        $y \geq 0$

Graph $x + 2y = 4$ and $2x - 3y = 6$ as solid lines. $x \geq 0$ and $y \geq 0$ restrict the region to quadrant I. Use (0, 0) as a test point to get $0 + 0 \leq 4$ and $0 + 0 \leq 6$, which are true. The region is all points on or below $x + 2y = 4$, and on or above $2x - 3y = 6$ in quadrant I.

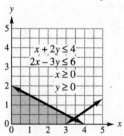

9. Let x = the number of batches of cakes and let y = the number of batches of cookies. Then we have the following inequalities:

$2x + \frac{3}{2}y \leq 15$ Oven time

$3x + \frac{2}{3}y \leq 13$ Decorating

$\quad\quad x \geq 0$

$\quad\quad y \geq 0.$

The solution of this system of inequalities is the graph of the feasible region.

(continued on next page)

(continued from page 245)

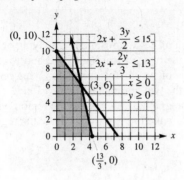

$(0, 10)$

$2x + \frac{3y}{2} \le 15$

$3x + \frac{2y}{3} \le 13$

$(3, 6)$ $x \ge 0$ $y \ge 0$

$\left(\frac{13}{3}, 0\right)$

10. Let x = the number of units of special and
let y = the number of basic
Then we have the following inequalities:
$$5x + 4y \le 100$$
$$2x + y \le 32$$
$$x \ge 6$$
$$y \ge 4$$

The solution of this system is the graph of the
feasible region.

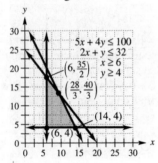

$5x + 4y \le 100$
$2x + y \le 32$
$x \ge 6$
$y \ge 4$

$\left(6, \frac{35}{2}\right)$

$\left(\frac{28}{3}, \frac{40}{3}\right)$

$(14, 4)$

$(6, 4)$

11.

Corner Point	Value of $z = 3x + 4y$
$(1, 6)$	27
$(6, 7)$	46 Maximum
$(7, 3)$	33
$\left(1, 2\frac{1}{2}\right)$	13
$(2, 1)$	10 Minimum

The maximum value of 46 occurs at $(6, 7)$ the
minimum value of 10 occurs at $(2, 1)$.

12.

Corner Point	Value of $z = 3x + 4y$
$(0, 8)$	32
$(8, 8)$	56 Maximum
$(5, 2)$	23
$(2, 0)$	6 Minimum

The maximum value is 56 at $(8, 8)$; the minimum
value is 6 at $(2, 0)$.

13. Maximize $z = 6x + 2y$
subject to $2x + 7y \le 14$
$2x + 3y \le 10$
$x \ge 0, \ y \ge 0.$
Graph the feasible region.

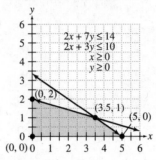

$2x + 7y \le 14$
$2x + 3y \le 10$
$x \ge 0$
$y \ge 0$

$(0, 2)$ $(3.5, 1)$ $(5, 0)$

$(0, 0)$

Corner Point	Value of $z = 6x + 2y$
$(0, 0)$	0
$(0, 2)$	4
$(3.5, 1)$	23
$(5, 0)$	30 Maximum

The maximum is 30 when $x = 5, y = 0$.

14. Find $x \ge 0$ and $y \ge 0$ such that
$$8x + 9y \ge 72$$
$$6x + 8y \ge 72$$
and $w = 2x + 10y$ is minimized.
Graph the feasible region.

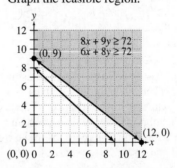

$8x + 9y \ge 72$
$6x + 8y \ge 72$

$(0, 9)$

$(12, 0)$

$(0, 0)$

Corner Point	Value of $z = 2x + 10y$
$(0, 9)$	90
$(12, 0)$	24 Minimum

The minimum is 24 when $x = 12, y = 0$.

15. Find $x \geq 0$ and $y \geq 0$ such that
$$x + y \leq 50$$
$$2x + y \geq 20$$
$$x + 2y \geq 30$$
and $w = 5x + 2y$ is minimized.
Graph the feasible region.

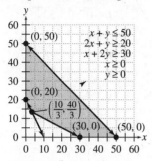

Corner Point	Value of $w = 5x + 2y$
(0, 20)	40 Minimum
(0, 50)	100
(50, 0)	250
(30, 0)	150
$\left(\frac{10}{3}, \frac{40}{3}\right)$	$\frac{130}{3} = 43\frac{1}{3}$

The minimum is 40 when $x = 0$, $y = 20$.

16. Maximize $z = 5x - 2y$
subject to: $3x + 2y \leq 12$
$5x + y \geq 5$
$x \geq 0$, $y \geq 0$.

Graph the feasible region.

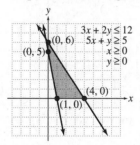

Corner Point	Value of $z = 5x - 2y$
(0, 5)	−10
(0, 6)	−12
(4, 0)	20 Maximum
(1, 0)	5

The maximum is 20 when $x = 4$, $y = 0$.

17. From the graph for Exercise 9, the corner points are (0, 10), (3, 6), $\left(\frac{13}{3}, 0\right)$, and (0, 0). Since x was the number of batches of cakes and y the number of batches of cookies, the revenue function is $z = 30x + 20y$. Evaluate this objective function at each corner point.

Corner Point	Value of $z = 30x + 20y$
(0, 10)	200
(3, 6)	210 Maximum
$\left(\frac{13}{3}, 0\right)$	130
(0, 0)	0

The bakery should make 3 batches of cakes and 6 batches of cookies to produce a maximum profit of $210.

18. The revenue function is $z = 20x + 18y$. Refer to the graph in Exercise 10, and evaluate at the corner points.

Corner Point	Value of $z = 20x + 18y$
(6, 4)	192
$\left(6, \frac{35}{2}\right)$	435 Maximum
$\left(\frac{28}{3}, \frac{40}{3}\right)$	$\frac{1280}{3} \approx 426.67$
(14, 4)	352

A maximum revenue of $435 is obtained by making 6 units of special pizza and $\frac{35}{2}$ $\left(\text{or } 17\frac{1}{2}\right)$ units of basic pizza.

19. Maximize $z = 5x_1 + 6x_2 + 3x_3$
subject to: $x_1 + x_2 + x_3 \leq 100$
$2x_1 + 3x_2 \leq 500$
$x_1 + 2x_3 \leq 350$
$x_1 \geq 0$, $x_2 \geq 0$, $x_3 \geq 0$.

a. $x_1 + x_2 + x_3 + s_1 = 100$
$2x_1 + 3x_2 + s_2 = 500$
$x_1 + 2x_3 + s_3 = 350$

b.

$$
\begin{array}{cccccc}
x_1 & x_2 & x_3 & s_1 & s_2 & s_3 \\
\end{array}
$$

$$
\left[\begin{array}{cccccc|c}
1 & 1 & 1 & 1 & 0 & 0 & 100 \\
2 & 3 & 0 & 0 & 1 & 0 & 500 \\
1 & 0 & 2 & 0 & 0 & 1 & 350 \\
\hline
-5 & -6 & -3 & 0 & 0 & 0 & 0
\end{array}\right]
$$

20. Maximize $z = 2x_1 + 9x_2$

subject to: $3x_1 + 5x_2 \le 47$
$\quad\quad\quad\quad x_1 + x_2 \le 25$
$\quad\quad\quad\quad 5x_1 + 2x_2 \le 35$
$\quad\quad\quad\quad 2x_1 + x_2 \le 30$
$\quad\quad\quad\quad x_1 \ge 0,\ x_2 \ge 0.$

a. $\quad 3x_1 + 5x_2 + s_1 \quad\quad\quad\quad\quad = 47$
$\quad\quad x_1 + x_2 \quad + s_2 \quad\quad\quad = 25$
$\quad\quad 5x_1 + 2x_2 \quad\quad + s_3 \quad\quad = 35$
$\quad\quad 2x_1 + x_2 \quad\quad\quad\quad + s_4 = 30$

b.

$$
\begin{array}{cccccc}
x_1 & x_2 & s_1 & s_2 & s_3 & s_4 \\
\end{array}
$$

$$
\left[\begin{array}{cccccc|c}
3 & 5 & 1 & 0 & 0 & 0 & 47 \\
1 & 1 & 0 & 1 & 0 & 0 & 25 \\
5 & 2 & 0 & 0 & 1 & 0 & 35 \\
2 & 1 & 0 & 0 & 0 & 1 & 30 \\
\hline
-2 & -9 & 0 & 0 & 0 & 0 & 0
\end{array}\right]
$$

21. Maximize $z = x_1 + 8x_2 + 2x_3$

subject to: $\quad x_1 + x_2 + x_3 \le 90$
$\quad\quad\quad 2x_1 + 5x_2 + x_3 \le 120$
$\quad\quad\quad x_1 + 3x_2 \quad\quad \le 80$
$\quad\quad x_1 \ge 0,\ x_2 \ge 0,\ x_3 \ge 0.$

a. $\quad x_1 + x_2 + x_3 + s_1 \quad\quad\quad = 90$
$\quad\quad 2x_1 + 5x_2 + x_3 \quad + s_2 \quad = 120$
$\quad\quad x_1 + 3x_2 \quad\quad\quad\quad + s_3 = 80$

b.

$$
\begin{array}{cccccc}
x_1 & x_2 & x_3 & s_1 & s_2 & s_3 \\
\end{array}
$$

$$
\left[\begin{array}{cccccc|c}
1 & 1 & 1 & 1 & 0 & 0 & 90 \\
2 & 5 & 1 & 0 & 1 & 0 & 120 \\
1 & 3 & 0 & 0 & 0 & 1 & 80 \\
\hline
-1 & -8 & -2 & 0 & 0 & 0 & 0
\end{array}\right]
$$

22. Maximize $z = 15x_1 + 12x_2$

subject to: $2x_1 + 5x_2 \le 50$
$\quad\quad\quad\quad x_1 + 3x_2 \le 25$
$\quad\quad\quad\quad 4x_1 + x_2 \le 18$
$\quad\quad\quad\quad x_1 + x_2 \le 12$
$\quad\quad\quad\quad x_1 \ge 0,\ x_2 \ge 0.$

a. $\quad 2x_1 + 5x_2 + s_1 \quad\quad\quad\quad\quad = 50$
$\quad\quad x_1 + 3x_2 \quad + s_2 \quad\quad\quad = 25$
$\quad\quad 4x_1 + x_2 \quad\quad + s_3 \quad\quad = 18$
$\quad\quad x_1 + x_2 \quad\quad\quad\quad + s_4 = 12$

b.

$$
\begin{array}{cccccc}
x_1 & x_2 & s_1 & s_2 & s_3 & s_4 \\
\end{array}
$$

$$
\left[\begin{array}{cccccc|c}
2 & 5 & 1 & 0 & 0 & 0 & 50 \\
1 & 3 & 0 & 1 & 0 & 0 & 25 \\
4 & 1 & 0 & 0 & 1 & 0 & 18 \\
1 & 1 & 0 & 0 & 0 & 1 & 12 \\
\hline
-15 & -12 & 0 & 0 & 0 & 0 & 0
\end{array}\right]
$$

23.

$$
\begin{array}{ccccc}
x_1 & x_2 & x_3 & s_1 & s_2 \\
\end{array}
$$

$$
\left[\begin{array}{ccccc|c}
1 & 2 & 3 & 1 & 0 & 28 \\
\boxed{2} & 4 & 8 & 0 & 1 & 32 \\
\hline
-5 & -2 & -3 & 0 & 0 & 0
\end{array}\right]
$$

Pivot on the 2 in row two, column one.

$$
\begin{array}{ccccc}
x_1 & x_2 & x_3 & s_1 & s_2 \\
\end{array}
$$

$$
\left[\begin{array}{ccccc|cl}
1 & 2 & 3 & 1 & 0 & 28 & \\
1 & 2 & 4 & 0 & \frac{1}{2} & 16 & \frac{1}{2}R_2 \\
\hline
-5 & -2 & -3 & 0 & 0 & 0 &
\end{array}\right]
$$

$$
\begin{array}{ccccc}
x_1 & x_2 & x_3 & s_1 & s_2 \\
\end{array}
$$

$$
\left[\begin{array}{ccccc|cl}
0 & 0 & -1 & 1 & -\frac{1}{2} & 12 & -R_2 + R_1 \\
1 & 2 & 4 & 0 & \frac{1}{2} & 16 & \\
\hline
0 & 8 & 17 & 0 & \frac{5}{2} & 80 & 5R_2 + R_3
\end{array}\right]
$$

The maximum is 80 when
$x_1 = 16,\ x_2 = 0,\ x_3 = 0,\ s_1 = 12,\ s_2 = 0.$

24.

$$
\begin{array}{cccc}
x_1 & x_2 & s_1 & s_2 \\
\end{array}
$$

$$
\left[\begin{array}{cccc|c}
2 & 1 & 1 & 0 & 10 \\
9 & \boxed{3} & 0 & 1 & 15 \\
\hline
-2 & -3 & 0 & 0 & 0
\end{array}\right]
$$

Pivot on the 3 in row two, column two.

$$
\begin{array}{cccc}
x_1 & x_2 & s_1 & s_2 \\
\end{array}
$$

$$
\left[\begin{array}{cccc|cl}
2 & 1 & 1 & 0 & 10 & \\
3 & 1 & 0 & \frac{1}{3} & 5 & \frac{1}{3}R_2 \\
\hline
-2 & -3 & 0 & 0 & 0 &
\end{array}\right]
$$

$$
\begin{array}{cccc}
x_1 & x_2 & s_1 & s_2 \\
\end{array}
$$

$$
\left[\begin{array}{cccc|cl}
-1 & 0 & 1 & -\frac{1}{3} & 5 & -R_2 + R_1 \\
3 & 1 & 0 & \frac{1}{3} & 5 & \\
\hline
7 & 0 & 0 & 1 & 15 & 3R_2 + R_3
\end{array}\right]
$$

The maximum is 15 when $x_1 = 0,\ x_2 = 5,\ s_1 = 5,$
and $s_2 = 0.$

25.

$$\begin{array}{cccccc} x_1 & x_2 & x_3 & s_1 & s_2 & s_3 \end{array}$$

$$\left[\begin{array}{cccccc|c} 1 & 2 & 2 & 1 & 0 & 0 & 50 \\ \boxed{4} & 24 & 0 & 0 & 1 & 0 & 20 \\ 1 & 0 & 2 & 0 & 0 & 1 & 15 \\ \hline -5 & -3 & -2 & 0 & 0 & 0 & 0 \end{array}\right]$$

Pivot on the 4 in row two, column one.

$$\begin{array}{cccccc} x_1 & x_2 & x_3 & s_1 & s_2 & s_3 \end{array}$$

$$\left[\begin{array}{cccccc|c} 1 & 2 & 2 & 1 & 0 & 0 & 50 \\ 1 & 6 & 0 & 0 & \frac{1}{4} & 0 & 5 \\ 1 & 0 & 2 & 0 & 0 & 1 & 15 \\ \hline -5 & -3 & -2 & 0 & 0 & 0 & 0 \end{array}\right] \begin{array}{c} \\ \frac{1}{4}R_2 \\ \\ \end{array}$$

$$\begin{array}{cccccc} x_1 & x_2 & x_3 & s_1 & s_2 & s_3 \end{array}$$

$$\left[\begin{array}{cccccc|c} 0 & -4 & 2 & 1 & -\frac{1}{4} & 0 & 45 \\ 1 & 6 & 0 & 0 & \frac{1}{4} & 0 & 5 \\ 0 & -6 & \boxed{2} & 0 & -\frac{1}{4} & 1 & 10 \\ \hline 0 & 27 & -2 & 0 & \frac{5}{4} & 0 & 25 \end{array}\right] \begin{array}{c} -R_2+R_1 \\ \\ -R_2+R_3 \\ 5R_2+R_4 \end{array}$$

Pivot on the 2 in row three, column three.

$$\begin{array}{cccccc} x_1 & x_2 & x_3 & s_1 & s_2 & s_3 \end{array}$$

$$\left[\begin{array}{cccccc|c} 0 & -4 & 2 & 1 & -\frac{1}{4} & 0 & 45 \\ 1 & 6 & 0 & 0 & \frac{1}{4} & 0 & 5 \\ 0 & -3 & 1 & 0 & -\frac{1}{8} & \frac{1}{2} & 5 \\ \hline 0 & 27 & -2 & 0 & \frac{5}{4} & 0 & 25 \end{array}\right] \begin{array}{c} \\ \\ \frac{1}{2}R_3 \\ \end{array}$$

$$\begin{array}{cccccc} x_1 & x_2 & x_3 & s_1 & s_2 & s_3 \end{array}$$

$$\left[\begin{array}{cccccc|c} 0 & 2 & 0 & 1 & 0 & -1 & 35 \\ 1 & 6 & 0 & 0 & \frac{1}{4} & 0 & 5 \\ 0 & -3 & 1 & 0 & -\frac{1}{8} & \frac{1}{2} & 5 \\ \hline 0 & 21 & 0 & 0 & 1 & 1 & 35 \end{array}\right] \begin{array}{c} -2R_3+R_1 \\ \\ \\ 2R_3+R_4 \end{array}$$

The maximum is 35 when
$x_1 = 5,\ x_2 = 0,\ x_3 = 5,\ s_1 = 35,\ s_2 = 0,$ and $s_3 = 0.$

26.

$$\begin{array}{ccccc} x_1 & x_2 & s_1 & s_2 & s_3 \end{array}$$

$$\left[\begin{array}{ccccc|c} 1 & -2 & 1 & 0 & 0 & 38 \\ 1 & -1 & 0 & 1 & 0 & 12 \\ 2 & \boxed{1} & 0 & 0 & 1 & 30 \\ \hline -1 & -2 & 0 & 0 & 0 & 0 \end{array}\right]$$

Pivot on the 1 in row three, column two.

$$\begin{array}{ccccc} x_1 & x_2 & s_1 & s_2 & s_3 \end{array}$$

$$\left[\begin{array}{ccccc|c} 5 & 0 & 1 & 0 & 2 & 98 \\ 3 & 0 & 0 & 1 & 1 & 42 \\ 2 & 1 & 0 & 0 & 1 & 30 \\ \hline 3 & 0 & 0 & 0 & 2 & 60 \end{array}\right] \begin{array}{c} 2R_3+R_1 \\ R_3+R_2 \\ \\ 2R_3+R_4 \end{array}$$

The maximum is 60 when
$x_1 = 0, x_2 = 30, s_1 = 98, s_2 = 42,$ and $s_3 = 0.$

27. From exercise 19, we have

$$\begin{array}{cccccc} x_1 & x_2 & x_3 & s_1 & s_2 & s_3 \end{array}$$

$$\left[\begin{array}{cccccc|c} 1 & 1 & 1 & 1 & 0 & 0 & 100 \\ 2 & 3 & 0 & 0 & 1 & 0 & 500 \\ 1 & 0 & 2 & 0 & 0 & 1 & 350 \\ \hline -5 & -6 & -3 & 0 & 0 & 0 & 0 \end{array}\right].$$

Pivot on the 1 in row 1 column 2.

$$\begin{array}{cccccc} x_1 & x_2 & x_3 & s_1 & s_2 & s_3 \end{array}$$

$$\left[\begin{array}{cccccc|c} 1 & \boxed{1} & 1 & 1 & 0 & 0 & 100 \\ -1 & 0 & -3 & -3 & 1 & 0 & 200 \\ 1 & 0 & 2 & 0 & 0 & 1 & 350 \\ \hline 1 & 0 & 3 & 6 & 0 & 0 & 600 \end{array}\right] \begin{array}{c} \\ -3R_1+R_2 \\ \\ 6R_1+R_4 \end{array}$$

The maximum is 600 when $x_1 = 0,\ x_2 = 100,$
and $x_3 = 0.$

28. From exercise 20, we have

$$\begin{array}{cccccc} x_1 & x_2 & s_1 & s_2 & s_3 & s_4 \end{array}$$

$$\left[\begin{array}{cccccc|c} 3 & 5 & 1 & 0 & 0 & 0 & 47 \\ 1 & 1 & 0 & 1 & 0 & 0 & 25 \\ 5 & 2 & 0 & 0 & 1 & 0 & 35 \\ 2 & 1 & 0 & 0 & 0 & 1 & 30 \\ \hline -2 & -9 & 0 & 0 & 0 & 0 & 0 \end{array}\right].$$

Pivot on the 5 in row 1 column 2. First divide row 1 by 5, then continue the process.

$$\begin{array}{cccccc} x_1 & x_2 & s_1 & s_2 & s_3 & s_4 \end{array}$$

$$\left[\begin{array}{cccccc|c} \frac{3}{5} & 1 & \frac{1}{5} & 0 & 0 & 0 & \frac{47}{5} \\ 1 & 1 & 0 & 1 & 0 & 0 & 25 \\ 5 & 2 & 0 & 0 & 1 & 0 & 35 \\ 2 & 1 & 0 & 0 & 0 & 1 & 30 \\ \hline -2 & -9 & 0 & 0 & 0 & 0 & 0 \end{array}\right] \begin{array}{c} \frac{1}{5}R_1 \\ \\ \\ \\ \end{array}$$

$$\begin{array}{cccccc} x_1 & x_2 & s_1 & s_2 & s_3 & s_4 \end{array}$$

$$\left[\begin{array}{cccccc|c} \frac{3}{5} & 1 & \frac{1}{5} & 0 & 0 & 0 & \frac{47}{5} \\ \frac{2}{5} & 0 & -\frac{1}{5} & 1 & 0 & 0 & \frac{78}{5} \\ \frac{19}{5} & 0 & -\frac{2}{5} & 0 & 1 & 0 & \frac{81}{5} \\ \frac{7}{5} & 0 & -\frac{1}{5} & 0 & 0 & 1 & \frac{103}{5} \\ \hline \frac{17}{5} & 0 & \frac{9}{5} & 0 & 0 & 0 & \frac{423}{5} \end{array}\right] \begin{array}{c} \\ -R_1+R_2 \\ -2R_1+R_3 \\ -R_1+R_4 \\ 9R_1+R_5 \end{array}$$

The maximum is $\frac{423}{5}$ when $x_1 = 0$ and
$x_2 = \frac{47}{5}.$

29. From exercise 21, we have

$$\begin{array}{cccccc} x_1 & x_2 & x_3 & s_1 & s_2 & s_3 \end{array}$$

$$\left[\begin{array}{cccccc|c} 1 & 1 & 1 & 1 & 0 & 0 & 90 \\ 2 & \boxed{5} & 1 & 0 & 1 & 0 & 120 \\ 1 & 3 & 0 & 0 & 0 & 1 & 80 \\ \hline -1 & -8 & -2 & 0 & 0 & 0 & 0 \end{array}\right].$$

Pivot on the 5 in row 2 column 2. First divide row 2 by 5, then continue the process.

$$\begin{array}{cccccc} x_1 & x_2 & x_3 & s_1 & s_2 & s_3 \end{array}$$

$$\left[\begin{array}{cccccc|c} 1 & 1 & 1 & 1 & 0 & 0 & 90 \\ \frac{2}{5} & 1 & \frac{1}{5} & 0 & \frac{1}{5} & 0 & 24 \\ 1 & 3 & 0 & 0 & 0 & 1 & 80 \\ \hline -1 & -8 & -2 & 0 & 0 & 0 & 0 \end{array}\right]\begin{array}{l} \frac{1}{5}R_1 \end{array}$$

$$\begin{array}{cccccc} x_1 & x_2 & x_3 & s_1 & s_2 & s_3 \end{array}$$

$$\left[\begin{array}{cccccc|c} \frac{3}{5} & 0 & \boxed{\frac{4}{5}} & 1 & -\frac{1}{5} & 0 & 66 \\ \frac{2}{5} & 1 & \frac{1}{5} & 0 & \frac{1}{5} & 0 & 24 \\ -\frac{1}{5} & 0 & -\frac{3}{5} & 0 & -\frac{3}{5} & 1 & 8 \\ \hline \frac{11}{5} & 0 & -\frac{2}{5} & 0 & \frac{8}{5} & 0 & 192 \end{array}\right]\begin{array}{l} -R_2+R_1 \\[6pt] \\ -3R_2+R_3 \end{array}$$

Now pivot on the $\frac{4}{5}$ in row 1 column 3. First multiply row 1 by $\frac{5}{4}$, then continue the process.

$$\begin{array}{cccccc} x_1 & x_2 & x_3 & s_1 & s_2 & s_3 \end{array}$$

$$\left[\begin{array}{cccccc|c} \frac{3}{4} & 0 & 1 & \frac{5}{4} & -\frac{1}{4} & 0 & \frac{165}{2} \\ \frac{2}{5} & 1 & \frac{1}{5} & 0 & \frac{1}{5} & 0 & 24 \\ -\frac{1}{5} & 0 & -\frac{3}{5} & 0 & -\frac{3}{5} & 1 & 8 \\ \hline \frac{11}{5} & 0 & -\frac{2}{5} & 0 & \frac{8}{5} & 0 & 192 \end{array}\right]\begin{array}{l} \frac{5}{4}R_1 \end{array}$$

$$\begin{array}{cccccc} x_1 & x_2 & x_3 & s_1 & s_2 & s_3 \end{array}$$

$$\left[\begin{array}{cccccc|c} \frac{3}{4} & 0 & 1 & \frac{5}{4} & -\frac{1}{4} & 0 & \frac{165}{2} \\ \frac{1}{4} & 1 & 0 & -\frac{1}{4} & \frac{1}{4} & 0 & \frac{15}{2} \\ \frac{1}{4} & 0 & 0 & \frac{3}{4} & -\frac{3}{4} & 1 & \frac{115}{2} \\ \hline \frac{5}{2} & 0 & 0 & \frac{1}{2} & \frac{3}{2} & 0 & 225 \end{array}\right]\begin{array}{l} \\ -\frac{1}{5}R_1+R_2 \\ \frac{3}{5}R_1+R_3 \\ \frac{2}{5}R_1+R_4 \end{array}$$

The maximum is 225 when $x_1 = 0$, $x_2 = \frac{15}{2}$, and $x_3 = \frac{165}{2}$.

30. From exercise 22, we have

$$\begin{array}{cccccc} x_1 & x_2 & s_1 & s_2 & s_3 & s_4 \end{array}$$

$$\left[\begin{array}{cccccc|c} 2 & 5 & 1 & 0 & 0 & 0 & 50 \\ 1 & 3 & 0 & 1 & 0 & 0 & 25 \\ \boxed{4} & 1 & 0 & 0 & 1 & 0 & 18 \\ 1 & 1 & 0 & 0 & 0 & 1 & 12 \\ \hline -15 & -12 & 0 & 0 & 0 & 0 & 0 \end{array}\right].$$

Pivot on the 4 in row 3 column 1. First divide row 3 by 4, then continue the process.

$$\begin{array}{cccccc} x_1 & x_2 & s_1 & s_2 & s_3 & s_4 \end{array}$$

$$\left[\begin{array}{cccccc|c} 2 & 5 & 1 & 0 & 0 & 0 & 50 \\ 1 & 3 & 0 & 1 & 0 & 0 & 25 \\ 1 & \frac{1}{4} & 0 & 0 & \frac{1}{4} & 0 & \frac{9}{2} \\ 1 & 1 & 0 & 0 & 0 & 1 & 12 \\ \hline -15 & -12 & 0 & 0 & 0 & 0 & 0 \end{array}\right]\begin{array}{l} \frac{1}{4}R_3 \end{array}$$

$$\begin{array}{cccccc} x_1 & x_2 & s_1 & s_2 & s_3 & s_4 \end{array}$$

$$\left[\begin{array}{cccccc|c} 0 & \frac{9}{2} & 1 & 0 & -\frac{1}{2} & 0 & 41 \\ 0 & \boxed{\frac{11}{4}} & 0 & 1 & -\frac{1}{4} & 0 & \frac{41}{2} \\ 1 & \frac{1}{4} & 0 & 0 & \frac{1}{4} & 0 & \frac{9}{2} \\ 0 & \frac{3}{4} & 0 & 0 & -\frac{1}{4} & 1 & \frac{15}{2} \\ \hline 0 & -\frac{33}{4} & 0 & 0 & \frac{15}{4} & 0 & \frac{135}{2} \end{array}\right]\begin{array}{l} -2R_3+R_1 \\ -R_3+R_2 \\ \\ -R_3+R_4 \\ 15R_3+R_5 \end{array}$$

Now pivot on the $\frac{11}{4}$ in row 2 column 2. First multiply row 2 by $\frac{4}{11}$, then continue the process.

$$\begin{array}{cccccc} x_1 & x_2 & s_1 & s_2 & s_3 & s_4 \end{array}$$

$$\left[\begin{array}{cccccc|c} 0 & \frac{9}{2} & 1 & 0 & -\frac{1}{2} & 0 & 41 \\ 0 & 1 & 0 & \frac{4}{11} & -\frac{1}{11} & 0 & \frac{82}{11} \\ 1 & \frac{1}{4} & 0 & 0 & \frac{1}{4} & 0 & \frac{9}{2} \\ 0 & \frac{3}{4} & 0 & 0 & -\frac{1}{4} & 1 & \frac{15}{2} \\ \hline 0 & -\frac{33}{4} & 0 & 0 & \frac{15}{4} & 0 & \frac{135}{2} \end{array}\right]\begin{array}{l} \\ \frac{11}{4}R_2 \end{array}$$

$$\begin{array}{cccccc} x_1 & x_2 & s_1 & s_2 & s_3 & s_4 \end{array}$$

$$\left[\begin{array}{cccccc|c} 0 & 0 & 1 & -\frac{18}{11} & -\frac{1}{11} & 0 & \frac{82}{11} \\ 0 & 1 & 0 & \frac{4}{11} & -\frac{1}{11} & 0 & \frac{82}{11} \\ 1 & 0 & 0 & -\frac{1}{11} & \frac{3}{11} & 0 & \frac{29}{11} \\ 0 & 0 & 0 & -\frac{3}{11} & -\frac{2}{11} & 1 & \frac{21}{11} \\ \hline 0 & 0 & 0 & 3 & 3 & 0 & 129 \end{array}\right]\begin{array}{l} -\frac{9}{2}R_2+R_1 \\ \\ -\frac{1}{4}R_2+R_3 \\ -\frac{3}{4}R_2+R_4 \\ \frac{33}{4}R_2+R_5 \end{array}$$

The maximum is 129 when $x_1 = \frac{29}{11}$ and $x_2 = \frac{82}{11}$.

31. a. Let x_1 = the number of item A, let x_2 = the number of item B, and x_3 = the number of item C.

   b. The objective function is
   $z = 4x_1 + 3x_2 + 3x_3$.

   c. The constraints are
   $$2x_1 + 3x_2 + 6x_3 \leq 1200$$
   $$x_1 + 2x_2 + 2x_3 \leq 800$$
   $$2x_1 + 2x_2 + 4x_3 \leq 500$$
   $$x_1 \geq 0, \ x_2 \geq 0, \ x_3 \geq 0.$$

32. a. Let x_1 = the amount invested in oil leases, x_2 = the amount in bonds, and x_3 = the amount invested in stock.

   b. We want to maximize the objective function
   $z = .15x_1 + .09x_2 + .05x_3$.

   c.
   $$x_1 + x_2 + x_3 \leq 50,000$$
   $$x_1 + x_2 \quad\ \leq 15,000$$
   $$x_1 \quad\ + x_3 \leq 25,000$$
   $$x_1 \geq 0, \ x_2 \geq 0, \ x_3 \geq 0$$

33. a. Let x_1 = the number of gallons of Fruity wine, and x_2 = the number of gallons of Crystal wine to be made.

   b. The objective function is $z = 12x_1 + 15x_2$.

   c. The ingredients available are the limitations. The constraints are
   $$2x_1 + x_2 \leq 110$$
   $$2x_1 + 3x_2 \leq 125$$
   $$2x_1 + x_2 \leq 90$$
   $$x_1 \geq 0, \ x_2 \geq 0.$$

34. a. Let x_1 = the number of 5-gallon bags, x_2 = the number of 10-gallon bags, and x_3 = the number of 20-gallon bags.

   b. The objective function is
   $z = x_1 + .9x_2 + .95x_3$.

   c.
   $$x_1 + 1.1x_2 + 1.5x_3 \leq 8$$
   $$x_1 + 1.2x_2 + 1.3x_3 \leq 8$$
   $$2x_1 + 3x_2 + 4x_3 \leq 8$$
   $$x_1 \geq 0, \ x_2 \geq 0, \ x_3 \geq 0$$

35. It is necessary to use the simplex method when there are more than two variables.

36. Problem with constraints involving "$\leq$" or "=" can be solved using slack variables, while those involving "$\geq$" or "=" can be solved using surplus variables.

37. Any standard minimization problem can be solved using the method of duals.

38. a. Maximize $z = 6x_1 + 7x_2 + 5x_3$
   subject to: $4x_1 + 2x_2 + 3x_3 \leq 9$
   $$5x_1 + 4x_2 + x_3 \leq 10$$
   $$x_1 \geq 0, \ x_2 \geq 0, \ x_3 \geq 0$$

   b. The first constraint would become
   $4x_1 + 2x_2 + 3x_3 \geq 9$

   c. $z = 22.7$ when $x_1 = 0$, $x_2 = 2.1$, $x_3 = 1.6$

   d. Minimize $w = 9y_1 + 10y_2$
   subject to: $4y_1 + 5y_2 \geq 6$
   $$2y_1 + 4y_2 \geq 7$$
   $$3y_1 + y_2 \geq 5$$
   $$y_1 \geq 0, \ y_2 \geq 0$$

   e. The minimum value is $w = 22.7$, when $y_1 = 1.3$, and $y_2 = 1.1$.

39. Using the method of duals,
$$\begin{bmatrix} 1 & 0 & 0 & 3 & 1 & 2 & | & 12 \\ 0 & 0 & 1 & 4 & 5 & 3 & | & 5 \\ 0 & 1 & 0 & -2 & 7 & -6 & | & 8 \\ \hline 0 & 0 & 0 & 5 & 7 & 3 & | & 172 \end{bmatrix}$$
indicates a minimum value of 172 at $(5, 7, 3, 0, 0, 0)$.

40. Using the method of duals,
$$\begin{bmatrix} 0 & 0 & 1 & 6 & 3 & 1 & | & 2 \\ 1 & 0 & 0 & 4 & -2 & 2 & | & 8 \\ 0 & 1 & 0 & 10 & 7 & 0 & | & 12 \\ \hline 0 & 0 & 0 & 9 & 5 & 8 & | & 62 \end{bmatrix}$$
indicates a minimum of 62 at $(9, 5, 8, 0, 0, 0)$.

41. Using the method of duals,

$$\left[\begin{array}{cccc|c} 1 & 0 & 7 & -1 & 100 \\ 0 & 1 & 1 & 3 & 27 \\ \hline 0 & 0 & 7 & 2 & 640 \end{array}\right]$$

indicates a minimum of 640 at $(7, 2, 0, 0)$.

42. Minimize $w = 5y_1 + 2y_2$

subject to $2y_1 + 3y_2 \geq 6$

$2y_1 + y_2 \geq 7$

$y_1 \geq 0, \ y_2 \geq 0$

Rewrite the objective function by forming the augmented matrix of the system and then form its transpose.

$$\left[\begin{array}{cc|c} 2 & 3 & 6 \\ 2 & 1 & 7 \\ \hline 5 & 2 & 0 \end{array}\right] \rightarrow \left[\begin{array}{cc|c} 2 & 2 & 5 \\ 3 & 1 & 2 \\ \hline 6 & 7 & 0 \end{array}\right].$$

This gives the problem

Maximize $z = 6x_1 + 7x_2$

subject to $2x_1 + 2x_2 \leq 5$

$3x_1 + x_2 \leq 2$

$x_1 \geq 0, \ x_2 \geq 0$

The initial simplex tableau is

$$\begin{array}{cccc} x_1 & x_2 & s_1 & s_2 \\ \end{array}$$
$$\left[\begin{array}{cccc|c} 2 & 2 & 1 & 0 & 5 \\ 3 & \boxed{1} & 0 & 1 & 2 \\ \hline -6 & -7 & 0 & 0 & 0 \end{array}\right].$$

Pivot on the 1 in row 2 column 2.

$$\begin{array}{cccc} x_1 & x_2 & s_1 & s_2 \\ \end{array}$$
$$\left[\begin{array}{cccc|c} -4 & 0 & 1 & -2 & 1 \\ 3 & 1 & 0 & 1 & 2 \\ \hline 15 & 0 & 0 & 7 & 14 \end{array}\right] \begin{array}{l} -2R_2 + R_1 \\ \\ 7R_2 + R_3 \end{array}$$

$$\qquad\quad \uparrow \quad\uparrow \quad\uparrow$$
$$\qquad\quad y_1 \quad y_2 \quad w$$

The minimum is 14 when $y_1 = 0$ and $y_2 = 7$.

43. Minimize $w = 18y_1 + 10y_2$

subject to: $y_1 + y_2 \geq 17$

$5y_1 + 8y_2 \geq 42$

$y_1 \geq 0, \ y_2 \geq 0.$

Rewrite the objective function by forming the augmented matrix of the system and then form its transpose.

$$\left[\begin{array}{cc|c} 1 & 1 & 17 \\ 5 & 8 & 42 \\ \hline 18 & 10 & 0 \end{array}\right] \rightarrow \left[\begin{array}{cc|c} 1 & 5 & 18 \\ 1 & 8 & 10 \\ \hline 17 & 42 & 0 \end{array}\right].$$

This gives the problem

Maximize $z = 17x_1 + 42x_2$

subject to $x_1 + 5x_2 \leq 18$.

$x_1 + 8x_2 \leq 10$

$x_1 \geq 0, \ x_2 \geq 0$

The initial simplex tableau is

$$\begin{array}{cccc} x_1 & x_2 & s_1 & s_2 \\ \end{array}$$
$$\left[\begin{array}{cccc|c} 1 & 5 & 1 & 0 & 18 \\ 1 & 8 & 0 & 1 & 10 \\ \hline -17 & -42 & 0 & 0 & 0 \end{array}\right].$$

Pivot on the 8 in row 2 column 2.

$$\begin{array}{cccc} x_1 & x_2 & s_1 & s_2 \\ \end{array}$$
$$\left[\begin{array}{cccc|c} 1 & 5 & 1 & 0 & 18 \\ \frac{1}{8} & 1 & 0 & \frac{1}{8} & \frac{5}{4} \\ \hline -17 & -42 & 0 & 0 & 0 \end{array}\right] \frac{1}{8}R_2$$

$$\begin{array}{cccc} x_1 & x_2 & s_1 & s_2 \\ \end{array}$$
$$\left[\begin{array}{cccc|c} \frac{3}{8} & 0 & 1 & -\frac{5}{8} & \frac{47}{4} \\ \frac{1}{8} & 1 & 0 & \frac{1}{8} & \frac{5}{4} \\ \hline -\frac{47}{4} & 0 & 0 & \frac{21}{4} & \frac{105}{2} \end{array}\right] \begin{array}{l} -5R_2 + R_1 \\ \\ 42R_2 + R_3 \end{array}$$

Now pivot on the $\frac{1}{8}$ in row 2 column 1.

$$\begin{array}{cccc} x_1 & x_2 & s_1 & s_2 \\ \end{array}$$
$$\left[\begin{array}{cccc|c} 0 & -3 & 1 & -1 & 8 \\ \frac{1}{8} & 1 & 0 & \frac{1}{8} & \frac{5}{4} \\ \hline 0 & 94 & 0 & 17 & 170 \end{array}\right] \begin{array}{l} -3R_2 + R_1 \\ \\ 94R_2 + R_3 \end{array}$$

$$\qquad\quad \uparrow \quad\uparrow \quad\uparrow$$
$$\qquad\quad y_1 \quad y_2 \quad w$$

The minimum is 170 when $y_1 = 0$ and $y_2 = 17$.

44. Minimize $w = 4y_1 + 5y_2$

subject to $10y_1 + 5y_2 \geq 100$

$20y_1 + 10y_2 \geq 150$

$y_1 \geq 0,\ y_2 \geq 0$

Rewrite the objective function by forming the augmented matrix of the system and then form its transpose.

$$\begin{bmatrix} 10 & 5 & | & 100 \\ 20 & 10 & | & 150 \\ \hline 4 & 5 & | & 0 \end{bmatrix} \rightarrow \begin{bmatrix} 10 & 20 & | & 4 \\ 5 & 10 & | & 5 \\ \hline 100 & 150 & | & 0 \end{bmatrix}.$$

This gives the problem

Maximize $z = 100x_1 + 150x_2$

subject to $10x_1 + 20x_2 \leq 4$

$5x_1 + 10x_2 \leq 5$

$x_1 \geq 0,\ x_2 \geq 0$

The initial simplex tableau is

$$\begin{array}{cccc} x_1 & x_2 & s_1 & s_2 \\ \end{array}$$
$$\begin{bmatrix} 10 & 20 & 1 & 0 & | & 4 \\ 5 & 10 & 0 & 1 & | & 5 \\ \hline -100 & -150 & 0 & 0 & | & 0 \end{bmatrix}.$$

Pivot on the 20 in row 1 column 2.

$$\begin{array}{cccc} x_1 & x_2 & s_1 & s_2 \\ \end{array}$$
$$\begin{bmatrix} \frac{1}{2} & 1 & \frac{1}{20} & 0 & | & \frac{1}{5} \\ 5 & 10 & 0 & 1 & | & 5 \\ \hline -100 & -150 & 0 & 0 & | & 0 \end{bmatrix} \frac{1}{20}R_1$$

$$\begin{array}{cccc} x_1 & x_2 & s_1 & s_2 \\ \end{array}$$
$$\begin{bmatrix} \frac{1}{2} & 1 & \frac{1}{20} & 0 & | & \frac{1}{5} \\ 0 & 0 & -\frac{1}{2} & 1 & | & 3 \\ \hline -25 & 0 & \frac{15}{2} & 0 & | & 30 \end{bmatrix} \begin{array}{l} -10R_1 + R_2 \\ 150R_1 + R_3 \end{array}$$

Now pivot on the $\frac{1}{2}$ in row 1 column 1.

$$\begin{array}{cccc} x_1 & x_2 & s_1 & s_2 \\ \end{array}$$
$$\begin{bmatrix} 1 & 2 & \frac{1}{10} & 0 & | & \frac{2}{5} \\ 0 & 0 & -\frac{1}{2} & 1 & | & 3 \\ \hline -25 & 0 & \frac{15}{2} & 0 & | & 30 \end{bmatrix} 2R_1$$

$$\begin{array}{cccc} x_1 & x_2 & s_1 & s_2 \\ \end{array}$$
$$\begin{bmatrix} 1 & 2 & \frac{1}{10} & 0 & | & \frac{2}{5} \\ 0 & 0 & -\frac{1}{2} & 1 & | & 3 \\ \hline 0 & 50 & 10 & 0 & | & 40 \end{bmatrix} \begin{array}{l} 2R_1 \\ 25R_1 + R_3 \end{array}$$

$$\quad\quad \uparrow \quad \uparrow \quad \uparrow$$
$$\quad\quad y_1 \quad y_2 \quad w$$

The minimum is 40 when $y_1 = 10$ and $y_2 = 0$.

45. Maximize $z = 20x_1 + 30x_2$

subject to $5x_1 + 10x_2 \leq 120$

$10x_1 + 15x_2 \geq 200$

$x_1 \geq 0,\ x_2 \geq 0$

The initial simplex tableau is

$$\begin{array}{cccc} x_1 & x_2 & s_1 & s_2 \\ \end{array}$$
$$\begin{bmatrix} 5 & 10 & 1 & 0 & | & 120 \\ 10 & 15 & 0 & -1 & | & 200 \\ \hline -20 & -30 & 0 & 0 & | & 0 \end{bmatrix}.$$

46. Minimize $w = 4y_1 + 2y_2$

subject to $y_1 + 3y_2 \geq 6$

$2y_1 + 8y_2 \leq 21$

$y_1 \geq 0,\ y_2 \geq 0$

The initial simplex tableau is

$$\begin{array}{cccc} y_1 & y_2 & s_1 & s_2 \\ \end{array}$$
$$\begin{bmatrix} 1 & 3 & -1 & 0 & | & 6 \\ 2 & 8 & 0 & 1 & | & 21 \\ \hline 4 & 2 & 0 & 0 & | & 0 \end{bmatrix}.$$

47. Minimize $w = 12y_1 + 20y_2 - 8y_3$

subject to: $y_1 + y_2 + 2y_3 \geq 48$

$y_1 + y_2 \quad\quad \leq 12$

$y_3 \geq 10$

$3y_1 \quad\quad + y_3 \geq 30$

$y_1 \geq 0,\ y_2 \geq 0,\ y_3 \geq 0.$

The initial simplex tableau is

$$\begin{array}{ccccccc} y_1 & y_2 & y_3 & s_1 & s_2 & s_3 & s_4 \\ \end{array}$$
$$\begin{bmatrix} 1 & 1 & 2 & -1 & 0 & 0 & 0 & | & 48 \\ 1 & 1 & 0 & 0 & 1 & 0 & 0 & | & 12 \\ 0 & 0 & 1 & 0 & 0 & -1 & 0 & | & 10 \\ 3 & 0 & 1 & 0 & 0 & 0 & -1 & | & 30 \\ \hline 12 & 20 & -8 & 0 & 0 & 0 & 0 & | & 0 \end{bmatrix}.$$

48. Maximize $w = 6x_1 - 3x_2 + 4x_3$

subject to: $2x_1 + x_2 + x_3 \leq 112$

$x_1 + x_2 + x_3 \geq 80$

$x_1 + x_2 \quad\quad \leq 45$

$x_1 \geq 0,\ x_2 \geq 0,\ x_3 \geq 0.$

The initial simplex tableau is

$$\begin{array}{cccccc} x_1 & x_2 & x_3 & s_1 & s_2 & s_3 \\ \end{array}$$
$$\begin{bmatrix} 2 & 1 & 1 & 1 & 0 & 0 & | & 112 \\ 1 & 1 & 1 & 0 & -1 & 0 & | & 80 \\ 1 & 1 & 0 & 0 & 0 & 1 & | & 45 \\ \hline -6 & 3 & -4 & 0 & 0 & 0 & | & 0 \end{bmatrix}.$$

49. If $w = -z$

$$\begin{bmatrix} 0 & 1 & 0 & 2 & 5 & 0 & | & 17 \\ 0 & 0 & 1 & 3 & 1 & 1 & | & 25 \\ 1 & 0 & 0 & 4 & 2 & \frac{1}{2} & | & 8 \\ \hline 0 & 0 & 0 & 2 & 5 & 0 & | & -427 \end{bmatrix}$$

indicates a minimum value of 427 at $(8, 17, 25, 0, 0, 0)$.

50.

$$\begin{bmatrix} 0 & 0 & 2 & 1 & 0 & 6 & 6 & | & 92 \\ 1 & 0 & 3 & 0 & 0 & 0 & 2 & | & 47 \\ 0 & 1 & 0 & 0 & 0 & 1 & 0 & | & 68 \\ 0 & 0 & 4 & 0 & 1 & 0 & 3 & | & 35 \\ \hline 0 & 0 & 5 & 0 & 0 & 2 & 9 & | & -1957 \end{bmatrix}$$

The minimum value is 1957 at $(47, 68, 0, 92, 35, 0, 0)$.

51. Maximize $z = 20x_1 + 30x_2$

subject to $5x_1 + 10x_2 \le 120$
$10x_1 + 15x_2 \ge 200$
$x_1 \ge 0, \; x_2 \ge 0$

The initial simplex tableau is

$$\begin{array}{cccc} x_1 & x_2 & s_1 & s_2 \end{array}$$
$$\begin{bmatrix} 5 & 10 & 1 & 0 & | & 120 \\ \boxed{10} & 15 & 0 & -1 & | & 200 \\ \hline -20 & -30 & 0 & 0 & | & 0 \end{bmatrix}.$$

Pivot on the 10 in row 2 column 1.

$$\begin{array}{cccc} x_1 & x_2 & s_1 & s_2 \end{array}$$
$$\begin{bmatrix} 0 & 5 & 2 & 1 & | & 40 \\ 10 & 15 & 0 & -1 & | & 200 \\ \hline 0 & 0 & 0 & -2 & | & 400 \end{bmatrix} \begin{array}{l} -R_2 + 2R_1 \\ \\ 2R_2 + R_3 \end{array}$$

This completes Stage I because the solution given in the usual way from the matrix has nonnegative values for all variables. Since there is a negative indicator in the objective row, we continue with Stage II. Pivot on the 1 in row 1 column 4.

$$\begin{array}{cccc} x_1 & x_2 & s_1 & s_2 \end{array}$$
$$\begin{bmatrix} 0 & 5 & 2 & \boxed{1} & | & 40 \\ 10 & 20 & 2 & 0 & | & 240 \\ \hline 0 & 10 & 4 & 0 & | & 480 \end{bmatrix} 2R_1 + R_3$$

$$\begin{array}{cccc} x_1 & x_2 & s_1 & s_2 \end{array}$$
$$\begin{bmatrix} 0 & 5 & 2 & 1 & | & 40 \\ 1 & 2 & \frac{1}{5} & 0 & | & 24 \\ \hline 0 & 10 & 4 & 0 & | & 480 \end{bmatrix} \frac{1}{10} R_2$$

The maximum is 480 when $x_1 = 24$ and $x_2 = 0$.

52. Minimize $w = 4y_1 + 2y_2$

subject to $y_1 + 3y_2 \ge 6$
$2y_1 + 8y_2 \le 21$
$y_1 \ge 0, \; y_2 \ge 0$

The initial simplex tableau is

$$\begin{array}{cccc} y_1 & y_2 & s_1 & s_2 \end{array}$$
$$\begin{bmatrix} \boxed{1} & 3 & -1 & 0 & | & 6 \\ 2 & 8 & 0 & 1 & | & 21 \\ \hline 4 & 2 & 0 & 0 & | & 0 \end{bmatrix}.$$

Pivot on the 1 in row 1 column 1.

$$\begin{array}{cccc} y_1 & y_2 & s_1 & s_2 \end{array}$$
$$\begin{bmatrix} 1 & 3 & -1 & 0 & | & 6 \\ 0 & 2 & 2 & 1 & | & 9 \\ \hline 0 & -10 & 4 & 0 & | & -24 \end{bmatrix} \begin{array}{l} -2R_1 + R_2 \\ -4R_1 + R_3 \end{array}$$

This completes Stage I because the solution given in the usual way from the matrix has nonnegative values for all variables. Since there is a negative indicator in the objective row, we continue with Stage II. Pivot on the 3 in row 1 column 2.

$$\begin{array}{cccc} y_1 & y_2 & s_1 & s_2 \end{array}$$
$$\begin{bmatrix} \frac{1}{3} & 1 & -\frac{1}{3} & 0 & | & 2 \\ 0 & 2 & 2 & 1 & | & 9 \\ \hline 0 & -10 & 4 & 0 & | & -24 \end{bmatrix} \frac{1}{3} R_1$$

$$\begin{array}{cccc} y_1 & y_2 & s_1 & s_2 \end{array}$$
$$\begin{bmatrix} \frac{1}{3} & 1 & -\frac{1}{3} & 0 & | & 2 \\ -\frac{2}{3} & 0 & \frac{8}{3} & 1 & | & 5 \\ \hline \frac{10}{3} & 0 & \frac{2}{3} & 0 & | & -4 \end{bmatrix} \begin{array}{l} -2R_1 + R_2 \\ 10R_1 + R_3 \end{array}$$

The minimum is 4 when $y_1 = 0$ and $y_2 = 2$.

53. Minimize $w = 4y_1 - 8y_2$

subject to: $y_1 + y_2 \le 50$
$2y_1 - 4y_2 \ge 20$
$y_1 - y_2 \le 22$
$y_1 \ge 0, \; y_2 \ge 0.$

The initial simplex tableau is

$$\begin{array}{ccccc} y_1 & y_2 & s_1 & s_2 & s_3 \end{array}$$
$$\begin{bmatrix} 1 & 1 & 1 & 0 & 0 & | & 50 \\ \boxed{2} & -4 & 0 & -1 & 0 & | & 20 \\ 1 & -1 & 0 & 0 & 1 & | & 22 \\ \hline 4 & -8 & 0 & 0 & 0 & | & 0 \end{bmatrix}.$$

(continued on next page)

(*continued from page 254*)

For Stage I, pivot on the 2 in row two, column one.

$$\begin{array}{ccccc} y_1 & y_2 & s_1 & s_2 & s_3 \\ \left[\begin{array}{ccccc|c} 1 & 1 & 1 & 0 & 0 & 50 \\ 1 & -2 & 0 & -\frac{1}{2} & 0 & 10 \\ 1 & -1 & 0 & 0 & 1 & 22 \\ \hline 4 & -8 & 0 & 0 & 0 & 0 \end{array}\right] & & & & \end{array} \frac{1}{2}R_2$$

$$\begin{array}{ccccc} y_1 & y_2 & s_1 & s_2 & s_3 \\ \left[\begin{array}{ccccc|c} 0 & 3 & 1 & \frac{1}{2} & 0 & 40 \\ 1 & -2 & 0 & -\frac{1}{2} & 0 & 10 \\ 0 & 1 & 0 & \frac{1}{2} & 1 & 12 \\ \hline 0 & 0 & 0 & 2 & 0 & -40 \end{array}\right] & \begin{array}{l} -R_2 + R_1 \\ \\ -R_2 + R_3 \\ -4R_2 + R_4 \end{array} \end{array}$$

The minimum is 40 when $y_1 = 10$ and $y_2 = 0$.

54. Maximize $z = 2x_1 + 4x_2$

subject to: $3x_1 + 2x_2 \le 12$
$5x_1 + x_2 \ge 5$
$x_1 \ge 0,\ x_2 \ge 0.$

The initial simplex tableau is

$$\begin{array}{cccc} x_1 & x_2 & s_1 & s_2 \\ \left[\begin{array}{cccc|c} 3 & 2 & 1 & 0 & 12 \\ 5 & \boxed{1} & 0 & -1 & 5 \\ \hline -2 & -4 & 0 & 0 & 0 \end{array}\right]. \end{array}$$

For Stage I, pivot on the 1 in row 2 column 1.

$$\begin{array}{cccc} x_1 & x_2 & s_1 & s_2 \\ \left[\begin{array}{cccc|c} -7 & 0 & 1 & \boxed{2} & 2 \\ 5 & 1 & 0 & -1 & 5 \\ \hline 18 & 0 & 0 & -4 & 20 \end{array}\right] & \begin{array}{l} -2R_2 + R_1 \\ \\ 4R_2 + R_3 \end{array} \end{array}$$

For Stage II, pivot on the 2 in row 1 column 4.

$$\begin{array}{cccc} x_1 & x_2 & s_1 & s_2 \\ \left[\begin{array}{cccc|c} -\frac{7}{2} & 0 & \frac{1}{2} & 1 & 1 \\ \frac{3}{2} & 1 & \frac{1}{2} & 0 & 6 \\ \hline 4 & 0 & 2 & 0 & 24 \end{array}\right] & \begin{array}{l} \frac{1}{2}R_1 \\ R_1 + R_2 \\ 4R_1 + R_3 \end{array} \end{array}$$

The maximum is 24 when $x_1 = 0$ and $x_2 = 6$.

55. From Exercise 31, the initial simplex tableau is

$$\begin{array}{cccccc} x_1 & x_2 & x_3 & s_1 & s_2 & s_3 \\ \left[\begin{array}{cccccc|c} 2 & 3 & 6 & 1 & 0 & 0 & 1200 \\ 1 & 2 & 2 & 0 & 1 & 0 & 800 \\ \boxed{2} & 2 & 4 & 0 & 0 & 1 & 500 \\ \hline -4 & -3 & -3 & 0 & 0 & 0 & 0 \end{array}\right]. \end{array}$$

Pivot on the 2 in row three, column one.

$$\begin{array}{cccccc} x_1 & x_2 & x_3 & s_1 & s_2 & s_3 \\ \left[\begin{array}{cccccc|c} 2 & 3 & 6 & 1 & 0 & 0 & 1200 \\ 1 & 2 & 2 & 0 & 1 & 0 & 800 \\ 1 & 1 & 2 & 0 & 0 & \frac{1}{2} & 250 \\ \hline -4 & -3 & -3 & 0 & 0 & 0 & 0 \end{array}\right] & \frac{1}{2}R_3 \end{array}$$

$$\begin{array}{cccccc} x_1 & x_2 & x_3 & s_1 & s_2 & s_3 \\ \left[\begin{array}{cccccc|c} 0 & 1 & 2 & 1 & 0 & -1 & 700 \\ 0 & 1 & 0 & 0 & 1 & -\frac{1}{2} & 550 \\ 1 & 1 & 2 & 0 & 0 & \frac{1}{2} & 250 \\ \hline 0 & 1 & 5 & 0 & 0 & 2 & 1000 \end{array}\right] & \begin{array}{l} -2R_3 + R_1 \\ -1R_3 + R_2 \\ \\ 4R_3 + R_4 \end{array} \end{array}$$

Roberta should get 250 units of item A, none of item B, and none of item C for a maximum profit of $1000.

56. From Exercise 32, we have the initial simplex tableau (after multiplying the fourth row by 100 to clear the decimals):

$$\begin{array}{cccccc} x_1 & x_2 & x_3 & s_1 & s_2 & s_3 \\ \left[\begin{array}{cccccc|c} 1 & 1 & 1 & 1 & 0 & 0 & 50,000 \\ \boxed{1} & 1 & 0 & 0 & 1 & 0 & 15,000 \\ 1 & 0 & 1 & 0 & 0 & 1 & 25,000 \\ \hline -15 & -9 & -5 & 0 & 0 & 0 & 0 \end{array}\right]. \end{array}$$

$$\begin{array}{cccccc} x_1 & x_2 & x_3 & s_1 & s_2 & s_3 \\ \left[\begin{array}{cccccc|c} 0 & 0 & 1 & 1 & -1 & 0 & 35,000 \\ 1 & 1 & 0 & 0 & 1 & 0 & 15,000 \\ 0 & -1 & \boxed{1} & 0 & -1 & 1 & 10,000 \\ \hline 0 & 6 & -5 & 0 & 15 & 0 & 225,000 \end{array}\right] & \begin{array}{l} -1R_2 + R_1 \\ \\ -1R_2 + R_3 \\ 15R_2 + R_4 \end{array} \end{array}$$

$$\begin{array}{cccccc} x_1 & x_2 & x_3 & s_1 & s_2 & s_3 \\ \left[\begin{array}{cccccc|c} 0 & 1 & 0 & 1 & 0 & -1 & 25,000 \\ 1 & 1 & 0 & 0 & 1 & 0 & 15,000 \\ 0 & -1 & 1 & 0 & -1 & 1 & 10,000 \\ \hline 0 & 1 & 0 & 0 & 10 & 5 & 275,000 \end{array}\right] & \begin{array}{l} -1R_3 + R_1 \\ \\ \\ 5R_3 + R_4 \end{array} \end{array}$$

Since $x_1 = 15,000$ and $x_3 = 10,000$, he should invest $15,000 in oil leases and $10,000 in stock for a maximum return of
$.01(275,000) = \$2750$.

57. From Exercise 33, we have the initial simplex tableau:

$$\begin{array}{ccccc} x_1 & x_2 & s_1 & s_2 & s_3 \\ \end{array}$$
$$\left[\begin{array}{ccccc|c} 2 & 1 & 1 & 0 & 0 & 110 \\ 2 & \boxed{3} & 0 & 1 & 0 & 125 \\ 2 & 1 & 0 & 0 & 1 & 90 \\ \hline -12 & -15 & 0 & 0 & 0 & 0 \end{array}\right]$$

$$\begin{array}{ccccc} x_1 & x_2 & s_1 & s_2 & s_3 \\ \end{array}$$
$$\left[\begin{array}{ccccc|c} 2 & 1 & 1 & 0 & 0 & 110 \\ \frac{2}{3} & \boxed{1} & 0 & \frac{1}{3} & 0 & \frac{125}{3} \\ 2 & 1 & 0 & 0 & 1 & 90 \\ \hline -12 & -15 & 0 & 0 & 0 & 0 \end{array}\right] \begin{array}{l} \\ \frac{1}{3}R_2 \\ \\ \\ \end{array}$$

$$\begin{array}{ccccc} x_1 & x_2 & s_1 & s_2 & s_3 \\ \end{array}$$
$$\left[\begin{array}{ccccc|c} \frac{4}{3} & 0 & 1 & -\frac{1}{3} & 0 & \frac{205}{3} \\ \frac{2}{3} & 1 & 0 & \frac{1}{3} & 0 & \frac{125}{3} \\ \boxed{\frac{4}{3}} & 0 & 0 & -\frac{1}{3} & 1 & \frac{145}{3} \\ \hline -2 & 0 & 0 & 5 & 0 & 625 \end{array}\right] \begin{array}{l} -R_2+R_1 \\ \\ -R_2+R_3 \\ 15R_2+R_4 \end{array}$$

$$\begin{array}{ccccc} x_1 & x_2 & s_1 & s_2 & s_3 \\ \end{array}$$
$$\left[\begin{array}{ccccc|c} \frac{4}{3} & 0 & 1 & -\frac{1}{3} & 0 & \frac{205}{3} \\ \frac{2}{3} & 1 & 0 & \frac{1}{3} & 0 & \frac{125}{3} \\ \boxed{1} & 0 & 0 & -\frac{1}{4} & \frac{3}{4} & \frac{145}{4} \\ \hline -2 & 0 & 0 & 5 & 0 & 625 \end{array}\right] \begin{array}{l} \\ \\ \frac{4}{3}R_3 \\ \\ \end{array}$$

$$\begin{array}{ccccc} x_1 & x_2 & s_1 & s_2 & s_3 \\ \end{array}$$
$$\left[\begin{array}{ccccc|c} 0 & 0 & 1 & 0 & -1 & 20 \\ 0 & 1 & 0 & \frac{1}{2} & -\frac{1}{2} & \frac{35}{2} \\ 1 & 0 & 0 & -\frac{1}{4} & \frac{3}{4} & \frac{145}{4} \\ \hline 0 & 0 & 0 & \frac{9}{2} & \frac{3}{2} & \frac{1395}{2} \end{array}\right] \begin{array}{l} -\frac{4}{3}R_3+R_1 \\ -\frac{2}{3}R_3+R_2 \\ \\ 2R_3+R_4 \end{array}$$

The winery should make 17.5 gallons of Crystal wine and 36.25 gallons of Fruity wine for a maximum profit of $697.50.

58. From Exercise 34, we have the initial tableau:

$$\begin{array}{cccccc} x_1 & x_2 & x_3 & s_1 & s_2 & s_3 \\ \end{array}$$
$$\left[\begin{array}{cccccc|c} 1 & 1.1 & 1.5 & 1 & 0 & 0 & 8 \\ 1 & 1.2 & 1.3 & 0 & 1 & 0 & 8 \\ \boxed{2} & 3 & 4 & 0 & 0 & 1 & 8 \\ \hline -1 & -.9 & -.95 & 0 & 0 & 0 & 0 \end{array}\right].$$

$$\begin{array}{cccccc} x_1 & x_2 & x_3 & s_1 & s_2 & s_3 \\ \end{array}$$
$$\left[\begin{array}{cccccc|c} 1 & 1.1 & 1.5 & 1 & 0 & 0 & 8 \\ 1 & 1.2 & 1.3 & 0 & 1 & 0 & 8 \\ \boxed{1} & 1.5 & 2 & 0 & 0 & .5 & 4 \\ \hline -1 & -.9 & -.95 & 0 & 0 & 0 & 0 \end{array}\right] \begin{array}{l} \\ \\ \frac{1}{2}R_3 \\ \\ \end{array}$$

$$\begin{array}{cccccc} x_1 & x_2 & x_3 & s_1 & s_2 & s_3 \\ \end{array}$$
$$\left[\begin{array}{cccccc|c} 0 & -.4 & -.5 & 1 & 0 & -.5 & 4 \\ 0 & -.3 & -.7 & 0 & 1 & -.5 & 4 \\ 1 & 1.5 & 2 & 0 & 0 & .5 & 4 \\ \hline 0 & .6 & 1.05 & 0 & 0 & .5 & 4 \end{array}\right] \begin{array}{l} -1R_3+R_1 \\ -1R_3+R_2 \\ \\ R_3+R_4 \end{array}$$

The final tableau gives $x_1 = 4$, $x_2 = 0$, $x_3 = 0$, and $z = 4$. Therefore, 4 units of 5-gallon bags (and none of the others) should be made for a maximum profit of $4 per unit.

59. Let y_1 = the number of cases of corn
y_2 = the number of cases of beans
y_3 = the number of cases of carrots

Minimize $w = 10y_1 + 15y_2 + 25y_3$

Maximize $z = -w = -10y_1 - 15y_2 - 25y_3$

subject to: $y_1 + y_2 + y_3 \geq 1000$
$\qquad\qquad y_1 - 2y_2 \qquad \geq 0$
$\qquad\qquad\qquad\qquad y_3 \geq 340$

$\qquad y_1 \geq 0,\ y_2 \geq 0,\ y_3 \geq 0$

The initial simplex tableau is

$$\begin{array}{cccccc} y_1 & y_2 & y_3 & s_1 & s_2 & s_3 \\ \end{array}$$
$$\left[\begin{array}{cccccc|c} 1 & 1 & 1 & -1 & 0 & 0 & 1000 \\ \boxed{1} & -2 & 0 & 0 & -1 & 0 & 0 \\ 0 & 0 & 1 & 0 & 0 & -1 & 340 \\ \hline 10 & 15 & 25 & 0 & 0 & 0 & 0 \end{array}\right]$$

For Stage I, pivot on the 1 in row two, column one.

$$\begin{array}{cccccc} y_1 & y_2 & y_3 & s_1 & s_2 & s_3 \\ \end{array}$$
$$\left[\begin{array}{cccccc|c} 0 & \boxed{3} & 1 & -1 & 1 & 0 & 1000 \\ 1 & -2 & 0 & 0 & -1 & 0 & 0 \\ 0 & 0 & 1 & 0 & 0 & -1 & 340 \\ \hline 0 & 35 & 25 & 0 & 10 & 0 & 0 \end{array}\right] \begin{array}{l} -R_2+R_1 \\ \\ \\ -10R_2+R_4 \end{array}$$

Continuing Stage I, pivot on the 3 in row one, column two.

$$\begin{array}{cccccc} y_1 & y_2 & y_3 & s_1 & s_2 & s_3 \\ \end{array}$$
$$\left[\begin{array}{cccccc|c} 0 & 1 & \frac{1}{3} & -\frac{1}{3} & \frac{1}{3} & 0 & \frac{1000}{3} \\ 1 & 0 & \frac{2}{3} & -\frac{2}{3} & -\frac{1}{3} & 0 & \frac{2000}{3} \\ 0 & 0 & \boxed{1} & 0 & 0 & -1 & 340 \\ \hline 0 & 0 & \frac{40}{3} & \frac{35}{3} & -\frac{5}{3} & 0 & -\frac{35,000}{3} \end{array}\right] \begin{array}{l} \frac{1}{3}R_1 \\ 2R_1+R_2 \\ \\ -35R_1+R_4 \end{array}$$

Pivot on the 1 in row three, column three.

(continued on next page)

(continued from page 256)

$$
\begin{array}{cccccc}
y_1 & y_2 & y_3 & s_1 & s_2 & s_3 \\
\end{array}
$$

$$
\left[\begin{array}{cccccc|c}
0 & 1 & 0 & -\frac{1}{3} & \boxed{\frac{1}{3}} & \frac{1}{3} & 220 \\
1 & 0 & 0 & -\frac{2}{3} & -\frac{1}{3} & \frac{2}{3} & 440 \\
0 & 0 & 1 & 0 & 0 & -1 & 340 \\
\hline
0 & 0 & 0 & \frac{35}{3} & -\frac{5}{3} & \frac{40}{3} & -16{,}200 \\
\end{array}\right]
\begin{array}{l}
-\frac{1}{3}R_3 + R_1 \\
-\frac{2}{3}R_3 + R_2 \\
\\
-\frac{40}{3}R_3 + R_4 \\
\end{array}
$$

For Stage II, pivot on the $\dfrac{1}{3}$ in row one, column five.

$$
\begin{array}{cccccc}
y_1 & y_2 & y_3 & s_1 & s_2 & s_3 \\
\end{array}
$$

$$
\left[\begin{array}{cccccc|c}
0 & 3 & 0 & -1 & 1 & 1 & 660 \\
1 & 1 & 0 & -1 & 0 & 1 & 660 \\
0 & 0 & 1 & 0 & 0 & -1 & 340 \\
\hline
0 & 5 & 0 & 10 & 0 & 15 & -15{,}100 \\
\end{array}\right]
\begin{array}{l}
3R_1 \\
\frac{1}{3}R_1 + R_2 \\
\\
\frac{5}{3}R_1 + R_4 \\
\end{array}
$$

The Cauchy Canners should produce 660 cases of corn, no cases of beans, and 340 cases of carrots for a minimum cost of $15,100.

60. First put the data into a table.

	Lumber	Concrete	Advertising	Total Spent
Atlantic	1000	3000	2000	$3000
Pacific	2000	3000	3000	$4000
Minimum Use	8000	18000	15000	

Let y_1 = the number of Atlantic boats and let y_2 = the number of Pacific boats.

The problem is to minimize $w = 3000y_1 + 4000y_2$

subject to: $1000y_1 + 2000y_2 \geq 8000$

$3000y_1 + 3000y_2 \geq 18{,}000$

$2000y_1 + 3000y_2 \geq 15{,}000$

$y_1 \geq 0,\ y_2 \geq 0.$

The matrix for this problem is
$$
\left[\begin{array}{cc|c}
1000 & 2000 & 8000 \\
3000 & 3000 & 18{,}000 \\
2000 & 3000 & 15{,}000 \\
\hline
3000 & 4000 & 0 \\
\end{array}\right].
$$

The transpose of this matrix is
$$
\left[\begin{array}{ccc|c}
1000 & 3000 & 2000 & 3000 \\
2000 & 3000 & 3000 & 4000 \\
\hline
8000 & 18{,}000 & 15{,}000 & 0 \\
\end{array}\right].
$$

The dual problem is as follows.

Minimize $z = 8000x_1 + 18{,}000x_2 + 15{,}000x_3$

subject to:

$1000x_1 + 3000x_2 + 2000x_3 \leq 3000$

$2000x_1 + 3000x_2 + 3000x_3 \leq 4000$

$x_1 \geq 0,\ x_2 \geq 0,\ x_3 \geq 0.$

(continued on next page)

(continued from page 257)

The initial simplex tableau is

$$
\begin{array}{ccccc}
x_1 & x_2 & x_3 & s_1 & s_2 \\
\end{array}
$$

$$
\left[
\begin{array}{ccccc|c}
1000 & \boxed{3000} & 2000 & 1 & 0 & 3000 \\
2000 & 3000 & 3000 & 0 & 1 & 4000 \\
\hline
-8000 & -18{,}000 & -15{,}000 & 0 & 0 & 0 \\
\end{array}
\right].
$$

Pivot on the 3000 in row one, column two.

$$
\begin{array}{ccccc}
x_1 & x_2 & x_3 & s_1 & s_2 \\
\end{array}
$$

$$
\left[
\begin{array}{ccccc|c}
\frac{1}{3} & \boxed{1} & \frac{2}{3} & \frac{1}{3000} & 0 & 1 \\
2000 & 3000 & 3000 & 0 & 1 & 4000 \\
\hline
-8000 & -18{,}000 & -15{,}000 & 0 & 0 & 0 \\
\end{array}
\right]
\begin{array}{l}
\frac{1}{3000}R_1 \\
\\
\\
\end{array}
$$

$$
\begin{array}{ccccc}
x_1 & x_2 & x_3 & s_1 & s_2 \\
\end{array}
$$

$$
\left[
\begin{array}{ccccc|c}
\frac{1}{3} & 1 & \frac{2}{3} & \frac{1}{3000} & 0 & 1 \\
1000 & 0 & \boxed{1000} & -1 & 1 & 1000 \\
\hline
-2000 & 0 & -3000 & 6 & 0 & 18{,}000 \\
\end{array}
\right]
\begin{array}{l}
\\
-3000R_1 + R_2 \\
18{,}000R_1 + R_3 \\
\end{array}
$$

Pivot on the 1000 in row two, column three.

$$
\begin{array}{ccccc}
x_1 & x_2 & x_3 & x_4 & x_5 \\
\end{array}
$$

$$
\left[
\begin{array}{ccccc|c}
\frac{1}{3} & 1 & \frac{2}{3} & \frac{1}{3000} & 0 & 1 \\
1 & 0 & 1 & -\frac{1}{1000} & \frac{1}{1000} & 1 \\
\hline
-2000 & 0 & -3000 & 6 & 0 & 18{,}000 \\
\end{array}
\right]
\begin{array}{l}
\\
\frac{1}{1000}R_2 \\
\\
\end{array}
$$

$$
\begin{array}{ccccc}
x_1 & x_2 & x_3 & x_4 & x_5 \\
\end{array}
$$

$$
\left[
\begin{array}{ccccc|c}
-\frac{1}{3} & 1 & 0 & \frac{1}{1000} & -\frac{1}{1500} & \frac{1}{3} \\
1 & 0 & 1 & -\frac{1}{1000} & \frac{1}{1000} & 1 \\
\hline
1000 & 0 & 0 & 3 & 3 & 21{,}000 \\
\end{array}
\right]
\begin{array}{l}
-\frac{2}{3}R_2 + R_1 \\
\\
3000R_2 + R_3 \\
\end{array}
$$

The contractor should build 3 Atlantic and 3 Pacific models for a minimum cost of $21,000.

61. Let $y_1 = $ kg of whole tomatoes and let $y_2 = $ kg of tomato sauce.

Minimize $w = 4y_1 + 3.25y_2$

subject to:
$$
\begin{aligned}
y_1 + \quad y_2 &\le 3{,}000{,}000 \\
y_1 \qquad\quad &\ge 800{,}000 \\
y_2 &\ge 80{,}000 \\
\frac{6}{60}y_1 + \frac{3}{60}y_2 &\ge 110{,}000 \\
y_1 \ge 0,\ y_2 &\ge 0
\end{aligned}
$$

The initial simplex tableau is

$$
\begin{array}{cccccc}
y_1 & y_2 & s_1 & s_2 & s_3 & s_4 \\
\end{array}
$$

$$
\left[
\begin{array}{cccccc|c}
1 & 1 & 1 & 0 & 0 & 0 & 3{,}000{,}000 \\
\boxed{1} & 0 & 0 & -1 & 0 & 0 & 800{,}000 \\
0 & 1 & 0 & 0 & -1 & 0 & 80{,}000 \\
.1 & .05 & 0 & 0 & 0 & -1 & 110{,}000 \\
\hline
4 & 3.25 & 0 & 0 & 0 & 0 & 0 \\
\end{array}
\right]
$$

(continued on next page)

(continued from page 258)

For Stage I, pivot on the 1 in row two, column one.

$$
\begin{array}{c}
\begin{array}{cccccc} y_1 & y_2 & s_1 & s_2 & s_3 & s_4 \end{array} \\
\left[\begin{array}{cccccc|c}
0 & 1 & 1 & 1 & 0 & 0 & 2,200,000 \\
1 & 0 & 0 & -1 & 0 & 0 & 800,000 \\
0 & \boxed{1} & 0 & 0 & -1 & 0 & 80,000 \\
0 & .05 & 0 & .1 & 0 & -1 & 30,000 \\
\hline
0 & 3.25 & 0 & 4 & 0 & 0 & -3,200,000
\end{array}\right]
\begin{array}{l}
-R_2 + R_1 \\
\\
\\
-.1R_2 + R_4 \\
-4R_2 + R_5
\end{array}
\end{array}
$$

Pivot on the 1 in row three, column two.

$$
\begin{array}{c}
\begin{array}{cccccc} y_1 & y_2 & s_1 & s_2 & s_3 & s_4 \end{array} \\
\left[\begin{array}{cccccc|c}
0 & 0 & 1 & 1 & 1 & 0 & 2,120,000 \\
1 & 0 & 0 & -1 & 0 & 0 & 800,000 \\
0 & 1 & 0 & 0 & -1 & 0 & 80,000 \\
0 & 0 & 0 & \boxed{.1} & .05 & -1 & 26,000 \\
\hline
0 & 0 & 0 & 4 & 3.25 & 0 & -3,460,000
\end{array}\right]
\begin{array}{l}
-R_3 + R_1 \\
\\
\\
-.05R_3 + R_4 \\
-3.25R_3 + R_5
\end{array}
\end{array}
$$

Pivot on the .1 in row four, column four.

$$
\begin{array}{c}
\begin{array}{cccccc} y_1 & y_2 & s_1 & s_2 & s_3 & s_4 \end{array} \\
\left[\begin{array}{cccccc|c}
0 & 0 & 1 & 0 & .5 & 10 & 1,860,000 \\
1 & 0 & 0 & 0 & .5 & -10 & 1,060,000 \\
0 & 1 & 0 & 0 & -1 & 0 & 80,000 \\
0 & 0 & 0 & 1 & .5 & -10 & 260,000 \\
\hline
0 & 0 & 0 & 0 & 1.25 & 40 & -4,500,000
\end{array}\right]
\begin{array}{l}
-R_4 + R_1 \\
R_4 + R_2 \\
\\
10R_4 \\
-4R_4 + R_5
\end{array}
\end{array}
$$

This program is optimal after Stage I.
1,060,000 kg of tomatoes should be used for canned whole tomatoes and 80,000 kg of tomatoes should be used for sauce at a minimum cost of $4,500,000.

62. Let y_1 = the number of runs of type I and let y_2 = the number of runs of type II.

Minimize $w = 15,000y_1 + 6000y_2$

Maximize $z = -w = -15,000y_1 - 6000y_2$

subject to: $3000y_1 + 3000y_2 = 18,000$
$\ \ 2000y_1 + 1000y_2 = 7000$
$\ \ 2000y_1 + 3000y_2 \geq 14,000$

$\ \ \ y_1 \geq 0,\ y_2 \geq 0$

The initial simplex tableau is

$$
\begin{array}{c}
\begin{array}{ccccccc} y_1 & y_2 & s_1 & s_2 & s_3 & s_4 & s_5 \end{array} \\
\left[\begin{array}{ccccccc|c}
3000 & 3000 & -1 & 0 & 0 & 0 & 0 & 18,000 \\
3000 & 3000 & 0 & 1 & 0 & 0 & 0 & 18,000 \\
\boxed{2000} & 1000 & 0 & 0 & -1 & 0 & 0 & 7000 \\
2000 & 1000 & 0 & 0 & 0 & 1 & 0 & 7000 \\
2000 & 3000 & 0 & 0 & 0 & 0 & -1 & 14,000 \\
\hline
15,000 & 6000 & 0 & 0 & 0 & 0 & 0 & 0
\end{array}\right]
\end{array}
$$

For Stage I, pivot on the 2000 in row three, column one.

(continued on next page)

(continued from page 259)

$$
\begin{array}{cccccccc}
y_1 & y_2 & s_1 & s_2 & s_3 & s_4 & s_5 & \\
\end{array}
$$

y_1	y_2	s_1	s_2	s_3	s_4	s_5		
0	1500	-1	0	1.5	0	0	7500	$-3000R_3 + R_1$
0	1500	0	1	1.5	0	0	7500	$-3000R_3 + R_2$
1	.5	0	0	$-.0005$	0	0	3.5	$.0005R_3$
0	0	0	0	1	1	0	0	$-2000R_3 + R_4$
0	$\boxed{2000}$	0	0	1	0	-1	7000	$-2000R_3 + R_5$
0	-1500	0	0	7.5	0	0	$-52,500$	$-15,000R_3 + R_6$

Pivot on the 2000 in row five, column two.

y_1	y_2	s_1	s_2	s_3	s_4	s_5		
0	0	-1	0	.75	0	$\boxed{.75}$	2250	$-1500R_5 + R_1$
0	0	0	1	.75	0	.75	2250	$-1500R_5 + R_2$
1	0	0	0	$-.00075$	0	.00025	1.75	$-.5R_5 + R_3$
0	0	0	0	1	1	0	0	
0	1	0	0	.0005	0	$-.0005$	3.5	$.0005R_5$
0	0	0	0	8.25	0	$-.75$	$-47,250$	$1500R_5 + R_6$

Pivot on the .75 in row one column seven.

y_1	y_2	s_1	s_2	s_3	s_4	s_5		
0	0	$-\frac{4}{3}$	0	1	0	1	3000	$\frac{4}{3}R_1$
0	0	$\boxed{1}$	1	0	0	0	0	$-.75R_1 + R_2$
1	0	$\frac{1}{3000}$	0	$-.001$	0	0	1	$-.00025R_1 + R_3$
0	0	0	0	1	1	0	0	
0	1	$-\frac{1}{1500}$	0	.001	0	0	5	$.0005R_1 + R_5$
0	0	-1	0	9	0	0	$-45,000$	$.75R_1 + R_6$

For Stage II, pivot on the 1 in row two, column three.

y_1	y_2	y_3	y_4	y_5	y_6	y_7		
0	0	0	$\frac{4}{3}$	1	0	1	3000	$\frac{4}{3}R_2 + R_1$
0	0	1	1	0	0	0	0	
1	0	0	$-\frac{1}{3000}$	$-.001$	0	0	1	$-\frac{1}{3000}R_2 + R_3$
0	0	0	0	1	1	0	0	
0	1	0	$\frac{1}{1500}$	.001	0	0	5	$\frac{1}{1500}R_2 + R_5$
0	0	0	1	9	0	0	$-45,000$	$R_2 + R_6$

The company should produce 1 run of type I and 5 runs of type II for a minimum cost of $45,000.

Case 7 Cooking with Linear Programming

1. Let x_1 = the number of 100 gram units of feta cheese, x_2 = the number of 100 gram units of lettuce, x_3 = the number of 100 gram units of salad dressing, and x_4 = the number of 100 gram units of tomato.

Maximize $z = 4.09x_1 + 2.37x_2 + 2.5x_3 + 4.64x_4$

subject to:

$$263x_1 + 14x_2 + 448.8x_3 + 21x_4 < 260$$
$$492.5x_1 + 36x_2 \qquad\qquad + 5x_4 > 210$$
$$10.33x_1 + 1.62x_2 \qquad\qquad + .85x_4 > 6$$
$$x_1 + x_2 \qquad + x_3 + x_4 < 4$$
$$x_3 \qquad\qquad \geq .3125$$

$x_1 \geq 0,\ x_2 \geq 0,\ x_3 \geq 0,\ x_4 \geq 0$

Using a computer with linear programming software, the optimal solution is:

$x_1 = .243037,\ x_2 = 2.35749,\ x_3 = .3125,$

$x_4 = 1.08698$

Converting to kitchen units gives approximately $\frac{1}{6}$ cup feta cheese $4\frac{1}{4}$ cups of lettuce, $\frac{1}{8}$ cup of salad dressing and $\frac{7}{8}$ of a tomato for a salad with a maximum of about 12.41 g of carbohydrates.

2. Let y_1 = the number of 100 gram units of beef, y_2 = the number of 100 gram units of oil, y_3 = the number of 100 gram units of onion, and y_4 = the number of 100 gram units of soy sauce.

Minimize $215y_1 + 884y_2 + 38y_3 + 60y_4$

subject to:

$$y_2 + 8.63y_3 + 5.57y_4 < 10$$
$$26y_1 \qquad + 1.16y_3 + 10.51y_4 > 50$$
$$6.4y_3 \qquad\qquad > 3.5$$
$$y_2 \qquad\qquad\qquad \geq .045$$

$y_1 \geq 0,\ y_2 \geq 0,\ y_3 \geq 0,\ y_4 \geq 0$

Using a computer with linear programming software, the optimal solution is

$y_1 = 1.51873,\ y_2 = .045,\ y_3 = .546875,$

$y_4 = .939941$

Converting to kitchen units gives approximately $5\frac{1}{3}$ ounces of beef, $\frac{1}{3}$ tablespoon of oil, $\frac{1}{2}$ an onion, and

$5\frac{1}{4}$ tablespoons of soy sauce for a stir-fry with a minimum 443.48 calories.

Chapter 8 Sets and Probability

Section 8.1 Sets

1. $3 \in \{2, 5, 7, 9, 10\}$
This statement is false. The number 3 is not an element of the set.

3. $9 \notin \{2, 1, 5, 8\}$
The statement is true. 9 is not an element of the set.

5. $\{2, 5, 8, 9\} = \{2, 5, 9, 8\}$
The statement is true. The sets contain exactly the same elements, so they are equal. The ordering of the elements in a set is unimportant.

7. {all whole numbers greater than 7 and less than 10} = $\{8, 9\}$
The statement is true. 8 and 9 are the only such numbers.

9. $\{x \mid x$ is an odd integer, $6 \le x \le 18\}$
$= \{7, 9, 11, 15, 17\}$
The statement is false. The number 13 should be included.

11. Answers vary.
Possible answer:
$\{0\}$ is a set containing 1 element, namely zero.
$\{\varnothing\}$ is a set containing 1 element, namely, a mathematical symbol representing the empty set.

13. Since every element of A is also an element of U, $A \subseteq U$.

15. $A \nsubseteq E$ since A contains elements that do not belong to E, namely –3, 0, 3.

17. $\varnothing \subseteq A$ since the empty set is a subset of every set.

19. $D \subseteq B$, since every element of D is also an element of B.

21. $\{A, B, C\}$ contains 3 elements. Therefore, it has $2^3 = 8$ subsets.

23. $\{x \mid x$ is an integer strictly between 0 and 8$\}$
$= \{1, 2, 3, 4, 5, 6, 7\}$. This set contains 7 elements. Therefore it has $2^7 = 128$ subsets.

25. $\left\{ x \middle| \begin{array}{l} x \text{ is an integer less than or equal to 0 or} \\ \qquad \text{greater than or equal to 8} \end{array} \right\}$

27. Answers vary.

29. $\{8, 11, 15\} \cap \{8, 11, 19, 20\} = \{8, 11\}$

31. $\{6, 12, 14, 16\} \cap \{6, 14, 19\} = \{6, 14\}$
$\{6, 14\}$ is the set of all elements belonging to both of the sets, so it is the intersection of those sets.

33. $\{3, 5, 9, 10\} \cup \varnothing = \{3, 5, 9, 10\}$.

35. $\{1, 2, 4\} \cup \{1, 2\} = \{1, 2, 4\}$
The answer set $\{1, 2, 4\}$ consists of all elements belonging to the first set, to the second set, or to both sets, and therefore it is the union of the first two sets.

37. $X \cap Y = \{b, 1, 3\}$ since only these elements are contained in both.

39. $X' = \{d, e, f, 4, 5, 6\}$ since these are the elements that are in U but not in X.

41. $X' \cap Y'$
$= \{d, e, f, 4, 5, 6\} \cap \{a, c, e, 2, 4, 6\}$
$= \{e, 4, 6\}$

43. $X \cup (Y \cap Z)$
$Y \cap Z = \{b, d, f, 1, 3, 5\} \cap \{b, d, 2, 3, 5\}$
$\qquad = \{b, d, 3, 5\}$
$X \cup (Y \cap Z) = \{a, b, c, 1, 2, 3\} \cup \{b, d, 3, 5\}$
$\qquad\qquad = \{a, b, c, d, 1, 2, 3, 5\}$

45. M' is the set of all students in this school not taking this course.

47. $N \cap P$ is the set of all students in this school taking both accounting and philosophy.

49. A pair of sets is disjoint if the two sets have no elements in common. The pairs of these sets that are disjoint are C and D, A and E, C and E, and D and E.

51. A' is the set of all stocks with a high price less than or equal to \$60. $A' = \{$Allstate, Microsoft$\}$

53. $A' \cap B'$ is the set of all stocks with a high price less than or equal to \$60 and a last price either less than or equal to \$25 or greater than or equal to \$55. $A' \cap B' = \{$Allstate, Microsoft$\}$.

55. $M \cap E$ is the set of all male employed applicants.

57. $M' \cup S'$ is the set of all female or married applicants.

59. {Internet}

61. F = {Comcast, Time Warner}

63. H = {Cablevision, Charter, Cox, Time Warner, Comcast Cable}

65. $H \cap F$ = {Cablevision, Charter, Cox, Time Warner, Comcast } $\cap$ {Comcast, Time Warner}
= {Comcast, Time Warner}

67. $U = \{s, d, c, g, i, m, h\}$
$O = \{i, m, h, g\}$
$O' = \{s, d, c\}$

69. $N = \{s, d, c, g\}$
$O = \{i, m, h, g\}$
$N \cap O = \{g\}$

Section 8.2 Applications of Venn Diagrams

1. $A \cap B'$
Shade the region inside B that is outside A.

$A \cap B'$

3. $B' \cup A'$
Shade the region outside B and the region outside of region A.

$B' \cup A'$

5. $B' \cup (A \cap B')$
First shade the common region that is in A and outside B. Then shade the rest of the region outside B.

$B' \cup (A \cap B')$

7. $U' = \varnothing$
There are no elements outside the universal set. Therefore, there is no region to be shaded.

9. Three sets divide the universal set into at most <u>8</u> regions.

11. $(A \cap C') \cup B$
First find $A \cap C'$, the region in A *and* not in C.

$A \cap C'$

For the union, we want the region in $(A \cap C')$ *or* in B, or both.

$(A \cap C') \cup B$

13. $A' \cap (B \cap C)$
First find $B \cap C$, the region in B *and* in C.

$B \cap C$

Now find A', the region not in A.

A'

For the intersection, we want the region in A' *and* in $(B \cap C)$.

$A' \cap (B \cap C)$

15. $(A \cap B') \cup C$

First find $A \cap B'$, the region in *A and* not in *B*.

$A \cap B'$

For the union, we want the region in $(A \cap B')$ *or* in *C*, or both.

$(A \cap B') \cup C$

17. Percentage of children under age 18 who lived with their father only:
$100\% - 70.7\% - 22.6\% - 3.5\% = 3.2\%$

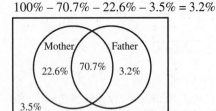

19. There are 26 applicants with sales experience and a college degree, of which 11 also have a real estate license. Therefore, 15 have only sales experience and a college degree. There are 16 applicants with sales experience and a real estate license, of which 11 also have a college degree. Therefore, there are 5 applicants with only sales experience and a real estate license. There are $66 - 15 - 11 - 5 = 35$ applicants with only sales experience. Using similar reasoning, we find that there are 10 applicants with only a college degree, 4 applicants with a only a college degree and a real estate license, and 3 applicants with only a real estate license. This is illustrated in the Venn diagram below.

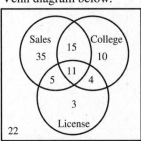

a. There were $35 + 15 + 11 + 5 + 10 + 4 + 3 + 22 = 105$ applicants.

b. $22 + 3 + 4 + 10 = 39$ applicants did not have sales experience.

c. 15 applicants had sales experience and a college degree.

d. Three applicants had only a real estate license.

21. a. The total number of people surveyed is the sum of all those who caught at least one fish minus the sum of those who caught at least two fish minus the number of those who caught all three fish.
$124 + 133 + 146 - (75 + 67 + 79 - 45)$
$= 277$

b. The total number of people who caught at least one walleye or at least one smallmouth bass equals the sum of the number of those who caught at least one walleye and the number of those who caught at least one smallmouth bass minus the number of those who caught at least one of each fish.
$124 + 133 - 75 = 182$

c. 45 people caught all three fish, while 75 people caught at least one walleye and at least one smallmouth bass, so 30 people caught at least one walleye and at least one smallmouth bass only. 67 people caught at least one walleye and at least one yellow perch, so $67 - 45 = 22$ people caught at least one walleye and at least one yellow perch only. Therefore, $124 - (30 + 45 + 22) = 27$ people caught at least one walleye only. The information in the exercise is summarized in the Venn diagram below.

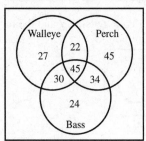

23. In the diagram, the regions are labeled (a) to (g). First, the number for each region is found by starting with innermost region (d). 15 had all three; then proceed to work outward as follows:
Region (a):
Since 25 had A, 25 − (2 + 15 + 1) = 7.
Region (b):
Since 17 had A and B, 17 − 15 = 2.
Region (c):
Since 27 had B, 27 − (2 + 15 + 7) = 3.
Region (d):
15 had all three.
Region (e):
Since 22 had B and Rh, 22 − 15 = 7.
Region (f):
Since 30 had Rh, 30 − (1 + 15 + 7) = 7.
Region (g):
12 had none.
Region with no label:
Since 16 had A and Rh, 16 − 15 = 1.

a. 7 + 2 + 3 + 15 + 7 + 7 + 12 + 1 = 54 patients were represented.

b. 7 + 3 + 7 = 17 patients had exactly one antigen.

c. 2 + 1 + 7 = 10 patients had exactly two antigens.

d. Since a person having only the Rh antigen has type O-positive blood, 7 had type O-positive blood.

e. Since a person having A, B, and Rh antigens is AB-positive, 15 had AB-positive blood.

f. Since a person having only the B antigen is B-negative, 3 had B-negative blood.

g. Since a person having neither A, B, nor Rh antigens is O-negative, 12 had O-negative blood.

h. Since a person having A and Rh antigens is A-positive, 1 had A-positive blood.

25. a. $n(A \cap M) = 1018$.

b. $n[C \cap (F \cup M)] = n(C \cap F) + n(C \cap M)$
$= 412 + 320 = 732$

c. $n(D \cup F) = n(D) + n(F) - n(D \cap F)$
$= 156 + 2506 - 95 = 2567$

d. $n(B' \cap E')$
$= 1018 + 1152 + 320 + 412 + 61 + 95$
$= 3058$

27. a. $n(C \cup D)$
$= (165,000 + 573,000) + (70,000 + 137,000)$
$= 945,000$

b. $n(B \cap G) = 3000$

c. $n[A \cap (E \cup F)] = n(A \cap E) + n(A \cap F)$
$= 18,000 + 18,000$
$= 36,000$

d. $n[F \cup G]' = n(C) + n(D) + n(E)$
$= 738,000 + 207,000 + 39,000$
$= 984,000$

e. $n(A' \cap C')$
$= n(D \cap B) + n(E \cap B)$
$+ n(F \cap B) + n(G \cap B)$
$= 137,000 + 21,000 + 18,000 + 3000$
$= 179,000$

29. Answers will vary.

31. $n(A \cup B) = n(A) + n(B) - n(A \cap B)$
$30 = 12 + 27 - n(A \cap B)$
$30 = 39 - n(A \cap B)$
$n(A \cap B) = 9$

33. $n(A \cup B) = n(A) + n(B) - n(A \cap B)$
$35 = 13 + n(B) - 5$
$n(B) = 27$

35. $n(A) = 28, n(B) = 12, n(A \cup B) = 30,$
$n(A') = 19$
This gives
$n(A \cup B) = n(A) + n(B) - n(A \cap B) \Rightarrow$
$30 = 28 + 12 - n(A \cap B)$
$n(A \cap B) = 28 + 12 - 30 = 10$
$n(A) = 28$ and $n(A \cap B) = 10$, so
$n(A \cap B') = 18$
$n(B) = 12$ and $n(A \cap B) = 10$, so
$n(B \cap A') = 2$
Since $n(A')=19$, 2 of which are accounted for in $B \cap A'$, 17 remain in $A' \cap B'$.

37. $n(A') = 28$, $n(B) = 25$, $n(A' \cup B') = 45$, $n(A \cap B) = 12$

$n(B) = 25$ and $n(A \cap B) = 12$, so
$n(B \cap A') = 13$
Since $n(A') = 28$, of which 13 are accounted for, 15 are in $A' \cap B'$.
$$n(A' \cup B') = n(A') + n(B') - n(A' \cap B')$$
$$45 = 28 + n(B') - 15$$
$$45 = 13 + n(B')$$
$$32 = n(B')$$

15 are in $A' \cap B'$, so the rest are in $A \cap B'$, and $n(A \cap B') = 17$.

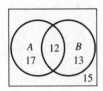

39.
$$n(A) = 54$$
$$n(A \cap B) = 22$$
$$n(A \cup B) = 85$$
$$n(A \cap B \cap C) = 4$$
$$n(A \cap C) = 15$$
$$n(B \cap C) = 16$$
$$n(C) = 44$$
$$n(B') = 63$$

Start with $A \cap B \cap C$. Now $n(A \cap C) = 15$, of which 4 are in $A \cap B \cap C$, so $n(A \cap B' \cap C) = 11$. $n(B \cap C) = 16$, of which 4 are in $A \cap B \cap C$, so $n(B \cap C \cap A') = 12$.
$n(C) = 44$, so 17 are in $C \cap A' \cap B'$.
$n(A \cap B) = 22$, so 18 are in $A \cap B \cap C'$.
$n(A) = 54$, so $54 - 11 - 18 - 4 = 21$ are in $A \cap B' \cap C'$.
$$n(A \cup B) = n(A) + n(B) - n(A \cap B)$$
$$85 = 54 + n(B) - 22$$
$$53 = n(B)$$

This leaves 19 in $B \cap A' \cap C'$. $n(B') = 63$, of which $21 + 11 + 17 = 49$ are accounted for, leaving 14 in $A' \cap B' \cap C'$.

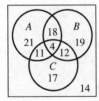

41. $(A \cap B)'$ is the complement of the intersection of A and B; hence it contains all elements not in $A \cap B$.

$(A \cap B)'$

$A' \cup B'$ is the union of the complements of A and B; hence it contains any element that is either not in A or not in B.

A'

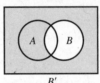

B'

Note that $(A \cap B)' = A' \cup B'$, as claimed.

$A' \cup B'$

43. $A \cup (B \cap C)$ contains all points in A and the points where B and C overlap.

$A \cup (B \cap C)$

$(A \cup B) \cap (A \cup C)$ contains the points where $A \cup B$ and $A \cup C$ overlap.

$A \cup B \qquad\qquad A \cup C$

Note that $A \cup (B \cap C) = (A \cup B) \cap (A \cup C)$, as claimed.

$(A \cup B) \cap (A \cup C)$

45. The complement of A and (intersection) B equals the union of the complement of A and the complement of B.

47. A union (B intersect C) equals (A union B) intersect (A union C).

Section 8.3 Introduction to Probability

1. Answers vary. Possible answer:
A coin or die is fair if the probability of any result is the same, that is, all results are equally likely.

3. There are 12 months in a year. The sample space is the set
{January, February, March, …, December}.

5. There are 80 points on the test. The sample space is the set {0, 1, 2, …, 80}.

7. There are only 2 choices the management can make. The sample space is the set
{go ahead, cancel}.

9. There are four possibilities, so the sample space is the set {Q1, Q2, Q3, Q4}.

11. Disjoint events are events that cannot occur at the same time.

13. Wearing a hat and wearing glasses are not disjoint, because it is possible to wear both at the same time.

15. Being a doctor and being under 5 years old are disjoint, since it is impossible to be under 5 years old and be a doctor.

17. Being a female and being a pilot are not disjoint, since there are many female pilots.

19. S = {(Connie, Kate), (Connie, Lindsey), (Connie, Jackie), (Connie, Taisa), (Connie, Nicole), (Kate, Lindsey), (Kate, Jackie), (Kate, Taisa), (Kate, Nicole), (Lindsey, Jackie), (Lindsey, Taisa), (Lindsey, Nicole), (Jackie, Taisa), (Jackie, Nicole), (Taisa, Nicole)}

  a. Taisa is selected.
  {(Connie, Taisa),(Kate, Taisa), (Lindsey, Taisa), (Jackie, Taisa), (Taisa, Nicole)}

  b. The two names selced have the same number of letters.
  {(Connie, Jackie), (Connie, Nicole), (Jackie, Nicole)}

21. The sample space is {(forest sage, rag painting), (forest sage, colorwash), (evergreen whisper, rag painting), (evergreen whisper, colorwash), (opaque emerald, rag painting), (opaque emerald, colorwash)}

  a. There are 3 combinations with colorwash: {(colorwash, forest sage), (colorwash, evergreen whisper), (colorwash, opaque emerald)}.
  $$P_C = \frac{n(C)}{n(S)} = \frac{3}{6} = \frac{1}{2}.$$

  b. There are 4 combinations with opaque emerald or rag painting: {(opaque emerald, rag painting), (opaque emerald, colorwash), (forest sage, rag painting), (evergreen whisper, rag painting)}.
  $$P_{O \cup R} = \frac{n(O \cup R)}{n(S)} = \frac{4}{6} = \frac{2}{3}$$

23. There are 2 possibilities for size and 3 possibilities for color. The sample space is {(10-foot, beige), (10-foot, forest green) (10-foot, rust), (12-foot, beige), (12-foot, forest green), (12-foot, rust)}

  a. Doug buys a 12-foot forest green umbrella. There is 1 possibility:
  {(12-foot, forest green)}.
  $$P_{12' \cap G} = \frac{n(12' \cap G)}{n(S)} = \frac{1}{6}$$

  b. Doug buys a 10-foot umbrella. There are 3 possibilities: {(10-foot, beige), (10-foot, forest green), (10-foot, rust)}
  $$P_{10'} = \frac{n(10')}{n(S)} = \frac{3}{6} = \frac{1}{2}$$

  c. Doug buys a rust-colored umbrella. There are 2 possibilities: {(10-foot, rust), (12-foot, rust)}
  $$P_R = \frac{n(R)}{n(S)} = \frac{2}{6} = \frac{1}{3}$$

25. Use the sample space S = {1, 2, 3, 4, 5, 6}. Let E be the event "getting a number less than 4." So, E = {1, 2, 3}.
Since S contains six elements,
$$P(E) = \frac{3}{6} = \frac{1}{2}.$$

27. Use the sample space
$S = \{1, 2, 3, 4, 5, 6\}$.
Let E be the event "getting a 2 or a 5." So,
$E = \{2, 5\}$. Since S contains 6 elements,
$$P(E) = \frac{2}{6} = \frac{1}{3}.$$

29. $S = \{1, 2, 3, 4, 5, 6\}$
Let E be the event "getting any number except
3." $E = \{1, 2, 4, 5, 6\}$
Since S contains 6 elements, $P(E) = \frac{5}{6}$.

31. There are four possibilities, {male beagle, male
boxer, male collie, male Labrador}.
$$P(E) = \frac{4}{8} = \frac{1}{2}.$$

33. There is one possibility, {female Labrador}.
$$P(E) = \frac{1}{8}.$$

35. There are six possibilities, {male beagle, male
boxer, male collie, female beagle, female boxer,
female collie}.
$$P(E) = \frac{6}{8} = \frac{3}{4}.$$

37. The total number of fatalities in 2007 is 5471.
$$P(E) = \frac{159}{5471} \approx .029$$

39. The total number of respondents is 1168

a. 18 or 19 when married:
$$P(E) = \frac{215}{1168} \approx .184$$

b. less than 22 when married:
$$P(E) = \frac{80 + 215 + 250}{1168} = \frac{545}{1168} \approx .467$$

c. over 25 when married:
$$P(E) = \frac{187 + 90 + 25}{1168} = \frac{302}{1168} \approx .259$$

41. a. $E \cup F$ = The person smokes or has a
family history of heart disease.

b. $E' \cap F$ = The person does not smoke and
has a family history of heart disease.

c. $F' \cup G'$ = The person does not have a
family history of heart disease or is not
overweight.

43. This experiment is possible, since all
probabilities are non-negative and
$.09 + .32 + .21 + .25 + .13 = 1$

45. This experiment is not possible, since the sum of
the probabilities is $\frac{39}{40}$, which is less than 1.

47. This experiment is not possible, since a
probability cannot be negative.

Section 8.4 Basic Concepts of Probability

1. Let G be the event of the marble landing in a
green slot, and let B be the event of the marble
landing in a black slot. There are 2 slots that are
green and 18 slots that are black, so
$$P(G \cup B) = P(G) + P(B) - P(G \cap B)$$
$$= \frac{2}{38} + \frac{18}{38} - \frac{0}{38} = \frac{20}{38} = \frac{10}{19}.$$

3. Let O be the event of the marble landing in an
odd slot, and let B be the event of the marble
landing in a black slot. There are 18 slots that are
odd, 18 slots that are black, and 8 slots that are
odd and black, so
$$P(O \cup B) = P(O) + P(B) - P(O \cap B)$$
$$= \frac{18}{38} + \frac{18}{38} - \frac{8}{38} = \frac{28}{38} = \frac{14}{19}.$$

5. Let E be the event of the marble landing in a slot
numbered 0, 00, 1, 2, or 3. There are 5 such
slots, so $P(E) = \frac{5}{38}$.

7. Let E be the event of the marble landing in a slot
numbered 25–36. There are 12 such slots, so
$$P(E) = \frac{12}{38} = \frac{6}{19}.$$

For Exercises 9–13, count outcomes by referring to
Figure 8.22 in the text.

9. a. $P(\text{sum is 8}) = \frac{5}{36}$

b. $P(\text{sum is 9}) = \frac{4}{36} = \frac{1}{9}$

c. $P(\text{sum is 10}) = \frac{3}{36} = \frac{1}{12}$

d. $P(\text{sum is 13}) = \frac{0}{36} = 0$

11. a. $P(\text{not more than 5}) = \dfrac{10}{36} = \dfrac{5}{18}$

b. $P(\text{not less than 8}) = \dfrac{15}{36} = \dfrac{5}{12}$

c. $P(\text{between 3 and 7 (exclusive)}) = \dfrac{12}{36} = \dfrac{1}{3}$

13. The shoes come in two shades of beige (light and dark) and black, so $P(\text{shoes are black}) = \dfrac{1}{3}$.

15. a. $P(\text{a sister or an aunt}) = \dfrac{2}{10} + \dfrac{3}{10} = \dfrac{5}{10} = \dfrac{1}{2}$

b. $P(\text{a sister or a cousin}) = \dfrac{2}{10} + \dfrac{2}{10} = \dfrac{4}{10} = \dfrac{2}{5}$

c. $P(\text{a sister or her mother}) = \dfrac{2}{10} + \dfrac{1}{10} = \dfrac{3}{10}$

17.

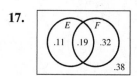

a. $P(E \cup F) = P(E) + P(F) - P(E \cap F)$
$= .30 + .51 - .19 = .62$

b. $P(E' \cap F) = P(F) - P(E \cap F)$
$= .51 - .19 = .32$

c. $P(E \cap F') = P(E) - P(E \cap F)$
$= .30 - .19 = .11$

d. $P(E' \cup F') = P(E') + P(F') - P(E' \cap F')$
$= .70 + .49 - .38 = .81$

19.

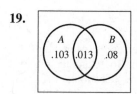

a. $P(A \cup B) = P(A) + P(B) - P(A \cap B)$
$= .116 + .093 - .013 = .196$

b. $P(A' \cap B) = P(B) - P(A \cap B)$
$= .093 - .013 = .08$

c. $P(A' \cap B') = P(A \cup B)' = 1 - P(A \cup B)$
$= 1 - .196 = .804$

d. $P(A' \cup B') = P(A') + P(B') - P(A' \cap B')$
$= (1 - .116) + (1 - .093) - .804$
$= .987$

21.

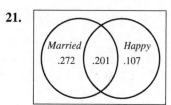

a. $P(M' \cap H')$
$= P(M') + P(H') - P(M' \cup H')$
$= [1 - P(M)] + [1 - P[H]]$
$\qquad - [1 - P(M \cap H)]$
$= (1 - .473) + (1 - .308) - (1 - .201) = .42$

b. $P(M \cup H') = P(M) + P(H') - P(M \cap H')$
$= .473 + (1 - .308) - .272 = .893$

c. $P(M' \cap H) = .107$

23. When rolling a die, there are 6 equally likely outcomes. The sample space is $S = \{1, 2, 3, 4, 5, 6\}$. Let E be the event "2 is rolled." Then $P(E) = \dfrac{1}{6}$ and $P(E') = \dfrac{5}{6}$. The odds in favor of rolling a 5 are $\dfrac{P(E)}{P(E')} = \dfrac{\frac{1}{6}}{\frac{5}{6}} = \dfrac{1}{5}$, written 1 to 5.

25. Let E be the event " 2, 3, 5, or 6 is rolled." Then $P(E) = \dfrac{4}{6} = \dfrac{2}{3}$ and $P(E') = \dfrac{1}{3}$. The odds in favor of rolling 2, 3, 5, or 6 are $\dfrac{P(E)}{P(E')} = \dfrac{\frac{2}{3}}{\frac{1}{3}} = 2$, written 2 to 1.

27. There are 3 yellow, 4 white, and 8 blue marbles.

a. Yellow: There are 3 ways to win and 12 ways to lose. The odds are 3 to 12 or 1 to 4.

b. Blue: There are 8 ways to win and 7 ways to lose; the odds are 8 to 7.

c. White: There are 4 ways to win and 11 ways to lose; the odds are 4 to 11.

29. Let E be the event "rolling a 7 or 11." Then

$P(E) = \dfrac{8}{36}$ and $P(E') = \dfrac{28}{36}$. The odds in favor

of rolling a 7 or 11 are $\dfrac{\frac{8}{36}}{\frac{28}{36}} = \dfrac{8}{28} = \dfrac{2}{7}$, written 2

to 7.

31. 21:79 **33.** 7:93

35. $\dfrac{2}{7}$ **37.** $\dfrac{1}{10}$

39. 1:9

41. Using a graphing calculator,

   a. P(the sum is not more than 5) $\approx .2778$

   b. P(the sum is not less than 8) $\approx .4167$

The probabilities compare very closely to the results in Exercise 11.

43. Using a graphing calculator,

   a. P(exactly 4 heads) $\approx .15625$

   b. P(2 heads and 3 tails) $\approx .3125$

45. $P(E) = \dfrac{1557}{5186} \approx .300$

47. $P(D') = 1 - P(D) = 1 - \dfrac{178}{5186} \approx .966$

49. $P(A \cup E) = P(A) + P(E) - P(A \cap E)$

$= \dfrac{2400}{5186} + \dfrac{1557}{5186} - \dfrac{872}{5186}$

$= \dfrac{3085}{5186} \approx .595$

51. $P(A) = \dfrac{763}{944} \approx .808$

53. $P(B \cup C) = \dfrac{99 + 257}{944} \approx .377$

55. $P(A \cup C) = \dfrac{763 + 257 - 234}{944} \approx .833$

In exercises 57–59, let A = the event "working full time," let B = the event "working part time," let C indicate "working 0–19 hours," let D indicate "working 40–49 hours," and let E indicate "working less than 30 hours".

57. $P(A) = \dfrac{2322}{2762} \approx .841$

59. $P(A \cap D) = \dfrac{1251}{2762} \approx .453$

61.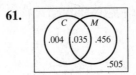

   a. $P(C') = .456 + .505 = .961$

   b. $P(M) = .035 + .456 = .491$

   c. $P(M') = .004 + .505 = .509$

   d. $P(M' \cap C') = 1 - (.004 + .035 + .456)$
                   $= .505$

   e. $P(C \cap M') = .004$ (inside C and outside M)

   f. $P(C \cup M') = .004 + .035 + .505 = .544$

63. a. Since red is no longer dominant, RW or WR results in pink.

$P(\text{red}) = P(RR) = \dfrac{1}{4}$

   b. Pink is produced by RW or WR, so

$P(\text{pink}) = \dfrac{2}{4} = \dfrac{1}{2}$.

   c. $P(\text{white}) = P(WW) = \dfrac{1}{4}$

65. The total number of men and women who answered the question is 2629, and the number of men who answered the question is 1124, so 2629 − 1124 = 1505 women answered the question. Of the 2177 men and women who believe in an afterlife, 891 are men, so 2177 − 891 = 1286 women believe in an afterlife. The number of men who do not believe in an afterlife is 1124 − 891 = 233. The number of women who do not believe in an afterlife is 1505 − 1286 = 219.

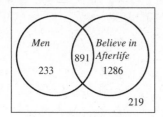

a. $P(\text{believes}) = \dfrac{2177}{2629} \approx .828$

b. $P(\text{male and believes}) = \dfrac{891}{2629} \approx .339$

c. $P(\text{female and does not believe}) = \dfrac{219}{2629}$
$\approx .083$

67. $P\left(\begin{array}{l}\text{valued at \$2 billion or more}\\ \text{and lost more than 5\%}\end{array}\right) = \dfrac{4}{203} \approx .020$

69. $P\left(\begin{array}{l}\text{0-5\% reduction or value of}\\ \text{\$300-499 million}\end{array}\right) = \dfrac{78+62-27}{203}$
$\approx .557$

Section 8.5 Conditional Probability and Independent Events

1.   Roll a fair die.

$P(3 \mid \text{odd}) = \dfrac{P(3 \cap \text{odd})}{P(\text{odd})} = \dfrac{\frac{1}{6}}{\frac{1}{2}} = \dfrac{1}{3}$

3. $P(\text{odd} \mid 3) = \dfrac{P(\text{odd} \cap 3)}{P(3)} = \dfrac{\frac{1}{6}}{\frac{1}{6}} = 1$

5.   There are 6 doubles, 1 of which has a sum of 6.

$P(\text{sum of 6} \mid \text{double}) = \dfrac{1}{6}$

7.   Since the first card is a heart, there are 51 cards remaining, 12 of them hearts.

$P(\text{second is heart} \mid \text{first is heart}) = \dfrac{12}{51} = \dfrac{4}{17}$

9. $P(\text{jack and 10}) = \dfrac{8 \cdot 4}{52 \cdot 51} \approx .012$

There are 8 possibilities for the first card (4 jacks and 4 tens), but for the second card there are only 4 (the 4 tens if a jack was picked or the 4 jacks if a 10 was picked).

11.   Answers vary.

13.   Answers vary.

15.   No. Knowledge that a college is a religiously affiliated school does affect the probability that four semesters of theology would be required to graduate from that college. So, the events are dependent.

17.   Yes. Knowledge that Tom Cruise's next movie grosses over \$200 million does not give any information about the occurrence or nonoccurrence of the event that the Republicans have a majority in Congress in 2016. So, these events are independent.

19.   No, for a two-child family, the knowledge that each child is the same sex influences the probability of the event that there is at most one male. Eliminating the "one of each" possibility lowers the probability of "at most one male" from .75 to .5. However, the events are independent for a three-child family because the first event does not influence the probability of the second event. The probability of at most one male is .5 whether male-female mixes are allowed or not.

21.   Let S be the event: cabinet made by Sitlington; C be the event: cabinet made by Capek; Y be the event: cabinet is satisfactory; and N be the event: cabinet is unsatisfactory. Construct a probability tree.

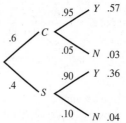

$P(C \mid N) = \dfrac{P(C \cap N)}{P(N)} = \dfrac{.03}{.03+.04} = \dfrac{.03}{.07}$
$= \dfrac{3}{7} \approx .43$

23. Let M represent "male" and L represent "not in labor force." We are seeking $P(L|M) = \dfrac{P(L \cap M)}{P(M)}$.

$$P(L \cap M) = P(L) + P(M) - P(L \cup M)$$
$$= .34 + .484 - .693 = .131$$

$$P(L|M) = \frac{P(L \cap M)}{P(M)} = \frac{.131}{.484} \approx .271.$$

The probability of not being in the labor force, given that the person is male is .271.

25. a. $P(\text{taller than } 6' \,|\, \text{male})$

$$= \frac{P(\text{taller than } 6' \cap \text{male})}{P(\text{male})}$$

$$= \frac{395}{2503} \approx .157$$

b. $P(\text{taller than } 6' \,|\, \text{female})$

$$= \frac{P(\text{taller than } 6' \cap \text{female})}{P(\text{female})}$$

$$= \frac{3}{2683} \approx .001$$

c. $P(\text{female} \,|\, \text{taller than } 6')$

$$= \frac{P(\text{female} \cap \text{taller than } 6')}{P(\text{taller than } 6')}$$

$$= \frac{3}{398} \approx .008$$

d. $P(\text{male} \,|\, \text{under } 6')$

$$= \frac{P(\text{male} \cap \text{under } 6')}{P(\text{under } 6')}$$

$$= \frac{2108}{4788} \approx .440$$

27. $P(\text{obese} \,|\, \text{female}) = \dfrac{P(\text{obese} \cap \text{female})}{P(\text{female})}$

$$= \frac{1003}{2683} \approx .374$$

29. $P(\text{female} \,|\, \text{underweight})$

$$= \frac{P(\text{female} \cap \text{underweight})}{P(\text{underweight})}$$

$$= \frac{93}{178} \approx .522$$

31. $P(\text{rain} \,|\, \text{rain forecast}) = \dfrac{66}{222} \approx .30$

33. Answers vary.

35. $P(C) = .049$ (directly from the chart)

37. $P(M \cup C) = P(M) + P(C) - P(M \cap C)$
$$= .527 + .049 - .042 = .534$$

39. $P(M'|C) = \dfrac{P(M' \cap C)}{P(C)} = \dfrac{.007}{.049}$

$$= \frac{1}{7} \approx .143$$

41. $P(M') = .473; \; P(M'|C) = .007$

Since $P(M') \neq P(M'|C)$, M' and C are dependent.

43. Complete the probability tree.

$P(\text{fails 1st and 2nd test})$
$= P(\text{fails 1st}) \cdot P(\text{fails 2nd} \,|\, \text{fails 1st})$
$= (.25)(.20) = .05$

45. $P(\text{requires at least 2 tries})$
$= P(\text{does not pass on 1st try}) = .25$

47. $P(0-2 \text{ cars} \,|\, \$0 - \$49{,}999) = \dfrac{504}{540} \approx .933$

49. $P(\$50{,}000 - \$99{,}999 \,|\, 0 - 2 \text{ cars})$

$$= \frac{199}{813} \approx .245$$

51. $P(\text{international flight is American})$

$$= \frac{147{,}768}{299{,}585} \approx .493$$

53. $P(\text{non-American flight is international})$

$$= \frac{87{,}656 + 64{,}161}{409{,}474 + 543{,}821} \approx .159$$

55. Once the singing group has a hit, subsequent records are also hits. The only way to have exactly one hit in their first three records is for the first two not to be hits while the third is a hit.
P(one hit in first three records) = (.68)(.84)(.08)
$$\approx .0457$$

57. P(have computer service)
= $1 - P$(no computer service)
= $1 - (.003)(.005)$
= $1 - .000015$
= $.999985$
Answers vary. It is fairly realistic to assume independence because the chance of a failure of one computer does not usually depend on the failure of another, so long as the cause of failure does not lie with something the two systems have in common, like a power source.

59. a. The probability of success with one component is
$$1 - .03 = .97;$$
with 2, $1 - (.03)^2 = .9991;$
with 3, $1 - (.03)^3 = .999973;$
with 4, $1 - (.03)^4 = .99999919.$
Therefore, 4 (the original and 3 backups) will do the job.

b. Answers vary. It is probably reasonable to assume independence here so long as the cause of failure is some internal defect and the components are from different manufacture lots.

61. Let A be the event "student studies" and B be the event "student gets a good grade." We are told that
$P(A) = .6$, $P(B) = .7$, and $P(A \cap B) = .52$.
$P(A) \cdot P(B) = (.6)(.7) = .42$
Since $P(A) \cdot P(B)$ is not equal to $P(A \cap B)$, A and B are not independent. Rather, they are dependent events.

Section 8.6 Bayes' Formula

1.

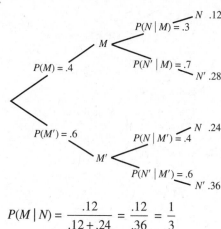

$$P(M \mid N) = \frac{.12}{.12 + .24} = \frac{.12}{.36} = \frac{1}{3}$$

For Exercises 3–5:

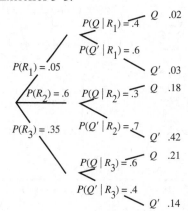

3. $P(R_1 \mid Q) = \dfrac{.02}{.02 + .18 + .21} = \dfrac{.02}{.41} = \dfrac{2}{41} \approx .0488$

5. $P(R_3 \mid Q) = \dfrac{.21}{.02 + .18 + .21} = \dfrac{.21}{.41} \approx .5122$

7.

jar 1 $\underline{P(\text{white} \mid \text{jar } 1) = 1/3}$ white 1/6

$P(\text{jar } 1) = 1/2$

$P(\text{jar } 2) = 1/3$ jar 2 $\underline{P(\text{white} \mid \text{jar } 2) = 2/3}$ white 2/9

$P(\text{jar } 3) = 1/6$

jar 3 $\underline{P(\text{white} \mid \text{jar } 3) = 1/2}$ white 1/12

$$P(\text{jar } 2 \mid \text{white}) = \frac{\dfrac{2}{9}}{\dfrac{1}{6} + \dfrac{2}{9} + \dfrac{1}{12}} = \frac{8}{17} \approx .4706$$

9.

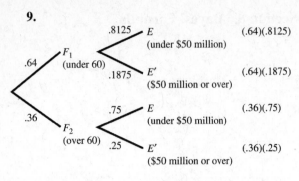

$$P(E' \mid F_1) = \frac{(.64)(.1875)}{(.64)(.1875) + (.36)(.25)} = .571$$

11.

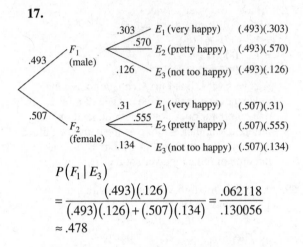

We are seeking $P(F_2 \mid E')$. Use Bayes' formula.

$$P(F_2 \mid E')$$
$$= \frac{(.279)(.056)}{(.382)(.084) + (.279)(.056) + (.339)(.037)}$$
$$= \frac{.015624}{.060255} \approx .259$$

13. We are seeking $P(F_1 \mid E)$. Use Bayes' formula.

$$P(F_1 \mid E)$$
$$= \frac{(.382)(.916)}{(.382)(.916) + (.279)(.944) + (.339)(.963)}$$
$$= \frac{.349912}{.939745} \approx .372$$

15.

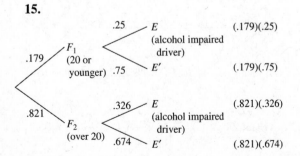

We are seeking $P(F_1 \mid E)$. Use Bayes' formula.

$$P(F_1 \mid E) = \frac{(.179)(.25)}{(.821)(.326) + (.179)(.25)}$$
$$= \frac{.04475}{.312396} \approx .143$$

17.

.493 F_1 (male) — .303 E_1 (very happy) $(.493)(.303)$
.570 E_2 (pretty happy) $(.493)(.570)$
.126 E_3 (not too happy) $(.493)(.126)$

.507 F_2 (female) — .31 E_1 (very happy) $(.507)(.31)$
.555 E_2 (pretty happy) $(.507)(.555)$
.134 E_3 (not too happy) $(.507)(.134)$

$$P(F_1 \mid E_3)$$
$$= \frac{(.493)(.126)}{(.493)(.126) + (.507)(.134)} = \frac{.062118}{.130056}$$
$$\approx .478$$

19. Let L be the event "the object was shipped by land," A be the event "the object was shipped by air," S be the event "the object was shipped by sea," and E be the event "an error occurred."

$$P(L \mid E)$$
$$= \frac{P(L) \cdot P(E \mid L)}{P(L) \cdot P(E \mid L) + P(A) \cdot P(E \mid A) + P(S) \cdot P(E \mid S)}$$
$$= \frac{(.50)(.02)}{(.50)(.02) + (.40)(.04) + (.10)(.14)}$$
$$= \frac{.0100}{.0400} = .25$$

The correct response is (c).

21.

.89 N — .1 T .089
.9 T' .801

.11 N' — .75 T .0825
.25 T' .0275

$$P(N' \mid T) = \frac{.0825}{.089 + .0825} = \frac{.0825}{.1715} \approx .481$$

23.

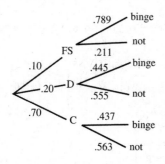

$$P(F_2 \mid E) = \frac{(.905)(.861)}{(.095)(.719) + (.905)(.861)} = \frac{.779205}{.84751} \approx .919$$

25. $P(18-39 \mid \text{high cholesterol}) = \dfrac{(.463)(.315)}{(.463)(.315) + (.349)(.504) + (.188)(.405)} = \dfrac{.145845}{.397881} \approx .367$

27. Let *FS* mean "lives in fraternity/sorority house," *D* mean "lives in dormitory," and *C* mean "lives off campus."

```
              .789   binge
         FS
              .211   not
   .10   .445         binge
    .20—D
              .555   not
   .70
         C    .437   binge
              .563   not
```

a. $P(\text{binge}) = .1(.789) + .2(.445) + .7(.437) = .4738$

b. $P(\text{FS} \mid \text{binge}) = \dfrac{.1(.789)}{.4738} = .1665$

29. F_1 means "fewer than 25,000,"
F_2 means "25,000–49,999," and F_3 means "50,000 or more."
$$P(F_3 \mid \text{decrease}) = \frac{(.043)(.117)}{(.91)(.479) + (.047)(.213) + (.043)(.117)} = \frac{.005031}{.450932} \approx .011$$

31. $P(45-64 \mid \text{lives alone})$
$$= \frac{(.318)(.144)}{(.177)(.038) + (.169)(.097) + (.179)(.086) + (.318)(.144) + (.157)(.287)} = \frac{.045792}{.129364} \approx .354$$

33. $P(\text{not living alone})$
$$= 1 - \big[(.177)(.038) + (.169)(.097) + (.179)(.086) + (.318)(.144) + (.157)(.287)\big] = 1 - .129364 \approx .871$$

35. $P(\text{not living alone} \mid 65 \text{ or over}) = 1 - .287 = .713$

Chapter 8 Review Exercises

1. $9 \in \{8, 4, -3, -9, 6\}$
Because 9 is not an element of the given set, the statement is false.

2. $4 \in \{3, 9, 7\}$
Because 4 is not an element of the given set, the statement is false.

3. $2 \notin \{0, 1, 2, 3, 4\}$
Because 2 is an element of the given set, the statement is false.

4. $0 \notin \{0, 1, 2, 3, 4\}$
Because 0 is an element of the given set, the statement is false.

5. $\{3, 4, 5\} \subseteq \{2, 3, 4, 5, 6\}$
The statement is true because every member of the first set is in the second set.

6. $(1, 2, 5, 8) \subseteq \{1, 2, 5, 10, 11\}$
This statement is false because 8 is an element of the first set but not of the second.

7. $\{1, 5, 9\} \subset \{1, 5, 6, 9, 10\}$
This statement is true because every member of the first set is a member of the second set and the second set contains at least one element not in the first set.

8. $0 \subseteq \varnothing$
This statement is false because the empty set has no subsets except itself. Also, 0 is an element, not a set, so it cannot be a subset.

9. $\{x \mid x \text{ is a national holiday}\} = \{$New Year's Day, Martin Luther King's Birthday, Presidents' Day, Memorial Day, Independence Day, Labor Day, Columbus Day, Veterans' Day, Thanksgiving, Christmas$\}$

10. $\{x \mid x \text{ is an integer}, -3 \le x < 1\} = \{-3, -2, -1, 0\}$

11. $\{$all counting numbers less than 5$\} = \{1, 2, 3, 4\}$

12. $\{x \mid x \text{ is a leap year between 1989 and 2006}\}$
$= \{1992, 1996, 2000, 2004\}$

13. M' contains all the elements of U not in M.
$M' = \{B_1, B_2, B_3, B_6, B_{12}\}$

14. N' contains all the elements of U not in N.
$N' = \{B_3, B_6, B_{12}, D\}$

15. $M \cap N$ contains all the elements that are common to M and N.
$M \cap N = \{A, C, E\}$

16. $M \cup N$ contains all the elements in either M or N or both.
$M \cup N = \{A, B_1, B_2, C, D, E\}$

17. $M \cup N'$ contains all the elements in either M or not in N, or both.
$M \cup N' = \{A, B_3, B_6, B_{12}, C, D, E\}$

18. $M' \cap N$ contains all the elements in N and not in M.
$M' \cap N = \{B_1, B_2\}$

19. $A \cap C$ is the set of all female students older than 22.

20. $B \cap D$ is the set of all finance majors with a GPA > 3.5.

21. $A \cup D$ is the set of females or students with a GPA > 3.5.

22. $A' \cap D$ is the set of male students with a GPA > 3.5.

23. $B' \cap C'$ is the set of non-finance majors who are 22 or younger

24. $B \cup A'$
Shade the region inside B as well as all of the region outside of A.

$B \cup A'$

25. $A' \cap B$
Shade all the region inside B that is also outside of A.

26. $A' \cap (B' \cap C)$
First choose the regions that are inside C and outside of B. From those regions, then shade the region outside of A.

27. $(A \cup B)' \cap C$

First choose the regions that are inside either *A* or *B*. Then choose all the regions outside those you have chosen. Now shade from those regions the region in *C*.

For Exercises 28–30, use the following Venn diagram. Since 20 movies were action, rated PG-13, and 2 hour or longer, and 28 were rated PG-13 and 2 hours or longer, 8 movies were rated PG-13 and 2 hours or longer only. Since 27 movies were action and 2 hours or longer, then 7 were action and 2 hours or longer only. 43 movies were action and PG-13, so 43 – 20 = 23 were action and PG-13 only. 56 movies were action, so 56 – (23 + 20 + 7) = 6 were action only. Similar reasoning shows that there were 36 movies that were PG-13 only and 9 movies that were 2 hours or longer only.

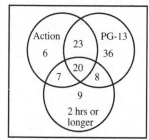

28. 6 + 23 = 29

29 movies were action movies and were not 2 hours or longer.

29. 56 + 87 – 43 = 100

100 movies were action or rated PG-13

30. There were 161 movies in total.

161 – (6 + 7 + 23 + 20 + 36 + 8 + 9) = 52

52 movies were neither action, PG-13, nor 2 hours or longer.

31. The sample space for rolling a die is {1, 2, 3, 4, 5, 6}.

32. The sample space for drawing a card from a deck containing only 4 aces is {ace of hearts, ace of diamonds, ace of spades, ace of clubs}.

33. The sample space for choosing a color and then a number is {(red, 10), (red, 20), (red, 30), (blue, 10), (blue, 20), (blue, 30), (green, 10), (green, 20), (green, 30)}.

34. {(rock, rock), (rock, pop), (rock, alternative), (pop, rock), (pop, pop), (pop, alternative), (alternative, rock), (alternative, pop), (alternative, alternative)}

35. No, the outcomes in the sample space are not equally likely because the probabilities for each type of music are not equal.

36. {(Dell, Epson), (Dell, HP), (Gateway, Epson), (Gateway, HP), (HP, Epson), (HP, HP)}

37. No, the outcomes in the sample space are not equally likely because the probabilities for each type of computer and printer are not equal.

38. "A customer buys neither" is written $E' \cap F'$. (This event can also be written as $(E \cup F)'$.)

39. "A customer buys at least one" is written $E \cup F$.

40. Answers vary. Possible answer:
This answer must be incorrect because any probability must be between 0 and 1 inclusive.

41. Answers vary. Possible answer:
Disjoint sets are sets which have no elements in common, for example, the set of all females and the set of all people who have been President of the United States.

42. Answers vary. Possible answer:
Two events are mutually exclusive if their intersection is the empty set. An example is the events "rolling a die and getting a 3" and "rolling a die and getting an even number."

43. Answers vary. Possible answer:
Disjoint sets and mutually exclusive events both have no elements in common, that is, their intersection is the empty set. Mutually exclusive events are in fact disjoint subsets of the sample space.

44. *P*(consumer discretionary or consumer staples)
=.084 + .1288 = .2128

45. *P*(information tech or telecommservices)
=.1527 + .0383 = .1910

46. $P(\$3 \text{ billion or greater}) = \dfrac{146+116}{8430} \approx .031$

47. $P\left(\$25\text{-}\$50 \text{ million} \cap \text{savings}\right) = \dfrac{115}{8430} \approx .014$

48. $P(\$1 \text{ billion} - \$3 \text{ billion} \cap \text{commercial})$
$$= \frac{310}{8430} \approx .037$$

49. $P(\text{less than } \$25 \text{ million} \cup \text{savings})$
$$= \frac{477}{8430} + \frac{1227}{8430} - \frac{55}{8430} = \frac{1649}{8430} \approx .196$$

50. $P(\$10 \text{ billion or greater} \cup \text{commercial})$
$$= \frac{116}{8430} + \frac{7203}{8430} - \frac{84}{8430} = \frac{7235}{8430} \approx .858$$

51. $P(\$500 \text{ million} - \$1 \text{ billion} \mid \text{savings})$
$$= \frac{P(\$500 \text{ million} - \$1 \text{ billion} \cap \text{savings})}{P(\text{savings})}$$
$$= \frac{159}{1227} \approx .130$$

52. $P(\$500 \text{ million} - \$1 \text{ billion} \mid \text{commercial})$
$$= \frac{P(\$500 \text{ million} - \$1 \text{ billion} \cap \text{commercial})}{P(\text{commercial})}$$
$$= \frac{548}{7203} \approx .076$$

53. $P(\text{savings} \mid \$10 \text{ billion or greater})$
$$= \frac{P(\text{savings} \cap \$10 \text{ billion or greater})}{P(\$100 \text{ billion or greater})}$$
$$= \frac{32}{116} \approx .276$$

54. $P(\text{commercial} \mid \text{less than } \$25 \text{ million})$
$$= \frac{P(\text{commercial} \cap \text{less than } \$25 \text{ million})}{P(\text{less than } \$25 \text{ million})}$$
$$= \frac{422}{477} \approx .885$$

55.

		2nd Parent	
		N_2	T_2
1st	N_1	N_1N_2	N_1T_2
Parent	T_1	T_1N_2	T_1T_2

56. $P(\text{child has disease}) = P(T_1T_2) = \frac{1}{4}$

57. There are 4 possible combinations, but only 2 have a normal cell combined with a trait cell (N_1T_2, T_1N_2).
$$P(\text{child is carrier}) = \frac{2}{4} = \frac{1}{2}$$

58. $P(\text{child is neither carrier nor has disease})$
$$= P(N_1N_2) = \frac{1}{4}$$

59. There are 36 possibilities, with 5 having a sum of 8:
$(4, 4), (3, 5), (5, 3), (2, 6)$ and $(6, 2)$.
$$P(8) = \frac{5}{36} \approx .139$$

60. $P(\text{no more than 4})$
$$= P(4) + P(3) + P(2)$$
$$= \frac{3}{36} + \frac{2}{36} + \frac{1}{36} = \frac{6}{36} \approx .167$$

61. $P(\text{at least 9})$
$$= P(9) + P(10) + P(11) + P(12)$$
$$= \frac{4}{36} + \frac{3}{36} + \frac{2}{36} + \frac{1}{36} = \frac{10}{36} \approx .278$$

62. $P(\text{odd and greater than 8})$
$$= P(9) + P(11)$$
$$= \frac{4}{36} + \frac{2}{36} = \frac{6}{36} \approx .167$$

63. A roll less than 4 means 3 or 2. There are 2 ways to get 3 and 1 way to get 2. Hence,
$$P(2 \mid \text{less than 4}) = \frac{1}{3}.$$

64. $P(7 \mid \text{at least one is a 4}) = \frac{2}{11} \approx .182$, since there are 11 possibilities with at least one of the dice being a 4 $\{(4, 1), (1, 4), (4, 2), (2, 4), (4, 3), (3, 4), (4, 4), (5, 4), (4, 5), (6, 4), (4, 6)\}$ with only $(4, 3)$ and $(3, 4)$ having a sum of 7.

For Exercises 65–68, draw a Venn diagram and use the given information to fill in the probabilities for each of the regions.

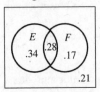

65. $P(E \cup F) = .34 + .28 + .17 = .79$

66. $P(E \cap F') = .34$

67. $P(E' \cup F) = .28 + .17 + .21 = .66$

68. $P(E' \cap F') = 1 - (.34 + .28 + .17) = .21$

69. Draw a probability tree.

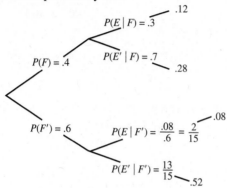

    a. $P(E' \mid F) = .7$

    b. $P(E \mid F') = \dfrac{2}{15} \approx .1333$

70. Answers vary.

71. Answers vary.

Use the following probability tree for Exercises 72–75.

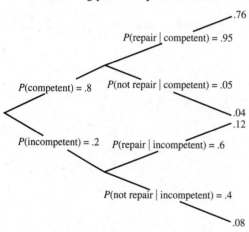

72. $P(\text{competent} \mid \text{repaired})$

$= \dfrac{.76}{.76 + .12} = \dfrac{.76}{.88} = \dfrac{19}{22}$

73. $P(\text{incompetent} \mid \text{repaired})$

$= \dfrac{.12}{.76 + .12} = \dfrac{12}{88} = \dfrac{3}{22}$

74. $P(\text{competent} \mid \text{not repaired})$

$= \dfrac{.04}{.04 + .08} = \dfrac{.04}{.12} = \dfrac{1}{3}$

75. $P(\text{incompetent} \mid \text{not repaired})$

$= \dfrac{.08}{.04 + .08} = \dfrac{8}{12} = \dfrac{2}{3}$

76. Let D mean defective.
$P(D) = .17(.04) + .39(.02) + .35(.07) + .09(.03)$
$= .0418$

    a. $P(4 \mid D) = \dfrac{P(4 \cap D)}{P(D)} = \dfrac{.09(.03)}{.0418}$

$= \dfrac{.0027}{.0418} \approx .0646$

    b. $P(2 \mid D) = \dfrac{P(2 \cap D)}{P(D)} = \dfrac{.39(.02)}{.0418}$

$= \dfrac{.0078}{.0418} \approx .1866$

77. **a.** $P(\text{second class}) = \dfrac{357}{1316} \approx .271$

    b. $P(\text{surviving}) = \dfrac{499}{1316} \approx .379$

    c. $P(\text{surviving} \mid \text{first class}) = \dfrac{203}{325} \approx .625$

    d. $P(\text{surviving} \mid \text{third class child}) = \dfrac{27}{79} \approx .342$

    e. $P(\text{female} \mid \text{first class survivor}) = \dfrac{140}{203} \approx .690$

    f. $P(\text{third class} \mid \text{male survivor}) = \dfrac{75}{146} \approx .514$

    g.   Answers vary. No, because third-class men had a slightly lower survival rate than men generally.

78. **a.**

	Too High	About Right	Too Low	Total
Male	479	364	8	851
Female	655	428	15	1098
Total	1134	792	23	1949

b. 1949 were surveyed.

c. 364 men think taxes are about right.

d. 655 women think taxes are too high.

e. 1098 women are in the survey

f. 479 of those who think taxes are too high are male.

g. Given that the respondent thought taxes are too high, how many are males?

h. $P(\text{males} \mid \text{taxes too high}) = \dfrac{479}{1134} \approx .422$

i. $P(\text{think taxes are about right} \mid \text{respondent is woman}) = \dfrac{428}{1098} \approx .390$

Additional Probability Review Exercises

1.

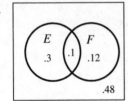

a. $P(E \cup F') = .3 + .1 + .48 = .88$

b. $P(E \cap F') = .30$

c. $P(E' \cup F) = .12 + .48 + .1 = .70$

2. There are a total of 18 (2 + 3 + 5 + 8) marbles in the jar.

a. $P(\text{white}) = \dfrac{2}{18} = \dfrac{1}{9}$

b. $P(\text{orange}) = \dfrac{3}{18} = \dfrac{1}{6}$

c. $P(\text{not black}) = \dfrac{18 - 8}{18} = \dfrac{10}{18} = \dfrac{5}{9}$

d. $P(\text{orange or yellow}) = \dfrac{3 + 5}{18} = \dfrac{8}{18} = \dfrac{4}{9}$

3. a. $P(\text{Consumer Discretionary}) = \dfrac{9.62}{100} = .0962$

b. $P(\text{Materials} \cup \text{Utilities}) = \dfrac{3.23 + 2.72}{100}$

$= \dfrac{5.95}{100} = .0595$

c. $P(\text{not Health}) = 1 - P(\text{Health})$

$= 1 - \dfrac{18.32}{100} = \dfrac{81.68}{100} = .8168$

4. a. $P(\text{Bermuda or U.S.})$

$= \dfrac{.86 + 95.35}{100} = \dfrac{96.21}{100} = .9621$

b. $P(\text{Europe})$

$= \dfrac{2.36 + 1.13}{100} = \dfrac{3.49}{100} = .0349$

c. $P(\text{not U.S.}) = 1 - \dfrac{95.35}{100} = \dfrac{4.65}{100} = .0465$

5. $S = \{1, 2, 3, 4, 5, 6\}$

a. $P(2 \mid \text{odd}) = 0$

b. $P(4 \mid \text{even}) = \dfrac{1}{3}$

c. $P(\text{even} \mid 6) = \dfrac{1}{1} = 1$

6. a. $P(\text{special education}) = \dfrac{11{,}543.0}{36{,}311.9} \approx .318$

b. $P(\text{Special Education or Vocational and Adult Education})$

$= \dfrac{11{,}543 + 2091.6}{36{,}311.9} = \dfrac{13{,}634.6}{36{,}311.9} \approx .375$

c. $P(\text{not from disadvantaged})$

$= 1 - P(\text{disadvantaged})$

$= 1 - \dfrac{14{,}842.9}{36{,}311.9} \approx .591$

7. $P(\text{age 18-44} \mid 2015) = \dfrac{116{,}686}{325{,}540} \approx .358$

8. $P(\text{age 18-44} \mid 2050) = \dfrac{150{,}400}{439{,}010} \approx .343$

9. $P(\text{65 or older} \mid 2015) = \dfrac{40{,}545 + 6292}{325{,}540} \approx .144$

10. $P(65 \text{ or older} \mid 2050) = \dfrac{69,506 + 19,041}{439,010} \approx .202$

11. No, because being a CPA increases one's likelihood of driving a luxury car.

12.

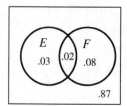

a. $P(E' \cap F) = .08$

b. $P(E' \cup F') = .98$

c. $P(E \cap F') = .03$

Use this diagram for Exercises 13–14.

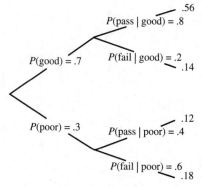

13. P(first box | orange) = $\dfrac{\frac{3}{40}}{\frac{3}{40}+\frac{3}{8}} = \dfrac{\frac{3}{40}}{\frac{18}{40}} = \dfrac{1}{6}$

14. $P(\text{second box} \mid \text{red}) = \dfrac{\frac{1}{4}}{\frac{3}{10}+\frac{1}{4}} = \dfrac{\frac{5}{20}}{\frac{11}{20}} = \dfrac{5}{11}$

15.

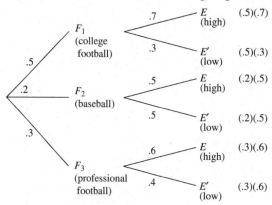

$P(\text{good} \mid \text{pass}) = \dfrac{.56}{.56 + .12} = .824 = 82.4\%$

For Exercises 16–19, use the following diagram.

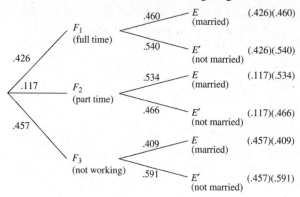

16. $P(\text{married})$
   $= .426(.460) + .117(.534) + .457(.409)$
   $\approx .445$

17. $P(\text{married} \cap \text{full time}) = .426(.460) \approx .196$

18. $P(\text{not married} \cap \text{part time}) = .117(.466) \approx .055$

19. $P(\text{not married} \cap \text{not working})$
   $= .457(.591) \approx .270$

For Exercises 20–21, use the following diagram.

20. $P(\text{college} \mid \text{high rating}) = \dfrac{.35}{.35 + .10 + .18} = \dfrac{.35}{.63}$
   $\approx .556$

21. $P(\text{professional football} \mid \text{high rating})$
   $= \dfrac{.18}{.35 + .10 + .18} = \dfrac{.18}{.63} \approx .286$

22. $P(\text{individual does not use a seat belt})$
   $= \dfrac{1}{1 + 26} = \dfrac{1}{27} = .037$

23. $P(\text{driver used a seatbelt}) = .635 \Rightarrow$
$P(\text{driver did not use a seatbelt}) = 1 - .635 = .365$
The odds that a driver used a seat belt are
$635{:}365 = 127{:}73$.

24. $P\left(\text{male} \cap 67'' - 72''\right) = \dfrac{1645}{5186} \approx .317$

25. $P\left(\text{female} \cup < 60''\right) = \dfrac{2683}{5186} + \dfrac{354}{5186} - \dfrac{316}{5186}$
$\qquad = \dfrac{2721}{5186} \approx .525$

26. $P\left(\le 72''\right) = \dfrac{2112}{5186} + \dfrac{2338}{5186} + \dfrac{354}{5186}$
$\qquad = \dfrac{4804}{5186} \approx .926$

27. $P\left(61'' - 66'' \mid \text{male}\right) = \dfrac{441}{2503} \approx .176$

28. $P\left(\text{female} \mid 67'' - 72''\right) = \dfrac{467}{2112} \approx .221$

Case 8 Medical Diagnosis

1. $P\left(T' \mid D'\right) = .999 \Rightarrow P\left(T \mid D'\right) = 1 - .999 = .001$

2. $P(D) = .005 \Rightarrow P(D') = 1 - .005 = .995$

$P(D \mid T) = \dfrac{P(D)P(T \mid D)}{P(D)P(T \mid D) + P(D')P(T \mid D')}$

$\qquad = \dfrac{.005(.99)}{.005(.99) + .995(.001)}$

$\qquad = \dfrac{.00495}{.005945} \approx .833$

3. $P(D \mid T) = \dfrac{P(D)P(T \mid D)}{P(D)P(T \mid D) + P(D')P(T \mid D')}$

$\qquad = \dfrac{.0005(.99)}{.0005(.99) + .9995(.001)}$

$\qquad = \dfrac{.000495}{.0014945} \approx .331$

Chapter 9 Counting, Probability Distributions, and Further Topics in Probability

Section 9.1 Probability Distributions and Expected Value

1. The number of possible samples is 16.

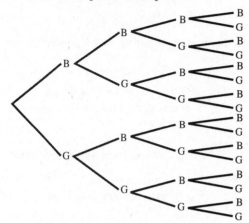

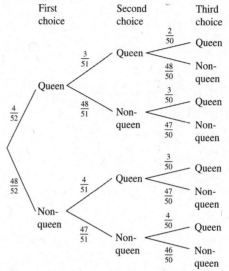

Number of Boys	$P(x)$
0	$\dfrac{1}{16} = .0625$
1	$\dfrac{4}{16} = .25$
2	$\dfrac{6}{16} = .375$
3	$\dfrac{4}{16} = .25$
4	$\dfrac{1}{16} = .0625$

3. Let x be the number of queens drawn. Then x can take on values 0, 1, 2, or 3.

Number of queens	$P(x)$
0	$\dfrac{48}{52} \cdot \dfrac{47}{51} \cdot \dfrac{46}{50} \approx .7826$
1	$\dfrac{4}{52} \cdot \dfrac{48}{51} \cdot \dfrac{47}{50} + \dfrac{48}{52} \cdot \dfrac{4}{51} \cdot \dfrac{47}{50}$ $+ \dfrac{48}{52} \cdot \dfrac{47}{51} \cdot \dfrac{4}{50} \approx .2042$
2	$\dfrac{4}{52} \cdot \dfrac{3}{51} \cdot \dfrac{48}{50} + \dfrac{4}{52} \cdot \dfrac{48}{51} \cdot \dfrac{3}{50}$ $+ \dfrac{48}{52} \cdot \dfrac{4}{51} \cdot \dfrac{3}{50} \approx .0130$
3	$\dfrac{4}{52} \cdot \dfrac{3}{51} \cdot \dfrac{2}{50} \approx .0002$

5.

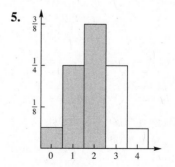

7. The histogram for Exercise 3 with P(at least one queens) follows.

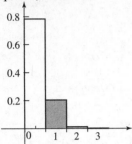

9. expected value $= 1(.1) + 3(.5) + 5(.2) + 7(.2)$
$= 4$

11. expected value
$= 0(.21) + 2(.24) + 4(.21) + 8(.17) + 16(.17)$
$= 5.4$

13. $E(x) = 1(.2) + 2(.3) + 3(.1) + 4(.4) = 2.7$

15. $E(x) = 1(.3) + 2(.25) + 3(.2) + 4(.15) + 5(.1)$
$= 2.5$

17. $E(x) = 1\left(\dfrac{18}{38}\right) - 1\left(\dfrac{20}{38}\right) = -\dfrac{2}{38} = -\dfrac{1}{19} \approx -0.05$

19. You have one chance in a thousand of winning $500 on a $1 bet for a net return of $499. In the 999 other outcomes you lose your dollar

$E(x) = 499\left(\dfrac{1}{1000}\right) + (-1)\left(\dfrac{999}{1000}\right)$

$= -\dfrac{500}{1000} = -\$.50 \text{ or } -50¢$

21. Let x denote the winnings.
$E(x)$

$= 49,999\left(\dfrac{1}{2,000,000}\right) + 9,999\left(\dfrac{2}{2,000,000}\right)$

$\qquad\qquad -1\left(\dfrac{1,999,997}{2,000,000}\right)$

$= -\$.965 = -96.5¢ \approx -97¢$

23. $E(x) = 25(-.87) = -21.75$

25. $E(x) = 0(.0004) + 1(.0100) + 2(.0901)$
$\qquad\qquad + 3(.3600) + 4(.5394)$
$\qquad = 3.4278$

27. The distribution is not valid since no probability can be less than 0.

29. The sum of the probabilities total 1.0, so this is a valid distribution.

31. Let x denote the missing probability. Then,
$.01 + .09 + .25 + .45 + .05 + x = 1.0 \Rightarrow$
$.85 + x = 1.0 \Rightarrow x = 1.0 - .85 = .15$

33. Let x denote the missing probability. Then
$.20 + x + .25 + .30 = 1.0 \Rightarrow x + .75 = 1.0 \Rightarrow$
$x = 1.0 - .75 = .25$

35. Let x and y denote the missing probabilities. Then
$.10 + .10 + .20 + .25 + .05 + x + y = 1.0 \Rightarrow$
$.70 + x + y = 1.0 \Rightarrow x + y = 1.0 - .7 = .3$
Answers may vary. One possible answer would be .1 and .2.

37. $E(x) = .0007\big[100(15,000) + 250(10,000)$
$\qquad\qquad\qquad\qquad + 500(5,000)\big]$
$\qquad = .0007(6,500,000) = \$4550$

39. $E(x) = 0(.595) + 1(.337) + 2(.064) + 3(.004)$
$\qquad = .477$

41.

Account Number	Expected value	Exist. vol. + exp. value	Class
3	2000	22,000	C
4	1000	51,000	B
5	25,000	30,000	C
6	60,000	60,000	A
7	16,000	46,000	B

43. **a.** Let x denote the cost of using each antibiotic. For amoxicillin,
$E(x) = .75(59.30) + .25(96.15) \approx \$68.51$.
For cefaclor,
$E(x) = .90(69.15) + .10(106.00) \approx \$72.84$.

b. Amoxicillin, since the total expected cost is less.

45. **a.** $E(x) = 750,000(.5) + 350,000(.5)$
$\qquad = 550,000 \text{ pounds}$

b. $E(x) = 750,000(.67) + 350,000(.33)$
$\qquad = 618,000 \text{ pounds}$

Section 9.2 The Multiplication Principle, Permutations and Combinations

1. $_4P_2 = \dfrac{4!}{(4-2)!} = \dfrac{4!}{2!} = 4 \cdot 3 = 12$

3. $_8C_5 = \dfrac{8!}{5!(8-5)!} = \dfrac{8!}{3!5!} = \dfrac{8 \cdot 7 \cdot 6}{3 \cdot 2 \cdot 1} = 56$

5. $_8P_1 = \dfrac{8!}{(8-1)!} = \dfrac{8!}{7!} = 8$

7. $4! = 4 \cdot 3 \cdot 2 \cdot 1 = 24$

9. $_9C_6 = \dfrac{9!}{6!(9-6)!}$

$= \dfrac{9!}{6!3!} = \dfrac{9 \cdot 8 \cdot 7}{3 \cdot 2 \cdot 1} = 84$

11. $_{13}P_3 = \dfrac{13!}{(13-3)!} = \dfrac{13!}{10!} = 13 \cdot 12 \cdot 11 = 1716$

13. $_{25}P_5 = 6,375,600$

15. $_{14}P_5 = 240,240$

17. $_{18}C_5 = \dfrac{18!}{5!13!} = \dfrac{18 \cdot 17 \cdot 16 \cdot 15 \cdot 14}{5 \cdot 4 \cdot 3 \cdot 2 \cdot 1} = 8568$

19. $_{28}C_{14} = 40,116,600$

21. If $0! = 0$, $_4P_4 = \dfrac{4!}{(4-4)!} = \dfrac{24}{0}$ = undefined

23. a. There are two possibilities for each line, and, by the multiplication principle, $2 \cdot 2 \cdot 2 = 8$ trigrams.

b. Since each hexagram is made up of two trigrams, and there are 8 possible trigrams, it follows that there are $8 \cdot 8 = 64$ possible hexagrams.

25. $6 \cdot 8 \cdot 4 \cdot 3 = 576$
There are 576 varieties of autos available.

27. Yes; Since a social security number has 9 digits with no restrictions, there are $10^9 = 1,000,000,000$ (1 billion) different social security numbers. This is enough for every one of the people in the United States to have a social security number.

29. Since a zip code has nine digits with no restrictions, there are 10^9 or 1,000,000,000 different 9-digit zip codes.

31. There are $12 \cdot 10 = 120$ different ways Sherri can buy 1 Pantene shampoo and 1 Pantene conditioner.

33. a. There are 8 possibilities for the first digit, 2 possibilities for the second digit, and 10 possibilities for the last digit. The total number of possible area codes is
$8 \cdot 2 \cdot 10 = 160$
There are 8 possibilities for the first digit and 10 possibilities for each of the next six digits. The total number of phone numbers is
$8 \cdot 10 \cdot 10 \cdot 10 \cdot 10 \cdot 10 \cdot 10 = 8 \cdot 10^6$
$= 8,000,000$

b. Some numbers, like 911, 800, and 900, are reserved for special purposes.

35. In this new plan, there would be 8 possibilities for the first digit, 2 possibilities for the second digit, and 10 possibilities for each of the last 2 digits. The total number of area codes is
$8 \cdot 2 \cdot 10 \cdot 10 = 1600$.

37. Answers vary. Possible answer:
A permutation of a elements ($a \geq 1$) from a set of b elements is any arrangement, without repetition, of the a elements.

39. Since order makes a difference, the number of arrangements is
$_{12}P_5 = \dfrac{12!}{(10-5)!} = \dfrac{12!}{5!} = 95,040.$

41. Order is important, and there are no repeats, so there are $_{24}P_3 = \dfrac{24!}{(24-3)!} = 24 \cdot 23 \cdot 22 = 12,144$
ways to name a fraternity using 3 Greek letters.

43. Four people are being selected. Since each receives a different job, order is important. The total number of different officer selections is
$_{32}P_4 = \dfrac{32!}{(32-4)!} = \dfrac{32!}{28!} = 32 \cdot 31 \cdot 30 \cdot 29$
$= 863,040$

45. $_{17}P_2 = \dfrac{17!}{(17-2)!} = \dfrac{17!}{15!} = 17 \cdot 16 = 272$

47. This is a combinations problem.

a. $_{10}C_4 = \dfrac{10!}{4!6!} = 210$

b. $_{10}C_6 = \dfrac{10!}{6!4!} = 210$

49. Order is not important, so use a combination.

$_{11}C_4 = \dfrac{11!}{(11-4)!4!} = 330$

There are 330 ways to select a group of 4 co-captains from a group of 11.

51. Answers vary. Possible answer: Combinations are not ordered, while permutations are ordered.

53. a. With repetition permitted, the tree diagram shows 9 different pairs.

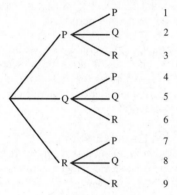

1ˢᵗ Choice   2ⁿᵈ Choice   Number of Ways

b. If repetition is not permitted, one branch is missing from each of the clusters of second branches, for a total of 6 different pairs.

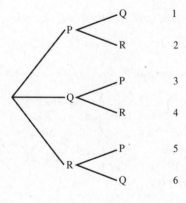

1ˢᵗ Choice   2ⁿᵈ Choice   Number of Ways

c. Find the number of combinations of 3 elements taken 2 at a time.

$_3C_2 = \dfrac{3!}{2!1!} = 3$

No repetitions are allowed, so the answer cannot equal that for part a. However, since order does not matter, our answer is only half of the answer for part b. For example PQ and QP are distinct in b. but not in c. Thus, the answer differs from both a. and b.

55. These are combination problems since order is not important.

a. $_{18}C_5 = 8568$ delegations

b. $_7C_5 = 21$ delegations

c. $_{11}C_3 \cdot _7C_2 = 165 \cdot 21 = 3465$ delegations

d. At least one means one or two or three or four or five.

$_{11}C_1 \cdot _7C_4 + _{11}C_2 \cdot _7C_3 + _{11}C_3 \cdot _7C_2$
$+ _{11}C_4 \cdot _7C_1 + _{11}C_5$
$= 385 + 1925 + 3465 + 2310 + 462$
$= 8547$

8547 delegations would have at least one Democrat.

57. There are no restrictions as to whether the scoops have to be different flavors.

a. The number of different double-scoops will be $_{21}C_2 = 210$.

b. The number of different triple-scoops will be $_{21}P_3 = 7980$ since order matters.

59. a. Since order is not important, this is a combination problem.
$_{99}C_6 = 1,120,529,256$

b. Since order is important, this is a permutation problem.
$_{99}P_6 = 806,781,064,300$

61. Using the multiplication principle, we have
$3 \cdot 3 \cdot 3 \cdot 3 = 81$

63. It is not possible since $26^3 = 17,576$ (the number of different 3-initial names) and the biologist needs 52,000 names. Actually, 4-initial names would do the biologist's job since $26^4 = 456,976$.

65. Since order is not important, these are combination problems.

   a. $_5C_3 = 10$

   b. Since you are taking 3 Diet Coke, and only 1 Diet Coke exists, this situation is impossible. The answer is 0.

   c. $_3C_3 = 1$

   d. $_5C_2 \cdot {_1C_1} = 10 \cdot 1 = 10$

   e. $_5C_2 \cdot {_3C_1} = 10 \cdot 3 = 30$

   f. $_3C_2 \cdot {_5C_1} = 3 \cdot 5 = 15$

   g. Again, since you are picking 2 Diet Coke and only 1 Diet Coke exists, this situation is impossible. The answer is 0.

67. Since order is not important, we use combinations. $_{35}C_{20} = 3,247,943,160$
There are 3,247,943,160 ways to select 20 volunteers from the group of 35.

69. Since the order in each column is important, we use permutations.
$$\left({_{15}P_5} \right)^4 \cdot \left({_{15}P_4} \right) = 5.524 \times 10^{26}$$

71. **a.** martini
There are 7 total letters, $1m$, $1a$, $1r$, $1t$, $2i$'s, and $1n$.
$$\frac{7!}{1!1!1!1!2!1!} = 2520$$

   b. nunnery
There are 7 letters, $3n$'s, $1u$, $1e$, $1r$, and $1y$.
$$\frac{7!}{3!1!1!1!1!} = 840$$

   c. grinding
There are 8 letters, $2g$'s, $1r$, $2i$'s, $2n$'s, and $1d$.
$$\frac{8!}{2!1!2!2!1!} = 5040$$

73. **a.** The key word is "distinguishable", so use permutations. Total of 12 dinners:
$_{12}P_{12} = 479,001,600$ ways

b. Since dinners of the same company are considered identical, the problem is to find the number of different arrangements of the three colors. $_3P_3 = 6$

c. $\dfrac{12!}{4!3!5!} = 27,720$

Section 9.3 Applications of Counting

1. There are $_{24}C_3 = 2024$ ways to select 3 tiles from the batch of 24. There are $_5C_1 = 5$ ways to select one defective tile from the batch, so there are $_{19}C_2 = 171$ ways to choose 2 good tiles from the 19 good tiles in the batch. Thus,
$$P(1 \text{ defective}) = \frac{{_5C_1} \cdot {_{19}C_2}}{_{24}C_3} = \frac{5 \cdot 171}{2024} \approx .422.$$

3. There are 8 computers, 5 non-defective, and a sample of size 1 is chosen.
$$P(\text{no defectives}) = \frac{_5C_1}{_8C_1} = \frac{5}{8}$$

5. There are 8 computers, 5 non-defective, and a sample of size 3 is chosen.
$$P(\text{no defectives}) = \frac{_5C_3}{_8C_3} = \frac{10}{56} = \frac{5}{28}$$

7. There are 42 prizes, 10 of which are $100 prizes; 3 are drawn.
$$P(\text{all are \$100 prizes}) = \frac{_{10}C_3}{_{242}C_3}$$
$$= \frac{120}{2,332,880} \approx .00005$$

9. There are 42 prizes, 20 of which are $25 prizes and 200 "dummy" tickets; 3 are drawn.
$$P(\text{two \$25 prizes}) = \frac{_{20}C_2 \cdot {_{200}C_1}}{_{242}C_3}$$
$$= \frac{190 \cdot 1 \cdot 200}{2,332,880} \approx .0163$$

11. There are 200 dummy tickets, and all 3 must come from that group.
$P(\text{no winning ticket})$
$$= \frac{_{200}C_3}{_{242}C_3} = \frac{1,313,400}{2,332,880} \approx .5630$$

13. Since order does not make a difference, the number of 2-card hands is $_{52}C_2 = 1326$.

15. There are 48 non-deuces (2's) in a deck of cards.

$$P(\text{no deuces}) = \frac{_{48}C_2}{_{52}C_2} = \frac{1128}{1326} \approx .851$$

17. From Exercise 13, we know there are 1326 different 2-card hands. To see how many hands have the same suit, there are $_{13}C_2 = 78$ ways to get a 2-card hand of one particular suit. Because there are 4 different suits, there are $4 \cdot 78 = 312$ hands of the same suit. Therefore, there are $1326 - 312 = 1014$ hands of different suits.

$$P(\text{different suits}) = \frac{1014}{1326} \approx .765$$

19. $P(\text{no more than 1 diamond})$
$= P(\text{no diamonds}) + P(\text{1 diamond})$
Because there are 39 non-diamonds in a deck,

$$P(\text{no more than 1 diamond}) = \frac{741}{1326} + \frac{39 \cdot 13}{1326}$$
$$= \frac{1248}{1326} \approx .941.$$

Alternate solution:
$P(\text{at least 1 black card}) = 1 - P(\text{2 diamonds})$

$$= 1 - \frac{_{13}C_2}{_{52}C_2} = \frac{78}{1326}$$
$$\approx .941$$

21. Answer varies. Possible answer:
The advantage of using this rule is that it is many times easier to calculate the probability of the complement of an event than of the given event. For example in dealing a hand of 5 cards, if E is the event to get at least one heart, it is much easier to calculate $P(E)'$, the probability of getting no hearts and subtracting it from 1.

23. There are 22 rocks in all, so there are $_{22}C_5 = 26,334$ ways to select 5 rocks. There are $_7C_3 = 35$ ways to select 3 sedimentary rocks and $_{15}C_2 = 105$ ways to select the remaining 2 rocks. Therefore, the probability of selecting 5 rocks of which 3 are sedimentary is

$$\frac{_7C_3 \cdot _{15}C_2}{_{22}C_5} = \frac{35 \cdot 105}{26,334} \approx .1396.$$

25. a. We must pick all 6 of our numbers from the 99 total numbers.

$$P(\text{all 6}) = \frac{_6C_6}{_{99}C_6} \approx 8.9 \times 10^{-10}$$

b. $P(\text{all 6}) = \dfrac{1}{_{99}P_6} \approx 1.2 \times 10^{-12}$

27. The probability of two individuals independently selecting the winning numbers is

$$\left(\frac{1}{195,249,054} \right)^2 \approx 2.62 \times 10^{-17}.$$

29. a. $P(\text{3 women and 1 man}) = \dfrac{_4C_3 \cdot _{11}C_1}{_{15}C_4}$
$$= \frac{4 \cdot 11}{1365} \approx .0322$$

b. $P(\text{all men}) = \dfrac{_{11}C_4}{_{15}C_4} = \dfrac{330}{1365} \approx .2418$

c. $P(\text{at least 1 woman})$
$= 1 - P(\text{no women})$
$= 1 - P(\text{all men})$
$= 1 - .2418 = .7582$

31. There are $_{20}C_7 = 77,520$ ways to select 7 volunteers from the pool of 20.

a. $P\left(\text{all 20-39}\right) = \dfrac{_{10}C_7}{_{20}C_7} = \dfrac{120}{77,520} \approx .0015$

b. $P\left(\text{5 20-39, 2 60 or older}\right) = \dfrac{_{10}C_5 \cdot _2C_2}{_{20}C_7}$
$$= \frac{252 \cdot 1}{77,520} \approx .0033$$

c. $P\left(\text{3 40-59, 4 other}\right) = \dfrac{_8C_3 \cdot _{12}C_4}{_{20}C_7}$
$$= \frac{56 \cdot 495}{77,520} \approx .3576$$

33. The probability that at least 2 of the 100 U.S. Senators have the same birthday is

$$1 - \frac{_{365}P_{100}}{(365)^{100}} \approx 1.$$

35. There are $_{20}C_4 = 4845$ ways to pick the correct 4 numbers out of the 20 the state picks. There are $_{80}C_4 = 1,581,580$ ways to pick 4 numbers out of 80. The probability of winning \$55 is

$$\frac{_{20}C_4}{_{80}C_4} = \frac{4845}{1581580} \approx .0031.$$

37. There are 60 losing numbers from which all 4 must be picked. There are 80 numbers from which 4 are picked. The probability of losing is

$$\frac{_{20}C_0 \cdot {}_{60}C_4}{_{80}C_4} = \frac{1 \cdot 487,635}{1,581,580} \approx .3083$$

39. Answers vary. Theoretical answers:
This exercise should be solved by computer methods. The solution will vary according to the computer program that is used. The answers are (a) .0399, (b) .5191, (c) .0226.

Section 9.4 Binomial Probability

1. $n = 10$, $p = .7$, $x = 6$, $1 - p = .3$
$P(\text{exactly } 6) = {}_{10}C_6(.7)^6(.3)^4 \approx .200$

3. $n = 10$, $p = .7$, $x = 0$, $1 - p = 1 - .7 = .3$
$P(\text{none}) = {}_{10}C_0(.7)^0(.3)^{10} \approx .000006$

5. $P(\text{at least } 1) = 1 - P(\text{none})$
From Exercise 3, we have $P(\text{none}) = .000006$, so
$P(\text{at least } 1) = 1 - .000006 = .999994$.

For exercises 7–11, we have $n = 10$, $p = .19$, and $1 - p = .81$.

7. $x = 2$; $P(\text{exactly } 2) = {}_{10}C_2(.19)^2(.81)^8 \approx .3010$.

9. $x = 0$; $P(\text{none}) = {}_{10}C_0(.19)^0(.81)^{10} \approx .1216$

11. $P(\text{at least } 1) = 1 - P(\text{none})$. From Exercise 9, we have $P(\text{none}) \approx .1216$, so
$P(\text{at least } 1) \approx 1 - .1216 \approx .8784$.

For exercises 13–15, we have $n = 5$, $p = \frac{1}{2}$, and $1 - p = \frac{1}{2}$.

13. $x = 5$; $P(\text{all heads}) = {}_5C_5\left(\frac{1}{2}\right)^5\left(\frac{1}{2}\right)^0 = \frac{1}{32}$

15. "No more than 3 heads" means 0 heads, 1 head, 2 heads, or 3 heads.

$P(0 \text{ heads}) = {}_5C_0\left(\frac{1}{2}\right)^0\left(\frac{1}{2}\right)^5 = \frac{1}{32}$

$P(1 \text{ head}) = {}_5C_1\left(\frac{1}{2}\right)^1\left(\frac{1}{2}\right)^4 = \frac{5}{32}$

$P(2 \text{ heads}) = {}_5C_2\left(\frac{1}{2}\right)^2\left(\frac{1}{2}\right)^3 = \frac{10}{32}$

$P(3 \text{ heads}) = {}_5C_3\left(\frac{1}{2}\right)^3\left(\frac{1}{2}\right)^2 = \frac{10}{32}$

$P(\text{no more than 3 heads})$
$= \frac{1}{32} + \frac{5}{32} + \frac{10}{32} + \frac{10}{32} = \frac{26}{32} = \frac{13}{16}$

17. Answer varies. Possible answer:
A problem involves a binomial experiment if the experiment is repeated several times, there are only two possible outcomes, and the repeated trials are independent.

For exercises 19–21, we have $n = 15$, $p = .34$, and $1 - p = .66$.

19. $P(3) = {}_{15}C_3(.34)^3(.66)^{12} \approx .1222$

21. $P(\text{at most } 2) = P(0) + P(1) + P(2)$
$= {}_{15}C_0(.34)^0(.66)^{15} + {}_{15}C_1(.34)^1(.66)^{14}$
$\qquad + {}_{15}C_2(.34)^2(.66)^{13}$
$\approx .0020 + .0152 + .0547 = .0719$

23. Since 34% of the mortgages were in California, $.34 \cdot 200 = 68$ of the 200 foreclosed mortgages would be expected to be from California.

For exercises 25–27, we have $n = 8$, $p = .56$, and $1 - p = .44$.

25. $P(\text{all but } 1) = P(7) = {}_8C_7(.56)^7(.44)^1 \approx .0608$

27. We use the result from exercise 24.
$P(\text{at most } 7) = 1 - P(8) = 1 - .0097 = .9903$

29. $n = 100$, $p = .027$, and $1 - p = .973$
$P(\text{exactly 2 sets of twins})$
$= {}_{100}C_2(.027)^2(.973)^{98} \approx .247$

31. $n = 9$, $p = .11$, $1 - p = .89$
$P(2) = {}_9C_2(.11)^2(.89)^7 \approx .1927$

33. $P(\text{none}) = {}_9C_0(.11)^0(.89)^9 \approx .3504$

35. Since 11% of Americans are left-handed, $.11 \times 35 = 3.85$ would be expected to be left-handed from the sample of 35.

37. $n = 16$, $p = .073$, and $1 - p = .927$.

$$P(\text{at most } 3) = P(0) + P(1) + P(2) + P(3)$$

$$= {}_{16}C_0 (.073)^0 (.927)^{16} + {}_{16}C_1 (.073)^1 (.927)^{15} + {}_{16}C_2 (.073)^2 (.927)^{14} + {}_{16}C_3 (.073)^3 (.927)^{13}$$

$$\approx .2974 + .3747 + .2213 + .0813 = .9747$$

39. The probability will be lower.

$$P(\text{at least } 2) = 1 - \left[P(0) + P(1) \right] = 1 - \left[{}_5C_0 (.158)^0 (.842)^5 + {}_5C_1 (.158)^1 (.842)^4 \right]$$

$$\approx 1 - (.4232 + .3971) = .1797$$

41. a. $P(\text{fewer than } 15) = {}_{40}C_{14}(.152)^{14}(.848)^{26} + {}_{40}C_{13}(.152)^{13}(.848)^{27} + {}_{40}C_{12}(.152)^{12}(.848)^{28}$

$$+ {}_{40}C_{11}(.152)^{11}(.848)^{29} + {}_{40}C_{10}(.152)^{10}(.848)^{30} + {}_{40}C_9(.152)^9(.848)^{31}$$

$$+ {}_{40}C_8(.152)^8(.848)^{32} + {}_{40}C_7(.152)^7(.848)^{33} + {}_{40}C_6(.152)^6(.848)^{34}$$

$$+ {}_{40}C_5(.152)^5(.848)^{35} + {}_{40}C_4(.152)^4(.848)^{36} + {}_{40}C_3(.152)^3(.848)^{37}$$

$$+ {}_{40}C_2(.152)^2(.848)^{38} + {}_{40}C_1(.152)^1(.848)^{39} + {}_{40}C_0(.152)^0(.848)^{40}$$

$$\approx .00112 + .00324 + .00840 + .01940 + .03968 + .07140 + .11204 + .15153 + .17405$$

$$+ .16646 + .12898 + .07779 + .03426 + .00980 + .00137$$

$$\approx .9995$$

b. Answers vary.

43. a. $P(\text{all 5 of the bands match}) = {}_5C_5(.25)^5 = \dfrac{1}{1024}$ or 1 chance in 1024

b. $P(\text{all 20 of the bands match}) = {}_{20}C_{20}(.25)^{20} \approx$ about 1 chance in 1.1×10^{12}

c. $P(\text{16 or more bands match}) = {}_{20}C_{16}(.25)^{16}(.75)^4 + {}_{20}C_{17}(.25)^{17}(.75)^3 + {}_{20}C_{18}(.25)^{18}(.75)^2$

$$+ {}_{20}C_{19}(.25)^{19}(.75) + {}_{20}C_{20}(.25)^{20}$$

$$\approx \text{about 1 chance in } 2.6 \times 10^6$$

d. Answers vary.

Section 9.5 Markov Chains

1. $\begin{bmatrix} \frac{1}{4} & \frac{3}{4} \end{bmatrix}$ could be a probability vector because it is a matrix with only one row containing only nonnegative entries whose sum is $\dfrac{1}{4} + \dfrac{3}{4} = 1$.

3. $\begin{bmatrix} 0 & 1 \end{bmatrix}$ could be a probability vector because it has only one row, the entries are nonnegative, and $0 + 1 = 1$.

5. $\begin{bmatrix} .3 & -.1 & .6 \end{bmatrix}$ cannot be a probability vector because it has a negative entry, $-.1$.

7. $\begin{bmatrix} .7 & .2 \\ .5 & .5 \end{bmatrix}$ cannot be a transition matrix because the sum of the entries in the first row is $.7 + .2 = .9 \neq 1$.

9. $\begin{bmatrix} \frac{4}{9} & \frac{1}{3} \\ \frac{1}{5} & \frac{7}{10} \end{bmatrix}$

This cannot be a transition matrix because the sum of the entries in row 1 is $\dfrac{4}{9} + \dfrac{1}{3} = \dfrac{7}{9} \neq 1$ and the sum of the entries in row 2 is $\dfrac{1}{5} + \dfrac{7}{10} = \dfrac{9}{10} \neq 1$.

11. $\begin{bmatrix} \frac{1}{2} & \frac{1}{4} & 1 \\ \frac{2}{3} & 0 & \frac{1}{3} \\ \frac{1}{3} & 1 & 0 \end{bmatrix}$

This could not be a transition matrix because the sum of the entries in row 1 is $\dfrac{1}{2} + \dfrac{1}{4} + 1 = \dfrac{7}{4} \neq 1$, and the sum of the entries in row 3 is

$\dfrac{1}{3} + 1 = \dfrac{4}{3} \neq 1.$

13. This is not a transition diagram because the sum of the probabilities for changing from state A to states A, B, and C is

$\dfrac{1}{3} + \dfrac{1}{2} + 1 \neq 1.$

15. This is a transition diagram. The information given in this diagram can also be given by the following matrix.

$\begin{array}{c} \\ A \\ B \\ C \end{array} \begin{array}{ccc} A & B & C \\ \begin{bmatrix} .6 & .2 & .2 \\ .9 & .02 & .08 \\ .4 & .0 & .6 \end{bmatrix} \end{array}$

17. Let $A = \begin{bmatrix} .2 & .8 \\ .9 & .1 \end{bmatrix}$

A is a regular transition matrix since A^2 contains all positive entries.

19. Let $P = \begin{bmatrix} 0 & 1 & 0 \\ .3 & .3 & .4 \\ 1 & 0 & 0 \end{bmatrix}$.

$P^2 = \begin{bmatrix} 0 & 1 & 0 \\ .3 & .3 & .4 \\ 1 & 0 & 0 \end{bmatrix}\begin{bmatrix} 0 & 1 & 0 \\ .3 & .3 & .4 \\ 1 & 0 & 0 \end{bmatrix}$

$= \begin{bmatrix} .3 & .3 & .4 \\ .49 & .39 & .12 \\ 0 & 1 & 0 \end{bmatrix}$

$P^3 = \begin{bmatrix} 0 & 1 & 0 \\ .3 & .3 & .4 \\ 1 & 0 & 0 \end{bmatrix}\begin{bmatrix} .3 & .3 & .4 \\ .49 & .39 & .12 \\ 0 & 1 & 0 \end{bmatrix}$

$= \begin{bmatrix} .49 & .39 & .12 \\ .237 & .607 & .156 \\ .3 & .3 & .4 \end{bmatrix}$

P is a regular transition matrix since P^3 contains all positive entries.

21. Let $A = \begin{bmatrix} .23 & .41 & 0 & .36 \\ 0 & .27 & .21 & .52 \\ 0 & 0 & 1 & 0 \\ .48 & 0 & .39 & .13 \end{bmatrix}$

$A^2 = \begin{bmatrix} .2257 & .205 & .2265 & .3428 \\ .2496 & .0729 & .4695 & .208 \\ 0 & 0 & 1 & 0 \\ .1728 & .1968 & .4407 & .1897 \end{bmatrix}$

A is not regular. Any power of A will have zero entries in the 1st, 2nd, and 4th column of row 3; thus, it cannot have all positive entries.

23. $\begin{bmatrix} v_1 & v_2 \end{bmatrix}\begin{bmatrix} .55 & .45 \\ .19 & .81 \end{bmatrix} = \begin{bmatrix} v_1 & v_2 \end{bmatrix}.$

We obtain the two equations

$.55v_1 + .19v_2 = v_1, \quad .45v_1 + .81v_2 = v_2.$

Simplify these equations to get the system

$-.45v_1 + .19v_2 = 0,$

$.45v_1 - .19v_2 = 0.$

These equations are dependent. Since $\mathbf{V}$ is a probability vector, $v_1 + v_2 = 1$.

Solve the system

$-.45v_1 + .19v_2 = 0.$

$v_1 + v_2 = 1$

by the substitution method.

$-.45(1 - v_2) + .19v_2 = 0 \Rightarrow$

$-.45 + .45v_2 + .19v_2 = 0 \Rightarrow .64v_2 = .45 \Rightarrow$

$v_2 = \dfrac{.45}{.64} = \dfrac{45}{64}$

$v_1 = 1 - \dfrac{45}{64} = \dfrac{19}{64}$

Thus, the equilibrium vector is

$\begin{bmatrix} \dfrac{19}{64}, & \dfrac{45}{64} \end{bmatrix}.$

25. $\begin{bmatrix} v_1 & v_2 \end{bmatrix} \begin{bmatrix} \frac{2}{3} & \frac{1}{3} \\ \frac{1}{8} & \frac{7}{8} \end{bmatrix} = \begin{bmatrix} v_1 & v_2 \end{bmatrix}.$

We obtain the two equations

$\frac{2}{3}v_1 + \frac{1}{8}v_2 = v_1$

$\frac{1}{3}v_1 + \frac{7}{8}v_2 = v_2.$

Multiply both equations by 24 to eliminate fractions.

$16v_1 + 3v_2 = 24v_1$

$8v_1 + 21v_2 = 24v_2$

Simplify both equations.

$-8v_1 + 3v_2 = 0$

$8v_1 - 3v_2 = 0$

These equations are dependent.

$v_1 + v_2 = 1 \Rightarrow v_1 = 1 - v_2.$

Substituting $1 - v_2$ for v_1 in the first equation gives

$-8(1 - v_2) + 3v_2 = 0 \Rightarrow -8 + 8v_2 + 3v_2 = 0 \Rightarrow$

$11v_2 = 8 \Rightarrow v_2 = \frac{8}{11}$ and $v_1 = 1 - v_2 = \frac{3}{11}.$

The equilibrium vector is

$\begin{bmatrix} \frac{3}{11} & \frac{8}{11} \end{bmatrix}.$

27. Let $P = \begin{bmatrix} .16 & .28 & .56 \\ .43 & .12 & .45 \\ .86 & .05 & .09 \end{bmatrix}$, and let **V** be the

probability vector $\begin{bmatrix} v_1 & v_2 & v_3 \end{bmatrix}.$

$\begin{bmatrix} v_1 & v_2 & v_3 \end{bmatrix} \begin{bmatrix} .16 & .28 & .56 \\ .43 & .12 & .45 \\ .86 & .05 & .09 \end{bmatrix} = \begin{bmatrix} v_1 & v_2 & v_3 \end{bmatrix}$

$.16v_1 + .43v_2 + .86v_3 = v_1$

$.28v_1 + .12v_2 + .05v_3 = v_2$

$.56v_1 + .45v_2 + .09v_3 = v_3$

Simplify these equations by the Gauss-Jordan method to obtain

$v_1 = \frac{7783}{16,799}, v_2 = \frac{2828}{16,799},$ and $v_3 = \frac{6188}{16,799}$

The equilibrium vector is

$\begin{bmatrix} \frac{7783}{16,799} & \frac{2828}{16,799} & \frac{6188}{16,799} \end{bmatrix},$

or $\begin{bmatrix} .4633 & .1683 & .3684 \end{bmatrix}$ in decimal form.

29. Let $P = \begin{bmatrix} .44 & .31 & .25 \\ .80 & .11 & .09 \\ .26 & .31 & .43 \end{bmatrix}$, and let **V** be the

probability vector $\begin{bmatrix} v_1 & v_2 & v_3 \end{bmatrix}.$

$\begin{bmatrix} v_1 & v_2 & v_3 \end{bmatrix} \begin{bmatrix} .44 & .31 & .25 \\ .80 & .11 & .09 \\ .26 & .31 & .43 \end{bmatrix} = \begin{bmatrix} v_1 & v_2 & v_3 \end{bmatrix}$

$.44v_1 + .80v_2 + .26v_3 = v_1$

$.31v_1 + .11v_2 + .31v_3 = v_2$

$.25v_1 + .09v_2 + .43v_3 = v_3$

Simplify these equations to get the system

$-.56v_1 + .80v_2 + .26v_3 = 0$

$.31v_1 - .89v_2 + .31v_3 = 0$

$.25v_1 + .09v_2 - .57v_3 = 0$

Since **V** is the probability vector,

$v_1 + v_2 + v_3 = 1.$

This gives us a system of four equations in three variables.

$v_1 + v_2 + v_3 = 1$

$-.56v_1 + .80v_2 + .26v_3 = 0$

$.31v_1 - .89v_2 + .31v_3 = 0$

$.25v_1 + .09v_2 - .57v_3 = 0$

This system can be solved by the Gauss-Jordan method. Start with the augmented matrix

$\begin{bmatrix} 1 & 1 & 1 \\ -.56 & .80 & .26 \\ .31 & -.89 & .31 \\ .25 & .09 & -.57 \end{bmatrix} \begin{bmatrix} 1 \\ 0 \\ 0 \\ 0 \end{bmatrix}.$

The solution of this system is

$v_1 = .4872, v_2 = .2583,$ and $v_3 = .2545,$ so the

equilibrium vector is $\begin{bmatrix} .4872 & .2583 & .2545 \end{bmatrix}.$

31. The following solution assumes the use of a TI-82 graphing calculator. Similar results can be obtained from other graphing calculators.

Store the given transition matrix as matrix $[A]$.

$$A = \begin{bmatrix} .3 & .2 & .3 & .1 & .1 \\ .4 & .2 & .1 & .2 & .1 \\ .1 & .3 & .2 & .2 & .2 \\ .2 & .1 & .3 & .2 & .2 \\ .1 & .1 & .4 & .2 & .2 \end{bmatrix}; \quad A^2 = \begin{bmatrix} .23 & .21 & .24 & .17 & .15 \\ .26 & .18 & .26 & .16 & .14 \\ .23 & .18 & .24 & .19 & .16 \\ .19 & .19 & .27 & .18 & .17 \\ .17 & .2 & .26 & .19 & .18 \end{bmatrix}$$

$$A^3 = \begin{bmatrix} .226 & .192 & .249 & .177 & .156 \\ .222 & .196 & .252 & .174 & .156 \\ .219 & .189 & .256 & .177 & .159 \\ .213 & .192 & .252 & .181 & .162 \\ .213 & .189 & .252 & .183 & .163 \end{bmatrix}$$

$$A^4 = \begin{bmatrix} .2205 & .1916 & .2523 & .1774 & .1582 \\ .2206 & .1922 & .2512 & .1778 & .1582 \\ .2182 & .1920 & .2525 & .1781 & .1592 \\ .2183 & .1909 & .2526 & .1787 & .1595 \\ .2176 & .1906 & .2533 & .1787 & .1598 \end{bmatrix}$$

$$A^5 = \begin{bmatrix} .21932 & .19167 & .25227 & .17795 & .15897 \\ .21956 & .19152 & .25226 & .17794 & .15872 \\ .21905 & .19152 & .25227 & .17818 & .15898 \\ .21880 & .19144 & .25251 & .17817 & .15908 \\ .21857 & .19148 & .25253 & .17824 & .15918 \end{bmatrix}$$

The entry in row 2, column 4 of A^5 is .17794, which gives the probability that state 2 changes to state 4 after 5 repetitions.

33. a. The transition matrix is

	Approve	Didn't Approve
Approve	.90	.10
Didn't Approve	.30	.70

b. $\begin{bmatrix} .35 & .65 \end{bmatrix} \begin{bmatrix} .90 & .10 \\ .30 & .70 \end{bmatrix} = \begin{bmatrix} .51 & .49 \end{bmatrix}$

c. $\begin{bmatrix} v_1 & v_2 \end{bmatrix} \begin{bmatrix} .9 & .1 \\ .3 & .7 \end{bmatrix} = \begin{bmatrix} .9v_1 + .3v_2 & .1v_1 + .7v_2 \end{bmatrix}$

We obtain the equations

$.9v_1 + .3v_2 = v_1$
$.1v_1 + .7v_2 = v_2$

which simplify to

$-.1v_1 + .3v_2 = 0$
$.1v_1 - .3v_2 = 0.$

$v_1 + v_2 = 1 \Rightarrow v_1 = 1 - v_2.$

Substituting $1 - v_2$ for v_1 in the second equation gives

$.1(1 - v_2) - .3v_2 = 0 \Rightarrow .1 - .4v_2 = 0 \Rightarrow$

$v_2 = \dfrac{.1}{.4} = .25$ and $v_1 = 1 - v_2 = .75.$

The long-range trend is that .75 of the voters will favor the initiative and .25 will not favor the initiative. This is represented as $[.75, .25].$

35. From the transition matrix, we obtain

$.81v_1 + .77v_2 = v_1$ and $.19v_1 + .23v_2 = v_2.$

Collecting like terms, we find that

$-.19v_1 + .77v_2 = 0$ or $v_1 = \dfrac{77}{19}v_2.$

Substitute this into $v_1 + v_2 = 1$, which yields

$\dfrac{96}{19}v_2 = 1.$

Therefore, $v_2 = \dfrac{19}{96} \approx .198$ and $v_1 = \dfrac{77}{96} \approx .802.$

The line works $\dfrac{77}{96}$ of the time.

37. Cross pink with color on left.

Resulting Color

$$
\begin{array}{c}
\quad\quad\quad\text{Red}\quad\text{Pink}\quad\text{White} \\
\begin{array}{c}\text{Red}\\[6pt]\text{Pink}\\[6pt]\text{White}\end{array}
\begin{bmatrix}
\dfrac{1}{2} & \dfrac{1}{2} & 0 \\[8pt]
\dfrac{1}{4} & \dfrac{1}{2} & \dfrac{1}{4} \\[8pt]
0 & \dfrac{1}{2} & \dfrac{1}{2}
\end{bmatrix}
\end{array}
$$

$$\frac{1}{2}v_1 + \frac{1}{4}v_2 = v_1$$
$$\frac{1}{2}v_1 + \frac{1}{2}v_2 + \frac{1}{2}v_3 = v_2$$
$$\frac{1}{4}v_2 + \frac{1}{2}v_3 = v_3$$

Also, $v_1 + v_2 + v_3 = 1$.

Solving this system, we obtain

$$v_1 = \frac{1}{4}, v_2 = \frac{1}{2}, v_3 = \frac{1}{4}.$$

The equilibrium vector is $\begin{bmatrix} \dfrac{1}{4} & \dfrac{1}{2} & \dfrac{1}{4} \end{bmatrix}$.

The long range prediction is $\dfrac{1}{4}$ red, $\dfrac{1}{2}$ pink, and $\dfrac{1}{4}$ white snapdragons.

39. a. $\begin{bmatrix} .6 & .395 & .005 \end{bmatrix}\begin{bmatrix} .90 & .10 & 0 \\ .09 & .909 & .001 \\ 0 & .34 & .66 \end{bmatrix}$

$$\approx \begin{bmatrix} .576 & .421 & .004 \end{bmatrix}$$

About 57.6% will own a home, 42.1% will rent, and .4% will be homeless.

b. $\begin{bmatrix} v_1 & v_2 & v_3 \end{bmatrix}\begin{bmatrix} .9 & .10 & 0 \\ .09 & .909 & .001 \\ 0 & .34 & .66 \end{bmatrix}$

$$= \begin{bmatrix} v_1 & v_2 & v_3 \end{bmatrix}$$

$$.90v_1 + .09v_2 + 0v_3 = v_1$$
$$.10v_1 + .909v_2 + .34v_3 = v_2$$
$$0v_1 + .001v_2 + .66v_3 = v_3.$$

Therefore, we have the system

$$v_1 + v_2 + v_3 = 1$$
$$.90v_1 + .09v_2 + 0v_3 = v_1$$
$$.10v_1 + .909v_2 + .34v_3 = v_2$$
$$0v_1 + .001v_2 + .66v_3 = v_3.$$

Solve this system by the Gauss-Jordan method to obtain

$v_1 \approx .473$, $v_2 = .526$, $v_3 = .002$. Therefore, 47.3% will be homeowners, 52.6% will be renters, and there will be 0.2% homeless in the long-range.

41. The transition matrix is

$$\begin{bmatrix} .85 & .10 & .05 \\ .15 & .75 & .10 \\ .10 & .30 & .60 \end{bmatrix}.$$

The square of the transition matrix is

$$\begin{bmatrix} .85 & .10 & .05 \\ .15 & .75 & .10 \\ .10 & .30 & .60 \end{bmatrix}\begin{bmatrix} .85 & .10 & .05 \\ .15 & .75 & .10 \\ .10 & .30 & .60 \end{bmatrix}$$

$$= \begin{bmatrix} .7425 & .175 & .0825 \\ .25 & .6075 & .1425 \\ .19 & .415 & .395 \end{bmatrix}.$$

The cube of the transition matrix is

$$\begin{bmatrix} .85 & .10 & .05 \\ .15 & .75 & .10 \\ .10 & .30 & .60 \end{bmatrix}\begin{bmatrix} .7425 & .175 & .0825 \\ .25 & .6075 & .1425 \\ .19 & .415 & .395 \end{bmatrix}$$

$$= \begin{bmatrix} .665625 & .23025 & .104125 \\ .317875 & .523375 & .15875 \\ .26325 & .44875 & .288 \end{bmatrix}.$$

a. $\begin{bmatrix} 50,000 & 0 & 0 \end{bmatrix}\begin{bmatrix} .85 & .10 & .05 \\ .15 & .75 & .10 \\ .10 & .30 & .60 \end{bmatrix}$

$$= \begin{bmatrix} 42,500 & 5000 & 2500 \end{bmatrix}$$

The numbers in the groups after 1 year are 42,500, 5000, and 2500.

b. $\begin{bmatrix} 50,000 & 0 & 0 \end{bmatrix}\begin{bmatrix} .7425 & .175 & .0825 \\ .25 & .6075 & .1425 \\ .19 & .415 & .395 \end{bmatrix}$

$$= \begin{bmatrix} 37,125 & 8750 & 4125 \end{bmatrix}$$

The numbers in the groups after 2 years are 37,125, 8750, and 4125.

c. $\begin{bmatrix} 50,000 & 0 & 0 \end{bmatrix}$

$$\begin{bmatrix} .665625 & .23025 & .104125 \\ .317875 & .523375 & .15875 \\ .26325 & .44875 & .288 \end{bmatrix}$$

$$= \begin{bmatrix} 33,281 & 11,513 & 5206 \end{bmatrix}$$

The numbers in the groups after 3 years are 33,281, 11,513, and 5206.

d. The system of equations is

$$v_1 + v_2 + v_3 = 1$$
$$.85v_1 + .15v_2 + .10v_3 = v_1$$
$$.10v_1 + .75v_2 + .30v_3 = v_2$$
$$.05v_1 + .10v_2 + .60v_3 = v_3$$

Solve this system by the Gauss-Jordan method to obtain

$$v_1 = \frac{28}{59} \approx .475, \; v_2 = \frac{22}{59} \approx .373, \text{ and }$$

$$v_3 = \frac{9}{59} \approx .152.$$

$$\mathbf{V} = \begin{bmatrix} \frac{28}{59} & \frac{22}{59} & \frac{9}{59} \end{bmatrix} \text{ or } \begin{bmatrix} .475 & .373 & .152 \end{bmatrix}.$$

In the long run, the probabilities of no accidents, one accident, and more than one accident are .475, .373, and .152, respectively.

43. a. $\begin{bmatrix} .232 & .150 & .159 & .459 \end{bmatrix} \begin{bmatrix} .85 & .04 & .05 & .06 \\ .02 & .91 & .03 & .04 \\ .01 & .01 & .95 & .03 \\ .03 & .02 & .06 & .89 \end{bmatrix}$

$$\approx \begin{bmatrix} .216 & .157 & .195 & .433 \end{bmatrix}.$$

The probability that a vehicle was purchased from Ford in the next year was about .157.

b. From the transition matrix we obtain

$$.85v_1 + .02v_2 + .01v_3 + .03v_4 = v_1$$
$$.04v_1 + .91v_2 + .01v_3 + .02v_4 = v_2$$
$$.05v_1 + .03v_2 + .95v_3 + .06v_4 = v_3$$
$$.06v_1 + .04v_2 + .03v_3 + .89v_4 = v_4$$
$$\text{and } v_1 + v_2 + v_3 + v_4 = 1.$$

Solve using the Gauss-Jordan method to obtain $v_1 = .103, v_2 = .156, \; v_3 = .494,$ and $v_4 = .248.$

Thus, the long-term probability that a vehicle is purchased from Toyota is .494.

45. a. If there is no one in line, then after 1 minute there will be either 0, 1, or 2 people in line with probabilities $p_{00} = .4, p_{01} = .3,$ and $p_{02} = .3.$ If there is one person in line, then that person will be served and either 0, 1 or 2 new people will join the line, with probabilities $p_{10} = .4, p_{11} = .3,$ and $p_{12} = .3.$ If there are two people in line, then one of them will be served and either 1 or 2 new people will join the line, with probabilities $p_{21} = .5$ and $p_{22} = .5;$ it is impossible for both people in line to be served, so $p_{20} = 0.$

Therefore, the transition matrix, A, is

$$\begin{matrix} & 0 & 1 & 2 \\ 0 & \\ 1 & \\ 2 & \end{matrix} \begin{bmatrix} .4 & .3 & .3 \\ .4 & .3 & .3 \\ 0 & .5 & .5 \end{bmatrix}.$$

b. The transition matrix for a two-minute period is

$$A^2 = \begin{bmatrix} .4 & .3 & .3 \\ .4 & .3 & .3 \\ 0 & .5 & .5 \end{bmatrix} \begin{bmatrix} .4 & .3 & .3 \\ .4 & .3 & .3 \\ 0 & .5 & .5 \end{bmatrix}$$

$$= \begin{bmatrix} .28 & .36 & .36 \\ .28 & .36 & .36 \\ .20 & .40 & .40 \end{bmatrix}.$$

c. The probability that a queue with no one in line has two people in line 2 min later is .36 since that is the entry in row 1, column 3 of A^2.

47. The following solution presupposes the use of a TI-82 graphics calculator. Similar results can be obtained from other graphics calculators. Store the given transition matrix as matrix $[A]$ and the probability vector $\begin{bmatrix} .5 & .5 & 0 & 0 \end{bmatrix}$ as matrix $[B]$. Multiply $[B]$ by successive powers of $[A]$ to obtain $\begin{bmatrix} 0 & 0 & .102273 & .897727 \end{bmatrix}.$

This gives the long-range prediction for the percent of employees in each state for the company training program.

Section 9.6 Decision Making

1. a. An optimist should choose the coast; $150,000 is the largest profit.

b. A pessimist should choose the highway; the worst case of $30,000 is better than −$40,000 if the coast is chosen.

c. If the possibility of heavy opposition is .8, the probability of light opposition is .2. Find his expected profit for each strategy.
Highway:
70,000(.2) + 30,000(.8) = $38,000
Coast:
150,000(.2) + −(40,000)(.8) = −$2000
He should choose the highway for an expected profit of $38,000

d. If the probability of heavy opposition is .4, the probability of light opposition is .6. Find his expected profit for each strategy.
Highway:
$70,000(.6) + 30,000(.4) = \$54,000$
Coast:
$150,000(.6) + (-40,000)(.4) = \$74,000$
He should choose the coast.

3. Note that the costs given in the payoff matrix are given in hundreds of dollars.

a. An optimist should make no upgrade; minimum cost is $2800.

b. A pessimist should make the upgrade; a worst case of $13,000 is better than a possible cost of $45,000 if no upgrade is made.

c. Find the expected cost of each strategy.
Make upgrade:
$.7(130) + .2(130) + .1(130) = 130,$
or $13,000
Make no upgrade:
$.7(28) + .2(180) + .1(450) = 100.6$
or $10,600
He should not upgrade. The expected cost to the company if this strategy is chosen is $10,060.

	Fails	Doesn't Fail
5. a. Overhaul $\begin{bmatrix} -\$8600 & -\$2600 \\ -\$6000 & \$0 \end{bmatrix}$
Don't Overhaul

b. Find the expected cost under each strategy.
Overhaul:
$.1(-8600) + .9(-2600) = -\$3200$
Don't Overhaul:
$.3(-6000) + .7(0) = -\$1800$
To minimize his expected costs, the business should not overhaul the machine before shipping.

	No rain	Rain
7. a. No rain $\begin{bmatrix} \$2500 & \$1500 \\ \$3000 & \$0 \end{bmatrix}$
Rain

b. Find her expected revenue for each scenario.
No rain:
$.6(2500) + .4(1500) = \$2100$
Rain
$.6(1500) + .4(0) = \$900$
She should rent the tent since the expected revenue is $2100.

9. Find the expected utility under each strategy.
Jobs:
$(.35)(40) + (.65)(-10) = 7.5$
Environment:
$(.35)(-12) + (.65)(30) = 15.3$
She should emphasize the environment. The expected utility of this strategy is 15.3.

Chapter 9 Review Exercises

1. $E(x) = 0(.22) + 1(.54) + 2(.16) + 3(.08)$
$= 1.1$

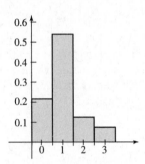

2. $E(x) = -3(.15) - 2(.20) - 1(.25) + 0(.18)$
$\quad\quad\quad + 1(.12) + 2(.06) + 3(.04)$
$= -.74$

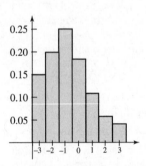

3. $E(x) = -10(.333) + 0(.333) + 10(.333) = 0$

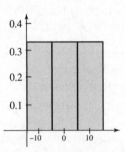

4. $E(x) = 0(.40898) + 1(.32788) + 2(.16616)$
$\quad\quad\quad + 3(.06281) + 4(.02110) + 5(.00771)$
$\quad\quad\quad + 6(.00321) + 7(.00132) + 8(.00061)$
$\quad\quad\quad + 9(.00023)$
$= 1.00703$

5. $E(x) = 0(.0138) + 1(.1049) + 2(.2722)$
    $\qquad + 3(.3909) + 4(.1748) + 5(.0434)$
    $\qquad = 2.7382$

6. **a.** In selecting 2 bouquets from a group of 10, there are $_{10}C_2 = 45$ ways.

    $P(0 \text{ roses}) = \dfrac{_3C_0 \cdot \, _7C_2}{_{10}C_2} = \dfrac{21}{45} = \dfrac{7}{15}$

    $P(1 \text{ rose}) = \dfrac{_3C_1 \cdot \, _7C_1}{_{10}C_2} = \dfrac{21}{45} = \dfrac{7}{15}$

    $P(2 \text{ roses}) = \dfrac{_3C_2 \cdot \, _7C_0}{_{10}C_2} = \dfrac{3}{45} = \dfrac{1}{15}$

x	0	1	2
$P(x)$	$\frac{7}{15}$	$\frac{7}{15}$	$\frac{1}{15}$

    b. $E(x) = 0\left(\frac{7}{15}\right) + 1\left(\frac{7}{15}\right) + 2\left(\frac{1}{15}\right) = .6$

7. **a.** Three of the 10 members did not do their homework, and 3 members of the 10 are selected. Three members of the 10 can be selected in $_{10}C_3 = 120$ ways. Let $P(n)$ represent the probability of selecting n students who did not do homework.

    $P(0) = \dfrac{_3C_0 \cdot \, _7C_3}{_{10}C_3} = \dfrac{35}{120} = .292$

    $P(1) = \dfrac{_3C_1 \cdot \, _7C_2}{_{10}C_3} = \dfrac{63}{120} = .525$

    $P(2) = \dfrac{_3C_2 \cdot \, _7C_1}{_{10}C_3} = \dfrac{21}{120} = .175$

    $P(3) = \dfrac{_3C_3 \cdot \, _7C_0}{_{10}C_3} = \dfrac{1}{120} = .008$

x	0	1	2	3
$P(x)$	.292	.525	.175	.008

    b. $E(x) = 0(.292) + 1(.525) + 2(.175) + 3(.008)$
    $\qquad = .899$

8. Probability of getting 3 hearts
    $= \dfrac{_{13}C_3}{_{52}C_3} = \dfrac{286}{22,100} \approx .0129$
    Probability of not getting 3 hearts
    $= 1 - .0129 = .9871.$
    Let $x =$ the amount to pay for the game.
    If you win, you get $100 - x$.
    If you lose, you get $-x$. The expected value must be 0.

$(100 - x)(.0129) + (-x)(.9871) = 0 \Rightarrow$
$1.29 - .0129x - .9871x = 0 \Rightarrow 1.29 = x$
You must pay $1.29.

9. The probability of winning if you bet "under" is $\dfrac{15}{36}$ and has a net return of $2. The probability of 7 is $\dfrac{6}{36}$ and has a net return of $4. Betting "over" has a probability of $\dfrac{15}{36}$ with a net return of $2.

    $P(\text{sum} < 7) = P(2) + P(3) + P(4)$
    $\qquad\qquad\qquad + P(5) + P(6)$
    $\qquad = \dfrac{1}{36} + \dfrac{2}{36} + \dfrac{3}{36} + \dfrac{4}{36} + \dfrac{5}{36} = \dfrac{15}{36}$

    $P(\text{sum} = 7) = \dfrac{6}{36}$

    $P(\text{sum} > 7) = P(8) + P(9) + P(10)$
    $\qquad\qquad\qquad + P(11) + P(12)$
    $\qquad = \dfrac{5}{36} + \dfrac{4}{36} + \dfrac{3}{36} + \dfrac{2}{36} + \dfrac{1}{36} = \dfrac{15}{36}$

    The expected return for each type of bet is as follows:
    Under:

    $E(x) = 2\left(\dfrac{15}{36}\right) - 2\left(\dfrac{21}{36}\right) = -\$.33$

    Exactly 7:

    $E(x) = 4\left(\dfrac{6}{36}\right) - 2\left(\dfrac{30}{36}\right) = -\$1.00$

    Over:

    $E(x) = 2\left(\dfrac{15}{36}\right) - 2\left(\dfrac{21}{36}\right) = -\$.33$

10. $E(x) = 9{,}998\left(\dfrac{1}{10{,}000}\right) + 998\left(\dfrac{2}{10{,}000}\right)$
    $\qquad + 98\left(\dfrac{2}{10{,}000}\right) - 2\left(\dfrac{9995}{10{,}000}\right)$
    $\qquad = -\$.78$

11. $E(x) = 0(.2373) + 1(.3955) + 2(.2637)$
    $\qquad + 3(.0879) + 4(.0146) + 5(.0010)$
    $\qquad = 1.25$

12. $E(x) = 0(.7915) + 1(.1925) + 2(.0156)$
    $\qquad + 3(.0004)$
    $\qquad = .2249$

13. Since order is important, this is a permutation. Eight taxis can line up in $_8P_8 = 8! = 40,320$ ways.

14. Since order is important, this is a permutation. If there are 8 finalists, there are

$$_8P_3 = \frac{8!}{(8-3)!} = 8 \cdot 7 \cdot 6 = 336 \text{ variations.}$$

15. Since order is not important, this is a combination. Three monitors can be selected from 12 monitors in

$$_{12}C_3 = \frac{12!}{3!9!} = \frac{12 \cdot 11 \cdot 10}{3 \cdot 2 \cdot 1} = 220 \text{ ways.}$$

16. a. One of the 4 broken monitors can be selected in $_4C_1 = 4$ ways. The remaining 2 must come from the 8 nonbroken monitors, and can be selected in $_8C_2 = 28$ ways. By the multiplication principle, the selection can be made in $112 = 4 \times 28$ ways.

b. All 3 monitors must come from the non-broken group of 8. This can be accomplished in $_8C_3 = 56$ ways.

c. At least one broken monitor can be accomplished by selecting 1, 2, or 3 defective monitors. If 1 monitor is broken, 2 must be non-broken. If 2 are broken, 1 must be non-broken. If 3 are broken, then 0 must be non-broken. The number of ways to select
1 broken: $_4C_1 \cdot _8C_2 = 4 \times 28 = 112$
2 broken: $_4C_2 \cdot _8C_1 = 6 \times 8 = 48$
3 broken: $_4C_3 \cdot _8C_0 = 4 \times 1 = 4$
is then $112 + 48 + 4 = 164$.

17. Since order is important, this is a permutation. There are 30 choices for the first seat, 29 for the second, 28 for the third, 27 for the fourth, 26 for the fifth, and 25 for the sixth, or $_{30}P_6 = 427,518,000$ ways.

18. The first seat will be occupied by the given student. That leaves 29 choices for the second seat, 28 choices for the third seat, 27 choices for the fourth seat, 26 choices for the fifth seat, and 25 for the sixth seat, or $_{29}P_5 = 14,250,600$ possible ways.

19. a. Since there are 15 students in each major, there are $_{15}P_3 = 2730$ ways to arrange the students within each major. There are also $_2P_2 = 2$ ways to arrange the two different groups. Thus, the total number of arrangements is
$2 \times 2730 \times 2730 = 14,905,800$.

b. Assume the odd seats are occupied by science majors, then because order is important, there are $_{15}P_3 = 2730$ possible arrangements for that group. Then the even seats would be occupied by the business majors with $_{15}P_3 = 2730$ possible arrangements By the multiplication principle, there would be $2730 \times 2730 = 7,452,900$ possible arrangements. However, if the odd seats were occupied by the business majors and the even by the science majors, there would also be $2730 \times 2730 = 7,452,900$ possible arrangements. So, there must be $7,452,900 + 7,452,900 = 14,905,800$ possibilities.

20. Answers vary.

21. Answers vary.

22. There are 26 black cards in the deck.
$$P(\text{both black}) = \frac{_{26}C_2}{_{52}C_2} = \frac{325}{1326} \approx .245$$

23. There are 13 hearts in a deck.
$$P(\text{both hearts}) = \frac{_{13}C_2}{_{52}C_2} = \frac{78}{1326} \approx .059$$

24. To get exactly one face card, you must have one non-face card. There are 12 face cards and 40 non-face cards in a deck.
$$P(\text{exactly one face card}) = \frac{_{12}C_1 \cdot _{40}C_1}{_{52}C_2}$$
$$= \frac{480}{1326} \approx .362$$

25. There are 4 aces in a deck of cards.
$$P(\text{at most one ace}) = \frac{_4C_0 \cdot _{48}C_2}{_{52}C_2} + \frac{_4C_1 \cdot _{48}C_1}{_{52}C_2}$$
$$= \frac{1128}{1326} + \frac{192}{1326}$$
$$= \frac{1320}{1326} \approx .995$$

26. There are 12 possible selections, 6 of them being ice cream, and 6 not ice cream.

$$P(3 \text{ ice cream}) = \frac{{}_6C_3 \cdot {}_6C_0}{{}_{12}C_3} = \frac{20 \cdot 1}{220} \approx .091$$

27. There are 12 possible selections, 4 custard and 8 non-custard.

$$P(4 \text{ custard}) = \frac{{}_4C_3 \cdot {}_8C_0}{{}_{12}C_3} = \frac{4 \cdot 1}{220} \approx .018$$

28. There are 12 possible selections, 2 frozen yogurt and 10 non-yogurt.

$$P(\text{at least 1 yogurt}) = 1 - P(\text{no yogurt})$$
$$= 1 - \frac{{}_2C_0 \cdot {}_{10}C_3}{{}_{12}C_3}$$
$$= 1 - \frac{1 \cdot 120}{220} = \frac{100}{220} \approx .455$$

29. There are 12 possible selections, 4 custard, 6 ice cream, and 2 frozen yogurt.

$$P(1 \text{ custard}, 1 \text{ ice cream}, 1 \text{ yogurt})$$
$$= \frac{{}_4C_1 \cdot {}_6C_1 \cdot {}_2C_1}{{}_{12}C_3} = \frac{4 \cdot 6 \cdot 2}{220} = \frac{48}{220} \approx .218$$

30. There are 12 possible selections, 6 ice cream, and 6 not ice cream.

$$P(\text{at most 1 ice cream}) = P(0) + P(1)$$
$$= \frac{{}_6C_0 \cdot {}_6C_3}{{}_{12}C_3} + \frac{{}_6C_1 \cdot {}_6C_2}{{}_{12}C_3}$$
$$= \frac{20}{220} + \frac{90}{220} = \frac{110}{220} = .5$$

31. a. The number of subsets of size 0 is ${}_nC_0$ or 1.
The number of subsets of size 1 is ${}_nC_1$ or n.
The number of subsets of size 2 is ${}_nC_2$ or $\frac{n(n-1)}{2}$.
The number of subsets of size n is ${}_nC_n$ or 1.

b. The total number of subsets of a set with n elements is ${}_nC_0 + {}_nC_1 + {}_nC_2 + \cdots + {}_nC_n$.

c. Answers vary.

d. Let $n = 4$.
$${}_4C_0 + {}_4C_1 + {}_4C_2 + {}_4C_3 + {}_4C_4$$
$$= 1 + 4 + 6 + 4 + 1 = 16$$
Since $2^4 = 16$, the equation from part c holds.

Let $n = 5$.
$${}_5C_0 + {}_5C_1 + {}_5C_2 + {}_5C_3 + {}_5C_4 + {}_5C_5$$
$$= 1 + 5 + 10 + 10 + 5 + 1 = 32$$
Since $2^5 = 32$, the equation from part c holds.

32. $n = 7, p = .36, x = 2, 1 - p = .64$
$$P(2) = {}_7C_2(.36)^2(.64)^5 \approx .292$$

33. $n = 7, p = .01, 1 - p = .99$
$$P(\text{at least 1}) = P(1) + P(2) + P(3) + P(4)$$
$$+ P(5) + P(6) + P(7).$$
Alternatively, $P(\text{at least 1}) = 1 - P(0)$.
$$P(0) = {}_7C_0(.36)^0(.64)^7 \approx .044$$
$$P(\text{at least 1}) = 1 - P(0) \approx .956$$

34. $n = 6, p = .78, 1 - p = .22$

x	$P(x)$	
0	.0001	$= {}_6C_0(.78)^0(.22)^6$
1	.0024	$= {}_6C_1(.78)^1(.22)^5$
2	.0214	$= {}_6C_2(.78)^2(.22)^4$
3	.1011	$= {}_6C_3(.78)^3(.22)^3$
4	.2687	$= {}_6C_4(.78)^4(.22)^2$
5	.3811	$= {}_6C_5(.78)^5(.22)^1$
6	.2252	$= {}_6C_6(.78)^6(.22)^0$

35. $E(x) = 0(.0001) + 1(.0024) + 2(.0214)$
$$+ 3(.1011) + 4(.2687)$$
$$+ 5(.3811) + 6(.2252)$$
$$= 4.68$$

36. a. $n = 4, p = .22, 1 - p = .78$
$$P(4) = {}_4C_4(.22)^4(.78)^0 \approx .002$$

b. $P(\text{at least 1}) = P(2) + P(3) + P(4)$
$$= 1 - P(0)$$
$$= 1 - {}_4C_0(.22)^0(.78)^4$$
$$\approx 1 - .370 = .630$$

c. $P(\text{at most 2})$
$$= P(0) + P(1) + P(2)$$
$$= {}_4C_0(.22)^0(.78)^4 + {}_4C_1(.22)^1(.78)^3$$
$$+ {}_4C_2(.22)^2(.78)^2$$
$$\approx .3701 + .4176 + .1767 \approx .964$$

37. a. $n = 5$, $p = .15$, $1 - p = .85$

x	$P(x)$	
0	.4437	$= {}_5C_0 (.15)^0 (.85)^5$
1	.3915	$= {}_5C_1 (.15)^1 (.85)^4$
2	.1382	$= {}_5C_2 (.15)^2 (.85)^3$
3	.0244	$= {}_5C_3 (.15)^3 (.85)^2$
4	.0022	$= {}_5C_4 (.15)^4 (.85)^1$
5	.0001	$= {}_5C_5 (.15)^5 (.85)^0$

b. $E(x) = 0(.4437) + 1(.3915) + 2(.1382)$
$\qquad + 3(.0244) + 4(.0022) + 5(.0001)$
$\qquad = .7504$

38. $\begin{bmatrix} 0 & 1 \\ .77 & .23 \end{bmatrix}$

This is a regular transition matrix because
$\begin{bmatrix} 0 & 1 \\ .77 & .23 \end{bmatrix}\begin{bmatrix} 0 & 1 \\ .77 & .23 \end{bmatrix} = \begin{bmatrix} .77 & .23 \\ .1771 & .8229 \end{bmatrix}$,
in which all entries are positive.

39. $\begin{bmatrix} -.2 & .4 \\ .3 & .7 \end{bmatrix}$

This is not a regular transition matrix because there is a negative entry in the first row and first column. In fact, this makes it not even a transition matrix.

40. $\begin{bmatrix} .21 & .15 & .64 \\ .50 & .12 & .38 \\ 1 & 0 & 0 \end{bmatrix}$

This is a regular transition matrix because
$\begin{bmatrix} .21 & .15 & .64 \\ .50 & .12 & .38 \\ 1 & 0 & 0 \end{bmatrix}\begin{bmatrix} .21 & .15 & .64 \\ .50 & .12 & .38 \\ 1 & 0 & 0 \end{bmatrix}$
$= \begin{bmatrix} .7591 & .0495 & .1914 \\ .545 & .0894 & .3656 \\ .21 & .15 & .64 \end{bmatrix}$,
in which all entries are positive.

41. $\begin{bmatrix} .22 & 0 & .78 \\ .40 & .33 & .27 \\ 0 & .61 & .39 \end{bmatrix}$

This is a regular transition matrix because
$\begin{bmatrix} .22 & 0 & .78 \\ .40 & .33 & .27 \\ 0 & .61 & .39 \end{bmatrix}\begin{bmatrix} .22 & 0 & .78 \\ .40 & .33 & .27 \\ 0 & .61 & .39 \end{bmatrix}$
$= \begin{bmatrix} .0484 & .4758 & .4758 \\ .220 & .2736 & .5064 \\ .244 & .4392 & .3168 \end{bmatrix}$,
in which all entries are positive.

42. $P = \begin{bmatrix} .35 & .15 & .50 \\ .30 & .35 & .35 \\ .15 & .30 & .55 \end{bmatrix}$
$I = \begin{bmatrix} .2 & .4 & .4 \end{bmatrix}$

a. The distribution after one month is
$\begin{bmatrix} .2 & .4 & .4 \end{bmatrix}\begin{bmatrix} .35 & .15 & .50 \\ .30 & .35 & .35 \\ .15 & .30 & .55 \end{bmatrix}$
$\qquad = \begin{bmatrix} .25 & .29 & .46 \end{bmatrix}.$

b. $I = \begin{bmatrix} .2 & .4 & .4 \end{bmatrix}$
$A = \begin{bmatrix} .35 & .15 & .50 \\ .30 & .35 & .35 \\ .15 & .30 & .55 \end{bmatrix}$
$A^2 = \begin{bmatrix} .2425 & .2550 & .5025 \\ .2625 & .2725 & .4650 \\ .2250 & .2925 & .4825 \end{bmatrix}$

The distribution after 2 months is given by
$IA^2 = \begin{bmatrix} .2435 & .2770 & .4795 \end{bmatrix}.$

c. To find the long-range distribution, we use the system
$\qquad v_1 + v_2 + v_3 = 1$
$\qquad .35v_1 + .3v_2 + .15v_3 = v_1$
$\qquad .15v_1 + .35v_2 + .3v_3 = v_2$
$\qquad .5v_1 + .35v_2 + .55v_3 = v_3.$
Simplify these equations to obtain the system
$\qquad v_1 + v_2 + v_3 = 1$
$\qquad -.65v_1 + .3v_2 + .15v_3 = 0$
$\qquad .15v_1 - .65v_2 + .3v_3 = 0$
$\qquad .5v_1 + .35v_2 - .45v_3 = 0.$

(*continued on next page*)

(*continued from page 300*)

Solve this system by the Gauss-Jordan method to obtain

$$v_1 = \frac{75}{313}, v_2 = \frac{87}{313}, \text{ and } v_3 \frac{151}{313}.$$

The long-range distribution is

$$\begin{bmatrix} .240 & .278 & .482 \end{bmatrix}.$$

43. a. $I = \begin{bmatrix} .15 & .60 & .25 \end{bmatrix}; \quad A = \begin{bmatrix} .80 & .14 & .06 \\ .04 & .85 & .11 \\ .03 & .13 & .84 \end{bmatrix}$

The distribution after one month is

$$\begin{bmatrix} .15 & .60 & .25 \end{bmatrix} \begin{bmatrix} .80 & .14 & .06 \\ .04 & .85 & .11 \\ .03 & .13 & .84 \end{bmatrix}$$
$$= \begin{bmatrix} .1515 & .5635 & .2850 \end{bmatrix}$$

b. $A^2 = \begin{bmatrix} .80 & .14 & .06 \\ .04 & .85 & .11 \\ .03 & .13 & .84 \end{bmatrix} \begin{bmatrix} .80 & .14 & .06 \\ .04 & .85 & .11 \\ .03 & .13 & .84 \end{bmatrix}$

$$= \begin{bmatrix} .6474 & .2388 & .1138 \\ .0693 & .7424 & .1883 \\ .0544 & .2239 & .7217 \end{bmatrix}$$

$$A^3 = \begin{bmatrix} .80 & .14 & .06 \\ .04 & .85 & .11 \\ .03 & .13 & .84 \end{bmatrix} \begin{bmatrix} .6474 & .2388 & .1138 \\ .0693 & .7424 & .1883 \\ .0544 & .2239 & .7217 \end{bmatrix}$$

$$= \begin{bmatrix} .530886 & .308410 & .160704 \\ .090785 & .665221 & .243994 \\ .074127 & .291752 & .634121 \end{bmatrix}$$

The distribution after 3 years is given by

$$IA^3 = \begin{bmatrix} .1526 & .5183 & .3290 \end{bmatrix}$$

c. To find the long range distribution, we use the system

$$v_1 + v_2 + v_3 = 1$$
$$.80v_1 + .04v_2 + .03v_3 = v_1$$
$$.14v_1 + .85v_2 + .13v_3 = v_3$$
$$.06v_1 + .11v_2 + .84v_3 = v_3$$

This system yields

$$v_1 + v_2 + v_3 = 1$$
$$-.20v_1 + .04v_2 + .03v_3 = 0$$
$$.14v_1 - .15v_2 + .13v_3 = 0$$
$$.06v_1 + .11v_2 - .16v_3 = 0$$

Solving this system yields $v_1 = \dfrac{97}{643}$,

$v_2 = \dfrac{302}{643}$ and $v_3 = \dfrac{244}{643}$. Thus the long range distribution is

$$\begin{bmatrix} .1509 & .4697 & .3795 \end{bmatrix}.$$

44. a. Since the candidate is an optimist, look for the biggest value in the matrix, which is 5000. Hence, she should oppose it.

b. A pessimistic candidate wants to find the best of the worst things that can happen. If she favors, the worst is –4000. If she waffles the worst is –500. If she opposes, then worst is 0. Since the best of these is 0, she should oppose.

c. Since there is a 40% chance the opponent favors the plant and a 35% that he will waffle, the chance he will oppose is
$1 - .4 - .35 = .25$
Expected gain if she favors:
$(0)(.4) + (-1000)(.35) + (-4000)(.25)$
$\qquad\qquad\qquad\qquad = -1350$
Expected gain if she waffles:
$(1000)(.4) + 0(.35) + (-500)(.25) = 275$
Expected gain if she opposes:
$(5000)(.4) + (2000)(.35) + (0)(.25) = 2700$
She should oppose and get 2700 additional votes.

d. Now the opponent has 0 probability of favoring, .7 of waffling and .3 of opposing. Expected gain if she favors:
$(0)(0) + (.7)(-1000) + (.3)(-4000) = -1900$
Expected gain if she waffles:
$(0)(1000) + (.7)(0) + (.3)(-500) = -150$
Expected gain if she opposes:
$(0)(5000) + (.7)(2000) + (.3)(0) = 1400$
She should oppose and get 1400 additional votes.

45. a. Since the department chair is an optimist, look for the biggest value in the matrix, which is 100. Hence, she should use active learning.

b. A pessimistic wants to find the best of the worst things that can happen. If she lectures, the worst is –80. If she uses active learning, the worst is –30. Since the best of these is –30, she should use active learning.

c. Since there is a 75% chance the class will prefer the lecture format, there is a 25% chance the class will support the active learning format.

Expected gain for lecture format:

$50(.75) - 80(.25) = 17.5$ points.

Expected gain for active learning format:

$-30(.75) + 100(.25) = 2.5$ points. She should use the lecture format where the expected gain is 17.5.

d. Since there is a 60% chance the class will prefer the active learning format, there is a 40% chance the class will support the lecture format.

Expected gain for lecture format:

$50(.4) - 80(.6) = -28$ points.

Expected gain for active learning format:

$-30(.4) + 100(.6) = 48$ points. She should use the active learning format where the expected gain is 48.

46. P(product is successful) = .5

P(product is unsuccessful) = .5

P(successful product passing quality control) = .8

P(unsuccessful product passing quality control)

$= .25$

P(successful product and passes quality control)

$= (.5)(.8) = .4$

P(unsuccessful product and passes quality control) = $(.5)(.25) = .125$

P(passes quality control) = $.4 + .125 = .525$

P(successful product passes quality control)

$$= \frac{.4}{.525} = .7619$$

P(unsuccessful product passes quality control)

$$= \frac{.125}{.525} = .2381$$

$E(x) = (40,000,000)(.7619)$

$\qquad + (-15,000,000)(.2381)$

$\qquad \approx 27,000,000$

The expected net profit is (e) $27 million.

47. If a box is good (probability .9) and the merchant samples an excellent piece of fruit from that box (probability .80), then he will accept the box and earn a $200 profit on it. If a box is bad (probability .1) and he samples an excellent piece of fruit from the box (probability .30), then he will accept the box and earn a –$1000 profit on it.

If the merchant ever samples a non-excellent piece of fruit, he will not accept the box. In this case he pays nothing and earns nothing, so the profit will be $0.

Let x denote the merchant's earnings.

Note that $\quad .9(.80) = .72,$

$\qquad\qquad .1(.30) = .03,$

and $1 - (.72 + .03) = .25.$

The probability distribution is as follows.

x	200	–1000	0
$P(x)$	.72	.03	.25

The expected value when the merchant samples the fruit is

$E(x) = 200(.72) + (-1000)(.03) + 0(.25)$

$\qquad = 144 - 30 + 0 = 114$

We must also consider the case in which the merchant does not sample the fruit. Let x again denote the merchant's earnings. The probability distribution is as follows.

x	200	–1000
$P(x)$	.9	.1

The expected value when the merchant does not sample the fruit is

$E(x) = 200(.9) + (-1000)(.1) = 180 - 100 = \$80.$

Combining these two results, the expected value of the right to sample is $114 – $80 = $34, which corresponds to choice (c).

48. Let $I(x)$ represent the airline's net income if x people show up.

$I(0) = 0; \ I(1) = 100$
$I(2) = 2 \cdot 100 = 200$
$I(3) = 3 \cdot 100 = 300$
$I(4) = 3 \cdot 100 - 100 = 200$
$I(5) = 3 \cdot 100 - 2 \cdot 100 = 100$
$I(6) = 3 \cdot 100 - 3 \cdot 100 = 0$

Let $P(x)$ represent the probability that x people will show up. Use the binomial probability formula to find the values of $P(x)$.

$P(0) = \binom{6}{0}(.6)^0(.4)^6 = .004$ $P(3) = \binom{6}{3}(.6)^3(.4)^3 = .276$ $P(6) = \binom{6}{6}(.6)^6(.4)^0 = .047$

$P(1) = \binom{6}{1}(.6)^1(.4)^5 = .037$ $P(4) = \binom{6}{4}(.6)^4(.4)^2 = .311$

$P(2) = \binom{6}{2}(.6)^2(.4)^4 = .138$ $P(5) = \binom{6}{5}(.6)^5(.4)^1 = .187$

On the basis of all calculations, the table given in the exercise is completed as follows.

x	0	1	2	3	4	5	6
Income	0	100	200	300	200	100	0
$P(x)$	.004	.037	.138	.276	.311	.187	.047

a. $E(I) = 0(.004) + 100(.037) + 200(.138) + 300(.276) + 200(.311) + 100(.187) + 0(.047) = \$195$

b. $n = 3$

x	0	1	2	3
Income	0	100	200	300
$P(x)$	$P(0) = \binom{3}{0}(.6)^0(.4)^3$ $= .064$	$P(0) = \binom{3}{1}(.6)^1(.4)^2$ $= .288$	$P(0) = \binom{3}{2}(.6)^2(.4)^1$ $= .432$	$P(0) = \binom{3}{3}(.6)^3(.4)^0$ $= ..216$

$E(I) = 0(.064) + 100(.288) + 200(.432) + 300(.216) = \$180$

$n = 4$

$P(0) = \binom{4}{0}(.6)^0(.4)^4 = .0256$ $P(1) = \binom{4}{1}(.6)^1(.4)^3 = .1536$ $P(2) = \binom{4}{2}(.6)^2(.4)^2 = .3456$

$P(3) = \binom{4}{3}(.6)^3(.4)^1 = .3456$ $P(4) = \binom{4}{4}(.6)^4(.4)^0 = .1296$

x	0	1	2	3	4
Income	0	100	200	300	200
$P(x)$	.0256	.1536	.3456	.3456	.1296

$E(I) = 0(.0256) + 100(.1536) + 200(.3456) + 300(.3456) + 200(.1296) = \$214.08$

(*continued on next page*)

(continued from page 303)

$n = 5$

$$P(0) = \binom{5}{0}(.6)^0(.4)^5 = .01024 \quad P(1) = \binom{5}{1}(.6)^1(.4)^4 = .0768 \quad P(2) = \binom{5}{2}(.6)^2(.4)^3 = .2304$$

$$P(3) = \binom{5}{3}(.6)^3(.4)^2 = .3456 \quad P(4) = \binom{5}{4}(.6)^4(.4)^1 = .2592 \quad P(5) = \binom{5}{5}(.6)^5(.4)^0 = .07776$$

x	0	1	2	3	4	5
Income	0	100	200	300	200	100
$P(x)$	.01024	.0768	.2304	.3456	.2592	.07776

$E(I) = 0(.01024) + 100(.0768) + 200(.2304) + 300(.3456) + 200(.2592) + 100(.07776) = \$217.06$

Since $E(I)$ is greatest when $n = 5$, the airlines should book 5 reservations to maximize revenue.

Case 9 Quick Draw® from the New York State Lottery

1. Using the probabilities determined in the case study, we have

x	Net winnings	$P(x)$
0	$2(0) - 1 = -\$1$	.16660
1	$2(0) - 1 = -\$1$	.36349
2	$2(0) - 1 = -\$1$	.30832
3	$2(1) - 1 = \$1$	.12982
4	$2(6) - 1 = \$11$	.02854
5	$2(55) - 1 = \$109$	.00310
6	$2(1000) - 1 = \$1999$	.00013

$$\begin{aligned} E(W) &= -1(.16660) - 1(.36349) - 1(.30832) \\ &\quad + 1(.12982) + 11(.02854) \\ &\quad + 109(.00310) + 1999(.00013) \\ &\approx .2031 \end{aligned}$$

2. It is not in the state's interest to offer such a promotion because the players win about 20 cents on average.

3.

x	$P(x)$
0	$\dfrac{_{20}C_0 \cdot {_{60}C_4}}{_{80}C_4} \approx .30832$
1	$\dfrac{_{20}C_1 \cdot {_{60}C_3}}{_{80}C_4} \approx .43273$
2	$\dfrac{_{20}C_2 \cdot {_{60}C_2}}{_{80}C_4} \approx .21264$

x	$P(x)$
3	$\dfrac{_{20}C_3 \cdot {_{60}C_1}}{_{80}C_4} \approx .04325$
4	$\dfrac{_{20}C_4 \cdot {_{60}C_0}}{_{80}C_4} \approx .00306$

For exercises 4 and 5, carry all decimal places in order to compute the expected winnings.

4.

x	Net winnings	$P(x)$
0	$-\$1$	.30832
1	$-\$1$	.43273
2	$\$0$	.21264
3	$\$4$	.04325
4	$\$54$	.00306

$$\begin{aligned} E(W) &= -1(.30832) - 1(.43273) + 0(.21264) \\ &\quad + 4(.04325) + 54(.00306) \\ &\approx -.40281 \end{aligned}$$

5.

x	Net winnings	$P(x)$
0	$2(0) - 1 = -\$1$	.30832
1	$2(0) - 1 = -\$1$	.43273
2	$2(1) - 1 = \$1$	.21264
3	$2(5) - 1 = \$9$	.04325
4	$2(55) - 1 = \$109$	.00306

$$\begin{aligned} E(W) &= -1(.30832) - 1(.43273) + 1(.21264) \\ &\quad + 9(.04325) + 109(.00306) \\ &\approx .19438 \end{aligned}$$

Chapter 10 Introduction to Statistics

Section 10.1 Frequency Distributions

1. a–b. Since 14–17 is to be the first interval and there are 4 numbers between 14 and 17 inclusive, we will let all six intervals be of size 4. The other five intervals are 14–17, 18–21, 22–25, 26–29, 30–33, 34–37, 38–41, and 42–45. Keeping a tally of how many data values lie in each interval leads to the following frequency distribution.

Interval	Frequency
14–17	2
18–21	11
22–25	12
26–29	11
30–33	7
34–37	5
38–41	1
42–45	1

c. Draw the histogram. It consists of 8 bars of equal width, having heights as determined by the frequency of each interval.

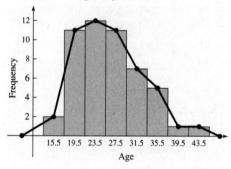

d. To construct the frequency polygon, join consecutive midpoints of the tops of the histogram bars with straight line segments. See the histogram in part (c).

3. a–b. Since 0–4 is to be the first interval, we let all the intervals be of size 5. The largest data value is 49, so the last interval that will be needed is 45–49. The frequency distribution is as follows.

Interval	Frequency
0–4	1
5–9	4
10–14	4
15–19	3
20–24	5
25–29	8
30–34	9
35–39	8
40–44	3
45–49	5

c. Draw the histogram. It consists of 8 bars of equal width and having heights as determined by the frequency of each interval.

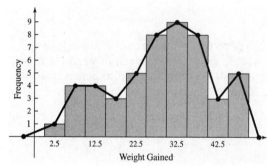

d. Construct the frequency polygon by joining consecutive midpoints of the tops of the histogram bars with straight line segments. See histogram in part (c).

5. The data ranges from a low of 9 to a high of 230, so we will use intervals of size 25 starting with 0–24. The frequency distribution is as follows.

Interval	Frequency
0–24	6
25–49	8
50–74	2
75–99	5
100–124	2
125–149	2
150–174	3
175–199	1
200–224	0
225–249	1

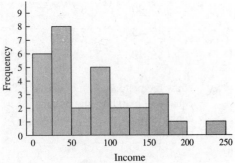

7. The data ranges from a low of 30 to a high of 500, so we will use intervals of size 50 starting with 0–49. The frequency distribution is as follows.

Interval	Frequency
0–49	1
50–99	10
100–149	6
150–199	5
200–249	3
250–299	1
300–349	1
350–399	1
400–449	1
450–499	0
500–549	1

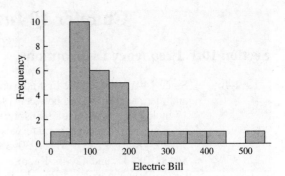

9. The data ranges from a low of 43 to a high of 73, so we will use intervals of size 5 starting with 40–44. The frequency distribution is as follows.

Interval	Frequency
40–44	1
45–49	2
50–54	6
55–59	9
60–64	6
65–69	4
70–74	2

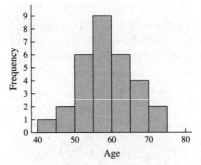

11. The data ranges from 15 to 44.

STEM	LEAVES
1	
1	55889
2	0000011223444444
2	55566677788899
3	00112334
3	55678
4	4

Units: 4|4 = 44 years

13. The data ranges from 0 to 50.

STEM	LEAVES
0	0
0	5578
1	0013
1	569
2	00022
2	55556889
3	000112223
3	55555679
4	012
4	55799
5	00

Units: 5|0 = 50 pounds

15. The data is given in thousands of dollars. Round the data to the nearest ten thousand. Then the data ranges from 10 to 230.

160	80	20	20	80	160
130	50	100	50	80	50
150	100	100	180	30	170
10	50	30	230	50	20
40	30	10	10	90	40

Since the stem is one digit and the leaves can represent one digit only, divide the by 10 to obtain a range of 1 to 23.

STEM	LEAVES
0	11122233344
0	555558889
1	0003
1	56678
2	3

Units: 2|3 = 230 thousand dollars

17. The data has been rounded to the nearest ten. It ranges from 30 to 520.

STEM	LEAVES
0	3
0	5556778889
1	011223
1	56899
2	003
2	5
3	1
3	7
4	0
4	
5	2

Units: 5|2 = 520 dollars

19. The data ranges from 79 to 91.

STEM	LEAVES
7	99
8	000011122333444
8	55666667777788888 9999999
9	000001111

Units: 9|1 = 91%

21. uniform

23. left skewed

25. right skewed

27. a. The distribution is normal.

b. 3 credit unions have loan-to-asset ratios of exactly 70%.

c. There are 19 credit unions with ratios in the 30s and 28 credit unions with ratios in the 40s, so there are more credit unions with ratios in the 40s.

29. a. The distribution is right skewed.

b. 8 states have an average expenditure of over $1000.

c. 5 states have an average expenditure of $800

Section 10.2 Measures of Central Tendency

1. $\Sigma x = 21,900 + 22,850 + 24,930$
$$+ 29,710 + 28,340 + 40,000$$
$$= 167,730$$

The mean of the 6 numbers is
$$\bar{x} = \frac{167,730}{6} = 27,955 \, .$$

3. $\Sigma x = 3.5 + 4.2 + 5.8 + 6.3 + 7.1$
$$+ 2.8 + 3.7 + 4.2 + 4.2 + 5.7$$
$$= 47.5$$
$$\bar{x} = \frac{47.5}{10} = 7.75 \approx 4.8$$

5. $\Sigma x = 9.2 + 10.4 + 13.5 + 8.7 + 9.7 = 51.5$
$$\bar{x} = \frac{51.5}{5} = 10.3$$

7.

Value	Frequency	Value × Frequency	
19	3	19·3	57
20	5	20·5	100
21	25	21·25	525
22	8	22·8	176
23	2	23·2	46
24	1	24·1	24
28	1	28·1	28
	Total: 45	Total:	956

The mean is $\bar{x} = \dfrac{956}{45} \approx 21.2$.

9.

x	f	xf
9	5	45
11	10	110
15	12	180
17	9	153
20	6	120
28	1	28
Totals	43	636

$$\bar{x} = \frac{636}{43} \approx 14.8$$

11. First arrange the numbers in numerical order, from smallest to largest.
$28458, \$29679, \$33679, \$38400, \$39720$
There are 5 numbers; the median is the middle term, in this case $33,679.

13. First arrange the numbers in numerical order, from smallest to largest.
94.1, 96.8, 97.4, 98.6, 98.4, 98.7, 99.2, 99.9
There are 8 numbers; the median is the mean of the 2 middle numbers, which is
$$\frac{98.4 + 98.6}{2} = \frac{197}{2} = 98.5 \, .$$

15. 1, 2, 2, 1, 2, 2, 1, 1, 2, 2, 3, 4, 2, 3, 4, 2, 3, 2, 3,
The mode is the number that occurs most often. The mode is 2.

17. 62, 65, 71, 74, 71, 76, 71, 63, 59, 65, 65, 64, 72, 71, 77, 63, 65
The mode is the number that occurs most often. There are two modes, 65 and 71, since they both appear four times.

19. 3.2, 2.7, 1.9, 3.7, 3.9
There is no one number that occurs more times than any other number; so, there is no mode.

21. Answers vary. Possible answer: The mode has advantages of being easily found and not being influenced by data that are very large or very small compared to the rest of the data. It is often used in samples where the data to be "averaged" are not numerical.

23.

Interval	Midpoint, x	f	xf
1000–1499	1249.5	1	1249.5
1500–1999	1749.5	3	5248.5
2000–2499	2249.5	2	4499.0
2500–2999	2749.5	9	24,745.5
3000–3499	3249.5	21	68,239.5
3500–3999	3749.5	13	48,743.5
4000–4499	4249.5	0	0
4500–4999	4749.5	1	4749.5
	Totals:	50	157,475

Mean $\bar{x} = \dfrac{157,475}{50} = 3149.5 \approx 3150$

The modal class is 3000–3499.

25. a. Mean $\bar{x} = \dfrac{35+37+37+49+57+64+67+72+77+80}{10} = \dfrac{575}{10} = 57.5$

b. The middle data when the data is in ascending order are 57 and 64. The median number of nations is $\dfrac{57+64}{2} = 60.5$.

c. The mode is 37, since this number occurs twice.

27. a. Mean $\bar{x} = \dfrac{601+533+461+436+435+431+423+404+380+377+373+370}{12}$

$= \dfrac{5224}{12} = \$435.333$ million

b. The two middle receipts are 431 million and 423 million. The median receipt is $\dfrac{431+423}{2} = \$427$ million

29. a. Mean $\bar{x} = \dfrac{1680.2+2169.2+2649.0+3288.9+4075.5+5294.3+6369.3+7786.9+9411.5+10,383.0}{12}$

$= \dfrac{53,107.8}{10} = \$5310.78$ million

The two middle data are 4075.5 and 5294.3.

Median $= \dfrac{4075.5+5294.3}{2} = \$4684.9$ million

b. The revenue in 2004 was closest to the mean.

31. The average monthly low temperatures in ascending order: 30, 32, 33, 40, 42, 48, 49, 57, 62, 65, 68, 69

$\bar{x} = \dfrac{\sum x}{n} = \dfrac{595}{12} \approx 49.6°F$

Median $= \dfrac{48+49}{2} = 48.5°F$

33. The distribution is right skewed, so the median is the better measure of center. There are 30 values, so the median is the average of the 15th and 16th values. These values fall in the 150–199.99 range, so the median is the midpoint of that range, $175.

35. a. The distribution is right skewed.

b. There are 51 values, so the median is the 26th value, 13%.

Section 10.3 Measures of Variation

Note: Answers may vary slightly throughout this section due to the number of decimal places carried throughout the computations.

1. Answers vary. Possible answer: The standard deviation of a sample of numbers is the square root of the variance of the sample.

3. 5, 10, 3, 11, 15

Range $= 15 - 3 = 12$

$\bar{x} = \dfrac{5+10+3+11+15}{5} = \dfrac{44}{5} = 8.8$

x	$x - \bar{x}$	$(x - \bar{x})^2$
5	−3.8	14.44
10	1.2	1.44
3	−5.8	33.64
11	2.2	4.84
15	6.2	38.44
Total		92.8

$s = \sqrt{\dfrac{92.8}{5-1}} \approx 4.8$

5. 1770, 39, 200, 77, 322

Range $= 1770 - 39 = 1731$

$$\bar{x} = \frac{1770 + 39 + 200 + 77 + 322}{5} = \frac{2408}{5} = 481.6$$

x	$x - \bar{x}$	$(x - \bar{x})^2$
1770	1288.4	1,659,974.56
39	−442.6	195,894.76
200	−281.6	79,298.56
77	−404.6	163,701.16
322	−159.6	25,472.16
	Total:	2,124,341.20

$$s = \sqrt{\frac{2,124,341.20}{5-1}} = \sqrt{531,085.3} \approx 728.8$$

7. 1118, 94, 363, 57, 270

Range $= 1118 - 57 = 1061$

$$\bar{x} = \frac{1118 + 94 + 363 + 57 + 270}{5} = \frac{1902}{5} = 380.4$$

x	$x - \bar{x}$	$(x - \bar{x})^2$
1118	737.6	544,053.76
94	−286.4	82,024.96
363	−17.4	302.76
57	−323.4	104,587.56
270	−110.4	12,188.16
	Total:	743,157.20

$$s = \sqrt{\frac{743,157.20}{5-1}} = \sqrt{185,789.30} \approx 431.0$$

9. 251, 136, 372, 103, 36

Range $= 372 - 36 = 336$

$$\bar{x} = \frac{251 + 136 + 372 + 103 + 36}{5} = \frac{898}{5} = 179.6$$

x	$x - \bar{x}$	$(x - \bar{x})^2$
251	71.4	5097.96
136	−43.6	1900.96
372	192.4	37,017.76
103	−76.6	5867.56
36	−143.6	20,620.96
	Total:	70,505.20

$$s = \sqrt{\frac{70,505.20}{5-1}} = \sqrt{17,626.30} \approx 132.8$$

11. Expand the table to include columns for the midpoint x of each interval, and for fx, x^2, and fx^2.

Interval	f	x	fx	x^2	fx^2
0–24	4	12	48	144	576
25–49	3	37	111	1369	4107
50–74	6	62	372	3844	23,064
75–99	3	87	261	7569	22,707
100–124	5	112	560	12,544	62,720
125–129	9	137	1233	18,769	168,921
Totals	30		2585		282,095

The mean of the grouped data is

$$\bar{x} = \frac{\Sigma fx}{n} = \frac{2585}{30} \approx 86.2 \;.$$

The standard deviation for the grouped data is

$$s = \sqrt{\frac{\Sigma fx^2 - n\bar{x}^2}{n-1}} = \sqrt{\frac{282,095 - 30(86.2)^2}{30-1}}$$
$$\approx \sqrt{2040.8} \approx 45.2$$

13. This exercise should be completed using a computer or calculator. The solution may vary according to the computer program or calculator that is used. Using a TI-84, we have $\bar{x} = 5.0876$ and $s \approx .1087$.

15. Use Chebyshev's theorem with $k = 2$.

$$1 - \frac{1}{2^2} = 1 - \frac{1}{4} = \frac{3}{4}$$

So, at least $\frac{3}{4}$ of the numbers lie within 2 standard deviations of the mean.

17. Use Chebyshev's theorem with $k = 1.5$.

$$1 - \frac{1}{(1.5)^2} = 1 - \frac{4}{9} = \frac{5}{9}, \text{ so at least } \frac{5}{9} \text{ of the}$$

numbers lie within 1.5 standard deviations of the mean.

19. Between 26 and 74, $\bar{x} = 50$ and $s = 6$; we have $26 = 50 - 4 \cdot 6$ and $74 = 50 + 4 \cdot 6$, so $k = 4$.

At least $1 - \frac{1}{k^2} = \frac{15}{16} = 93.75\%$ of the numbers lie between 26 and 74.

21. Less than 32 or more than 68
From exercise 18, 88.9% of the data lie between 32 and 68. So, no more than
100% − 88.9% = 11.1% of the data are less then 32 or more than 68.

23. Aerobic shoe sales
281, 239, 222, 237, 261, 262, 268

$$\overline{x} = \frac{281 + 239 + 222 + 237 + 261 + 262 + 268}{7}$$

$$= \frac{1770}{7} \approx 252.9$$

The mean sales for aerobic shoes is about $252.9 million.

x	$x - \overline{x}$	$(x - \overline{x})^2$
281	28.1	789.61
239	−13.9	193.21
222	−30.9	954.81
237	−15.9	252.81
261	8.1	65.61
262	9.1	82.81
268	15.1	228.01
	Total:	2566.87

$$s = \sqrt{\frac{2566.87}{7-1}} \approx 20.7$$

The standard deviation is about $20.7 million.

25. Cross-training shoe sales
1476, 1421, 1407, 1327, 1437, 1516, 1561

$$\overline{x} = \frac{1476 + 1421 + 1407 + 1327 + 1437 + 1516 + 1561}{7}$$

$$= \frac{10,145}{7} \approx 1449.3$$

The mean sales for aerobic shoes is about $1449.3 million.

x	$x - \overline{x}$	$(x - \overline{x})^2$
1476	26.7	712.89
1421	−28.3	800.89
1407	−42.3	1789.29
1327	−122.3	14,957.29
1437	−12.3	151.29
1516	66.7	4448.89
1561	111.7	12,476.89
	Total:	35,337.43

$$s = \sqrt{\frac{35,337.43}{7-1}} \approx 76.7$$

The standard deviation is about $76.7 million.

27. a. $\overline{x} = \dfrac{16 + 12 + 11 + 9 + 8 + 7 + 7 + 6 + 5 + 5 + 4 + 4 + 2}{13}$

$$= \frac{96}{13} \approx 7.4$$

x	$x - \overline{x}$	$(x - \overline{x})^2$
16	8.6	73.96
12	4.6	21.16
11	3.6	12.96
9	1.6	2.56
8	.6	.36
7	−.4	.16
7	−.4	.16
6	−1.4	1.96
5	−2.4	5.76
5	−2.4	5.76
4	−3.4	11.56
4	−3.4	11.56
2	−5.4	29.16
	Total:	177.08

$$s = \sqrt{\frac{177.08}{13-1}} \approx \sqrt{14.8} \approx 3.8$$

b. One standard deviation from the mean consists of the values 3.58 to 11.18. Ten animals have blood types within 1 standard deviation of the mean.

29. a. $\bar{x} = \dfrac{84+91+128+131+143+153+164}{7}$

$= \dfrac{894}{7} \approx 127.71$

x	$x - \bar{x}$	$(x - \bar{x})^2$
84	–43.71	1910.564
91	–36.71	1347.624
128	.29	.0841
131	3.29	10.8241
143	15.29	233.7841
153	25.29	639.5841
164	36.29	1316.964
	Total:	5459.429

$s = \sqrt{\dfrac{5459.429}{7-1}} \approx \sqrt{909.9} \approx 30.2$

b. Doubling times within 2 standard deviations of the mean range from 67.33 to 188.09 days. All of these cancers have doubling times within 2 standard deviations of the mean.

c. Answers vary. Possible answer: A doubling time of 200 days is more than 2 standard deviations from the mean of these slow growing cancers. The tumor is not growing as expected for these slow-growing cancers.

31. a, b. 1: $\bar{x} = \dfrac{2-2+1}{3} = \dfrac{1}{3}$

$s = \sqrt{\dfrac{\left(2-\frac{1}{3}\right)^2 + \left(-2-\frac{1}{3}\right)^2 + \left(1-\frac{1}{3}\right)^2}{3-1}}$

$= \sqrt{\dfrac{26}{6}} \approx 2.1$

2: $\bar{x} = \dfrac{3-1+4}{3} = 2$

$s = \sqrt{\dfrac{(3-2)^2 + (-1-2)^2 + (4-2)^2}{3-1}}$

$= \sqrt{\dfrac{14}{2}} \approx 2.6$

3: $\bar{x} = \dfrac{-2+0+1}{3} = -\dfrac{1}{3}$

$s = \sqrt{\dfrac{\left(-2+\frac{1}{3}\right)^2 + \left(0+\frac{1}{3}\right)^2 + \left(1+\frac{1}{3}\right)^2}{3-1}}$

$= \sqrt{\dfrac{14}{6}} \approx 1.5$

4: $\bar{x} = \dfrac{-3+1+2}{3} = 0$

$s = \sqrt{\dfrac{(-3-0)^2 + (1-0)^2 + (2-0)^2}{3-1}}$

$= \sqrt{\dfrac{14}{2}} \approx 2.6$

5: $\bar{x} = \dfrac{-1+2+4}{3} = \dfrac{5}{3}$

$s = \sqrt{\dfrac{\left(-1-\frac{5}{3}\right)^2 + \left(2-\frac{5}{3}\right)^2 + \left(4-\frac{5}{3}\right)^2}{3-1}}$

$= \sqrt{\dfrac{38}{6}} \approx 2.5$

6: $\bar{x} = \dfrac{3+2+2}{3} = \dfrac{7}{3}$

$s = \sqrt{\dfrac{\left(3-\frac{7}{3}\right)^2 + \left(2-\frac{7}{3}\right)^2 + \left(2-\frac{7}{3}\right)^2}{3-1}}$

$= \sqrt{\dfrac{2}{6}} \approx .6$

7: $\bar{x} = \dfrac{0+1+2}{3} = 1$

$s = \sqrt{\dfrac{(0-1)^2 + (1-1)^2 + (2-1)^2}{3-1}}$

$= \sqrt{\dfrac{2}{2}} = 1$

8: $\bar{x} = \dfrac{-1+2+3}{3} = \dfrac{4}{3}$

$s = \sqrt{\dfrac{\left(-1-\frac{4}{3}\right)^2 + \left(2-\frac{4}{3}\right)^2 + \left(3-\frac{4}{3}\right)^2}{3-1}}$

$= \sqrt{\dfrac{26}{6}} \approx 2.1$

9: $\bar{x} = \dfrac{2+3+2}{3} = \dfrac{7}{3}$

$s = \sqrt{\dfrac{\left(2-\frac{7}{3}\right)^2 + \left(3-\frac{7}{3}\right)^2 + \left(2-\frac{7}{3}\right)^2}{3-1}}$

$= \sqrt{\dfrac{2}{6}} \approx .6$

(continued on next page)

(*continued from page 312*)

10: $\bar{x} = \dfrac{0+0+2}{3} = \dfrac{2}{3}$

$s = \sqrt{\dfrac{\left(0-\frac{2}{3}\right)^2 + \left(0-\frac{2}{3}\right)^2 + \left(2-\frac{2}{3}\right)^2}{3-1}}$

$= \sqrt{\dfrac{8}{6}} \approx 1.2$

c. $\bar{X} = \dfrac{\Sigma x}{n} \approx \dfrac{11.3}{10} = 1.13$

d. $\bar{s} = \dfrac{\Sigma s}{n} = \dfrac{16.8}{10} = 1.68$

e.   The upper control limit for the sample means is
$\bar{x} + 1.954\bar{s} = 1.13 + (1.954)(1.68) \approx 4.41$
The lower control limit for the sample means is
$\bar{x} - 1.954\bar{s} = 1.13 - (1.954)(1.68) \approx -2.15$
One of the measurements, -3, is outside of these limits, so the process is out of control.

33.

Interval	f	x	xf	x^2	x^2f
50–59	2	54.5	109	2970.25	5940.5
60–69	4	64.5	258	4160.25	16,641
70–79	7	74.5	521.5	5550.25	38,851.75
80–89	9	84.5	760.5	7140.25	64,262.25
90–99	8	94.5	756	8930.25	71,442
Totals:	30		2405		197,137.5

$\bar{x} = \dfrac{2405}{30} \approx 80.17$

$s = \sqrt{\dfrac{197{,}137.5 - 30(80.17)^2}{30-1}} \approx 12.2$

For individualized instruction, the mean is 80.17 and the standard deviation is 12.2.

35. Answers vary. Possible answer: You would expect the mean of the traditional instruction to be smaller, since 13 of the 34 students lie in the first two intervals while only 6 of 30 students lie in the same intervals for individualized instruction. As far as standard deviation is concerned, there is a small difference between the two, 1.1. This gives the impression that the data is pretty much spread out in nearly the same fashion in both types of instruction.

Section 10.4   Normal Distributions and Boxplots

1. The peak in a normal curve occurs directly above the mean.

3. Answers vary. Possible answer:
   If a normal distribution has mean μ and standard deviation σ, then the z-score for the number x is
   $z = \dfrac{x-\mu}{\sigma}$.

5. By looking up 1.75 in Table 2, we get $.4599 = 45.99\%$. Thus, the percent of the total area between the mean and 1.75 standard deviations from the mean is 45.99%.

7. $-.43$ indicates that we are below the mean. Looking up .43 in Table 2, we get $.1664 = 16.64\%$. Thus, the percent of the total area between the mean and .43 standard deviations from the mean is 16.64%.

9. The entry corresponding to $z = 1.41$ is .4207, and the entry for $z = 2.83$ is .4977, so the area between $z = 1.41$ and $z = 2.83$ is $.4977 - .4207 = .0770$ or 7.7%.

11. For $z = 2.48$, the entry is .4934, and for $z = .05$, the entry is .0199. The area between $z = -2.48$ and $z = -.05$ is $.4934 - .0199 = .4735$ or 47.35%.

13. To find the area between $z = -3.05$ and $z = 1.36$, add the area between $z = -3.05$ and $z = 0$ to the area between $z = 1.36$ and $z = 0$. The area is $.4989 + .4131 = .9120 = 91.20\%$.

15. 5% of the total area is to the right of z. The mean divides the area in half or .5. The area from the mean to z standard deviations is $.5 - .05 = .45$. Using Table 2 backwards, we get the z-score corresponding to the area .45 as 1.64 or 1.65 (approximately).

17. 15% of the total area is to the left of z. The area from the mean to z standard deviations is $.5 - .15 = .35$. The z-score from the table is 1.04. As the area is to the left of the mean, $z = -1.04$.

19. To find $P(x \le \mu)$ and $P(x \ge \mu)$ consider the normal curve. The curve is symmetric about the vertical line through μ, so $P(x \le \mu)$ represents half the area under the curve. Since the area under the curve is 1, $P(x \le \mu) = .5$. Similarly, $P(x \ge \mu) = .5$.

21. Using Chebyshev's theorem with $k = 3$:

$$1 - \frac{1}{3^2} = 1 - \frac{1}{9} = \frac{8}{9} \approx .889$$

Using the normal distribution, the area between $z = -3$ and $z = 3$ is $.4987 + .4987 = .9974$ The probability a number will lie within 3 standard deviations of the mean is greater as indicated by the normal distribution at .9974 than Chebyshev's theorem shows at .889.

23. Let x represent the number of grams of peanut butter in a jar.

$\mu = 453$, $\sigma = 10.1$

Find the z-score for $x = 450$.

$$z = \frac{x - \mu}{\sigma} = \frac{450 - 453}{10.1} \approx -.30$$

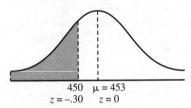

To find the area to the left of $z = -.30$, find the area between $z = 0$ and $z = -.30$ and subtract that answer from .5. The area is $.5 - .1179 = .3821$. Thus, the probability that a jar will contain less than 450 grams is .3821.

25. Let x represent the number of lumens.

$\mu = 1640$, $\sigma = 62$

For $x = 1600$, $z = \dfrac{x - \mu}{\sigma} = \dfrac{1600 - 1640}{62} \approx -.65.$

For $x = 1700$, $z = \dfrac{x - \mu}{\sigma} = \dfrac{1700 - 1640}{62} \approx .97.$

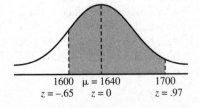

To find the area between $z = -.65$ and $z = .97$, find the area between $z = 0$ and $z = -.65$. Then find the area between $z = 0$ and $z = .97$. Add these two answers together. The area is $.2422 + .3340 = .5762$. Thus, the probability that a 100-watt bulb will have a brightness between 1600 and 1700 lumens is .5762.

27. Let x represent HDL cholesterol level.

$\mu = 51.6$, $\sigma = 14.3$

For $x = 60$, $z = \dfrac{60 - 51.6}{14.3} \approx .59$

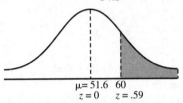

To find the area to the right of $z = .59$, find the area between $z = .59$ and $z = 0$ and subtract that from 0.5. The area is $0.5 - .2224 = .2776$. Thus, the probability that an individual will have an HDL cholesterol level greater than 60 mg/dL is .2776.

29. Let x represent the starting salaries for accounting majors.

$\mu = 45,000$, $\sigma = 3200$

For $x = 53,000$, $z = \dfrac{53,000 - 45,000}{3200} = 2.5.$

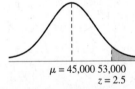

To find the area to the right of $z = 2.5$, find the area between $z = 0$ and $z = 2.5$ and subtract it from .5. The area is $.5 - .4938 = .0062$. Thus, the probability that an individual will have a starting salary above \$53,000 is .0062.

For exercises 31–37, $\mu = 540$ and $\sigma = 100$. Refer to Table 2 in the back of the text.

31. For $x = 700$, $z = \dfrac{700 - 540}{100} = 1.6.$

The area between $z = 0$ and $z = 1.6$ is .4452. Thus, the probability that a GMAT test taker earns a score in the range 540–700 is .4452.

33. From exercise 31, we know that $z = 1.6$ for $x = 700$ and $P = .4452$. From exercise 32, we know that $z = -2.4$ for $x = 300$ and $P = .4918$. Therefore, the probability that a GMAT test taker earns a score in the range 300–700 is $.4452 + .4918 = .9370$.

35. For $x = 750$, $z = \dfrac{750 - 540}{100} = 2.1$.

The area between $z = 0$ and $z = 2.1$ is .4821. The area to the right of $z = 0$ is .5, so the area to the right of $z = 2.1$ is $.5 - .4821 = .0179$. Thus, the probability that a GMAT test taker earns a score greater than 750 is .0179.

37. For $x = 300$, $z = \dfrac{300 - 540}{100} = -2.4$.

For $x = 400$, $z = \dfrac{400 - 540}{100} = -1.4$.

The area between $z = 0$ and $z = -2.4$ is .4918, while the area between $z = 0$ and $z = -1.4$ is .4192. Therefore, the area between $z = -2.4$ and $z = -1.4$ is $.4918 - .4192 = .0726$, which is the probability that a GMAT test taker earns a score in the range 300–400.

39. If 85% of the total area is to the left of z, then $z > 0$ and 35% of the area is between z and the mean. Look for a value of z in Table 2 with an area of .35. The closest match is $z = 1.04$. Let x stand for the speed.

$$\dfrac{x - 40}{5} = 1.04 \Rightarrow x - 40 = 5.2 \Rightarrow x = 45.2$$

The 85th percentile speed is 45.2 mph.

41.

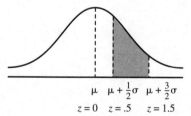

To find the area between $z = .5$ and $z = 1.5$, first find the area between $z = 1.5$ and $z = 0$. Then subtract the area between $z = 0$ and $z = .5$. The area is $.4332 - .1915 = .2417 = 24.17\%$, which means 24.17% of the students receive a B.

43. Answers vary. Possible answer: This system would be more fair in a large freshman class in psychology than in a graduate seminar of five students since the large class is more apt to have grades ranging the entire spectrum. The graduate students might have all grades in the 90's which would mean that the graduate student with the lowest score in the 90's would get an F.

45. $\mu = 550$ units; $\sigma = 45$ units
The recommended daily allowance is
$$\mu + 2.5\sigma = 550 + (2.5)(46) = 665 \text{ units}$$

47. $\mu = 155$ units; $\sigma = 14$ units
The recommended daily allowance is
$$\mu + 2.5\sigma = 155 + (2.5)(14) = 190 \text{ units.}$$

49. To find the area to the right of $z = 1$, find the area between $z = 0$ and $z = 1$, and subtract that from .5. The area is $.5 - .3413 = .1587$, which means 15.87% of the students had scores more than 1 standard deviation above the mean.

51. Less than $400
$\mu = 500$, $\sigma = 65$

For $x = 400$, $z = \dfrac{400 - 500}{65} \approx -1.54$

To find the area to the left of $z = -1.54$, subtract the area between $z = -1.54$ and $z = 0$ from .5. The area is $.5 - .4382 = .0618$. Thus, the probability that a woman will pay less than $400 additional is .0618.

53. Between $350 and $600

For $x = 350$, $z = \dfrac{350 - 500}{65} = -2.31$.

For $x = 600$, $z = \dfrac{600 - 500}{65} = 1.54$.

To find the area between $z = -2.31$ and $z = 1.54$, find the area between $z = 0$ and $z = 1.54$ and add it to the area between $z = -2.31$ and $z = 0$. The area is $.4382 + .4896 = .9278$. Thus, the probability that a woman will pay between $350 and $600 additional is .9278.

55. Arrange the data in ascending order:
19.6, 19.8, 20.1, 20.5, 21.0, 22.0, 23.1, 24.1, 28.9, 31.9
Minimum = 19.6; maximum = 31.9

Median $= Q_2 = \dfrac{21.0 + 22.0}{2} = 21.5.$

$n = 10$, so $.25 \cdot 10 = 2.5$, which rounds up to 3. Count to the third data point to obtain $Q_1 = 20.1.$

$.75 \cdot 10 = 7.5$, which rounds up to 8, so count to the eighth data point to obtain $Q_3 = 24.1.$

57.

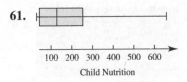

Coca-Cola

59. Arrange the child nutrition data in ascending order:
33, 35, 40, 52, 86, 109, 144, 214, 254, 449, 473, 661
Minimum = 33; maximum = 661

Median $= Q_2 = \dfrac{109 + 144}{2} = 126.5.$

$n = 12$, so $.25 \cdot 12 = 3$. Count to the third data point to obtain $Q_1 = 40$. $.75 \cdot 12 = 9$, so count to the ninth data point to obtain $Q_3 = 254.$

61.

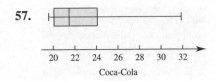

Child Nutrition

63. Child nutrition has higher variability.

Section 10.5 Normal Approximation to the Binomial Distribution

1. To find the mean and standard deviation of a binomial distribution, you must know n, the number of independent repeated trials, and p, the probability of a success in a single trial.

3. $n = 16, p = .5, (1 - p) = .5$

   a. Using the binomial distribution
$$P(x = 8) =_{16} C_8 (.5)^8 (.5)^8 \approx .1964$$

   b. Since we are using a normal curve approximation.

$$\mu = np = 16\left(\frac{1}{2}\right) = 8 \text{ and}$$

$$\sigma = \sqrt{np(1 - p)} = \sqrt{16\left(\frac{1}{2}\right)\left(1 - \frac{1}{2}\right)} = 2.$$

The required probability is the area of the region corresponding to the x-values of 7.5 and 8.5.

$$z = \frac{x - \mu}{\sigma} = \frac{8.5 - 8}{2} = .25$$

From Table 2, the corresponding area is .0987. Therefore, the total area is $2(.0987) = .1974$. This is the probability of getting exactly 8 heads.

5. $n = 16, p = .5, (1 - p) = .5$

   a. Using the binomial distribution,
$$P(x > 13)$$
$$= P(x = 14) + P(x = 15) + P(x = 16)$$
$$P(x = 14) =_{16} C_{14} (.5)^{14} (.5)^2 \approx .00183$$
$$P(x = 15) =_{16} C_{15} (.5)^{15} (.5)^1 \approx .00024$$
$$P(x = 16) =_{16} C_{16} (.5)^{16} (.5)^0 \approx .00002$$
$$P(x > 13) \approx .00183 + .00024 + .00002$$
$$\approx .00209 \approx .0021$$

   b. For more than 13 tails, the desired area is to the right of 13.5.
$$\mu = np = 16(.5) = 8$$
$$\sigma = \sqrt{16(.5)(.5)} = 2$$

For $x = 13.5$, $z = \dfrac{13.5 - 8}{2} = 2.75$ and

$A = .4970.$
$$P(x > 13) = .5 - .4970 = .0030$$

For exercises 7–9, $n = 1000$, $p = \dfrac{1}{2}$,

$$\mu = np = 1000\left(\frac{1}{2}\right) = 500$$

$$\sigma = \sqrt{np(1 - p)} = \sqrt{1000\left(\frac{1}{2}\right)\left(\frac{1}{2}\right)} = 15.8$$

7. For 500 heads, we need to find the area of the region corresponding to the x-values of 499.5 and 500.5.
$$z = \frac{500.5 - 500}{15.8} = .03$$
The area is .0120. Since the region is symmetric about the mean $\mu = 500$, the probability (area) is $2(.0120) = .0240.$

9. 475 heads or more

$p = \dfrac{1}{2}$, $\mu = 500$, and $\sigma = 15.8$.

The required area is to the right of the x-value 474.5.

$z = \dfrac{474.5 - 500}{15.8} = -1.61$

The area is .4463. This area is to the left of the mean. So, the required area (probability) is .4463 + .5 = .9463.

11. Exactly 20 fives

The probability of a five is $p = \dfrac{1}{6}$, and $n = 120$.

$\mu = np = 120\left(\dfrac{1}{6}\right) = 20$

$\sigma = \sqrt{120\left(\dfrac{1}{6}\right)\left(\dfrac{5}{6}\right)} = 4.08$

$z = \dfrac{20.5 - 20}{4.08} = .12$

The area is .0478. So, the area between the x-values 19.5 and 20.5 is 2(.0478) = .0956. This is the required probability.

13. More than 17 threes

$p = \dfrac{1}{6}$, and the required area is to the right of $x = 17.5$.

$z = \dfrac{17.5 - 20}{4.08} = -.61$

The area is .2291. So, for more than 17 threes, the probability is .2291 + .5 = .7291.

15. $n = 130$, $p = \dfrac{1}{6}$, $1 - p = \dfrac{5}{6}$

$\mu = np = 130\left(\dfrac{1}{6}\right) \approx 21.6667$

$\sigma = \sqrt{np(1-p)} = \sqrt{130\left(\dfrac{1}{6}\right)\left(\dfrac{5}{6}\right)} \approx 4.2492$

The required area is to the right of $x = 25.5$.

$z = \dfrac{25.5 - 21.6667}{4.2492} \approx .90$

The area for $z = .90$ is .3159, so the probability is .5 − .3159 = .1841.

17. $\mu = np = 120(.6) = 72$

$\sigma = \sqrt{np(1-p)} = \sqrt{120(.6)(.4)} \approx 5.37$

a. For exactly 80 units of food, the area is between 79.5 and 80.5. For $x = 79.5$,

$z = \dfrac{79.5 - 72}{5.37} \approx 1.40$, which gives an area of .4192. For $x = 80.5$,

$z = \dfrac{80.5 - 72}{5.37} \approx 1.58$, which gives an area of .4430.

The probability is .4430 − .4192 = .0238.

b. For at least 70 units, the area is to the right of 69.5. For $x = 69.5$, $z = \dfrac{69.5 - 72}{5.37} = -.47$ which gives an area of .1808.

The probability is .5 + .1808 = .6808.

19. $p = .27$, $1 - p = .73$, and $n = 500$

$\mu = np = 500(.27) = 135$

$\sigma = \sqrt{np(1-p)} = \sqrt{135(.73)} \approx 9.9272$

For $x = 150$, $z = \dfrac{149.5 - 135}{9.9272} \approx 1.46$.

The area between $z = 0$ and $z = 1.46$ is .4279, so the probability of $z \geq 1.46$ is .5 − .4279 = .0721.

21. $p = .30$, $1 - p = .70$, $n = 150$

$\mu = np = 150(.30) = 45$

$\sigma = \sqrt{np(1-p)} = \sqrt{45(.7)} \approx 5.6125$

For 40 or less, let $x = 40.5$

$z = \dfrac{40.5 - 45}{5.6125} \approx -.80$

The area is .2881, so the probability is .5 − .2881 = .2119.

23. $\mu = np = (134)(.20) = 26.8$

$\sigma = \sqrt{np(1-p)} = \sqrt{(134)(.20)(.80)} = 4.63$

a. $P(x = 12) = P(11.5 < x < 12.5)$

If $x = 11.5$, $z = \dfrac{11.5 - 26.8}{4.63} \approx -3.31$ and $A = .4995$.

If $x = 12.5$, $z = \dfrac{12.5 - 26.8}{4.63} = -3.09$ and $A = .4990$

$P(11.5 < x < 12.5) = .4995 - .4990 = .0005$

b. $P(\text{no more than } 12) = P(x < 12.5)$

If $x = 12.5$, $z = -3.09$ and $A = .4990$ from (a). $P(x < 12.5) = .5 - .4990 = .001$

c. $P(x = 0) = P(x < .5)$

If $x = .5$, $z = \dfrac{.5 - 26.8}{4.63} = -5.680$ and $A = .5$.

$P(x \leq .5) = P(z < -5.680) = .5 - .5 = .0000$.

25. $n = 700$, $p = .164$, $1 - p = .836$

$\mu = np = (700)(.164) = 114.8$

$\sigma = \sqrt{np(1-p)} = \sqrt{114.8(.836)} \approx 9.7966$

$P(\text{at least } 100) = P(x \geq 100) = P(x > 99.5)$.

If $x = 99.5$, $z = \dfrac{99.5 - 114.8}{9.7966} \approx -1.56$ and

$A = .4406$. Thus,

$P(\text{at least } 100) = .5 + .4406 = .9406$

27. a. $n = 1000$, $p = .006$, $1 - p = .994$

$\mu = np = (1000)(.006) = 6$

$\sigma = \sqrt{np(1-p)} = \sqrt{1000(.006)(.994)}$

≈ 2.4421

The required area is to the right of $x = 9.5$.

$z = \dfrac{9.5 - 6}{2.4421} \approx 1.43$

The area for $z = 1.43$ is $.4236$, so the probability is $.5 - .4236 = .0764$.

b. $n = 1000$, $p = .015$, $1 - p = .985$

$\mu = np = (1000)(.015) = 15$

$\sigma = \sqrt{np(1-p)} = \sqrt{1000(.015)(.985)}$

≈ 3.8438

The area required is between $x = 19.5$ and $x = 40.5$.

If $x = 19.5$, $z = \dfrac{19.5 - 15}{3.8438} \approx 1.17$ and

$A = .3790$

If $x = 40.5$, $z = \dfrac{40.5 - 15}{3.8438} \approx 6.63$ and

$A = .5000$.

$P(20 \leq x \leq 40) = .5000 - .3790 = .121$

c. $n = 500$, $p = .015$, $1 - p = .985$

$\mu = np = 7.5$

$\sigma = \sqrt{np(1-p)} = \sqrt{500(.015)(.985)}$

≈ 2.7180

For $x = 14.5$, $z = \dfrac{14.5 - 7.5}{2.7180} \approx 2.58$ and

$A = .4951$, so the probability is $.5 - .4951 = .0049$.

Yes, this town does appear to have a higher than normal number of B– donors, since the probability of getting these kinds of results from a normal town is only .49%.

29. $n = 500$, $p = .421$, $1 - p = .579$

$\mu = np = 500(.421) = 210.5$

$\sigma = \sqrt{np(1-p)} = \sqrt{210.5(.579)} \approx 11.0399$

$P(\text{at least } 220) = P(x \geq 220) = P(x > 219.5)$

For $x = 219.5$, $z = \dfrac{219.5 - 210.5}{11.0399} \approx .82$ and

$A = .2939$

$P(\text{at least } 220) = .5 - .2939 = .2061$

31. a. Using the binomial distribution to find the probability of 4 or more holes in one, $P(x > 3.5)$, is equivalent to finding the probability $1 - P(x < 3.5)$.

$1 - P(x < 3.5)$

$\quad = 1 - (P(0) + P(1) + P(2) + P(3))$

$P(x = 0) = \binom{156}{0}\left(\dfrac{1}{3709}\right)^0\left(\dfrac{3708}{3709}\right)^{156}$

$\quad \approx .95881$

$P(x = 1) = \binom{156}{1}\left(\dfrac{1}{3709}\right)^1\left(\dfrac{3708}{3709}\right)^{155}$

$\quad \approx .04034$

$P(x = 2) = \binom{156}{2}\left(\dfrac{1}{3709}\right)^2\left(\dfrac{3708}{3709}\right)^{154}$

$\quad \approx .00084$

$P(x = 3) = \binom{156}{3}\left(\dfrac{1}{3709}\right)^3\left(\dfrac{3708}{3709}\right)^{153}$

$\quad \approx .00001$

$1 - (.95881 + .04034 + .00084 + .00001)$

$\quad\quad\quad\quad\quad \approx 1.2139 \times 10^{-7}$

The probability that 4 or more of 156 golf pros shoot a hole in one is 1.2139×10^{-7}.

b. $n = 156$, $p = \dfrac{1}{3709}$, $1 - p = \dfrac{3708}{3709}$

$\mu = np = \dfrac{156}{3709} \approx .04206$

$\sigma = \sqrt{np(1-p)} = \sqrt{156\left(\dfrac{1}{3709}\right)\left(\dfrac{3708}{3709}\right)}$

$\approx .20506$

The required area is to the right of $x = 3.5$.

$z = \dfrac{3.5 - .04206}{.20506} \approx 16.86$

The area for $z = 16.86$ is essentially $.5000$. The probability is then $.5 - .5000 = 0$. (The probability is actually very small.) We must be cautious in using this approximation in this application because the data doesn't follow the rule of thumb that $np \geq 5$.

c. $n = 20{,}000, \ p = \dfrac{1}{3709}, \ 1 - p = \dfrac{3708}{3709}$

$\mu = np = 20{,}000\left(\dfrac{1}{3709}\right) \approx 5.3923$

$\sigma = \sqrt{np(1-p)} = \sqrt{20{,}000\left(\dfrac{1}{3709}\right)\left(\dfrac{3708}{3709}\right)}$

≈ 2.3218

The required area is to the right of $x = 3.5$.

$z = \dfrac{3.5 - 5.3923}{2.3218} = -.82$

The area for $z = -.82$ is .2939, so the probability is $.5 + .2939 = .7939$.
Discussion answers vary.

Chapter 10 Review Exercises

1. Answers vary. Possible answer: Some reasons for organizing data into a grouped frequency distribution are that
   a. it organizes data so that one can visually understand data;
   b. it allows for graphical representation like histogram;
   c. it provides easier computation of mean and standard deviation when there is a large number of data.

2. Answers vary. Possible answer: In a grouped frequency distribution, there should be from 6 to 15 intervals.

3. **a.**

Interval	Frequency
40–49	2
50–59	2
60–69	3
70–79	8
80–89	9
90–99	6

   b.

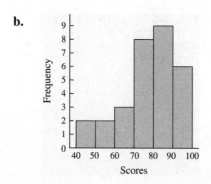

c.

STEM	LEAVES
4	25
5	35
6	389
7	11667777
8	122444899
9	111249

Units: 9|9 = 99 points

4. **a.**

Interval	Frequency
9–10	3
11–12	6
13–14	6
15–16	7

b–c.

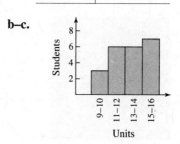

d.

STEM	LEAVES
0	9
1	00
1	122222
1	333344
1	5555566

Units: 1|6 = 16 units

5. $\Sigma x = 68 + 45 + 71 + 77 + 82 + 94 + 89 + 63 + 99$
   $+ 55 + 76 + 77 + 82 + 92 + 77 + 76 + 53$
   $+ 84 + 91 + 81 + 71 + 69 + 42 + 88 + 91$
   $+ 84 + 77 + 84 + 89 + 91 = 2318$
   The mean of the 30 numbers is
   $\bar{x} = \dfrac{\Sigma x}{30} = \dfrac{2318}{30} = 77.27$.

6. $\Sigma x = 10 + 9 + 16 + 12 + 13 + 15 + 13 + 16 + 15$
   $+ 11 + 13 + 12 + 12 + 15 + 12 + 14 + 10$
   $+ 12 + 14 + 15 + 15 + 13 = 287$
   $\bar{x} = \dfrac{\Sigma x}{n} = \dfrac{287}{22} \approx 13.0$

7.

Interval	Midpoint, x	Frequency, f	Product, xf
60–69	64.5	10	645
70–79	74.5	24	1788
80–89	84.5	6	507
90–99	94.5	3	283.5
100–109	104.5	1	104.5
	Total:	44	3328

The mean of this collection of grouped data is

$$\overline{x} = \frac{3328}{44} = 75.6 \text{ cm}.$$

8.

Interval	Midpoint, x	Freq, f	Product xf
0–999	499.5	1	499.5
1000–1999	1499.5	12	17,994
2000–2999	2499.5	14	34,993
3000–3999	3499.5	11	38,494.5
4000–4999	4499.5	5	22,497.5
5000–5999	5499.5	1	5499.5
	Total:	44	119,978

Use the formula for the mean of a grouped frequency distribution.

$$\overline{x} = \frac{\Sigma xf}{n} = \frac{119,978}{44} \approx 2726.8$$

9. Answers vary. Possible answer: Mean, median, and mode are all types of averages. The mode is the value that occurs the most. The median is the middle value when the data are ranked from highest to lowest. The mean is the sum of the data divided by the number of data items.

10. 65, 68, 71, 72, 72, 73, 73, 73, 78, 80, 84, 89
There are 12 numbers here; the median is the mean of the 2 middle numbers, which is
$$\frac{73 + 73}{2} = 73.$$
The mode = 73, which occurs 3 times.

11. Arrange the numbers in numerical order, from smallest to largest.
66, 67, 68, 70, 71, 72, 72, 72, 73, 74, 76, 77, 80
The median is the middle number; in this case it is 72.
The mode is the number that occurs most often; in this case it is 72.

12. The modal class is the interval with the greatest frequency. For the distribution of Exercise 7, the modal class is 70–79.

13. The modal class for the distribution of Exercise 8 is the interval 2000–2999, since it contains more data values than any of the other intervals.

14. right skewed

15. uniform

16. normal

17. left skewed

18. Answers vary. Possible answer: The range of a distribution is the difference between the largest and smallest data values.

19. Answers vary. Possible answer: The standard deviation is the square root of the variance. The standard deviation measures how spread out the data are from the mean.

20. 14, 17, 18, 19, 30
The range is the difference between the largest and smallest numbers. For this distribution, the range is $30 - 14 = 16$.
To find the standard deviation, the first step is to find the mean.
$$\overline{x} = \frac{\Sigma x}{n} = \frac{14 + 17 + 18 + 19 + 30}{5} = \frac{98}{5} = 19.6$$
Now complete the following chart.

x	x^2
14	196
17	289
18	324
19	361
30	900

Total: 2070

$$s = \sqrt{\frac{\Sigma x^2 - n\overline{x}^2}{n-1}} = \sqrt{\frac{2070 - 5(19.6)^2}{4}}$$
$$= \sqrt{37.3} \approx 6.11$$

21. The range is $54 - 17 = 37$, the difference of the highest and lowest numbers in the distribution.

The mean is $\bar{x} = \dfrac{\Sigma x}{n} = \dfrac{289}{10} = 28.9$.

Construct a table with the values of x, $x - \bar{x}$, and $(x - \bar{x})^2$.

x	$x - \bar{x}$	$(x - \bar{x})^2$
26	−2.9	8.41
43	14.1	198.81
17	−11.9	141.61
20	−8.9	79.21
25	−3.9	15.21
37	8.1	65.61
54	25.1	630.01
28	−.9	.81
20	−8.9	79.21
19	−9.9	98.01
Totals: 289		1316.90

The standard deviation is $s = \sqrt{\dfrac{1316.90}{10-1}} \approx \sqrt{146.3} \approx 12.10$.

22. Recall that when working with grouped data, x represents the midpoint of each interval. Complete the following table, which extends the table from Exercise 7.

Interval	f	x	xf	x^2	fx^2
60–69	10	64.5	645	4160.25	41,602.50
70–79	24	74.5	1788	5550.25	133,206.00
80–89	6	84.5	507	7140.25	42,841.50
90–99	3	94.5	283.5	8930.25	26,790.75
100–09	1	104.5	104.5	10,920.25	10,920.25
Totals:	44		3328		255,361.00

Use the formulas for grouped frequency distributions to find the mean and then the standard deviation. (The mean was also calculated in Exercise 7.)

$$\bar{x} = \frac{\Sigma xf}{n} = \frac{3328}{44} = 75.6$$

$$s = \sqrt{\frac{\Sigma fx^2 - n\bar{x}^2}{n-1}} = \sqrt{\frac{255,361 - 44(75.6)^2}{43}} \approx 9.51$$

23. Start with the frequency distribution that was the answer to Exercise 8, and expand the table to include columns for the midpoint x of each interval, and for xf, x^2, and fx^2.

Interval	f	x	xf	x^2	fx^2
0–999	1	499.5	499.50	249,500.25	249,500.25
1000–1999	12	1499.5	17,994.00	2,248,500.25	26,982,003.00
2000–2999	14	2499.5	34,993.00	6,247,500.25	87,465,003.50
3000–3999	11	3499.5	38,494.50	12,246,500.25	134,711,502.75
4000–4999	5	4499.5	22,497.50	20,245,500.25	101,227,501.25
5000–5999	1	5499.5	5,499.50	30,244,500.25	30,244,500.25
Totals:	44		119,978.00		380,880,011.00

The mean of the grouped data is
$$\bar{x} = \frac{\Sigma xf}{n} = \frac{119,978}{44} \approx 2726.8.$$
The standard deviation for the grouped data is
$$s = \sqrt{\frac{\Sigma fx^2 - n\bar{x}^2}{n-1}} = \sqrt{\frac{380,880,011 - 44(2726.8)^2}{44-1}} \approx \sqrt{1,249,319} \approx 1117.7$$

24. Answers vary. Possible answer: A normal distribution is a continuous distribution with the following properties:
- The highest frequency is at the mean;
- The graph is symmetric about a vertical line through the mean; and
- The total area under the curve, above the x-axis, is 1.

25. Answers vary. Possible answer: A distribution in which the peak is not at the center, or mean, is called skewed.

26. Between $z = 0$ and $z = 1.35$
By Table 2, the area between $z = 0$ and $z = 1.35$ is .4115.

27. To the left of $z = .38$
The area between $z = 0$ and $z = .38$ is .1480, so the area to the left of $z = .38$ is
.5 + .1480 = .6480.

28. Between $z = -1.88$ and $z = 2.41$
The area between $z = -1.88$ and $z = 0$ is .4700.
The area between $z = 0$ and $z = 2.41$ is .4920.
The total area between $z = -1.88$ and $z = 2.41$ is
.4700 + .4920 = .9620.

29. Between $z = 1.53$ and $z = 2.82$
The area between $z = 2.82$ and $z = 1.53$ is
.4976 − .4370 = .0606.

30. Since 8% of the area is to the right of the z-score, 42% or .4200 is between $z = 0$ and the appropriate z-score. Use Table 2 to find an appropriate z-score whose area is .4200. The closest approximation is $z = 1.41$.

31. Answers vary. Possible answer: The normal distribution is not a good approximation of a binomial distribution that has a value of p close to 0 or 1 because the histogram of such a binomial distribution is skewed and therefore not close to the shape of a normal distribution.

32.

Interval	Frequency
0–499	28
500–999	15
1000–1499	3
1500–1999	2
2000–2499	1
2500–2999	0
3000–3499	0
3500–3999	1

33. a.

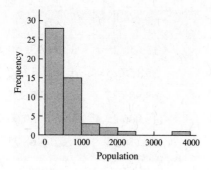

b.   The distribution is right skewed.

34. $\bar{x} = \dfrac{\Sigma x}{n} = \dfrac{6.6 + 4.7 + (-11.7) + (-30.4)}{4}$

$= -\dfrac{30.8}{4} = -7.7$

$s = \sqrt{\dfrac{\Sigma x^2 - n\bar{x}^2}{n-1}}$

$= \sqrt{\dfrac{6.6^2 + 4.7^2 + (-11.7)^2 + (-30.4)^2 - 4(-7.7)^2}{3}}$

$= \sqrt{\dfrac{1126.7 - 4(-7.7)^2}{3}} = \sqrt{\dfrac{889.54}{3}} \approx 17.2$

35. $\bar{x} = \dfrac{\Sigma x}{n} = \dfrac{-10.3 + .1 + 4.9 + 20.0}{4}$

$= \dfrac{14.7}{4} = 3.675$

$s = \sqrt{\dfrac{\Sigma x^2 - n\bar{x}^2}{n-1}}$

$= \sqrt{\dfrac{(-10.3)^2 + .1^2 + 4.9^2 + 20.0^2 - 4(3.675)^2}{3}}$

$= \sqrt{\dfrac{530.11 - 4(3.675)^2}{3}} = \sqrt{\dfrac{476.0875}{3}} \approx 12.6$

36.   Diet A:

$\bar{x} = \dfrac{1 + 0 + 3 + 7 + 1 + 1 + 5 + 4 + 1 + 4}{10} = \dfrac{27}{10} = 2.7$

$\Sigma x^2 = 1^2 + 0^2 + 3^2 + 7^2 + 1^2 + 1^2 + 5^2$
$\qquad\qquad + 4^2 + 1^2 + 4^2$

$= 119$

$s = \sqrt{\dfrac{\Sigma x^2 - n\bar{x}^2}{n-1}} = \sqrt{\dfrac{119 - 10(2.7)^2}{10-1}}$

$= \sqrt{\dfrac{46.1}{9}} \approx 2.26$

Diet B:

$\bar{x} = \dfrac{2 + 1 + 1 + 2 + 3 + 2 + 1 + 0 + 1 + 0}{10} = \dfrac{13}{10} = 1.3$

$\Sigma x^2 = 2^2 + 1^2 + 1^2 + 2^2 + 3^2 + 2^2 + 1^2$
$\qquad\qquad + 0^2 + 1^2 + 0^2$

$= 25$

$s = \sqrt{\dfrac{\Sigma x^2 - n\bar{x}^2}{n-1}} = \sqrt{\dfrac{25 - 10(1.3)^2}{10-1}} = \sqrt{\dfrac{8.1}{9}} \approx .95$

a.   Diet A produced the greater mean gain.

b.   "Most consistent" means least variable. Diet B has the smaller standard deviation and so produced the most consistent gain.

37. a.   Arrange the data for diet A in ascending order: 0, 1, 1, 1, 1, 3, 4, 4, 5, 7

Minimum = 0

Maximum = 7

Median $= \dfrac{1+3}{2} = 2$

$n = 10$

$.25 \cdot 10 = 2.5$ rounds up to 3.

$Q_1 = 1$

$.75 \cdot 10 = 7.5$ rounds up to 8.

$Q_3 = 4$

Diet A

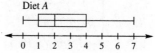

Arrange the data for diet B in ascending order: 0, 0, 1, 1, 1, 1, 2, 2, 2, 3

Minimum = 0

Maximum = 3

Median = 1

$n = 10$

$.25 \cdot 10 = 2.5$ rounds up to 3.

$Q_1 = 1$

$.75 \cdot 10 = 7.5$ rounds up to 8.

$Q_3 = 2$

Diet B

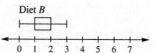

b.   Answers vary. Diet A had a greater mean gain but Diet B had more consistent results.

38.

Interval	f	x	xf	x^2	fx^2
75–149.999999	102	112.5	11,475.0	12,656.25	1,290,937.50
150–224.999999	34	187.5	6375.0	35,156.25	1,195,312.50
225–299.999999	13	262.5	3412.5	68,906.25	895,751.25
300–374.999999	8	337.5	2700.0	113,906.25	911,250.00
375–449.999999	3	412.5	1237.5	170,156.25	510,468.75
450–524.999999	0	487.5	0	237,656.25	0
525–599.999999	1	562.5	562.5	316,406.25	316,406.25
Totals:	161		25,762.5		5,120,126.25

$$\bar{x} = \frac{25,762.50}{161} \approx \$160.0155 \text{ million}$$

$$s = \sqrt{\frac{\Sigma fx^2 - n\bar{x}^2}{n-1}} = \sqrt{\frac{5,120,126.25 - 161(160.0155)^2}{160}} \approx 78.967$$

39. a. March 2008   March 2009

x	x^2
115	13,225
73	5,329
47	2209
130	16,900
82	6724
66	4356
31	961
22	484
8	64
14	196
19	361
7	49
614	50,858

x	x^2
68	4624
46	2116
24	576
81	6561
55	3025
43	1849
32	1024
15	225
3	9
12	144
13	169
5	25
397	20,347

$$\bar{x} = \frac{614}{12} \approx 51.17$$

$$s = \sqrt{\frac{\Sigma x^2 - n\bar{x}^2}{n-1}} = \sqrt{\frac{50,858 - 12(51.17)^2}{12-1}}$$
$$\approx 42.04$$

$$\bar{x} = \frac{397}{12} \approx 33.08$$

$$s = \sqrt{\frac{\Sigma x^2 - n\bar{x}^2}{n-1}} = \sqrt{\frac{20,347 - 12(33.08)^2}{12-1}}$$
$$\approx 25.61$$

b. Chrysler LLC is closest to the mean sales in 2008. Hyundai Motor America is closest to the mean sales in 2009.

40. March 2008:

Arrange the data in ascending order:

7, 8, 14, 19, 22, 31, 47, 66, 73, 82, 115, 130

Minimum: 7, maximum: 130

$$Q_2 = \frac{31+47}{2} = \frac{78}{2} = 39$$

$n = 12$, so $Q_1 =$ the third data entry, 14, and

$Q_3 =$ the ninth data entry, 73.

March 2009:

Arrange the data in ascending order:

3, 5, 12, 13, 15, 24, 32, 43, 46, 55, 68, 81

Minimum: 3, maximum: 81

$$Q_2 = \frac{24+32}{2} = \frac{56}{2} = 28$$

$n = 12$, so $Q_1 =$ the third data entry, 12, and

$Q_3 =$ the ninth data entry, 46.

41. a.

b. It appears that sales decreased. Explanations will vary.

42. a. $\mu = 100$ and $\sigma = 15$

More than 130

We need to find the area to the right of $x = 130$.

$$z = \frac{130-100}{15} = 2$$

The area is .4773. The area to the right of $x = 130$ is $.5 - .4773 = .0227$.

Therefore, 2.27% of the people score more than 130.

b. Less than 85

For $x = 85$, $z = \frac{85-100}{15} = -1$.

The area to the left of $z = -1$ is equal to the area to the right of $z = 1$, which is $.5 - .3413 = .1587$.

Therefore, 15.87% of the people score less than 85.

c. Between 85 and 115

For $x = 115$, $z = \frac{115-100}{15} = 1$.

The area is .3413. Since 85 and 115 are equidistant from the mean, the total required area is $2(.3413) = .6826$. Thus, 68.26% of the people score between 85 and 115.

43. Let x represent the number of ounces of juice in a can.

$\mu = 32.1$, $\sigma = .1$

Find the z-score for $x = 32$.

$$z = \frac{x-\mu}{\sigma} = \frac{32-32.1}{.1} = -1$$

As shown in the solution for Exercise 42(b), the area to the left of $z = -1$ is $.5 - .3413 = .1587$. Therefore, 15.87% of the cartons contain less than a quart.

44. $n = 1000$, $p = .164$, $1 - p = .836$

$$\mu = np = 1000(.164) = 164$$

$$\sigma = \sqrt{np(1-p)} = \sqrt{164(.836)} \approx 11.7091$$

$P(\text{less than } 150) = P(x \le 150) = P(x < 149.5)$

For $x = 149.5$, $z = \frac{149.5-164}{11.7091} \approx -1.24$ and

$A = .3925$

$P(\text{less than } 150) = .5 - .3925 = .1075$

45. $n = 1000$, $p = .017$, $1 - p = .983$

$$\mu = np = 1000(.017) = 17$$

$$\sigma = \sqrt{np(1-p)} = \sqrt{17(.983)} \approx 4.0879$$

$P(\text{between 12 and 25}) = P(12 \le x \le 25)$
$= P(11.5 < x < 25.5)$

For $x = 11.5$, $z = \frac{11.5-17}{4.0879} \approx -1.35$ and

$A = .4115$

For $x = 25.5$, $z = \frac{25.5-17}{4.0879} \approx 2.08$ and

$A = .4812$

$P(12 \le x \le 25) = .4115 + .4812 = .8927$

46. $n = 1000$, $p = .06$, $1 - p = .94$

$$\mu = np = 1000(.06) = 60$$

$$\sigma = \sqrt{np(1-p)} = \sqrt{60(.94)} \approx 7.5100$$

$P(\text{more than } 50) = P(x > 50) = P(x > 50.5)$

For $x = 50.5$, $z = \frac{50.5-60}{7.5100} \approx -1.26$ and

$A = .3962$

$P(\text{more than } 50) = .5 + .3962 = .8962$

Case 10 Statistics in the Law: The Castañeda Decision

1. $n = 870, p = .791, 1 - p = .209$

   $\mu = np = 870(.791) = 688.17$

   $\sigma = \sqrt{np(1-p)} = \sqrt{870(.791)(.209)} \approx 11.9928$

   $z = \dfrac{339 - 688.17}{11.9928} \approx -29.1$

2. Answers vary. Possible answer: The courts' figure of 1 in 10^{140} probably comes from the normal approximation to the binomial distribution.

3. **a.** $n = 220, p = .791$

   $\mu = np = 220(.791) \approx 174$

   Of 220 jurors from this population, we should expect 174 Mexican-American jurors.

   b. $n = 220, p = .791, 1 - p = .209$

   $\sigma = \sqrt{np(1-p)} = \sqrt{220(.791)(.209)} \approx 6.03$

   c. $z = \dfrac{100 - 174}{6.03} \approx -12.3$

   d. The probability at -12.3 standard deviations from the mean is less than .004, the smallest probability in the table.

4. **a.** $n = 112, \; p = \dfrac{88}{294}, \; (1 - p) = \dfrac{206}{294}$

   $\mu = np = 112\left(\dfrac{88}{294}\right) \approx 33.5238$

   $\sigma = \sqrt{np(1-p)} = \sqrt{112\left(\dfrac{88}{294}\right)\left(\dfrac{206}{294}\right)}$

   ≈ 4.8466

   $z = \dfrac{6 - 33.5238}{4.8466} \approx -5.7$

   The expected number of women in management is about -5.7 standard deviations from the mean.

   b. Answers vary. Possible answer: Yes, it appears to be purposeful discrimination. The women in management do not represent the women in the employee body.

Chapter 11 Differential Calculus

Section 11.1 Limits

1. a. By reading the graph, as x gets closer to 3 from the left or the right, $f(x)$ approaches 8, so $\lim_{x \to 3} f(x) = 8$.

b. As x gets closer to -1 from the left or right, $f(x)$ approaches 0, so $\lim_{x \to -1} f(x) = 0$.

3. a. By reading the graph, as x gets closer to -3 from the left $f(x)$ approaches -1. As x gets closer to -3 from the right, $f(x)$ approaches 2. Since these two values of $f(x)$ are not equal, $\lim_{x \to -3} f(x)$ does not exist.

b. By reading the graph, as x gets closer to 2 from the left or right, $f(x)$ approaches 2 so $\lim_{x \to 2} f(x) = 2$.

5. a. By reading the graph, as x gets closer to 3 from the left, $f(x)$ becomes infinitely small, and from the right, $f(x)$ becomes infinitely large. Since $f(x)$ has no limit as x approaches 3 from both the right and left, $\lim_{x \to 3} f(x)$ does not exist.

b. By reading the graph, as x gets closer to 2 from the left or the right, $f(x)$ gets approaches -4 so $\lim_{x \to 2} f(x) = -4$.

7. a. By reading the graph, as x gets closer to 1 from the left or right, $g(x)$ approaches 6, so $\lim_{x \to 1} g(x) = 6$.

b. By reading the graph, as x approaches 0 from the left or right, $g(x)$ approaches 0, so $\lim_{x \to 0} g(x) = 0$.

9. $\lim_{x \to 1} F(x)$ in Exercise 2(a) exists because, as x gets closer to 1 from the left or the right, $F(x)$ gets closer to 2, a single number. On the other hand, $\lim_{x \to -3} f(x)$ in Exercise 3(a) does not exist because, as x approaches -3 from the left, $f(x)$ gets closer to -1, but, as x approaches -3 from the right, $f(x)$ approaches another number, 2.

11. $\lim_{x \to 1} \dfrac{\ln x}{x - 1}$

x	$\dfrac{\ln x}{x-1}$
.9	1.0536
.99	1.0050
.999	1.0005
1	
1.001	.9995
1.01	.9950
1.1	.9531

As x gets closer to 1 from the left or right, the value of $\dfrac{\ln x}{x - 1}$ gets closer to 1, so $\lim_{x \to 1} \dfrac{\ln x}{x - 1} = 1$. Verify this by examining the graph of the function.

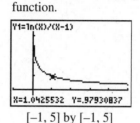

$[-1, 5]$ by $[-1, 5]$

13. $\lim_{x \to 0} \dfrac{e^{3x} - 1}{x}$

x	$\dfrac{e^{3x} - 1}{x}$
$-.1$	2.5918
$-.01$	2.9554
$-.001$	2.9955
0	
.001	3.0045
.01	3.0455
.1	3.4986

As x gets closer to 0 from the left or right, the value of $\dfrac{e^{3x} - 1}{x}$ gets closer to 3, so $\lim_{x \to 0} \dfrac{e^{3x} - 1}{x} = 3$. Verify this by examining the graph of the function.

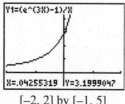

$[-2, 2]$ by $[-1, 5]$

15. $\lim\limits_{x\to 0}\left(x\cdot\ln|x|\right)$

x	$x\cdot\ln|x|$
−.1	.2303
−.01	.0461
−.001	.0069
0	
.001	−.0069
.01	−.0461
.1	−.2303

As x gets closer to 0 from the left or right, the value of $x\cdot\ln|x|$ gets closer to 0, so

$\lim\limits_{x\to 0}\left(x\cdot\ln|x|\right)=0.$ Verify this by examining the graph of the function.

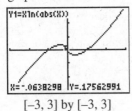

[−3, 3] by [−3, 3]

17. $\lim\limits_{x\to 3}\dfrac{x^3-3x^2-4x+12}{x-3}$

x	$\dfrac{x^3-3x^2-4x+12}{x-3}$
2.9	4.4100
2.99	4.9401
2.999	4.9940
3	
3.001	5.0060
3.01	5.0601
3.1	5.6100

As x gets closer to 3 from the left or right, the

value of $\dfrac{x^3-3x^2-4x+12}{x-3}$ gets closer to 5, so

$\lim\limits_{x\to 3}\dfrac{x^3-3x^2-4x+12}{x-3}=5.$ Verify this by examining the graph of the function.

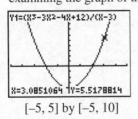

[−5, 5] by [−5, 10]

19. $\lim\limits_{x\to -2}\dfrac{x^4+2x^3-x^2+3x+1}{x+2}$

x	$\dfrac{x^4+2x^3-x^2+3x+1}{x+2}$
−2.1	87.839
−2.01	898.8894
−2.001	8998.9889
−2	
−1.999	−9000.989
−1.99	−900.8906
−1.9	−89.959

As x gets closer to −2 from the left, the value of $\dfrac{x^4+2x^3-x^2+3x+1}{x+2}$ becomes infinitely large, and from the right, the value becomes infinitely small, so

$\lim\limits_{x\to -2}\dfrac{x^4+2x^3-x^2+3x+1}{x+2}$ does not exist.

Verify this by examining the graph of the function.

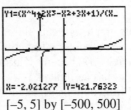

[−5, 5] by [−500, 500]

For exercises 21–27, $\lim\limits_{x\to 4}f(x)=25$ and $\lim\limits_{x\to 4}g(x)=10.$

21. $\lim\limits_{x\to 4}[f(x)-g(x)]$

Use property 2.

$\lim\limits_{x\to 4}[f(x)-g(x)]=\lim\limits_{x\to 4}f(x)-\lim\limits_{x\to 4}g(x)$
$=25-10=15$

23. $\lim\limits_{x\to 4}\dfrac{f(x)}{g(x)}$

Use property 4.

$\lim\limits_{x\to 4}\dfrac{f(x)}{g(x)}=\dfrac{\lim\limits_{x\to 4}f(x)}{\lim\limits_{x\to 4}g(x)}=\dfrac{25}{10}=\dfrac{5}{2}$

25. $\lim\limits_{x\to 4}\sqrt{f(x)}=\lim\limits_{x\to 4}[f(x)]^{1/2}$

Use property 5.

$\lim\limits_{x\to 4}\sqrt{f(x)}=\left[\lim\limits_{x\to 4}f(x)\right]^{1/2}=25^{1/2}=5$

27. $\lim\limits_{x\to 4}\dfrac{f(x)+g(x)}{2g(x)}$

$=\dfrac{\lim\limits_{x\to 4}[f(x)+g(x)]}{\lim\limits_{x\to 4}[2g(x)]}$ Property 4

$=\dfrac{\lim\limits_{x\to 4}f(x)+\lim\limits_{x\to 4}g(x)}{2\lim\limits_{x\to 4}g(x)}$ Property 1 and L.C.F.

$=\dfrac{25+10}{2(10)}=\dfrac{35}{20}=\dfrac{7}{4}$

29. a. $f(x)\begin{cases}3-x & \text{if } x<-2\\ x+2 & \text{if } -2\le x<2\\ 1 & \text{if } x\ge 2\end{cases}$

The graph of $f(x)=3-x$ if $x<-2$ is the ray through $(-4, 7)$ and $(-2, 5)$, with excluded (open) endpoint $(-2, 5)$. The graph of $f(x)=x+2$ if $-2\le x<2$ is the segment with endpoint $(-2, 0)$ and open endpoint $(2, 4)$. The graph of $f(x)$ if $x\ge 2$ is the horizontal ray to the right with endpoint $(2, 1)$.

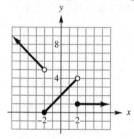

b. $\lim\limits_{x\to -2}f(x)$

As x gets closer to -2 from the left, $f(x)=3-x$ gets closer to $3-(-2)=5$ because $f(x)=3-x$ if $x<-2$. As x gets closer to -2 from the right, $f(x+2)$ gets closer to 0 because $f(x)=x+2$ if $-2\le x<2$. $f(x)$ does not get closer to a single real number. Therefore $\lim\limits_{x\to -2}f(x)$

does not exist.

c. $\lim\limits_{x\to 1}f(x)$

$f(x)$ is the polynomial function $f(x)=x+2$ if $-2\le x<2$, so $\lim\limits_{x\to 1}f(x)=\lim\limits_{x\to 1}f(1)=1+2=3.$

d. $\lim\limits_{x\to 2}f(x)$

As x gets closer to 2 from the left, $f(x)=x+2$ gets closer to $2+2=4$ because $f(x)=x+2$ if $-2\le x<2$. As x gets closer to 2 from the right, $f(x)=1$ gets closer to 1 because $f(x)=1$ if $x\ge 2$. $f(x)$ does not get closer to a single real number. Therefore, $\lim\limits_{x\to 2}f(x)$ does not exist.

31. $\lim\limits_{x\to 2}\left(3x^3-4x^2-5x+2\right)$

$=3(2)^3-4(2)^2-5(2)+2=0$

33. $\lim\limits_{x\to 3}\dfrac{4x+7}{10x+1}=\dfrac{\lim\limits_{x\to 3}(4x+7)}{\lim\limits_{x\to 3}(10x+1)}=\dfrac{4(3)+7}{10(3)+1}=\dfrac{19}{31}$

35. $\lim\limits_{x\to 5}\dfrac{x^2-25}{x-5}=\lim\limits_{x\to 5}\dfrac{(x-5)(x+5)}{x-5}$

$=\lim\limits_{x\to 5}(x+5)=5+5=10$

37. $\lim\limits_{x\to 4}\dfrac{x^2-x-12}{x-4}=\lim\limits_{x\to 4}\dfrac{(x-4)(x+3)}{x-4}$

$=\lim\limits_{x\to 4}(x+3)=4+3=7$

39. $\lim\limits_{x\to 2}\dfrac{x^2-5x+6}{x^2-6x+8}=\lim\limits_{x\to 2}\dfrac{(x-2)(x-3)}{(x-2)(x-4)}$

$=\lim\limits_{x\to 2}\dfrac{x-3}{x-4}=\dfrac{\lim\limits_{x\to 2}(x-3)}{\lim\limits_{x\to 2}(x-4)}$

$=\dfrac{2-3}{2-4}=\dfrac{1}{2}$

41. $\lim\limits_{x\to 4}\dfrac{(x+4)^2(x-5)}{(x-4)(x+4)^2}=\lim\limits_{x\to 4}\dfrac{x-5}{x-4}$

$=\dfrac{-1}{0}$ Undefined

The limit does not exist.

43. $\lim\limits_{x\to 3}\sqrt{x^2-3}=\sqrt{\lim\limits_{x\to 3}(x^2-3)}=\sqrt{3^2-3}=\sqrt{6}$

45. $\lim\limits_{x\to 4}\dfrac{-6}{(x-4)^2}=\dfrac{\lim\limits_{x\to 4}(-6)}{\lim\limits_{x\to 4}(x-4)^2}$

$=\dfrac{-6}{0}$ Undefined

The limit does not exist.

47. $\lim\limits_{x\to 0}\dfrac{\frac{1}{x+3}-\frac{1}{3}}{x}=\lim\limits_{x\to 0}\left(\dfrac{1}{x+3}-\dfrac{1}{3}\right)\left(\dfrac{1}{x}\right)$

$\qquad =\lim\limits_{x\to 0}\dfrac{3-x-3}{3(x+3)(x)}=\lim\limits_{x\to 0}\dfrac{-x}{3(x+3)x}$

$\qquad =\lim\limits_{x\to 0}\dfrac{-1}{3(x+3)}=-\dfrac{1}{9}$

49. $\lim\limits_{x\to 25}\dfrac{\sqrt{x}-5}{x-25}=\lim\limits_{x\to 25}\dfrac{\sqrt{x}-5}{x-25}\cdot\dfrac{\sqrt{x}+5}{\sqrt{x}+5}$

$\qquad =\lim\limits_{x\to 25}\dfrac{x-25}{(x-25)\left(\sqrt{x}+5\right)}$

$\qquad =\lim\limits_{x\to 25}\dfrac{1}{\sqrt{x}+5}=\dfrac{1}{\sqrt{25}+5}$

$\qquad =\dfrac{1}{5+5}=\dfrac{1}{10}$

51. $\lim\limits_{x\to 5}\dfrac{\sqrt{x}-\sqrt{5}}{x-5}=\lim\limits_{x\to 5}\dfrac{\sqrt{x}-\sqrt{5}}{x-5}\cdot\dfrac{\sqrt{x}+\sqrt{5}}{\sqrt{x}+\sqrt{5}}$

$\qquad =\lim\limits_{x\to 5}\dfrac{x-5}{(x-5)(\sqrt{x}+\sqrt{5})}$

$\qquad =\lim\limits_{x\to 5}\dfrac{1}{\sqrt{x}+\sqrt{5}}$

$\qquad =\dfrac{1}{2\sqrt{5}}$ or $\dfrac{\sqrt{5}}{10}$

53. $P(s)=\dfrac{105s}{s+7}$

a. $P(1)=\dfrac{105(1)}{1+7}=\dfrac{105}{8}=13.125$

b. $P(13)=\dfrac{105(13)}{13+7}=\dfrac{1365}{20}=68.25$

c. $\lim\limits_{s\to 13}P(s)=\lim\limits_{s\to 13}\dfrac{105s}{s+7}=\dfrac{105(13)}{13+7}=68.25$

55. $c(x)=150,000+3x$

$\overline{c}(x)=\dfrac{c(x)}{x}$

a. $\overline{c}(1000)=\dfrac{150,000+3(1000)}{1000}$

$\qquad =\dfrac{153,000}{1000}=\$153$

b. $\overline{c}(10,000)=\dfrac{150,000+3(10,000)}{10,000}$

$\qquad =\dfrac{180,000}{10,000}=\$18$

c. $\lim\limits_{x\to 100,000}\overline{c}(x)=\lim\limits_{x\to 100,000}\dfrac{150,000+3x}{x}$

$\qquad =\dfrac{150,000+3(100,000)}{100,000}$

$\qquad =\dfrac{450,000}{100,000}=\$4.50$

57. The curves are continuous.

a. $\lim\limits_{x\to 2030}C(x)=C(2030)=1.5$

b. $\lim\limits_{x\to 2015}I(x)=I(2015)=1.2$

c. $\lim\limits_{x\to 2045}C(x)-I(x)$

$\qquad =\lim\limits_{x\to 2045}C(x)-\lim\limits_{x\to 2045}I(x)$

$\qquad =C(2045)-I(2045)=1.5-1.5=0$

d. $\lim\limits_{x\to 2045}C(x)+I(x)$

$\qquad =\lim\limits_{x\to 2045}C(x)+\lim\limits_{x\to 2045}I(x)$

$\qquad =C(2045)+I(2045)=1.5+1.5=3$

59. a. $P(x)=\begin{cases}.99 & \text{if}\quad 0\le x\le 20\\ 1.06 & \text{if}\quad 20<x\le 21\\ 1.13 & \text{if}\quad 21<x\le 22\\ 1.20 & \text{if}\quad 22<x\le 23\end{cases}$

b.

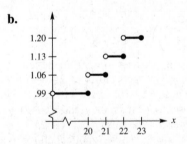

c. The $\lim\limits_{x\to 10}P(x)=.99$ since 10 is between 0 and 20.

d. The $\lim\limits_{x\to 20}P(x)$ does not exist because the limit as x approaches 20 from the left is not equal to the limit as x approaches 20 from the right.

e. The $\lim\limits_{x\to 22.5}P(x)=1.20$ since 22.5 is between 22 and 23.

Section 11.2 One-Sided Limits and Limits Involving Infinity

1. As x approaches 2 from the left, $f(x)$ approaches 3. Thus, $\lim\limits_{x \to 2^-} f(x) = 3$.

3. As x approaches -3 from the left, $f(x)$ approaches -2. Thus, $\lim\limits_{x \to -3^-} f(x) = -2$.

5. As x approaches -1 from the left, $f(x)$ approaches 6. Thus, $\lim\limits_{x \to -1^-} f(x) = 6$.

7. As x approaches -4 from the left, $f(x)$ approaches 0. Thus, $\lim\limits_{x \to -4^-} f(x) = 0$.

9. a. As x approaches -2 from the left, $f(x)$ approaches 0. Thus $\lim\limits_{x \to -2^-} f(x) = 0$.

  b. As x approaches 0 from the right, $f(x)$ approaches 1. Thus, $\lim\limits_{x \to 0^+} f(x) = 1$.

  c. As x approaches 3 from the left, $f(x)$ approaches -1. Thus, $\lim\limits_{x \to 3^-} f(x) = -1$.

  d. As x approaches 3 from the right, $f(x)$ approaches -1. Thus, $\lim\limits_{x \to 3^+} f(x) = -1$.

11. a. As x approaches -2 from the left, $f(x)$ approaches 2. Thus $\lim\limits_{x \to -2^-} f(x) = 2$.

  b. As x approaches 0 from the right, $f(x)$ is undefined. Thus, $\lim\limits_{x \to 0^+} f(x)$ is undefined.

  c. As x approaches 3 from the left, $f(x)$ approaches 0. Thus, $\lim\limits_{x \to 3^-} f(x) = 0$.

  d. As x approaches 3 from the right, $f(x)$ approaches 2. Thus, $\lim\limits_{x \to 3^+} f(x) = 2$.

13. $\lim\limits_{x \to 2^+} \sqrt{x^2 - 4} = \lim\limits_{x \to 2^+} \left(x^2 - 4\right)^{1/2}$
$$= \left[\lim\limits_{x \to 2^+} \left(x^2 - 4\right)\right]^{1/2} = 0^{1/2} = 0$$

15. $\lim\limits_{x \to 0^+} \sqrt{x} + x + 1$
$$= \lim\limits_{x \to 0^+} \sqrt{x} + \lim\limits_{x \to 0^+} x + \lim\limits_{x \to 0^+} 1$$
$$= 0 + 0 + 1 = 1$$

17. $\lim\limits_{x \to -2^+} \left(x^3 - x^2 - x + 1\right)$
$$= \lim\limits_{x \to -2^+} x^3 - \lim\limits_{x \to -2^+} x^2 - \lim\limits_{x \to -2^+} x + \lim\limits_{x \to -2^+} 1$$
$$= -8 - 4 + 2 + 1 = -9$$

19. $\lim\limits_{x \to -5^+} \dfrac{\sqrt{x+5} + 5}{x^2 - 5}$
$$= \lim\limits_{x \to -5^+} \left[\frac{\sqrt{x+5} + 5}{x^2 - 5} \cdot \frac{\sqrt{x+5} - 5}{\sqrt{x+5} - 5}\right]$$
$$= \lim\limits_{x \to -5^+} \left[\frac{x - 20}{x^2\sqrt{x+5} - 5\sqrt{x+5} - 5x^2 + 25}\right]$$
$$= \lim\limits_{x \to -5^+} \left[\frac{\dfrac{x - 20}{x^2}}{\sqrt{x+5} - \dfrac{5\sqrt{x+5}}{x^2} - 5 + \dfrac{25}{x^2}}\right]$$
$$= \frac{-1}{-5 + 1} = \frac{1}{4}$$

21.

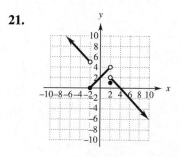

  a. As x approaches -2 from the left, $f(x)$ approaches $3 - (-2) = 5$. Thus,
  $$\lim\limits_{x \to -2^-} f(x) = 5.$$

  b. As x approaches -2 from the right, $f(x)$ approaches $-2 + 2 = 0$. Thus,
  $$\lim\limits_{x \to -2^+} f(x) = 0.$$

  c. Since $f(x)$ does not approach the same value as x approaches -2 from the left and right, $\lim\limits_{x \to -2} f(x)$ does not exist.

23. a. As x approaches 2 from the left and the right, $f(x)$ approaches positive infinity. Thus, $\lim\limits_{x \to 2} f(x) = \infty$.

b. As x approaches -2 from the right, $f(x)$ approaches positive infinity. Thus,
$$\lim_{x \to -2^+} f(x) = \infty.$$

c. As x approaches -2 from the left, $f(x)$ approaches negative infinity. Thus,
$$\lim_{x \to -2^-} f(x) = -\infty.$$

25. $\lim_{x \to 4} \dfrac{-6}{(x-4)^2} = -\infty$

27. $\lim_{x \to -1^+} \dfrac{2}{1+x} = \infty$

29. $\lim_{x \to \infty} f(x) = \infty$ and $\lim_{x \to -\infty} f(x) = 0$

31. $\lim_{x \to \infty} f(x) = 2$ and $\lim_{x \to -\infty} f(x) = -1$

33. $\lim_{x \to \infty} f(x) = \infty$ and $\lim_{x \to -\infty} f(x) = -\infty$

35. $\lim_{x \to \infty} \left[\sqrt{x^2 + 1} - (x+1) \right] = -1$

37. $\lim_{x \to -\infty} \dfrac{x^{2/3} - x^{4/3}}{x^3} = 0$

39. $\lim_{x \to -\infty} e^{1/x} = 1$

41. $\lim_{x \to \infty} \dfrac{\ln x}{x} = 0$

43. $\lim_{x \to \infty} \dfrac{10x^2 - 5x + 8}{3x^2 + 8x - 29} = \lim_{x \to \infty} \dfrac{10 - \dfrac{5}{x} + \dfrac{8}{x^2}}{3 + \dfrac{8}{x} - \dfrac{29}{x^2}}$
$$= \dfrac{10 - 0 + 0}{3 + 0 - 0} = \dfrac{10}{3}$$

45. $\lim_{x \to -\infty} \dfrac{4x^2 + 11x + 21}{7 + 67x - x^2} = \lim_{x \to -\infty} \dfrac{4 + \dfrac{11}{x} + \dfrac{21}{x^2}}{\dfrac{7}{x^2} + \dfrac{67}{x} - 1}$
$$= \dfrac{4 + 0 + 0}{0 + 0 - 1} = -4$$

47. $\lim_{x \to -\infty} \dfrac{8x^5 + 7x^4 - 10x^3}{25 - 4x^5} = \lim_{x \to -\infty} \dfrac{8 + \dfrac{7}{x} - \dfrac{10}{x^2}}{\dfrac{25}{x^5} - 4}$
$$= \dfrac{8 + 0 - 0}{0 - 4} = -2$$

49. $\lim_{x \to -\infty} \dfrac{(4x-1)(x+2)}{5x^2 - 3x + 1} = \lim_{x \to -\infty} \dfrac{4x^2 + 7x - 2}{5x^2 - 3x + 1}$
$$= \lim_{x \to -\infty} \dfrac{4 + \dfrac{7}{x} - \dfrac{2}{x^2}}{5 - \dfrac{3}{x} + \dfrac{1}{x^2}}$$
$$= \dfrac{4 + 0 - 0}{5 - 0 + 0} = \dfrac{4}{5}$$

51. $\lim_{x \to \infty} \left(5x - \dfrac{1}{4x^2} \right) = \lim_{x \to \infty} 5x - \lim_{x \to \infty} \dfrac{1}{4x^2}$
$$= \infty - 0 = \infty$$

53. $\lim_{x \to -\infty} \left(\dfrac{18x}{x-9} - \dfrac{3x}{x+4} \right)$
$$= \lim_{x \to -\infty} \left(\dfrac{18x}{x-9} \right) - \lim_{x \to -\infty} \left(\dfrac{3x}{x+4} \right)$$
$$= \lim_{x \to -\infty} \dfrac{18}{1 - \dfrac{9}{x}} - \lim_{x \to -\infty} \dfrac{3}{1 + \dfrac{4}{x}}$$
$$= \dfrac{18}{1 - 0} - \dfrac{3}{1 + 0} = 15$$

55. $\lim_{x \to \infty} \dfrac{x}{|x|}$

Since x approaches positive infinity, $|x| = x$.

Thus, $\lim_{x \to \infty} \dfrac{x}{|x|} = \lim_{x \to \infty} \dfrac{x}{x} = \lim_{x \to \infty} 1 = 1$.

57. $\lim_{x \to -\infty} \dfrac{x}{|x| + 1}$

Since x approaches negative infinity, $|x| = -x$.

Thus, $\lim_{x \to -\infty} \dfrac{x}{|x| + 1} = \lim_{x \to -\infty} \dfrac{x}{-x + 1}$
$$= \lim_{x \to \infty} \dfrac{1}{-1 + \dfrac{1}{x}} = \dfrac{1}{-1} = -1$$

59. a. $N(t) = 71.8e^{-8.96e^{-.0685t}}$

$N(65) = 71.8e^{-8.96e^{-.0685(65)}}$

$= 71.8e^{-8.96e^{-4.4525}}$

≈ 64.68

b. $\lim\limits_{t \to \infty} N(t) =$

$\lim\limits_{t \to \infty} 71.8e^{-8.96e^{-.0685t}}$

$= \lim\limits_{t \to \infty} 71.8e^{-8.96e^{-\infty}}$

$= \lim\limits_{t \to \infty} 71.8e^{-8.96(0)} = \lim\limits_{t \to \infty} 71.8e^{0}$

$= \lim\limits_{t \to \infty} 71.8 \approx 71$ teeth

61. $\lim\limits_{h \to \infty} A(h) = \lim\limits_{h \to \infty} \dfrac{.17h}{h^2 + 2} = \lim\limits_{h \to \infty} \dfrac{.17}{h + \dfrac{2}{h}}$

$= \dfrac{.17}{\infty + 0} = \dfrac{.17}{\infty} = 0$

The concentration of the drug approaches 0 as time increases.

Section 11.3 Rates of Change

1.   The average rate of change for $f(x) = x^2 + 2x$

between $x = 0$ and $x = 6$ is $\dfrac{f(6) - f(0)}{6 - 0}$.

$f(6) = 6^2 + 2(6) = 36 + 12 = 48$

$f(0) = 0^2 + 2(0) = 0$

The average rate of change is $\dfrac{48 - 0}{6 - 0} = \dfrac{48}{6} = 8$.

3. $f(x) = 2x^3 - 4x^2 + 6$

between $x = -1$ and $x = 2$.
The average rate of change is

$\dfrac{f(2) - f(-1)}{2 - (-1)} = \dfrac{6 - (-0)}{3} = \dfrac{6}{3} = 2$.

5. $f(x) = \sqrt{x}$ between $x = 1$ and $x = 9$.
The average rate of change is

$\dfrac{f(9) - f(1)}{9 - 1} = \dfrac{3 - 1}{9 - 1} = \dfrac{2}{8} = \dfrac{1}{4}$.

7. $f(x) = \dfrac{1}{x - 1}$ between $x = -2$ and $x = 0$
The average rate of change is

$\dfrac{f(0) - f(-2)}{0 - (-2)} = \dfrac{-1 - \left(-\frac{1}{3}\right)}{2} = \dfrac{-1 + \frac{1}{3}}{2}$

$= \dfrac{-\frac{2}{3}}{2} = -\dfrac{1}{3}$

9.   Let $S(x)$ be the function representing sales in thousands of dollars of x thousand catalogs.

a. $S(20) = 40$
$S(10) = 30$

Average rate of change $= \dfrac{S(20) - S(10)}{20 - 10}$

$= \dfrac{40 - 30}{10} = 1$

As catalog distribution changes from 10,000 to 20,000, sales will have an average increase of $1000 for each additional 1000 catalogs distributed.

b. $S(30) = 46$
$S(20) = 40$

Average rate of change $= \dfrac{S(30) - S(20)}{30 - 20}$

$= \dfrac{46 - 40}{10} = \dfrac{6}{10} = \dfrac{3}{5}$

As catalog distribution changes from 20,000 to 30,000, sales will have an average

increase of $\left(\dfrac{3}{5}\right)(1000)$ or $600 for each

additional 1000 catalogs distributed.

c. $S(40) = 50$
$S(30) = 46$

Average rate of change $= \dfrac{S(40) - S(30)}{40 - 30}$

$= \dfrac{50 - 46}{10} = \dfrac{4}{10} = \dfrac{2}{5}$

As catalog distribution changes from 30,000 to 40,000, sales will have an average

increase of $\left(\dfrac{2}{5}\right)(1000)$ or $400 for each

additional 1000 catalogs distributed.

d.   As more catalogs are distributed, sales increase at a smaller and smaller rate.

e.   It might be that the market for items in the catalog is becoming saturated.

11. Let $R(t)$ represent the revenue in year t.

    a. Average rate of change in revenue from 1996 to 2002 is

$$\frac{R(2002) - R(1996)}{2002 - 1996} = \frac{17 - 14}{6} = \$.5 \text{ billion}$$

    b. Average rate of change in revenue from 2002 to 2007 is

$$\frac{R(2007) - R(2002)}{2007 - 2002} = \frac{21 - 17}{5} = \$.8 \text{ billion}$$

13. Let $C(t)$ represent the carbon monoxide concentration in parts per million for 8 hours of exposure in year t.

    a. Average rate of change in carbon monoxide levels from 1992 to 1999 is

$$\frac{C(1999) - C(1992)}{1999 - 1992} = \frac{3.9 - 5.4}{7} \approx -.214$$

    The average rate of change in carbon monoxide levels from 1992 to 1999 was about $-.214\%$.

    b. Average rate of change in carbon monoxide levels from 2004 to 2007 is

$$\frac{C(2007) - C(2004)}{2007 - 2004} = \frac{2.0 - 2.5}{3} \approx -.167$$

    The average rate of change in carbon monoxide levels from 2004 to 2007 was about $-.167\%$.

15. Let $f(t)$ represent the value of the S&P index for t months.

    a. Average rate of change of the S&P index from month 1 to month 10:

$$\frac{f(10) - f(1)}{10 - 1} = \frac{701 - 1386}{9} \approx -76.11$$

    The average rate of change of the S&P index from month 1 to month 10 was about -76.11.

    b. Average rate of change of the S&P index from month 4 to month 12:

$$\frac{f(12) - f(4)}{12 - 4} = \frac{874 - 1278}{8} = -50.5$$

    The average rate of change of the S&P index from month 4 to month 12 was -50.5.

    c. Average rate of change of the S&P index from month 10 to month 12:

$$\frac{f(12) - f(10)}{12 - 10} = \frac{874 - 701}{2} = 86.5$$

    The average rate of change of the S&P index from month 10 to month 12 was 86.5.

17. The average rate of change of y as x changes from a to b is found by $\dfrac{y(b) - y(a)}{b - a}$.

The instantaneous rate of change of y as $x = a$ is found by $\lim\limits_{h \to 0} \dfrac{y(a + h) - y(a)}{h}$.

19. $s(t) = 2.2t^2$

Time Interval	Average Speed
$t = 5$ to $t = 5.1$	$\frac{s(5.1) - s(5)}{5.1 - 5} = \frac{57.22 - 55}{.1}$ $= 22.2$
$t = 5$ to $t = 5.01$	$\frac{s(5.01) - s(5)}{5.01 - 5} = 22.022$
$t = 5$ to $t = 5.001$	$\frac{s(5.001) - s(5)}{5.001 - 5} = 22.0022$

This chart suggests that the instantaneous velocity at $t = 5$ is 22 ft/sec.

21. $s(t) = 2.2t^2$

$$\text{Average speed} = \frac{s(30) - s(0)}{30 - 0}$$
$$= \frac{1980 - 0}{30} = 66 \text{ ft/sec.}$$

23. $s(t) = t^2 + 4t + 3$

For $t = 1$, the instantaneous velocity is

$$\lim_{h \to 0} \frac{s(1 + h) - s(1)}{h}.$$

We have

$$s(1 + h) = (1 + h)^2 + 4(1 + h) + 3$$
$$= 1 + 2h + h^2 + 4 + 4h + 3$$
$$= h^2 + 6h + 8,$$

and $s(1) = 1^2 + 4(1) + 3 = 8$.

Thus,

$$s(1 + h) - s(1) = (h^2 + 6h + 8) - 8$$
$$= h^2 + 6h,$$

and the instantaneous velocity at $t = 1$ is

$$\lim_{h \to 0} \frac{h^2 + 6h}{h} = \lim_{h \to 0} \frac{h(h + 6)}{h} = \lim_{h \to 0} (h + 6)$$
$$= 6 \text{ ft/sec.}$$

25. a. $\text{Average velocity} = \dfrac{s(2) - s(0)}{2 - 0} = \dfrac{10 - 0}{2}$
$$= 5 \text{ ft/sec}$$

    b. $\text{Average velocity} = \dfrac{s(4) - s(2)}{4 - 2} = \dfrac{14 - 10}{2}$
$$= 2 \text{ ft/sec}$$

c.   Average velocity $= \dfrac{s(6)-s(4)}{6-4} = \dfrac{20-14}{2} = 3$ ft/sec

d.   Average velocity $= \dfrac{s(8)-s(6)}{8-6} = \dfrac{30-20}{2} = 5$ ft/sec

e.   **i.** Let $t = 4$ and $h = 2$

   instantaneous velocity $\approx \dfrac{s(4+2)-s(4)}{2} = \dfrac{20-14}{2} = 3$ ft/sec

   ii. instantaneous velocity $\approx \dfrac{2+3}{2} = 2.5$ ft/sec

f.   **i.** Let $t = 6$ and $h = 2$

   instantaneous velocity $\approx \dfrac{s(6+2)-s(6)}{2} = \dfrac{30-20}{2} = 5$ ft/sec

   ii. instantaneous velocity $\approx \dfrac{5+3}{2} = 4$ ft/sec

27. $f(x) = x^2 - x - 1$

a. $f(a+h) = (a+h)^2 - (a+h) - 1 = a^2 + 2ah + h^2 - a - h - 1$

b. $\dfrac{f(a+h)-f(a)}{h} = \dfrac{a^2 + 2ah + h^2 - a - h - 1 - (a^2 - a - 1)}{h} = \dfrac{2ah + h^2 - h}{h} = \dfrac{h(2a+h-1)}{h} = 2a + h - 1$

c.   The instantaneous rate of change is $\displaystyle\lim_{h\to 0} \dfrac{f(a+h)-f(a)}{h} = \lim_{h\to 0} 2a + h - 1$
   $= 2a - 1,$

   so the instantaneous rate of change when
   $a = 5$ is $2(5) - 1 = 9$.

29. $f(x) = x^3$

a. $f(a+h) = (a+h)^3 = a^3 + 3a^2h + 3ah^2 + h^3$

b. $\dfrac{f(a+h)-f(a)}{h} = \dfrac{a^3 + 3a^2h + 3ah^2 + h^3 - a^3}{h} = \dfrac{h(3a^2 + 3ah + h^2)}{h} = 3a^2 + 3ah + h^2$

c.   The instantaneous rate of change is $\displaystyle\lim_{h\to 0} \dfrac{f(a+h)-f(a)}{h} = \lim_{h\to 0} 3a^2 + 3ah + h^2 = 3a^2,$

   the instantaneous rate of change when $a = 5$ is $3(5)^2 = 75$.

31. $R = 10x - .002x^2$
   $R(1000) = 10(1000) - .002(1000)^2 = 8000$
   $R(1001) = 10(1001) - .002(1001)^2 = 8005.998$

a.   Average rate of change $= \dfrac{R(1001)-R(1000)}{1001-1000} = \dfrac{8005.998-8000}{1} = 5.998$
   Since the revenue is given in thousands of dollars, the average rate of change is \$5998 per unit.

b. Marginal revenue $= \lim\limits_{h\to 0} \dfrac{R(1000+h) - R(1000)}{h} = \lim\limits_{h\to 0} \dfrac{10(1000+h) - .002(1000+h)^2 - 8000}{h}$

$$= \lim\limits_{h\to 0} \frac{10,000 + 10h - .002(1,000,000 + 2000h + h^2) - 8000}{h}$$

$$= \lim\limits_{h\to 0} \frac{6h - .002h^2}{h} = \lim\limits_{h\to 0} \frac{h(6 - .002h)}{h} = \lim\limits_{h\to 0} 6 - .002h = 6$$

The marginal revenue is $6000 per unit.

c. Additional revenue $= R(1001) - R(1000) = 8005.998 - 8000 = 5.998$

The additional revenue is $5998.

d. The answers in (a) and (c) are the same.

33. $p(t) = t^2 + t$ for $0 \le t \le 5$

a. Average rate of change $= \dfrac{p(4) - p(1)}{4 - 1} = \dfrac{(4^2 + 4) - (1^2 + 1)}{3} = \dfrac{18}{3} = 6\%$ per day

b. Instantaneous rate of change $= \lim\limits_{h\to 0} \dfrac{p(t+h) - p(t)}{h} = \lim\limits_{h\to 0} \dfrac{(t+h)^2 + t + h - t^2 - t}{h}$

$$= \lim\limits_{h\to 0} \frac{t^2 + 2th + h^2 + t + h - t^2 - t}{h} = \lim\limits_{h\to 0} \frac{2th + h^2 + h}{h}$$

$$= \lim\limits_{h\to 0} 2t + h + 1 = 2t + 1$$

When $t = 3$, the instantaneous rate of change is $2(3) + 1 = 7\%$ per day.

35. a. Since $x = 0$ corresponds to 1990, the year 2006 corresponds to $x = 16$.

$$\lim\limits_{h\to 0} \frac{g(h+16) - g(16)}{h} = \lim\limits_{h\to 0} \frac{268(1.0846)^{h+16} - 268(1.0846)^{16}}{h}$$

X	Y₁
-.001	79.808
-1E⁻⁴	79.811
-1E⁻⁶	79.811
0	ERROR
1E⁻⁶	79.811
1E⁻⁴	79.812
.001	79.815

X=1E⁻6

For 2006, the revolving consumer credit is increasing at a rate of about $79.81 billion per year.

b. Since $x = 0$ corresponds to 1990, the year 2010 corresponds to $x = 20$.

$$\lim\limits_{h\to 0} \frac{g(h+20) - g(20)}{h} = \lim\limits_{h\to 0} \frac{268(1.0846)^{h+20} - 268(1.0846)^{20}}{h}$$

X	Y₁
-.001	110.44
-1E⁻⁴	110.44
-1E⁻⁶	110.44
0	ERROR
1E⁻⁶	110.44
1E⁻⁴	110.44
.001	110.45

X=1E⁻6

For 2010, the revolving consumer credit is increasing at a rate of about $110.44 billion per year.

37. a. Since $x = 1$ corresponds to 1991, the year 2002 corresponds to $x = 12$.

$$\lim_{h \to 0} \frac{g(12+h) - g(12)}{h} = \lim_{h \to 0} \frac{\left(5.78 + 8.59\ln(12+h)\right) - \left(5.78 + 8.59\ln(12)\right)}{h}$$

$$= \lim_{h \to 0} \frac{8.59\ln(12+h) - 8.59\ln(12)}{h}$$

X	Y₁	
-.001	.71586	
-1E-4	.71584	
-1E-6	.71583	
0	ERROR	
1E-5	.71583	
1E-4	.71583	
.001	.7158	

X=1E-5

In 2002, the rate of change of revenue was $.716 billion or $716 million per year.

b. Since $x = 1$ corresponds to 1991, the year 2007 corresponds to $x = 172$.

$$\lim_{h \to 0} \frac{g(17+h) - g(17)}{h} = \lim_{h \to 0} \frac{\left(5.78 + 8.59\ln(17+h)\right) - \left(5.78 + 8.59\ln(17)\right)}{h}$$

$$= \lim_{h \to 0} \frac{8.59\ln(17+h) - 8.59\ln(17)}{h}$$

X	Y₁	
-.001	.50531	
-1E-4	.5053	
-1E-6	.5053	
0	ERROR	
1E-5	.50529	
1E-4	.50529	
.001	.50528	

X=1E-5

In 2002, the rate of change of revenue was $.505 billion or $505 million per year.

39. Let $F(t)$ represent the thermal effect of food in kJ/hr, where t is the hours elapsed since eating a meal.

a. $F(t) = -10.28 + 175.9te^{-t/1.3}$

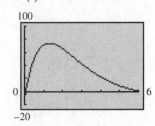

b. Average rate of change of the thermal effect for $F(t) = -10.28 + 175.9te^{-t/1.3}$ between 0 and 1 hour is

$$\frac{F(1) - F(0)}{1 - 0}.$$

$$F(1) = -10.28 + 175.9(1)e^{-1/1.3} = 71.23$$

$$F(0) = -10.28 + 175.9(0)e^{0/1.3} = -10.28$$

The average rate of change is $\dfrac{71.23 - (-10.28)}{1 - 0} = \dfrac{81.51}{1} = 81.51$ kJ/hr/hr.

c. $\lim_{h \to 0} \dfrac{F(1+h) - F(1)}{h} = \lim_{h \to h} \dfrac{-10.28 + 175.9(1+h)e^{-(1+h)/1.3} - \left(-10.28 + 175.9(1)e^{-1/1.3}\right)}{h}$

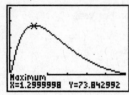

At exactly 1 hour after eating, the rate of change for the thermal effect is 18.81 kJ/hr/hr.

d. The function stops increasing and begins to decrease at approximately $t = 1.3$ hours.

Section 11.4 Tangent Lines and Derivatives

1. $f'(x) = 3x^2 - 4$

$f'(2) = 3(2)^2 - 4(1) = 12 - 4 = 8$

The slope of the tangent line at $(2, f(2)) = (2, 0)$ is 8.

$y - y_1 = m(x - x_1) \Rightarrow y - (0) = 8(x - 2) \Rightarrow$

$y = 8x - 16$

3. $f'(x) = -\dfrac{1}{x^2}$

$f'(-2) = -\dfrac{1}{(-2)^2} = -\dfrac{1}{4}$

The slope of the tangent line at

$(-2, f(-2)) = \left(-2, -\dfrac{1}{2}\right)$ is $-\dfrac{1}{4}$.

$y - y_1 = m(x - x_1) \Rightarrow y + \dfrac{1}{2} = -\dfrac{1}{4}(x + 2) \Rightarrow$

$y = -\dfrac{1}{4}x - 1$

5. $f(x) = x^2 + 3$ at $x = 2$.

a. $f(2 + h) = (2 + h)^2 + 3 = 7 + 4h + h^2$

$f(2) = 2^2 + 3 = 7$

$f(2 + h) - f(2) = 7 + 4h + h^2 - 7 = 4h + h^2$

$\dfrac{f(2 + h) - f(2)}{h} = \dfrac{4h + h^2}{h} = 4 + h$

Letting $h \to 0$, slope of tangent = 4

b. The slope of the tangent line at $(2, f(2)) = (2, 7)$ is 4.

$y - y_1 = m(x - x_1) \Rightarrow y - 7 = 4(x - 2) \Rightarrow$

$y = 4x - 1$

7. $h(x) = \dfrac{7}{x}$ at $x = 3$.

a. $h(3 + h) = \dfrac{7}{3 + h}$

$h(3 + h) - h(3) = \dfrac{7}{3 + h} - \dfrac{7}{3} = \dfrac{21 - (21 + 7h)}{3(3 + h)}$

$= \dfrac{-7h}{3(3 + h)}$

$\dfrac{h(3 + h) - h(3)}{h} = \dfrac{\frac{-7h}{3(3 + h)}}{h} = \dfrac{-7}{3(3 + h)}$

Letting $h \to 0$,

slope of tangent $= \dfrac{-7}{3(3)} = -\dfrac{7}{9}$

b. The slope of the tangent line at

$(3, h(3)) = \left(3, \dfrac{7}{3}\right)$ is $-\dfrac{7}{9}$.

$y - \dfrac{7}{3} = -\dfrac{7}{9}(x - 3) \Rightarrow y = -\dfrac{7}{9}x + \dfrac{14}{3}$

9. $g(x) = 5\sqrt{x}$ at $x = 9$.

    a. $g(9+h) = 5\sqrt{9+h}$

        $g(9) = 5\sqrt{9} = 15$

        $g(9+h) - g(9) = 5\sqrt{9+h} - 15$

                          $= 5\left(\sqrt{9+h} - 3\right)$

        $\dfrac{g(9+h) - g(9)}{h}$

        $= \dfrac{5\left(\sqrt{9+h} - 3\right)}{h}$

        $= 5 \cdot \dfrac{\sqrt{9+h} - 3}{h} \cdot \dfrac{\left(\sqrt{9+h} + 3\right)}{\left(\sqrt{9+h} + 3\right)}$

        $= \dfrac{5\left[(9+h) - 9\right]}{h\left(\sqrt{9+h} + 3\right)} = \dfrac{5h}{h\left(\sqrt{9+h} + 3\right)}$

        $= \dfrac{5}{\sqrt{9+h} + 3}$

    Letting $h \to 0$,

        slope of tangent $= \dfrac{5}{\sqrt{9} + 3} = \dfrac{5}{6}$

    b. The slope of the tangent line at

        $(9, g(9)) = (9, 15)$ is $\dfrac{5}{6}$.

        $y - 15 = \dfrac{5}{6}(x - 9) \Rightarrow y = \dfrac{5}{6}x + \dfrac{15}{2}$

11. a. From the graph, $f(x)$ is largest at $x = x_5$.

    b. $f(x)$ is smallest at $x = x_4$.

    c. $f'(x)$ is smallest at $x = x_3$ because x_3 is the only labeled value at which the tangent line has a negative slope.

    d. $f'(x)$ is closest to 0 at $x = x_2$ because the tangent line is closer to being horizontal (where the slope is 0) at x_2 than at any other labeled value.

13. If $g'(x) > 0$ for $x < 0$, the slope of the tangent line must be positive, so $g(x)$ must increase for $x < 0$. If $g'(x) < 0$ for $x > 0$, the slope of the tangent line must be negative, so $g(x)$ must decrease for $x > 0$. One of the many possible graphs follows.

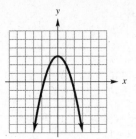

15. a. The derivative is positive because the tangent line is rising from left to right at $x = 100$ and thus has a positive slope.

    b. The derivative is negative because the tangent line is falling from left to right at $x = 200$ and thus has a negative slope.

17. Since velocity is the rate of change of distance from a starting point, we must decide which graph represents the function (distance over time) and which represents the derivative of that function (velocity over time). For $t > 0$, the graph of (a) is positive. If (a) is the derivative, the graph of (b) would have no high or low points when (a) would be 0, which is not the case. Furthermore, assuming (b) is the derivative of (a), we see that (b) is positive when (a) is increasing ($0 < t < 2$ and $t > 4$) and (b) is negative when (a) is decreasing ($2 < t < 4$). Therefore, graph (a) is distance and graph (b) is velocity.

19. $f(x) = -4x^2 + 11x$

    Step 1

    $f(x+h) = -4(x+h)^2 + 11(x+h)$

                $= -4(x^2 + 2xh + h^2) + 11x + 11h$

                $= -4x^2 - 8xh - 4h^2 + 11x + 11h$

    Step 2

    $f(x+h) - f(x)$

        $= (-4x^2 - 8xh - 4h^2 + 11x + 11h)$

                        $- (-4x^2 + 11x)$

        $= -8xh - 4h^2 + 11h$

    Step 3

    $\dfrac{f(x+h) - f(x)}{h} = \dfrac{-8xh - 4h^2 + 11h}{h}$

                    $= \dfrac{h(-8x - 4h + 11)}{h}$

                    $= -8x - 4h + 11$

(continued on next page)

(continued from page 339)

Step 4

$$f'(x) = \lim_{h \to 0} \frac{f(x+h) - f(x)}{h}$$
$$= \lim_{h \to 0} (-8x - 4h + 11) = -8x + 11$$
$$f'(2) = -8(2) + 11 = -5$$
$$f'(0) = -8(0) + 11 = 11$$
$$f'(-3) = -8(-3) + 11 = 35$$

21. $f(x) = 8x + 6$

We condense the steps as follows

$$\frac{f(x+h) - f(x)}{h} = \frac{[8(x+h) + 6] - (8x + 6)}{h}$$
$$= \frac{(8x + 8h + 6) - (8x + 6)}{h}$$
$$= \frac{8h}{h} = 8$$

$$f'(x) = \lim_{h \to 0} 8 = 8$$
$$f'(2) = 8; \quad f'(0) = 8; \quad f(-3) = 8$$

23. $f(x) = -\dfrac{2}{x}$

$$\frac{f(x+h) - f(x)}{h} = \frac{\frac{-2}{x+h} - \left(\frac{-2}{x}\right)}{h} = \frac{\frac{-2x + 2(x+h)}{(x+h)x}}{h}$$
$$= \frac{2h}{h(x+h)(x)} = \frac{2}{(x+h)(x)}$$

$$f'(x) = \lim_{h \to 0} \frac{2}{(x+h)(x)} = \frac{2}{x^2}$$

$$f'(2) = \frac{2}{2^2} = \frac{1}{2}$$

$$f'(0) = \frac{2}{0^2} \text{ Undefined}$$

The derivative does not exist at $x = 0$.

$$f'(-3) = \frac{2}{(-3)^2} = \frac{2}{9}$$

25. $f(x) = \dfrac{4}{x-1}$

$$\frac{f(x+h) - f(x)}{h}$$
$$= \frac{\frac{4}{x+h-1} - \frac{4}{x-1}}{h} = \frac{\frac{4(x-1) - 4(x+h-1)}{(x+h-1)(x-1)}}{h}$$
$$= \frac{-4h}{h(x-1+h)(x-1)}$$
$$= \frac{-4}{(x-1+h)(x-1)}$$

$$f'(x) = \lim_{h \to 0} \frac{-4}{(x-1+h)(x-1)} = \frac{-4}{(x-1)^2}$$

$$f'(2) = \frac{-4}{(2-1)^2} = -4$$

$$f'(0) = \frac{-4}{(0-1)^2} = -4$$

$$f'(-3) = \frac{-4}{(-3-1)^2} = \frac{-4}{16} = -\frac{1}{4}$$

27. For $x = \pm 6$, the graph of $f(x)$ has sharp points. Therefore, there is no derivative for $x = 6$ or $x = -6$.

29. The derivative does not exist at the following x-values.

x	Reason
-5	Function is not defined.
-3	Function has a sharp point.
0	Function is not defined.
2	Function has a sharp point.
4	Function has a vertical tangent

31. $R(x) = 20x - \dfrac{x^2}{500}$

a. The marginal revenue is the rate of change of the revenue, or $R'(x)$, so find $R'(x)$.

$$\frac{R(x+h) - R(x)}{h}$$
$$= \frac{20(x+h) - \frac{(x+h)^2}{500} - 20x + \frac{x^2}{500}}{h}$$
$$= \frac{20h - \frac{2xh + h^2}{500}}{h} = 20 - \frac{2x+h}{500}$$

$$R'(x) = \lim_{h \to 0} \frac{R(x+h) - R(x)}{h}$$
$$= \lim_{h \to 0} \left(20 - \frac{2x+h}{500}\right) = 20 - \frac{2x}{500}$$

At $x = 1000$, $R'(x) = 20 - \dfrac{2(1000)}{500} = 16$

The marginal revenue is \$16/table.

b.   The actual revenue from the sale of the 1001st item is

$$R(1001) - R(1000) = 20(1001) - \frac{1001^2}{500} - \left[20(1000) - \frac{1000^2}{500} \right] = 18,015.998 - 18,000 = 15.998 \text{ or } 16.$$

The actual revenue is \$15.998 or \$16.

c.   The marginal revenue found in part (a) approximates the actual revenue from the sale of the 1001st item found in part (b).

33. a.   Given that the demand for the items is $D(p) = -p^2 + 5p + 1,$ where p represents the price.

$$D'(p) = \lim_{h \to 0} \frac{D(p+h) - D(p)}{h} = \lim_{h \to 0} \frac{\left[-(p+h)^2 + 5(p+h) + 1\right] - \left[-p^2 + 5p + 1\right]}{h}$$

$$= \lim_{h \to 0} \frac{\left[-p^2 - 2ph - h^2 + 5p + 5h + 1\right] - \left[-p^2 + 5p + 1\right]}{h} = \lim_{h \to 0} \frac{-2ph - h^2 + 5h}{h}$$

$$= \lim_{h \to 0} (-2p - h + 5) = -2p + 5$$

The rate of change of demand with respect to price is $D'(p)$ or $-2p + 5.$

b.   When $p = 12,$ $D'(12) = -2(12) + 5 = -19.$

This means that the demand is decreasing at the rate of about 19 items for each increase in price of \$1.

35.   At $t = 2,$ the slope of the tangent line is 1000.
Thus, the population is increasing at a rate of
1000 shellfish per time unit.
At $t = 10,$ the slope of the tangent line is 570.
Thus, the population is increasing more slowly at
570 shellfish per time unit.
At $t = 13,$ the slope of the tangent line is 250; the
population is increasing at a much slower rate of
250 shellfish per time unit.

37. $f(x) = x^2 - 5x + 2$
$f'(x) = 2x - 5$

Numerical derivative:

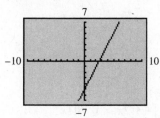

39. $f(x) = \ln x + x$
$f'(x) = \frac{1}{x} + 1$

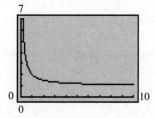

41. $f(t) = 1.704 \left(1.223^t \right)$

Set $Y_1 = f(t).$

a.

```
nDeriv(Y1,X,8)
        1.716879509
nDeriv(Y1,X,15)
        7.026272945
nDeriv(Y1,X,20)
        19.22459953
```

1998: 1.72
2005: 7.03
2010: 19.22

b.   The rate is increasing.

43. $g(x) = 145.033x^{.119}$

Set $Y_1 = g(x)$.

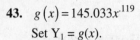

$g'(20) \approx 1.23$

The rate of increase is 1.23 million drivers in 2010.

45. a. $f(x) = .5x^5 - 2x^3 + x^2 - 3x + 2;\ -3 \le x \le 3$

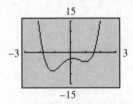

b. $g(x) = 2.5x^4 - 6x^2 + 2x - 3$

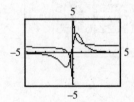

c. The graphs appear identical, which suggests that $f'(x) = g(x)$.

47. $y = \dfrac{4x^2 + x}{x^2 + 1}$

Graph the derivative of y and each possible derivative function on the same screen.

a. $f(x) = \dfrac{2x+1}{2x}$

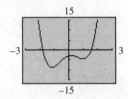

b. $g(x) = \dfrac{x^2 + x}{2x}$

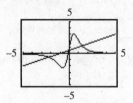

c. $h(x) = \dfrac{2x+1}{x^2+1}$

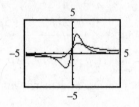

d. $k(x) = \dfrac{-x^2 + 8x + 1}{\left(x^2+1\right)^2}$

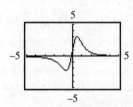

The derivative of y could possibly be

$$k(x) = \dfrac{-x^2 + 8x + 1}{\left(x^2+1\right)^2}.$$

Section 11.5 Techniques for Finding Derivatives

1. $f(x) = 4x^2 - 3x + 5$

$f'(x) = 4\left(2x^{2-1}\right) - 3\left(1x^{1-1}\right) + 0$

$\quad = 8x - 3$

3. $y = 2x^3 + 3x^2 - 6x + 2$

$y' = 2\left(3x^{3-1}\right) + 3\left(2x^{2-1}\right) - 6\left(1x^{1-1}\right) + 0$

$\quad = 6x^2 + 6x - 6$

5. $g(x) = x^4 + 3x^3 - 8x - 7$

$g'(x) = 4x^{4-1} + 3\left(3x^{3-1}\right) - 8\left(1x^{1-1}\right) - 0$

$\quad = 4x^3 + 9x^2 - 8$

7. $f(x) = 6x^{1.5} - 4x^{.5}$

$f'(x) = 6\left(1.5x^{1.5-1}\right) - 4\left(.5x^{-.5}\right)$

$\quad = 9x^{.5} - 2x^{-.5} = 9x^{.5} - \dfrac{2}{x^{.5}}$

9. $y = -15x^{3/2} + 2x^{1.9} + x$

$y' = -15\left(\dfrac{3}{2}x^{3/2-1}\right) + 2\left(1.9x^{1.9-1}\right) + 1x^{1-1}$

$\quad = -22.5x^{1/2} + 3.8x^{.9} + 1$

11. $y = 24t^{3/2} + 4t^{1/2}$

$y' = 24\left(\dfrac{3}{2}t^{1/2}\right) + 4\left(\dfrac{1}{2}t^{-1/2}\right)$

$\quad = 36t^{1/2} + 2t^{-1/2} = 36t^{1/2} + \dfrac{2}{t^{1/2}}$

13. $y = 8\sqrt{x} + 6x^{3/4}$

$\quad = 8x^{1/2} + 6x^{3/4}$

$y' = 8\left(\dfrac{1}{2}x^{-1/2}\right) + 6\left(\dfrac{3}{4}x^{-1/4}\right)$

$\quad = 4x^{-1/2} + \dfrac{9}{2}x^{-1/4} = \dfrac{4}{x^{1/2}} + \dfrac{9}{2x^{1/4}}$

15. $g(x) = 6x^{-5} - 2x^{-1}$

$g'(x) = 6\left(-5x^{-6}\right) - 2(-1)x^{-2}$

$\quad = -30x^{-6} + 2x^{-2} = -\dfrac{30}{x^6} + \dfrac{2}{x^2}$

17. $y = 10x^{-2} + 6x^{-4} + 3x$

$y' = 10(-2x^{-3}) + 6(-4x^{-5}) + 3$

$\quad = -20x^{-3} - 24x^{-5} + 3 = -\dfrac{20}{x^3} - \dfrac{24}{x^5} + 3$

19. $f(t) = \dfrac{-4}{t} + \dfrac{8}{t^2} = -4t^{-1} + 8t^{-2}$

$f'(t) = 4t^{-2} - 16t^{-3} = \dfrac{4}{t^2} - \dfrac{16}{t^3}$

21. $y = \dfrac{12 - 7x + 6x^3}{x^4} = \dfrac{12}{x^4} - \dfrac{7x}{x^4} + \dfrac{6x^3}{x^4}$

$\quad = 12x^{-4} - 7x^{-3} + 6x^{-1}$

$y' = -48x^{-5} + 21x^{-4} - 6x^{-2}$

$\quad = -\dfrac{48}{x^5} + \dfrac{21}{x^4} - \dfrac{6}{x^2}$

23. $g(x) = 5x^{-1/2} - 7x^{1/2} + 101x$

$g'(x) = 5\left(-\dfrac{1}{2}x^{-3/2}\right) - 7\left(\dfrac{1}{2}x^{-1/2}\right) + 101$

$\quad = -\dfrac{5}{2}x^{-3/2} - \dfrac{7}{2}x^{-1/2} + 101$

$\quad = -\dfrac{5}{2x^{3/2}} - \dfrac{7}{2x^{1/2}} + 101$

25. $y = 7x^{-3/2} + 8x^{-1/2} + x^3 - 9$

$y' = 7\left(-\dfrac{3}{2}x^{-5/2}\right) + 8\left(-\dfrac{1}{2}x^{-3/2}\right) + 3x^2 + 0$

$\quad = -\dfrac{21}{2}x^{-5/2} - 4x^{-3/2} + 3x^2$

$\quad = -\dfrac{21}{2x^{5/2}} - \dfrac{4}{x^{3/2}} + 3x^2$

27. $y = \dfrac{19}{\sqrt[4]{x}} = 19x^{-1/4}$

$y' = 19\left(-\dfrac{1}{4}x^{-5/4}\right) = -\dfrac{19}{4}x^{-5/4} = -\dfrac{19}{4x^{5/4}}$

29. $y = \dfrac{-8t}{\sqrt[3]{t^2}} = -8t\left(t^{-2/3}\right) = -8t^{1/3}$

$y' = -8\left(\dfrac{1}{3}t^{-2/3}\right) = -\dfrac{8}{3}t^{-2/3} = -\dfrac{8}{3t^{2/3}}$

31. $y = 8x^{-5} - 9x^{-4} + 9x^4$

$\dfrac{dy}{dx} = 8\left(-5x^{-6}\right) - 9\left(-4x^{-5}\right) + 9\left(4x^{4-1}\right)$

$\quad = -40x^{-6} + 36x^{-5} + 36x^3$

$\quad = -\dfrac{40}{x^6} + \dfrac{36}{x^5} + 36x^3$

33. $D_x\left(9x^{-1/2} + \dfrac{2}{x^{3/2}}\right)$

$\quad = D_x\left(9x^{-1/2} + 2x^{-3/2}\right)$

$\quad = 9\left(-\dfrac{1}{2}x^{-3/2}\right) + 2\left(-\dfrac{3}{2}x^{-5/2}\right)$

$\quad = -\dfrac{9}{2}x^{-3/2} - 3x^{-5/2} = -\dfrac{9}{2x^{3/2}} - \dfrac{3}{x^{5/2}}$

35. $f(x) = 6x^2 - 2x$

$f'(x) = 12x - 2$

$f'(-2) = 12(-2) - 2 = -24 - 2 = -26$

37. $f(t) = 2\sqrt{t} - \dfrac{3}{\sqrt{t}} = 2t^{1/2} - 3t^{-1/2}$

$f'(t) = 2\left(\dfrac{1}{2}t^{-1/2}\right) - 3\left(-\dfrac{1}{2}t^{-3/2}\right)$

$= t^{-1/2} + \dfrac{3}{2}t^{-3/2} = \dfrac{1}{t^{1/2}} + \dfrac{3}{2t^{3/2}}$

$f'(4) = \dfrac{1}{4^{1/2}} + \dfrac{3}{2(4^{3/2})} = \dfrac{1}{2} + \dfrac{3}{16} = \dfrac{11}{16}$

39. $f(x) = -\dfrac{(3x^2 + x)^2}{7} = -\dfrac{9x^4 + 6x^3 + x^2}{7}$

$= -\dfrac{9}{7}x^4 - \dfrac{6}{7}x^3 - \dfrac{1}{7}x^2$

$f'(x) = -\dfrac{9}{7}(4x^3) - \dfrac{6}{7}(3x^2) - \dfrac{1}{7}(2x)$

$= -\dfrac{36}{7}x^3 - \dfrac{18}{7}x^2 - \dfrac{2}{7}x$

$f'(1) = -\dfrac{36}{7} - \dfrac{18}{7} - \dfrac{2}{7} = -\dfrac{56}{7} = -8$

Of the given choices, –9 is closest to –8, so the answer is (b).

41. $f(x) = x^4 - 2x^2 + 1; x = 1$

$f'(x) = 4x^3 - 4x$

The slope of the tangent line at $x = 1$ is

$f'(1) = 4(1^3) - 4(1) = 4 - 4 = 0.$

The tangent line at $x = 1$ is the horizontal line through $f(1) = 1^4 - 2(1^2) + 1 = 0.$

The equation of the tangent line is $y = 0.$

43. $y = 4x^{1/2} + 2x^{3/2} + 1; x = 4.$

$y' = 4\left(\dfrac{1}{2}x^{-1/2}\right) + 2\left(\dfrac{3}{2}x^{1/2}\right)$

$= \dfrac{2}{x^{1/2}} + 3x^{1/2}$

The slope of the tangent line at $x = 4$ is

$y'(4) = \dfrac{2}{4^{1/2}} + 3(4^{1/2}) = 7$

Because $y(4) = 4(4^{1/2}) + 2(4^{3/2}) + 1 = 25,$

the tangent line $x = 4$ goes through (4, 25). Using the point-slope form, the equation of the tangent line at $x = 4$ is

$y - 25 = 7(x - 4) \Rightarrow y = 7x - 28 + 25 \Rightarrow$

$y = 7x - 3.$

45. $P(x) = .03x^2 - 4x - 3x^8 + -5000$

Marginal profit

$= P'(x) = .03(2x) - 4 + 3(.8x^{-.2})$

$= .06x - 4 + 2.4x^{-.2}$

a. $P'(100) = .06(100) - 4 + 2.4(100)^{-.2} = \$2.96$

b. $P'(1000) = .06(1000) - 4 + 2.4(1000)^{-.2}$
$= \$56.60$

c. $P'(5000) = .06(5000) - 4 + 2.4(5000)^{-.2}$
$= \$296.44$

d. $P'(10,000) = .06(10,000) - 4 + 2.4(10,000)^{-.2}$
$= \$596.38$

47. $S(t) = 10,000 - 10,000t^{-.2} + 100t^{.1}$

$S'(t) = -10,000(-.2t^{-1.2}) + 100\left(.1t^{-.9}\right)$

$= \dfrac{2000}{t^{1.2}} + \dfrac{2010}{t^{.9}}$

a. $S'(1) = \dfrac{2000}{(1)^{1.2}} + \dfrac{10}{1^{.9}} = \dfrac{100}{1} = 2010$

b. $S'(10) = \dfrac{2000}{(10)^{1.2}} + \dfrac{10}{10^{.9}} = 127.45$

49. $R(x) = .0608x^2 + 1.4284x + 2.3418$

a. $R'(x) = .1216x + 1.4284$

b. $R'(5) = .1216(5) + 1.4284 = 2.0364$

The marginal revenue for 5 hundred million bushels is $2.0364 million.

c. $R'(7) = .1216(7) + 1.4284 = 2.2796$

The marginal revenue for 7 hundred million bushels is $2.2796 million.

51. Demand is $x = 5000 - 100p \Rightarrow p = \dfrac{5000 - x}{100}$

$$R(x) = x\left(\dfrac{5000 - x}{100}\right) = \dfrac{5000x - x^2}{100}$$

$$R'(x) = \dfrac{5000 - 2x}{100}$$

a. $R'(1000) = \dfrac{5000 - 2000}{100} = 30$

b. $R'(2500) = \dfrac{5000 - 2(2500)}{100} = 0$

c. $R'(3000) = \dfrac{5000 - 2(3000)}{100} = -10$

53. $V = \pi r^2 h$

Since $h = 80$ mm, $V = \pi r^2(80) = 80\pi r^2$.

a. $\dfrac{dV}{dr} = 80\pi(2r) = 160\pi r$

b. When $r = 4$, $\dfrac{dV}{dr} = 160\pi(4) = 640\pi$.

The volume of blood is increasing at a rate of 640π mm^3 / mm of change in the radius.

c. When $r = 6$, $\dfrac{dV}{dr} = 160\pi(6) = 960\pi$.

The volume of blood is increasing at a rate of 960π mm^3/mm of change in the radius.

d. When $r = 8$, $\dfrac{dV}{dr} = 160\pi(8) = 1280\pi$.

The volume of blood is increasing at a rate of 1280π mm^3/mm of change in the radius.

55. $f(x) = .00036x^4 - .015x^3 + .197x^2$
$$- .257x + 28.31$$
$f'(x) = .00144x^3 - .045x^2 + .394x - .257$

a. In 2000, $x = 10$

$$f'(10) = .00144(10)^3 - .045(10)^2$$
$$+ .394(10) - .257$$
$$= .623$$

Living standards were increasing by \$.623 million per year in 2000.

b. In 2009, $x = 19$

$$f'(19) = .00144(19)^3 - .045(19)^2$$
$$+ .394(19) - .257$$
$$= .86096$$

Living standards were increasing by about \$.861 million per year in 2009.

c. The rate of increase in living standards was higher in 2009 than in 2000.

57. $y_1 = 4.13x + 14.63$
$y_2 = -.033x^2 + 4.647x + 13.347$

a. When $x = 5$
$y_1 = 4.13(5) + 14.63 \approx 35$
$y_2 = -.033(5)^2 + 4.647(5) + 13.347 \approx 36$

b. $\dfrac{dy_1}{dx} = 4.13$; when $x = 5$, $\dfrac{dy_1}{dx} = 4.13$

$\dfrac{dy_2}{dx} = -.033(2x) + 4.647 = -.066x + 4.647$

When $x = 5$,

$\dfrac{dy_2}{dx} = -.066(5) + 4.647 = 4.317$

c. Use points $(15, 76)$ and $(5, 36)$ to find the slope.

$$m = \dfrac{76 - 36}{15 - 5} = \dfrac{40}{10} = 4$$

$y - 36 = 4(x - 5) \Rightarrow y = 4x + 16$

(Actually, all points except the first lie on this line.)

d. Answers vary.

59. a. $V = C(R_0 - R)R^2 = CR_0R^2 - CR^3$

$$\dfrac{dV}{dR} = 2CR_0R - 3CR^2$$

b. Let $\dfrac{dV}{dR} = 0$.

$2CR_0R - 3CR^2 = 0 \Rightarrow CR(2R_0 - 3R) = 0 \Rightarrow$

$2R_0 - 3R = 0 \Rightarrow 2R_0 = 3R \Rightarrow \dfrac{2}{3}R_0 = R$

61. $s(t) = 8t^2 + 3t + 1$

a. $v(t) = 8(2t) + 3 = 16t + 3$

b. $v(0) = 16(0) + 3 = 3$
$v(5) = 16(5) + 3 = 83$
$v(10) = 16(10) + 3 = 163$

63. $s(t) = 2t^3 + 6t^2$

a. $v(t) = 2(3t^2) + 6(2t) = 6t^2 + 12t$

b. $v(0) = 6(0^2) + 12(0) = 0$
$v(5) = 6(5^2) + 12(5) = 210$
$v(10) = 6(10^2) + 12(10) = 720$

65. $s(t) = -16t^2 + 144$
velocity $= s'(t) = -32t$

a. $s'(1) = -32$ ft/sec
$s'(2) = -32 \cdot 2 = -64$ ft/sec

b. The rock will hit the ground when $s(t) = 0$.
$-16t^2 + 144 = 0 \Rightarrow t = \sqrt{\dfrac{144}{16}} = 3$ sec

c. The velocity at impact is the velocity at 3 sec.
$v = -32 \cdot 3 = -96$ ft/sec

67. Proof that if $y = x^n$, $y' = n \cdot x^{n-1}$:

a. $y = (x+h)^n$
By the binomial theorem,
$y = x^n + n \cdot x^{n-1}h$
$\qquad + \dfrac{n(n-1)}{2!}x^{n-2}h^2 + \ldots + h^n$

b. $\dfrac{(x+h)^n - x^n}{h}$
$\qquad = n \cdot x^{n-1} + \dfrac{n(n-1)}{2!}x^{n-2}h + \ldots + h^{n-1}$

c. $\lim\limits_{h \to 0} \dfrac{(x+h)^n - x^n}{h} = n \cdot x^{n-1} = y'$

69. $g(x) = 6 - 4x + 3x^2 - x^3$
$g'(x) = -4 + 6x - 3x^2$

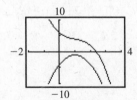

a. $g'(x) > 0$ for no value of x.

b. $g'(x) = 0$ for no value of x.

c. $g'(x) < 0$ on $(-\infty, \infty)$.

d. The derivative is always negative, so the graph of $g(x)$ is always decreasing.

Section 11.6 Derivatives of Products and Quotients

1. $y = (x^2 - 2)(3x + 2)$
$y' = (x^2 - 2)D_x(3x + 2)$
$\qquad + (3x + 2)D_x(x^2 - 2)$
$= (x^2 - 2)(3) + (3x + 2)(2x)$
$= 3x^2 - 6 + 6x^2 + 4x$
$= 9x^2 + 4x - 6$

3. $y = (6x^3 + 2)(5x - 3)$
$y' = (6x^3 + 2)D_x(5x - 3)$
$\qquad + (5x - 3)D_x(6x^3 + 2)$
$= (6x^3 + 2)(5) + (5x - 3)(18x^2)$
$= 30x^3 + 10 + 90x^3 - 54x^2$
$= 120x^3 - 54x^2 + 10$

5. $y = (x^4 - 2x^3 + 2x)(4x^2 + x - 3)$
$y' = (x^4 - 2x^3 + 2x)[D_x(4x^2 + x - 3)]$
$\qquad + (4x^2 + x - 3)[D_x(x^4 - 2x^3 + 2x)]$
$= (x^4 - 2x^3 + 2x)(8x + 1)$
$\qquad + (4x^2 + x - 3)(4x^3 - 6x^2 + 2)$
$= 8x^5 - 16x^4 + 16x^2 + x^4 - 2x^3 + 2x$
$\qquad + 16x^5 - 20x^4 - 18x^3 + 26x^2 + 2x - 6$
$= 24x^5 - 35x^4 - 20x^3 + 42x^2 + 4x - 6$

7. $y = \left(5x^2 + 12x\right)^2$
$y' = (5x^2 + 12x)[D_x(5x^2 + 12x)]$
$\qquad + (5x^2 + 12x)[D_x(5x^2 + 12x)]$
$= (5x^2 + 12x)(10x + 12)$
$\qquad + (5x^2 + 12x)(10x + 12)$
$= 50x^3 + 180x^2 + 144x$
$\qquad + 50x^3 + 180x^2 + 144x$
$= 100x^3 + 360x^2 + 288x$

9. $y = (4x^3 + 6x^2)^2$

$$y' = (4x^3 + 6x^2)[D_x(4x^3 + 6x^2)]$$
$$+ (4x^3 + 6x^2)[D_x(4x^3 + 6x^2)]$$
$$= (4x^3 + 6x^2)(12x^2 + 12x)$$
$$+ (4x^3 + 6x^2)(12x^2 + 12x)$$
$$= 48x^5 + 48x^4 + 72x^4 + 72x^3$$
$$+ 48x^5 + 48x^4 + 72x^4 + 72x^3$$
$$= 96x^5 + 240x^4 + 144x^3$$

11. $y = \dfrac{3x - 5}{x - 3}$

$$y' = \frac{(x-3)D_x(3x-5) - (3x-5)D_x(x-3)}{(x-3)^2}$$
$$= \frac{(x-3)(3) - (3x-5)(1)}{(x-3)^2}$$
$$= \frac{3x - 9 - 3x + 5}{(x-3)^2} = \frac{-4}{(x-3)^2}$$

13. $f(t) = \dfrac{t^2 - 4t}{t + 3}$

$D_t(t^2 - 4t) = 2t - 4$ and $D_t(t + 3) = 1$

$$f'(t) = \frac{(t+3)(2t-4) - (t^2-4t)(1)}{(t+3)^2}$$
$$= \frac{2t^2 - 4t + 6t - 12 - t^2 + 4t}{(t+3)^2}$$
$$= \frac{t^2 + 6t - 12}{(t+3)^2}$$

15. $g(x) = \dfrac{3x^2 + x}{2x^3 - 2}$

$D_x(3x^2 + x) = 6x + 1$ and $D_x(2x^3 - 2) = 6x^2$

$$g'(x) = \frac{(2x^3-2)(6x+1)}{(2x^3-1)^2} - \frac{(3x^2+x)(6x^2)}{(2x^3-2)^2}$$
$$= \frac{12x^4 + 2x^3 - 12x - 2 - 18x^4 - 6x^3}{(2x^3-2)^2}$$
$$= \frac{-6x^4 - 4x^3 - 12x - 2}{(2x^3-2)^2}$$

17. $y = \dfrac{x^2 - 4x + 2}{x + 3}$

$D_x(x^2 - 4x + 2) = 2x - 4$ and $D_x(x + 3) = 1$

$$y' = \frac{(x+3)(2x-4)}{(x+3)^2} - \frac{(x^2-4x+2)(1)}{(x+3)^2}$$
$$= \frac{2x^2 - 4x + 6x - 12 - x^2 + 4x - 2}{(x+3)^2}$$
$$= \frac{x^2 + 6x - 14}{(x+3)^2}$$

19. $r(t) = \dfrac{\sqrt{t}}{3t + 4} = \dfrac{t^{1/2}}{3t + 4}$

$D_t(t^{1/2}) = \dfrac{1}{2}t^{-1/2}$ and $D_t(3t + 4) = 3$

$$r'(t) = \frac{(3t+4)\left(\frac{1}{2}t^{-1/2}\right) - (t^{1/2})(3)}{(3t+4)^2}$$
$$= \frac{\frac{3}{2}t^{1/2} + 2t^{-1/2} - 3t^{1/2}}{(3t+4)^2}$$
$$= \frac{-\frac{3}{2}t^{1/2} + \frac{2}{t^{1/2}}}{(3t+4)^2}$$
$$= \frac{-\frac{3}{2}\sqrt{t} + \frac{2}{\sqrt{t}}}{(3t+4)^2} = \frac{-3t + 4}{2\sqrt{t}(3t+4)^2}$$

21. $y = \dfrac{9x - 8}{\sqrt{x}} = \dfrac{9x - 8}{x^{1/2}}$

$D_x(9x - 8) = 9$ and $D_x(x^{1/2}) = \dfrac{1}{2}x^{-1/2}$

$$y' = \frac{x^{1/2}(9) - (9x-8)\left(\frac{1}{2}x^{-1/2}\right)}{(x^{1/2})^2}$$
$$= \frac{9x^{1/2} - \frac{9}{2}x^{1/2} + 4x^{-1/2}}{x}$$
$$= \frac{\frac{9x^{1/2}}{2} + 4x^{-1/2}}{x}$$
$$= \frac{\frac{9x^{1/2}}{2} + \frac{4}{x^{1/2}}}{x} = \frac{\frac{9x+8}{2x^{1/2}}}{x} = \frac{9x+8}{2x^{3/2}} = \frac{9x+8}{2x\sqrt{x}}$$

23. $y = \dfrac{2 - 7x}{5 - x}$

$D_x(2 - 7x) = -7$ and $D_x(5 - x) = -1$

$$y' = \frac{(5-x)(-7) - (2-7x)(-1)}{(5-x)^2}$$
$$= \frac{-35 + 7x + 2 - 7x}{(5-x)^2} = \frac{-33}{(5-x)^2}$$

25. $f(p) = \dfrac{(6p-7)(11p-1)}{4p+3}$

$D_p\left[(6p-7)(11p-1)\right] = (6p-7)(11) + (11p-1)(6) = 66p - 77 + 66p - 6 = 132p - 83$

$D_p(4p+3) = 4$

$f'(p) = \dfrac{(4p+3)D_p\left[(6p-7)(11p-1)\right] - (6p-7)(11p-1)D_p(4p+3)}{(4p+3)^2}$

$ = \dfrac{(4p+3)(132p-83) - (6p-7)(11p-1)4}{(4p+3)^2} = \dfrac{\left(528p^2 + 64p - 249\right) - \left(264p^2 - 332p + 28\right)}{(4p+3)^2}$

$ = \dfrac{264p^2 + 396p - 277}{(4p+3)^2}$

27. $g(x) = \dfrac{x^3 - 8}{(3x+9)(2x-1)}$

$D_x\left(x^3 - 8\right) = 3x^2$

$D_x\left[(3x+9)(2x-1)\right] = (3x+9)(2) + (2x-1)(3) = 6x + 18 + 6x - 3 = 12x + 15$

$g'(x) = \dfrac{\left[(3x+9)(2x-1)\right]D_x\left(x^3-8\right) - \left(x^3-8\right)D_x\left[(3x+9)(2x-1)\right]}{\left[(3x+9)(2x-1)\right]^2}$

$ = \dfrac{\left(6x^2 + 15x - 9\right)\left(3x^2\right) - \left(x^3 - 8\right)(12x + 15)}{(3x+9)^2(2x-1)^2}$

$ = \dfrac{\left(18x^4 + 45x^3 - 27x^2\right) - \left(12x^4 + 15x^3 - 96x - 120\right)}{(3x+9)^2(2x-1)^2}$

$ = \dfrac{6x^4 + 30x^3 - 27x^2 + 96x + 120}{(3x+9)^2(2x-1)^2}$

29. In the first step, the numerator should be
$(x^2 - 1)(2) - (2x + 5)(2x).$

31. $f(x) = \dfrac{x}{x-2}$

$f'(x) = \dfrac{(x-2)[D_x(x)] - (x)[D_x(x-2)]}{(x-2)^2} = \dfrac{(x-2)(1) - (x)(1)}{(x-2)^2} = \dfrac{x-2-x}{(x-2)^2} = \dfrac{-2}{(x-2)^2}$

$f'(3) = \dfrac{-2}{(3-2)^2} = -2$

The slope is equal to $f'(3) = -2.$

$(y-3) = -2(x-3) \Rightarrow y = -2x + 9$

33. $f(x) = \dfrac{(x-1)(2x+3)}{x-5}$

$f(x) = \dfrac{2x^2 + x - 3}{x - 5}$

$f'(x) = \dfrac{(x-5)[D_x(2x^2 + x - 3)] - (2x^2 + x - 3)[D_x(x-5)]}{(x-5)^2}$

$f'(x) = \dfrac{(x-5)(4x+1) - (2x^2 + x - 3)(1)}{(x-5)^2} = \dfrac{4x^2 - 19x - 5 - 2x^2 - x + 3}{(x-5)^2} = \dfrac{2x^2 - 20x - 2}{(x-5)^2}$

$f'(9) = \dfrac{2(9)^2 - 20(9) - 2}{(9-5)^2} = -1.25$

The slope is equal to $f'(9) = -1.25$.

$y - 42 = -1.25(x-9) \Rightarrow y = -1.25x + 53.25$

35. $C(x) = \dfrac{4x+3}{x+5}$

a. $C(15) = \dfrac{4(15)+3}{15+5} = \dfrac{63}{20} = \$3.15 \text{ hundred}$

$\dfrac{C(15)}{15} = \dfrac{3.15}{15} = \$.21 \text{ hundred per toaster}$

$= \$21 \text{ per toaster}$

b. $C(25) = \dfrac{4(25)+3}{25+5} = \dfrac{103}{30} = \$3.433 \text{ hundred}$

$\dfrac{C(25)}{25} = \dfrac{3.433}{25} = \$.1373 \text{ hundred per toaster}$

$= \$13.73 \text{ per toaster}$

c. $\dfrac{C(x)}{x} = \dfrac{\frac{4x+3}{x+5}}{x} = \dfrac{4x+3}{x(x+5)}$

$= \dfrac{4x+3}{x^2 + 5x} \text{ hundred dollars per toaster}$

d. $\bar{C}(x) = \dfrac{4x+3}{x^2 + 5x}$

Marginal average cost function:

$\bar{C}'(x) = \dfrac{d}{dx}\left(\dfrac{4x+3}{x^2+5x}\right)$

$= \dfrac{(x^2+5x)(4) - (4x+3)(2x+5)}{(x^2+5x)^2}$

$= \dfrac{4x^2 + 20x - 8x^2 - 26x - 15}{(x^2+5x)^2}$

$= \dfrac{-4x^2 - 6x - 15}{(x^2+5x)^2}$

37. a. Let $x = 60$

$C(60) = \dfrac{175,000(60)}{100 - 60} = \$262,500$

b. $C'(x) = \dfrac{(100-x)(175,000) - (175,000x)(-1)}{(100-x)^2}$

$= \dfrac{17,500,000 - 175,000x + 175,000x}{(100-x)^2}$

$= \dfrac{17,500,000}{(100-x)^2}$

c. $C'(80) = \dfrac{17,500,000}{(100-80)^2} = \$43,750 \text{ for each}$

1% removed

d. $C'(90) = \dfrac{17,500,000}{(100-90)^2} = \$175,000 \text{ for each}$

1% removed

e. $C'(95) = \dfrac{17,500,000}{(100-95)^2} = \$700,000 \text{ for each}$

1% removed

39. $T(x) = \dfrac{10x}{x^2+5} + 98.6$

a. $\dfrac{dT}{dx} = \dfrac{(x^2+5)[D_x(10x)] - (10x)[D_x(x^2+5)]}{(x^2+5)^2}$

$= \dfrac{(x^2+5)(10) - (10x)(2x)}{(x^2+5)^2}$

$= \dfrac{10x^2 + 50 - 20x^2}{(x^2+5)^2} = \dfrac{-10x^2 + 50}{(x^2+5)^2}$

b. $T'(0) = \dfrac{-10(0)^2 + 50}{((0)^2 + 5)^2} = 2$

The temperature is increasing at 2 degrees per hour.

c. $T'(1) = \dfrac{-10(1)^2 + 50}{((1)^2 + 5)^2} \approx 1.1111$

The temperature is increasing at about 1.1111 degrees per hour.

d. $T'(3) = \dfrac{-10(3)^2 + 50}{((3)^2 + 5)^2} \approx -.2041$

The temperature is decreasing at about .2041 degrees per hour.

e. $T'(9) = \dfrac{-10(9)^2 + 50}{((9)^2 + 5)^2} \approx -.1028$

The temperature is decreasing at about .1028 degrees per hour.

41. $R(w) = 30\dfrac{w - 4}{w - 1.5}$

a. $R(5) = 30\left(\dfrac{5 - 4}{5 - 1.5}\right) \approx 8.57$ minutes

b. $R(7) = 30\left(\dfrac{7 - 4}{7 - 1.5}\right) \approx 16.36$ minutes

c. $R'(x) = 30\left(\dfrac{(w - 1.5)(1) - (w - 4)(1)}{(w - 1.5)^2}\right)$

$= 30\left(\dfrac{w - 1.5 - w + 4}{(w - 1.5)^2}\right)$

$= \dfrac{75}{(w - 1.5)^2}$

$R'(5) = \dfrac{75}{(5 - 1.5)^2} \approx 6.12$ min/(kcal/min)

$R'(7) = \dfrac{75}{(7 - 1.5)^2} \approx 2.48$ min/(kcal/min)

43. $f(x) = \dfrac{x^2}{2(1 - x)}$

The rate of change of the waiting time is

$f'(x) = \dfrac{2(1 - x)(2x) - x^2(-2)}{[2(1 - x)]^2}$

$= \dfrac{(2 - 2x)(2x) + 2x^2}{(2 - 2x)^2} = \dfrac{4x - 4x^2 + 2x^2}{(2 - 2x)^2}$

$= \dfrac{-2x^2 + 4x}{(2 - 2x)^2}$

a. $f'(.1) = \dfrac{-2(.1)^2 + 4(.1)}{[2 - 2(.1)]^2} = \dfrac{.38}{3.24} \approx .1173$

The rate of change of the number of vehicles waiting when the traffic intensity is .1 is .1173.

b. $f'(.6) = \dfrac{-2(.6)^2 + 4(.6)}{[2 - 2(.6)]^2} = \dfrac{1.68}{.64} = 2.625$

The rate of change of the number of vehicles waiting when the traffic intensity is .6 is 2.625.

45. $N(t) = \dfrac{70t^2}{30 + t^2}$

a. $N'(t) = \dfrac{(30 + t^2)(140t) - (70t^2)(2t)}{(30 + t^2)^2}$

$= \dfrac{4200t + 140t^3 - 140t^3}{(30 + t^2)^2}$

$= \dfrac{4200t}{(30 + t^2)^2}$

b. $N'(3) = \dfrac{4200(3)}{(30 + 3^2)^2}$

≈ 8.28 wpm/hr of instruction

$N'(5) = \dfrac{4200(5)}{(30 + 5^2)^2}$

≈ 6.94 wpm/hr of instruction

$N'(7) = \dfrac{4200(7)}{(30 + 7^2)^2}$

≈ 4.71 wpm/hr of instruction

$N'(10) = \dfrac{4200(10)}{(30 + 10^2)^2}$

≈ 2.49 wpm/hr of instruction

$N'(15) = \dfrac{4200(15)}{(30 + 15^2)^2}$

$\approx .97$ wpm/hr of instruction

c. The rate of improvement decreases as the time increases.

Section 11.7 The Chain Rule

In Exercises 1–3, $f(x) = 2x^2 + 3x$ and $g(x) = 4x - 1$.

1. $f[g(5)]$

Find $g(5)$ first.
$$g(5) = 4(5) - 1 = 19$$
$$f[g(5)] = f(19)$$
$$= 2(19^2) + 3(19) = 779$$

3. $g[f(5)]$

Find $f(5)$ first.
$$f(5) = 2(5^2) + 3(5) = 65$$
$$g[f(5)] = g(65)$$
$$= 4(65) - 1 = 259$$

5. $f(x) = 8x + 12; g(x) = 2x + 3$
$$f[g(x)] = 8(2x + 3) + 12 = 16x + 36$$
$$g[f(x)] = 2(8x + 12) + 3 = 16x + 27$$

7. $f(x) = -x^3 + 2; g(x) = 4x + 2$
$$f[g(x)] = -(4x + 2)^3 + 2$$
$$= -64x^3 - 96x^2 - 48x - 6$$
$$g[f(x)] = 4(-x^3 + 2) + 2 = -4x^3 + 10$$

9. $f(x) = \dfrac{1}{x}; g(x) = x^2$
$$f[g(x)] = \frac{1}{x^2}$$
$$g[f(x)] = \left(\frac{1}{x}\right)^2 = \frac{1}{x^2}$$

11. $f(x) = \sqrt{x + 2}; g(x) = 8x^2 - 6x$
$$f[g(x)] = \sqrt{8x^2 - 6x + 2}$$
$$g[f(x)] = 8(\sqrt{x + 2})^2 - 6\sqrt{x + 2}$$
$$= 8x + 16 - 6\sqrt{x + 2}$$

13. $y = (4x + 3)^5$

If $f(x) = x^5$ and $g(x) = 4x + 3$, then
$$y = f[g(x)] = (4x + 3)^5.$$

15. $y = \sqrt{6 + 3x^2}$

If $f(x) = \sqrt{x}$ and $g(x) = 6 + 3x^2$, then
$$y = f[g(x)] = \sqrt{6 + 3x^2}.$$

17. $y = \dfrac{\sqrt{x} + 3}{\sqrt{x} - 3}$

If $f(x) = \dfrac{x + 3}{x - 3}$ and $g(x) = \sqrt{x}$, then
$$y = f[g(x)] = \frac{\sqrt{x} + 3}{\sqrt{x} - 3}.$$

19. $y = (x^{1/2} - 3)^3 + (x^{1/2} - 3) + 5$

If $f(x) = x^3 + x + 5$ and
$g(x) = x^{1/2} - 3$, then
$$y = f[g(x)] = (x^{1/2} - 3)^3 + (x^{1/2} - 3) + 5.$$

21. $y = (7x - 12)^5$
$$y' = 5(7x - 12)^{5-1}\left[\frac{d}{dx}(7x - 12)\right]$$
$$= 5(7x - 12)^4(7) = 35(7x - 12)^4$$

23. $y = 7(4x + 3)^4$
$$y' = 7\left[\frac{d}{dx}\left((4x + 3)^4\right)\right] + \frac{d}{dx}(7) \cdot (4x + 3)^4$$
$$= 7\left[4(4x + 3)^3\right]\left[\frac{d}{dx}(4x + 3)\right] + 0$$
$$= 28(4x + 3)^3(4) = 112(4x + 3)^3$$

25. $y = 11(3x + 5)^{3/2}$
$$y' = 11\left[\frac{d}{dx}\left((3x + 5)^{3/2}\right)\right] + \frac{d}{dx}(11) \cdot (3x + 5)^{3/2}$$
$$= \frac{3}{2}\left[11(3x + 5)^{1/2}\right]\left[\frac{d}{dx}(3x + 5)\right] + 0$$
$$= \frac{33}{2}(3x + 5)^{1/2}(3) = \frac{99}{2}(3x + 5)^{1/2}$$

27. $y = -4(10x^2 + 8)^5$
$$y' = -4\left[\frac{d}{dx}\left((10x^2 + 8)^5\right)\right]$$
$$\quad + \frac{d}{dx}(-4) \cdot (10x^2 + 8)^5$$
$$= -4\left[5(10x^2 + 8)^4\right]\left[\frac{d}{dx}(10x^2 + 8)\right] + 0$$
$$= -20(10x^2 + 8)^4(20x) = -400x(10x^2 + 8)^4$$

29. $y = -7(4x^2 + 9x)^{3/2}$

$y' = -7\left(\dfrac{3}{2}\right)(4x^2 + 9x)^{1/2}(8x + 9)$

$\quad = -\dfrac{21}{2}(8x + 9)(4x^2 + 9x)^{1/2}$

31. $y = 8\sqrt{4x + 7} = 8(4x + 7)^{1/2}$

$y' = 8\left(\dfrac{1}{2}\right)(4x + 7)^{-1/2}(4)$

$\quad = 16(4x + 7)^{-1/2} = \dfrac{16}{\sqrt{4x + 7}}$

33. $y = -2\sqrt{x^2 + 4x} = -2(x^2 + 4x)^{1/2}$

$y' = -2\left(\dfrac{1}{2}\right)(x^2 + 4x)^{-1/2}(2x + 4)$

$\quad = -(2x + 4)(x^2 + 4x)^{-1/2} = \dfrac{-2x - 4}{\sqrt{x^2 + 4x}}$

35. $y = (x + 1)(x - 3)^2$

$y' = (x + 1)D_x(x - 3)^2 + (x - 3)^2 D_x(x + 1)$

$\quad = (x + 1)(2)(x - 3)(1) + (x - 3)^2(1)$

$\quad = 2(x + 1)(x - 3) + (x - 3)^2$

37. $y = 5(x + 3)^2(2x - 1)^5$

$y' = 5(x + 3)^2(5)(2x - 1)^4(2)$

$\qquad\qquad + (5)(2x - 1)^5(2)(x + 3)$

$\quad = 50(x + 3)^2(2x - 1)^4 + 10(2x - 1)^5(x + 3)$

$\quad = 10(x + 3)(2x - 1)^4 \cdot [5(x + 3) + (2x - 1)]$

$\quad = 10(x + 3)(2x - 1)^4(7x + 14)$

$\quad = 70(x + 3)(2x - 1)^4(x + 2)$

39. $y = (3x + 1)^3 \sqrt{x} = (3x + 1)^3(x^{1/2})$

$y' = (3x + 1)^3\left(\dfrac{1}{2}\right)(x^{-1/2})$

$\qquad\quad + (x^{1/2})(3)(3x + 1)^2(3)$

$\quad = \dfrac{1}{2}(x^{-1/2})(3x + 1)^3 + 9(x^{1/2})(3x + 1)^2$

$\quad = (x^{-1/2})(3x + 1)^2\left[\dfrac{1}{2}(3x + 1) + 9x\right]$

$\quad = (x^{-1/2})(3x + 1)^2\left(\dfrac{3x + 1 + 18x}{2}\right)$

$\quad = (x^{-1/2})(3x + 1)^2\left(\dfrac{21x + 1}{2}\right)$

$\quad = \dfrac{(3x + 1)^2(21x + 1)}{2\sqrt{x}}$

41. $y = \dfrac{1}{(x - 2)^3} = (x - 2)^{-3}$

$y' = -3(x - 2)^{-4} = \dfrac{-3}{(x - 2)^4}$

43. $y = \dfrac{(2x - 7)^2}{4x - 1}$

$y' = \dfrac{(4x - 1)(2)(2x - 7)(2) - (2x - 7)^2(4)}{(4x - 1)^2}$

$\quad = \dfrac{(16x - 4)(2x - 7) - (4)(2x - 7)^2}{(4x - 1)^2}$

$\quad = \dfrac{(2x - 7)[(16x - 4) - 4(2x - 7)]}{(4x - 1)^2}$

$\quad = \dfrac{(2x - 7)(16x - 4 - 8x + 28)}{(4x - 1)^2}$

$\quad = \dfrac{(2x - 7)(8x + 24)}{(4x - 1)^2}$

45. $y = \dfrac{x^2 + 5x}{(3x + 1)^3}$

$y' = \dfrac{(3x + 1)^3(2x + 5) - \left(x^2 + 5x\right)(3)(3x + 1)^2(3)}{(3x + 1)^6}$

$\quad = \dfrac{(3x + 1)^3(2x + 5) - \left(9x^2 + 45x\right)(3x + 1)^2}{(3x + 1)^6}$

$\quad = \dfrac{(3x + 1)^2\left[(3x + 1)(2x + 5) - \left(9x^2 + 45x\right)\right]}{(3x + 1)^6}$

$\quad = \dfrac{(3x + 1)^2\left(6x^2 + 17x + 5 - 9x^2 - 45x\right)}{(3x + 1)^6}$

$\quad = \dfrac{-3x^2 - 28x + 5}{(3x + 1)^4}$

47. $y = \left(x^{1/2} + 2\right)(2x + 3)$

$y' = \left(x^{1/2} + 2\right)(2) + (2x + 3)\left(\dfrac{1}{2}x^{-1/2}\right)$

$\quad = 2\sqrt{x} + 4 + \dfrac{2x + 3}{2\sqrt{x}}$

$\quad = \dfrac{4x + 8\sqrt{x} + 2x + 3}{2\sqrt{x}} = \dfrac{6x + 8\sqrt{x} + 3}{2\sqrt{x}}$

49. Use the table of values in the textbook.

    a. $x = 1$

$$D_x(f[g(x)]) = f'[g(x)] \cdot g'(x)$$
$$= f'[g(1)] \cdot g'(1)$$
$$= [f'(2)] \cdot \left(\frac{2}{7}\right)$$
$$= (-7)\left(\frac{2}{7}\right) = -2$$

    b. $x = 2$

$$D_x(f[g(x)]) = f'[g(x)] \cdot g'(x)$$
$$= f'[g(2)] \cdot g'(2)$$
$$= [f'(3)] \cdot \frac{3}{7}$$
$$= (-8)\left(\frac{3}{7}\right) = -\frac{24}{7}$$

51. $f(x) = (2x^2 + 3x + 1)^{50}$

$$f'(x) = 50(2x^2 + 3x + 1)^{49}(4x + 3)$$
$$f'(0) = 50(2 \cdot 0^2 + 3 \cdot 0 + 1)^{49}(4 \cdot 0 + 3)$$
$$= 50(1)(3) = 150$$

Since (d) is 150, the correct choice is (d).

53. $C(x) = 600 + \sqrt{200 + 25x^2 - x}$

$$= 600 + \left(200 + 25x^2 - x\right)^{1/2}$$

The marginal cost function is

$$C'(x) = \left(\frac{1}{2}\right)\left(200 + 25x^2 - x\right)^{-1/2}(50x - 1)$$

$$= \frac{50x - 1}{2\sqrt{200 + 25x^2 - x}}.$$

55. $R(x) = 15\sqrt{500x - 2x^2}, \quad 0 \le x \le 200$

$$= 15\left(500x - 2x^2\right)^{1/2}$$

    a.   The marginal revenue function is

$$R'(x) = 15\left(\frac{1}{2}\right)\left(500x - 2x^2\right)^{-1/2} \cdot (500 - 4x)$$

$$= \frac{15(500 - 4x)}{2\sqrt{500x - 2x^2}} = \frac{15(250 - 2x)}{\sqrt{500x - 2x^2}}$$

    b. $R'(40) = \dfrac{15(250 - 2 \cdot 40)}{\sqrt{500 \cdot 40 - 2 \cdot 40^2}} \approx \$19.67$

$$R'(80) = \frac{15(250 - 2 \cdot 80)}{\sqrt{500 \cdot 80 - 2 \cdot 80^2}} \approx \$8.19$$

$$R'(120) = \frac{15(250 - 2 \cdot 120)}{\sqrt{500 \cdot 120 - 2 \cdot 120^2}} \approx \$0.85$$

$$R'(160) = \frac{15(250 - 2 \cdot 160)}{\sqrt{500 \cdot 160 - 2 \cdot 160^2}} \approx -\$6.19$$

    c.   As the number of items sold increases, the rate of change of revenue per item sold decreases.

57. $A = 15{,}000\left(1 + \dfrac{r}{1200}\right)^{96}$

$$A' = 15{,}000(96)\left(1 + \frac{r}{1200}\right)^{95}\left(\frac{1}{1200}\right)$$

$$= 1200\left(1 + \frac{r}{1200}\right)^{95}$$

    a. $A'(2.5) = 1200\left(1 + \dfrac{2.5}{1200}\right)^{95}$

$$= \$1462.33 \text{ per percentage point}$$

    b. $A'(3.25) = 1200\left(1 + \dfrac{3.25}{1200}\right)^{95}$

$$= \$1551.57 \text{ per percentage point}$$

    c. $A'(4) = 1200\left(1 + \dfrac{4}{1200}\right)^{95}$

$$= \$1646.19 \text{ per percentage point}$$

59. $p = \dfrac{350}{x^{1/2}}; \quad x = 10n$

The marginal revenue product is

$$\frac{dR}{dn} = \left(p + x\frac{dp}{dx}\right)\frac{dx}{dn}$$

$$= \left[350x^{-1/2} + 10n\frac{d}{dx}\left(350x^{-1/2}\right)\right]10$$

$$= \left[350x^{-1/2} + 10n(350)\left(-\frac{1}{2}\right)\left(x^{-3/2}\right)\right]10$$

$$= \left[350x^{-1/2} + 10n\left(-175x^{-3/2}\right)\right]10$$

$$= \frac{3500}{x^{1/2}} - \frac{17{,}500n}{x^{3/2}}$$

If $n = 10$, then $x = 100$ and

$$\frac{dR}{dn} = \left[\frac{3500}{\sqrt{100}} - \frac{17{,}500(10)}{100^{3/2}}\right] = 350 - 175 = 175$$

The marginal revenue product is \$175/additional worker.

61. $r(t) = 2t; \ A(r) = \pi r^2$

$A[r(t)] = A(2t) = \pi(2t)^2 = 4\pi t^2$

$A = 4\pi t^2$ gives the area of the pollution in terms of the time since the pollutants were first emitted.

63. $D(p) = \dfrac{-p^2}{100} + 500; \ p(c) = 2c - 10$

The demand in terms of the cost is

$D(c) = D[p(c)] = \dfrac{-(2c-10)^2}{100} + 500$

$= \dfrac{-4(c-5)^2}{100} + 500$

$= \dfrac{-c^2 + 10c - 25}{25} + 500$

65. a. $x = \sqrt{15{,}000 - 1.5p}$ Solve for p.

$x^2 = 15{,}000 - 1.5p \Rightarrow$

$x^2 - 15{,}000 = -1.5p \Rightarrow \dfrac{x^2 - 150{,}000}{-1.5} = p \Rightarrow$

$\dfrac{-2(x^2 - 15{,}000)}{3} = p \Rightarrow \dfrac{30{,}000 - 2x^2}{3} = p$

Use the Revenue formula, $R = px$.

$R = \left(\dfrac{30{,}000 - 2x^2}{3}\right)x = \dfrac{30{,}000x - 2x^3}{3}$

b. $P(x) = R(x) - C(x)$

$= \left(\dfrac{30{,}000x - 2x^3}{3}\right) - (2000x + 3500)$

$= 10{,}000x - \dfrac{2x^3}{3} - 2000x - 3500$

$= 8000x - \dfrac{2x^3}{3} - 3500$

c. $P'(x) = 8000 - \dfrac{2}{3}\left(3x^2\right) = 8000 - 2x^2$

d. Find x by plugging $p = 25$ into the demand equation.

$x = \sqrt{15{,}000 - 1.5(25)} \approx 122.3213$

$P'(122.3213) \approx 8000 - 2(122.3213)^2$

$\approx -\$21{,}925$

67. a. To find the life expectancy of a "jawbreaker," set $r(t) = 0$ and solve for t.

$6 - \dfrac{3}{17}t = 0 \Rightarrow 6 = \dfrac{3}{17}t \Rightarrow 34 = t$

The life expectancy is 34 minutes.

b. Replace r with $6 - \dfrac{3}{17}t$.

$V(t) = \dfrac{4}{3}\pi\left(6 - \dfrac{3}{17}t\right)^3$

$V'(t) = 3\left[\dfrac{4}{3}\pi\left(6 - \dfrac{3}{17}t\right)^2\right]\left(-\dfrac{3}{17}\right)$

$= -\dfrac{12\pi}{17}\left(6 - \dfrac{3}{17}t\right)^2$

$V'(17) = -\dfrac{12\pi}{17}\left(6 - \dfrac{3}{17}(17)\right)^2 = -\dfrac{12\pi}{17}(3)^2$

$= -\dfrac{108\pi}{17} \ \text{mm}^3/\text{min}$

Replace r with $6 - \dfrac{3}{17}t$.

$S(t) = 4\pi\left(6 - \dfrac{3}{17}t\right)^2$

$S'(t) = 2\left[4\pi\left(6 - \dfrac{3}{17}t\right)\right]\left(-\dfrac{3}{17}\right)$

$= -\dfrac{24\pi}{17}\left(6 - \dfrac{3}{17}t\right)$

$S'(17) = -\dfrac{24\pi}{17}\left(6 - \dfrac{3}{17}(17)\right) = -\dfrac{24\pi}{17}(3)$

$= -\dfrac{72\pi}{17} \ \text{mm}^2/\text{min}$

In a person's mouth, at the 17-minute mark a jawbreaker's volume is decreasing at $\dfrac{108\pi}{17} \approx 20 \ \text{mm}^3/\text{min}$ and its surface area is decreasing at $\dfrac{72\pi}{17} \approx 13 \ \text{mm}^2/\text{min}$.

c. Answers vary.

69. $F(t) = 60 - \dfrac{150}{\sqrt{8+t^2}} = 60 - 150(8+t^2)^{-1/2}$

a. $F'(t) = -150\left(-\dfrac{1}{2}\right)(8+t^2)^{-3/2}(2t)$

$= 150t(8+t^2)^{-3/2} = \dfrac{150t}{(8+t^2)^{3/2}}$

b. The rate at which the cashier's speed is increasing after x hr is $F'(x)$.

After 5 hr,

$F'(5) = \dfrac{150(5)}{(8+5^2)^{3/2}}$

≈ 3.96 items/min per hr

After 10 hr,

$F'(10) = \dfrac{150(10)}{(8+10^2)^{3/2}}$

≈ 1.34 items/min per hr

After 20 hr,

$F'(20) = \dfrac{150(20)}{(8+20^2)^{3/2}}$

$\approx .36$ items/min per hr

After 40 hr,

$F'(40) = \dfrac{150(40)}{(8+40^2)^{3/2}}$

$\approx .09$ items/min per hr

c. The answers are decreasing with time. This is reasonable because there is an upper limit to how fast a cashier can work, so $F(t)$ has to level off as t increases.

71. The following solution presupposes the use of a TI-83 graphics calculator. Similar results can be obtained from other graphics calculators.

Enter $K(x) = \sqrt[3]{(2x-1)^2}$ as y_1 and

$K'(x) = \dfrac{4}{3\sqrt[3]{2x-1}}$ as y_2. Use $-3 \le x \le 3$ and $-3 \le y \le 3$. Graph the functions.

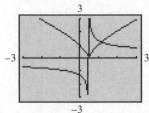

a. From the graph, $K'(x)$ is positive on the interval $(.5, \infty)$.

b. From the graph, $K'(x) = 0$ for no values of x, i.e., the derivative is never zero.

c. From the graph, $K'(x)$ is negative on the interval $(-\infty, .5)$.

d. The derivative does not exist at $x = .5$, which corresponds to a sharp point on the graph of $K(x)$. The derivative is positive when $K(x)$ is increasing and negative when $K(x)$ is decreasing.

Section 11.8 Derivatives of Exponential and Logarithmic Functions

1. $y = e^{5x}$

Let $g(x) = 5x$. Then $y = e^{g(x)}$, so $y' = g'(x) \cdot e^{g(x)}$. Since $g'(x) = 5$, $y' = 5e^{5x}$.

3. $f(x) = 5e^{2x}$

$f'(x) = 5(2e^{2x}) = 10e^{2x}$

5. $g(x) = -4e^{-7x}$

$g'(x) = -4(-7)e^{-7x} = 28e^{-7x}$

7. $y = e^{x^2}$

$y' = 2xe^{x^2}$

9. $f(x) = e^{x^3/3}$

$f'(x) = \dfrac{1}{3}\left(3x^2\right)e^{x^3/3} = x^2 e^{x^3/3}$

11. $y = -3e^{3x^2+5}$

$y' = (-3)(6x)e^{3x^2+5} = -18xe^{3x^2+5}$

13. $y = \ln\left(-10x^2 + 7x\right)$

$y' = \dfrac{D_x(-10x^2 + 7x)}{-10x^2 + 7x} = \dfrac{-20x + 7}{-10x^2 + 7x}$

15. $y = \ln\sqrt{5x+1} = \ln(5x+1)^{1/2}$

Use a property of logarithms.

$y = \dfrac{1}{2}\ln(5x+1)$

$y' = \dfrac{1}{2}\cdot\dfrac{D_x(5x+1)}{5x+1} = \dfrac{1}{2}\cdot\dfrac{5}{5x+1} = \dfrac{5}{2(5x+1)}$

17. $f(x) = \ln\left[(5x+9)(x^2+1)\right]$

$$f'(x) = \frac{(5x+9)(2x)+(x^2+1)(5)}{(x^2+1)(5x+9)}$$

$$= \frac{10x^2+18x+5x^2+5}{(5x+9)(x^2+1)} = \frac{15x^2+18x+5}{(x^2+1)(5x+9)}$$

19. $y = x^4 e^{-3x}$
$y' = 4x^3 e^{-3x} + x^4(-3)e^{-3x}$
$= 4x^3 e^{-3x} - 3x^4 e^{-3x} = \left(4x^3 - 3x^4\right)e^{-3x}$

21. $y = \left(6x^2 - 5x\right)e^{-3x}$

$y' = \left(6x^2 - 5x\right)(-3)e^{-3x} + e^{-3x}(12x-5)$
$= e^{-3x}\left(-18x^2 + 15x + 12x - 5\right)$
$= e^{-3x}\left(-18x^2 + 27x - 5\right)$

23. $y = \ln\left(\dfrac{4-x}{3x+8}\right)$

Use a property of logarithms.
$y = \ln(4-x) - \ln(3x+8)$
$y' = \dfrac{-1}{4-x} - \dfrac{3}{3x+8} = \dfrac{-(3x+8)-3(4-x)}{(4-x)(3x+8)}$
$= \dfrac{-3x-8-12+3x}{(4-x)(3x+8)} = \dfrac{-20}{(4-x)(3x+8)}$

25. $y = \ln\left(8x^3 - 3x\right)^{1/2} = \dfrac{1}{2}\ln\left(8x^3 - 3x\right)$

$= \dfrac{1}{2}\ln\left[x\left(8x^2 - 3\right)\right] = \dfrac{1}{2}\left[\ln x + \ln\left(8x^2 - 3\right)\right]$

$y' = \dfrac{1}{2}\left[\dfrac{1}{x} + \dfrac{16x}{8x^2 - 3}\right] = \dfrac{1}{2}\left[\dfrac{8x^2 - 3 + 16x^2}{x\left(8x^2 - 3\right)}\right]$

$= \dfrac{24x^2 - 3}{2x\left(8x^2 - 3\right)} = \dfrac{3\left(8x^2 - 1\right)}{2x\left(8x^2 - 3\right)}$

27. $y = x\ln\left(9 - x^4\right)$

$y' = x \cdot \dfrac{-4x^3}{9 - x^4} + \ln\left(9 - x^4\right) \cdot 1$
$= \dfrac{-4x^4}{9 - x^4} + \ln\left(9 - x^4\right)$

29. $y = \dfrac{\ln|x|}{x^4}$

$y' = \dfrac{x^4\left(\frac{1}{x}\right) - (\ln|x|)(4x^3)}{(x^4)^2} = \dfrac{x^3 - 4x^3\ln|x|}{x^8}$

$= \dfrac{x^3(1 - 4\ln|x|)}{x^8} = \dfrac{1 - 4\ln|x|}{x^5}$

31. $y = \dfrac{-5\ln|x|}{5 - 2x}$

$y' = \dfrac{(5-2x)\left(\frac{-5}{x}\right) - (-5\ln|x|)(-2)}{(5-2x)^2}$

$= \dfrac{\frac{-25}{x} + 10 - 10\ln|x|}{(5-2x)^2}$

33. $y = \dfrac{x^3 - 1}{2\ln|x|}$

$y' = \dfrac{2\ln|x|(3x^2) - (x^3-1)\left(\frac{2}{x}\right)}{4[\ln|x|]^2}$

$= \dfrac{6x^2\ln|x| - \frac{2(x^3-1)}{x}}{4(\ln|x|)^2}$

$= \dfrac{6x^3\ln|x| - 2(x^3-1)}{4x(\ln|x|)^2}$

$= \dfrac{3x^3\ln|x| - (x^3-1)}{2x(\ln|x|)^2}$

35. $y = \sqrt{\ln(x-3)} = [\ln(x-3)]^{1/2}$

$y' = \dfrac{1}{2}[\ln(x-3)]^{-1/2} \cdot D_x[\ln(x-3)]$

$= \dfrac{1}{2}[\ln(x-3)]^{-1/2}\dfrac{1}{(x-3)}$

$= \dfrac{1}{2(x-3)\sqrt{\ln(x-3)}}$

37. $y = \dfrac{e^x - 1}{\ln|x|}$

$y' = \dfrac{\ln|x| \cdot e^x - (e^x - 1)\frac{1}{x}}{(\ln|x|)^2}$

$= \dfrac{xe^x\ln|x| - e^x + 1}{x(\ln|x|)^2}$

39. $y = \dfrac{e^x - e^{-x}}{x}$

$$y' = \dfrac{x\left[e^x - (-1)e^{-x}\right] - \left(e^x - e^{-x}\right)(1)}{x^2}$$

$$= \dfrac{xe^x + xe^{-x} - e^x + e^{-x}}{x^2}$$

$$\text{or } \dfrac{e^x(x-1) + e^{-x}(x+1)}{x^2}$$

41. $f(x) = e^{3x+2}\ln(4x-5)$

$$f'(x) = e^{3x+2}\left(\dfrac{4}{4x-5}\right)$$

$$+ \left[\ln(4x-5)\right](e^{3x+2})(3)$$

$$= \dfrac{4e^{3x+2}}{4x-5} + 3e^{3x+2}\ln(4x-5)$$

43. $y = \dfrac{700}{7 - 10e^{.4x}} = 700(7 - 10e^{.4x})^{-1}$

$$y' = -700(7 - 10e^{.4x})^{-2}[-10e^{.4x}(.4)]$$

$$= \dfrac{2800e^{.4x}}{(7 - 10^{.4x})^2}$$

45. $y = \dfrac{500}{12 + 5e^{-.5x}}$

$$= 500(12 + 5e^{-.5x})^{-1}$$

$$y' = -500\left(12 + 5e^{-.5x}\right)^{-2}\left[(5)e^{-.5x}(-.5)\right]$$

$$= -500\left(12 + 5e^{-.5x}\right)^{-2}\left(-2.5e^{-.5x}\right)$$

$$= 1250e^{-.5x}\left(12 + 5e^{-.5x}\right)^{-2} = \dfrac{1250e^{-.5x}}{(12 + 5e^{-.5x})^2}$$

47. $y = 8^x$

$$y = \left(e^{\ln 8}\right)^x = e^{(\ln 8)x}$$

$$y' = (\ln 8)e^{(\ln 8)x} = (\ln 8)8^x$$

49. $y = 15^{2x}$

$$y = \left(e^{\ln 15}\right)^{2x} = e^{(\ln 15)(2x)}$$

$$y' = 2(\ln 15)e^{(\ln 15)(2x)} = 2(\ln 15) \cdot 15^{2x}$$

51. $g(x) = \log 6x = \dfrac{\ln 6x}{\ln 10} = \dfrac{1}{\ln 10} \cdot \ln 6x$

$$g'(x) = \dfrac{1}{\ln 10} \cdot \dfrac{d}{dx}(\ln 6x)$$

$$= \dfrac{1}{\ln 10} \cdot \left(\dfrac{1}{6x}\right) \cdot (6) = \dfrac{1}{(\ln 10)x}$$

53. $f(x) = \dfrac{e^x}{2x+1}$

$$f'(x) = \dfrac{(2x+1)(e^x) - (e^x)(2)}{(2x+1)^2} = \dfrac{e^x(2x-1)}{(2x+1)^2}$$

$$f'(0) = \dfrac{e^0(2(0)-1)}{(2(0)+1)^2} = \dfrac{-1}{1} = -1$$

The slope of the tangent line to the graph of f is $f'(0) = -1$.

$$(y - 1) = -1(x - 0) \Rightarrow y = -x + 1$$

55. $f(x) = \dfrac{x^2}{e^x}$

$$f'(x) = \dfrac{(e^x)(2x) - (x^2)(e^x)}{(e^x)^2} = \dfrac{-x^2 + 2x}{e^x}$$

$$f'(1) = \dfrac{-(1)^2 + 2(1)}{e^{(1)}} = \dfrac{1}{e}$$

The slope of the tangent line to the graph of f is

$$f'(1) = \dfrac{1}{e}. \text{ When } x = 1, \ f(1) = \dfrac{(1)^2}{e^1} = \dfrac{1}{e}.$$

$$\left(y - \dfrac{1}{e}\right) = \dfrac{1}{e}(x - 1) \Rightarrow y = \dfrac{1}{e}x$$

57. $f(x) = e^{2x}$

$$f'(x) = 2e^{2x}$$

$$f'\left[\ln\left(\dfrac{1}{4}\right)\right] = 2e^{2\ln(1/4)} = 2e^{\ln(1/4)^2}$$

$$= 2\left(\dfrac{1}{4}\right)^2 = 2\left(\dfrac{1}{16}\right) = \dfrac{1}{8}$$

59. $f(x) = 175e^{.014x}$

$$f'(x) = 175(.014)e^{.014x} = 2.45e^{.014x}$$

a. The year 2010 corresponds to $x = 20$.

$$f(20) = 175e^{.014(20)} \approx 231.5$$

$$f'(20) = 2.45e^{.014(20)} \approx 3.24$$

In 2010 there are about 231.5 million Internet users, increasing at about 3.24 million per year.

b. The year 2012 corresponds to $x = 22$.

$$f(20) = 175e^{.014(22)} \approx 238.1$$

$$f'(20) = 2.45e^{.014(22)} \approx 3.33$$

In 2012 there will be about 238.1 million Internet users, increasing at about 3.33 million per year.

61. $g(x) = .965e^{.287x}$

$g'(x) = (.287)(.965)e^{.287x} = .276955e^{.287x}$

a. For 2009, $x = 9$.

$g(9) = .965e^{.287(9)} \approx 12.77$

$g'(9) = .276955e^{.287(9)} \approx 3.67$

In 2009, the amount spent on broadcast and satellite ration was about $12.77, increasing at about $3.67 per year.

b. For 2010, $x = 10$.

$g(10) = .965e^{.287(10)} \approx 17.02$

$g'(10) = .276955e^{.287(10)} \approx 4.88$

In 2010, the amount spent on broadcast and satellite radio is about $17.02, increasing at about $4.88 per year.

c. For 2011, $x = 11$.

$g(11) = .965e^{.287(11)} \approx 22.68$

$g'(11) = .276955e^{.287(11)} \approx 6.51$

In 2011, the amount spent on broadcast and satellite radio will be about $22.68, increasing at about $6.51 per year.

63. $g(x) = 17.6e^{.043x}$

$g'(x) = (.043)(17.6e^{.043x}) = .7568e^{.043x}$

a. $g(10) = 17.6e^{.043(10)} \approx 27.1$

b. $g'(10) = .7568e^{.043(10)} \approx 1.2$

c. In 2010, about 27,100 doctorates were awarded in science in the U.S., increasing at about 1200 per year.

65. a. $M(200) = 3102e^{-e^{-.022(200-56)}} \approx 2974.15$ g

b. $M'(t) = 3102\left(-e^{-.022(t-56)}\right)(-.022)$

$\cdot \left(e^{-e^{-.022(t-56)}}\right)$

$= 68.244e^{-.022(t-56)}\left(e^{-e^{-.022(t-56)}}\right)$

$M'(200) = 68.244e^{-.022(200-56)}e^{-e^{-.022(200-56)}}$

≈ 2.75 g/day

c.

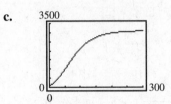

The growth pattern for $M(t)$ is increasing for 0 to 200 days and staying approximately the same for t greater than 200.

d.

Weight	Day	Rate
990.9797	50	24.87793
2121.673	100	17.72981
2733.571	150	7.603823
2974.153	200	2.753855
3058.845	250	.942782
3087.568	300	.316772

67. $f(x) = 1272e^{.0671x}$

$f'(x) = .0671\left(1272e^{.0671x}\right)$

The year 2010 corresponds to $x = 30$.

$f'(30) = .0671\left(1272e^{.0671(30)}\right) \approx 638.9$

Health care expenditures are changing at about $638.90 per capita in 2010.

69. a. $V(240) = \dfrac{1100}{(1+1023e^{-.02415(240)})^4} \approx 3.857 \text{ cm}^3$

b. $\dfrac{4}{3}\pi r^3 = 3.857 \Rightarrow r^3 \approx .92079 \Rightarrow r \approx .973 \text{ cm}$

c. $\dfrac{1100}{\left(1+1023e^{-.02415t}\right)^4} = .5 \Rightarrow \dfrac{\left(1+1023e^{-.02415t}\right)^4}{1100} = \dfrac{1}{.5} \Rightarrow \left(1+1023e^{-.02415t}\right)^4 = 2200 \Rightarrow$

$1+1023e^{-.02415t} \approx 6.84866 \Rightarrow 1023e^{-.02415t} \approx 5.84866 \Rightarrow e^{-.02415t} \approx .00572 \Rightarrow$

$\ln e^{-.02415t} \approx \ln .00572 \Rightarrow -.02415t \approx -5.16428 \Rightarrow t \approx 214$

According to the formula, a tumor of size 0.5 cm^3 has been growing for about 214 months.

d. $V'(t) = \dfrac{\left(1+1023e^{-.02415t}\right)^4 (0) - (1100)(4)\left(1+1023e^{-.02415t}\right)^3 (1023)(-.02415)\left(e^{-.02415t}\right)}{\left(1+1023e^{-.02415t}\right)^8}$

$= \dfrac{108,703.98\left(1+1023e^{-.02415t}\right)^3 \left(e^{-.02415t}\right)}{\left(1+1023e^{-.02415t}\right)^8} = \dfrac{108,703.98e^{-.02415t}}{\left(1+1023e^{-.02415t}\right)^5}$

$V'(240) = \dfrac{108,703.98e^{-.02415(240)}}{\left(1+1023e^{-.02415(240)}\right)^5} \approx .282 \text{ cm}^3 / \text{month}$

When the tumor is 240 months old, it is increasing in volume at the instantaneous rate of $.282 \text{ cm}^3 / \text{month}$.

71. Let $p = 100 + \dfrac{50}{\ln x}$.

a.   The revenue function is given by
$R(x) = xp = x\left(100 + \dfrac{50}{\ln x}\right)$. The marginal revenue is

$R'(x) = x\dfrac{d}{dx}\left(100 + \dfrac{50}{\ln x}\right) + \left(100 + \dfrac{50}{\ln x}\right)\dfrac{d}{dx}x$

$= x\left(-\dfrac{50}{x(\ln x)^2}\right) + \left(100 + \dfrac{50}{\ln x}\right)$

$= \dfrac{-50}{(\ln x)^2} + \dfrac{50}{\ln x} + 100$

$= \dfrac{50(\ln x - 1)}{(\ln x)^2} + 100$

b.   The revenue from the next thousand items when $x = 8$ is
$R'(8) = \dfrac{50(\ln 8 - 1)}{(\ln 8)^2} + 100 = \$112.48$

73. $f(x) = \dfrac{67,338 - 12,595\ln x}{x}$

$f'(x) = \dfrac{x\left(-12,595 \cdot \frac{1}{x}\right) - (67,338 - 12,595\ln x)}{x^2}$

$= \dfrac{-12,595 - 67,338 + 12,595\ln x}{x^2}$

$= \dfrac{-79,933 + 12,595\ln x}{x^2}$

a. $f(30) = \dfrac{67,338 - 12,595\ln(30)}{30} \approx 816.66$

$f'(30) = \dfrac{-79,933 + 12,595\ln(30)}{30^2} \approx -41.2$

When the street is 30 feet wide, the maximum traffic flow is about 817 cars per hour, and the rate of change of the maximum traffic flow is about –41.2 cars per hour per foot.

b. $f(40) = \dfrac{67,338 - 12,595\ln(40)}{40} \approx 521.91$

$f'(40) = \dfrac{-79,933 + 12,595\ln(40)}{40^2} \approx -20.9$

When the street is 40 feet wide, the maximum traffic flow is about 522 cars per hour, and the rate of change of the maximum traffic flow is about –20.9 cars per hour per foot.

75. $g(x) = -232.3 + 2403\ln x, \quad (x \ge 5)$

$g'(x) = \dfrac{2403}{x}$

a. For 2009, $x = 19$.

$g(19) = -232.3 + 2403\ln 19 \approx 6843$

There were about 6843 liver transplants in the U.S. in 2009.

b. $g'(19) = \dfrac{2403}{19} \approx 126$

In 2009, liver transplants were increasing at about 126 per year.

77. a. $\dfrac{2\ln E - 2\ln.007}{3\ln 10} = 8.9$

$2\ln E - 2\ln.007 \approx 61.479$

$2\ln E \approx 51.555$

$\ln E \approx 25.778$

$E \approx 1.567 \times 10^{11}$ kWh

b. $247 \times 10,000,000 = 2,470,000,000$

10 million households use 2,470,000,000 kWh/month.

$\dfrac{1.567 \times 10^{11}}{2,470,000,000} \approx 63.4$

The energy released by this earthquake would power 10 million households for about 63.4 months.

c. $M'(E) = \dfrac{(3\ln 10)\left(\frac{2}{E}\right) - (0)(2\ln E - 2\ln.007)}{(3\ln 10)^2}$

$= \dfrac{\frac{6\ln 10}{E}}{(3\ln 10)^2} = \dfrac{2}{E(3\ln 10)}$

$M'(70,000) = \dfrac{2}{70,000(3\ln 10)}$

$\approx 4.14 \times 10^{-6}$ Munit/kWh

d. $\displaystyle\lim_{E \to \infty} \dfrac{2}{E(3\ln 10)} = 0$

As E increases, $\dfrac{dM}{dE}$ approaches zero.

Section 11.9 Continuity and Differentiability

1. $\displaystyle\lim_{x \to 0} f(x)$ does not exist and $f(x)$ is not defined at $x = 0$. The open circle at $x = 2$ means $f(2)$ does not exist. Thus, the function is discontinuous at $x = 0$ and $x = 2$.

3. The limit of $h(x)$ as x approaches 1 is –2.5. The value of $h(x)$ at $x = 1$ is 2 (solid circle). Since the limit of $h(x)$ and the value of $h(x)$ are not equal, $h(x)$ is discontinuous at $x = 1$.

5. $\displaystyle\lim_{x \to -3} y$ does not exist because y approaches 2 from the left and 3 from the right at $x = -3$, so y is not continuous at $x = -3$. Likewise, $\displaystyle\lim_{x \to 0} y$ and $\displaystyle\lim_{x \to 3} y$ do not exist because y approaches different values from the left and right at both $x = 0$ and $x = 3$. Therefore, y is discontinuous at $x = -3$, $x = 0$, and $x = 3$.

7. $\displaystyle\lim_{x \to 0} y$ and $\displaystyle\lim_{x \to 6} y$ do not exist because y approaches different values from the left and from the right at both $x = 0$ and $x = 6$. Thus, y is discontinuous at $x = 0$ and $x = 6$.

In Exercises 9–19, the value of the limit of a function is found by the methods established earlier in the chapter. Details are omitted here.

9. $f(x) = \dfrac{4}{x - 2}; x = 0, x = 2$

$f(0) = \dfrac{4}{-2} = -2$

$f(x)$ is continuous at $x = 0$ because $f(0)$ is defined, $\displaystyle\lim_{x \to 0} f(x)$ exists, and $\displaystyle\lim_{x \to 0} f(x) = f(0)$.

$f(2) = \dfrac{4}{2 - 2} = \dfrac{4}{0}$, so $f(x)$ is not defined at $x = 2$. Hence, $f(x)$ is not continuous at $x = 2$.

11. $h(x) = \dfrac{1}{x(x-3)}; x = 0; x = 3; x = 5$

$h(0) = \dfrac{1}{0(0-3)} = \dfrac{1}{0}$

so $h(x)$ is not continuous at $x = 0$.

$h(3) = \dfrac{1}{3(3-3)} = \dfrac{1}{0}$

so $h(x)$ is not continuous at $x = 3$.

$h(5) = \dfrac{1}{5(5-3)} = \dfrac{1}{10}$

$\displaystyle\lim_{x \to 5} \dfrac{1}{x(x-3)} = \dfrac{1}{10}$, so $h(x)$ is continuous at

$x = 5$.

13. $g(x) = \dfrac{x+2}{x^2 - x - 2} = \dfrac{x+2}{(x-2)(x+1)}$;

$x = 1, x = 2, x = -2$

$g(1) = \dfrac{1+2}{(1-2)(1+1)} = -\dfrac{3}{2}$

$\displaystyle\lim_{x \to 1} \dfrac{x+2}{x^2 - x - 2} = -\dfrac{3}{2}$, so $g(x)$ is continuous at

$x = 1$.

$g(2) = \dfrac{2+2}{(2-2)(2+1)} = \dfrac{4}{0(3)} = \dfrac{4}{0}$

$g(x)$ is not defined at $x = 2$, so $g(x)$ is not continuous at $x = 2$.

$g(-2) = \dfrac{-2+2}{(-2-2)(-2+1)} = \dfrac{0}{-4(-1)}$

$\quad\quad = 0$

$\displaystyle\lim_{x \to -2} \dfrac{x+2}{x^2 - x - 2} = 0$, so $g(x)$ is continuous at

$x = -2$.

15. $g(x) = \dfrac{x^2 - 4}{x - 2}; x = 0, x = 2, x = -2$

$g(0) = \dfrac{0-4}{0-2} = 2$

$\displaystyle\lim_{x \to 0} g(x) = 2$, so $g(x)$ is continuous at $x = 0$.

$g(2) = \dfrac{4-4}{2-2} = \dfrac{0}{0}$

so $g(x)$ is not continuous at $x = 2$.

$g(-2) = \dfrac{4-4}{-4} = 0$

$\displaystyle\lim_{x \to -2} g(x) = 0$, so $g(x)$ is continuous at $x = -2$.

17. $f(x) = \begin{cases} x - 2 & \text{if } x \le 3 \\ 2 - x & \text{if } x > 3 \end{cases}; x = 2, x = 3$

$f(2) = 2 - 2 = 0$

$\displaystyle\lim_{x \to 2} f(x) = 0$

so $f(x)$ is continuous at $x = 2$.

$f(3) = 3 - 2 = 1$

$\displaystyle\lim_{x \to 3} f(x)$ does not exist because the limit approaches different values when approached from the left and right so, $f(x)$ is not continuous at $x = 3$.

19. $f(x) = \dfrac{7+x}{x(x-3)}$ $f(x)$ is not defined at $x = 0$,

so $f(x)$ is not continuous at $x = 0$. $f(x)$ is not defined at $x = 3$; so $f(x)$ is not continuous at $x = 3$. Thus, the function is discontinuous at $x = 0$ and $x = 3$.

21. $g(x) = x^2 - 5x + 12$

$g(x)$ is continuous everywhere.

23. $k(x) = e^{\sqrt{x-2}}$

Since $\sqrt{x-2}$ is not defined for $x - 2 < 0$, or $x < 2$, $k(x)$ is not defined for $x < 2$. Thus, $k(x)$ is not continuous for $x < 2$.

25. a.

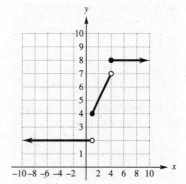

b. Since $\displaystyle\lim_{x \to 1} f(x)$ and $\displaystyle\lim_{x \to 4} f(x)$ do not exist, $f(x)$ is discontinuous at $x = 1$ and $x = 4$.

27. $f(x) = \begin{cases} x+k & \text{if } x \le 2 \\ 5-x & \text{if } x > 2 \end{cases}$

We wish to find k, such that

$2 + k = 5 - 2$,

in order for $\lim\limits_{x \to 2} f(x)$ to be the same number

from the left or right.

$2 + k = 5 - 2 \Rightarrow 2 + k = 3 \Rightarrow k = 1$

29. a. $\lim\limits_{x \to 82} T(x) = 4.75$ cents

b. $\lim\limits_{x \to 04^-} T(x) = 6$ cents

c. $\lim\limits_{x \to 04^+} T(x) = 6.25$ cents

d. $\lim\limits_{x \to 04} T(x)$ does not exists since $T(x)$ does

not have the same value as x approaches
from the left and right.

f. The graph of $T(x)$ is discontinuous for
years 1991, 2000, 2001, 2004 and 2009.

31. The function is discontinuous only at $t = m$,
where $\lim\limits_{t \to m} f(t) = 45\%$ from the left and

$\lim\limits_{t \to m} f(t) = 70\%$ from the right. The function is

differentiable everywhere except at $t = m$ and at
the endpoints.

33. a. $\lim\limits_{x \to 5^-} C(x) = \$4.60$

b. $\lim\limits_{x \to 5^+} C(x) = \$5.44$

c. $\lim\limits_{x \to 5} C(x)$ does not exists since $C(x)$ does

not have the same value as x approaches
from the left and right.

d. $C(5) = \$4.60$

e. $C(x)$ is discontinuous on 1, 2, 3, 4, 5, 6,
and 7.

f.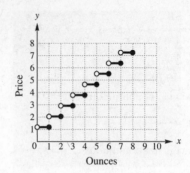

35. a. $C(5) = 5(120) = 600$
The cost for 5 days is $600.

b. $C(6) = C(5) = 5(120) = 600$
The cost for 6 days is $600.

c. $C(7) = C(5) = 600$
The cost for 7 days is $600.

d. Because 6 and 7 are free, 8 days is
equivalent to 6 days for which a fee is
charged.
$C(8) = 6(120) = 720$
The cost for 8 days is $720.

e. $\lim\limits_{t \to 5} C(t) = C(5) = \$600$

f. $\lim\limits_{t \to 6} C(t) = C(6) = \$600$

37. $[-3, 16]$ since x must be greater than or equal to
-3 and less than or equal to 16.

39. $(12, \infty)$ since x must be greater than 12.

41. From the graph in Exercise 6, the function is
continuous on $(-6, 0)$, $(4, 8)$. Because the
function is not continuous as x approaches 0
from the right, the function is not continuous on
$[0, 3]$.

Chapter 11 Review Exercises

1. The graph of $f(x)$ approaches 4 from the left and
from the right of $x = -3$. Therefore, $\lim\limits_{x \to -3} f(x) = 4$.

2. $\lim\limits_{x \to -1} g(x)$ does not exist, since the graph shows
that $g(x)$ approaches 2 as x approaches -1 from
the right but $g(x)$ approaches -2 as x approaches
-1 from the left.

3. $\lim\limits_{x\to 1}\dfrac{x^3-1.1x^2-2x+2.1}{x-1}$

x	$\dfrac{x^3-1.1x^2-2x+2.1}{x-1}$
.9	−1.38
.99	−1.2189
.999	−1.2019
1	
1.001	−1.1981
1.01	−1.1809
1.1	−1

As x gets closer to 1 from the left or right, the value of $\dfrac{x^3-1.1x^2-2x+2.1}{x-1}$ gets closer to −1.2, so $\lim\limits_{x\to 1}\dfrac{x^3-1.1x^2-2x+2.1}{x-1}=-1.2$.

4. $\lim\limits_{x\to 2}\dfrac{x^4+.5x^3-4.5x^2-2.5x+3}{x-2}$

x	$\dfrac{x^4+.5x^3-4.5x^2-2.5x+3}{x-2}$
1.9	15.334
1.99	17.2758
1.999	17.4775
2	
2.001	17.5225
2.01	17.7259
2.1	19.836

As x gets closer to 2 from the left or right, the value of $\dfrac{x^4+.5x^3-4.5x^2-2.5x+3}{x-2}$ gets closer to 17.5, so $\lim\limits_{x\to 2}\dfrac{x^4+.5x^3-4.5x^2-2.5x+3}{x-2}=17.5$.

5. $\lim\limits_{x\to 0}\dfrac{\sqrt{2-x}-\sqrt{2}}{x}$

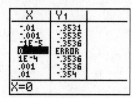

As x gets closer to 0 from the left or right, $\dfrac{\sqrt{2-x}-\sqrt{2}}{x}$ gets closer to −.3536, so $\lim\limits_{x\to 0}\dfrac{\sqrt{2-x}-\sqrt{2}}{x}=-.3536$.

6. $\lim\limits_{x\to -1}\dfrac{10^x-.1}{x+1}$

X	Y₁	
-1.1	.20567	
-1.01	.22763	
-1.001	.22999	
	ERROR	
-.9999	.23029	
-.999	.23052	
-.99	.23293	
X= -1		

As x approaches −1 from the left or right, $\dfrac{10^x-.1}{x+1}$ approaches .23, so $\lim\limits_{x\to -1}\dfrac{10^x-.1}{x+1}=.23.$

7. x^2-5x+2 is a polynomial, so $\lim\limits_{x\to 3}\left(x^2-5x+2\right)=3^2-5(3)+2=-4.$

8. $-3x^2+x+4$ is a polynomial, so $\lim\limits_{x\to -2}\left(-3x^2+x+4\right)=-3(-2)^2+(-2)+4=-10.$

9. $\dfrac{5x+1}{x-4}$ is continuous except at 4, where the expression is not defined. We must investigate the expression as x approaches 4 from the left and right.

X	Y₁	
4.1	215	
4.01	2105	
4.001	21005	
	ERROR	
3.999	-20995	
3.99	-2095	
3.9	-205	
X=4		

(*continued on next page*)

(continued from page 363)

Because the expression approaches values that are not the same (21,000) and (–21,000) as x approaches 4 from the left and from the right, $\lim\limits_{x \to 4} \dfrac{5x+1}{x-4}$ does not exist.

10. $\dfrac{5x-3}{x+1}$ is continuous except at –1, where the expression is not defined. The function is defined for $x = 5$, so $\lim\limits_{x \to 5} \dfrac{5x-3}{x+1} = \dfrac{5(5)-3}{5+1} = \dfrac{22}{6} = \dfrac{11}{3}$.

11. $\lim\limits_{x \to 5} \dfrac{x^2-25}{x+5}$

Although $f(x) = \dfrac{x^2-25}{x+5}$ is discontinuous at $x = -5$, this limit exists. We rewrite the fraction and use the Limit Theorem from Section 11.1.

$$\lim_{x \to 5} \frac{x^2-25}{x+5} = \lim_{x \to 5} \frac{(x+5)(x-5)}{x+5}$$
$$= \lim_{x \to 5}(x-5) = 5 - 5 = 0$$

12. $\lim\limits_{x \to 5} \dfrac{x^2-2x-3}{x+1}$

Although the function is discontinuous at $x = -1$, this limit exists.

$$\lim_{x \to 5} \frac{x^2-2x-3}{x+1} = \lim_{x \to 5} \frac{(x-3)(x+1)}{x+1}$$
$$= \lim_{x \to 5}(x-3) = 5 - 3 = 2$$

13. $\lim\limits_{x \to -4} \dfrac{2x^2+3x-20}{x+4} = \lim\limits_{x \to -4} \dfrac{(x+4)(2x-5)}{x+4}$
$$= \lim_{x \to -4}(2x-5)$$
$$= 2(-4) - 5 = -13$$

14. $\lim\limits_{x \to 3} \dfrac{3x^2-2x-21}{x-3} = \lim\limits_{x \to 3} \dfrac{(x-3)(3x+7)}{x-3}$
$$= \lim_{x \to 3}(3x+7) = 3(3)+7$$
$$= 16$$

15. $\lim\limits_{x \to 9} \dfrac{\sqrt{x}-3}{x-9} = \lim\limits_{x \to 9} \dfrac{(\sqrt{x}-3)(\sqrt{x}+3)}{(x-9)(\sqrt{x}+3)}$
$$= \lim_{x \to 9} \frac{x-9}{(x-9)(\sqrt{x}+3)}$$
$$= \lim_{x \to 9} \frac{1}{\sqrt{x}+3} = \frac{1}{\sqrt{9}+3} = \frac{1}{6}$$

16. $\lim\limits_{x \to 16} \dfrac{\sqrt{x}-4}{x-16} = \lim\limits_{x \to 16} \dfrac{(\sqrt{x}-4)(\sqrt{x}+4)}{(x-16)(\sqrt{x}+4)}$
$$= \lim_{x \to 16} \frac{x-16}{(x-16)(\sqrt{x}+4)}$$
$$= \lim_{x \to 16} \frac{1}{\sqrt{x}+4} = \frac{1}{\sqrt{16}+4} = \frac{1}{8}$$

17. $\lim\limits_{x \to -1^+} \sqrt{9+8x-x^2}$
$$= \lim_{x \to -1^+} \sqrt{(9-x)(1+x)}$$
$$= \left(\lim_{x \to -1^+} \sqrt{9-x}\right)\left(\lim_{x \to -1^+} \sqrt{1+x}\right)$$
$$= \sqrt{10} \cdot 0 = 0$$

18. $\lim\limits_{x \to 8^+} \dfrac{x^2-64}{x-8} = \lim\limits_{x \to 8^+} \dfrac{(x-8)(x+8)}{x-8}$
$$= \lim_{x \to 8^+}(x+8) = \lim_{x \to 8^+} x + \lim_{x \to 8^+} 8$$
$$= 8 + 8 = 16$$

19. $\lim\limits_{x \to -5^+} \dfrac{|x+5|}{x+5}$

As x approaches –5 from the right, $|x+5| = (x+5)$, so

$$\lim_{x \to -5^+} \frac{|x+5|}{x+5} = \lim_{x \to -5^+} \frac{(x+5)}{x+5} = \lim_{x \to -5^+} 1 = 1$$

20. $\lim\limits_{x \to 7^-} \left(\sqrt{7-x^2+6x}+2\right)$
$$= \lim_{x \to 7^-} \sqrt{7-x^2+6x} + \lim_{x \to 7^-} 2 = 0 + 2 = 2$$

21. $\lim\limits_{x \to \infty} \dfrac{2x^3-3x^2+5x-1}{4x^3+2x^2-x+10} = \lim\limits_{x \to \infty} \dfrac{2-\dfrac{3}{x}+\dfrac{5}{x^2}-\dfrac{1}{x^3}}{4+\dfrac{2}{x}-\dfrac{1}{x^2}+\dfrac{10}{x^3}}$

$$= \frac{\lim\limits_{x \to \infty} 2-\dfrac{3}{x}+\dfrac{5}{x^2}-\dfrac{1}{x^3}}{\lim\limits_{x \to \infty} 4+\dfrac{2}{x}-\dfrac{1}{x^2}+\dfrac{10}{x^3}}$$

$$= \frac{2}{4} = \frac{1}{2}$$

22. $\displaystyle\lim_{x \to -\infty} \frac{4 - 3x - 2x^2}{x^3 + 2x + 5} = \lim_{x \to -\infty} \frac{\dfrac{4}{x^3} - \dfrac{3}{x^2} - \dfrac{2}{x}}{1 + \dfrac{2}{x^2} + \dfrac{5}{x^3}}$

$\qquad = \dfrac{\displaystyle\lim_{x \to -\infty} \dfrac{4}{x^3} - \dfrac{3}{x^2} - \dfrac{2}{x}}{\displaystyle\lim_{x \to -\infty} 1 + \dfrac{2}{x^2} + \dfrac{5}{x^3}}$

$\qquad = \dfrac{0}{1} = 0$

23. $\displaystyle\lim_{x \to -\infty} \left(\frac{2x+1}{x-3} + \frac{4x-1}{3x} \right)$

$\qquad = \displaystyle\lim_{x \to -\infty} \left(\frac{6x^2 + 3x + 4x^2 - 13x + 3}{3x^2 - 9x} \right)$

$\qquad = \displaystyle\lim_{x \to -\infty} \frac{10x^2 - 10x + 3}{3x^2 - 9x} = \lim_{x \to -\infty} \frac{10 - \dfrac{10}{x} + \dfrac{3}{x^2}}{3 - \dfrac{9}{x}}$

$\qquad = \dfrac{\displaystyle\lim_{x \to -\infty} 10 - \dfrac{10}{x} + \dfrac{3}{x^2}}{\displaystyle\lim_{x \to -\infty} 3 - \dfrac{9}{x}} = \dfrac{10}{3}$

24. $\displaystyle\lim_{x \to \infty} \frac{x}{(\ln x)^2}$

X	Y1	
5	1.9303	
50	3.2671	
500	12.946	
5000	68.925	
50000	427.1	
500000	2903.7	
5E6	21015	

X=5000000

By constructing a table using a calculator, it is obvious that as x approaches infinity, $\dfrac{x}{(\ln x)^2}$

gets larger and larger. Thus, $\displaystyle\lim_{x \to \infty} \frac{x}{(\ln x)^2} = \infty$.

25. $x = 0$ to $x = 4$

From the graph, $f(0) = 0$ and $f(4) = 1$.

Average rate of change

$= \dfrac{f(4) - f(0)}{4 - 0} = \dfrac{1 - 0}{4 - 0} = \dfrac{1}{4}$

26. $x = 2$ to $x = 8$

From the graph, $f(8) = 4$ and $f(2) = 4$.

Average rate of change

$= \dfrac{f(8) - f(2)}{8 - 2} = \dfrac{4 - 4}{8 - 2} = 0$

27. $f(x) = 3x^2 - 5$, from $x = 1$ to $x = 5$

$f(1) = 3(1^2) - 5 = -2$

$f(5) = 3(5^2) - 5 = 70$

Average rate of change

$= \dfrac{f(5) - f(1)}{5 - 1} = \dfrac{70 - (-2)}{4} = 18$

28. $g(x) = -x^3 + 2x^2 + 1$,

from $x = -2$ to $x = 3$

$g(-2) = -(-2)^3 + 2(-2)^2 + 1 = 17$

$g(3) = -3^3 + 2(3)^2 + 1 = -8$

Average rate of change

$= \dfrac{g(3) - g(-2)}{3 - (-2)} = \dfrac{-8 - 17}{5} = \dfrac{-25}{5} = -5$

29. $h(x) = \dfrac{6 - x}{2x + 3}$, from $x = 0$ to $x = 6$

$h(0) = \dfrac{6 - 0}{2(0) + 3} = \dfrac{6}{3} = 2$

$h(6) = \dfrac{6 - 6}{2(6) + 3} = \dfrac{0}{15} = 0$

Average rate of change

$= \dfrac{h(6) - h(0)}{6 - 0} = \dfrac{0 - 2}{6} = -\dfrac{1}{3}$

30. $f(x) = e^{2x} + 5 \ln x$

from $x = 1$ to $x = 7$

$f(1) = e^2 + 5 \ln 1 \approx 7.389056099$

$f(7) = e^{14} + 5 \ln 7 \approx 1,206,614$

Average rate of change

$= \dfrac{f(7) - f(1)}{7 - 1} = \dfrac{1,206,206.6}{6}$

$\approx 200,434.44$

31. $y = 2x + 3 = f(x)$

$f(x + h) = 2(x + h) + 3 = 2x + 2h + 3$

$f(x + h) - f(x) = (2x + 2h + 3) - (2x + 3) = 2h$

$\dfrac{f(x + h) - f(x)}{h} = \dfrac{2h}{h} = 2$

$f'(x) = y' = \displaystyle\lim_{h \to 0} \frac{f(x + h) - f(x)}{h} = \lim_{h \to 0} 2 = 2$

32. $y = x^2 + 2x$

$$y' = \lim_{h \to 0} \frac{[(x+h)^2 + 2(x+h)] - (x^2 + 2x)}{h}$$

$$= \lim_{h \to 0} \frac{x^2 + 2xh + h^2 + 2x + 2h - x^2 - 2x}{h}$$

$$= \lim_{h \to 0} \frac{2xh + h^2 + 2h}{h} = \lim_{h \to 0} (2x + h + 2)$$

$$= 2x + 2$$

33. $y = 2x^2 - x - 1$

$$y' = \lim_{h \to 0} \frac{[2(x+h)^2 - (x+h) - 1] - (2x^2 - x - 1)}{h}$$

$$= \lim_{h \to 0} \frac{2x^2 + 4xh + 2h^2 - x - h - 1 - 2x^2 + x + 1}{h}$$

$$= \lim_{h \to 0} \frac{4xh + 2h^2 - h}{h} = \lim_{h \to 0} (4x + 2h - 1)$$

$$= 4x - 1$$

34. $y = x^3 + 5$

$$y' = \lim_{h \to 0} \frac{[(x+h)^3 + 5] - (x^3 + 5)}{h}$$

$$= \lim_{h \to 0} \frac{x^3 + 3x^2h + 3xh^2 + h^3 + 5 - x^3 - 5}{h}$$

$$= \lim_{h \to 0} \frac{3x^2h + 3xh^2 + h^3}{h}$$

$$= \lim_{h \to 0} (3x^2 + 3xh + h^2) = 3x^2$$

35. $y = x^2 - 6x$; at $x = 2$

Let $f(x) = y$.

$$f(2) = 2^2 - 6(2) = -8$$

$$f'(x) = 2x - 6$$

The slope of the tangent line at $(2, -8)$ is

$$f'(2) = 2(2) - 6 = -2.$$

Use point-slope form with $x_1 = 2$, $y_1 = -8$ and $m = -2$ to find equation of tangent line.

$$y - y_1 = m(x - x_1)$$

$$y - (-8) = -2(x - 2) \Rightarrow y + 8 = -2x + 4 \Rightarrow$$

$$y = -2x - 4$$

36. $y = 8 - x^2$; at $x = 1$

Let $f(x) = y$.

$$f(1) = 8 - 1^2 = 7$$

$$f'(x) = -2x$$

The slope of the tangent line at $(1, 7)$ is

$$f'(1) = -2(1) = -2$$

The equation is

$$y - y_1 = m(x - x_1)$$

$$y - 7 = -2(x - 1) \Rightarrow y - 7 = -2x + 2 \Rightarrow$$

$$y = -2x + 9.$$

37. $y = \dfrac{-2}{x+5}$; at $x = -2$

Let $f(x) = y$.

$$f(-2) = \frac{-2}{-2+5} = -\frac{2}{3}$$

$$f(x) = -2(x+5)^{-1}$$

$$f'(x) = -2[-1(x+5)^{-2}(1)] = \frac{2}{(x+5)^2}$$

$$\text{slope} = f'(-2) = \frac{2}{(-2+5)^2} = \frac{2}{(3)^2} = \frac{2}{9}$$

The equation is

$$y - y_1 = m(x - x_1)$$

$$y + \frac{2}{3} = \frac{2}{9}(x+2) \Rightarrow 9y + 6 = 2(x+2) \Rightarrow$$

$$9y + 6 = 2x + 4 \Rightarrow y = \frac{2}{9}x - \frac{2}{9}.$$

38. $y = \sqrt{6x - 2}$; at $x = 3$

Let $f(x) = y$.

$$f(3) = \sqrt{6(3) - 2} = 4$$

$$f(x) = (6x - 2)^{1/2}$$

$$f'(x) = \frac{1}{2}(6x-2)^{-1/2}(6) = 3(6x-2)^{-1/2}$$

$$\text{slope} = y'(3) = 3(6 \cdot 3 - 2)^{-1/2}$$

$$= 3(16)^{-1/2} = \frac{3}{16^{1/2}} = \frac{3}{4}$$

The equation is

$$y - 4 = \frac{3}{4}(x - 3) \Rightarrow 4y - 16 = 3x - 9 \Rightarrow$$

$$4y = 3x + 7 \Rightarrow y = \frac{3}{4}x + \frac{7}{4}$$

39. $T(p) = .06p^4 - 1.25p^3 + 6.5p^2 - 18p + 200$

  $(0 < p \le 11)$

  $T(8) = .06(8)^4 - 1.25(8)^3 + 6.5(8)^2 - 18(8) + 200$
  $\quad = 77.76$

  $T(5) = .06(5)^4 - 1.25(5)^3 + 6.5(5)^2 - 18(5) + 200$
  $\quad = 153.75$

  a. Average rate of change in demand for price change from \$5 to \$8 is

  $\dfrac{T(8) - T(5)}{8 - 5} = \dfrac{77.76 - 153.75}{3}$
  $\qquad\qquad = -25.33$ per dollar.

  b. $T'(p) = .24p^3 - 3.75p^2 + 13p - 18$

  Instantaneous rate of change in demand when price is \$5 is

  $T'(5) = .24(5)^3 - 3.75(5)^2 + 13(5) - 18$
  $\quad = -16.75$ per dollar

  c. Instantaneous rate of change in demand when price is \$8 is

  $T'(8) = .24(8)^3 - 3.75(8)^2 + 13(8) - 18$
  $\quad = -31.12$ per dollar

40. The average rate of change of a function $f(x)$

  from $x = 0$ to $x = 4$ is $\dfrac{f(4) - f(0)}{4 - 0}$.

  If this quotient equals 0, it means that
  $f(4) - f(0) = 0$ or $f(4) = f(0)$.

  It is not necessary for f to be constant between $x = 0$ and $x = 4$, although that could be the case. For example, suppose we have the quadratic function $f(x) = (x - 2)^2$. Then $f(0) = 4$ and

  $f(4) = 4$, and $\dfrac{f(4) - f(0)}{4 - 0} = \dfrac{4 - 4}{4} = 0$, although

  $f(x)$ is not constant from $x = 0$ to $x = 4$.

41. $y = 6x^2 - 7x + 3$
  $y' = 6(2x) - 7 = 12x - 7$

42. $y = x^3 - 5x^2 + 1$
  $y' = 3x^2 - 2(5x) = 3x^2 - 10x$

43. $y = 4x^{9/2}$

  $y' = 4\left(\dfrac{9}{2}\right)\left(x^{7/2}\right) = 18x^{7/2}$

44. $y = -4x^{-5}$

  $y' = (-5)(-4)x^{-6} = 20x^{-6}$ or $\dfrac{20}{x^6}$

45. $f(x) = x^{-6} + \sqrt{x} = x^{-6} + x^{1/2}$

  $f'(x) = -6x^{-7} + \left(\dfrac{1}{2}\right)x^{-1/2} = -\dfrac{6}{x^7} + \dfrac{1}{2x^{1/2}}$

46. $f(x) = 8x^{-3} - 5\sqrt{x} = 8x^{-3} - 5(x)^{1/2}$

  $f'(x) = 8(-3)x^{-4} - 5\left(\dfrac{1}{2}\right)x^{-1/2}$

  $\quad = -24x^{-4} - \dfrac{5}{2}x^{-1/2} = -\dfrac{24}{x^4} - \dfrac{5}{2x^{1/2}}$

47. $y = \left(4t^2 + 9\right)\left(t^3 - 5\right)$

  Use the product rule.

  $y' = \left(4t^2 + 9\right)\left(3t^2\right) + \left(t^3 - 5\right)(8t)$
  $\quad = 12t^4 + 27t^2 + 8t^4 - 40t$
  $\quad = 20t^4 + 27t^2 - 40t$

48. $y = \left(-5t + 3\right)\left(t^3 - 4t\right)$

  Use the product rule.

  $y' = \left(-5t + 3\right)\left(3t^2 - 4\right) + \left(t^3 - 4t\right)(-5)$
  $\quad = -15t^3 + 9t^2 + 20t - 12 - 5t^3 + 20t$
  $\quad = -20t^3 + 9t^2 + 40t - 12$

49. $y = 8x^{3/4}\left(5x + 1\right)$

  Use the product rule.

  $y' = 8x^{3/4}(5) + (5x + 1)\left(\dfrac{3}{4}\right)8x^{-1/4}$

  $\quad = 40x^{3/4} + 6(5x + 1)x^{-1/4}$

50. $y = 40x^{-1/4}\left(x^2 + 1\right)$

  Use the product rule.

  $y' = 40x^{-1/4}(2x) + \left(x^2 + 1\right)(40)\left(-\dfrac{1}{4}\right)x^{-5/4}$

  $\quad = 80x^{3/4} - 10x^{-5/4}\left(x^2 + 1\right)$

51. $f(x) = \dfrac{4x}{x^2 - 8}$

  Use the quotient rule.

  $f'(x) = \dfrac{\left(x^2 - 8\right)(4) - 4x(2x)}{\left(x^2 - 8\right)^2}$

  $\quad = \dfrac{4x^2 - 32 - 8x^2}{\left(x^2 - 8\right)^2} = \dfrac{-4x^2 - 32}{\left(x^2 - 8\right)^2}$

52. $g(x) = \dfrac{-5x^2}{3x+1}$

Use the quotient rule.

$$g'(x) = \frac{(3x+1)(-10x)-\left(-5x^2\right)(3)}{(3x+1)^2}$$

$$= \frac{-30x^2-10x+15x^2}{(3x+1)^2} = \frac{-15x^2-10x}{(3x+1)^2}$$

53. $y = \dfrac{\sqrt{x}-6}{3x+7} = \dfrac{x^{1/2}-6}{3x+7}$

Use the quotient rule.

$$y' = \frac{(3x+7)\left(\frac{1}{2}x^{-1/2}\right)-\left(x^{1/2}-6\right)(3)}{(3x+7)^2}$$

$$= \frac{\frac{3}{2}x^{1/2}+\frac{7}{2}x^{-1/2}-3x^{1/2}+18}{(3x+7)^2}$$

$$= \frac{-\frac{3}{2}x^{1/2}+\frac{7}{2}x^{-1/2}+18}{(3x+7)^2}\cdot\frac{2}{2}$$

$$= \frac{-3x^{1/2}+7x^{-1/2}+36}{2(3x+7)^2}\cdot\frac{x^{1/2}}{x^{1/2}}$$

$$= \frac{-3x+7+36x^{1/2}}{2x^{1/2}(3x+7)^2} = \frac{-3x+36\sqrt{x}+7}{2\sqrt{x}(3x+7)^2}$$

54. $y = \dfrac{\sqrt{x}+9}{x-4} = \dfrac{x^{1/2}+9}{x-4}$

Use the quotient rule.

$$y' = \frac{(x-4)\left(\frac{1}{2}x^{-1/2}\right)-\left(x^{1/2}+9\right)(1)}{(x-4)^2}$$

$$= \frac{\frac{1}{2}x^{1/2}-2x^{-1/2}-x^{1/2}-9}{(x-4)^2}$$

$$= \frac{-\frac{1}{2}x^{1/2}-2x^{-1/2}-9}{(x-4)^2}\cdot\frac{2x^{1/2}}{2x^{1/2}}$$

$$= \frac{-x-4-18x^{1/2}}{2x^{1/2}(x-4)^2} = \frac{-x-18\sqrt{x}-4}{2\sqrt{x}(x-4)^2}$$

55. $y = \dfrac{x^2-x+1}{x-1}$

Use the quotient rule.

$$y' = \frac{(x-1)(2x-1)-(x^2-x+1)(1)}{(x-1)^2}$$

$$= \frac{2x^2-3x+1-x^2+x-1}{(x-1)^2} = \frac{x^2-2x}{(x-1)^2}$$

56. $y = \dfrac{2x^3-5x^2}{x+2}$

Use the quotient rule.

$$y' = \frac{(x+2)(6x^2-10x)-(2x^3-5x^2)(1)}{(x+2)^2}$$

$$= \frac{6x^3-10x^2+12x^2-20x-2x^3+5x^2}{(x+2)^2}$$

$$= \frac{4x^3+7x^2-20x}{(x+2)^2}$$

57. $f(x) = (4x-2)^4$

Use the chain rule.

$$f'(x) = 4(4x-2)^3(4) = 16(4x-2)^3$$

58. $k(x) = (5x-1)^6$

Use the chain rule.

$$k'(x) = 6(5x-1)^5(5) = 30(5x-1)^5$$

59. $y = \sqrt{2t-5} = (2t-5)^{1/2}$

Use the chain rule.

$$y' = \frac{1}{2}(2t-5)^{-1/2}(2)$$

$$= (2t-5)^{-1/2} = \frac{1}{(2t-5)^{1/2}} = \frac{1}{\sqrt{2t-5}}$$

60. $y = -3\sqrt{8t-1} = -3(8t-1)^{1/2}$

$$y' = -3\left[\frac{1}{2}(8t-1)^{-1/2}\right](8)$$

$$= -12(8t-1)^{-1/2}$$

$$= \frac{-12}{(8t-1)^{1/2}} = \frac{-12}{\sqrt{8t-1}}$$

61. $y = 2x(3x-4)^3$

Use the product rule and the chain rule.

$$y' = 2x(3)(3x-4)^2(3)+(3x-4)^3(2)$$

$$= 18x(3x-4)^2+2(3x-4)^3$$

62. $y = 5x^2(2x+3)^5$

$$y' = 5x^2(5)(2x+3)^4(2)+(2x+3)^5(10x)$$

$$= 50x^2(2x+3)^4+10x(2x+3)^5$$

63. $f(u) = \dfrac{3u^2 - 4u}{(2u+3)^3}$

Use the quotient rule and the chain rule.

$f'(u)$

$= \dfrac{(2u+3)^3(6u-4) - (3u^2-4u)(3)(2u+3)^2(2)}{(2u+3)^6}$

$= \dfrac{(2u+3)^3(6u-4) - 6(3u^2-4u)(2u+3)^2}{(2u+3)^6}$

$= \dfrac{(2u+3)(6u-4) - 6(3u^2-4u)}{(2u+3)^4}$

64. $g(t) = \dfrac{t^3 + t - 2}{(2t-1)^5}$

$g'(t)$

$= \dfrac{(2t-1)^5(3t^2+1) - (t^3+t-2)(5)(2t-1)^4(2)}{(2t-1)^{10}}$

$= \dfrac{(2t-1)^4\left[(2t-1)(3t^2+1) - 10(t^3+t-2)\right]}{(2t-1)^{10}}$

$= \dfrac{(2t-1)(3t^2+1) - 10(t^3+t-2)}{(2t-1)^6}$

65. $y = e^{-2x^3}$

$y' = -6x^2 e^{-2x^3}$

66. $y = -5e^{x^2}$

$y' = -5(2xe^{x^2}) = -10xe^{x^2}$

67. $y = 5x \cdot e^{2x}$

$y' = 5x(2)e^{2x} + e^{2x}(5) = 10xe^{2x} + 5e^{2x}$

68. $y = -7x^2 \cdot e^{-3x}$

$y' = -7x^2(-3e^{-3x}) + e^{-3x}(-14x)$

$= 21x^2 e^{-3x} - 14xe^{-3x}$

69. $y = \ln(x^2 + 4x - 1)$

$y' = \dfrac{D_x(x^2+4x+1)}{x^2+4x-1} = \dfrac{2x+4}{x^2+4x-1}$

70. $y = \ln(4x^3 + 2x)$

$y' = \dfrac{12x^2 + 2}{4x^3 + 2x} = \dfrac{2(6x^2+1)}{2(2x^3+x)} = \dfrac{6x^2+1}{2x^3+x}$

71. $y = \dfrac{\ln 6x}{x^2 - 1}$

$y' = \dfrac{(x^2-1)\left(\dfrac{6}{6x}\right) - (\ln 6x)(2x)}{(x^2-1)^2}$

$= \dfrac{x - \dfrac{1}{x} - 2x(\ln 6x)}{(x^2-1)^2}$

72. $y = \dfrac{\ln(3x+5)}{x^2 + 5x}$

$y' = \dfrac{(x^2+5x)\left(\dfrac{3}{3x+5}\right) - [\ln(3x+5)](2x+5)}{(x^2+5x)^2}$

$= \dfrac{3}{(x^2+5x)(3x+5)} - \dfrac{(2x+5)\ln(3x+5)}{(x^2+5x)^2}$

73. $y = \dfrac{x^2 + 3x - 10}{x - 3}$

$y' = \dfrac{(x-3)(2x+3) - (x^2+3x-10)(1)}{(x-3)^2}$

$= \dfrac{2x^2 - 3x - 9 - x^2 - 3x + 10}{(x-3)^2}$

$= \dfrac{x^2 - 6x + 1}{(x-3)^2}$

74. $y = \dfrac{x^2 - x - 6}{x - 2}$

$y' = \dfrac{(x-2)(2x-1) - (x^2-x-6)(1)}{(x-2)^2}$

$= \dfrac{2x^2 - 5x + 2 - x^2 + x + 6}{(x-2)^2}$

$= \dfrac{x^2 - 4x + 8}{(x-2)^2}.$

75. $y = -6e^{2x}$

$y' = -6(2)e^{2x} = -12e^{2x}$

76. $y = 8e^{5x}$

$y' = 8(.5e^{5x}) = 4e^{5x}$

77. $D_x\left(\dfrac{\sqrt{x}+1}{\sqrt{x}-1}\right) = D_x\left(\dfrac{x^{1/2}+1}{x^{1/2}-1}\right)$

$= \dfrac{(x^{1/2}-1)\left(\frac{1}{2}x^{-1/2}\right)-(x^{1/2}+1)\left(\frac{1}{2}x^{-1/2}\right)}{(x^{1/2}-1)^2}$

$= \dfrac{(x^{1/2}-1-x^{1/2}-1)\left(\frac{1}{2}x^{-1/2}\right)}{(x^{1/2}-1)^2}$

$= \dfrac{-2\left(\frac{1}{2}\right)x^{-1/2}}{(x^{1/2}-1)^2} = \dfrac{-1}{x^{1/2}(x^{1/2}-1)^2}$

78. $D_x\left(\dfrac{2x+\sqrt{x}}{1-x}\right) = D_x\left(\dfrac{2x+x^{1/2}}{1-x}\right)$

$= \dfrac{(1-x)\left(2+\frac{1}{2}x^{-1/2}\right)-(2x+x^{1/2})(-1)}{(1-x)^2}$

$= \dfrac{(1-x)\left(2+\frac{1}{2x^{1/2}}\right)+(2x+x^{1/2})}{(1-x)^2}$

$= \dfrac{(1-x)\left(\frac{4x^{1/2}+1}{2x^{1/2}}\right)+(2x+x^{1/2})}{(1-x)^2}$

$= \dfrac{(1-x)(4x^{1/2}+1)+2x^{1/2}(2x+x^{1/2})}{2x^{1/2}(1-x)^2}$

$= \dfrac{4x^{1/2}+1-4x^{3/2}-x+4x^{3/2}+2x}{2x^{1/2}(1-x)^2}$

$= \dfrac{4x^{1/2}+x+1}{2x^{1/2}(1-x)^2}$

79. $y = \sqrt{t^{1/2}+t} = (t^{1/2}+t)^{1/2}$

$\dfrac{dy}{dt} = \dfrac{1}{2}(t^{1/2}+t)^{-1/2}\left(\dfrac{1}{2}t^{-1/2}+1\right)$

$= \dfrac{\left(\frac{1}{2}t^{-1/2}+1\right)(2t^{1/2})}{2(t^{1/2}+t)^{1/2}(2t^{1/2})}$

$= \dfrac{1+2t^{1/2}}{4t^{1/2}(t^{1/2}+t)^{1/2}}$

80. $y = \dfrac{\sqrt{x-1}}{x} = \dfrac{(x-1)^{1/2}}{x}$

$\dfrac{dy}{dx} = \dfrac{x\left[\frac{1}{2}(x-1)^{-1/2}(1)\right]-(x-1)^{1/2}(1)}{x^2}$

$= \dfrac{\frac{x}{2(x-1)^{1/2}}-(x-1)^{1/2}}{x^2} = \dfrac{x-2(x-1)}{2x^2(x-1)^{1/2}}$

$= \dfrac{2-x}{2x^2(x-1)^{1/2}}$

81. $f(x) = \dfrac{\sqrt{8+x}}{x+1} = \dfrac{(8+x)^{1/2}}{x+1}$

$f'(x) = \dfrac{(x+1)\left(\frac{1}{2}\right)(8+x)^{-1/2}(1)}{(x+1)^2}$

$\quad - \dfrac{(8+x)^{1/2}(1)}{(x+1)^2}$

$= \dfrac{\frac{1}{2}(x+1)(8+x)^{-1/2}-(8+x)^{1/2}}{(x+1)^2}$

$f'(1) = \dfrac{\frac{1}{2}(1+1)(8+1)^{-1/2}-(8+1)^{1/2}}{(1+1)^2}$

$= \dfrac{\frac{1}{2}(2)(9)^{-1/2}-(9)^{1/2}}{2^2}$

$= \dfrac{\frac{1}{3}-3}{4} = \dfrac{-\frac{8}{3}}{4} = -\dfrac{2}{3}$

82. $f(t) = \dfrac{2-3t}{\sqrt{2+t}}$

$f'(-2)$ does not exist since $f(-2)$ is undefined.

83. The graph is continuous throughout, because the function is defined for every x and the limit of the function is equal to the value of the function for every x. Thus, there is no point of discontinuity for this function.

84. The graph is discontinuous at $x = -4$ and $x = 2$, since the limit of the function fails to exist at each point.

85. $f(x) = \dfrac{2x-3}{2x+3}; x = -\dfrac{3}{2}, x = 0, x = \dfrac{3}{2}$

$f\left(-\dfrac{3}{2}\right) = \dfrac{2\left(-\frac{3}{2}\right)-3}{2\left(-\frac{3}{2}\right)+3} = \dfrac{-6}{0}$,

so $f(x)$ is not defined at $x = -\dfrac{3}{2}$. Thus, $f(x)$ is not

continuous at $x = -\dfrac{3}{2}$.

$f(0) = \dfrac{2(0)-3}{2(0)+3} = -1$

$\lim_{x\to 0} f(x) = -1$, so $f(x)$ is continuous at $x = 0$.

$f\left(\dfrac{3}{2}\right) = \dfrac{2\left(\frac{3}{2}\right)-3}{2\left(\frac{3}{2}\right)+3} = \dfrac{0}{6} = 0$

$\lim_{x\to 3/2} f(x) = 0$, so $f(x)$ is continuous at $x = \dfrac{3}{2}$.

86. $g(x) = \dfrac{2x-1}{x^3+x^2}; x = -1, x = 0, x = \dfrac{1}{2}$

$g(-1) = \dfrac{2(-1)-1}{(-1)^3+(-1)^2} = \dfrac{-3}{0},$ so $g(x)$ is not

continuous at $x = -1$.

$g(0) = \dfrac{2(0)-1}{0^3+0^2} = \dfrac{-1}{0},$ so $g(x)$ is not continuous

at $x = 0$.

$g\left(\dfrac{1}{2}\right) = \dfrac{2\left(\frac{1}{2}\right)-1}{\left(\frac{1}{2}\right)^3+\left(\frac{1}{2}\right)^2} = \dfrac{0}{\frac{3}{8}} = 0$

$\lim\limits_{x\to 1/2} g(x) = 0,$ so $g(x)$ is continuous at $x = \dfrac{1}{2}$.

87. $h(x) = \dfrac{2-3x}{2-x-x^2}; \ x = -2, \ x = \dfrac{2}{3}, \ x = 1$

$h(-2) = \dfrac{2-3(-2)}{2-(-2)-(-2)^2} = \dfrac{8}{0},$ so $h(x)$ is not

continuous at $x = -2$.

$h\left(\dfrac{2}{3}\right) = \dfrac{2-3\left(\frac{2}{3}\right)}{2-\frac{2}{3}-\left(\frac{2}{3}\right)^2} = \dfrac{0}{\frac{8}{9}} = 0$

$\lim\limits_{x\to 2/3} h(x) = 0,$ so $h(x)$ is continuous at $x = \dfrac{2}{3}$.

$h(1) = \dfrac{2-3\cdot 1}{2-1-1^2} = \dfrac{-1}{0},$ so $h(x)$ is not continuous

at $x = 1$.

88. $f(x) = \dfrac{x^2-4}{x^2-x-6}; x = 2, x = 3, x = 4$

$f(2) = \dfrac{2^2-4}{2^2-2-6} = \dfrac{0}{-4} = 0,$

$\lim\limits_{x\to 2} f(x) = 0,$ so $f(x)$ is continuous at $x = 2$.

$f(3) = \dfrac{3^2-4}{3^2-3-6} = \dfrac{5}{0},$ so $f(x)$ is not continuous

at $x = 3$.

$f(4) = \dfrac{4^2-4}{4^2-4-6} = \dfrac{12}{6} = 2$

$\lim\limits_{x\to 4} f(x) = 2,$ so $f(x)$ is continuous at $x = 4$.

89. $f(x) = \dfrac{x-6}{x+5}; x = 6, x = -5, x = 0$

Since $f(x)$ is a rational function, it will not be continuous at any x for which $x + 5 = 0$. Then, if $x = -5, f(x)$ will not be continuous, but it is continuous at all other points, including $x = 6$ and $x = 0$.

90. $f(x) = \dfrac{x^2-9}{x+3}; x = 3, x = -3, x = 0$

$f(-3)$ is not defined, but $f(x)$ will be continuous for all other values of x. Thus, $f(x)$ is continuous at $x = 3$ and $x = 0$ and discontinuous at $x = -3$.

91. $L(x) = .286x^2 - 3.669x + 24.475$

a. In 2005, $x = 55$.

$L(55) = .286(55)^2 - 3.669(55) + 24.475$
$= 687.83$

The amount of consumer loan for 2005 was $687.83 billion.

b. In 2009, $x = 59$.

$L(59) = -.286(59)^2 - 3.669(59) + 24.475$
$= 803.57$

The average rate of change from 2005 to 2009 is

$\dfrac{L(59)-L(55)}{4} = \dfrac{803.57-687.83}{4} \approx 28.935$

The average rate of change from 2005 to 2009 is about $28.9 billion.

c. In 2010, $x = 60$.

$L'(x) = .572x - 3.669$

$L'(60) = .572(60) - 3.669 = 30.651$

In 2010, the rate of change for the amount of consumer loans is $30.651 billion per year.

92. $f(x) = -.000144x^3 + .014151x^2 + .1388x + 23.35$

a. For 1990, $x = 20$.

$f(20) \approx 30.63$ quadrillion Btu's

For 2000, $x = 30$.

$f(30) \approx 36.36$ quadrillion Btu's

For 2008, $x = 38$

$f(38) \approx 41.16$ quadrillion Btu's

b. The average rate change in energy consumption between 2000 and 2008 is

$\dfrac{f(38)-f(30)}{38-30} = \dfrac{41.16-36.36}{8}$

$\approx .60$ quadrillion Btu's/yr

c. $f'(x) = -.000432x^2 + .028302x + .1388$

$f'(38) = .59$

Energy consumption was changing by .59 quadrillion Btu's per year in 2008.

93. $f(x) = 69.52e^{.018x}$

 a. For 2005, $x = 25$.

 $f(25) = 69.52e^{.018(25)} \approx 109.03$ million kilowatts.

 b. For 2010, $x = 30$.

 $f'(x) = 69.52(.018)e^{.018x}$

 $f'(30) = 69.52(.018)e^{.018(30)} \approx 2.15$

 In 2010, the rate of change was about 2.15 million kilowatts per year.

94. $g(x) = -.033741x^4 + 1.62176x^3 - 28.4297x^2 + 216.603x - 599.806$

 a. For 2006, $x = 16$.

 $g(16) \approx \$19.3$ billion

 b. For 2008, $x = 18$.

 $g(18) \approx \$3.93$ billion

 c. For 2007, $x = 17$.

 $g'(x) = -.134964x^3 + 4.86528x^2 - 56.8594x + 216.603$

 $g'(17) \approx -7.02$

 In 2007, revenue was decreasing by about $\$7.02$ billion per year.

95. $h(x) = -13.93 + 17.25 \ln x$

 a. For 2007, $x = 7$.

 $h(7) = -13.93 + 17.25 \ln 7 \approx 19.64$

 About $\$19.64$ billion was spent in 2007.

 b. For 2008, $x = 8$.

 $h'(x) = \dfrac{17.25}{x}$

 $h'(8) = \dfrac{17.25}{8} \approx 2.16$

 In 2008, spending was increasing by $\$2.16$ billion per year.

96. $f(x) = .0079x^3 - .3180x^2 + 4.1636x - 3.9136$

 a. In 2000, $x = 10$.

 $f(10) \approx 13.82$

 There were about 13,820 facilities in 2000.
 In 2007, $x = 17$.

 $f(17) \approx 13.78$

 There were about 13,780 facilities in 2007.

 b. $f'(x) = .0237x^2 - .636x + 4.1636$

 $f'(17) \approx .2$

 In 2007, the number of facilities was changing at about 200 per year.

97. $g(x) = \dfrac{1660}{1 + 36.08e^{-.247x}}$

 a. In 2000, $x = 20$.

 $g(20) = \dfrac{1660}{1 + 36.08e^{-.247(20)}} \approx 1319.4$

 About $\$1319.4$ billion in foreign bonds were purchased in 2000.
 In 2006, $x = 26$

 $g(26) = \dfrac{1660}{1 + 36.08e^{-.247(26)}} \approx 1568.0$

 About $\$1568.0$ billion in foreign bonds were purchased in 2006.

 b. $g'(x) = \dfrac{-1660(-.247)\left(36.08e^{-.247x}\right)}{\left(1 + 36.08e^{-.247x}\right)^2}$

 $g'(26) \approx 21.5$

 The rate of change in 2006 was about $\$21.5$ billion per year.

98. $h(x) = 1252.3e^{.025x}$

 a. For 2005, $x = 25$.

 $h(25) \approx 2339.6$

 About 2339.6 thousand barrels were produced in 2005.

 b. $h'(x) = 1252.3(.025)e^{.025x}$

 $h'(25) \approx 58.5$

 In 2005, production was increasing by about 58.5 thousand barrels per year.

99. From the graph, the slope of the tangent line to the Hands curve where it crosses the Bat curve is 0, and the slope of the tangent line to the Bat curve where it crosses the Hands curve is approximately $\dfrac{40 - 15}{.15 - .1} \approx 500$. So when the Hands and Bat functions are equal, the derivative of the Hands function is 0 mph per sec, and the derivative of the Bat function is approximately 500 mph per sec. This represents the acceleration of the hands and the bat at the moment when their velocities are equal.

100. $L = 71.5(1 - e^{-.1t})$

$W = .01289 \cdot L^{2.9}$

a. The approximate length of a 5-year-old monkeyface is

$L(5) = 71.5(1 - e^{-.1(5)}) = 28.1$ cm.

b. The rate at which the length of a 5-year-old is increasing is $L'(5)$.

$L'(t) = 71.5(.1e^{-.1t}) = 7.15e^{-.1t}$.

$L'(5) = 7.15e^{-.1(5)} = 4.34$ cm/year.

c. The approximate weight of a 5-year-old monkeyface is

$W(L(5)) = W(28.1) = .01289 \cdot 28.1^{2.9}$

≈ 205 g.

d. The rate of change of the weight with respect to length for a 5-year-old monkeyface is $W'(L(5))$.

$W' = .01289(2.9)L^{1.9} = .037381 \cdot L^{1.9}$

$W'(L(5)) = W'(28.1) = .037381\left(28.1^{1.9}\right)$

$= 21.14$ g/cm.

e. The rate at which the weight of a 5-year-old monkeyface is increasing is $W'(5)$.

$W'(t) = W'(t) \cdot L'(t)$.

So $W'(5) = W'(L(5)) \cdot L'(5)$

$= 21.1$ g/cm $\cdot$ 4.34 cm/year

$= 91.6$ g/year.

101. a. $\lim\limits_{x \to 6^-} C(x) = \$54.75$

b. $\lim\limits_{x \to 6^+} C(x) = \$59.00$

c. Since $\lim\limits_{x \to 6^-} C(x) \neq \lim\limits_{x \to 6^+} C(x)$, $\lim\limits_{x \to 6} C(x)$ does not exist.

d. $C(6) = \$54.75$

e. C is discontinuous at .5, 1, 2, 3, 4, 5, and 6.

f.

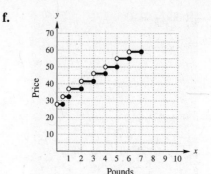

102. a.

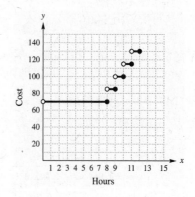

b. $\lim\limits_{x \to 10^-} C(x) = \$100$

c. $\lim\limits_{x \to 10^+} C(x) = \$115$

d. Since $\lim\limits_{x \to 10^-} C(x) \neq \lim\limits_{x \to 10^+} C(x)$, $\lim\limits_{x \to 10} C(x)$ does not exist.

e. $C(10) = \$100$

Case 11 Price Elasticity of Demand

1. $q = -3.003p + 675.23$

$\dfrac{dq}{dp} = -3.003$

$E = -\dfrac{p}{q} \cdot \dfrac{dq}{dp} = \dfrac{3.003p}{-3.003p + 675.23}$

For $p = 70$,

$E = \dfrac{3.003(70)}{-3.003(70) + 675.23} \approx .452$

2. $q = -1000p + 70,000$

$$\frac{dq}{dp} = -1000$$

$$E = -\frac{p}{q} \cdot \frac{dq}{dp} = \frac{1000p}{-1000p + 70,000}$$

a. For $p = 30$,

$$E = \frac{1000(30)}{-1000(30) + 70,000} = .75$$

b. For $p = 40$,

$$E = \frac{1000(40)}{-1000(40) + 70,000} \approx 1.33$$

Since the demand is elastic at $p = \$40$, a smaller price increase would be better.

3. $q = -2481.52p + 472,191.2$

$$\frac{dq}{dp} = -2481.52$$

$$E = -\frac{p}{q} \cdot \frac{dq}{dp} = \frac{2481.52p}{-2481.52p + 472,191.2}$$

a. For $p = 100$,

$$E = \frac{2481.52(100)}{-2481.52(100) + 472,191.2} \approx 1.1$$

For $p = 75$,

$$E = \frac{2481.52(75)}{-2481.52(75) + 472,191.2} \approx .65$$

b. Let $1 = \dfrac{2481.52p}{-2481.52p + 472,191.2}$

$$-2481.52p + 472,191.2 = 2481.52p$$
$$472,191.2 = 4963.04p$$
$$p \approx 95.14$$

The demand has unit elasticity at $p \approx \$95.14$.

4. $q = -2.35p + 28.26$

$$\frac{dq}{dp} = -2.35$$

$$E = -\frac{p}{q} \cdot \frac{dq}{dp} = \frac{2.35p}{-2.35p + 28.26}$$

For $p = 3.00$,

$$E = \frac{2.35(3)}{-2.35(3) + 28.26} \approx .33$$

Demand is inelastic at a price of $3.00. Even though demand drops as price rises, at this price level a small price increase leads to an overall revenue increase.

5. $E = 0$ means that $-\dfrac{p}{q} \cdot \dfrac{dq}{dp} = 0$.

Because the price $p \neq 0$, $\dfrac{dq}{dp} = 0$. Changes in price produce no change at all in demand, i.e., the demand is constant.

Chapter 12 Applications of the Derivative

Section 12.1 Derivatives and Graphs

1. The function is increasing on $(1, \infty)$, and decreasing on $(-\infty, 1)$. The lowest point on the graph has coordinates $(1, -4)$, so the local minimum of -4 occurs at $x = 1$.

3. The function is increasing on $(-\infty, -2)$ and decreasing on $(-2, \infty)$. The highest point on the graph has coordinates $(-2, 3)$, so a local maximum of 3 occurs at $x = -2$.

5. The function is increasing on $(-\infty, -4)$ and $(-2, \infty)$, and decreasing on $(-4, -2)$. A high point occurs at $(-4, 3)$ and a low point occurs at $(-2, 1)$, so a local maximum of 3 occurs at $x = -4$ and a local minimum of 1 occurs at $x = -2$.

7. The function is increasing on $(-7, -4)$ and $(-2, \infty)$ and decreasing on $(-\infty, -7)$ and $(-4, -2)$. Low points occur at $(-7, -2)$ and $(-2, -2)$, and a high point occurs at $(-4, 3)$. Thus a local minimum of -2 occurs at $x = -7$ and $x = -2$, and a local maximum of 3 occurs at $x = -4$.

9. $f(x) = 2x^3 - 5x^2 - 4x + 2$
   $f'(x) = 6x^2 - 10x - 4$
   $6x^2 - 10x - 4 = 0$
   $2(3x^2 - 5x - 2) = 0$
   $2(3x + 1)(x - 2) = 0$
   $x = -\dfrac{1}{3}$ or $x = 2$
   Test $f'(x)$ at $x = -2, x = 0, x = 3$.
   $f'(-2) = 6(-2)^2 - 10(-2) - 4 = 40 > 0$
   $f'(0) = 6(0)^2 - 10(0) - 4 = -4 < 0$
   $f'(3) = 6(3)^2 - 10(3) - 4 = 20 > 0$
   $f'(x)$ is positive on $\left(-\infty, -\dfrac{1}{3}\right)$, so $f(x)$ is increasing.
   $f'(x)$ is negative on $\left(-\dfrac{1}{3}, 2\right)$, so $f(x)$ is decreasing.
   $f'(x)$ is positive on $(2, \infty)$, so $f(x)$ is increasing.

11. $f(x) = \dfrac{x+1}{x+3}$
    $f'(x) = \dfrac{(x+3)(1) - (x+1)(1)}{(x+3)^2} = \dfrac{2}{(x+3)^2}$
    so $f'(x) = \dfrac{2}{(x+3)^2} = 0$ has no solution.
    $f'(x)$ does not exist if $(x+3)^2 = 0 \Rightarrow x = -3$
    Test $f'(x)$ at $x = -4$ and $x = 0$.
    $f'(-4) = \dfrac{2}{(-4+3)^2} = 2$
    $f'(0) = \dfrac{2}{(0+3)^2} = \dfrac{2}{9}$
    $f'(x)$ is positive on $(-\infty, -3)$ and $(-3, \infty)$, so $f(x)$ is increasing on those intervals.

13. $f(x) = \sqrt{6-x} = (6-x)^{\frac{1}{2}}$, so $f(x)$ is defined on $x \le 6$.
    $f'(x) = \dfrac{1}{2}(6-x)^{-\frac{1}{2}}(-1) = \dfrac{-1}{2\sqrt{6-x}}$
    so $f'(x) = \dfrac{-1}{2\sqrt{6-x}} = 0$ has no solution.
    $f'(x)$ does not exist if $2\sqrt{6-x} = 0 \Rightarrow$
    $6 - x = 0 \Rightarrow x = 6$. Test $f'(x)$ at $x = 0$,
    $f'(0) = \dfrac{-1}{2\sqrt{6-0}} = \dfrac{-1}{2\sqrt{6}} < 0$
    $f'(x)$ is negative on $(-\infty, 6)$, so $f(x)$ is decreasing.

15. $f(x) = 2x^3 + 3x^2 - 12x + 5$
    $f'(x) = 6x^2 + 6x - 12$
    $6x^2 + 6x - 12 = 0$
    $6(x^2 + x - 2) = 0$
    $6(x + 2)(x - 1) = 0 \Rightarrow x = -2, 1$
    Test $f'(x)$ at $x = -3, x = 0, x = 3$
    $f'(-3) = 6(-3)^2 + 6(-3) - 12 = 24 > 0$
    $f'(0) = 6(0)^2 + 6(0) - 12 = -12 < 0$
    $f'(3) = 6(3)^2 + 6(3) - 12 = 60 > 0$
    $f'(x)$ is positive on $(-\infty, -2)$ and $(1, \infty)$, so $f(x)$ is increasing.
    $f'(x)$ is negative on $(-2, 1)$, so $f(x)$ is decreasing.

17. The graph of f' is 0 when $x = -2$, $x = -1$, $x = 2$, so these are the critical numbers of f.

19. $f(x) = x^3 + 3x^2 - 3$

$f'(x) = 3x^2 + 6x = 3x(x + 2)$

Find the critical numbers by solving the equation $f'(x) = 0$

$3x(x + 2) = 0$

$3x(x + 2) = 0 \Rightarrow x = 0, -2$

Use the first derivative test by testing $f'(x)$ at $x = -3$, $x = 1$, and $x = 3$.

$f'(-3) = 3(-3)^2 + 6(-3) = 9 > 0$

$f'(-1) = 3(-1)^2 + 6(-1) = -3 < 0$

$f'(3) = 3(3)^2 + 6(3) = 45 > 0$

$f'(x)$ is positive on $(-\infty, -2)$ and negative on $(-2, 0)$, so a local maximum occurs at $x = -2$.

$f'(x)$ is negative on $(-2, 0)$ and positive on $(0, \infty)$, so a local minimum occurs at $x = 0$.

Values of extrema:

$f(0) = 0^3 + 3(0)^2 - 3 = -3$

$f(-2) = (-2)^3 + 3(-2)^2 - 3 = 1$

A local maximum of 1 occurs at $x = -2$ and a local minimum of -3 occurs at $x = 0$.

21. $f(x) = x^3 + 6x^2 + 9x + 2$

$f'(x) = 3x^2 + 12x + 9$

Find critical numbers by solving $f'(x) = 0$.

$f'(x) = 3x^2 + 12x + 9 = 3(x^2 + 4x + 3)$

$= 3(x + 3)(x + 1) = 0 \Rightarrow x = -3, -1$

Test $f'(x)$ at $x = -4$, $x = -2$, and $x = 0$.

$f'(-4) = 3(-4)^2 + 12(-4) + 9 = 9 > 0$

$f'(-2) = 3(-2)^2 + 12(-2) + 9 = -3 < 0$

$f'(0) = 3(0)^2 + 12(0) + 9 = 9 > 0$

$f'(x)$ is positive on $(-\infty, -3)$ and negative on $(-3, -1)$, so a local maximum occurs at $x = -3$.

$f'(x)$ is negative on $(-3, -1)$ and positive on $(-1, \infty)$, so a local minimum occurs at $x = -1$.

$f(-3) = (-3)^3 + 6(-3)^2 + 9(-3) + 2 = 2$

$f(-1) = (-1)^3 + 6(-1)^2 + 9(-1) + 2 = -2$

Thus, f has a local maximum of 2 at $x = -3$ and a local minimum of -2 at $x = -1$.

23. $f(x) = \frac{4}{3}x^3 - \frac{21}{2}x^2 + 5x + 3$

$f'(x) = 4x^2 - 21x + 5$

Find critical numbers by solving $f'(x) = 0$.

$4x^2 - 21x + 5 = 0 \Rightarrow (4x - 1)(x - 5) = 0 \Rightarrow$

$x = \frac{1}{4}, 5$

Test $f'(x)$ at 0, 1, and 6.

$f'(0) = 5 > 0$; $f'(1) = -12 < 0$;

$f'(6) = 23 > 0$

$f'(x)$ is positive on $(-\infty, \frac{1}{4})$ and positive on

$(5, \infty)$. $f'(x)$ is negative on $\left(\frac{1}{4}, 5\right)$.

By the first derivative test, $f(x)$ has a local

maximum at $x = \frac{1}{4}$, and a local minimum at

$x = 5$.

$f(5) = -\frac{407}{6}$; $f\left(-\frac{1}{4}\right) = \frac{347}{96}$

Thus, f has a local maximum of $\frac{347}{96}$ at $x = -\frac{1}{4}$

and a local minimum of $-\frac{407}{6}$ at $x = 5$.

25. $f(x) = \frac{2}{3}x^3 - x^2 - 12x + 2$

$f'(x) = 2x^2 - 2x - 12$

Find critical numbers by solving $f'(x) = 0$.

$2(x^2 - x - 6) = 0 \Rightarrow 2(x + 2)(x - 3) = 0 \Rightarrow$

$x = -2, 3$

Test $f'(x)$ at $x = -3$, $x = 0$, and $x = 4$.

$f'(-3) = 2(-3)^2 - 2(-3) - 12 = 12 > 0$

$f'(0) = 2(0)^2 - 2(0) - 12 = -12 < 0$

$f'(4) = 2(4)^2 - 2(4) - 12 = 12 > 0$

$f'(x)$ is positive on $(-\infty, -2)$ and negative on $(-2, 3)$, so $f(x)$ has a local maximum at $x = -2$.

$f'(x)$ is negative on $(-2, 3)$ and positive on $(3, \infty)$, so $f(x)$ has a local minimum at $x = 3$.

$f(-2) = \frac{2}{3}(-2)^3 - (-2)^2 - 12(-2) + 2 = \frac{50}{3}$

$f(3) = \frac{2}{3}(3)^3 - 3^2 - 12(3) + 2 = -25$

Thus, f has a local maximum of $\frac{50}{3}$ at $x = -2$

and a local minimum of -25 at $x = 3$.

27. $f(x) = x^5 + 20x^2 + 8$

$f'(x) = 5x^4 + 40x$

Find critical numbers by solving $f'(x) = 0$.

$5x^4 + 40x = 0 \Rightarrow 5x(x^3 + 8) = 0 \Rightarrow$

$5x(x+2)(x^2 - 2x + 4) = 0 \Rightarrow x = 0, -2$

Test $f'(x)$ at $x = -3$, $x = -1$, and $x = 3$.

$f'(-3) = 5(-3)^4 + 40(-3) = 285 > 0$

$f'(-1) = 5(-1)^4 + 40(-1) = -35 < 0$

$f'(3) = 5(3)^4 + 40(3) = 585 > 0$

$f'(x)$ is positive on $(-\infty, -2)$ and negative on $(-2, 0)$, so $f(x)$ has a local maximum at $x = -2$. $f'(x)$ is negative on $(-2, 0)$ and positive on $(0, \infty)$, so $f(x)$ has a local minimum at $x = 0$.

$f(-2) = (-2)^5 + 20(-2)^2 + 8 = 56$

$f(0) = 0^5 + 20(0)^2 + 8 = 8$

Thus, f has a local maximum of 56 at $x = -2$ and a local minimum of 8 at $x = 0$.

29. $f(x) = x^{11/5} - x^{6/5} + 1$

$f'(x) = \frac{11}{5}x^{6/5} - \frac{6}{5}x^{1/5} = \frac{1}{5}x^{1/5}(11x - 6)$

Find critical numbers by solving $f'(x) = 0$.

$\frac{1}{5}x^{1/5}(11x - 6) = 0 \Rightarrow x = 0, \frac{6}{11}$

Test $f'(x)$ at $x = -1$, $x = \frac{1}{2}$, and $x = 1$.

$f'(-1) = \frac{11}{5}(-1)^{6/5} - \frac{6}{5}(-1)^{1/5} = \frac{17}{5} > 0$

$f'\left(\frac{1}{2}\right) = \frac{11}{5}\left(\frac{1}{2}\right)^{6/5} - \frac{6}{5}\left(\frac{1}{2}\right)^{1/5} \approx -.087 < 0$

$f'(1) = \frac{11}{5}(1)^{6/5} - \frac{6}{5}(1)^{1/5} = 1 > 0$

$f'(x)$ is positive on $(-\infty, 0)$ and negative on $\left(0, \frac{6}{11}\right)$, so a local maximum occurs at $x = 0$.

$f'(x)$ is negative on $\left(0, \frac{6}{11}\right)$ and positive on $\left(\frac{6}{11}, \infty\right)$, so a local minimum occurs at $x = \frac{6}{11}$.

$f(0) = 0^{11/5} - 0^{6/5} + 1 = 1$

$f\left(\frac{6}{11}\right) = \left(\frac{6}{11}\right)^{11/5} - \left(\frac{6}{11}\right)^{6/5} + 1 \approx .7804$

A local maximum of 1 occurs at $x = 0$, and a local minimum of approximately .7804 occurs at $x = \frac{6}{11}$.

31. $f(x) = -(3 - 4x)^{2/5} + 4$

$f'(x) = -\frac{2}{5}(3 - 4x)^{-3/5}(-4) = \frac{8}{5}(3 - 4x)^{-3/5}$

$= \frac{8}{5(3 - 4x)^{3/5}}$

$f'(x)$ is never zero, but $f'(x)$ does not exist if $3 - 4x = 0$ or $x = \frac{3}{4}$. Test $f'(x)$ at $x = 0$ and $x = 1$.

$f'(0) = \frac{8}{5(3 - 4 \cdot 0)^{3/5}} \approx .828 > 0$

$f'(1) = \frac{8}{5(3 - 4 \cdot 1)^{3/5}} \approx -\frac{8}{5} < 0$

$f'(x)$ is positive on $\left(-\infty, \frac{3}{4}\right)$ and negative on $\left(\frac{3}{4}, \infty\right)$, so a local maximum occurs at $x = \frac{3}{4}$.

$f\left(\frac{3}{4}\right) = -\left[3 - 4\left(\frac{3}{4}\right)\right]^{2/5} + 4 = 4$

A local maximum of 4 occurs at $x = \frac{3}{4}$. There is no local minimum.

33. $f(x) = \frac{x^3}{x^3 + 1}$

$f'(x) = \frac{(x^3 + 1)3x^2 - x^3(3x^2)}{(x^3 + 1)^2} = \frac{3x^2}{(x^3 + 1)^2}$

$f'(x) = 0$ when $x = 0$.

Note that both $f(x)$ and $f'(x)$ do not exist at $x = -1$, so -1 is not a critical number. Test $f'(x)$ at $x = -2$ and $x = 1$.

$f'(-2) = \frac{12}{49} > 0$; $f'(1) = \frac{3}{4} > 0$

$f'(x)$ is positive on $(-\infty, 0)$ and $(0, \infty)$.

f has no local extrema.

35. $f(x) = -xe^x$

$$f'(x) = (-1)\left(e^x\right) + (-x)\left(e^x\right)$$

$$= -e^x - xe^x = -e^x(1+x)$$

$f'(x) = 0$ when

$$-e^x = 0 \quad \text{or} \quad 1+x = 0 \Rightarrow x = -1$$

No solution

The only critical number is –1. Test $f'(x)$ at $x = -2$ and $x = 0$.

$$f'(-2) = -e^{-2}(1-2) = -e^{-2}(-1)$$

$$= e^{-2} \approx .135 > 0$$

$$f'(0) = -e^0(1) = -1 < 0$$

$f'(x)$ is positive on $(-\infty, -1)$ and negative on $(-1, \infty)$.

$$f(-1) = -(-1)e^{-1} = \frac{1}{e}$$

f has a local maximum of $\frac{1}{e}$ at $x = -1$.

37. $f(x) = x \cdot \ln|x|$

$$f'(x) = (1)\ln|x| + x\left(\frac{1}{x}\right) = \ln|x| + 1$$

Find critical numbers by solving $f'(x) = 0$.

$$\ln|x| + 1 = 0 \Rightarrow \ln|x| = -1$$

If $x > 0$, $x = e^{-1} = \frac{1}{e} \approx .3679$

Test $f'(x)$ at $x = .2$ and $x = 1$.

$$f'(.2) = \ln.2 + 1 \approx -.6094 < 0$$

$$f'(1) = \ln 1 + 1 = 1 > 0$$

$f'(x)$ is negative on $\left(-\infty, \frac{1}{e}\right)$ and positive on $\left(\frac{1}{e}, \infty\right)$.

$$f\left(\frac{1}{e}\right) = \frac{1}{e}\ln\left(\frac{1}{e}\right) = \frac{1}{e}(-1) = -\frac{1}{e}$$

f has a local minimum of $-\frac{1}{e}$ at $x = \frac{1}{e}$. By symmetry and absolute value, there is a local maximum of $\frac{1}{e}$ at $x = -\frac{1}{e}$.

39. $f(x) = xe^{3x} - 2$

$$f'(x) = (1)e^{3x} + (3)xe^{3x} = e^{3x}(1+3x)$$

Find critical numbers by solving $f'(x) = 0$.

$$e^{3x}(1+3x) = 0$$

$$e^{3x} = 0 \quad \text{or} \quad 1+3x = 0 \Rightarrow x = -\frac{1}{3}$$

No solution

The critical number is $x = -\frac{1}{3}$.

$$f'(-1) = e^{-3}(1-3) = -2e^{-3} < 0$$

$$f'(0) = e^{3(0)}(1+3(0)) = 1 > 0$$

Values of extremum:

$$f\left(-\frac{1}{3}\right) = \frac{-1}{3}e^{3(-1/3)} - 2 = \frac{-1}{3e} - 2 \approx -2.1226$$

A local minimum of –2.1226 occurs at $x = -\frac{1}{3}$.

Solutions 41–43 presuppose the use of a TI-83 graphics calculator. Similar results can be obtained from other graphics calculators.

41. Graph the function

$$f(x) = .2x^4 - x^3 - 12x^2 + 99x - 5$$

by entering it as y_1.

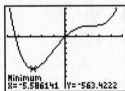

[–10, 10] by [–900, 500]

Under the CALC menu use, "minimum" and "maximum" to find a local minimum of –563.42 at $x = -5.5861$.

43. Graph the function

$$f(x) = .01x^5 + .2x^4 - x^3 - 6x^2 + 5x + 40$$

by entering it as y_1.

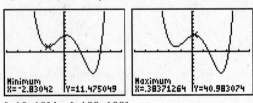

[–10, 10] by [–100, 100]

(continued on next page)

(continued from page 378)

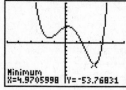 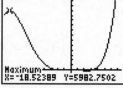

[−20, 10] by [−1000, 7000]

Under the CALC menu, use "minimum" and "maximum" to find local maxima of 5982.75 at $x = -18.5239$ and 40.9831 at $x = .3837$, and local minima of 11.4750 at $x = -2.8304$, and −53.7683 at $x = 4.9706$. Note that two different viewing windows were used.

45. The slope is 0 during 2001 and 2004.

47. a. The revenue of both industries is increasing on (2004, 2006].

b. Programming revenue is constant in 2003.

c. Design revenue is decreasing more rapidly. The slope of its graph is steeper than the slope of the graph for programming revenue.

49. $f(x)$
$$= -.547x^3 + 16.072x^2 - 102.955x + 658.046$$
$f'(x) = -1.641x^2 + 32.144x - 102.955$

Find critical numbers by solving $f'(x) = 0$.

$$x = \frac{-32.144 \pm \sqrt{32.144^2 - 4(-1.641)(-102.955)}}{2(-1.641)}$$

$$\approx 15.55, 4.03$$

Test f' at 0, 10, and 20.

$f'(0) = -102.955 < 0$
$f'(10) = 54.385 > 0$
$f'(20) = -116.475 < 0$

Thus, there is a local maximum at about 15.55 and a local minimum at about 4.03.

$f(15.55) \approx 886.61$; $f(4.03) = 468.36$

Local maximum: (15.55, 886.61)
Local minimum: (4.03, 468.36)

51. $f(x) = \dfrac{1}{\sqrt{2\pi}} e^{-x^2/2}$

$f'(x) = \dfrac{-x}{\sqrt{2\pi}} e^{-x^2/2}$

$\dfrac{-x}{\sqrt{2\pi}} e^{-x^2/2} = 0$ when $x = 0$.

Test $f'(x)$ at $x = -1$, $x = 1$.

$$f'(-1) = \frac{-(-1)}{\sqrt{2\pi}} e^{-(-1)^2/2} = \frac{1}{\sqrt{2\pi}} e^{-1/2} > 0$$

$$f'(1) = \frac{-1}{\sqrt{2\pi}} e^{-1^2/2} = \frac{-1}{\sqrt{2\pi}} e^{-1/2} < 0$$

$f'(x)$ is positive on $(-\infty, 0)$, so $f(x)$ is increasing. $f'(x)$ is negative on $(0, \infty)$, so $f(x)$ is decreasing.

53. $C(x) = -.0021x^3 + .0159x^2 + 3.1442x + 24.1502$
$R(x) = -.0090x^3 + .1335x^2 + 3.8922x + 28.9465$
Profit function,
$P(x) = R(x) - C(x)$
$\quad = -.0069x^3 + .1176x^2 + .748x + 4.7963$
$P'(x) = -.0207x^2 + .2352x + .748$

Find critical numbers by solving $P'(x) = 0$..

$$x = \frac{-.2352 \pm \sqrt{.2352^2 - 4(-.0207)(.748)}}{2(-.0207)}$$

$$\approx -2.59, 13.95$$

Neither solution is in the given interval (0, 8), so there are no critical points in the interval . Therefore, the function is either increasing or decreasing over the entire interval. We check two values, 1 and 2. $P(1) \approx 5.66$ and $P(2) \approx 6.71$, so the function is increasing over (0, 8).

55. $A(x) = .004x^3 - .05x^2 + .16x + .05$, $0 \le x \le 6$
$A'(x) = .012x^2 - .1x + .16$
Set the derivative equal to zero and solve for x.
$.012x^2 - .1x + .16 = 0$
$x \approx 2.16$ or $x \approx 6.17$
since $0 \le x \le 6$, $x \approx 2.16$ is the only critical number. Test $A'(x)$ for $x = 0$, and $x = 3$.

$A'(0) = .012(0)^2 - .1(0) + .16 = .16 > 0$

$A'(3) = .012(3)^2 - .1(3) + .16 = -.032 < 0$

a. $A'(x)$ is positive on [0, 2.16), so $A(x)$, the alcohol concentration, is increasing.

b. $A'(x)$ is negative on (2.16, 6], so $A(x)$ is decreasing.

57. a. The average number of calls is increasing on (5, 16) and (18, 19).

b. The average number of calls is decreasing on (0, 5), (16, 18), and (19, 23).

c. Local maxima occur at $t = 16$ and $t = 19$, which correspond to 4:00 pm and 7:00 pm. Local minima occur at $t = 5$ and $t = 18$, which correspond to 5:00 am and 6:00 pm.

59. a. 1980: $7.75; 1990: $6.10; 2000: $6.25

b. Between 1973 and 2004 there are 5 local maxima; they correspond approximately to the years 1974, 1976, 1978, 1991, and 1997.

c. Between 1980 and 2003 local minima occur around the years 1989 and 1995.

d. Between 1986 and 2007, there were local minima in 1989, 1995, and 2006.

61. $g(x) = 4.63 + 60.66x - 33.98x^2 + 8.686x^3$
$\qquad - 1.1526x^4 + .08246x^5 - .0030185x^6$
$\qquad + .00004435x^7$

a.

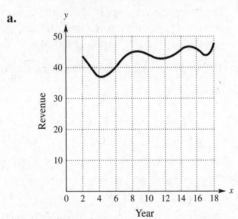

b. There appear to be 5 local extrema, 2 local maxima and 3 local minima.

c.

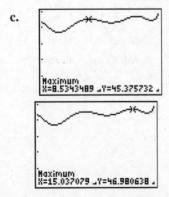

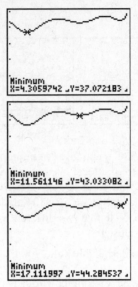

The local maxima are (8.534, 45.376) and (15.037, 46.981). The local minima are (4.306, 37.072), (11.561, 43.033), and (17.112, 44.285).

63. $h(x) = -.1351x^3 + 4.472x^2 - 38.864x + 332.459$

a.

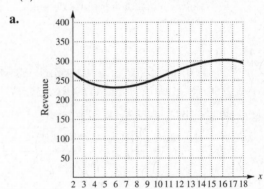

b. There appear to be two local extrema, one local maximum and one local minimum.

c.

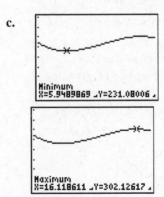

The local minimum is (5.95, 231.08) and the local maximum is (16.12, 302.13).

65. $f(x) = -.000164x^2 + .014532x + 1.1385$

$f'(x) = -.000328x + .014532$. $f'(x) = 0$ when

$x \approx 44.3$. $f'(50) = -.001868 < 0$,

$f'(40) = .001412 > 0$. Hence, $x \approx 44.3$ is where a local maximum occurs. This corresponds to mid-2034, and the population will be $f(44.3) \approx 1.46$ billion. .

67. $g(x) = -.0084x^4 + .4298x^3 - 7.9062x^2$
$\qquad + 61.4056x - 165.102$

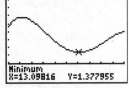
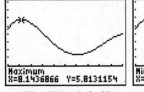

[7, 17] by [−2, 8]

There is a local maximum at about (8.14, 5.81) and a local minimum at about (13.10, 1.38). This means that the bond rate was a maximum of 5.81% in 1998 and a minimum of 1.38% in 2003.

69. $R(t) = \dfrac{20t}{t^2 + 100}$,

$R'(t) = \dfrac{\left(t^2 + 100\right)(20) - 20t(2t)}{\left(t^2 + 100\right)^2}$

$\qquad = \dfrac{20t^2 - 2000 - 40t^2}{\left(t^2 + 100\right)^2} = \dfrac{-20\left(t^2 - 100\right)}{\left(t^2 + 100\right)^2}$

$R'(t) = 0$ when $t = 10$. This is the only critical number. $R'(9) \approx .0116 > 0$ and

$R'(11) \approx -.0086 < 0$. Hence a local maximum occurs at $t = 10$. The film length that received the highest rating is 10 minutes.

Section 12.2 The Second Derivative

1. $f(x) = x^3 - 6x^2 + 1$

$f'(x) = 3x^2 - 12x$

$f''(x)$ is the derivative of $f'(x)$.

$\quad f''(x) = 6x - 12$

$\quad f''(0) = 6(0) - 12 = -12$

$\quad f''(2) = 6(2) - 12 = 0$

$\quad f''(-3) = 6(-3) - 12 = -30$

3. $f(x) = (x+3)^4$

$f'(x) = 4(x+3)^3(1) = 4(x+3)^3$

$f''(x) = 12(x+3)^2$

$f''(0) = 12(0+3)^2 = 108$

$f''(2) = 12(2+3)^2 = 300$

$f''(-3) = 12(-3+3)^2 = 0$

5. $f(x) = \dfrac{x^2}{1+x}$

$f'(x) = \dfrac{(1+x)(2x) - x^2(1)}{(1+x)^2} = \dfrac{2x + x^2}{(1+x)^2}$

$f''(x) = \dfrac{(1+x)^2(2+2x) - \left(2x+x^2\right)(2)(1+x)}{(1+x)^4}$

$\qquad = \dfrac{(1+x)(2+2x) - \left(2x+x^2\right)(2)}{(1+x)^3}$

$\qquad = \dfrac{2}{(1+x)^3}$

$f''(0) = \dfrac{2}{(1+0)^3} = 2$

$f''(2) = \dfrac{2}{(1+2)^3} = \dfrac{2}{27}$

$f''(-3) = \dfrac{2}{(1-3)^3} = -\dfrac{1}{4}$

7. $f(x) = \sqrt{x+4} = (x+4)^{1/2}$

$f'(x) = \dfrac{1}{2}(x+4)^{-1/2}$

$f''(x) = \left(-\dfrac{1}{2}\right)\dfrac{1}{2}(x+4)^{-3/2} = -\dfrac{(x+4)^{-3/2}}{4}$

or $-\dfrac{1}{4(x+4)^{3/2}}$

$f''(0) = -\dfrac{1}{4(0+4)^{3/2}} = -\dfrac{1}{4(4)^{3/2}} = -\dfrac{1}{32}$

$f''(2) = -\dfrac{1}{4(2+4)^{3/2}} = -\dfrac{1}{4(6)^{3/2}}$

$f''(-3) = -\dfrac{1}{4(-3+4)^{3/2}} = -\dfrac{1}{4(1)^{3/2}} = -\dfrac{1}{4}$

9. $f(x) = 5x^{4/5}$

$f'(x) = 4x^{-1/5}$

$f''(x) = -\frac{4}{5}x^{-6/5}$ or $-\frac{4}{5x^{6/5}}$

$f''(0)$ does not exist.

$f''(2) = -\frac{4}{5\left(2^{6/5}\right)} = -\frac{4}{5(2)\left(2^{1/5}\right)} = -\frac{2^{4/5}}{5}$

$f''(-3) = -\frac{4}{5}(-3)^{-6/5} = -\frac{4}{5(-3)(-3)^{1/5}} = \frac{4}{15(-3)^{1/5}}$

11. $f(x) = 2e^x$; $f'(x) = 2e^x$; $f''(x) = 2e^x$

$f''(0) = 2e^0 = 2(1) = 2$

$f''(2) = 2e^2$

$f''(-3) = 2e^{-3}$ or $\frac{2}{e^3}$

13. $f(x) = 6e^{2x}$

$f'(x) = 6e^{2x}(2) = 12e^{2x}$

$f''(x) = 12e^{2x}(2) = 24e^{2x}$

$f''(0) = 24e^0 = 24$

$f''(2) = 24e^4$

$f''(-3) = 24e^{-6} = \frac{24}{e^6}$

15. $f(x) = \ln|x|$

$f'(x) = \frac{1}{x} = x^{-1}$

$f''(x) = -x^{-2} = -\frac{1}{x^2}$

$f''(0) = -\frac{1}{0^2}$ does not exist

$f''(2) = -\frac{1}{2^2} = -\frac{1}{4}$

$f''(-3) = -\frac{1}{(-3)^2} = -\frac{1}{9}$

17. $f(x) = x\ln|x|$

$f'(x) = (1)\ln|x| + x\left(\frac{1}{x}\right) = \ln|x| + 1$

$f''(x) = \frac{1}{x}$

$f''(0) = \frac{1}{0}$ does not exist

$f''(2) = \frac{1}{2}$

$f''(-3) = -\frac{1}{3}$

19. As the price $P(t)$ decreases, the slope of the graph is negative, so $P'(t)$ is negative. Since the price is decreasing faster and faster, the graph resembles the right-hand side of the graph in Figure 12.23(a). Here the slope is decreasing, so the second derivative $P''(t)$ is negative.

21. $s(t) = 6t^2 + 5t$

$v(t) = s'(t) = 12t + 5$

$a(t) = v'(t) = s''(t) = 12$

$v(0) = 12(0) + 5 = 5$ cm/sec

$v(4) = 12(4) + 5 = 53$ cm/sec

$a(0) = 12$ cm/sec^2

$a(4) = 12$ cm/sec^2

23. $s(t) = 3t^3 - 4t^2 + 8t - 9$

$v(t) = s'(t) = 9t^2 - 8t + 8$

$a(t) = v'(t) = s''(t) = 18t - 8$

$v(0) = 9(0)^2 - 8(0) + 8 = 8$ cm/sec

$v(4) = 9(4)^2 - 8(4) + 8 = 120$ cm/sec

$a(0) = 18(0) - 8 = -8$ cm/sec^2

$a(4) = 18(4) - 8 = 64$ cm/sec^2

25. $f(x) = x^3 + 3x - 5$

$f'(x) = 3x^2 + 3$

$f''(x) = 6x$

$f''(x) > 0$ if $x > 0$

$f''(x) < 0$ if $x < 0$

so $f''(x) > 0$ on $(0, \infty)$ and $f(x)$ is concave upward. $f''(x) < 0$ on $(-\infty, 0)$ and $f(x)$ is concave downward. $f''(x) = 0$ when $x = 0$ and so there is a point of inflection at $(0, -5)$.

27. $f(x) = x^3 + 4x^2 - 6x + 3$

$f'(x) = 3x^2 + 8x - 6$

$f''(x) = 6x + 8$

f is concave upward when

$f''(x) = 6x + 8 > 0 \Rightarrow x > -\dfrac{4}{3}$

f is concave downward when

$f''(x) = 6x + 8 < 0 \Rightarrow x < -\dfrac{4}{3}$

point of inflection at $x = -\dfrac{4}{3}$

$f\left(-\dfrac{4}{3}\right) = \left(-\dfrac{4}{3}\right)^3 + 4\left(-\dfrac{4}{3}\right)^2 - 6\left(-\dfrac{4}{3}\right) + 3$

$\qquad = \dfrac{425}{27}$

so f is concave upward on $\left(-\dfrac{4}{3}, \infty\right)$, concave

downward on $\left(-\infty, -\dfrac{4}{3}\right)$, and has a point of

inflection at $\left(-\dfrac{4}{3}, \dfrac{425}{27}\right)$.

29. $f(x) = \dfrac{2}{x-4}; \quad f'(x) = \dfrac{-2}{(x-4)^2};$

$f''(x) = \dfrac{4}{(x-4)^3}$

$\dfrac{4}{(x-4)^3} > 0$ when $(x-4)^3 > 0 \Rightarrow x > 4.$

$\dfrac{4}{(x-4)^3} < 0$ when $(x-4)^3 < 0 \Rightarrow x < 4.$

$f(4) = \dfrac{2}{4-4} = \dfrac{2}{0}$ which does not exist.

So f is concave upward on $(4, \infty)$, concave downward on $(-\infty, 4)$, and has no points of inflection.

31. $f(x) = x^4 + 8x^3 - 30x^2 + 24x - 3$

$f'(x) = 4x^3 + 24x^2 - 60x + 24$

$f''(x) = 12x^2 + 48x - 60$

$f''(x) = 12x^2 + 48x - 60 = 0 \Rightarrow x = -5, 1$

$f''(x) > 0$ when $12x^2 + 48x - 60 > 0 \Rightarrow$

$12\left(x^2 + 4 - 5\right) > 0 \Rightarrow 12(x+5)(x-1) > 0 \Rightarrow$

$x + 5 > 0$ and $x - 1 > 0$, or $x + 5 < 0$ and

$x - 1 < 0 \Rightarrow x > -5$ and $x > 1$ (which is $x > 1$), or

$x < -5$ and $x < 1$ (which is $x < -5$).

$f''(x) < 0$ when $12x^2 + 48x - 60 < 0 \Rightarrow$

$12\left(x^2 + 4 - 5\right) < 0 \Rightarrow 12(x+5)(x-1) < 0 \Rightarrow$

$x + 5 > 0$ and $x - 1 < 0$, or $x + 5 < 0$ and

$x - 1 > 0 \Rightarrow x > -5$ and $x < 1$, or $x < -5$ and

$x > 1 \Rightarrow -5 < x < 1.$

$f(-5) = (-5)^4 + 8(-5)^3 - 30(-5)^2 + 24(-5) - 3$

$\qquad = -1248$

$f(1) = (1)^4 + 8(1)^3 - 30(1)^2 + 24(1) - 3 = 0$

So f is concave upward on $(-\infty, -5)$ and $(1, \infty)$, concave downward on $(-5, 1)$, and has points of inflection at $(-5, -1248)$ and $(1, 0)$.

33. $R(x) = 10,000 - x^3 + 42x^2 + 800x, \ 0 \le x \le 20$

$R'(x) = -3x^2 + 84x + 800$

$R''(x) = -6x + 84$

$R''(x) = 0$ when $-6x + 84 = 0 \Rightarrow x = 14.$

$R(14) = 10,000 - (14)^3 + 42(14)^2 + 800(14)$

$\qquad = 26,688$

So the point of diminishing returns is $(14, 26{,}688)$.

35. $f(x) = -2x^3 - 3x^2 - 72x + 1$

$f'(x) = -6x^2 - 6x - 72$

$f''(x) = -12x - 6$

$f'(x) = 0 \Rightarrow -6x^2 - 6x - 72 = 0 \Rightarrow$

$-6\left(x^2 + x + 12\right) = 0$

the solutions to $x^2 + x + 12 = 0$ are not real numbers. Therefore, there are no critical numbers, and there are no local maxima or minima.

37. $f(x) = x^3 + \dfrac{3}{2}x^2 - 60x + 100$

$f'(x) = 3x^2 + 3x - 60$

Solve $f'(x) = 0$.

$3x^2 + 3x - 60 = 0 \Rightarrow 3\left(x^2 + x - 20\right) = 0 \Rightarrow$

$3(x+5)(x-4) = 0 \Rightarrow x = -5, 4$

Use the second derivative test.

$f''(x) = 6x + 3$

$f''(-5) = 6(-5) + 3 = -27 < 0$

$f''(4) = 6(4) + 3 = 27 > 0$

Thus, $x = -5$ gives a local maximum and $x = 4$ gives a local minimum.

39. $f(x) = x^4 - 8x^2$

$f'(x) = 4x^3 - 16x$

Solve for $f'(x) = 0$.

$4x^3 - 16x = 0 \Rightarrow 4x(x^2 - 4) = 0 \Rightarrow x = 0, \pm 2$

$x = 0$ or $x = 2$ or $x = -2$

Use the second derivative test.

$f''(x) = 12x^2 - 16$

$f''(0) = -16 < 0$

$f''(2) = 48 - 16 = 32 > 0$

$f''(-2) = 48 - 16 = 32 > 0$

Thus, $x = 0$ gives a local maximum, and $x = 2$ and $x = -2$ give local minima.

41. $f(x) = x + \dfrac{4}{x} = x + 4x^{-1}$

$f'(x) = 1 - 4x^{-2}$

Solve $f'(x) = 0$.

$1 - 4x^{-2} = 0 \Rightarrow 1 - \dfrac{4}{x^2} = 0 \Rightarrow 1 = \dfrac{4}{x^2} \Rightarrow$

$x^2 = 4 \Rightarrow x = \pm 2$

Use the second derivative test.

$f''(x) = 8x^{-3} = \dfrac{8}{x^3}$

$f''(2) = \dfrac{8}{(2)^3} = \dfrac{8}{8} = 1 > 0$

$f''(-2) = \dfrac{8}{(-2)^3} = \dfrac{8}{-8} = -1 < 0$

Thus, $x = 2$ gives a local minimum and $x = -2$ gives a local maximum.

43. $f(x) = \dfrac{x^2 + 9}{2x}$

$f'(x) = \dfrac{(2x)(2x) - (2)(x^2 + 9)}{(2x)^2} = \dfrac{4x^2 - 2x^2 - 18}{4x^2}$

$= \dfrac{2x^2 - 18}{4x^2} = \dfrac{x^2 - 9}{2x^2}$

Solve $f'(x) = 0$.

$\dfrac{x^2 - 9}{2x^2} = 0 \Rightarrow \dfrac{(x+3)(x-3)}{2x^2} = 0 \Rightarrow$

$(x+3)(x-3) = 0 \Rightarrow x = -3, 3$

Use the second derivative test.

$f''(x) = \dfrac{(4x)(4x^2) - (8x)(2x^2 - 18)}{(4x^2)^2}$

$= \dfrac{144x}{16x^4} = \dfrac{9}{x^3}$

$f''(3) = \dfrac{9}{(3)^3} = \dfrac{9}{27} = \dfrac{1}{3} > 0$

$f''(-3) = \dfrac{9}{(-3)^3} = -\dfrac{9}{27} = -\dfrac{1}{3} < 0$

Thus $x = 3$ gives a local minimum and $x = -3$ gives a local maximum.

45. $f(x) = \dfrac{2-x}{2+x}$

$f'(x) = \dfrac{(-1)(2+x) - (1)(2-x)}{(2+x)^2}$

$= \dfrac{-2 - x - 2 + x}{(2+x)^2} = \dfrac{-4}{(2+x)^2}$

$f'(x)$ does not equal 0 for any value. $f'(-2)$ does not exist, but -2 is not a critical number because $f(-2)$ is undefined. Thus, there are no critical numbers and, consequently, there can be no local maxima or minima.

47. $f'(x) = (x-1)(x-2)(x-4)$

$= x^3 - 7x^2 + 14x - 8$

$f'(x) = 0$ when $x = 1$ or $x = 2$ or $x = 4$

$f''(x) = 3x^2 - 14x + 14$

$f''(1) = 3(1)^2 - 14(1) + 14 = 3 > 0$

$f''(2) = 3(2)^2 - 14(2) + 14 = -2 < 0$

$f''(4) = 3(4)^2 - 14(4) + 14 = 6 > 0$

$f''(x) = 0$ when $3x^2 - 14x + 14 = 0$

$x = \dfrac{-(-14) \pm \sqrt{(-14^2) - 4(3)(14)}}{2(3)}$

$= \dfrac{14 \pm \sqrt{28}}{6} = \dfrac{7 \pm \sqrt{7}}{3}$

So f has a local maximum at $x = 2$, local minimum at $x = 1$ and $x = 4$, and points of inflection at $x = \dfrac{7 + \sqrt{7}}{3}$ and $x = \dfrac{7 - \sqrt{7}}{3}$

49. $f'(x) = (x-2)^2(x-1) = x^3 - 5x^2 + 8x - 4$

$f'(x) = 0$ when $x = 2$ or $x = 1$

$f''(x) = 3x^2 - 10x + 8$

$f''(2) = 3(2)^2 - 10(2) + 8 = 0$

$f''(1) = 3(1)^2 - 10(1) + 8 = 1 > 0$

$f''(x) = 0$ when

$3x^2 - 10x + 8 = 0 \Rightarrow (3x-4)(x-2) = 0 \Rightarrow$

$x = \dfrac{4}{3}, 2$

So f has a local minimum at $x = 1$, and points of

inflection at $x = \dfrac{4}{3}$ and $x = 2$.

51. a. $f'(x) > 0$ and $f''(x) > 0$

These conditions indicate a point where the function is increasing and the graph is concave up. Point E satisfies these conditions.

b. $f'(x) < 0$ and $f''(x) > 0$

These conditions indicate a point where the function is decreasing and the graph concave up. Point C satisfies these conditions.

c. $f'(x) = 0$ and $f''(x) < 0$

These conditions indicate a critical point of the function where the graph is concave down. Points A and F satisfy these conditions.

d. $f'(x) = 0$ and $f''(x) > 0$

These conditions indicate a critical point of the function where the graph is concave up. Point D satisfies these conditions.

e. $f'(x) < 0$ and $f''(x) = 0$

These conditions indicate a point where the function is decreasing and the same point is an inflection point. Point B satisfies these conditions.

53. $f(x) = 2.5x^{1/2} + 13, \quad 1 \le x \le 15$

$f'(x) = 1.25x^{-1/2} = \dfrac{1.25}{x^{1/2}}$

a. $f'(x)$ is always positive, so the function is increasing over the defined interval, [1, 15].

b. $f'(10) = 1.25(10)^{-1/2} \approx .40$

The stock is increasing at about 40 cents per week at the beginning of the 10th week.

55. $s(t) = -16t^2$

$v(t) = s'(t) = -32t$

a. $s'(3) = -32(3) = -96$ ft/sec

b. $s'(5) = -32(5) = -160$ ft/sec

c. $s'(8) = -32(8) = -256$ ft/sec

d. $a(t) = s''(t) = -32$ ft/sec^2

57. $R(x) = -3.21x^3 + 62.14x^2 + 172.92x + 48.65,$
$0 \le x \le 14$

The point of diminishing returns occurs at a point of inflection.

$R'(x) = -9.63x^2 + 124.28x + 172.92$

$R''(x) = -19.26x + 124.28$

$R''(x) = -19.26x + 124.28 = 0 \Rightarrow x \approx 6.5$

Verify that this is a point of inflection:

$R''(7) = -19.26(7) + 124.28 = -10.54 < 0$

$R''(6) = -19.26(6) + 124.28 = 8.72 > 0$

The point of diminishing returns is about 6.5 million farms.

59. $f(x) = .0024x^3 - .0633x^2 + .994x + 3.67,$
$5 \le x \le 17$

$f'(x) = .0072x^2 - .1266x + .994$

$f''(x) = .0144x - .1266$

$f''(x) = .0144x - .1266 = 0 \Rightarrow x \approx 8.8$

$f(8.8) \approx 9.2$

The point of inflection is (8.8, 9.2). This means that the GDP began increasing in late 1998.

61. $f(x) = .067x^3 - 2.495x^2 + 24.932x + 273.9,$
$0 \le x \le 29$

a. $f'(x) = .201x^2 - 4.99x + 24.932$

$.201x^2 - 4.99x + 24.932 = 0$

$x = \dfrac{-(-4.99) \pm \sqrt{(-4.99)^2 - 4(.201)(24.932)}}{2(.201)}$

$\approx \dfrac{4.99 \pm \sqrt{4.8548}}{.402} \approx 6.93, 17.89$

$x \approx 6.93$ or $x \approx 17.89$ are the critical numbers.

b. Test $f'(x)$ at $x = 0$, $x = 10$, and $x = 20$.

$$f'(0) = .201(0)^2 - 4.99(0) + 24.932$$
$$= 24.932 > 0$$
$$f'(10) = .201(10)^2 - 4.99(10) + 24.932$$
$$= -4.868 < 0$$
$$f'(20) = .201(20)^2 - 4.99(20) + 24.932$$
$$= 5.532 > 0$$

Defense spending was at a local maximum in late-1986 ($x \approx 6.93$) and at a local minimum in late 1997 ($x \approx 17.89$).

c. $f''(x) = .402x - 4.99$

$$f''(x) = .402x - 4.99 = 0 \Rightarrow x \approx 12.41$$
$$f(12.41) \approx 327.11$$

$f(x)$ has an inflection point at $(12.41, 327.11)$. This indicates that the rate of change in spending began to increase in mid-1992.

63. $h(x) = .108x^3 - 6.135x^2 + 99.581x - 57.14$, $5 \leq x \leq 27$

a. $h'(x) = .324x^2 - 12.27x + 99.581$

Solve $f'(x) = 0$.

$$.324x^2 - 12.27x + 99.581 = 0 \Rightarrow$$
$$x = \frac{-(-12.27) \pm \sqrt{(-12.27)^2 - 4(.324)(99.581)}}{2(.324)}$$
$$\approx \frac{-(-12.27) \pm \sqrt{21.4959}}{2(.324)} \approx 11.78, 26.09$$

Test $h'(x)$ at $x = 10, 20, 27$.

$$h'(10) = .324(10)^2 - 12.27(10) + 99.581$$
$$\approx 9.281 > 0$$
$$h'(20) = .324(20)^2 - 12.27(20) + 99.581$$
$$\approx -16.219 < 0$$
$$h'(30) = .324(30)^2 - 12.27(30) + 99.581$$
$$\approx 23.081 > 0$$

A local maximum occurs at $x \approx 11.78$, so assaults peaked in 1991.

b. $h''(x) = .648x - 12.27$

$$h''(x) = 0 \Rightarrow .648x - 12.27 = 0 \Rightarrow x \approx 18.93$$

There is a point of inflection at $x \approx 18.93$, so there is a point of inflection in late 1998.

65. $f(x) = .864x^3 - 32.10x^2 + 478.04x + 429.6$, $5 \leq x \leq 17$

a. $f'(x) = 2.592x^2 - 64.2x + 478.04$

Find the critical numbers by solving $f'(x) = 0$.

$$x = \frac{-(-64.2) \pm \sqrt{(-64.2)^2 - 4(2.592)(478.04)}}{2(2.592)}$$
$$= \frac{64.2 \pm \sqrt{-834.679}}{5.184}$$

There are no real solutions, so there are no critical numbers. Therefore, there are no local extrema in the interval.

b. $f''(x) = 5.184x - 64.2$

$$f''(x) = 0 \Rightarrow 5.184x - 64.2 = 0 \Rightarrow x \approx 12.38$$
$$f''(12) \approx -1.992 < 0 \text{ and}$$
$$f''(13) \approx 3.192 > 0, \text{ so there is a point of}$$
inflection at $x \approx 12.38$.

$$f(12.38) = .864(12.38)^3 - 32.10(12.38)^2$$
$$+ 478.04(12.38) + 429.6$$
$$\approx 3067.31$$

The point of inflection is $(12.38, 3067.31)$.

67. $p(t) = \frac{30t^3 - t^4}{1200}$ $(0 \leq t \leq 30)$

$$p'(t) = \frac{90t^2 - 4t^3}{1200}$$

a. $p'(t) = 0$ when

$$90t^2 - 4t^3 = 0 \Rightarrow 4t^2(22.5 - t) = 0 \Rightarrow$$
$$t = 0, 22.5$$
$$p''(t) = \frac{180t - 12t^2}{1200}$$
$$p''(0) = \frac{180(0) - 12(0)^2}{1200} = 0$$
$$p''(22.5) = \frac{180(22.5) - 12(22.5)^2}{1200}$$
$$= -1.6875 < 0$$

So p has a local maximum at 22.5, thus the percent of the population infected is a maximum at 22.5 days.

b. $p(22.5) = \frac{30(22.5)^3 - (22.5)^4}{1200} \approx 71.19$

so the maximum percent of the population infected is about 71.19%.

Section 12.3 Optimization Applications

1. On [0, 4] f has an absolute maximum at $x = 4$ and an absolute minimum at $x = 1$.

3. On [–4, 2] f has an absolute maximum at $x = 2$ and an absolute minimum at $x = -2$.

5. On [–8, 0] f has an absolute maximum at $x = -4$, and an absolute minimum at $x = -7$ and $x = -2$.

7. $f(x) = x^4 - 32x^2 - 7$ on [–5, 6]

$f'(x) = 4x^3 - 64x = 0 \Rightarrow 4x\left(x^2 - 16\right) = 0 \Rightarrow$
$4x(x-4)(x+4) = 0 \Rightarrow x = 0,\ 4,\ -4$

x	$f(x)$
–5	–182
–4	–263
0	–7
4	–263
6	137

So on [–5, 6] f has an absolute maximum at $x = 6$, and an absolute minimum at $x = -4$ and $x = 4$.

9. $f(x) = \dfrac{8+x}{8-x}$ on [4, 6]

$f'(x) = \dfrac{(8-x)(1)-(8+x)(-1)}{(8-x)^2} = \dfrac{16}{(8-x)^2}$

$f'(x)$ is never 0, but is undefined when $x = 8$. Since 8 is not [4, 6], check only 4 and 6.

x	$f(x)$
4	3
6	7

So on [4, 6], f has an absolute maximum at $x = 6$ and an absolute minimum at $x = 4$.

11. $f(x) = \dfrac{x}{x^2 + 2}$ on [–1, 4]

$f'(x) = \dfrac{\left(x^2+2\right)(1)-x(2x)}{\left(x^2+2\right)^2} = \dfrac{-x^2+2}{\left(x^2+2\right)^2} = 0$

when $x = \sqrt{2}$ or $-\sqrt{2}$.
$-\sqrt{2}$ is not in [0, 4], so check $x = -1$, $\sqrt{2}$, 4.

x	$f(x)$
–1	$-\dfrac{1}{3}$
$\sqrt{2}$	$\dfrac{\sqrt{2}}{4}$
4	$\dfrac{2}{9}$

So on [–1, 4] f has an absolute maximum at $x = \sqrt{2}$ and an absolute minimum at $x = -1$.

13. $f(x) = \left(x^2 + 18\right)^{2/3}$ on [–3, 2]

$f'(x) = \dfrac{2}{3}\left(x^2+18\right)^{-1/3}(2x) = \dfrac{4x}{3\left(x^2+18\right)^{1/3}}$

$f'(x) = 0$ when $x = 0$.

x	$f(x)$
–3	9
0	$3\sqrt[3]{12} \approx 6.87$
2	$22^{2/3} \approx 7.85$

So on [–3, 2] f has an absolute maximum at $x = -3$ and an absolute minimum at $x = 0$.

15. $f(x) = \dfrac{1}{\sqrt{x^2 + 1}}$ on [–1, 1]

$f(x) = \left(x^2 + 1\right)^{-1/2}$

$f'(x) = -\dfrac{1}{2}\left(x^2+1\right)^{-3/2}(2x) = \dfrac{-x}{\left(x^2+1\right)^{3/2}} = 0$

when $x = 0$.

x	$f(x)$
–1	$\dfrac{1}{\sqrt{2}}$
0	1
1	$\dfrac{1}{\sqrt{2}}$

So on [–1, 1] f has an absolute maximum at $x = 0$, and an absolute minimum at $x = -1$ and $x = 1$.

17. $g(x) = 2x^3 - 3x^2 - 12x + 1$ on $(0, 4)$

$g'(x) = 6x^2 - 6x - 12 = 0$. Solve $g'(x) = 0$.

$6(x-2)(x+1) = 0 \Rightarrow x = 2, -1$. $x = 2$ is the only critical number that lies in the given interval.

$g''(x) = 12x - 6$

$g''(2) = 12 > 0$

Hence, f has a local minimum at $x = 2$. By the Critical Point Theorem, the absolute minimum of f on the interval $(0, 4)$ occurs at $x = 2$.

19. $g(x) = \dfrac{1}{x} = x^{-1}$ on $(0, \infty)$

$g'(x) = -x^{-2} = -\dfrac{1}{x^2}$

Solve $g'(x) = 0$. $g'(x) = \dfrac{1}{x^2} = 0$ Since $g'(x)$ is

never equal to 0 and $g'(x)$ is defined on $(0, \infty)$ there are no absolute extrema.

21. $g(x) = 6x^{2/3} - 4x$ on $(0, \infty)$

$g'(x) = \dfrac{4}{x^{1/3}} - 4$

Solve $g'(x) = 0$. $g' = \dfrac{4}{x^{1/3}} - 4 = 0 \Rightarrow$

$\dfrac{4 - 4x^{1/3}}{x^{1/3}} = 0 \Rightarrow 4 = 4x^{1/3} \Rightarrow x = 1$ is the only critical number that lies in the given interval.

$g''(x) = \dfrac{-4}{3x^{4/3}}$

$g''(1) = \dfrac{-4}{3} < 0$

Hence, g has a local maximum at $x = 1$. By the Critical Point Theorem, the absolute maximum of g on the interval $(0, \infty)$ occurs at $x = 1$.

23. The average cost per stereo receiver is

$\bar{C}(x) = \dfrac{C(x)}{x}$.

$\dfrac{C(x)}{x} = \dfrac{.28x^3 - 100.5x^2 + 9500x}{x}$

$= .28x^2 - 100.5x + 9500$

$\left(\dfrac{C(x)}{x}\right)' = .56x - 100.5$

$.56x - 100.5 = 0 \Rightarrow x \approx 179$

This is the minimum in the interval $[0, 250]$.

$\bar{C}(179) = \dfrac{C(179)}{179}$

$= \dfrac{.28(179)^3 - 100.5(179)^2 + 9500(179)}{179}$

$= 481.98$

A production level of 179 receivers will give the minimum average cost of $481.98.

25. a. The revenue function is $R(x) = 400x$.

b. The profit function is

$P(x) = R(x) - C(x)$

$= 400x - \left(525,000 - 30x + .012x^2\right)$

$= -.012x^2 + 430x - 525,000$

c. $P'(x) = -.024x + 430 = 0 \Rightarrow x \approx 17,917$

$P(17,917)$

$= -.012(17,917)^2 + 430(17,917) - 525,000$

$\approx \$3,327,083$

17,917 grtills should be made to generate the maximum profit of $3,327,083.

27. $p(t) = 10te^{-t/8}$, $(0 \le t \le 40)$

a. $p'(t) = 10t\left(-\dfrac{1}{8}\right)e^{-t/8} + 10e^{-t/8}$

$= e^{-t/8}\left(-\dfrac{5}{4}t + 10\right)$

$0 = e^{-t/8}\left(-\dfrac{5}{4}t + 10\right) \Rightarrow 0 = -\dfrac{5}{4}t + 10 \Rightarrow$

$8 = t$

Test $p(t)$ at $x = 0$, $x = 8$, and $x = 40$.

$p(0) = 0$; $p(8) \approx 29.43$; $p(40) \approx 2.70$

The maximum number of people are infected after 8 days.

b. $p(8) = 10(8)e^{-8/8} = 29.43$

The maximum percent of people infected is 29.43%.

29. $b(x) = -.0245x^3 + .216x^2 - .095x + 1.06$,
$0 \le x \le 7$

Solve $b'(x) = -.0735x^2 + .432x - .095 = 0$.

$$x = \frac{-.432 \pm \sqrt{.432^2 - 4(-.0735)(-.095)}}{2(-.0735)}$$

$\approx .23, 5.65$

Test $b'(x)$ at $x = 0, 2, 10$.

$b'(0) = -.095$, $b'(2) = .475$, $b'(10) = -3.125$

Using the First Derivative test, there is a local maximum at $x \approx 5.65$, so the number of injections was highest during 2005.

31. $f(x) = 2.96x^2 - 78.63x + 672.4$, $9 \le x \le 16$

$f'(x) = 5.92x - 78.63 = 0 \Rightarrow x \approx 13.3$

This is the only critical point in the given interval. Confirm that this is a minimum by testing $f'(x)$ at $x = 9$ and $x = 16$.

$f'(9) = -25.35$; $f'(17) = 16.09$

Therefore, the minimum occurs at $x \approx 13$, so the number of workers was at a minimum in 2003.

33. Let x = the length of a side of the square that is cut from each corner, in feet. Then the width of the box is $3 - 2x$, the length is $8 - 2x$, and the height is x. The height must not be more than $\frac{3}{2}$ ft because the width of the cardboard is 3 ft.

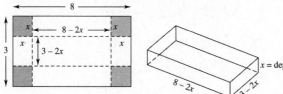

$V(x) = (8 - 2x)(3 - 2x)x = 24x - 22x^2 + 4x^3$

The domain is $\left(0, \frac{3}{2}\right)$.

$V'(x) = 24 - 44x + 12x^2$

Solve $V'(x) = 0$.

$24 - 44x + 12x^2 = 0 \Rightarrow 4(6 - 11x + 3x^2) = 0 \Rightarrow$

$(3 - x)(2 - 3x) = 0 \Rightarrow x = 3, \frac{2}{3}$

Because 3 is not in the domain, the only critical number of interest is $x = \frac{2}{3}$. Evaluate $V(x)$ at

$x = \frac{2}{3}$ and at the endpoints of the domain.

x	$V(x)$
0	0
$\frac{2}{3}$	$\frac{200}{27} = 7\frac{11}{27}$ ← Maximum
$\frac{3}{2}$	0

The square should be $\frac{2}{3}$ ft by $\frac{2}{3}$ ft or 8 inches by 8 inches.

35. Let x = a side of the square base. The volume is $16{,}000$ cm^3 and $V = LWH$, or $H = \frac{V}{LW}$, so the height of the box is $\frac{16{,}000}{x^2}$. The area of each base (top or bottom) is x^2. The area of each of the four sides is $x\left(\frac{16{,}000}{x^2}\right)$. The cost of both the top and bottom is $3 \cdot 2x^2$. The cost of the other four sides is $1.50 \cdot 4x\left(\frac{16{,}000}{x^2}\right)$.

The total cost is

$$C(x) = 3 \cdot 2x^2 + 1.50 \cdot 4x\left(\frac{16{,}000}{x^2}\right)$$

$$= 6x^2 + \frac{96{,}000}{x}$$

$$C'(x) = 12x - \frac{96{,}000}{x^2} = \frac{12x^3 - 96{,}000}{x^2}$$

When $C'(x) = 0$, $12x^3 - 96{,}000 = 0 \Rightarrow x = 20$.

$$C''(x) = 12 + \frac{192{,}000}{x^3} > 0 \text{ if } x > 0.$$

So $x = 20$ leads to a minimum cost. The dimensions of the box of minimum total cost are

20 cm by 20 cm by $\frac{16{,}000}{20^2} = 40$ cm.

$$C(20) = 6 \cdot 20^2 + \frac{96{,}000}{20} = 7200$$

The minimum total cost is 7200 cents or \$72.

37. Let x = width. Then $2x$ = length, and h = height. An equation for the volume is

$$36 = (2x)(x)h \Rightarrow 36 = 2x^2 h \Rightarrow \frac{18}{x^2} = h.$$

The surface area is

$$S(x) = 2x(x) + 2xh + 2(2x)h = 2x^2 + 6xh$$

$$= 2x^2 + 6x\left(\frac{18}{x^2}\right) = 2x^2 + \frac{108}{x}.$$

$$S'(x) = 4x - \frac{108}{x^2}$$

When $S'(x) = 0$,

$$\frac{4x^3 - 108}{x^2} = 0 \Rightarrow 4\left(x^3 - 27\right) = 0 \Rightarrow x = 3.$$

$$S''(x) = 4 + \frac{108(2)}{x^3} = 4 + \frac{216}{x^3} > 0$$

since $x > 0$. So $x = 3$ minimizes the volume.

If $x = 3$, $h = \frac{18}{x^2} = \frac{18}{9} = 2$.

The dimensions are 3 ft by 6 ft by 2 ft.

39. a. The length of the field in meters is $1200 - 2x$.

b. $A(x) = x(1200 - 2x) = 1200x - 2x^2$

c. $A'(x) = 1200 - 4x$

$1200 - 4x = 0 \Rightarrow 4(300 - x) = 0 \Rightarrow x = 300.$

$A''(x) = -4 < 0$ for all x.

The maximum area occurs at $x = 300$ m.

d. $A(300) = 1200(300) - 2(300)^2 = 180,000$

The maximum area is 180,000 sq m.

41. Let x = length of sides that cost $6 per foot.
y = length of sides that cost $3 per foot.
An equation for the cost of the fencing is
$2(3y) + 2(6x) = 2400 \Rightarrow 6y = 2400 - 12x \Rightarrow$
$y = 400 - 2x.$

$A = xy \Rightarrow A(x) = x(400 - 2x) = 400x - 2x^2$
$A'(x) = 400 - 4x$
$400 - 4x = 0, \Rightarrow x = 100.$
$A''(x) = -4 < 0$ for all x.

$A(x)$ is maximum when $x = 100$.
If $x = 100$, $y = 400 - 2(100) = 200.$
$A(100) = 100[400 - 2(100)] = 20,000$

The maximum area is 20,000 sq ft with 200 ft on the $3 sides and 100 ft on the $6 sides.

43. Let x = the length of the side opposite the existing fence.

The area is 15,625 m^2, so $\frac{15,625}{x}$ = the length

of each end. The cost of each end is $2\left(\frac{15,625}{x}\right)$,

and the cost of the side opposite the existing fence is $4x$. The total cost of the new fence for three sides of the rectangular area is

$$C(x) = 4x + 2(2)\left(\frac{15,625}{x}\right) = 4x + \frac{62,500}{x}$$

$$C'(x) = 4 - \frac{62,500}{x^2}$$

$$4 - \frac{62,500}{x^2} = 0 \Rightarrow 4x^2 = 62,500 \Rightarrow x = \pm 125.$$

Only $x = 125$ is in the domain.

$$C''(x) = \frac{125,000}{x^3} > 0, \text{ so } x = 125 \text{ gives a}$$

minimum cost.

$$C(125) = 4(125) + \frac{62,500}{125} = 1000$$

The least expensive fence costs $1000.

45. a. Price in cents for x thousand bars is

$$P(x) = 100 - \frac{x}{10}.$$

Revenue (in cents) is

$$1000x \cdot p(x) = 1000x\left(100 - \frac{x}{10}\right) \Rightarrow$$

$$R(x) = 100,000x - 100x^2$$

b. $R'(x) = 100,000 - 200x$

When $100,000 - 200x = 0$,

$$x = 500.$$

$R''(x) = -200 < 0$ for all x.

The maximum revenue is attained when $x = 500$.

c. $R(500) = 100,000(500) - 100(500)^2$

$$= 25,000,000$$

The maximum revenue is 25,000,000 cents, or $250,000.

47. $f(x) = 55.44(54.3x - 143.9)e^{-.15x}$
$$= 3010.392xe^{-.15x} - 7977.816e^{-.15x}$$
$$f'(x) = -451.5588xe^{-.15x} + 3010.392e^{-.15x}$$
$$+ 1196.6724e^{-.15x}$$
$$= -451.5588xe^{-.15x} + 4207.0644e^{-.15x}$$
$$-451.5588xe^{-.15x} + 4207.0644e^{-.15x} = 0 \Rightarrow$$
$$4207.0644e^{-.15x} = 451.5588xe^{-.15x} \Rightarrow$$
$$x = \frac{4207.0644}{451.5588} \approx 9.3$$

Verify that this is a maximum by testing $f'(0)$ and $f'(10)$. $f'(0) \approx 4207$; $f'(10) \approx -69$

According to the model, revenue reaches its maximum in 2009.

49. Distance on shore: $9 - x$ miles
Cost on shore: $400 per mile

Distance underwater: $\sqrt{x^2 + 36}$
Cost underwater: $500 per mile
Find the distance from A, that is, $(9 - x)$, to minimize cost, $C(x)$.
$$C(x) = (9-x)400 + \left(\sqrt{x^2+36}\right)500$$
$$= 3600 - 400x + 500\left(x^2+36\right)^{1/2}$$
$$C'(x) = -400 + 500\left(\frac{1}{2}\right)\left(x^2+36\right)^{-1/2}(2x)$$
$$= -400 + \frac{500x}{\sqrt{x^2+36}}$$
$$C'(x) = 0 \Rightarrow \frac{500x}{\sqrt{x^2+36}} = 400 \Rightarrow$$
$$\frac{5x}{4} = \sqrt{x^2+36} \Rightarrow \frac{25}{16}x^2 = x^2+36 \Rightarrow$$
$$\left(\frac{25}{16}-1\right)x^2 = 36 \Rightarrow \frac{9}{16}x^2 = 36 \Rightarrow x^2 = 64$$
$$x = \pm 8$$

A distance cannot be negative, so we discard -8. Use the second derivative test to determine whether the critical number 8 yields a maximum or minimum.
$$C''(x)$$
$$= \frac{\left(x^2+36\right)^{1/2}(500) - 500x\left(\frac{1}{2}\right)\left(x^2+36\right)^{-1/2}(2x)}{x^2+36}$$
$$= \frac{500\left(x^2+36\right)^{1/2} - 500x^2\left(x^2+36\right)^{-1/2}}{x^2+36}$$

$C''(8) > 0$, so $x = 8$ produces the minimum total cost. The cost is a minimum when the distance is $9 - x = 9 - 8 = 1$ mile from point A.

51. x = number of batches per year
M = 12,300 sinks annually
k = $2.25, cost to store one unit for one year
a = $350, cost to set up, or fixed cost
From Example 9, the annual number of batches that gives the minimu total production cost is
$$x = \sqrt{\frac{kM}{2a}} = \sqrt{\frac{(2.25)(12,300)}{2(350)}}$$
$$= \sqrt{39.54} \approx 6$$

Each year, 6 batches should be produced.

53. x = number of orders per year
M = 15,900 cases of beer annually
k = $3.25, cost to store one case for one year
a = $5.50, ordering cost
$$x = \sqrt{\frac{kM}{2a}} = \sqrt{\frac{(3.25)(15,900)}{2(5.50)}}$$
$$= \sqrt{4697.73} \approx 69$$

Each year, 69 batches should be produced.

55. x = number of booklets per year
k = $.75, cost to store for one year
M = 50,000, annual demand
a = 80, cost to place an order (fixed cost)
$$x = \sqrt{\frac{kM}{2a}} = \sqrt{\frac{.75(50,000)}{2(80)}} \approx 15 \text{ orders}$$

Since the annual demand is 50,000, and the number of orders is 15, the optimum number of copies per order is $\frac{50,000}{15} \approx 3333$ booklets per order.

57. The formula for the economic order quantity (which is another name for the economic lot size) finds the number of batches to be ordered depending on the cost of storing one unit per year, a fixed cost of placing an order, and a known number of units demanded. The best answer is (c). Periodic demand for the goods is known.

Section 12.4 Curve Sketching

1. $f(x) = -x^2 - 10x - 25$

The y-intercept is $f(0) = -25$, but $(0, -25)$ is not a convenient point to plot. To find any x-intercepts, let $y = 0$.

$0 = -x^2 - 10x - 25 \Rightarrow 0 = -1\left(x^2 + 10x + 25\right) \Rightarrow$

$0 = -1(x+5)^2 \Rightarrow 0 = (x+5)^2 \Rightarrow 0 = x + 5 \Rightarrow$
$x = -5$

The only x-intercept is -5. This is a polynomial function, so there are no asymptotes.

Find $f'(x)$ and set it equal to zero to find critical numbers.

$f'(x) = -2x - 10 = -2(x + 5) = 0$

Critical number: $x = -5$

$f''(x) = -2 < 0$ for all x

$f(-5) = 0$

The graph is concave downward on $(-\infty, \infty)$. Hence there is a local maximum of 0 at $x = -5$. There are no points of inflection. The graph is a parabola opening downward. The local maximum (which is also the absolute maximum) occurs at the vertex $(-5, 0)$. Plot some additional points and sketch the graph.

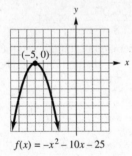

$f(x) = -x^2 - 10x - 25$

3. $f(x) = 3x^3 - 3x^2 + 1$

The y-intercept is $f(0) = 1$. It is not convenient to solve $f(x) = 0$ to find any x-intercepts.

$f'(x) = 9x^2 - 6x = 3x(3x - 2) = 0$

Critical numbers: $x = 0$ or $x = \dfrac{2}{3}$

$f''(x) = 18x - 6$

$f''(0) = -6 < 0$

$f''\left(\dfrac{2}{3}\right) = 6 > 0$

There is a local maximum of 1 at 0 and a local minimum of $\dfrac{5}{9}$ at $\dfrac{2}{3}$.

$f''(x) = 18x - 6 = 0 \Rightarrow x = \dfrac{1}{3}$

$f''(0) = -6 < 0$
$f''(1) = 12 > 0$

There is a point of inflection at $\left(\dfrac{1}{3}, \dfrac{7}{9}\right)$.

Plot the critical points, point of inflection, and a few additional points to sketch the graph.

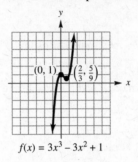

$f(x) = 3x^3 - 3x^2 + 1$

5. $f(x) = -2x^3 - 9x^2 + 108x - 10$

The y-intercept is $f(0) = -10$. It is not convenient to find the x-intercepts.

$f'(x) = -6x^2 - 18x + 108 = 0 \Rightarrow$

$-6\left(x^2 + 3x - 18\right) = 0 \Rightarrow (x + 6)(x - 3) = 0 \Rightarrow$

$x = -6, 3$ (critical numbers)

$f''(x) = -12x - 18$

$f''(-6) = 54 > 0$

$f''(3) = -54 < 0$

There is a local maximum of 179 at 3 and a local minimum of -550 at -6.

$f''(x) = -12x - 18 = 0 \Rightarrow x = -\dfrac{3}{2}$

$f''(-2) = 6 > 0$
$f''(-1) = -6 < 0$

There is a point of inflection at $\left(-\dfrac{3}{2}, -\dfrac{371}{2}\right)$.

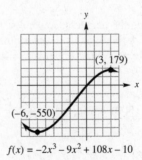

$f(x) = -2x^3 - 9x^2 + 108x - 10$

7. $f(x) = 2x^3 + \dfrac{7}{2}x^2 - 5x + 3$

The y-intercept is $f(0) = 3$. It is not convenient to find any possible x-intercepts.

$f'(x) = 6x^2 + 7x - 5 = 0 \Rightarrow$

$(2x - 1)(3x + 5) = 0 \Rightarrow x = \dfrac{1}{2}, \ -\dfrac{5}{3}$ (critical

numbers)

$f''(x) = 12x + 7$

$f''\left(\dfrac{1}{2}\right) = 13 > 0$

$f''\left(-\dfrac{5}{3}\right) = -13 < 0$

There is a local maximum of $\dfrac{637}{54}$ at $-\dfrac{5}{3}$ and a

local minimum of $\dfrac{13}{8}$ and $\dfrac{1}{2}$.

$f''(x) = 12x + 7 = 0 \Rightarrow x = -\dfrac{7}{12}$

$f''(-1) = -5 < 0$

$f''(0) = 7 > 0$

There is a point of inflection at $\left(-\dfrac{7}{12}, \dfrac{2899}{432}\right)$.

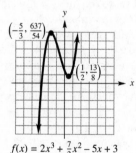

$f(x) = 2x^3 + \frac{7}{2}x^2 - 5x + 3$

9. $f(x) = (x + 3)^4$

The y-intercept is $f(0) = 3^4 = 81$, but $(0, 81)$ is not a convenient point to plot. Because the right-hand side is in factored form, it is easy to find any x-intercepts.

$0 = (x + 3)^4 \Rightarrow x = -3$

The only x-intercept is -3.

$f'(x) = 4(x + 3)^3 (1) = 4(x + 3)^3 = 0 \Rightarrow$

$x = -3$ (critical number)

$f''(x) = 12(x + 3)^2$

$f''(-3) = 12(-3 + 3)^2 = 0$

The second derivative test fails, so use the first derivative test.

$f'(-2) = 4(-2 + 3)^3 = 4(1)^3 = 4 > 0$

$f'(-4) = 4(-4 + 3)^3 = 4(-1)^3 = -4 < 0$

$f(x)$ is increasing on $(-3, \infty)$ and decreasing on $(-\infty, -3)$. Thus, there is a local minimum at $x = -3$, which is also an absolute minimum.

$f''(x) = 12(x + 3)^2 > 0$ for all $x \neq -3$, so the graph is always concave upward and there is no point of inflection.

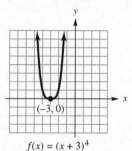

$f(x) = (x + 3)^4$

11. $f(x) = x^4 - 18x^2 + 5$

The y-intercept is $f(0) = 5$. It is not convenient to find the x-intercepts.

$f'(x) = 4x^3 - 36x = 4x\left(x^2 - 9\right) = 0 \Rightarrow$

$x = 0, \ -3, \ 3$ (critical numbers)

$f''(x) = 12x^2 - 36$

$f''(0) = -36 < 0$

$f''(3) = 72 > 0$

$f''(-3) = 72 > 0$

There is a local maximum of 5 at $x = 0$ and a local minimum of -76 at $x = 3$ and $x = -3$.

$f''(x) = 12x^2 - 36 = 0 \Rightarrow x^2 = 3 \Rightarrow x = \pm\sqrt{3}$

$f''(-2) = 12 > 0$

$f''(-1) = -24 < 0$

There is a point of inflection at $\left(-\sqrt{3}, -40\right)$.

$f''(1) = -24 < 0$

$f''(2) = 12 > 0$

There is another point of inflection at $\left(\sqrt{3}, -40\right)$.

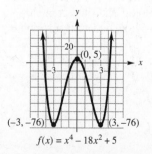

$f(x) = x^4 - 18x^2 + 5$

13. $f(x) = x - \dfrac{1}{x}$

Step 1: There is no y-intercept since $f(0)$ is undefined. Find any x-intercepts by solving the equation $f(x) = 0$.

$$f(x) = x - \frac{1}{x} = 0 \Rightarrow x\left(\frac{x^2 - 1}{x}\right) = x \cdot 0 \Rightarrow$$

$$x^2 - 1 = 0 \Rightarrow (x+1)(x-1) = 0 \Rightarrow x = \pm 1$$

The x-intercepts are -1 and 1.

Step 2: f is a rational function. To find any vertical asymptotes, rewrite the function rule so that we have a single fraction.

$$f(x) = x - \frac{1}{x} = \frac{x^2 - 1}{x}$$

When $x = 0$, the denominator is 0 but the numerator is not, so the line $x = 0$ (the y-axis) is a vertical asymptote.

Step 3: Find the first derivative and use it to look for critical numbers.

$$f(x) = x - \frac{1}{x} = x - x^{-1}$$

$$f'(x) = 1 - (-1)x^{-2} = 1 + x^{-2} = 1 + \frac{1}{x^2}$$

The equation $f'(x) = 1 + \dfrac{1}{x^2} = 0$ has no real

solution. Although $f'(0)$ does not exist, 0 is not a critical number because $f(0)$ also does not exist. Thus, there are no critical numbers and consequently no local extrema. The vertical asymptote at $x = 0$ (the y-axis) separates the graph into two branches. Note that

$f'(x) = 1 + \dfrac{1}{x^2}$ is always positive, so the graph

is increasing on both $(-\infty, 0)$ and $(0, \infty)$.

Step 4 We now consider the second derivative.

$$f''(x) = -2x^{-3} = -\frac{2}{x^3}$$

The equation $f''(x) = -\dfrac{2}{x^3} = 0$ has no solution.

Although $f''(x)$ does not exist at $x = 0$, there cannot be a point of inflection because $f(0)$ does not exist. (Recall that $x = 0$ is a vertical asymptote.) Thus, the graph has no point of inflection. Although there is no point of inflection, concavity may also change at a vertical asymptote, so we examine the second derivative on the two open intervals determined by the asymptote. We will test $x = -1$ in $(-\infty, 0)$ and $x = 1$ in $(0, \infty)$.

$$f''(-1) = -\frac{2}{(-1)^3} = 2 > 0$$

$$f''(1) = -\frac{2}{1^3} = -2 < 0$$

Thus, the graph is concave upward on $(-\infty, 0)$ and concave downward on $(0, \infty)$.

Step 5: To find any horizontal asymptote, consider the following limit.

$$\lim_{x \to \infty} x - \frac{1}{x} = \lim_{x \to \infty} \frac{x^2 - 1}{x} = \lim_{x \to \infty} \frac{\dfrac{x^2}{x^2} - \dfrac{1}{x^2}}{\dfrac{x}{x^2}}$$

$$= \lim_{x \to \infty} \frac{1 - \dfrac{1}{x^2}}{\dfrac{1}{x}} = \frac{1 - 0}{0} = \frac{1}{0}$$

Because division by 0 is undefined, this limit does not exist. Similarly, $\lim_{x \to \infty} f(x)$ does not exist. Therefore, the graph has no horizontal asymptote. Notice that as $x \to \infty$, $\dfrac{1}{x} \to 0$, so

that $f(x) = x - \dfrac{1}{x} \approx x$. Likewise, as $x \to -\infty$,

$f(x) = x - \dfrac{1}{x} \approx x$. This means that as x gets

larger and larger in absolute value, the graph gets closer and closer to the line $y = x$, which we call an oblique or slant asymptote.

Step 6: All of the information we have gathered about this function will guide us in sketching the graph. Because there are no local extrema or points of inflection and also no y-intercept, the only points we have found are the x-intercepts -1 and 1. A few additional points should be plotted on each branch of the curve. The asymptotes are very helpful because they serve as guidelines for the graph.

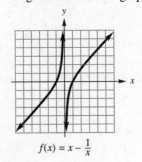

$$f(x) = x - \frac{1}{x}$$

15. $f(x) = \dfrac{x^2 + 25}{x}$

$f(0)$ is undefined, so there is no y-intercept.

$f(x) = 0$ has no real solution because $x^2 + 25$ is always positive. Thus, there are no x-intercepts. When $x = 0$, the denominator is 0 but the numerator is not, so the line $x = 0$ (the y-axis) is a vertical asymptote. We now consider the first derivative and look for critical numbers.

$$f'(x) = \frac{x(2x) - \left(x^2 + 25\right)}{x^2} = \frac{x^2 - 25}{x^2}$$

$$f'(x) = \frac{(x+5)(x-5)}{x^2} = 0 \Rightarrow$$

$(x+5)(x-5) = 0 \Rightarrow x = \pm 5$

Although $f'(0)$ does not exist, 0 is not a critical number because $f(0)$ also does not exist. Use the second derivative test to determine if there are local maxima or minima at the critical numbers -5 and 5.

$$f''(x) = \frac{x^2(2x) - 2x\left(x^2 - 25\right)}{x^4} = \frac{50}{x^3}$$

$$f''(-5) = -\frac{2}{5} < 0; \quad f''(5) = \frac{2}{5} > 0$$

Thus, there is a local maximum of -10 at $x = -5$ and a local minimum of 10 at $x = 5$. f is increasing on $(-\infty, -5)$ and $(5, \infty)$ and decreasing on $(-5, 0)$ and $(0, 5)$.

$f''(x) = \dfrac{50}{x^3} \neq 0$ for any x, so there are no points of inflection. This graph changes concavity at the vertical asymptote $x = 0$. Earlier we found that $f''(-5) < 0$ and $f''(5) > 0$. This tells us that the graph is concave downward on $(-\infty, 0)$ and concave upward on $(0, \infty)$. Following the method shown in Example 1 and the solution for exercise 13, we determine that there is no horizontal asymptote. However, if we rewrite the function rule as $f(x) = x + \dfrac{25}{x}$,

we see that as x gets very large in absolute value, $\dfrac{25}{x} \to 0$, and the graph approaches the line $y = x$, which is an oblique or slant asymptote.

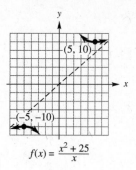

$f(x) = \dfrac{x^2 + 25}{x}$

17. $f(x) = \dfrac{x-1}{x+1}$

$f(0) = -1$, so the y-intercept is -1. $f(x) = 0$ when $x - 1 = 0$ or $x = 1$, so there is one x-intercept, 1. When $x = -1$, the denominator is 0 but the numerator is not, so there is one vertical asymptote, the line $x = -1$.

$$f'(x) = \frac{(x+1) - (x-1)}{(x+1)^2} = \frac{2}{(x+1)^2}$$

$f'(x)$ is never zero.

$f'(x)$ fails to exist for $x = -1$, but -1 is not a critical number because $f(-1)$ does not exist. Thus, there are no critical points and there can be no local extrema. Because $f'(x)$ is always positive where it exists, the function is increasing on $(-\infty, -1)$ and $(-1, \infty)$.

$$f''(x) = \frac{(x+1)^2(0) - 2(2)(x+1)^3}{(x+1)^4}$$

$$= \frac{-4(x+1)^3}{(x+1)^4} = \frac{-4}{x+1}$$

$f''(x)$ is never equal to 0, so there are no points of inflection. The concavity of the graph may change at the vertical asymptote, so we test numbers in the intervals $(-\infty, -1)$ and $(-1, \infty)$.

$$f''(-2) = \frac{-4}{-2+1} = 4 > 0$$

$$f''(0) = \frac{-4}{0+1} = -4 < 0$$

Thus, f is concave upward on $(-\infty, -1)$ and concave downward on $(-1, \infty)$. Because this is a rational function, we now look for a horizontal asymptote.

$$\lim_{x \to \infty} \frac{x-1}{x+1} = \lim_{x \to \infty} \frac{\frac{x}{x} - \frac{1}{x}}{\frac{x}{x} + \frac{1}{x}} = \lim_{x \to \infty} \frac{1 - \frac{1}{x}}{1 + \frac{1}{x}} = \frac{1-0}{1+0} = 1$$

Likewise, $\lim_{x \to -\infty} f(x) = 1$. Thus, there is a horizontal asymptote at $y = 1$. The vertical and horizontal asymptotes act as guidelines for the graph.

(continued on next page)

(continued from page 395)

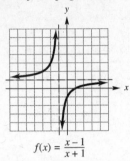

$$f(x) = \frac{x-1}{x+1}$$

19. $y = x - \ln|x|$

$$y' = 1 - \frac{1}{x} = \frac{x-1}{x}$$

$$y'' = \frac{1}{x^2}$$

$y' = 0$ when $x = 1$. There is a vertical asymptote at $x = 0$. The critical numbers are 0 and 1. Test y' at $x = -1, \frac{1}{2}$, and $\frac{3}{2}$.

$$y'(-1) = 2, \quad y'\left(\frac{1}{2}\right) = -1, \quad y'\left(\frac{3}{2}\right) = \frac{1}{3}.$$

So y is increasing on $(-\infty, 0)$ and $(1, \infty)$, and decreasing on $(0, 1)$. Since $y'' \neq 0$, there are no points of inflection. Also, y'' is always positive. Hence, the graph is always concave upward.

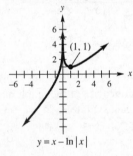

$$y = x - \ln|x|$$

21. $y = xe^{-x}$

$$y' = (1-x)e^{-x}$$

$$y'' = (x-2)e^{-x}$$

Set $y' = 0$ and solve for x to get the critical numbers. Since $e^{-x} \neq 0$, the only critical number is $x = 1$. Test y' at $x = 0$ and $x = 2$.

$$y'(0) = 1; \quad y'(2) = -e^{-2}$$

y is increasing on $(-\infty, 1)$ and decreasing on $(1, \infty)$. Set $y'' = 0$ to find points of inflection.

$y'' = 0 \Rightarrow (x-2)e^{-x} = 0$. An inflection point occurs at $x = 2$.

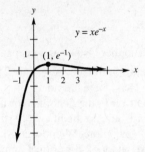

$$y = xe^{-x}$$

For Exercises 23–27, many correct answers are possible, including the following:

23.

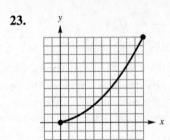

25.

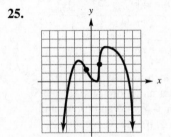

27.

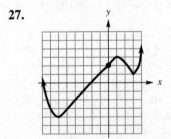

29. a. The first derivative is always positive because the function is increasing.

b. The second derivative is positive because the graph is concave upward. This means that the risk is increasing at a faster and faster rate.

The solutions given for exercises 31–39 were done using a TI-84 Plus graphing calculator. Similar results can be obtained using other graphing calculators.

31. Plot

$$R(x) = .0725x^4 - 3.9421x^3 + 79.159x^2 - 693.2x + 2315.4$$

over the interval [9, 18], then find the maxima and minima.

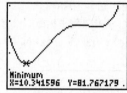 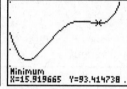

[9, 18] by [75, 100]

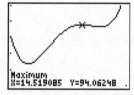

Local minima: (10.3, 81.8), (15.9, 93.4)
Local maximum: (14.5, 94.1)
Now plot

$$R''(x) = .87x^3 - 23.6526x^2 + 158.318x \text{ over}$$

[9, 18] and determine the zeros.

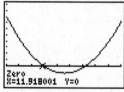 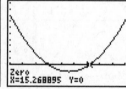

[9, 18] by [–6, 20]

Thus, there are inflection points at $x \approx 11.9$ and $x \approx 15.3$

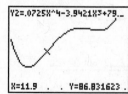 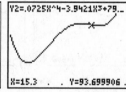

The inflection points are (11.9, 86.8) and (15.3, 93.7).

33. Plot $R(x) = -.0046x^3 + .090x^2 - .101x + 2.1$

over the interval [0, 9], then find the maxima and minima.

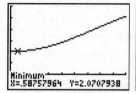 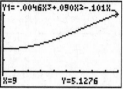

[0, 9] by [–1, 6]

Local minimum: (.6, 2.1)
No local maximum
Now plot $R''(x) = -.0276x + .18$ over

[0, 9] and determine the zero.

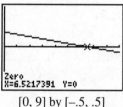

[0, 9] by [–.5, .5]

Thus, there is an inflection point at $x \approx 6.5$.

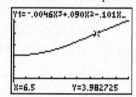

The inflection point is (6.5, 4.0).

35. Plot

$$R(x) = -.0432x^4 + .8586x^3 - 5.226x^2 + 7.58x + 31.4$$

over the interval [0, 9], then find the maxima and minima.

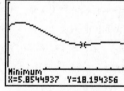 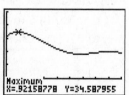

[0, 9] by [–10, 50]

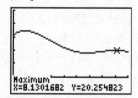

Local minimum: (5.9, 18.2)
Local maxima: (.9, 34.6), (8.1, 20.3)

(*continued on next page*)

(*continued from page 397*)

Now plot $R''(x) = -.5184x^2 + 5.1516x - 10.452$
over [0, 9] and determine the zeros.

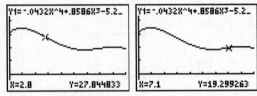

[0, 9] by [−20, 10]

Thus, there are inflection points at $x \approx 2.8$ and $x \approx 7.1$.

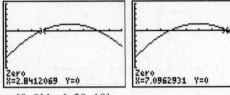

The inflection points are (2.8, 27.8) and (7.1, 19.3).

37. Plot
$$f(x) = .000000143x^4 + .0000178x^3$$
$$- .00145x^2 - .9279x + 77.8$$
over the interval [0, 110], then find the maxima and minima.

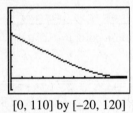

[0, 110] by [−20, 120]

Since the curve decreases steadily, there are no local extrema. Now plot
$$f''(x) = .000001716x^2 + .0001068x - .0029$$
over [0, 110] and determine any zeros.

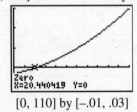

[0, 110] by [−.01, .03]

Thus, there is an inflection point at $x = 20.4$.

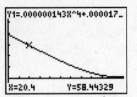

The inflection point is (20.4, 58.4).

39. Plot
$$g(x) = -.000039x^4 + .002393x^3 - .03311x^2$$
$$- .0845x + 14.3$$
over the interval [0, 35], then find the maxima and minima.

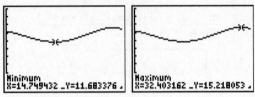

[0, 35] by [0, 20]

Local minimum: (14.7, 11.7)
Local maximum: (32.4, 15.2)
Now plot
$$g''(x) = -.000468x^2 + .014358x - .06622 \text{ over}$$
[0, 35] and determine the zeros.

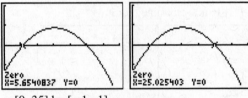

[0, 35] by [−.1, .1]

Thus, there are inflection points at $x \approx 5.7$ and $x \approx 25.0$.

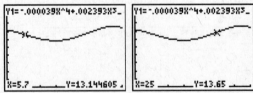

The inflection points are (5.7, 13.1) and (25.0, 13.7).

Chapter 12 Review Exercises

1. If $f'(x) > 0$ for each x in an interval, then f is increasing on the interval; if $f'(x) < 0$ for each x in an interval, then f is decreasing on the interval.

2. If either $f'(c) = 0$ or $f'(c)$ does not exist, there may be a local extremum at c.
   The first derivative test says that $f(c)$ is a local maximum if $f'(a) > 0$ and $f'(b) < 0$ and $a < c < b$. $f(c)$ is a local minimum of $f'(a) < 0$ and $f'(b) > 0$. If $f''(c)$ exists, the second derivative test says that if $f''(c) < 0$, then $f(c)$ is a local maximum. If $f''(c) > 0$, then $f(c)$ is a local minimum.

3. A local extremum is a maximum or minimum value of the function on an open interval. An absolute extremum is the largest or smallest possible value of the function. A local extremum can be an absolute extremum, but not necessarily.

4. The first derivative can be used to determine where a graph is increasing or decreasing. The second derivative can be used to determine where a graph is concave upward or concave downward.

5. $f(x) = x^2 + 9x - 9$; $f'(x) = 2x + 9$

   $f'(x) = 0 \Rightarrow 2x + 9 = 0 \Rightarrow x = -\dfrac{9}{2}$

   $f''(x) = 2 > 0$ for all x. A local minimum occurs

   at $x = -\dfrac{9}{2}$. f is increasing on $\left(-\dfrac{9}{2}, \infty\right)$ and

   decreasing on $\left(-\infty, -\dfrac{9}{2}\right)$.

6. $f(x) = -3x^2 - 3x + 11$
   $f'(x) = -6x - 3$

   $f'(x) = 0 \Rightarrow -6x - 3 = 0 \Rightarrow x = -\dfrac{1}{2}$

   $f''(x) = -6 < 0$ for x. A local maximum occurs

   at $x = -\dfrac{1}{2}$. f is increasing on $\left(-\infty, -\dfrac{1}{2}\right)$ and

   decreasing on $\left(-\dfrac{1}{2}, \infty\right)$.

7. $g(x) = 2x^3 - x^2 - 4x + 7$
   $g'(x) = 6x^2 - 2x - 4$
   $g'(x) = 0 \Rightarrow 6x^2 - 2x - 4 = 0 \Rightarrow$
   $2(3x + 2)(x - 1) = 0 \Rightarrow x = -\dfrac{2}{3}, 1$

   Use the second derivative test.
   $g''(x) = 12x - 2$
   $g''\left(-\dfrac{2}{3}\right) = -10 < 0$; $g''(1) = 10 > 0$

   A local maximum occurs at $x = -\dfrac{2}{3}$. A local

   minimum occurs at $x = 1$. f is increasing on

   $\left(-\infty, -\dfrac{2}{3}\right)$ and $(1, \infty)$, and decreasing on

   $\left(-\dfrac{2}{3}, 1\right)$.

8. $g(x) = -4x^3 - 5x^2 + 8x + 1$
   $g'(x) = -12x^2 - 10x + 8$
   $g'(x) = 0 \Rightarrow -12x^2 - 10x + 8 = 0 \Rightarrow$
   $-2(3x + 4)(2x - 1) = 0 \Rightarrow x = -\dfrac{4}{3}, \dfrac{1}{2}$

   Use the second derivative test.
   $g''(x) = -24x - 10$
   $g''\left(-\dfrac{4}{3}\right) = 22 > 0$; $g''\left(\dfrac{1}{2}\right) = -22 < 0$

   A local maximum occurs at $x = \dfrac{1}{2}$.

   A local minimum occurs at $x = -\dfrac{4}{3}$.

   f is increasing on $\left(-\dfrac{4}{3}, \dfrac{1}{2}\right)$, and decreasing on

   both $\left(-\infty, -\dfrac{4}{3}\right)$ and $\left(\dfrac{1}{2}, \infty\right)$.

9. $f(x) = \dfrac{4}{x-4} = 4(x-4)^{-1}$

$f'(x) = -4(x-4)^{-2} = \dfrac{-4}{(x-4)^2}$

$f'(x) = 0$ has no solution.

$f'(4)$ does not exist. Although 4 is not a critical number because $f(4)$ is undefined, this number determines the open intervals $(-\infty, 4)$ and $(4, \infty)$ we need to examine. (The graph will have a vertical asymptote at $x = 4$.)

$f'(2) = -1 < 0; \ f'(5) = -4 < 0$

f is never increasing. It is decreasing on $(-\infty, 4)$ and $(4, \infty)$.

10. $f(x) = \dfrac{-6}{3x-5} = -6(3x-5)^{-1}$

$f'(x) = -6(-1)(3x-5)^{-2}(3)$

$\quad = 18(3x-5)^{-2} = \dfrac{18}{(3x-5)^2}$

$f'(x) = 0$ has no solution.

$f'(x)$ does not exist if $x = \dfrac{5}{3}$, but this is not a

critical number because $f\left(\dfrac{5}{3}\right)$ is undefined.

There are no critical numbers and there is no local extrema.

Test -1 in $\left(-\infty, \dfrac{5}{3}\right)$ and 2 in $\left(\dfrac{5}{3}, \infty\right)$.

$f'(-1) = \dfrac{9}{32} > 0; \ f'(2) = 18 > 0$

f is never decreasing. It is increasing on

$\left(-\infty, \dfrac{5}{3}\right)$ and $\left(\dfrac{5}{3}, \infty\right)$.

11. $f(x) = 2x^3 - 3x^2 - 36x + 10$

$f'(x) = 6x^2 - 6x - 36 = 0 \Rightarrow$

$6\left(x^2 - x - 6\right) = 0 \Rightarrow (x-3)(x+2) = 0 \Rightarrow$

$x = 3, -2$

Critical numbers: 3, –2

$f''(x) = 12x - 6$

$f''(3) = 30 > 0$, so there is a local minimum at $x = 3$.

$f''(-2) = -30 < 0$, so there is a local maximum at $x = -2$.

$f(-2) = 54; \ f(3) = -71$

f has a local maximum of 54 at $x = -2$ and a local minimum of –71 at $x = 3$.

12. $f(x) = 2x^3 - 3x^2 - 12x + 2$

$f'(x) = 6x^2 - 6x - 12 = 0 \Rightarrow x^2 - x - 2 = 0 \Rightarrow$

$(x-2)(x+1) = 0 \Rightarrow x = 2, -1$

–1 and 2 are the critical numbers.

$f''(x) = 12x - 6$

$f''(-1) = -18 < 0$

$f''(2) = 18 > 0$

$f(-1) = 9; \ f(2) = -18$

f has a local maximum of 9 at $x = -1$ and a local minimum of –18 at $x = 2$.

13. $f(x) = x^4 - \dfrac{8}{3}x^3 - 6x^2 + 2$

$f'(x) = 4x^3 - 8x^2 - 12x = 4x\left(x^2 - 2x - 3\right)$

$\quad = 4x(x-3)(x+1)$

$f'(x) = 0$ when $x = -1, 0$, or 3, so these are the critical numbers.

$f(-1) = -\dfrac{1}{3}; f(0) = 2; \ f(3) = -43$

$f''(x) = 12x^2 - 16x - 12$

$f''(3) = 48 > 0$

$f''(0) = -12 < 0$

$f''(-1) = 16 > 0$

A local minimum of $-\dfrac{1}{3}$ occurs at $x = -1$.

A local maximum of 2 occurs at $x = 0$.

A local minimum of –43 occurs at $x = 3$.

14. $f(x) = x \cdot e^x$

Find the derivative and set it equal to 0.

$f'(x) = xe^x + e^x = 0 \Rightarrow e^x(x+1) = 0 \Rightarrow x = -1$

The only critical number is –1.

$f''(x) = xe^x + e^x + e^x = xe^x + 2e^x$

$f''(-1) = -1e^{-1} + 2e^{-1} = -\dfrac{1}{e} + \dfrac{2}{e} = \dfrac{1}{e} > 0$

$f(-1) = -1e^{-1} = -\dfrac{1}{e} \approx -.368$

f has a local minimum of $-\dfrac{1}{e}$ at $x = -1$.

15. $f(x) = 3x \cdot e^{-x}$

$f'(x) = 3x\left(-e^{-x}\right) + 3e^{-x} = (3 - 3x)e^{-x}$

$f''(x) = (3 - 3x)\left(-e^{-x}\right) + (-3)e^{-x} = (3x - 6)e^{-x}$

$f'(x) = 0 \Rightarrow (3 - 3x)e^{-x} = 0 \Rightarrow 3 - 3x = 0 \Rightarrow$
$x = 1$

$x = 1$ is the critical number. $f(1) = 3e^{-1} = \dfrac{3}{e}$

$f''(1) = -3e^{-1} < 0$, so $x = 1$ is a local maximum.

A local maximum of $\dfrac{3}{e}$ occurs at

$x = 1$.

16. $f(x) = \dfrac{e^x}{x - 1}$

Find the first derivative and set it equal to zero.

$f'(x) = \dfrac{(x - 1)e^x - e^x}{(x - 1)^2} = 0 \Rightarrow$

$\dfrac{e^x(x - 1 - 1)}{(x - 1)^2} = 0 \Rightarrow \dfrac{e^x(x - 2)}{(x - 1)^2} = 0 \Rightarrow$

$x - 2 = 0 \Rightarrow x = 2$

$f''(x) = \dfrac{(x - 1)^2\left[e^x + (x - 2)e^x\right]}{(x - 1)^4}$

$\qquad - \dfrac{e^x(x - 2)(2)(x - 1)}{(x - 1)^4}$

$f''(2) = \dfrac{1^2\left(e^2\right) - 0}{1^4} = e^2 > 0$

$f(2) = \dfrac{e^2}{1} = e^2$

A local minimum of e^2 occurs at $x = 2$.

17. $f(x) = 2x^5 - 5x^3 + 3x - 1$

$f'(x) = 10x^4 - 15x^2 + 3$

$f''(x) = 40x^3 - 30x$

$f''(1) = 40(1)^3 - 30(1) = 10$

$f''(-2) = 40(-2)^3 - 30(-2) = -260$

18. $f(x) = \dfrac{3 - 2x}{x + 3}$

$f'(x) = \dfrac{(x + 3)(-2) - (3 - 2x)(1)}{(x + 3)^2} = \dfrac{-9}{(x + 3)^2}$

$\qquad = -9(x + 3)^{-2}$

$f''(x) = -9(-2)(x + 3)^{-3} = \dfrac{18}{(x + 3)^3}$

$f''(1) = \dfrac{18}{(1 + 3)^3} = \dfrac{18}{64} = \dfrac{9}{32}$

$f''(-2) = \dfrac{18}{(-2 + 3)^3} = \dfrac{18}{1} = 18$

19. $f(x) = -5e^{2x}$

$f'(x) = -5e^{2x}(2) = -10e^{2x}$

$f''(x) = -10e^{2x}(2) = -20e^{2x}$

$f''(1) = -20e^{2(1)} = -20e^2$

$f''(-2) = -20e^{2(-2)} = -20e^{-4}$

20. $f(x) = \ln|5x + 2|$

$f'(x) = \dfrac{1}{5x + 2}(5) = 5(5x + 2)^{-1}$

$f''(x) = 5(-1)(5x + 2)^{-2}(5) = -\dfrac{25}{(5x + 2)^2}$

$f''(1) = -\dfrac{25}{[5(1) + 2]^2} = -\dfrac{25}{49}$

$f''(-2) = -\dfrac{25}{[5(-2) + 2]^2} = -\dfrac{25}{64}$

21. $f(x) = -2x^3 - \dfrac{1}{2}x^2 - x - 3$

$f'(x) = -6x^2 - x - 1$

Solve $-6x^2 - x - 1 = 0$ for x:

$x = \dfrac{-(-1) \pm \sqrt{(-1)^2 - 4(-6)(-1)}}{2(-6)} = \dfrac{1 \pm \sqrt{-23}}{-12}$,

which has no real-number solution. Therefore there are no local extrema.

$f''(x) = -12x - 1$

$f''(x) = 0 \Rightarrow -12x - 1 = 0 \Rightarrow x = -\dfrac{1}{12}$

$f(x)$ is concave upward on $\left(-\infty, -\dfrac{1}{12}\right)$ and

concave downward on $\left(-\dfrac{1}{12}, \infty\right)$.

$f\left(-\dfrac{1}{12}\right) = -2\left(-\dfrac{1}{12}\right)^3 - \dfrac{1}{2}\left(-\dfrac{1}{12}\right)^2 - \left(-\dfrac{1}{12}\right) - 3$

$\qquad \approx -2.92$

$\left(-\dfrac{1}{12}, -2.92\right)$ is an inflection point.

(*continued on next page*)

(continued from page 401)

$$f(0) = -2(0)^3 - \frac{1}{2}(0)^2 - 0 - 3 = -3, \text{ so the}$$

y-intercept is –3. It is not convenient to find the x-intercept. The function is a polynomial, so there are no asymptotes. Use all of the above information and plot some additional points to sketch the graph.

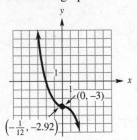

22. $f(x) = -\frac{4}{3}x^3 + x^2 + 30x - 7$

$f'(x) = -4x^2 + 2x + 30$

When ,

$f'(x) = 0 \Rightarrow -2\left(2x^2 - x - 15\right) = 0 \Rightarrow$

$(2x + 5)(x - 3) = 0 \Rightarrow x = -\frac{5}{2}, 3$

$$f\left(-\frac{5}{2}\right) = -\frac{4}{3}\left(-\frac{5}{2}\right)^3 + \left(-\frac{5}{2}\right)^2 + 30\left(-\frac{5}{2}\right) - 7$$
$$\approx -54.9$$

$f(3) = -\frac{4}{3}(3)^3 + 3^2 + 30(3) - 7 = 56$

$\left(-\frac{5}{2}, -54.9\right)$ and (3, 56) are critical points.

$f''(x) = -8x + 2; \quad f''\left(-\frac{5}{2}\right) = 22 > 0$

There is a local minimum at $x = -\frac{5}{2}$.

$f''(3) = -22 < 0$

There is a local maximum at x = 3. f(x) is

increasing on $\left(-\frac{5}{2}, 3\right)$ and decreasing on

$\left(-\infty, -\frac{5}{2}\right)$ and (3, ∞).

$f''(x) = 0 \Rightarrow -8x + 2 = 0 \Rightarrow x = \frac{1}{4}$.

f(x) is concave upward on $\left(-\infty, \frac{1}{4}\right)$ and concave

downward on $\left(\frac{1}{4}, \infty\right)$.

$f\left(\frac{1}{4}\right) = -\frac{4}{3}\left(\frac{1}{4}\right)^3 + \left(\frac{1}{4}\right)^2 + 30\left(\frac{1}{4}\right) - 7 \approx .54$

$\left(\frac{1}{4}, .54\right)$ is an inflection point. f(0) = –7, so the

y-intercept is 0. It is not convenient to find the x-intercept.

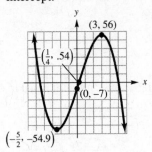

23. $f(x) = x^4 - \frac{4}{3}x^3 - 4x^2 + 1$

$f'(x) = 4x^3 - 4x^2 - 8x$

$f'(x) = 0 \Rightarrow 4x\left(x^2 - x - 2\right) = 0 \Rightarrow$

$x(x + 1)(x - 2) = 0 \Rightarrow x = 0, -1, 2$

–1, 0, and 2 are critical numbers.

$f(-1) = (-1)^4 - \frac{4}{3}(-1)^3 - 4(-1)^2 + 1 = -\frac{2}{3}$

$f(0) = 1$

$f(2) = 2^4 - \frac{4}{3}(2)^3 - 4(2)^2 + 1 = -\frac{29}{3}$

$\left(-1, -\frac{2}{3}\right)$, (0, 1), and $\left(2, -\frac{29}{3}\right)$ are critical

points.

$f''(x) = 12x^2 - 8x - 8$

$f''(-1) = 12 > 0$

There is a local minimum at x = –1.

$f''(0) = -8 < 0$

There is a local maximum at x = 0.

$f''(2) = 24 > 0$

There is a local minimum at x = 2.

f(x) is increasing on (–1, 0) and (2, ∞) and

decreasing on (–∞, –1) and (0, 2).

(continued on next page)

(continued from page 402)

$$f''(x) = 0 \Rightarrow 4\left(3x^2 - 2x - 2\right) = 0 \Rightarrow$$

$$x = \frac{2 \pm \sqrt{(-2)^2 + 24}}{6} = \frac{2 \pm 2\sqrt{7}}{6} = \frac{1 \pm \sqrt{7}}{3}$$

$f(x)$ is concave upward on

$\left(-\infty, \dfrac{1-\sqrt{7}}{3}\right)$ and $\left(\dfrac{1+\sqrt{7}}{3}, \infty\right)$ and concave

downward on $\left(\dfrac{1-\sqrt{7}}{3}, \dfrac{1+\sqrt{7}}{3}\right)$.

$$f\left(\frac{1-\sqrt{7}}{3}\right) \approx .11$$

$$f\left(\frac{1+\sqrt{7}}{3}\right) \approx -5.12$$

$\left(\dfrac{1-\sqrt{7}}{3}, .11\right)$ and $\left(\dfrac{1+\sqrt{7}}{3}, -5.12\right)$ are

inflection points.
$f(0) = 1$, so the y-intercept is 1. The x-intercepts
are not convenient to find but can be estimated
from the graph. This is a polynomial function, so
there are no asymptotes.

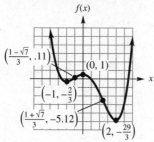

24. $f(x) = -\dfrac{2}{3}x^3 + \dfrac{9}{2}x^2 + 5x + 1$

$f'(x) = -2x^2 + 9x + 5$

$f'(x) = 0 \Rightarrow 2x^2 - 9x - 5 = 0 \Rightarrow$

$(2x+1)(x-5) = 0 \Rightarrow x = -\dfrac{1}{2}, 5$

$-\dfrac{1}{2}$ and 5 are critical numbers.

$f\left(-\dfrac{1}{2}\right) \approx -.29; f(5) \approx 55.17$

$\left(-\dfrac{1}{2}, -.29\right)$ and $(5, 55.17)$ are critical points.

$$f''(x) = -4x + 9; \quad f''\left(-\frac{1}{2}\right) = 11 > 0$$

There is a local minimum at $x = -\dfrac{1}{2}$.

$$f''(5) = -11 < 0$$

There is a local maximum at $x = 5$.

$f(x)$ is increasing on $\left(-\dfrac{1}{2}, 5\right)$ and decreasing on

$\left(-\infty, -\dfrac{1}{2}\right)$ and $(5, \infty)$.

$$f''(x) = 0 \Rightarrow -4x + 9 = 0 \Rightarrow x = \frac{9}{4},$$

$f(x)$ is concave upward on $\left(-\infty, \dfrac{9}{4}\right)$ and

concave downward on $\left(\dfrac{9}{4}, \infty\right)$.

$$f\left(\frac{9}{4}\right) \approx 27.44$$

$\left(\dfrac{9}{4}, 27.44\right)$ is an inflection point. $f(0) = 1$, so

the y-intercept is 1. It is not convenient to find
the x-intercepts.

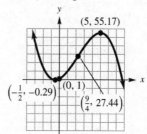

25. $f(x) = \dfrac{x-1}{2x+1}$

$f'(x) = \dfrac{(2x+1) - (x-1)(2)}{(2x+1)^2} = \dfrac{3}{(2x+1)^2}$

$\quad\ = 3(2x+1)^{-2}$

$f'(x)$ is never zero. $f'\left(-\dfrac{1}{2}\right)$ does not exist,

but since $f\left(-\dfrac{1}{2}\right)$ is undefined, $-\dfrac{1}{2}$ is not a

critical number. (There is a vertical asymptote at

$x = -\dfrac{1}{2}$.) This function has no critical numbers,

so there can be no local extrema.

(continued on next page)

(continued from page 403)

To determine where the function is increasing and decreasing, test points in the intervals

$$\left(-\infty, -\frac{1}{2}\right) \text{ and } \left(-\frac{1}{2}, \infty\right).$$

$$f'(-1) = \frac{2}{[2(-1)+1]^2} = 2 > 0$$

$$f'(0) = \frac{2}{[2(0)+1]^2} = 2 > 0$$

$f(x)$ is increasing on $\left(-\infty, -\frac{1}{2}\right)$ and on $\left(-\frac{1}{2}, \infty\right).$

$$f''(x) = 3(-2)(2x+1)^{-3}(2) = -\frac{12}{(2x+1)^3}$$

Test $x = -1$ and $x = 0$ in $f''(x)$.

$$f''(-1) = -\frac{12}{[2(-1)+1]^3} = 12 > 0$$

$f(x)$ is concave upward on $\left(-\infty, -\frac{1}{2}\right).$

$$f''(0) = -\frac{12}{[2(0)+1]^3} = -12 < 0$$

$f(x)$ is concave downward on $\left(-\frac{1}{2}, \infty\right).$

$f''(x)$ is never zero, so there are no inflection points.

When $f(x) = 0$,

$$f(x) = 0 \Rightarrow \frac{x-1}{2x+1} = 0 \Rightarrow x-1 = 0 \Rightarrow x = 1$$

The only x-intercept is 1.

$$f(0) = \frac{0-1}{2(0)+1} = -1, \text{ so the } y\text{-intercept is } -1.$$

This rational function has a vertical asymptote where the denominator is 0, that is, at $x = -\frac{1}{2}$.

To find the horizontal asymptote, find the limit at infinity.

$$\lim_{x\to\infty} f(x) = \lim_{x\to\infty} \frac{x-1}{2x+1} = \lim_{x\to\infty} \frac{\frac{x}{x}-\frac{1}{x}}{\frac{2x}{x}+\frac{1}{x}}$$

$$= \lim_{x\to\infty} \frac{1-\frac{1}{x}}{2+\frac{1}{x}} = \frac{1-0}{2+0} = \frac{1}{2}$$

$y = \frac{1}{2}$ is a horizontal asymptote.

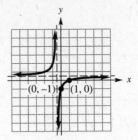

26. $f(x) = \dfrac{2x-5}{x+3}$

$$f'(x) = \frac{(x+3)(2)-(2x-5)}{(x+3)^2} = \frac{11}{(x+3)^2}$$

$$= 11(x+3)^{-2}$$

$f'(x)$ is never zero. -3 is not a critical number because $f(x)$ does not exist at $x = -3$. There are no local extrema. Test $x = -4$ and $x = -2$ in $f'(x)$.

$$f'(-4) = \frac{11}{(-4+3)^2} = 11 > 0$$

$$f(-2) = \frac{11}{(-2+3)^2} = 11 > 0$$

$f(x)$ is increasing on $(-\infty, -3)$ and on $(-3, \infty)$.

$$f''(x) = 11(-2)(x+3)^{-3} = -22(x+3)^{-3}$$

Test $x = -4$ and $x = -2$ in $f''(x)$.

$$f''(-4) = -22(-4+3)^{-3} = 22 > 0$$

$f(x)$ is concave upward on $(-\infty, -3)$.

$$f''(-2) = -22(-2+3)^{-3} = -22 < 0$$

$f(x)$ is concave downward on $(-3, \infty)$.

$f''(x)$ is never zero, so there are no inflection points.

If $f(x) = 0$,

$$f(x) = 0 \Rightarrow \frac{2x-5}{x+3} = 0 \Rightarrow 2x-5 = 0 \Rightarrow x = \frac{5}{2}$$

The x-intercept is $\frac{5}{2}$.

$$f(0) = \frac{2\cdot 0 - 5}{0+3} = -\frac{5}{3}$$

The y-intercept is $-\frac{5}{3}$.

This rational function has a vertical asymptote at $x = -3$.

$$\lim_{x\to\infty} f(x) = \lim_{x\to\infty} \frac{2x-5}{x+3} = \lim_{x\to\infty} \frac{\frac{2x}{x}-\frac{5}{x}}{\frac{x}{x}+\frac{3}{x}}$$

$$= \lim_{x\to\infty} \frac{2-\frac{5}{x}}{1+\frac{3}{x}} = \frac{2-0}{1+0} = 2$$

$y = 2$ is a horizontal asymptote.

(continued on next page)

(*continued from page 404*)

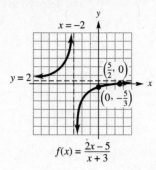

$$f(x) = \frac{2x-5}{x+3}$$

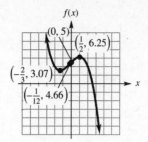

27. $f(x) = -4x^3 - x^2 + 4x + 5$

$f'(x) = -12x^2 - 2x + 4$

$f'(x) = 0 \Rightarrow -2\left(6x^2 + x - 2\right) = 0 \Rightarrow$

$(3x+2)(2x-1) = 0$

$-\dfrac{2}{3}$ and $\dfrac{1}{2}$ are critical numbers.

$f\left(-\dfrac{2}{3}\right) \approx 3.07;\ f\left(\dfrac{1}{2}\right) = 6.25$

$\left(-\dfrac{2}{3}, 3.07\right)$ and $\left(\dfrac{1}{2}, 6.25\right)$ are critical points.

$f''(x) = -24x - 2$

$f''\left(-\dfrac{2}{3}\right) = 16 > 0$

There is a local minimum at $x = -\dfrac{2}{3}$.

$f''\left(\dfrac{1}{2}\right) = -14 < 0$

There is a local maximum at $x = \dfrac{1}{2}$.

f is increasing on $\left(-\dfrac{2}{3}, \dfrac{1}{2}\right)$ and decreasing on

$\left(-\infty, -\dfrac{2}{3}\right)$ and $\left(\dfrac{1}{2}, \infty\right)$.

$f''(x) = 0 \Rightarrow -24x - 2 = 0 \Rightarrow x = -\dfrac{1}{12}$,

f is concave upward on $\left(-\infty, -\dfrac{1}{12}\right)$ and

concave downward on $\left(-\dfrac{1}{12}, \infty\right)$.

$f\left(-\dfrac{1}{12}\right) \approx 4.66,\ \text{so}\ \left(-\dfrac{1}{12}, 4.66\right)$ is an

inflection point. $f(0) = 5$, so the *y*-intercept is 5.
It is not convenient to find the *x*-intercept.

28. $f(x) = x^3 + \dfrac{5}{2}x^2 - 2x - 3$

$f'(x) = 3x^2 + 5x - 2 = (3x-1)(x+2) = 0 \Rightarrow$

$x = \dfrac{1}{3}, -2$

$\dfrac{1}{3}$ and -2 are critical numbers.

$f(-2) = 3;\ f\left(\dfrac{1}{3}\right) \approx -3.35$

$(-2, 3)$ and $\left(\dfrac{1}{3}, -3.35\right)$ are critical points.

$f''(x) = 6x + 5 = 0 \Rightarrow x = -\dfrac{5}{6}$

$f''\left(\dfrac{1}{3}\right) = 7 > 0$, so $f(x)$ has a local minimum at

$x = \dfrac{1}{3}$.

$f''(-2) = -7 < 0$, so $f(x)$ has a local maximum
at $x = -2$.

f is increasing on $(-\infty, -2)$ and $\left(\dfrac{1}{3}, \infty\right)$ and

decreasing on $\left(-2, \dfrac{1}{3}\right)$. Thus, f is concave

downward on $\left(-\infty, -\dfrac{5}{6}\right)$ and concave upward

on $\left(-\dfrac{5}{6}, \infty\right)$. $f\left(-\dfrac{5}{6}\right) \approx -.18$

There is a point of inflection at $\left(-\dfrac{5}{6}, -.18\right)$.

The *y*-intercept is -3. It is not convenient to find
the *x*-intercepts.

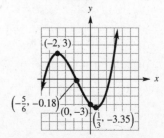

29. $f(x) = x^4 + 2x^2$

$f'(x) = 4x^3 + 4x = 0 \Rightarrow 4x(x^2 + 1) = 0$

$x^2 + 1 = 0$ has no real solution, so $x = 0$ is the only critical number. $f(0) = 0$, so $(0, 0)$ is a critical number.

$f''(x) = 12x^2 + 4$

$f''(0) = 4 > 0$

There is a local minimum at $x = 0$. $f(x)$ is increasing on $(0, \infty)$ and decreasing on $(-\infty, 0)$.

$f''(x) = 12x^2 + 4 > 0$ for all x, so $f(x)$ is concave upward on $(-\infty, \infty)$.

$f''(x)$ is never zero, so there are no inflection points. $f(0) = 0$, so the y-intercept is 0. To find any x-intercepts, let $f(x) = 0$.

$f(x) = x^4 + 2x^2 = 0 \Rightarrow x^2(x^2 + 2) = 0$

The only real solution is $x = 0$, so the only x-intercept is 0.

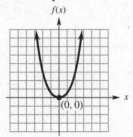

30. $f(x) = 6x^3 - x^4$

$f'(x) = 18x^2 - 4x^3 = 0 \Rightarrow 2x^2(9 - 2x) = 0 \Rightarrow$

$x = 0, \dfrac{9}{2}$

0 and $\dfrac{9}{2}$ are critical numbers.

$f(0) = 0; \quad f\left(\dfrac{9}{2}\right) \approx 136.7$

$(0, 0)$ and $\left(\dfrac{9}{2}, 136.7\right)$ are critical points.

$f''(x) = 36x - 12x^2$

$f''(0) = 0$

There is neither a local minimum nor a local maximum at $x = 0$.

$f''\left(\dfrac{9}{2}\right) = -81 < 0$

There is a local maximum at $x = \dfrac{9}{2}$.

f is increasing on $\left(-\infty, \dfrac{9}{2}\right)$ and decreasing on

$\left(\dfrac{9}{2}, \infty\right)$.

$f''(x) = 0 \Rightarrow 36x - 12x^2 = 0 \Rightarrow$

$12x(3 - x) = 0 \Rightarrow x = 0, 3$

Test $f''(x)$ at $x = -1$, $x = 1$, and $x = 4$.

$f''(-1) = -48 < 0$

f is concave downward on $(-\infty, 0)$.

$f''(1) = 24 > 0$

f is concave upward on $(0, 3)$.

$f''(4) = -48 < 0$

f is concave downward on $(3, \infty)$.

$f(3) = 81$

There are inflection points at $(0, 0)$ and $(3, 81)$.

$f(0) = 0$, so the y-intercept is 0.

To find any x-intercepts, let $f(x) = 0$.

$f(x) = 6x^3 - x^4 = 0 \Rightarrow x^3(6 - x) = 0 \Rightarrow$

$x = 0, 6$

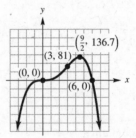

31. $f(x) = \dfrac{x^2 + 4}{x} = x + \dfrac{4}{x}$

$f'(x) = 1 - \dfrac{4}{x^2} = 0 \Rightarrow \dfrac{4}{x^2} = 1 \Rightarrow 4 = x^2 \Rightarrow$

$x = \pm 2$

-2 and 2 are critical numbers. Notice that $f(x)$ does not exist at $x = 0$. The line $x = 0$ is a vertical asymptote.

$f(-2) = -4; \quad f(2) = 4$

$(-2, -4)$ and $(2, 4)$ are critical points.

$f''(x) = \dfrac{8}{x^3}$

$f''(-2) = -1 < 0$

There is a local maximum at $x = -2$.

$f''(2) = 1 > 0$

There is a local minimum at $x = 2$. $f(x)$ is increasing on $(-\infty, -2)$ and $(2, \infty)$ and decreasing on $(-2, 0)$ and $(0, 2)$. $f''(x)$ is never zero, so there are no points of inflection.

(continued on next page)

(continued from page 406)

$f(x)$ is concave upward on $(0, \infty)$ and concave downward on $(-\infty, 0)$. Because $f(0)$ does not exist, there is no y-intercept. To find any x-intercepts, let $f(x) = 0$.

$$f(x) = \frac{x^2 + 4}{x} = 0 \Rightarrow x^2 + 4 = 0$$

This equation has no real solution, so there are no x-intercepts. When $x = 0$, the denominator of $\dfrac{x^2 + 4}{x}$ is equal to 0 but the numerator is not, so $x = 0$ (the y-axis) is a vertical asymptote.

We now determine whether this rational function has a horizontal asymptote.

$$\lim_{x \to \infty} \frac{x^2 + 4}{x} = \lim_{x \to \infty} \frac{\frac{x^2}{x^2} + \frac{4}{x^2}}{\frac{x}{x^2}} = \lim_{x \to \infty} \frac{1 + \frac{4}{x^2}}{\frac{1}{x}}$$

$$= \frac{1 + 0}{0} = \frac{1}{0}$$

This expression is undefined, which indicates that there is no horizontal asymptote. Rewrite the function rule as

$$f(x) = \frac{x^2 + 4}{x} = x + \frac{4}{x}.$$

As x gets larger and larger in absolute value, $\dfrac{4}{x} \to 0$, and the graph approaches the line $y = x$, which is an oblique or slant asymptote.

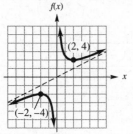

32. $f(x) = x + \dfrac{8}{x} = x + 8x^{-1}$

$f'(x) = 1 - 8x^{-2} = 1 - \dfrac{8}{x^2} = 0 \Rightarrow x^2 = 8 \Rightarrow$

$x = \pm 2\sqrt{2}$ Critical numbers

Notice that $f(x)$ does not exist at $x = 0$. The line $x = 0$ is a vertical asymptote.

$f\left(2\sqrt{2}\right) = 2\sqrt{2} + \dfrac{8}{2\sqrt{2}} = 2\sqrt{2} + 2\sqrt{2} = 4\sqrt{2}$

$f\left(-2\sqrt{2}\right) = -4\sqrt{2}$

Thus, $\left(2\sqrt{2},\, 4\sqrt{2}\right)$ and $\left(-2\sqrt{2},\, -4\sqrt{2}\right)$ are critical points.

$$f''(x) = \frac{16}{x^3}$$

$f''\left(2\sqrt{2}\right) > 0$, so $f(x)$ has a local minimum at $x = 2\sqrt{2}$.

$f''\left(-2\sqrt{2}\right) < 0$, so $f(x)$ has a local maximum at $x = -2\sqrt{2}$.

$f(x)$ is increasing on $\left(-\infty,\, -2\sqrt{2}\right)$ and $\left(2\sqrt{2},\, \infty\right)$ and decreasing on $\left(-2\sqrt{2},\, 0\right)$ and $\left(0,\, 2\sqrt{2}\right)$.

$f''(x) > 0$ when $x > 0$. $f''(x) < 0$ when $x < 0$.

Thus, $f(x)$ is concave upward on $(0, \infty)$ and concave downward on $(-\infty, 0)$. There is no point of inflection at $x = 0$, even though $f(x)$ changes concavity because $f(0)$ is undefined. $f(0)$ is undefined, so there is no y-intercept.

To any x-intercepts, let $f(x) = 0$.

$$f(x) = x + \frac{8}{x} = 0 \Rightarrow x\left(x + \frac{8}{x}\right) = x(0) \Rightarrow$$

$x^2 + 8 = 0$

This equation has no real solution, so there are no x-intercepts. To look for vertical and horizontal asymptotes, we rewrite the function rule so that the right-hand side is a single fraction.

$$f(x) = x + \frac{8}{x} = \frac{x^2 + 8}{x}$$

When $x = 0$, the denominator is 0 but the numerator is not, so $x = 0$ (the y-axis) is a vertical asymptote. Using the method shown in the solution for Exercise 31, we find that

$\lim\limits_{x \to \infty} \dfrac{x^2 + 8}{x} = \dfrac{1}{0}$, which is undefined. This indicates that there is no horizontal asymptote. To find a possible slant asymptote, use the function rule in its original form. As $x \to \infty$ or

$x \to -\infty$, $\dfrac{8}{x} \to 0$, so the graph will get closer and closer to the line $y = x$, which is a slant asymptote.

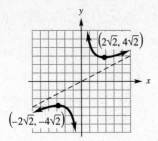

33. $f(x) = -x^2 + 6x + 1$; [2, 4]

$f'(x) = -2x + 6 = 0$ when $x = 3$

The absolute extrema must occur at the critical number 3 or at the endpoints.

$f(2) = 9$ Absolute minimum

$f(3) = 10$ Absolute maximum

$f(4) = 9$ Absolute minimum

The absolute maximum of 10 occurs at $x = 3$. The absolute minimum of 9 occurs at $x = 2$ and $x = 4$.

34. $f(x) = 4x^2 - 4x - 7$; [−1, 3]

$f'(x) = 8x - 4 = 0 \Rightarrow x = \dfrac{1}{2}$

x	−1	$\frac{1}{2}$	3
$f(x)$	1	−8	17

The absolute maximum of 17 occurs at $x = 3$ and the absolute minimum of −8 occurs at

$x = \dfrac{1}{2}$.

35. $f(x) = x^3 + 2x^2 - 15x + 3$; [−.5, 3.3]

$f'(x) = 3x^2 + 4x - 15 = 0 \Rightarrow$

$(3x - 5)(x + 3) = 0 \Rightarrow x = \dfrac{5}{3}, -3$

Find the values of the function at the critical numbers and at the endpoints.

$f(-.5) = 10.875$

$f(3.3) = 11.217$ Absolute maximum

$f\left(\dfrac{5}{3}\right) = -\dfrac{319}{27}$ Absolute minimum

$f(2) = -11$

The absolute maximum of 11.217 occurs at $x = 3.3$.

The absolute minimum of $-\dfrac{319}{27}$ occurs at

$x = \dfrac{5}{3}$.

36. $f(x) = -2x^3 - x^2 + 4x - 1$; [−3, 1]

$f'(x) = -6x^2 - 2x + 4 = -2\left(3x^2 + x - 2\right)$

$\quad\quad = -2(3x - 2)(x + 1)$

$f'(x) = 0$ when $x = -1$ or $x = \dfrac{2}{3}$,

so the critical numbers are −1 and $\dfrac{2}{3}$. Find the values of the function at the critical numbers and at the endpoint.

x	−3	−1	$\frac{2}{3}$	1
$f(x)$	32	−4	$\frac{17}{27}$	0

The absolute maximum of 32 occurs at $x = -3$ and the absolute minimum of −4 occurs at $x = -1$.

37. $S(x) = -x^3 + 3x^2 + 360x + 5000, \ 6 \le x \le 20$

$S'(x) = -3x^2 + 6x + 360 = 0 \Rightarrow$

$x = \dfrac{-6 \pm \sqrt{6^2 - 4(-3)(360)}}{2(-3)} = \dfrac{-6 \pm \sqrt{4356}}{-6} \Rightarrow$

$x = -10$ or $x = 12$

The point $x = -10$ is out of range. Test $S(x)$ at $x = 6$, $x = 12$, and $x = 20$.

$S(6) = 7052; \ S(12) = 8024; \ S(20) = 5400$

A temperature of 12°C produces the maximum number of bacteria.

38. $f(x) = -.000092x^3 + .011x^2 - .537x + 11.61,$
$\ 5 \le x \le 27$

$f'(x) = -.000276x^2 + .022x - .537 = 0 \Rightarrow$

$x = \dfrac{-.022 \pm \sqrt{.022^2 - 4(-.000276)(-.537)}}{2(-.000276)}$

$\quad = \dfrac{-.022 \pm \sqrt{-.000108848}}{2(-.000276)}$

Since the discriminant is negative, there are no real solutions, so there are no critical points.

$f(5) \approx 9.2, \ f(27) \approx 3.3$

The function decreases over the interval [5, 27] and there are no local extrema.

39. $g(x) = -.038x^3 + 1.48x^2 - 17.6x + 78.0,$
$9 \le x \le 18$

$g'(x) = -.114x^2 + 2.96x - 17.6 = 0 \Rightarrow$

$$x = \frac{-2.96 \pm \sqrt{2.96^2 - 4(-.114)(-17.6)}}{2(-.114)}$$

$$= \frac{-2.96 \pm \sqrt{.736}}{-.228} \approx 9.22, 16.75$$

$g''(x) = -.228x + 2.96$

$g''(9.22) \approx .86 > 0 \Rightarrow 9.22,$ local minimum

$g''(16.75) \approx -.86 < 0 \Rightarrow 16.75,$ local maximum

$g(9.22) \approx 11.76;\ g(16.75) \approx 19.85$

The minimum earnings of about \$11.76 billion occurred in 1999. The maximum earnings of about \$19.85 billion occurred in 2006.

40. $f(x) = .043x^3 - 1.297x^2 + 8.412x + 52.4,$
$0 \le x \le 17$

$f'(x) = .129x^2 - 2.594x + 8.412$
$f''(x) = .258x - 2.594$

$f'(x) = .129x^2 - 2.594x + 8.412 = 0 \Rightarrow$

$$x = \frac{2.594 \pm \sqrt{(-2.594)^2 - 4(.129)(8.412)}}{2(.129)}$$

$$= \frac{2.594 \pm \sqrt{2.388244}}{.258} \approx 4.06, 16.04$$

$f''(4.06) \approx -1.5 < 0 \Rightarrow 4.06,$ local max

$f''(16.04) \approx 1.5 > 0 \Rightarrow 16.04,$ local min

$f(4.06) \approx 68$

The most heart and lung transplants were performed in 2004. There were about 68 transplants.

41. $R(x) = .0913x^3 - 3.52x^2 + 44.9x - 167.47,$
$9 \le x \le 18$

a. $R'(x) = .2739x^2 - 7.04x + 44.9 = 0 \Rightarrow$

$$x = \frac{7.04 \pm \sqrt{(-7.04)^2 - 4(.2739)(44.9)}}{2(.2739)}$$

$$= \frac{7.04 \pm \sqrt{.36916}}{.5478} \approx 11.74, 13.96$$

$R''(x) = .5478x - 7.04$

$R''(11.74) \approx -.61 < 0 \Rightarrow 11.74,$ local max

$R''(13.96) \approx .61 > 0 \Rightarrow 13.96,$ local min

$R(11.74) \approx 22.24;\ R(13.96) \approx 21.74$

There is a local maximum at (11.74, 22.24). There is a local minimum at (13.96, 21.74).

b. Evaluate $R(x)$ at the endpoints of the interval. $R(9) \approx 18.07;\ R(18) \approx 32.71$

The absolute minimum revenue is about \$18.07 billion in 1999.

c. From part (b), the absolute maximum revenue is about \$32.71 billion in 2008.

42. $g(x) = -.0019x^3 + .111x^2 - 1.93x + 88.9$
$3 \le x \le 39$

a. $g'(x) = -.0057x^2 + .222x - 1.93$
$g''(x) = -.0114x + .222$
$g'(x) = 0 \Rightarrow$

$$x = \frac{-.222 \pm \sqrt{.222^2 - 4(-.0057)(-1.93)}}{2(-.0057)}$$

$$= \frac{-.222 \pm \sqrt{.00528}}{2(-.0057)} \approx 13.10, 25.85$$

$g''(13.10) \approx .073 > 0 \Rightarrow 13.10,$ local min

$g''(25.85) \approx -.073 > 0 \Rightarrow 25.85,$ local max

$g(3) \approx 84.06,\ g(13.10) \approx 78.39,$
$g(25.85) \approx 80.36,\ g(39) \approx 69.75$

Local minimum: (13.10, 78.39)
Local maximum: (25.85, 80.36)
Absolute minimum: (39, 69.75)
Absolute maximum: (3, 84.06)

43. $f(x) = -.00035x^3 + .0266x^2 - .123x + 20.7$
$0 \le x \le 58$

a. $f'(x) = -.00105x^2 + .0532x - .123 = 0 \Rightarrow$

$$x = \frac{-.0532 \pm \sqrt{.0532^2 - 4(-.00105)(-.123)}}{2(-.00105)}$$

$$= \frac{-.0532 \pm \sqrt{.00231364}}{2(-.00105)} \approx 2.43, 48.24$$

$f''(x) = -.0021x + .0532$

$f''(2.43) \approx .05 > 0 \Rightarrow 2.43,$ local min

$f''(48.24) \approx -.05 < 0 \Rightarrow 48.24,$ local max

$f(48.24) \approx 37.4$

The percentage was at its maximum in 1998. It was about 37.4%

b. $f''(x) = -.0021x + .0532 = 0 \Rightarrow x \approx 25.33$
$f(25.33) \approx 28.96$

The point of inflection is (25.33, 28.96).

44. Let x = length and width of a side of base and let h = height.

Volume = 27 cubic meters with a square base and top. Find height, length, and width for minimum surface area.

$$\text{Volume} = x^2 h = 27 \Rightarrow h = \frac{27}{x^2}$$

Surface area = $A(x) = 2x^2 + 4xh$

$$A(x) = 2x^2 + 4x\left(\frac{27}{x^2}\right) = 2x^2 + 108x^{-1}$$

$$A'(x) = 4x - 108x^{-2}$$

$$A' = 0 \Rightarrow 4x - \frac{108}{x^2} = 0 \Rightarrow \frac{4x^3 - 108}{x^2} = 0 \Rightarrow$$

$$4x^3 = 108 \Rightarrow x^3 = 27 \Rightarrow x = 3$$

$$A''(x) = 4 + 2(108)x^{-3} = 4 + \frac{216}{x^3}$$

$$A''(3) = 12 > 0$$

The minimum occurs at $x = 3$ where $h = \frac{27}{3^2} = 3$.

The dimensions are 3 m by 3 m by 3 m.

45. Let x = width of fence and let y = length of fence. An equation describing the amount of fencing is

$$500 = 2x + y \Rightarrow y = 500 - 2x$$

$$A = xy \Rightarrow A(x) = x(500 - 2x) = 500x - 2x^2$$

$$A'(x) = 500 - 4x = 0 \Rightarrow x = 125$$

Then $y = 500 - 2(125) = 250$.

$A''(x) = -4 < 0$, so $A(125)$ is a maximum.

Dimensions for maximum area are 125 m by 250 m.

46. $500 = 2x + 2y \Rightarrow y = 250 - x$

$$A = xy \Rightarrow A(x) = x(250 - x) = 250x - x^2$$

$$A'(x) = 250 - 2x = 0 \Rightarrow x = 125$$

Then $y = 250 - 125 = 125$.

$A''(x) = -2 < 0$, so $A(125)$ is a maximum.

Dimensions for maximum area are 125 m by 125 m.

47. Volume of open cylinder = 27π cubic inches. Find radius of bottom to minimize cost of material.

Volume of cylinder $= \pi r^2 h$

Surface area of cylinder open at one end $= 2\pi rh + \pi r^2$,

$$V = \pi r^2 h = 27\pi \Rightarrow h = \frac{27\pi}{\pi r^2}$$

$$A = 2\pi r \left(\frac{27\pi}{\pi r^2}\right) + \pi r^2 = 54\pi r^{-1} + \pi r^2$$

$$A' = -54\pi r^{-2} + 2\pi r = 0 \Rightarrow 2\pi r = \frac{54\pi}{r^2} \Rightarrow$$

$$r^3 = 27 \Rightarrow r = 3$$

$A'' = 108\pi r^{-3} + 2\pi > 0$, so for minimum cost, the radius should be 3 inches.

48. x = number of batches per year
M = 145,000 phones produced annually
k = $3.25, cost to store one phone for one year
a = $230, cost to produce each batch

$$x = \sqrt{\frac{kM}{2a}} = \sqrt{\frac{(3.25)(145,000)}{2(230)}} \approx 32$$

32 batches should be produced annually.

49. $\dfrac{145,000}{32} \approx 4531$

4531 phones should be produced for each batch.

50. a. x = number of batches per year
M = 250,000 cases sold per year
k = 1.25, cost to store 1 pound for 1 year
a = 350, cost to produce each batch

$$x = \sqrt{\frac{kM}{2a}} = \sqrt{\frac{(1.25)(250,000)}{2(350)}} \approx 21$$

The company should produce 21 batches annually.

b. Each batch should contain $\dfrac{250,000}{21} \approx 11,905$ pounds.

Case 12: A Total Cost Model for a Training Program

1. $Z(m) = \dfrac{C_1}{m} + DtC_2 + DC_3\left(\dfrac{m-1}{2}\right)$

$$= C_1 m^{-1} + DtC_2 + \frac{DC_3 m}{2} - \frac{DC_3}{2}$$

$$Z'(m) = -C_1 m^{-2} + \frac{DC_3}{2} = -\frac{C_1}{m^2} + \frac{DC_3}{2}$$

2. $Z'(m) = 0 \Rightarrow -\dfrac{C_1}{m^2} + \dfrac{DC_3}{2} = 0 \Rightarrow$

$$-\frac{C_1}{m^2} = -\frac{DC_3}{2} \Rightarrow -\frac{C_1}{m^2} = -\frac{DC_3}{2} \Rightarrow$$

$$2C_1 = DC_3 m^2 \Rightarrow \sqrt{\frac{2C_1}{DC_3}} = m \ (m \text{ is positive})$$

3. Let $D = 3$, $t = 12$, $C_1 = 15,000$, $C_3 = 900$.

$$m = \sqrt{\frac{2(15,000)}{(3)(900)}} = \sqrt{\frac{100}{9}} = \frac{10}{3}$$

4. $m^+ = 4$ Whole number larger than $\dfrac{10}{3}$

$m^- = 3$ Whole number smaller than $\dfrac{10}{3}$

5. Let $C_1 = 15,000$, $m^+ = 4$, $m^- = 3$, $D = 3$,
$C_2 = 100$, $C_3 = 900$, $t = 12$.

$$Z(m) = \frac{C_1}{m} + DtC_2 + DC_3 \left(\frac{m-1}{2} \right)$$

$$Z\left(m^+\right) = Z(4)$$
$$= \frac{15,000}{4} + (3)(12)(100) + (3)(900)\left(\frac{4-1}{2} \right)$$
$$= 3750 + 3600 + 4050 = \$11,400$$

$$Z\left(m^-\right) = Z(3)$$
$$= \frac{15,000}{3} + (3)(12)(100) + (3)(900)\left(\frac{3-1}{2} \right)$$
$$= 5000 + 3600 + 2700 = \$11,300$$

6. Since $Z\left(m^-\right)$ is smaller than $Z\left(m^+\right)$, the optimum value of Z is \$11,300. This occurs when $t = 3$ months. Now $N = mD = (3)(3) = 9$ so the number of trainees in a batch is 9.

Chapter 13 Integral Calculus

Section 13.1 Antiderivatives

1. If $F(x)$ and $G(x)$ are both antiderivatives of $f(x)$, then they differ only by a constant.

3. Answers vary.

5. $\int 12x\,dx = 12\int x\,dx = 12\left(\dfrac{x^2}{2}\right) + C = 6x^2 + C$

7. $\int 8p^3\,dp = 8\int p^3\,dp = 8\left(\dfrac{p^4}{4}\right) + C = 2p^4 + C$

9. $\int 105\,dx = 105\int 1\,dx = 105\int x^0\,dx$
$= 105\left(\dfrac{x^1}{1}\right) + C = 105x + C$

11. $\int (5z-1)\,dz = 5\int z\,dz - \int 1\,dz = 5\left(\dfrac{z^2}{2}\right) - \int z^0\,dz$
$= \dfrac{5z^2}{2} - \dfrac{z^1}{1} + C = \dfrac{5z^2}{2} - z + C$

13. $\int (z^2 - 4z + 6)\,dz = \int z^2\,dz - 4\int z\,dz + 6\int 1\,dz$
$= \dfrac{z^3}{3} - 4\left(\dfrac{z^2}{2}\right) + 6\int z^0\,dz$
$= \dfrac{z^3}{3} - 2z^2 + 6\left(\dfrac{z^1}{1}\right) + C$
$= \dfrac{z^3}{3} - 2z^2 + 6z + C$

15. $\int (x^3 - 14x^2 + 22x + 8)\,dx$
$= \int x^3\,dx - 14\int x^2\,dx + 22\int x\,dx + 8\int 1\,dx$
$= \dfrac{x^4}{4} - 14\left(\dfrac{x^3}{3}\right) + 22\left(\dfrac{x^2}{2}\right) + 8\int x^0\,dx$
$= \dfrac{x^4}{4} - \dfrac{14x^3}{3} + 11x^2 + 8\left(\dfrac{x^1}{1}\right) + C$
$= \dfrac{x^4}{4} - \dfrac{14x^3}{3} + 11x^2 + 8x + C$

17. $\int 6\sqrt{y}\,dy = 6\int y^{1/2}\,dy = 6\left(\dfrac{y^{3/2}}{\frac{3}{2}}\right) + C$
$= 6\left(\dfrac{2}{3}y^{3/2}\right) + C = 4y^{3/2} + C$

19. $\int \left(6t\sqrt{t} + 3\sqrt[4]{t}\right)dt = 6\int t \cdot t^{1/2}\,dt + 3\int t^{1/4}\,dt$
$= 6\int t^{3/2}\,dt + 3\int t^{1/4}\,dt$
$= 6\left(\dfrac{t^{5/2}}{\frac{5}{2}}\right) + 3\left(\dfrac{t^{5/4}}{\frac{5}{4}}\right) + C$
$= 6\left(\dfrac{2}{5}t^{5/2}\right) + 3\left(\dfrac{4}{5}t^{5/4}\right) + C$
$= \dfrac{12t^{5/2}}{5} + \dfrac{12t^{5/4}}{5} + C$

21. $\int \left(56t^{1/2} + 18t^{7/2}\right)dt = 56\int t^{1/2}\,dt + 18\int t^{7/2}\,dt$
$= 56\left(\dfrac{t^{3/2}}{\frac{3}{2}}\right) + 18\left(\dfrac{t^{9/2}}{\frac{9}{2}}\right) + C$
$= 56\left(\dfrac{2}{3}t^{3/2}\right) + 18\left(\dfrac{2}{9}t^{9/2}\right) + C$
$= \dfrac{112t^{3/2}}{3} + 4t^{9/2} + C$

23. $\int \dfrac{24}{x^3}\,dx = 24\int \dfrac{1}{x^3}\,dx = 24\int x^{-3}\,dx = 24\left(\dfrac{x^{-2}}{-2}\right) + C$
$= -12x^{-2} + C = -12\left(\dfrac{1}{x^2}\right) + C = -\dfrac{12}{x^2} + C$

25. $\int \left(\dfrac{1}{y^3} - \dfrac{2}{\sqrt{y}}\right)dy = \int \left(y^{-3} - 2y^{-1/2}\right)dy$
$= \int y^{-3}\,dy - 2\int y^{-1/2}\,dy$
$= \dfrac{y^{-2}}{-2} - 2\left(\dfrac{y^{1/2}}{\frac{1}{2}}\right) + C$
$= \dfrac{1}{-2y^2} - 2\left(2y^{1/2}\right) + C$
$= -\dfrac{1}{2y^2} - 4y^{1/2} + C$
$= -\dfrac{1}{2y^2} - 4\sqrt{y} + C$

27. $\int \left(6x^{-3} + 2x^{-1}\right)dx = 6\int x^{-3}dx + 2\int x^{-1}dx$

$$= 6\left(\frac{x^{-2}}{-2}\right) + 2\ln|x| + C$$

$$= -3x^{-2} + 2\ln|x| + C$$

29. $\int 4e^{3u}du = 4\int e^{3u}du = 4\left(\frac{1}{3}e^{3u}\right) + C$

$$= \frac{4e^{3u}}{3} + C$$

31. $\int 3e^{-.8x}dx = \frac{3\left(e^{-.8x}\right)}{-.8} + C = \frac{-15e^{-.8x}}{4} + C$

33. $\int \left(\frac{6}{x} + 4e^{-.5x}\right)dx = 6\ln|x| + \frac{4e^{-.5x}}{-.5} + C$

$$= 6\ln|x| - 8e^{-.5x} + C$$

35. $\int \frac{1 + 2t^4}{t}dt = \int\left(\frac{1}{t} + \frac{2t^4}{t}\right)dt = \int\left(\frac{1}{t} + 2t^3\right)dt$

$$= \ln|t| + \frac{2t^4}{4} + C = \ln|t| + \frac{t^4}{2} + C$$

37. $\int \left(e^{2u} + \frac{4}{u}\right)du = \int e^{2u}du + \int \frac{4}{u}du$

$$= \frac{1}{2}e^{2u} + 4\ln|u| + C$$

39. $\int (5x+1)^2 dx = \int \left(25x^2 + 10x + 1\right)dx$

$$= 25\int x^2dx + 10\int xdx + \int 1dx$$

$$= 25\left(\frac{x^3}{3}\right) + 10\left(\frac{x^2}{2}\right) + x + C$$

$$= 25\left(\frac{x^3}{3}\right) + 5x^2 + x + C$$

41. $\int \frac{\sqrt{x}+1}{\sqrt[3]{x}}dx = \int\left(\frac{\sqrt{x}}{\sqrt[3]{x}} + \frac{1}{\sqrt[3]{x}}\right)dx$

$$= \int\left(x^{(1/2 - 1/3)} + x^{-1/3}\right)dx$$

$$= \int\left(x^{1/6} + x^{-1/3}\right)dx$$

$$= \frac{x^{7/6}}{\frac{7}{6}} + \frac{x^{2/3}}{\frac{2}{3}} + C$$

$$= \frac{6x^{7/6}}{7} + \frac{3x^{2/3}}{2} + C$$

43. $f'(x) = 6x^2 - 4x + 3$

$$f(x) = \int\left(6x^2 - 4x + 3\right)dx$$

$$= 6\left(\frac{x^3}{3}\right) - 4\left(\frac{x^2}{2}\right) + 3x + C$$

$$= 2x^3 - 2x^2 + 3x + C$$

Since $(0, 1)$ is on the curve, $f(0) = 1$. Thus,
$2(0)^3 - 2(0)^2 + 3(0) + C = 1$ and $C = 1$.
Therefore, the equation of the curve is
$f(x) = 2x^3 - 2x^2 + 3x + 1$.

45. a. $G'(x) = g(x)$ and $G(1) = 216.63$

$$G(x) = \int\left(.21x^2 - .36x + 11.38\right)dx$$

$$= .21\int x^2dx - .36\int xdx + 11.38\int 1dx$$

$$= .21\left(\frac{x^3}{3}\right) - .36\left(\frac{x^2}{2}\right) + 11.38x + C$$

$$= .07x^3 - .18x^2 + 11.38x + C$$

$G(x) = .07x^3 - .18x^2 + 11.38x + C$
$216.63 = .07(1)^3 - .18(1)^2 + 11.38(1) + C$
$205.36 = C$
$G(x) = .07x^3 - .18x^2 + 11.38x + 205.36$

b. $x = 2007 - 2000 = 7$
$G(7) = .07(7)^3 - .18(7)^2 + 11.38(7) + 205.36$
$\quad \approx 300.21$
There were about \$300.21 billion in imports from Canada in 2007.

47. a. $F'(x) = f(x) = 9.42x + 87.14$ and
$F(2) = 1589.92$

$$F(x) = \int(9.42x + 87.14)dx$$

$$= 9.42\int xdx + 87.14\int dx$$

$$= 9.42\left(\frac{x^2}{2}\right) + 87.14x + C$$

$$= 4.71x^2 + 87.14x + C$$

$F(2) = 1589.92 = 4.71(2)^2 + 87.14(2) + C \Rightarrow$
$C = 1396.8$
$F(x) = 4.71x^2 + 87.14x + 1396.8$

b. $x = 2014 - 2000 = 14$
$F(14) = 4.71(14)^2 + 87.14(14) + 1396.8$
$\quad = 3539.92$
Total health care expenditures will be
\$3539.92 billion in 2014.

49. $P'(x) = 4 - 6x + 3x^2$ and $P(0) = -40$

$$P(x) = \int \left(4 - 6x + 3x^2\right) dx$$

$$= 4\int 1\, dx - 6\int x\, dx + 3\int x^2\, dx$$

$$= 4x - 6\left(\frac{x^2}{2}\right) + 3\left(\frac{x^3}{3}\right) + C$$

$$P(x) = 4x - 3x^2 + x^3 + C$$

$$-40 = 4(0) - 3(0)^2 + (0)^3 + C \Rightarrow -40 = C$$

Therefore,

$$P(x) = x^3 - 3x^2 + 4x - 40.$$

51. $C'(x) = x^{2/3} + 5$, $C(8) = 58$

$$C(x) = \int \left(x^{2/3} + 5\right) dx = \frac{3x^{5/3}}{5} + 5x + C$$

$$58 = \frac{3(8)^{5/3}}{5} + 5(8) + C \Rightarrow 58 = \frac{3(32)}{5} + 40 + C \Rightarrow$$

$$58 = \frac{96}{5} + 40 + C \Rightarrow \frac{-6}{5} = -1.2 = C$$

So, $C(x) = \frac{3x^{5/3}}{5} + 5x - 1.2$.

53. $C'(x) = .2x^2 + .4x + .8$, $C(6) = 32.50$

$$C(x) = \int (.2x^2 + .4x + .8)\, dx$$

$$= \frac{x^3}{15} + \frac{x^2}{5} + .8x + k$$

$$C(6) = \frac{6^3}{15} + \frac{6^2}{5} + .8(6) + k$$

Since $C(6) = 32.50$,

$$32.50 = \frac{216}{15} + \frac{36}{5} + 4.8 + k \Rightarrow k = 6.1.$$

$$C(x) = \frac{x^3}{15} + \frac{x^2}{5} + .8x + 6.1$$

55. $C'(x) = -\dfrac{40}{e^{.05x}} + 100$, $C(5) = 1400$

$$C(x) = \int \left(-\frac{40}{e^{.05x}} + 100\right) dx$$

$$= \frac{800}{e^{.05x}} + 100x + k$$

$$C(5) = \frac{800}{e^{.05(5)}} + 100(5) + k$$

Since $C(5) = 1400$

$$1400 = \frac{800}{e^{.25}} + 500 + k \Rightarrow k \approx 276.96$$

$$C(x) \approx \frac{800}{e^{.05x}} + 100x + 276.96$$

57. a. $F'(t) = f(t)$, $F(0) = 5.282$

$$F(t) = \int .06750396 e^{.01278t}\, dt$$

$$= .06750396 \int e^{.01278t}\, dt$$

$$= .06750396 \cdot \frac{1}{.01278} e^{.01278t} + C$$

$$= 5.282 e^{.01278t} + C$$

$$F(0) = 5.282 = 5.282 e^{.01278(0)} + C \Rightarrow C = 0$$

$$F(t) = 5.282 e^{.01278t}$$

b. $t = 2015 - 1990 = 25$

$$F(25) = 5.282 e^{.01278(25)} \approx 7.2704$$

The function estimates the world population in 2015 to be 7.2704 billion.

59. a. $G'(x) = g(x)$, $G(16) = 20$

$$G(x) = \int -\frac{10.2}{x}\, dx = -10.2 \int \frac{1}{x}\, dx$$

$$= -10.2 \ln x + C$$

$$G(16) = 20 = -10.2 \ln 16 + C \Rightarrow C \approx 48.28$$

$$G(x) = -10.2 \ln x + 48.28$$

b. $G(18) = -10.2 \ln 18 + 48.28 \approx 18.8$
$G(20) = -10.2 \ln 20 + 48.28 \approx 17.7$
$G(50) = -10.2 \ln 50 + 48.28 \approx 8.4$
$G(70) = -10.2 \ln 70 + 48.28 \approx 4.9$

Drivers of age 18 had about 18.8 accidents per hundred drivers.
Drivers of age 20 had about 17.7 accidents per hundred drivers.
Drivers of age 50 had about 8.4 accidents per hundred drivers.
Drivers of age 70 had about 4.9 accidents per hundred drivers.

c. Answers will vary.

Section 13.2 Integration by Substitution

1. Integration by substitution is related to the chain rule for derivatives, but in reverse. Difficult integrals in which u replaces an expression in the integrand and the derivative of u also appears to suggest using integration by substitution.

3. $\int 3(12x-1)^2\,dx$

Let $u = 12x - 1$. Then $du = 12\,dx$.

$$\int 3(12x-1)^2\,dx = 3\int (12x-1)^2\,dx$$

$$= 3\cdot\frac{1}{12}\cdot 12\int (12x-1)^2\,dx$$

$$= \frac{3}{12}\int (12x-1)^2\,12\,dx$$

$$= \frac{1}{4}\int u^2\,du = \frac{1}{4}\left(\frac{u^3}{3}\right)+C$$

$$= \frac{u^3}{12}+C = \frac{(12x-1)^3}{12}+C$$

5. $\int \dfrac{5}{(3t-6)^2}\,dt$

Let $u = 3t - 6$. Then $du = 3\,dt$.

$$\int \frac{5}{(3t-6)^2}\,dt = 5\int (3t-6)^{-2}\,dt$$

$$= 5\left(\frac{1}{3}\right)\int (3t-6)^{-2}\,(3)\,dt$$

$$= \frac{5}{3}\int u^{-2}\,du = \frac{5}{3}\left(\frac{u^{-1}}{-1}\right)+C$$

$$= -\frac{5}{3u}+C = -\frac{5}{3(3t-6)}+C$$

7. $\int \dfrac{x+1}{\left(x^2+2x-4\right)^{3/2}}\,dx$

Let $u = x^2 + 2x - 4$. Then $du = (2x + 2)\,dx$.

$$\int \frac{(x+1)}{\left(x^2+2x-4\right)^{3/2}}\,dx$$

$$= \int \left(x^2+2x-4\right)^{-3/2}(x+1)\,dx$$

$$= \frac{1}{2}\cdot 2\int \left(x^2+2x-4\right)^{-3/2}(x+1)\,dx$$

$$= \frac{1}{2}\int \left(x^2+2x-4\right)^{-3/2}(2x+2)\,dx$$

$$= \frac{1}{2}\int u^{-3/2}\,du = \frac{1}{2}\left(\frac{u^{-1/2}}{-\frac{1}{2}}\right)+C = -\frac{1}{u^{1/2}}+C$$

$$= -\frac{1}{\left(x^2+2x-4\right)^{1/2}}+C = -\frac{1}{\sqrt{x^2+2x-4}}+C$$

9. $\int r^2\sqrt{r^3+3}\,dr$

Let $u = r^3 + 3$. Then $du = 3r^2\,dr$.

$$\int r^2\sqrt{r^3+3}\,dr = \int \left(r^3+3\right)^{1/2}r^2\,dr$$

$$= \frac{1}{3}\cdot 3\int \left(r^3+3\right)^{1/2}r^2\,dr$$

$$= \frac{1}{3}\int \left(r^3+3\right)^{1/2}3r^2\,dr$$

$$= \frac{1}{3}\int u^{1/2}\,du = \frac{1}{3}\left(\frac{u^{3/2}}{\frac{3}{2}}\right)+C$$

$$= \frac{1}{3}\left(\frac{2}{3}u^{3/2}\right)+C = \frac{2}{9}\left(r^3+3\right)^{3/2}+C$$

$$= \frac{2\left(r^3+3\right)^{3/2}}{9}+C$$

11. $\int \left(-4e^{5k}\right)dk$

Let $u = 5k$. Then $du = 5\,dk$.

$$\int \left(-4e^{5k}\right)dk = -4\int e^{5k}\,dk$$

$$= -4\cdot\frac{1}{5}\cdot 5\int e^{5k}\,dk = -\frac{4}{5}\int e^{5k}\,5\,dk$$

$$= -\frac{4}{5}\int e^{u}\,du = -\frac{4}{5}e^{u}+C$$

$$= -\frac{4}{5}e^{5k}+C$$

13. $\int 4w^3 e^{2w^4}\,dw$

Let $u = 2w^4$. Then $du = 8w^3\,dw$.

$$\int 4w^3 e^{2w^4}\,dw = 4\int e^{2w^4}\cdot w^3\,dw$$

$$= 4\cdot\frac{1}{8}\cdot 8\int e^{2w^4}\cdot w^3\,dw$$

$$= \frac{4}{8}\int e^{2w^4}\cdot 8w^3\,dw = \frac{1}{2}\int e^{u}\,du$$

$$= \frac{1}{2}e^{u}+C = \frac{e^{2w^4}}{2}+C$$

15. $\int (2-t)e^{4t-t^2}\,dt$

Let $u = 4t - t^2$.

Then $du = (4 - 2t)\,dt = 2(2-t)\,dt$.

$\int (2-t)e^{4t-t^2}\,dt = \dfrac{1}{2}\cdot 2\int e^{4t-t^2}\cdot(2-t)\,dt$

$= \dfrac{1}{2}\int e^{4t-t^2}(4-2t)\,dt$

$= \dfrac{1}{2}\int e^u\,du = \dfrac{1}{2}e^u + C$

$= \dfrac{e^{4t-t^2}}{2} + C$

17. $\int \dfrac{e^{\sqrt{y}}}{\sqrt{y}}\,dy = \int e^{y^{1/2}}\cdot y^{-1/2}\,dy$

Let $u = y^{1/2}$. Then $du = \dfrac{1}{2}y^{-1/2}\,dy$.

$\int \dfrac{e^{\sqrt{y}}}{\sqrt{y}}\,dy = 2\cdot\dfrac{1}{2}\int e^{y^{1/2}}\cdot y^{-1/2}\,dy$

$= 2\int e^{y^{1/2}}\cdot\dfrac{1}{2}y^{-1/2}\,dy = 2\int e^u\,du$

$= 2e^u + C = 2e^{\sqrt{y}} + C$

19. $\int \dfrac{-5}{12+6x}\,dx$

Let $u = 12 + 6x$. Then $du = 6\,dx$.

$\int \dfrac{-5}{12+6x}\,dx = -5\int \dfrac{1}{12+6x}\,dx$

$= -5\cdot\dfrac{1}{6}\cdot 6\int \dfrac{1}{12+6x}\,dx$

$= -\dfrac{5}{6}\int \dfrac{1}{12+6x}(6)\,dx = -\dfrac{5}{6}\int \dfrac{1}{u}\,du$

$= -\dfrac{5}{6}\ln|u| + C = -\dfrac{5\ln|12+6x|}{6} + C$

21. $\int \dfrac{e^{2t}}{e^{2t}+1}\,dt$

Let $u = e^{2t} + 1$. Then $du = e^{2t}(2)\,dt$.

$\int \dfrac{e^{2t}}{e^{2t}+1}\,dt = \dfrac{1}{2}\cdot 2\int \dfrac{1}{e^{2t}+1}\cdot e^{2t}\,dt$

$= \dfrac{1}{2}\int \dfrac{1}{e^{2t}+1}\cdot 2e^{2t}\,dt = \dfrac{1}{2}\int \dfrac{1}{u}\,du$

$= \dfrac{1}{2}\ln|u| + C = \dfrac{1}{2}\ln\left|e^{2t}+1\right| + C$

23. $\int \dfrac{x+2}{\left(2x^2+8x\right)^3}\,dx = \int \left(2x^2+8x\right)^{-3}(x+2)\,dx$

Let $u = 2x^2 + 8x$. Then $du = (4x+8)\,dx$.

$\int \dfrac{x+2}{\left(2x^2+8x\right)^3}\,dx = \dfrac{1}{4}\cdot 4\int \left(2x^2+8x\right)^{-3}(x+2)\,dx$

$= \dfrac{1}{4}\int \left(2x^2+8x\right)^{-3}(4x+8)\,dx$

$= \dfrac{1}{4}\int u^{-3}\,du = \dfrac{1}{4}\cdot\dfrac{u^{-2}}{-2} + C$

$= -\dfrac{1}{8u^2} + C$

$= -\dfrac{1}{8\left(2x^2+8x\right)^2} + C$

25. $\int 5\left(\dfrac{1}{r}+r\right)\left(1-\dfrac{1}{r^2}\right)\,dr$

Let $u = \dfrac{1}{r} + r$. Then $du = \left(-\dfrac{1}{r^2}+1\right)dr$.

$\int 5\left(\dfrac{1}{r}+r\right)\left(1-\dfrac{1}{r^2}\right)\,dr = 5\int u\,du = \dfrac{5u^2}{2} + C$

$= \dfrac{5\left(\frac{1}{r}+r\right)^2}{2} + C$

27. $\int \dfrac{x^2+1}{\left(x^3+3x\right)^{\frac{2}{3}}}\,dx = \dfrac{1}{3}\int \dfrac{3\left(x^2+1\right)dx}{\left(x^3+3x\right)^{\frac{2}{3}}}$

Let $u = x^3 + 3x$. Then

$du = \left(3x^2+3\right)dx = 3\left(x^2+1\right)dx$

$\int \dfrac{x^2+1}{\left(x^3+3x\right)^{2/3}}\,dx = \dfrac{1}{3}\int \dfrac{du}{u^{2/3}} = \dfrac{1}{3}\int u^{-2/3}\,du$

$= \dfrac{1}{3}\left(\dfrac{u^{1/3}}{\frac{1}{3}}\right) + C = u^{1/3} + C$

$= \left(x^3+3x\right)^{1/3} + C$

29. $\int \dfrac{6x+7}{3x^2+7x+8}\,dx = \int \dfrac{1}{3x^2+7x+8}(6x+7)\,dx$

Let $u = 3x^2 + 7x + 8$. Then $du = (6x+7)\,dx$.

$\int \dfrac{6x+7}{3x^2+7x+8}\,dx = \int \dfrac{1}{u}\,du = \ln|u| + C$

$= \ln\left|3x^2+7x+8\right| + C$

31. $\int 2x\left(x^2+5\right)^3 dx$

Let $u = x^2 + 5$. Then $du = 2x\, dx$.

$\int 2x\left(x^2+5\right)^3 dx = \int u^3 du = \dfrac{u^4}{4} + C$

$\qquad\qquad = \dfrac{\left(x^2+5\right)^4}{4} + C$

33. $\int \left(\sqrt{x^2+12x}\right)(x+6)dx$

$\qquad = \int \left(x^2+12x\right)^{1/2}(x+6)dx$

Let $u = x^2 + 12x$. Then $du = 2x + 12\,dx$ or

$(x+6)dx = \dfrac{du}{2}$.

$\int \left(\sqrt{x^2+12x}\right)(x+6)dx = \dfrac{1}{2}\int u^{1/2}du$

$\qquad\qquad = \dfrac{1}{2}\left(\dfrac{2}{3}\right)u^{3/2} + C$

$\qquad\qquad = \dfrac{\left(x^2+12x\right)^{3/2}}{3} + C$

35. $\int \dfrac{(10+\ln x)^2}{x} dx = \int (10+\ln x)^2 \cdot \dfrac{1}{x} dx$

Let $u = 10 + \ln x$. Then $du = \dfrac{1}{x} dx$.

$\int \dfrac{(10+\ln x)^2}{x} dx = \int u^2 du = \dfrac{u^3}{3} + C$

$\qquad\qquad = \dfrac{(10+\ln x)^3}{3} + C$

37. $\int \dfrac{5u}{\sqrt{u-1}} du = \int 5u(u-1)^{-\frac{1}{2}} du$

Let $w = u - 1$. Then $dw = du$ and $u = w + 1$.

$\int \dfrac{5u}{\sqrt{u-1}} du = 5\int (w+1)w^{-1/2}dw$

$\qquad\qquad = 5\int w^{1/2} + w^{-1/2}dw$

$\qquad\qquad = \dfrac{5w^{3/2}}{\frac{3}{2}} + \dfrac{5w^{1/2}}{\frac{1}{2}} + C$

$\qquad\qquad = \dfrac{10}{3} w^{3/2} + 10w^{1/2} + C$

$\qquad\qquad = \dfrac{10}{3}(u-1)^{3/2} + 10(u-1)^{1/2} + C$

39. $\int t\sqrt{5t-1}dt = \dfrac{1}{5}\int 5t(5t-1)^{1/2} dt$

Let $u = 5t - 1$. Then $du = 5\, dt$ and $t = \dfrac{u+1}{5}$.

$\int t\sqrt{5t-1}dt = \dfrac{1}{5}\int \left(\dfrac{u+1}{5}\right)u^{1/2}du$

$\qquad = \dfrac{1}{25}\int \left(u^{3/2} + u^{1/2}\right)du$

$\qquad = \dfrac{1}{25}\left[\dfrac{u^{5/2}}{\frac{5}{2}} + \dfrac{u^{3/2}}{\frac{3}{2}}\right] + C$

$\qquad = \dfrac{1}{25}\left[\dfrac{2}{5}(5t-1)^{5/2} + \dfrac{2}{3}(5t-1)^{3/2}\right] + C$

$\qquad = \dfrac{2(5t-1)^{5/2}}{125} + \dfrac{2(5t-1)^{3/2}}{75} + C$

41. a. $C(x) = \int C'(x)dx = \int \dfrac{60x}{5x^2+1} dx$

Let $u = 5x^2 + 1$. Then $du = 10x\, dx$.

$C(x) = \int (5x^2+1)^{-1}6\cdot 10x\, dx = 6\int u^{-1}du$

$\qquad = 6\ln|u| + k$

Notice that for any x, $u \geq 0$.

$C(x) = 6\ln(5x^2+1) + k$

$C(0) = 6\ln(5(0)^2+1) + k \Rightarrow 10 = k$

$C(x) = 6\ln(5x^2+1) + 10$

b. $C(5) = 6\ln(5(5)^2+1) + 10 \approx 39.02$

Since this function is always increasing, only the cost at the 5th year is about \$39.02 thousand, so, yes, they should add the new line.

43. a. $G'(x) = g(x)$, $G(3) = 8.7$

$G(x) = \int 1.1128806e^{.2483x}dx$

$\qquad = 1.1128806\left(\dfrac{1}{.2483}\right)e^{.2483x} + C$

$\qquad = 4.482e^{.2483x} + C$

Since $G(3) = 8.7$, we have

$8.7 = 4.482e^{.2483(3)} + C \Rightarrow C \approx -.74$

$G(x) = 4.482e^{.2483x} - .74$

b. $x = 2007 - 2000 = 7$

$G(7) = 4.482e^{.2483(7)} - .74 \approx 24.747$

According to the model, the revenue was about \$24.747 billion in 2007.

45. a. $R'(x) = MR(x) = 1.8x\left(x^2 + 27,000\right)^{-2/3}$, $R(150) = 32$

$R(x) = \int 1.8x\left(x^2 + 27,000\right)^{-2/3} dx$

Let $u = x^2 + 27,000$. Then $du = 2x\,dx$.

$R(x) = \int 1.8x\left(x^2 + 27,000\right)^{-2/3} dx = 1.8 \cdot \frac{1}{2}\int 2x\left(x^2 + 27,000\right)^{-2/3} dx = .9\int u^{-2/3} du = .9\left(\frac{u^{1/3}}{1/3}\right) + C$

$\qquad = 2.7u^{1/3} + C = 2.7\left(x^2 + 27,000\right)^{1/3} + C$

Since $R(150) = 32$, we have $32 = 2.7\left(150^2 + 27,000\right)^{1/3} + C \Rightarrow C \approx -67.14$.

Thus, $R(x) = 2.7\left(x^2 + 27,000\right)^{1/3} - 67.14$.

b. $R(250) = 2.7\left(250^2 + 27,000\right)^{1/3} - 67.14 \approx 53.633$

The revenue from selling 250 players is about \$53,633.

c. $2.7\left(x^2 + 27,000\right)^{1/3} - 67.14 = 100 \Rightarrow 2.7\left(x^2 + 27,000\right)^{1/3} = 167.14 \Rightarrow \left(x^2 + 27,000\right)^{1/3} \approx 61.90 \Rightarrow$

$x^2 + 27,000 \approx 237,177 \Rightarrow x^2 \approx 210,177 \Rightarrow x \approx 458$

About 458 players must be sold to produce revenue of at least \$100,000.

47. a. $G'(x) = g(x)$, $G(1970) = 61.298$

$G(x) = \int .00040674x(x - 1970)^4 dx$

Let $u = x - 1970$. Then $x = u + 1970$ and $du = dx$.

$G(x) = \int .00040674x(x - 1970)^4 dx = .00040674\int (u + 1970)u^4 du = .00040674\int \left(u^{1.4} + 1970u^4\right) du$

$\qquad = .00040674\left(\int u^{1.4} du + 1970\int u^4 du\right) = .00040674\left[\frac{u^{2.4}}{2.4} + 1970\left(\frac{u^{1.4}}{1.4}\right)\right] + C$

$\qquad = .00040674\left[\frac{(x - 1970)^{2.4}}{2.4} + 1970\left(\frac{(x - 1970)^{1.4}}{1.4}\right)\right] + C$

Since $G(1970) = 61.298$, we have

$61.298 = .00040674\left[\frac{(1970 - 1970)^{2.4}}{2.4} + 1970\left(\frac{(1970 - 1970)^{1.4}}{1.4}\right)\right] + C \Rightarrow 61.298 = C$

Thus, $G(x) = .00040674\left[\frac{(x - 1970)^{2.4}}{2.4} + 1970\left(\frac{(x - 1970)^{1.4}}{1.4}\right)\right] + 61.298$

b. $G(2015) = .00040674\left[\frac{(2015 - 1970)^{2.4}}{2.4} + 1970\left(\frac{(2015 - 1970)^{1.4}}{1.4}\right)\right] + 61.298 \approx 180.945$

According to the model, in 2015, there will be about 180,945 urban transit vehicles.

Section 13.3 Area and the Definite Integral

1. The total usage of electricity is the total area between the graph of the rate function and the x-axis from $x = 0$ (midnight) to $x = 24$ (the next midnight). Approximate this area by using 12 rectangles each with base of length 2 and heights determined by the graph.

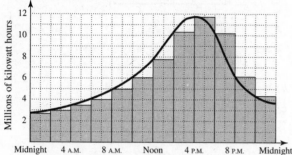

Estimate the area to be the sum
$2 \cdot 3 + 2 \cdot 3 + 2 \cdot 3.5 + 2 \cdot 4 + 2 \cdot 5 + 2 \cdot 6 + 2 \cdot 8 + 2 \cdot 11 + 2 \cdot 11.5 + 2 \cdot 10 + 2 \cdot 6 + 2 \cdot 4.5 = 151$
The total usage of electricity is about 151 kWh. (Your answer may vary depending on how you interpreted the height of each rectangle from the graph.)

3. Approximate coal consumption by finding the sum of the areas of 5 rectangles, each with base of length 2 and heights determined by the graph.

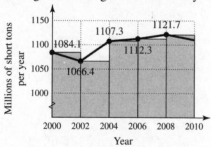

$2 \cdot 1084.1 + 2 \cdot 1066.4 + 2 \cdot 1107.3 + 2 \cdot 1112.3 + 2 \cdot 1121.7 = 10{,}983.6$
Coal consumption fro 2000–2010 was about 10,983.6 million short tons.

5. The indefinite integral $\int f(x)dx$ denotes a set of functions, whereas the definite integral represents a number.

7. **a.** $f(x) = 3x + 8;\ [0, 4]$

    two rectangles: $\Delta x = \dfrac{4 - 0}{2} = 2$

i	x_i	$f(x_i)$
1	0	8
2	2	14

    $$\sum_{i=1}^{2} f(x_i)\Delta x = 8(2) + 14(2) = 16 + 28 = 44$$

    b. four rectangles: $\Delta x = \dfrac{4 - 0}{4} = 1$

i	x_i	$f(x_i)$
1	0	8
2	1	11
3	2	14
4	3	17

    $$\sum_{i=1}^{4} f(x_i)\Delta x = 8(1) + 11(1) + 14(1) + 17(1)$$
    $$= 8 + 11 + 14 + 17 = 50$$

    c. 40 rectangles: $\Delta x = \dfrac{4 - 0}{40} = .1$

    ```
    sum(seq(3X+8,X,0
    ,3.9,.1))
                  554
    Ans*.1
                 55.4
    ```

    $$\sum_{i=1}^{40} f(x_i)\Delta x = 55.4$$

9. **a.** $f(x) = 4 - x^2;\ [-2, 2]$

    two rectangles: $\Delta x = \dfrac{2 - (-2)}{2} = 2$

i	x_i	$f(x_i)$
1	–2	0
2	0	4

    $$\sum_{x=1}^{2} f(x_i)\Delta x = 0(2) + 4(2) = 8$$

b. four rectangles: $\Delta x = \dfrac{2-(-2)}{4} = 1$

i	x_i	$f(x_i)$
1	-2	0
2	-1	3
3	0	4
4	1	3

$$\sum_{i=1}^{4} f(x_i)\Delta x = 0(1)+3(1)+4(1)+3(1) = 10$$

c. 40 rectangles: $\Delta x = \dfrac{2-(-2)}{40} = .1$

```
sum(seq(4-X²,X,-
2,1.9,.1))
            106.6
Ans*.1
            10.66
```

$$\sum_{i=1}^{40} f(x_i)\Delta x = 10.66$$

11. a. $f(x) = e^{2x} - .5$; [0, 2]

two rectangles: $\Delta x = \dfrac{2-0}{2} = 1$

i	x_i	$f(x_i)$
1	0	.5
2	1	6.89

$$\sum_{i=1}^{2} f(x_i)\Delta x = .5(1)+6.89(1) = 7.39$$

b. four rectangles: $\Delta x = \dfrac{2-0}{4} = 0.5$

i	x_i	$f(x_i)$
1	0	.5
2	.5	2.22
3	1	6.89
4	1.5	19.59

$$\sum_{i=1}^{4} f(x_i)\Delta x$$
$$= .25+1.110+3.445+9.795 = 14.6$$

c. 40 rectangles: $\Delta x = \dfrac{2-0}{40} = .05$

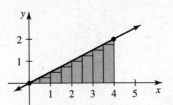

```
sum(seq(e^(2X)-.
5,X,0,1.95,.05))
          489.6290021
Ans*.05
          24.48145011
```

$$\sum_{i=1}^{40} f(x_i)\Delta x = 24.48$$

13. $f(x) = \dfrac{x}{2}$ between $x = 0$ and $x = 4$

a. four rectangles: $\Delta x = \dfrac{4-0}{4} = 1$

i	x_i	$f(x_i)$
1	0	0
2	1	.5
3	2	1
4	3	1.5

$$\sum_{i=1}^{4} f(x_i)\Delta x = 0+.5+1+1.5 = 3$$

b. eight rectangles: $\Delta x = \dfrac{4-0}{8} = .5$

i	x_i	$f(x_i)$
1	0	0
2	.5	.25
3	1.0	.50
4	1.5	.75
5	2.0	1.00
6	2.5	1.25
7	3.0	1.50
8	3.5	1.75

(continued on next page)

(*continued from page 421*)

$$\sum_{i=1}^{8} f(x_i)\Delta x$$
$$= 0(.5) + .25(.5) + .5(.5) + .75(.5)$$
$$+ 1(.5) + 1.25(.5) + 1.5(.5) + 1.75(.5)$$
$$= 3.5$$

c. $\int_0^4 f(x)dx = \int_0^4 \frac{x}{2} dx = \frac{1}{2}$ (base)(height)
$$= \frac{1}{2}(4)(2) = 4$$

Exercises 15–19 were worked using a TI-84 Plus graphing calculator. Similar results can be obtained using other calculators.

15. $\int_{-5}^{0} \left(x^3 + 6x^2 - 10x + 2\right) dx$

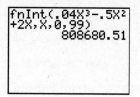

```
fnInt(X³+6X²-10X
+2,X,-5,0)
             228.75
```

17. $\int_2^7 5\ln\left(2x^2+1\right) dx$

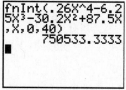

```
fnInt(5ln(2X²+1)
,X,2,7)
         90.54808787
```

19. $\int_0^3 4x^2 e^{-3x} dx$

```
fnInt(4X²e^(-3X)
,X,0,3)
         .29444972
```

Exercises 21–25 were worked using a TI-84 Plus graphing calculator. Similar results can be obtained using other calculators.

21. $MR(x) = .04x^3 - .5x^2 + 2x,\ [0, 99]$

Total revenue $= \int_0^{99} (.04x^3 - .5x^2 + 2x)dx$

```
fnInt(.04X³-.5X²
+2X,X,0,99)
         808680.51
```

The total revenue over the period is $808,680.51.

23. $MR(x) = .26x^4 - 6.25x^3 - 30.2x^2 + 87.5x,\ [0, 40]$
Total revenue
$$= \int_0^{40} \left(.26x^4 - 6.25x^3 - 30.2x^2 + 87.5x\right)dx$$

```
fnInt(.26X^4-6.2
5X³-30.2X²+87.5X
,X,0,40)
         750533.3333
```

The total revenue over the period is $750,533.33.

25. $R(x) = 927.7e^{.0244x}$

The year 2004 corresponds to $x = 14$, and the year 2007 corresponds to $x = 17$.

Total electricity consumption $= \int_{14}^{17} 927.7e^{.0244x}$.

```
fnInt(927.7e^(.0
244X),X,14,17)
         4063.272423
```

From 2004 to 2007, about 4063.27 billion kWh were consumed.

27. Read values of the function on the graph for every 5 sec from $t = 0$ to $t = 25$.

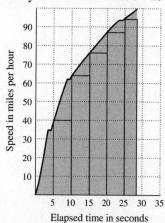

(*continued on next page*)

(continued from page 421)

These are speeds in miles per hour, so multiply these values by $\dfrac{5280}{3600}$ to get the speeds in feet per second. Then to estimate the distance traveled, find the sum of areas of rectangles with heights the sides of the rectangles and with widths of 5 sec. The last rectangle has a width of 3 sec. The total distance traveled is approximately

$$\dfrac{5280}{3600}\left[\begin{array}{l}0(5)+40(5)+64(5)\\ \qquad +77(5)+88(5)+95(3)\end{array}\right]$$

$$=\dfrac{5280}{3600}(1630)\approx 2391.$$

The total distance traveled is approximately 2400 ft.

29. a. At 2 seconds, the area under the curve for car A is $\dfrac{1}{2}(1)(6)+1(6)=9$, so car A has traveled 9 ft.

b. Car A is furthest ahead of car B when the graphs intersect, at 2 sec.

c. At 2 seconds, car B has traveled approximately $.5(.3+1+2.9+5)=4.6$ So car A is $9-4.6\approx 4$ feet ahead of car B.

d. Car B catches up with car A when the areas under the curves are equal, which is roughly somewhere between 3 and 3.5 seconds.

Section 13.4 The Fundamental Theorem of Calculus

1. $\displaystyle\int_{-1}^{3}\left(6x^2-7x+8\right)dx$

$$=\left(2x^3-\dfrac{7x^2}{2}+8x\right)\Bigg|_{-1}^{3}$$

$$=\left[2(3)^3-\dfrac{7(3)^2}{2}+8(3)\right]$$

$$\qquad -\left[2(-1)^3-\dfrac{7(-1)^2}{2}+8(-1)\right]$$

$$=\left(54-\dfrac{63}{2}+24\right)-\left(-2-\dfrac{7}{2}-8\right)$$

$$=\dfrac{93}{2}-\left(-\dfrac{27}{2}\right)=\dfrac{120}{2}=60$$

3. $\displaystyle\int_{0}^{2}3\sqrt{4u+1}\,du$

Let $w=4u+1$. Then $dw=4\,du$. If $u=2$, $w=9$. If $u=0$, $w=1$.

$$\int_{0}^{2}3\sqrt{4u+1}\,du=\dfrac{3}{4}\int_{0}^{2}(4u+1)^{1/2}4\,du$$

$$=\dfrac{3}{4}\int_{1}^{9}w^{1/2}\,dw=\dfrac{3}{4}\left(\dfrac{w^{3/2}}{\frac{3}{2}}\right)\Bigg|_{1}^{9}$$

$$=\dfrac{3}{4}\cdot\dfrac{2}{3}w^{3/2}\Bigg|_{1}^{9}$$

$$=\dfrac{1}{2}(9)^{3/2}-\dfrac{1}{2}(1)^{3/2}$$

$$=\dfrac{27}{2}-\dfrac{1}{2}=13$$

5. $\displaystyle\int_{0}^{1}3\left(t^{1/2}+4t\right)dt=3\left(\dfrac{t^{3/2}}{3/2}+2t^2\right)\Bigg|_{0}^{1}$

$$=3\left(\dfrac{1^{3/2}}{3/2}+2(1)^2\right)$$

$$\qquad -3\left(\dfrac{0^{3/2}}{3/2}+2(0)^2\right)$$

$$=3\left(\dfrac{2}{3}+2\right)-0=8$$

7. $\displaystyle\int_{1}^{4}\left(10y\sqrt{y}+3\sqrt{y}\right)dy=\int_{1}^{4}\left(10y^{3/2}+3y^{1/2}\right)dy$

$$=\left(10\cdot\dfrac{y^{5/2}}{\frac{5}{2}}+3\cdot\dfrac{y^{3/2}}{\frac{3}{2}}\right)\Bigg|_{1}^{4}$$

$$=\left(4y^{5/2}+2y^{3/2}\right)\Bigg|_{1}^{4}$$

$$=\left[4(4)^{5/2}+2(4)^{3/2}\right]$$

$$\qquad -\left[4(1)^{5/2}+2(1)^{3/2}\right]$$

$$=(128+16)-(4+2)=138$$

9. $\displaystyle\int_{4}^{7}\dfrac{11}{(x-3)^2}\,dx=\int_{4}^{7}11(x-3)^{-2}\,dx$

Let $u=x-3$. Then $du=dx$. If $x=7$, $u=4$. If $x=4$, $u=1$.

$$\int_{4}^{7}\dfrac{11}{(x-3)^2}\,dx=\int_{1}^{4}11u^{-2}\,du=11\cdot\dfrac{u^{-1}}{-1}\Bigg|_{1}^{4}$$

$$=-\dfrac{11}{u}\Bigg|_{1}^{4}=-\dfrac{11}{4}-\left(-\dfrac{11}{1}\right)$$

$$=-\dfrac{11}{4}+\dfrac{44}{4}=\dfrac{33}{4}=8\dfrac{1}{4}$$

11. $\displaystyle\int_1^5 \left(6n^{-1} + n^{-4}\right)dn = \int_1^5 6n^{-1}dn + \int_1^5 n^{-4}dn$

$\qquad = 6\int_1^5 n^{-1}dn + \int_1^5 n^{-4}dn$

$\qquad = 6\left(\ln n\right)\Big|_1^5 + \dfrac{n^{-3}}{-3}\Big|_1^5$

$\qquad = 6\left(\ln 5 - \ln 1\right) - \dfrac{1}{3}\left(5^{-3} - 1^{-3}\right)$

$\qquad = 6\ln 5 - \dfrac{1}{3}\left(\dfrac{1}{125} - 1\right) \approx 9.9873$

13. $\displaystyle\int_2^3 \left(.1e^{-.1A} + \dfrac{3}{A}\right)dA$

$\qquad = 0.1\int_2^3 e^{-.1A}dA + 3\int_2^3 \dfrac{1}{A}dA$

$\qquad = 0.1 \cdot \dfrac{e^{-.1A}}{-.1}\Big|_2^3 + 3\ln|A|\Big|_2^3 = -e^{-.1A}\Big|_2^3 + 3\ln|A|\Big|_2^3$

$\qquad = -e^{-.3} - \left(-e^{-.2}\right) + 3\ln 3 - 3\ln 2$

$\qquad = e^{-.2} - e^{-.3} + 3\ln 3 - 3\ln 2 \approx 1.294$

15. $\displaystyle\int_1^2 \left(e^{6u} - \dfrac{1}{u^2}\right)du = \int_1^2 e^{6u}du - \int_1^2 \dfrac{1}{u^2}du$

$\qquad = \dfrac{e^{6u}}{6}\Big|_1^2 + \dfrac{1}{u}\Big|_1^2$

$\qquad = \dfrac{e^{12}}{6} - \dfrac{e^6}{6} + \dfrac{1}{2} - 1$

$\qquad = \dfrac{e^{12}}{6} - \dfrac{e^6}{6} - \dfrac{1}{2} \approx 27{,}058.06$

17. $\displaystyle\int_{-1}^0 y\left(2y^2 - 3\right)^5 dy$

Let $u = 2y^2 - 3$. Then $du = 4y\,dy$. If $y = 0$, then
$u = -3$. If $y = -1$, then $u = -1$.

$\displaystyle\int_{-1}^0 y\left(2y^2 - 3\right)^5 dy = \dfrac{1}{4}\int_{-1}^0 \left(2y^2 - 3\right)^5 4y\,dy$

$\qquad = \dfrac{1}{4}\int_{-1}^{-3} u^5 du = \dfrac{1}{4} \cdot \dfrac{u^6}{6}\Big|_{-1}^{-3}$

$\qquad = \dfrac{1}{24}(-3)^6 - \dfrac{1}{24}(-1)^6$

$\qquad = \dfrac{729}{24} - \dfrac{1}{24} = \dfrac{728}{24} = \dfrac{91}{3}$

19. $\displaystyle\int_1^{64} \dfrac{\sqrt{z} - 2}{\sqrt[3]{z}}\,dz$

$\qquad = \int_1^{64} \left(\dfrac{z^{1/2}}{z^{1/3}} - 2z^{-1/3}\right)dz$

$\qquad = \int_1^{64} z^{1/6}dz - 2\int_1^{64} z^{-1/3}dz$

$\qquad = \dfrac{z^{7/6}}{\frac{7}{6}}\Big|_1^{64} - 2 \cdot \dfrac{z^{2/3}}{\frac{2}{3}}\Big|_1^{64} = \dfrac{6z^{7/6}}{7}\Big|_1^{64} - 3z^{2/3}\Big|_1^{64}$

$\qquad = \dfrac{6(64)^{7/6}}{7} - \dfrac{6(1)^{7/6}}{7} - 3\left(64^{2/3} - 1^{2/3}\right)$

$\qquad = \dfrac{6(128)}{7} - \dfrac{6}{7} - 3(16 - 1) = \dfrac{447}{7} \approx 63.857$

21. $\displaystyle\int_1^3 \dfrac{\ln x}{4x}\,dx = \dfrac{1}{4}\int_1^3 \ln x \cdot \dfrac{1}{x}\,dx$

Let $u = \ln x$. Then $du = \dfrac{1}{x}dx$. If $x = 3$, then
$u = \ln 3$. If $x = 1$, $u = \ln 1 = 0$.

$\displaystyle\int_1^3 \dfrac{\ln x}{4x}\,dx = \dfrac{1}{4}\int_0^{\ln 3} u\,du = \left(\dfrac{1}{4}\right) \cdot \dfrac{u^2}{2}\Big|_0^{\ln 3}$

$\qquad = \dfrac{(\ln 3)^2}{8} - \dfrac{0^2}{8} = \dfrac{(\ln 3)^2}{8} \approx .1509$

23. $\displaystyle\int_0^8 x^{1/3}\sqrt{x^{4/3} + 9}\,dx$

Let $u = x^{4/3} + 9$. Then $du = \dfrac{4}{3}x^{1/3}dx$. If $x = 8$,
then $u = 25$. If $x = 0$, then $u = 9$.

$\displaystyle\int_0^8 x^{1/3}\sqrt{x^{4/3} + 9}\,dx$

$\qquad = \dfrac{3}{4}\int_0^8 \left(x^{4/3} + 9\right)^{1/2} \cdot \dfrac{4}{3}x^{1/3}dx$

$\qquad = \dfrac{3}{4}\int_9^{25} u^{1/2}du = \dfrac{3}{4} \cdot \dfrac{u^{3/2}}{\frac{3}{2}}\Big|_9^{25}$

$\qquad = \dfrac{3}{4} \cdot \dfrac{2}{3}u^{3/2}\Big|_9^{25} = \dfrac{1}{2}(25)^{3/2} - \dfrac{1}{2}(9)^{3/2} = 49$

25. $\displaystyle\int_0^1 \frac{4e^t}{\left(3+e^t\right)^2}\,dt$

Let $u = 3+e^t$. Then $du = e^t\,dt$. If $t = 1$, then $u = 3 + e$. If $t = 0$, $u = 4$.

$$\int_0^1 \frac{4e^t}{\left(3+e^t\right)^2}\,dt = 4\int_0^1 \left(3+e^t\right)^{-2} e^t\,dt$$

$$= 4\int_4^{3+e} u^{-2}\,du = \left.\frac{4u^{-1}}{-1}\right|_4^{3+e}$$

$$= \left.-\frac{4}{u}\right|_4^{3+e} = -\frac{4}{3+e} - \left(-\frac{4}{4}\right)$$

$$\approx .3005$$

27. $\displaystyle\int_1^{49} \frac{\left(1+\sqrt{x}\right)^{4/3}}{\sqrt{x}}\,dx = \int_1^{49} \left(1+x^{1/2}\right)^{4/3} x^{-1/2}\,dx$

Let $u = 1+x^{1/2}$. Then $du = \frac{1}{2}x^{-1/2}\,dx$.

If $x = 49$, $u = 8$. If $x = 1$, $u = 2$.

$$\int_1^{49} \frac{\left(1+\sqrt{x}\right)^{4/3}}{\sqrt{x}}\,dx = \int_1^{49} \left(1+x^{1/2}\right)^{4/3} x^{-1/2}\,dx$$

$$= 2\int_1^{49} \left(1+x^{1/2}\right)^{4/3} \frac{1}{2} x^{-1/2}\,dx$$

$$= 2\int_2^8 u^{4/3}\,du = \left.2\cdot\frac{u^{7/3}}{\frac{7}{3}}\right|_2^8$$

$$= \left.2\cdot\frac{3}{7}u^{7/3}\right|_2^8$$

$$= \frac{6}{7}(8)^{7/3} - \frac{6}{7}(2)^{7/3}$$

$$\approx 105.3946$$

29. $\displaystyle\int_2^3 \frac{4x^3+6}{x^4+6x+9}\,dx$

Let $u = x^4+6x+9$. Then $du = \left(4x^3+6\right)dx$.

If $x = 3$, then $u = 3^4+6(3)+9 = 108$. If $x = 2$, then $u = 2^4+6(2)+9 = 37$.

$$\int_2^3 \frac{4x^3+6}{x^4+6x+9}\,dx = \int_{37}^{108} \frac{1}{u}\,du = \left.\ln|u|\right\|_{37}^{108}$$

$$= \ln 108 - \ln 37 \approx 1.0712$$

31. A negative definite integral for the first year and a half would indicate a loss for that period. The overall profit for the two-year period is represented by $\displaystyle\int_0^2 \left(6x^2-7x-3\right)dx$.

33. $f(x) = 9-x^2$; $[0, 4]$

The graph of f is above the x-axis on $[0, 3)$ and below the x-axis on $(3, 4)$.

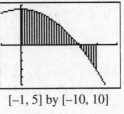

[–1, 5] by [–10, 10]

$$A = \int_0^3 \left(9-x^2\right)dx + \left|\int_3^4 \left(9-x^2\right)dx\right|$$

$$= \left.\left(9x-\frac{x^3}{3}\right)\right|_0^3 + \left|\left.\left(9x-\frac{x^3}{3}\right)\right|_3^4\right|$$

$$= (27-9)-0 + \left|(36-\frac{64}{3})-(27-9)\right|$$

$$= 18 + \left|\frac{44}{3}-18\right| = \frac{64}{3} = 21.3\overline{3}$$

35. $f(x) = x^3-1$; $[-1, 2]$

The graph of f is below the x-axis on $[-1, 1)$ and above the x-axis on $(1, 2]$.

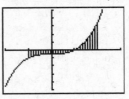

[–2, 3] by [–10, 10]

$$A = \left|\int_{-1}^1 \left(x^3-1\right)dx\right| + \int_1^2 \left(x^3-1\right)dx$$

$$= \left|\left.\left(\frac{x^4}{4}-x\right)\right|_{-1}^1\right| + \left.\left(\frac{x^4}{4}-x\right)\right|_1^2$$

$$= \left|\left(\frac{1}{4}-1\right)-(\frac{1}{4}+1)\right| + (4-2)-\left(\frac{1}{4}-1\right)$$

$$= \left|-\frac{3}{4}-\frac{5}{4}\right| + 2 - \left(-\frac{3}{4}\right) = 4.75$$

37. $f(x) = e^{2x} - 1$; $[-2, 1]$

Solve $f(x) = 0$ to determine where the graph crosses the x-axis.

$e^{2x} - 1 = 0 \Rightarrow e^{2x} = 1 \Rightarrow 2x \ln e = \ln 1 \Rightarrow$
$2x = 0 \Rightarrow x = 0$

The graph crosses the x-axis at 0 in the given interval $[-2, 1]$. The total area is

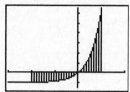

$[-3, 2]$ by $[-1.5, 6.5]$

$$\left| \int_{-2}^{0} \left(e^{2x} - 1\right) dx \right| + \int_{0}^{1} \left(e^{2x} - 1\right) dx$$

$$= \left| \left(\frac{e^{2x}}{2} - x\right)\Big|_{-2}^{0} \right| + \left(\frac{e^{2x}}{2} - x\right)\Big|_{0}^{1}$$

$$= \left| \left(\frac{1}{2} - 0\right) - \left(\frac{e^{-4}}{2} + 2\right) \right| + \left(\frac{e^2}{2} - 1\right) - \left(\frac{1}{2} - 0\right)$$

$$= \left| -\frac{3}{2} - \frac{e^{-4}}{2} \right| + \frac{e^2}{2} - 1 - \frac{1}{2}$$

$$= \left| -\frac{3}{2} - \frac{e^{-4}}{2} \right| + \frac{e^2}{2} - \frac{3}{2} \approx 3.7037$$

39. $f(x) = x^2 e^{-x^3/2} dx$ on $[0, 3]$

The graph of f is above the x-axis on $[0, 3]$.

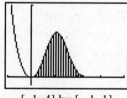

$[-1, 4]$ by $[-.1, 1]$

Let $u = -\frac{1}{2}x^3$. Then $du = -\frac{3}{2}x^2 dx$. If $x = 3$,

then $u = -\frac{27}{2}$. If $x = 0$, then $u = 0$.

$$\int_{0}^{3} x^2 e^{-x^3/2} dx = -\frac{2}{3} \int_{0}^{3} x^2 e^{-x^3/2} \left(-\frac{3}{2}\right) dx$$

$$= -\frac{2}{3} \int_{0}^{-27/2} e^u du = -\frac{2}{3} e^u \Big|_{0}^{-27/2}$$

$$= -\frac{2}{3}\left(e^{-27/2} - 1\right) \approx .6667$$

41. $f(x) = \frac{1}{x}$; $[1, e]$

$\frac{1}{x} = 0$ has no solution, so the graph does not cross the x-axis in the given interval $[1, e]$.

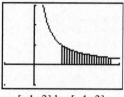

$[-1, 3]$ by $[-1, 3]$

$$\int_{1}^{e} \frac{1}{x} dx = \ln x \Big|_{1}^{e} = \ln e - \ln 1 = 1$$

43. $f(x) = \frac{12(\ln x)^3}{x}$ on $[1, 4]$

The graph of f is above the x-axis on $[0, 3]$.

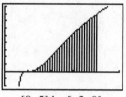

$[0, 5]$ by $[-2, 9]$

$$\int_{1}^{4} \frac{12(\ln x)^3}{x} dx$$

Let $u = \ln x$. Then $du = \frac{1}{x} dx$. If $x = 4$, then
$u = \ln 4$. If $x = 1$, then $u = 0$.

$$\int_{1}^{4} \frac{12(\ln x)^3}{x} dx = 12 \int_{0}^{\ln 4} u^3 du = 12 \frac{u^4}{4}\Big|_{0}^{\ln 4}$$

$$= 3\left((\ln 4)^4 - 0\right) \approx 11.08$$

45. $A = \int_{0}^{2} \left(2 - .5x^2\right) dx + \left| \int_{2}^{3} \left(2 - .5x^2\right) dx \right|$

$$= \left(2x - \frac{.5x^3}{3}\right)\Big|_{0}^{2} + \left| \left(2x - \frac{.5x^3}{3}\right)\Big|_{2}^{3} \right|$$

$$= \left(4 - \frac{4}{3}\right) - 0 + \left| \left(6 - \frac{9}{2}\right) - \left(4 - \frac{4}{3}\right) \right|$$

$$= \frac{8}{3} + \left| \frac{3}{2} - \frac{8}{3} \right| = \frac{23}{6}$$

47. $A = \int_{-1}^{0} (x^2 - 2x) dx + \left| \int_{0}^{2} (x^2 - 2x) dx \right|$

$= \left(\frac{x^3}{3} - x^2 \right) \Big|_{-1}^{0} + \left| \left(\frac{x^3}{3} - x^2 \right) \Big|_{0}^{2} \right|$

$= 0 - \left(-\frac{1}{3} - 1 \right) + \left| \left(\frac{8}{3} - 4 \right) - 0 \right|$

$= \frac{4}{3} + \left| -\frac{4}{3} \right| = \frac{8}{3} \approx 2.67$

49. $f(x) = \begin{cases} 2x + 3 & \text{if } x \le 2 \\ -.5x + 8 & \text{if } x > 2 \end{cases}$

$\int_{1}^{4} f(x) dx = \int_{1}^{2} f(x) dx + \int_{2}^{4} f(x) dx$

$= \int_{1}^{2} (2x + 3) dx + \int_{2}^{4} (-.5x + 8) dx$

$= \left(x^2 + 3x \right) \Big|_{1}^{2} + \left(-.5 \frac{x^2}{2} + 8x \right) \Big|_{2}^{4}$

$= (4 + 6) - (1 + 3) + (-4 + 32)$
$\qquad\qquad\qquad - (-1 + 16)$

$= 19$

51. $E(x) = 4x + 2$ is the rate of expenditure per day.

a. The total expenditure is hundreds of dollars in 10 days is

$\int_{0}^{10} (4x + 2) dx = \left(\frac{4x^2}{2} + 2x \right) \Big|_{0}^{10}$

$= 2(100) + 20 - 0 = 220$

Therefore, since $220(100) = 22{,}000$, the total expenditure is $22,000.

b. From the tenth to the twenty-fifth day:

$\int_{10}^{25} (4x + 2) dx = \left(\frac{4x^2}{2} + 2x \right) \Big|_{10}^{25}$

$= [2(625) + 50] - [2(100) + 20]$

$= 1080$

That is, $108,000 is spent.

c. If $76,000, or $760(100)$, is spent,

$\int_{0}^{a} (4x + 2) dx = 760.$

$\int_{0}^{a} (4x + 2) dx = \left(2x^2 + 2x \right) \Big|_{0}^{a} = 2a^2 + 2a$

Solve $760 = 2a^2 + 2a$ by the quadratic formula.

$2a^2 + 2a - 760 = 0 \Rightarrow a^2 + a - 380 = 0 \Rightarrow$

$a = \frac{-1 \pm \sqrt{1 - 4(-380)}}{2}$

Since the number of days must be positive,

$a = \frac{-1 \pm \sqrt{1521}}{2} = 19$ days.

53. $f(t) = \frac{6000(.3 + .28t)}{\left(1 + .3t + .14t^2 \right)^2}$

a. $\int_{0}^{3} \frac{6000(.3 + .28t)}{\left(1 + .3t + .14t^2 \right)^2}$

Let $u = 1 + .3t + .14t^2$ and $du = (.3 + .28t) dt$.

$\int \frac{6000(.3 + .28t)}{\left(1 + .3t + .14t^2 \right)^2} = 6000 \int \frac{du}{u^2}$

$= 6000 \left(\frac{u^{-1}}{-1} \right) + C$

$= -\frac{6000}{u} + C = -\frac{6000}{1 + .3t + .14t^2} + C$

$\int_{0}^{3} \frac{6000(.3 + .28t)}{(1 + .3t + .14t^2)^2} dt$

$= -\frac{6000}{1 + .3t + .14t^2} \Big|_{0}^{3}$

$= -\frac{6000}{1 + .3(3) + .14(3)^2}$

$\qquad - \left(-\frac{6000}{1 + .3(0) + .14(0)^2} \right)$

$\approx -1898.734 - (-6000) \approx 4101.27$

Total depreciation at the end of 3 years is about $4101.27.

b. Let b be the year the total depreciation will be at least $3000.

$3000 = \int_{0}^{b} \frac{6000(.3 + .28t)}{(1 + .3t + .14t^2)^2} dt$

$3000 = -6000 \frac{1}{1 + .3b + .14b^2} - (-6000)$

$.5 = \frac{1}{1 + .3b + .14b^2}$

$1 = .5(1 + .3b + .14b^2)$

$0 = -.5 + .15b + .07b^2$

Using the quadratic formula, $b \approx 1.8079$.
Total depreciation will reach $3000 in year 1.81.

55. $\int_{4}^{20} (-.0028x^3 + .168x^2 + 7.8x + 404)dx$

$$= \left(-.0028\frac{x^4}{4} + .168\frac{x^3}{3} + 7.8\frac{x^2}{2} + 404x \right)\Big|_{4}^{20}$$

$$= -.0028\frac{20^4}{4} + .168\frac{20^3}{3} + 7.8\frac{20^2}{2} + 404(20) - \left(-.0028\frac{4^4}{4} + .168\frac{4^3}{3} + 7.8\frac{4^2}{2} + 404(4) \right)$$

$$= 9976 - 1681.8048 = 8294.1952$$

World energy consumption from 2004 to 2020 is projected to be about 8294.1952 quadrillion BTUs.

57. $L'(t) = \dfrac{70\ln(t+1)}{t+1}$

Consider $\int \dfrac{70\ln(t+1)}{t+1}\,dt$. Let $u = \ln(t+1)$. Then $du = \dfrac{1}{t+1}\,dt$.

$$\int \frac{70\ln(t+1)}{(t+1)} = 70\int \ln(t+1)\frac{1}{t+1}\,dt = 70\int u\,du = 70\left(\frac{u^2}{2} \right) + C = 35u^2 + C = 35[\ln(t+1)]^2 + C$$

a. The total number of barrels leaked on the first day is

$$\int_{0}^{24} \frac{70\ln(t+1)}{t+1} = 35[\ln(t+1)]^2\Big|_{0}^{24} = 35(\ln 25)^2 - 35(\ln 1)^2 = 35(\ln 25)^2 \approx 363 \text{ barrels.}$$

b. The total number of barrels leaked on the second day is

$$\int_{24}^{48} \frac{70\ln(t+1)}{t+1}\,dt = 35[\ln(t+1)]^2\Big|_{24}^{48} = 35(\ln 49)^2 - 35(\ln 25)^2 \approx 167 \text{ barrels.}$$

c. The amount of oil leaked from one day to the next is given by

$$\int_{a}^{a+24} \frac{70\ln(t+1)}{t+1}\,dt = [35\ln(t+1)]^2\Big|_{a}^{a+24} = 35[\ln(a+25)]^2 - 35[\ln(a+1)]^2$$

$$= 35[\ln(a+25) + \ln(a+1)] \cdot [\ln(a+25) - \ln(a+1)]$$

$$= 35[\ln(a+25) + \ln(a+1)] \cdot \ln\left(\frac{a+25}{a+1} \right).$$

As a gets larger, $\ln\left(\dfrac{a+25}{a+1} \right) \approx \ln 1 = 0$.

In the long run, the amount of oil leaked per day is decreasing to 0.

59. **a.** $\int_{0}^{9} f(x)dx = \int_{0}^{9} (-.77x^2 + 2.5x + 40)dx = \left(-.77\frac{x^3}{3} + 2.5\frac{x^2}{2} + 40x \right)\Big|_{0}^{9}$

$$= -.77\frac{9^3}{3} + 2.5\frac{9^2}{2} + 40(9) - 0 = 274.14$$

This integral represents the total number of people from age 0 to 90. There are about 274.14 million people aged 0 to 90 according to the 2000 census.

b. $\int_{3.5}^{5.5} f(x)dx = \left(-.77\frac{x^3}{3} + 2.5\frac{x^2}{2} + 40x \right)\Big|_{3.5}^{5.5}$

$$= -.77\frac{5.5^3}{3} + 2.5\frac{5.5^2}{2} + 40(5.5) - \left(-.77\frac{3.5^3}{3} + 2.5\frac{3.5^2}{2} + 40(3.5) \right) \approx 70.8$$

There are about 70.8 million baby boomers.

61. a. $y = x(\ln|x| - 1)$

$$y' = x\left(\frac{1}{x}\right) + (\ln|x| - 1)(1)$$

$$= 1 + \ln|x| - 1 = \ln|x|$$

This gives

$$y = \int \ln|x|\,dx = x(\ln|x| - 1)\,dx$$

$$= (x\ln|x| - x)\,dx.$$

b. $\int (3.2 + 28\ln x)\,dx$

$$= \int 3.2\,dx + 28\int (\ln x)\,dx$$

$$= 3.2x + 28(x\ln|x| - x) + C$$

$$= 28x\ln x - 24.8x + C$$

$$\int_4^9 (3.2 + 28\ln x)\,dx = (28x\ln x - 24.8x)\Big|_4^9$$

$$\approx 330.5006 - 56.0650$$

$$\approx 274.4356$$

Dell's net revenue from January 1, 2004 to January 1, 2009 was about $274.4356 billion.

63. $\int_0^5 \frac{1}{3}e^{-x/3}\,dx = -e^{-x/3}\Big|_0^5 = -e^{-5/3} - \left(-e^{-0/3}\right)$

$$= -e^{-1} + 1 \approx -.189 + 1 = .811$$

The probability that the first customer arrives in the first 5 minutes is about .811.

65. Since the function computes the probability after 3 hours, when we compute the probability that the battery will last between 5 and 8 hours, we are actually computing the probability that the battery will last between 2 and 5 hours after the first 3 hours.

$$\int_2^5 \frac{1}{8}e^{-x/8}\,dx = -e^{-x/8}\Big|_2^5 = -e^{-5/8} - \left(-e^{-2/8}\right)$$

$$\approx .244$$

The probability that a battery will last between 5 and 8 hours is about .244.

Section 13.5 Applications of Integrals

1. $M(x) = 60(1 + x^2)$

The total maintenance charge on a two-year lease is given by

$$\int_0^2 60(1 + x^2)\,dx$$

$$= 60\int_0^2 (1 + x^2)\,dx = 60\left(x + \frac{x^3}{3}\right)\Big|_0^2$$

$$= 60\left(2 + \frac{2^3}{3} - 0\right) = 60\left(2 + \frac{8}{3}\right) = \$280$$

Monthly addition for maintenance:

$$\frac{280}{24} = \$11.67$$

3. $MR(x) = R'(x) = .56x + 3.1$

$$R = \int_0^{13} (.56x + 3.1)\,dx = (.28x^2 + 3.1x)\Big|_0^{13}$$

$$= \left[.28(13^2) + 3.1(13)\right] - 0 = 87.62$$

Total revenue from selling 13 billion tickets was $87.62 billion.

5. Rate of savings: $S(t) = 1000(t + 2)$

During the first year, total savings were

$$\int_0^1 1000(t + 2)\,dt = 1000\int_0^1 (t + 2)\,dt$$

$$= 1000\left(\frac{t^2}{2} + 2t\right)\Big|_0^1$$

$$= 1000\left(\frac{1}{2} + 2 - 0\right) = \$2500$$

During the first 6 years total savings were

$$\int_0^6 1000(t + 2)\,dt = 100\int_0^6 (t + 2)\,dt$$

$$= 1000\left(\frac{t^2}{2} + 2t\right)\Big|_0^6$$

$$= 1000\left(\frac{36}{2} + 12 - 0\right)$$

$$= \$30,000$$

7. The rate of production is given by the function $P(x) = 1000e^{.2x}$.

In the first 4 years, the total production will be

$$\int_0^4 1000e^{.2x}\,dx = 1000\int_0^4 e^{.2x}\,dx = 1000\left(\frac{e^{.2x}}{.2}\right)\Big|_0^4$$

$$= \frac{1000}{.2}\left(e^{.8} - e^0\right) \approx 6127.7$$

so a production of 20,000 units in the first 4 years will not be met.

9. $y = 3x$ and $y = x^2 - 4$ from $x = -1$ to $x = 3$

On the interval, $3x \geq x^2 - 4$

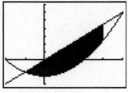

[–2, 4] by [–6, 14]

$$\int_{-1}^{3}\left(3x - (x^2 - 4)\right)dx = \int_{-1}^{3}\left(-x^2 + 3x + 4\right)dx = -\frac{x^3}{3} + \frac{3x^2}{2} + 4x\Big|_{-1}^{3}$$

$$= -\frac{(3)^3}{3} + \frac{3(3)^2}{2} + 4(3) - \left(-\frac{(-1)^3}{3} + \frac{3(-1)^2}{2} + 4(-1)\right) = \frac{33}{2} - \left(-\frac{13}{6}\right) = \frac{56}{3}$$

11. $y = x^2$ and $y = x^3$ from $x = 0$ to $x = 1/2$

On the interval, $x^2 \geq x^3$

[–.5, 1] by [–.25, .5]

$$\int_{0}^{1/2}(x^2 - x^3)dx = \left(\frac{x^3}{3} - \frac{x^4}{4}\right)\Big|_{0}^{1/2} = \left(\frac{1}{24} - \frac{1}{64}\right) - 0 = \frac{5}{192}$$

Exercises 13–15 were done using a TI-84 Plus graphing calculator. Similar results can be obtained using other graphing calculators.

13. $y = \ln x$ and $y = 2xe^x$, $[1, 4]$

$$A = \int_{1}^{4}\left(2xe^x - \ln x\right)dx$$

The area is approximately 325.04.

15. $y = \sqrt{9 - x^2}$ and $y = \sqrt{x+1}$, $[-1, 2]$

$$A = \int_{-1}^{2} \left(\sqrt{9 - x^2} - \sqrt{x+1} \right) dx$$

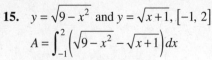

```
fnInt(√(9-X²)-√(
X+1),X,-1,2)
         4.999220084
```

The area is approximately 4.999.

17. $S'(x) = 150 - x^2$; $C'(x) = x^2 + \dfrac{11}{4}x$

a. Let a be the last year it will be profitable to use this new machine. Then,

$$\int_{a}^{a+1} \left(150 - x^2 \right) dx - \int_{a}^{a+1} \left(x^2 + \frac{11}{4}x \right) dx < 0$$

$$\left(150x - \frac{x^3}{3} \right) \Bigg|_{a}^{a+1} - \left(\frac{x^3}{3} + \frac{11}{4} \cdot \frac{x^2}{2} \right) \Bigg|_{a}^{a+1} < 0$$

Multiply by 24 to clear fractions.

$$\left(3600x - 8x^3 \right) \Bigg|_{a}^{a+1} - \left(8x^3 + 33x^2 \right) \Bigg|_{a}^{a+1} < 0$$

$$\left[3600(a+1) - 8(a+1)^3 \right] - \left(3600a - 8a^3 \right) - \left[8(a+1)^3 + 33(a+1)^2 \right] + \left(8a^3 + 33a^2 \right) < 0$$

$$3600a + 3600 - 8\left(a^3 + 3a^2 + 3a + 1 \right) - 3600a + 8a^3$$

$$- \left[8\left(a^3 + 3a^2 + 3a + 1 \right) + 33\left(a^2 + 2a + 1 \right) \right] + 8a^3 + 33a^2 < 0$$

$$3600a + 3600 - 8a^3 - 24a^2 - 24a - 8 - 3600a + 8a^3 - 8a^3 - 24a^2 - 24a - 8 - 33a^2$$

$$-66a - 33 + 8a^3 + 33a^2 < 0$$

$-48a^2 - 114a + 3351 < 0 \Rightarrow 48a^2 + 114a - 3351 > 0$

Solving the equation $48a^2 + 114a - 3351 = 0$, we find $a \approx -9.63$ or $a \approx 7.25$.

The solution of $48a^2 + 114a - 3351 > 0$ is $(-\infty, -9.63)$ or $(7.25, \infty)$.

Thus, the last year it will be profitable is during the 8th yr. (7.25 yr occurs during year 8.)

b. The total net savings during the first year is given by

$$\int_{0}^{1} \left(150 - x^2 \right) dx - \int_{0}^{1} \left(x^2 + \frac{11}{4}x \right) dx = \left(150x - \frac{x^3}{3} \right) \Bigg|_{0}^{1} - \left(\frac{x^3}{3} + \frac{11}{4} \cdot \frac{x^2}{2} \right) \Bigg|_{0}^{1}$$

$$= \left(150 - \frac{1}{3} \right) - 0 - \left[\left(\frac{1}{3} + \frac{11}{8} \right) - 0 \right] \approx 147.96$$

The total is about $147.96.

c. The total net savings over the entire period of use is given by

$$\int_{0}^{8} \left(150 - x^2 \right) dx - \int_{0}^{8} \left(x^2 + \frac{11}{4}x \right) dx = \left(150x - \frac{x^3}{3} \right) \Bigg|_{0}^{8} - \left(\frac{x^3}{3} + \frac{11}{4} \cdot \frac{x^2}{2} \right) \Bigg|_{0}^{8}$$

$$= \left(1200 - \frac{512}{3} \right) - 0 - \left[\left(\frac{512}{3} + \frac{11}{4} \cdot \frac{64}{2} \right) - 0 \right] \approx 770.6667$$

The total is about $771.

19. $E(x) = e^{.15x}$, $I(x) = 120.3 - e^{.15x}$

a. To find the optimum number of days, solve the equation $E(x) = I(x)$. The solution will give the value of x where the two curves meet.

$e^{.15x} = 120.3 - e^{.15x} \Rightarrow 2e^{.15x} = 120.3 \Rightarrow$

$e^{.15x} = 60.15 \Rightarrow .15x = \ln 60.15 \Rightarrow$

$x = \dfrac{\ln 60.15}{.15} \approx 27$

The optimum number of days is 27.

b. The total income for the optimum number of days is given by

$\displaystyle\int_0^{27} \left(120.3 - e^{.15x}\right) dx$

$= 120.3x - \dfrac{1}{.15} e^{.15x} \Big|_0^{27}$

$= (3248.1 - 382.65) - (0 - 6.67)$

$= 2872.12$

The total is $2872.12.

c. The total expenditure for the optimum number of days is given by

$\displaystyle\int_0^{27} e^{.15x} dx = \dfrac{1}{.15} e^{.15x} \Big|_0^{27}$

$= 382.65 - 6.67 = 375.98$.

The total is $375.98.

d. The maximum profit is
$2872.12 - $375.98 = $2496.14.

21. a. The rate of consumption will equal the rate of production when $\dfrac{20}{1.2t + 1.6} = t + .8$.

$\dfrac{20}{1.2t + 1.6} = t + .8 \Rightarrow$

$20 = (1.2t + 1.6)(t + .8) \Rightarrow$

$20 = 1.2t^2 + 2.56t + 1.28 \Rightarrow$

$0 = 1.2t^2 + 2.56t - 18.72$

Applying the quadratic formula we have, $t = 3.02$ or $t = -5.16$. Reject $t = -5.16$. In about 3.02 years the rates will be equal.

b. The total excess production before consumption and production are equal is given by

$\displaystyle\int_0^{3.02} \dfrac{20}{1.2t + 1.6} dt - \int_0^{3.02} (t + .8) dt$

$= \dfrac{20}{1.2} \ln|1.2t + 1.6| \Big|_0^{3.02} - \left(\dfrac{t^2}{2} + .8t\right)\Big|_0^{3.02}$

$= \left[\dfrac{20}{1.2} \ln 5.224 - \dfrac{20}{1.2} \ln 1.6\right]$

$\qquad - \left[\dfrac{(3.02)^2}{2} + .8(3.02) - 0\right]$

≈ 12.74

The total is about 12.74 trillion gallons.

23. a. The total pollution in the lake after 20 hours is given by $\displaystyle\int_0^{20} 8\left(1 - e^{-.25t}\right) dt - \int_0^{20} .2t\, dt$

$= \displaystyle\int_0^{20} \left[8\left(1 - e^{-.25t}\right) - .2t\right] dt$.

```
fnInt(8(1-e^(-.2
5X))-.2X,X,0,20)
        88.2156143
■
```

The total pollution in the lake after 20 hours was about 88.22 gallons.

b. The rate at which pollution enters the lake will be equal to the rate at which pollution is being removed when $f(t) = g(t)$.

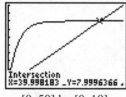

```
Intersection
X=39.998183  Y=7.9996366
```
[0, 50] by [0, 10]

The rate at which pollution enters the lake will be equal to the rate at which pollution is being removed at about 40 hours.

c. The total pollution in the lake after 40 hours is given by $\displaystyle\int_0^{40} 8\left(1 - e^{-.25t}\right) dt - \int_0^{40} .2t\, dt$

$= \displaystyle\int_0^{40} \left[8\left(1 - e^{-.25t}\right) - .2t\right] dt$.

```
fnInt(8(1-e^(-.2
5X))-.2X,X,0,40)
        128.0014528
```

The total pollution in the lake when the rate at which pollution enters the lake will be equal to the rate at which pollution is being removed (40 hours) is about 128 gallons.

d. All the pollution will be removed from the lake when $\int_0^t 8\left(1-e^{-.25t}\right)dt - \int_0^t .2t\,dt = 0$.

$$\int_0^x 8\left(1-e^{-.25t}\right)dt - \int_0^x .2t\,dt = 0$$

$$\int_0^x 8\left(1-e^{-.25t}\right)dt = \int_0^x .2t\,dt$$

$$\left(32e^{-.25t}+8t\right)\Big|_0^x = .1t^2\Big|_0^x$$

$$\left(32e^{-.25x}+8x-32\right) = .1x^2$$

Using a graphing calculator, we have

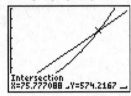

```
Intersection
X=75.777088  Y=574.2167
```

All pollution will be removed from the lake in about 75.78 hours.

25. a. We graph the linear functions

$$S(q)=\frac{7}{5}q \text{ and } D(q)=-\frac{3}{5}q+10$$

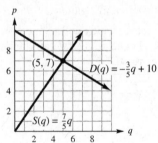

b. $S(q)=D(q) \Rightarrow \dfrac{7}{5}q = -\dfrac{3}{5}q+10 \Rightarrow$

$7q = -3q+50 \Rightarrow q = 5$

$S(5)=7$

The equilibrium point is $(5, 7)$.

c. The consumers' surplus is given by

$$\int_0^5 \left[\left(-\frac{3}{5}q+10\right)-7\right]dq = \int_0^5 \left(-\frac{3}{5}q+3\right)dq$$

$$= \left(-\frac{3}{5}\cdot\frac{q^2}{2}+3q\right)\Bigg|_0^5$$

$$= \left(-\frac{15}{2}+15\right)-0$$

$$= 7.5$$

The consumers' surplus is $7.50.

d. The producers' surplus is given by

$$\int_0^5 \left(7-\frac{7}{5}q\right)dq = \left(7q-\frac{7}{5}\cdot\frac{q^2}{2}\right)\Bigg|_0^5$$

$$= \left(35-\frac{35}{2}\right)-0 = 17.5$$

The producers' surplus is $17.50.

27. If $S(q)=100+3q^{3/2}+q^{5/2}$,

$S(9)=100+81+243=424$.

Therefore, the producers' surplus is given by

$$\int_0^9 \left[424-\left(100+3q^{3/2}+q^{5/2}\right)\right]dq$$

$$= \int_0^9 \left(324-3q^{3/2}-q^{5/2}\right)dq$$

$$= \left(324q-3\cdot\frac{q^{5/2}}{\frac{5}{2}}-\frac{q^{7/2}}{\frac{7}{2}}\right)\Bigg|_0^9$$

$$= \left(324q-\frac{6}{5}q^{5/2}-\frac{2}{7}q^{7/2}\right)\Bigg|_0^9$$

$$\approx (2916-291.6-624.86)-0 = 1999.54$$

The producers' surplus is $1999.54.

29. If $D(q)=\dfrac{15,500}{(3.2q+7)^3}$, then

$$D(5)=\frac{15,500}{(16+7)^3}=1.2739.$$

Therefore, consumers' surplus is given by

$$\int_0^5 \left[\frac{15,500}{(3.2q+7)^3}-1.2739\right]dq.$$

Let $u = 3.2q+7$. Then $du = 3.2\,dq$. If $q = 5$, $u = 23$. If $q = 0$, $u = 7$.

$$\int_0^5 \left[\frac{15,500}{(3.2q+7)^3}-1.2739\right]dq$$

$$= \frac{1}{3.2}\int_0^5 \left[15,500(3.2q+7)^{-3}-1.2739\right]3.2\,dq$$

$$= \frac{1}{3.2}\int_7^{23} \left(15,500u^{-3}-1.2739\right)du$$

$$= \frac{1}{3.2}\left(15,500\cdot\frac{u^{-2}}{-2}-1.2739u\right)\Bigg|_7^{23}$$

$$= \frac{1}{3.2}\left(-\frac{7750}{u^2}-1.2739u\right)\Bigg|_7^{23}$$

$$= \frac{1}{3.2}\left(-\frac{7750}{529}-29.2997\right)$$

$$\quad -\frac{1}{3.2}\left(-\frac{7750}{49}-8.9173\right) = 38.50$$

The consumers' surplus is $38.50.

31. a. We graph the quadratic functions
$$S(q) = q^2 + \frac{11}{4}q \text{ and } D(q) = 150 - q^2.$$

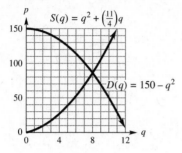

b. $S(q) = D(q)$

$$q^2 + \frac{11}{4}q = 150 - q^2$$

$$2q^2 + \frac{11}{4}q - 150 = 0$$

$$8q^2 + 11q - 600 = 0$$

$$(8q + 75)(q - 8) = 0 \Rightarrow q = -\frac{75}{8} \text{ or } q = 8$$

We discard the negative value.

$$S(8) = 8^2 + \frac{11}{4} \cdot 8 = 86$$

The equilibrium point is (8, 86).

c. The consumers' surplus is given by

$$\int_0^8 \left[(150 - q^2) - 86 \right] dq = \int_0^8 (64 - q^2) dq$$

$$= \left(64q - \frac{q^3}{3} \right) \Big|_0^8$$

$$= \left(512 - \frac{512}{3} \right) - 0$$

$$\approx 341.33.$$

The consumers' surplus is \$341.33.

d. The producers' surplus is given by

$$\int_0^8 \left[86 - \left(q^2 + \frac{11}{4}q \right) \right] dq$$

$$= \left(86q - \frac{q^3}{3} - \frac{11}{4} \cdot \frac{q^2}{2} \right) \Big|_0^8$$

$$\approx (688 - 170.67 - 88) - 0 = 429.33$$

The producers' surplus is \$429.33.

33. $y = 3\sqrt{x}$ and $y = 2x$

Determine the points of intersection

$$3\sqrt{x} = 2x \Rightarrow 9x = 4x^2 \Rightarrow 4x^2 - 9x = 0 \Rightarrow$$

$$x(4x - 9) = 0 \Rightarrow x = 0 \text{ or } x = \frac{9}{4}$$

On the interval $\left[0, \frac{9}{4} \right]$, $3\sqrt{x} \geq 2x$.

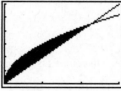

[0, 3] by [0, 6]

$$\int_0^{9/4} (3\sqrt{x} - 2x) dx = \left(\frac{3x^{3/2}}{\frac{3}{2}} - x^2 \right) \Big|_0^{9/4}$$

$$= \left(2x^{3/2} - x^2 \right) \Big|_0^{9/4}$$

$$= 2 \left(\frac{9}{4} \right)^{3/2} - \left(\frac{9}{4} \right)^2 - 0$$

$$= 2 \left(\frac{27}{8} \right) - \left(\frac{81}{16} \right) = \frac{27}{16}$$

The area of the poster board is $\frac{27}{16}$ square units.

35. $f(x) = 70.71e^{.43x}$

The total increase is given by

$$\int_0^4 70.71e^{.43x} dx = 70.71 \cdot \frac{e^{.43x}}{.43} \Big|_0^4$$

$$= 70.71\frac{e^{.43(4)}}{.43} - \frac{70.71}{.43}$$

$$= 753.89.$$

The total increase in costs is \$753.89.

37. $I(x) = .9x^2 + .1x$

a. $I(.1) = .9(.1)^2 + .1(.1) = .019$

The lower 10% of income producers earn 1.9% of the total income of the population.

b. $I(.5) = .9(.5)^2 + .1(.5) = .275$

The lower 50% of income producers earn 27.5% of the total income of the population.

c. $I(.9) = .9(.9)^2 + .1(.9) = .819$

The lower 90% of income producers earn 81.9% of the total income of the population.

d. Graph $I(x) = x$ and $I(x) = .9x^2 + .1x$ for $0 \le x \le 1$ on the same set of axes.

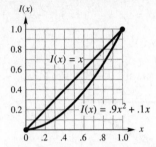

e. The area between the curves is given by

$$\int_0^1 \left[x - \left(.9x^2 + .1x \right) \right] dx$$

$$= \int_0^1 \left(.9x - .9x^2 \right) dx = \left(.9 \cdot \frac{x^2}{2} - .9 \cdot \frac{x^3}{3} \right) \Big|_0^1$$

$$= \left(.45x^2 - .3x^3 \right) \Big|_0^1 = (.45 - .3) - 0 = .15$$

This represents the amount of inequality of income distribution.

39. a. $\int_0^T H(x)dx$ will be a maximum when its derivative, $H(x)$, is equal to 0. This occurs when $H(x) = 20 - 2x = 0 \Rightarrow x = 10$.

Thus, to achieve the maximum production rate, 10 items must be made.

b. $\int_0^T H(x)dx = \int_0^T (20 - 2x)dx = \left(20x - x^2 \right) \Big|_0^T$

$$= 20T - T^2$$

This will be a maximum when its derivative, $H(T)$, is equal to 0.

$H(T) = 20 - 2T = 0 \Rightarrow T = 10$

Thus, the maximum production rate per item is 10 hours per item.

Section 13.6 Tables of Integrals (Optional)

1. $\int \dfrac{-7}{\sqrt{x^2 + 36}}\, dx = -7 \int \dfrac{dx}{\sqrt{x^2 + 36}}$

Use entry 5 from the table with $a = 6$.

$-7 \int \dfrac{dx}{\sqrt{x^2 + 36}} = -7 \ln \left| x + \sqrt{x^2 + 36} \right| + C$

3. $\int \dfrac{6}{x^2 - 9}\, dx = 6 \int \dfrac{1}{x^2 - 9}\, dx$

Use entry 8 from the table with $a = 3$. Note that $x^2 > 3^2$.

$6 \int \dfrac{1}{x^2 - 9}\, dx = 6 \left(\dfrac{1}{2(3)} \ln \left| \dfrac{x - 3}{x + 3} \right| \right) + C$

$$= \ln \left| \dfrac{x - 3}{x + 3} \right| + C, \left(x^2 > 9 \right)$$

5. $\int \dfrac{-4}{x\sqrt{9 - x^2}}\, dx = -4 \int \dfrac{dx}{x\sqrt{9 - x^2}}$

Use entry 9 from the table with $a = 3$. Note that $0 < x < 3$.

$-4 \int \dfrac{dx}{x\sqrt{9 - x^2}} = \dfrac{4}{3} \ln \left| \dfrac{3 + \sqrt{9 - x^2}}{x} \right| + C,$

$(0 < x < 3)$

7. $\int \dfrac{-5x}{3x + 1}\, dx = -5 \int \dfrac{x}{3x + 1}\, dx$

Use entry 11 from table with $a = 3, b = 1$.

$-5 \int \dfrac{x}{3x + 1}\, dx = -5 \left(\dfrac{x}{3} - \dfrac{1}{9} \ln |3x + 1| \right) + C$

$$= -\dfrac{5x}{3} + \dfrac{5}{9} \ln |3x + 1| + C$$

9. $\int \dfrac{13}{3x(3x - 5)}\, dx = \dfrac{13}{3} \int \dfrac{1}{x(3x - 5)}\, dx$

Use entry 13 from the table with $a = 3, b = -5$.

$\dfrac{13}{3} \int \dfrac{1}{x(3x - 5)}\, dx = \dfrac{13}{3} \left(-\dfrac{1}{5} \ln \left| \dfrac{x}{3x - 5} \right| \right) + C$

$$= -\dfrac{13}{15} \ln \left| \dfrac{x}{3x - 5} \right| + C$$

11. $\int \dfrac{4}{4x^2 - 1}\, dx = 4 \int \dfrac{dx}{4x^2 - 1} = 2 \int \dfrac{2\, dx}{4x^2 - 1}$

Let $u = 2x$. Then $du = 2\, dx$.

$\int \dfrac{4}{4x^2 - 1}\, dx = 2 \int \dfrac{du}{u^2 - 1}$

Use entry 8 from the table with $a = 1, u^2 > 1$.

$2 \int \dfrac{du}{u^2 - 1} = 2 \cdot \dfrac{1}{2} \ln \left| \dfrac{u - 1}{u + 1} \right| + C, u^2 > 1,$

Substitute $2x$ for u. Since

$u^2 > 1, 4x^2 > 1,$ or $x^2 > \dfrac{1}{4}$.

$\int \dfrac{4}{4x^2 - 1}\, dx = \ln \left| \dfrac{2x - 1}{2x + 1} \right| + C, \left(x^2 > \dfrac{1}{4} \right)$

13. $\displaystyle\int \frac{3}{x\sqrt{1-9x^2}}\, dx = 3\int \frac{3\, dx}{3x\sqrt{1-9x^2}}$

Let $u = 3x$. Then $du = 3\, dx$.

$\displaystyle\int \frac{3}{x\sqrt{1-9x^2}}\, dx = 3\int \frac{du}{u\sqrt{1-u^2}}$

Use entry 9 from the table with $a = 1$.
Note $0 < u < 1$.

$\displaystyle 3\int \frac{du}{u\sqrt{1-u^2}} = -3\ln\left|\frac{1+\sqrt{1-u^2}}{u}\right| + C$

$\displaystyle\qquad = -3\ln\left|\frac{1+\sqrt{1-9x^2}}{3x}\right| + C,$

$\displaystyle\left(0 < 3x < 1,\ \text{or}\ 0 < x < \frac{1}{3}\right)$

15. $\displaystyle\int \frac{15x}{2x+3}\, dx = 15\int \frac{x}{2x+3}\, dx$

Use entry 11 from the table with $a = 2$,
$b = 3$.

$\displaystyle 15\int \frac{x}{2x+3}\, dx = 15\left(\frac{x}{2} - \frac{3}{4}\ln|2x+3|\right) + C$

$\displaystyle\qquad = \frac{15x}{2} - \frac{45}{4}\ln|2x+3| + C$

17. $\displaystyle\int \frac{-x}{(5x-1)^2}\, dx = -\int \frac{x\, dx}{(5x-1)^2}$

Use entry 12 form the table with $a = 5$, $b = -1$.

$\displaystyle -\int \frac{x\, dx}{(5x-1)^2} = -\left(\frac{-1}{25(5x-1)} + \frac{1}{25}\ln|5x-1|\right) + C$

$\displaystyle\qquad = \frac{1}{25(5x-1)} - \frac{\ln|5x-1|}{25} + C$

19. $\displaystyle\int \frac{3x^4\ln|x|}{4}\, dx$

Use entry 16, with $n = 4$.

$\displaystyle\int \frac{3x^4\ln|x|}{4}\, dx = \frac{3}{4}x^5\left(\frac{\ln|x|}{5} - \frac{1}{25}\right) + C$

21. $\displaystyle\int \frac{7\ln|x|}{x^2}\, dx = 7\int x^{-2}\ln|x|\, dx$

Use entry 16 from the table with $n = -2$.

$\displaystyle 7\int x^{-2}\ln|x|\, dx = 7x^{-1}\left(\frac{\ln|x|}{-1} - \frac{1}{1}\right) + C$

$\displaystyle\qquad = \frac{7}{x}\left(-\ln|x| - 1\right) + C$

23. $\displaystyle\int xe^{-2x}\, dx$

Use entry 17 from the table with $n = 1$ and $a = -2$.

$\displaystyle\int xe^{-2x}\, dx = \frac{xe^{-2x}}{-2} - \frac{1}{-2}\int x^0 e^{-2x}\, dx$

$\displaystyle\qquad = -\frac{1}{2}xe^{-2x} + \frac{1}{2}\int e^{-2x}\, dx$

$\displaystyle\qquad = -\frac{1}{2}xe^{-2x} + \frac{1}{2}\left(\frac{e^{-2x}}{-2}\right) + C$

$\displaystyle\qquad = -\frac{1}{2}xe^{-2x} - \frac{1}{4}e^{-2x} + C$

$\displaystyle\qquad = -\frac{xe^{-2x}}{2} - \frac{e^{-2x}}{4} + C$

25. $\displaystyle R'(x) = \frac{1000}{\sqrt{x^2+25}}$

The total revenue is given by

$\displaystyle\int_0^{20} \frac{1000}{\sqrt{x^2+25}}\, dx = 1000\int_0^{20} \frac{1}{\sqrt{x^2+5^2}}\, dx.$

We use entry 5 in the table with $a = 5$.

$\displaystyle\int_0^{20} \frac{1000}{\sqrt{x^2+25}}\, dx$

$\displaystyle\qquad = 1000\ln\left|x + \sqrt{x^2+25}\right|\Big|_0^{20}$

$\displaystyle\qquad = 1000\left[\ln\left(20 + \sqrt{425}\right) - \ln\left(0 + \sqrt{25}\right)\right]$

$\displaystyle\qquad \approx 2094.71$

The total revenue is approximately \$2094.71.

27. Rate of growth: $m'(x) = 25xe^{2x}$

The total growth is given by $\displaystyle\int_0^3 25xe^{2x}\, dx$.

Use entry 17 with $n = 1$ and $a = 2$.

$\displaystyle\int_0^3 25xe^{2x}\, dx = 25\left(\frac{xe^{2x}}{2}\Big|_0^3 - \frac{1}{2}\int_0^3 e^{2x}\, dx\right)$

$\displaystyle\qquad = 25\left(\frac{3e^6}{2} - \frac{e^{2x}}{4}\Big|_0^3\right)$

$\displaystyle\qquad = 25\left[\frac{3e^6}{2} - \frac{1}{4}\left(e^6 - e^0\right)\right]$

$\displaystyle\qquad = \frac{75}{2}e^6 - \frac{25}{4}\left(e^6 - 1\right)$

$\displaystyle\qquad = 15{,}128.58 - 2515.18 \approx 12{,}613$

The total accumulated growth after 3 days is about 12,613 microbes.

Section 13.7 Differential Equations

1. $\dfrac{dy}{dx} = -3x^2 + 7x$

$dy = \left(-3x^2 + 7x\right)dx$

$\displaystyle\int dy = \int\left(-3x^2 + 7x\right)dx$

$y = -x^3 + \dfrac{7}{2}x^2 + C$

3. $3x^3 - 2\dfrac{dy}{dx} = 0 \Rightarrow -2\dfrac{dy}{dx} = -3x^3 \Rightarrow$

$\dfrac{dy}{dx} = \dfrac{3}{2}x^3 \Rightarrow y = \displaystyle\int\dfrac{3}{2}x^3\,dx \Rightarrow y = \dfrac{3}{2}\int x^3\,dx \Rightarrow$

$y = \dfrac{3}{2}\cdot\dfrac{x^4}{4} + C \Rightarrow y = \dfrac{3x^4}{8} + C$

5. $y\dfrac{dy}{dx} = x \Rightarrow y\,dy = x\,dx \Rightarrow$

$\dfrac{y^2}{2} = \dfrac{x^2}{2} + C_1 \Rightarrow y^2 = x^2 + C$

7. $\dfrac{dy}{dx} = 4xy$

$\dfrac{dy}{y} = 4x\,dx$

$\displaystyle\int\dfrac{dy}{y} = 4\int x\,dx$

$\ln|y| = 4\dfrac{x^2}{2} + C$

$\ln|y| = 2x^2 + C$

$|y| = e^{2x^2 + C} = e^{2x^2}e^c$

$y = Me^{2x^2}$

9. $\dfrac{dy}{dx} = 4x^3 y - 3x^2 y \Rightarrow \dfrac{dy}{dx} = y\left(4x^3 - 3x^2\right) \Rightarrow$

$\dfrac{dy}{y} = \left(4x^3 - 3x^2\right)dx$

$\displaystyle\int\dfrac{dy}{y} = \int\left(4x^3 - 3x^2\right)dx$

$\ln|y| = \dfrac{4x^4}{4} - \dfrac{3x^3}{3} + C$

$\ln|y| = x^4 - x^3 + C$

$|y| = e^{x^4 - x^3 + C} \Rightarrow |y| = e^{x^4 - x^3}e^C \Rightarrow$

$y = Me^{x^4 - x^3}$

11. $\dfrac{dy}{dx} = \dfrac{y}{x},\ x > 0 \Rightarrow \dfrac{dy}{y} = \dfrac{dx}{x} \Rightarrow$

$\displaystyle\int\dfrac{dy}{y} = \int\dfrac{dx}{x} \Rightarrow \ln|y| = \ln|x| + C \Rightarrow$

$|y| = e^{\ln|x| + C} \Rightarrow y = e^{\ln|x|}e^C = Mx$

13. $\dfrac{dy}{dx} = y - 7$

$\dfrac{dy}{y - 7} = dx$

$\ln|y - 7| = x + C$

$|y - 7| = e^{x + C} = e^x e^C$

$y - 7 = Me^x \Rightarrow y = 7 + Me^x$

15. $\dfrac{dy}{dx} = y^2 e^x \Rightarrow \dfrac{dy}{y^2} = e^x\,dx \Rightarrow \displaystyle\int y^{-2}\,dy = \int e^x\,dx \Rightarrow$

$-y^{-1} = e^x + C \Rightarrow -\dfrac{1}{y} = e^x + C \Rightarrow y = -\dfrac{1}{e^x + C}$

17. $\dfrac{dy}{dx} + 2x = 3x^2$; $y = 4$ when $x = 0$.

$\dfrac{dy}{dx} = 3x^2 - 2x \Rightarrow y = \displaystyle\int\left(3x^2 - 2x\right)dx \Rightarrow$

$y = y = x^3 - x^2 + C$

The general solution is

$y = x^3 - x^2 + C$.

When $x = 0$, $y = 4$,

$4 = 0^3 - 0^2 + C \Rightarrow C = 4$

Thus, the particular solution is

$y = x^3 - x^2 + 4$.

19. $\dfrac{dy}{dx}\left(x^3 + 28\right) = \dfrac{6x^2}{y}$; $y^2 = 6$ when $x = -3$

$y\,dy = \dfrac{6x^2}{x^3 + 28}\,dx \Rightarrow$

$\displaystyle\int y\,dy = \int\dfrac{6x^2}{x^3 + 28}\,dx = \int\dfrac{2(3)x^2}{x^3 + 28}\,dx \Rightarrow$

$\dfrac{y^2}{2} = 3\ln\left|x^3 + 28\right| + C \Rightarrow y^2 = 4\ln\left|x^3 + 28\right| + C$

$y^2 = 6$ when $x = -3$, so

$6 = 4\ln\left|(-3)^3 + 28\right| + C \Rightarrow C = 6$

Thus, the particular solution is

$y^2 = 4\ln\left|x^3 + 28\right| + 6$

21. $\dfrac{dy}{dx} = \dfrac{x^2}{y}$; $y = 4$ when $x = 0$.

$y\,dy = x^2\,dx \Rightarrow \displaystyle\int y\,dy = \int x^2\,dx \Rightarrow \dfrac{y^2}{2} = \dfrac{x^3}{3} + C$

Let $x = 0$ and $y = 4$.

$\dfrac{16}{2} = \dfrac{0}{3} + C \Rightarrow C = 8$

The particular solution is $y^2 = \dfrac{2}{3}x^3 + 8$.

23. $(5x+3)y = \dfrac{dy}{dx}$; $y = 1$ when $x = 0$.

$(5x+3)dx = \dfrac{dy}{y} \Rightarrow \displaystyle\int (5x+3)dx = \int \dfrac{dy}{y} \Rightarrow$

$\dfrac{5x^2}{2} + 3x = \ln|y| + C$

If $x = 0$ and $y = 1$, $0 = \ln|1| + C \Rightarrow 0 = C$.

We now have $\dfrac{5x^2}{2} + 3x = \ln|y|$.

Rewrite this equation in exponential form. The particular solution is $y = e^{2.5x^2 + 3x}$.

25. $\dfrac{dy}{dx} = \dfrac{7x+1}{y-3}$; $y = 4$ when $x = 0$.

$(y-3)dy = (7x+1)dx$

$\displaystyle\int (y-3)dy = \int (7x+1)dx$

$\dfrac{y^2}{2} - 3y = \dfrac{7x^2}{2} + x + C$

If $x = 0$ and $y = 4$, $\dfrac{16}{2} - 12 = C \Rightarrow -4 = C$.

The particular solution is $\dfrac{y^2}{2} - 3y = \dfrac{7x^2}{2} + x - 4$

or $y^2 - 6y = 7x^2 + 2x - 8$.

27. Answers vary.

29. $\dfrac{dy}{dx} = \dfrac{3x^2 - 4x}{25} \Rightarrow$

$y = \displaystyle\int \dfrac{3x^2 - 4x}{25}\,dx = \dfrac{1}{25}\int (3x^2 - 4x)dx \Rightarrow$

$y = \dfrac{1}{25}(x^3 - 2x^2) + C$

Since production is 2300 units when investment is \$5000, we have

$23 = \dfrac{1}{25}(5^3 - 2(5)^2) + C \Rightarrow C = 20$.

Thus, $y = \dfrac{1}{25}(x^3 - 2x^2) + 20$.

a. $y = \dfrac{1}{25}(8^3 - 2(8)^2) + 20 = 35.36$

The production level is 3536 units if the investment is \$8000.

b. $y = \dfrac{1}{25}(11^3 - 2(11)^2) - 20 = 63.56$

The production level is 6356 units if the investment is \$11,000.

31. $\dfrac{dy}{dt} = kt$; $k = 8$ and $y = 50$ when $t = 0$.

a. $\dfrac{dy}{dt} = 8t \Rightarrow dy = 8t\,dt \Rightarrow$

$y = \displaystyle\int 8t\,dt = 4t^2 + C$

When $t = 0$, $y = 50$, so

$50 = 4(0)^2 + C \Rightarrow C = 50$, and

$y = 4t^2 + 50$.

b. If $y = 550$, then

$550 = 4t^2 + 50 \Rightarrow 4t^2 = 500 \Rightarrow t^2 = 125 \Rightarrow$
$t \approx \pm 11$.

The product should be dated 11 days from time $t = 0$.

33. a. $\dfrac{dy}{dt} = ky \Rightarrow \displaystyle\int \dfrac{dy}{y} = \int k\,dt \Rightarrow$

$\ln|y| = kt + C \Rightarrow e^{\ln|y|} = e^{kt+C} \Rightarrow$

$y = \pm(e^{kt})(e^C) = Me^{kt}$

If $y = 1$ when $t = 0$ and $y = 5$ when $t = 2$, we have the following system of equations.

$1 = Me^{k(0)}$ (1)

$5 = Me^{2k}$ (2)

From equation (1),

$1 = M(1) \Rightarrow M = 1$.

Substitute 1 for M in equation (2) to find k.

$5 = (1)e^{2k} \Rightarrow e^{2k} = 5 \Rightarrow 2k\ln e = \ln 5 \Rightarrow$

$k = \dfrac{\ln 5}{2} \approx .8$

b. If $k = .8$ and $M = 1$, then $y = e^{.8t}$.

When $t = 3$, $y = e^{.8(3)} \approx 11$.

In 3 days, 11 people heard the rumor.

c. When $t = 5$, $y = e^{.8(5)} \approx 55$.

In 5 days, 55 people heard the rumor.

d. When $t = 10$, $y = e^{.8(10)} \approx 2981$.

In 10 days, 2981 people heard the rumor.

35. a. $\dfrac{dy}{dt} = -.15y$

b. $\dfrac{dy}{dt} = -.15y \Rightarrow \dfrac{dy}{y} = -.15dt \Rightarrow$

$\displaystyle\int \dfrac{dy}{y} = \int -.15dt \Rightarrow \ln y = -.15t + C \Rightarrow$

$y = e^{-.15t+C} = e^{-.15t}e^{C} \Rightarrow$

$y = Me^{-.15t}$

c. At $t = 0$, $M = 1$ because when $t = 0$,

$e^{-.15t} = 1$. So we have

$.25 = e^{-.15t} \Rightarrow \ln(.25) = -.15t \Rightarrow$

$t = \dfrac{\ln(.25)}{-.15} \approx 9.2$

Sales will decrease to 25% of their original level in about 9.2 years.

37. From examples 5 and 6, we have $y = Me^{-.03t}$.

When $t = 0$, $y = 6$, so $6 = Me^{-.03(0)} \Rightarrow M = 6$.

$y = 6e^{-.03(8)} \approx 4.7$

About 4.7 cc of dye are present after 8 minutes.

39. $\dfrac{dy}{dx} = 30 - y$; $y = 1000$ when $x = 0$

$\dfrac{dy}{30-y} = dx \Rightarrow \displaystyle\int \dfrac{dy}{30-y} = \int dx \Rightarrow$

$-\ln|30 - y| = x + C \Rightarrow$

$\ln|30 - y| = -x + C \Rightarrow |30 - y| = e^{-x+C} \Rightarrow$

$30 - y = Me^{-x} \Rightarrow y = 30 - Me^{-x} \Rightarrow$

$1000 = 30 - M \Rightarrow M = -970$

$y = 30 + 970e^{-x}$

a. If $x = 2$, $y = 30 + 970e^{-2} \approx 161.28$.

About 161.28 thousand bacteria are present.

b. If $x = 5$, $y = 30 + 970e^{-5} \approx 36.54$.

About 36.54 thousand bacteria are present.

c. If $x = 10$, $y = 30 + 970e^{-10} \approx 30.04$.

About 30.44 thousand bacteria are present.

41. $\dfrac{dT}{dt} = -k(T - C) = -k(T - 68) \Rightarrow$

$\dfrac{1}{T-68}dT = -k\,dt \Rightarrow \displaystyle\int \dfrac{1}{T-68}dT = \int -k\,dt \Rightarrow$

$\ln|T - 68| = -kt \Rightarrow T - 68 = Me^{-kt} \Rightarrow$

$T = Me^{-kt} + 68$

$M = 98.6 - 68 = 30.6$, so we have

$T = 30.6e^{-kt} + 68$. When $t = 1$, $T = 90$, so

$90 = 30.6e^{-.33t} + 68$, $k = .33$.

a. $T = 68 + 30.6e^{-.33t}$

b. After two hours,

$T = 68 + 30.6e^{-.33(2)} = 83.82$, so the temperature of the body is about 83.8°.

c. When $T = 75$, we have $75 = 68 + 30.6e^{-.33t}$ which gives $t = 4.46$ when solved, so the body will be 75° in about 4.5 hours.

d. The body will be within .01° of the surrounding air when

$68.01 = 68 + 30.6e^{-.33t} \Rightarrow t = 24.32$, so the body will be within .01° in about 24.3 hours.

43. $E = 2$

$-\dfrac{p}{q} \cdot \dfrac{dq}{dp} = 2$

$\dfrac{1}{q}dq = -\dfrac{2}{p}dp$

$\displaystyle\int \dfrac{1}{q}dq = -2\int \dfrac{1}{p}dp$

$\ln q = -2\ln p + \ln C$

$\ln q = \ln p^{-2} + \ln C$

$e^{\ln q} = e^{\ln p^{-2} + \ln C} = e^{\ln p^{-2}}e^{\ln C}$

$q = Cp^{-2}$

$q = \dfrac{C}{p^2}$

Chapter 13 Review Exercises

1. $\displaystyle\int \left(x^2 - 8x - 7\right) dx = \frac{x^3}{3} - 8 \cdot \frac{x^2}{2} - 7x + C$

$\qquad = \dfrac{x^3}{3} - 4x^2 - 7x + C$

2. $\displaystyle\int \left(6 - x^2\right) dx = 6x - \frac{x^3}{3} + C$

3. $\displaystyle\int 7\sqrt{x}\,dx = 7\int x^{1/2}\,dx = 7 \cdot \frac{x^{3/2}}{\frac{3}{2}} + C$

$\qquad = 7 \cdot \dfrac{2}{3} x^{3/2} + C = \dfrac{14x^{3/2}}{3} + C$

4. $\displaystyle\int \frac{\sqrt{x}}{8}\,dx = \frac{1}{8}\int x^{1/2} + C = \frac{1}{8} \cdot \frac{x^{3/2}}{\frac{3}{2}} + C$

$\qquad = \dfrac{1}{8} \cdot \dfrac{2}{3} x^{3/2} + C = \dfrac{x^{3/2}}{12} + C$

5. $\displaystyle\int \left(x^{1/2} + 3x^{-2/3}\right) dx = \frac{x^{3/2}}{\frac{3}{2}} + 3 \cdot \frac{x^{1/3}}{\frac{1}{3}} + C$

$\qquad = \dfrac{2}{3} x^{3/2} + 3 \cdot \dfrac{3}{1} x^{1/3} + C$

$\qquad = \dfrac{2x^{3/2}}{3} + 9x^{1/3} + C$

6. $\displaystyle\int \left(8x^{4/3} + x^{-1/2}\right) dx = 8 \cdot \frac{x^{7/3}}{\frac{7}{3}} + \frac{x^{1/2}}{\frac{1}{2}} + C$

$\qquad = 8 \cdot \dfrac{3}{7} x^{7/3} + 2x^{1/2} + C$

$\qquad = \dfrac{24x^{7/3}}{7} + 2x^{1/2} + C$

7. $\displaystyle\int \frac{-6}{x^3}\,dx = -6\int x^{-3}\,dx = -6 \cdot \frac{x^{-2}}{-2} + C$

$\qquad = 3x^{-2} + C$

8. $\displaystyle\int \frac{9}{x^4}\,dx = 9\int x^{-4}\,dx = 9 \cdot \frac{x^{-3}}{-3} + C = -\frac{3}{x^3} + C$

9. $\displaystyle\int -3e^{2x}\,dx = -3\int e^{2x}\,dx = -3 \cdot \frac{1}{2}e^{2x} + C$

$\qquad = -\dfrac{3e^{2x}}{2} + C$

10. $\displaystyle\int 8e^{-x}\,dx = 8\int e^{-x}\,dx = 8 \cdot \frac{1}{-1}e^{-x} + C$

$\qquad = -8e^{-x} + C$

11. $\displaystyle\int \frac{10}{x-3}\,dx = 10\int \frac{1}{x-3}\,dx = 10\ln|x-3| + C$

12. $\displaystyle\int \frac{-14}{x+2}\,dx = -14\int \frac{1}{x+2}\,dx = -14\ln|x+2| + C$

13. $\displaystyle\int 5xe^{3x^2}\,dx$

Let $u = 3x^2$. Then $du = 6x\,dx$.

$\displaystyle\int 5xe^{3x^2}\,dx = \frac{1}{6} \cdot 5\int e^{3x^2} \cdot 6x\,dx = \frac{5}{6}\int e^u\,du$

$\qquad = \dfrac{5}{6}e^u + C = \dfrac{5}{6}e^{3x^2} + C = \dfrac{5e^{3x^2}}{6} + C$

14. $\displaystyle\int 4xe^{x^2}\,dx$

Let $u = x^2$. Then $du = 2x\,dx$.

$\displaystyle\int 4xe^{x^2}\,dx = 2\int e^{x^2} \cdot 2x\,dx = 2\int e^u\,du$

$\qquad = 2e^u + C = 2e^{x^2} + C$

15. $\displaystyle\int \frac{3x}{x^2-1}\,dx$

Let $u = x^2 - 1$. Then $du = 2x\,dx$.

$\displaystyle\int \frac{3x}{x^2-1}\,dx = 3 \cdot \frac{1}{2}\int \frac{1}{x^2-1}\,2x\,dx$

$\qquad = \dfrac{3}{2}\int \dfrac{1}{u}\,du = \dfrac{3}{2}\ln|u| + C$

$\qquad = \dfrac{3\ln|x^2-1|}{2} + C$

16. $\displaystyle\int \frac{-6x}{4-x^2}\,dx$

Let $u = 4 - x^2$. Then $du = -2x\,dx$.

$\displaystyle\int \frac{-6x}{4-x^2}\,dx = 6 \cdot \frac{1}{2}\int \frac{1}{4-x^2}(-2x)\,dx = 3\int \frac{1}{u}\,du$

$\qquad = 3\ln|u| + C = 3\ln|4-x^2| + C$

17. $\displaystyle\int \frac{12x^2 dx}{\left(x^3+5\right)^4} = 12 \cdot \frac{1}{3}\int \frac{3x^2 dx}{\left(x^3+5\right)^4}$

Let $u = x^3 + 5$. Then $du = 3x^2 dx$.

$\displaystyle\int \frac{12x^2 dx}{\left(x^3+5\right)^4} = 12 \cdot \frac{1}{3}\int \frac{du}{u^4} = 4\int u^{-4}du$

$\displaystyle\qquad = 4\left(\frac{u^{-3}}{-3}\right) + C = -\frac{4}{3}\left(x^3+5\right)^{-3} + C$

18. $\displaystyle\int \left(x^2 - 5x\right)^4 (2x-5)dx$

Let $u = x^2 - 5x$. Then $du = (2x-5)dx$.

$\displaystyle\int \left(x^2 - 5x\right)^4 (2x-5)dx = \int u^4 du = \frac{u^5}{5} + C$

$\displaystyle\qquad = \frac{\left(x^2-5x\right)^5}{5} + C$

19. $\displaystyle\int \frac{4x-5}{2x^2-5x}dx$

Let $u = 2x^2 - 5x$. Then $du = (4x-5)dx$.

$\displaystyle\int \frac{4x-5}{2x^2-5x}dx = \int \frac{du}{u} = \ln|u| + C$

$\displaystyle\qquad = \ln\left|2x^2-5x\right| + C$

20. $\displaystyle\int \frac{8(2x+9)}{x^2+9x+1}dx$

Let $u = x^2 + 9x + 1$. Then $du = (2x+9)\,dx$.

$\displaystyle\int \frac{8(2x+9)}{x^2+9x+1}dx = 8\int \frac{du}{u} = 8\ln|u| + C$

$\displaystyle\qquad = 8\ln\left|x^2+9x+1\right| + C$

21. $\displaystyle\int \frac{x^2}{e^{3x^3}}dx = \int x^2 e^{-3x^3} = -\frac{1}{9}\int -9x^2 e^{-3x^3}dx$

Let $u = -3x^3$. Then $du = -9x^2 dx$.

$\displaystyle\int \frac{x^2}{e^{3x^3}}dx = -\frac{1}{9}\int -9x^2 e^{-3x^3}dx = -\frac{1}{9}\int e^u du$

$\displaystyle\qquad = -\frac{1}{9}e^u + C = -\frac{1}{9}e^{-3x^3} + C$

22. $\displaystyle\int 2e^{3x^2+4}x\,dx$

Let $u = 3x^2 + 4$. Then $du = 6x\,dx$.

$\displaystyle 2\int e^{3x^2+4}x\,dx = 2\cdot\frac{1}{6}\int e^{3x^2+4}\cdot 6x\,dx = \frac{1}{3}\int e^u du$

$\displaystyle\qquad = \frac{1}{3}e^u + C = \frac{1}{3}e^{3x^2+4} + C$

$\displaystyle\qquad = \frac{e^{3x^2+4}}{3} + C$

23. $\displaystyle\int -2e^{-5x}dx = \frac{-2}{-5}\int -5e^{-5x}dx$

Let $u = -5x$. Then $du = -5\,dx$.

$\displaystyle\int -2e^{-5x}dx = \frac{2}{5}\int e^u du = \frac{2}{5}e^u + C$

$\displaystyle\qquad = \frac{2e^{-5x}}{5} + C$

24. $\displaystyle\int 11e^{-4x}dx = \frac{11e^{-4x}}{-4} + C = -\frac{11e^{-4x}}{4} + C$

25. $\displaystyle\int \frac{3(\ln x)^5}{x}dx = 3\int \frac{(\ln x)^5}{x}dx$

Let $u = \ln x$. Then $du = \frac{1}{x}dx$.

$\displaystyle\int \frac{3(\ln x)^5}{x}dx = 3\int u^5 du = 3\left(\frac{u^6}{6}\right) + C$

$\displaystyle\qquad = \frac{u^6}{2} + C = \frac{(\ln x)^6}{2} + C$

26. $\displaystyle\int \frac{10\left(\ln(5x+3)\right)^2}{5x+3}dx = 10\int \frac{\left(\ln(5x+3)\right)^2}{5x+2}dx$

Let $u = \ln(5x+3)$. Then $du = \left(\frac{1}{5x+3}\cdot 5\right)dx$.

$\displaystyle\int \frac{10\left(\ln(5x+3)\right)^2}{5x+3}dx = 10\cdot\frac{1}{5}\int u^2 du = 2\frac{u^3}{3} + C$

$\displaystyle\qquad = \frac{2\left[\ln(5x+3)\right]^3}{3} + C$

27. $\displaystyle\int 25e^{-50x}dx = 25\int e^{-50x}dx$

Let $u = -50x$. Then $du = -50dx$.

$\displaystyle\int 25e^{-50x}dx = 25\left(-\frac{1}{50}\right)\int e^u du = -\frac{1}{2}e^u + C$

$\displaystyle\qquad = -\frac{1}{2}e^{-50x} + C$

28. $\int 4xe^{-3x^2+7}\,dx = 4\int xe^{-3x^2+7}\,dx$

Let $u = -3x^2 + 7$. Then $du = -6x\,dx$.

$\int 4xe^{-3x^2+7}\,dx = 4\left(-\dfrac{1}{6}\right)\int e^u\,du$

$= -\dfrac{2}{3}e^u + C = -\dfrac{2}{3}e^{-3x^2+7} + C$

29. Answers vary. Possible answer: By dividing the area under the curve into rectangles and then finding the sum of the areas of the rectangles.

30. a. $f(x) = 16x^2 - x^4 + 2$ from $x = -2$ to $x = 3$

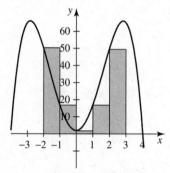

Area $= f(-2) + f(-1) + f(0) + f(1) + f(2)$

$= 50 + 17 + 2 + 17 + 50 = 136$

b.

The area is approximately 141.667.

31. a. $g(x) = -x^4 + 12x^2 + x + 5$ from $x = -3$ to $x = 3$

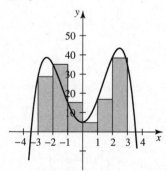

Area $= f(-3) + f(-2) + f(-1)$
    $+ f(0) + f(1) + f(2)$
$= 29 + 35 + 15 + 5 + 17 + 39 = 140$

b.
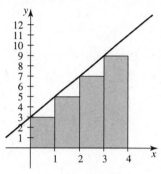

The area is approximately 148.8.

32. $f(x) = 2x + 3$ from $x = 0$ to $x = 4$

$\Delta x = \dfrac{4-0}{4} = 1$

i	x_i	$f(x_i)$
1	0	3
2	1	5
3	2	7
4	3	9

$\displaystyle\sum_{i=1}^{4} f(x_i)\Delta x = 3(1) + 5(1) + 7(1) + 9(1) = 24$

33. $\displaystyle\int_0^4 (2x+3)\,dx$

$A = \dfrac{1}{2}(B+b)h$

$h = 4 - 0 = 4$

$B = f(4) = 2(4) + 3 = 11$

$b = f(0) = 2(0) + 3 = 3$

$\displaystyle\int_0^4 (2x+3) = \dfrac{1}{2}(11+3)(4) = 28$

The area formula gives the exact answer, 28. The answer obtained in Exercise 32, which is 24, is an approximation of the area. Because it uses left endpoints and the line $y = 2x + 3$ has a positive slope, the area obtained using the four rectangles will be less than the exact area.

34. Answers vary. Possible answer: Substitution is useful in integration with complicated functions when some expression in the integral can be replaced by u along with the derivative of u, du.

35. $\int_0^4 \left(x^3 - x^2\right) dx = \left(\frac{x^4}{4} - \frac{x^3}{3}\right)\Big|_0^4$

$= \left(\frac{256}{4} - \frac{64}{3}\right) - 0 = \frac{128}{3}$

36. $\int_0^1 e^{3t} dt = \frac{1}{3} e^{3t}\Big|_0^1 = \frac{1}{3} e^{3(1)} - \frac{1}{3} e^{3(0)}$

$= \frac{1}{3}(e^3 - 1) \approx 6.362$

37. $\int_2^5 \left(6x^{-2} + x^{-3}\right) dx = \left(\frac{6x^{-1}}{-1} + \frac{x^{-2}}{-2}\right)\Big|_2^5$

$= \left(-\frac{6}{x} - \frac{1}{2x^2}\right)\Big|_2^5$

$= \left(-\frac{6}{5} - \frac{1}{50}\right) - \left(-3 - \frac{1}{8}\right)$

$= \frac{381}{200}$

38. $\int_2^3 \left(5x^{-2} + 7x^{-4}\right) dx$

$= \left(\frac{5x^{-1}}{-1} + \frac{7x^{-3}}{-3}\right)\Big|_2^3 = \left(\frac{-5}{x} - \frac{7}{3x^3}\right)\Big|_2^3$

$= \left(\frac{-5}{3} - \frac{7}{81}\right) - \left(\frac{-5}{2} - \frac{7}{24}\right) = \frac{673}{648} \approx 1.039$

39. $\int_1^5 15x^{-1} dx = 15\int_1^5 \frac{dx}{x} = 15\ln|x|\Big\|_1^5$

$= 15\ln 5 - 15\ln 1 = 15\ln 5 \approx 24.1416$

40. $\int_1^6 \frac{8x^{-1}}{3} dx = \frac{8}{3}\int_1^6 \frac{1}{x} dx = \frac{8}{3}(\ln x)\Big|_1^6$

$= \frac{8}{3}(\ln 6 - \ln 1) = \frac{8}{3}\ln 6 \approx 4.778$

41. $\int_0^4 2e^{-5x} dx = \frac{2e^{-5x}}{-5}\Big|_0^4 = -\frac{2}{5} e^{-20} - \frac{2}{5} e^0$

$= -\frac{2}{5} e^{-20} - \frac{2}{5} \approx .40$

42. $\int_1^2 \frac{7}{2} e^{4x} dx = \frac{7}{2} \cdot \frac{e^{4x}}{4}\Big|_1^2 = \frac{7e^{4x}}{8}\Big|_1^2$

$= \frac{7\left(e^8 - e^4\right)}{8} \approx 2560.5649$

43. $\int_{\sqrt{5}}^5 2x\sqrt{x^2 - 3}\, dx$

Let $u = x^2 - 3$. Then $du = 2x\, dx$.

If $x = 5$, $u = 22$. If $x = \sqrt{5}$, $u = 2$.

$\int_{\sqrt{5}}^5 2x\sqrt{x^2 - 3}\, dx = \int_{\sqrt{5}}^5 \sqrt{x^2 - 3} \cdot 2x\, dx$

$= \int_2^{22} u^{1/2}\, du = \frac{u^{3/2}}{\frac{3}{2}}\Big|_2^{22}$

$= \frac{2}{3}\left(22^{3/2} - 2^{3/2}\right) \approx 66.907$

44. $\int_0^1 x\sqrt{5x^2 + 4}\, dx$

Let $u = 5x^2 + 4$. Then $du = 10x\, dx$.

If $x = 1$, $u = 9$. If $x = 0$, $u = 4$.

$\int_0^1 x\sqrt{5x^2 + 4}\, dx = \frac{1}{10}\int_0^1 \left(5x^2 + 4\right)^{1/2} \cdot 10x\, dx$

$= \frac{1}{10}\int_4^9 u^{1/2}\, du = \frac{1}{10} \cdot \frac{u^{3/2}}{\frac{3}{2}}\Big|_4^9$

$= \frac{2}{30}\left(9^{3/2} - 4^{3/2}\right) = \frac{19}{15}$

45. $\int_1^6 \frac{8x + 12}{x^2 + 3x + 9}\, dx = 4\int_1^6 \frac{2x + 3}{x^2 + 3x + 9}\, dx$

Let $u = x^2 + 3x + 9$. Then $du = (2x + 3)\, dx$.

If $x = 6$, then $u = 63$. If $x = 1$, then $u = 13$.

$4\int_1^6 \frac{2x + 3}{x^2 + 3x + 9}\, dx = 4\int_{13}^{63} \frac{1}{u}\, du = 4\ln|u|\Big\|_{13}^{63}$

$= 4(\ln 63 - \ln 13)$

$= 4\ln \frac{63}{13} \approx 6.313$

46. $\int_1^4 \frac{3(\ln x)^5}{x}\, dx = 3\int_1^4 \frac{(\ln x)^5}{x}\, dx$

Let $u = \ln x$. Then $du = \frac{1}{x}\, dx$.

If $x = 4$, the $u = \ln 4$. If $x = 1$, then $u = 0$.

$\int_1^4 \frac{3(\ln x)^5}{x}\, dx = 3\int_0^{\ln 4} u^5\, du = \frac{3u^6}{6}\Big|_0^{\ln 4}$

$= \frac{1}{2}\left[(\ln 4)^6 - 0\right]$

$= \frac{(\ln 4)^6}{2} \approx 3.549$

47. $f(x) = e^x$; $[0, 2]$

$$A = \int_0^2 e^x dx = e^x \Big|_0^2 = e^2 - e^0 = e^2 - 1 \approx 6.3891$$

48. $f(x) = 1 + e^{-x}$; $[0, 4]$

$$A = \int_0^4 \left(1 + e^{-x}\right) dx = \left(x + \frac{1}{-1}e^{-x}\right)\Big|_0^4$$

$$= \left(4 - e^{-4}\right) - (0 - 1) = 5 - e^{-4} \approx 4.982$$

49. $C'(x) = 10 - 5x$; fixed cost is $4. The fixed cost tells us that $C(0) = 4$.

$$C(x) = \int (10 - 5x)dx$$

$$C(x) = 10x - \frac{5x^2}{2} + k$$

$$4 = 10(0) - \frac{5}{2}(0)^2 + k \Rightarrow k = 4$$

Thus, the cost function is

$$C(x) = 10x - \frac{5x^2}{2} + 4 .$$

50. $C'(x) = 2x + 3x^2$; 2 units cost $12. Since, 2 units cost $12, we have $C(2) = 12$.

$$C(x) = \int \left(2x + 3x^2\right)dx$$

$$C(x) = x^2 + x^3 + k \Rightarrow 12 = 2^2 + 2^3 + k \Rightarrow$$

$$12 = 4 + 8 + k \Rightarrow k = 0$$

Thus, the cost function is $C(x) = x^2 + x^3$.

51. $C'(x) = 3\sqrt{2x - 1}$; 13 units cost $270. Since 13 units cost $270, we have $C(13) = 270$.

$$C(x) = \int 3\sqrt{2x - 1}dx$$

Let $u = 2x - 1$. Then $du = 2\,dx$.

$$C(x) = 3 \cdot \frac{1}{2}\int \sqrt{2x - 1} \cdot 2\,dx = \frac{3}{2}\int u^{1/2}du \Rightarrow$$

$$C(x) = \frac{3}{2} \cdot \frac{u^{3/2}}{\frac{3}{2}} + k \Rightarrow C(x) = u^{3/2} + k \Rightarrow$$

$$C(x) = (2x - 1)^{3/2} + k \Rightarrow 270 = 25^{3/2} + k \Rightarrow$$

$$k = 145$$

Thus, the cost function is

$$C(x) = (2x - 1)^{3/2} + 145 .$$

52. $C'(x) = \dfrac{6}{x + 1}$; fixed cost is $18. Since the fixed cost is $18, $C(0) = 18$.

$$C(x) = \int \frac{6}{x + 1}dx = 6\ln|x + 1| + k \Rightarrow$$

$$18 = 6\ln 1 + k \Rightarrow k = 18$$

Thus, the cost function is $C(x) = 6\ln|x + 1| + 18$.

53. $S' = \sqrt{x} + 3$

$$S(x) = \int_0^9 \left(x^{1/2} + 3\right)dx = \left(\frac{x^{3/2}}{\frac{3}{2}} + 3x\right)\Big|_0^9$$

$$= \frac{2}{3}(9)^{3/2} + 27 = 45$$

Total sales are 45(1000), or 45,000 units.

54.

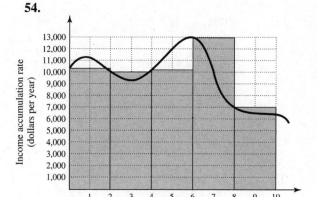

Left Endpoint	Height of rectangle (function values)
0	10,500
2	10,000
4	10,200
6	13,000
8	7000

The total area is approximately
10,500(2) + 10,000(2) + 10,200(2)
  + 13,000(2) + 7000(2) = 101,400.
The total income is approximately $101,400.

55. $\int_0^T 100,000e^{.03t}\,dt = 4,000,000$

$$100,000 \cdot \frac{1}{.03}e^{.03t}\Big|_0^T = 4,000,000$$

$$\frac{1}{.03}\left(e^{.03T}-1\right) = 40$$

$$e^{.03T}-1 = 1.2$$

$$e^{.03T} = 2.2$$

$$.03T = \ln 2.2$$

$$T = \frac{\ln 2.2}{.03} \approx 26.28$$

The supply will be used up in about 26.3 years.

56. $\int_0^{10}\left(150 - \sqrt{3.2t+4}\right)dt$

$$= \int_0^{10} 150\,dt - \int_0^{10}\sqrt{3.2t+4}\,dt$$

Let $u = 3.2t + 4$ and $du = 3.2\,dt$ or $\dfrac{1}{3.2}du = dt$.

$$\int_0^{10} 150\,dt - \int_0^{10}\sqrt{3.2t+4}\,dt$$

$$= 150t\Big|_0^{10} - \int_0^{10}\sqrt{u}\,\frac{1}{3.2}\,du$$

$$= 1500 - \frac{1}{3.2}\frac{u^{1.5}}{1.5}$$

$$\approx 1500 - .2083(3.2t+4)^{1.5}\Big|_0^{10}$$

$$\approx 1500$$

$$\quad -\left(.2083(3.2\cdot 10 + 4)^{1.5} - .2083(3.2\cdot 0 + 4)^{1.5}\right)$$

$$\approx 1456.66$$

In the first 10 months there were about 1457 spiders.

57. $f(t) = .222t^3 + 2.179t^2 - 27.23t + 478.1$

Since we are seeking the number of births from the beginning of 2000 to the end of 2005, we use the interval [0, 6).

$$\int_0^6\left(.222t^3 + 2.179t^2 - 27.23t + 478.1\right)dt$$

$$= \left[\begin{array}{l}.222\left(\dfrac{t^4}{4}\right) + 2.179\left(\dfrac{t^3}{3}\right) \\[2mm] \quad - 27.23\left(\dfrac{t^2}{2}\right) + 478.1t\end{array}\right]\Bigg|_0^6$$

$$= 2607.28 - 0 = 2607.28$$

The total number of births from the beginning of 2000 to the end of 2005 was about 2,607,280.

58. $g(t) = 123.88e^{.0511t}$

$$\int_0^6 123.88e^{.0511t} = \frac{123.88}{.0511}e^{.0511t}\Big|_0^6$$

$$\approx 3294.008 - 2424.27$$

$$= 869.82$$

From 2000 to 2005 about \$869.82 billion was paid out in life insurance premiums.

59. $h(t) = -.0471t + 2.6391$

$$\int_5^{19} -.0471t + 2.6391 = -\frac{.0471t^2}{2} + 2.6391t\Big|_5^{19}$$

$$= 41.64135 - 12.60675$$

$$= 29.0346$$

Total oil production from the beginning of 1995 to the end of 2008 was about 29.0346 billion barrels.

60. $S'(x) = 225 - x^2$

$$C'(x) = x^2 + 25x + 150$$

$$S'(x) = C'(x) \Rightarrow 225 - x^2 = x^2 + 25x + 150 \Rightarrow$$

$$2x^2 + 25x - 75 = 0 \Rightarrow (2x-5)(x+15) = 0 \Rightarrow$$

$$x = \frac{5}{2}, -15$$

Discard the negative solution. The company should use the machinery for 2.5 years.

$$\int_0^{2.5}\left[\left(225-x^2\right) - \left(x^2 + 25x + 150\right)\right]dx$$

$$= \int_0^{2.5}\left(-2x^2 - 25x + 75\right)dx$$

$$= \left(\frac{-2x^3}{3} - \frac{25x^2}{2} + 75x\right)\Bigg|_0^{2.5}$$

$$= \frac{-2(2.5)^3}{3} - \frac{25(2.5)^2}{2} + 75(2.5)$$

$$\approx 98.95833$$

The net savings are 98,958.33, or about \$99,000.

61. Answers vary. Possible answer: Consumers' surplus is the total of all differences between the equilibrium price on an item and the higher prices individuals will be willing to pay and is thought of as savings realized by those individuals. Producers' surplus is the total of all differences between the equilibrium price and the lower prices at which the manufacturers would sell the product and is considered added income for the manufacturers.

62. $S(q) = q^2 + 5q + 100$

$D(q) = 350 - q^2$

$S(q) = D(q)$

$q^2 + 5q + 100 = 350 - q^2$

$2q^2 + 5q - 250 = 0$

$(2q + 25)(q - 10) = 0$

$q = \dfrac{-25}{2}$ or $q = 10$

Discard the negative solution.

$S(10) = 10^2 + 5(10) + 100 = 250$

The equilibrium point is (10, 250).

a. The producers' surplus is given by

$\displaystyle\int_0^{10}\left[250 - \left(q^2 + 5q + 100\right)\right]dq$

$= \displaystyle\int_0^{10}\left(150 - q^2 - 5q\right)dq$

$= \left(150q - \dfrac{q^3}{3} - \dfrac{5q^2}{2}\right)\Bigg|_0^{10} \approx 916.67$

The producers' surplus is $916.67.

b. The consumers' surplus is given by

$\displaystyle\int_0^{10}\left[\left(350 - q^2\right) - 250\right]dq$

$= \displaystyle\int_0^{10}\left(100 - q^2\right)dq = \left(100q - \dfrac{q^3}{3}\right)\Bigg|_0^{10}$

$= \left[100(10) - \dfrac{10^3}{3}\right] - 0 \approx 666.67$

The consumers' surplus is $666.67.

63. $I'(t) = \dfrac{100t}{t^2 + 2}$

The total number of infected people is given by

$\displaystyle\int_0^4 \dfrac{100t}{t^2 + 2}\,dt$.

Let $u = t^2 + 2$. Then $du = 2t\,dt$. If $t = 4$, $u = 18$.
If $t = 0$, $u = 2$.

$\displaystyle\int_0^4 \dfrac{100t}{t^2 + 2}\,dt = 50\int_0^4 \dfrac{2t\,dt}{t^2 + 2} = 50\int_2^{18} \dfrac{du}{u}$

$= 50 \ln|u|\Big\|_2^{18} = 50\ln 18 - 50\ln 2$

$= 50\ln\left(\dfrac{18}{2}\right) \approx 109.86$

Approximately 110 people will be infected.

64. a. $\displaystyle\int_0^2 \left(\dfrac{1}{2}e^{-x/2}\right)dx = -e^{-x/2}\Big|_0^2 = -e^{-1} - \left(-e^0\right)$

$\approx .6321$

The probability that a customer in line waits less than 2 minutes is about .6321.

b. $1 - \displaystyle\int_0^3 \left(\dfrac{1}{2}e^{-x/2}\right)dx = 1 - \left[-e^{-x/2}\Big|_0^3\right]$

$= 1 - \left[-e^{-3/2} - \left(-e^0\right)\right]$

$\approx .2231$

The probability that a customer in line waits more than 3 minutes is about .2231.

65. a. $\displaystyle\int_1^2 .44e^{-.44t}\,dt = -e^{-.44t}\Big|_1^2 = -e^{-.88} - \left(-e^{-.44}\right)$

$\approx .2293$

The probability that the person will take between 1 and 2 seconds between blinks is about .2293.

b. $1 - \displaystyle\int_0^4 .44e^{-.44t}\,dt = 1 - \left[-e^{-.44t}\Big|_0^4\right]$

$= 1 - \left[-e^{-1.76} - \left(-e^0\right)\right]$

$\approx .1720$

The probability that the person will take more than 4 seconds between blinks is about .1720.

66. $\displaystyle\int \dfrac{1}{\sqrt{x^2 - 64}}\,dx$

Use entry 6 from the table of integrals with $a = 8$.

$\displaystyle\int \dfrac{1}{\sqrt{x^2 - 64}}\,dx = \ln\left|x + \sqrt{x^2 - 64}\right| + C$

67. $\displaystyle\int \dfrac{10}{x\sqrt{25 + x^2}}\,dx = 10\int \dfrac{1}{x\sqrt{25 + x^2}}\,dx$

Use entry 10 in the table with $a = 5$.

$10\displaystyle\int \dfrac{1}{x\sqrt{25 + x^2}}\,dx$

$= -10\left[\dfrac{1}{5}\ln\left|\dfrac{5 + \sqrt{5^2 + x^2}}{x}\right|\right] + C$

$= -2\ln\left|\dfrac{5 + \sqrt{25 + x^2}}{x}\right| + C$

68. $\int \dfrac{18}{x^2-9}dx = 18\int \dfrac{dx}{x^2-9}$

Use entry 8 from the table with $a = 3$.

$18\int \dfrac{dx}{x^2-9} = 18\left(\dfrac{1}{6}\ln\left|\dfrac{x-3}{x+3}\right|\right)+C$

$\qquad = 3\ln\left|\dfrac{x-3}{x+3}\right|+C \ \left(x^2 > 9\right)$

69. $\int \dfrac{15x}{2x-5}dx = 15\int \dfrac{x}{2x-5}dx$

This matches entry 11 in the table with $a = 2$ and $b = -5$.

$15\int \dfrac{x}{2x-5}dx = 15\left[\dfrac{x}{2}-\dfrac{-5}{2^2}\ln|2x-5|\right]+C$

$\qquad = 15\left[\dfrac{x}{2}+\dfrac{5\ln|2x-5|}{4}\right]+C$

$\qquad$ or $\dfrac{15x}{2}+\dfrac{75}{4}\ln|2x-5|+C$

70. Answers vary.

71. $\dfrac{dy}{dx} = 2x^3+6x+5$

$dy = \left(2x^3+6x+5\right)dx$

$\int dy = \int\left(2x^3+6x+5\right)dx$

$y = \dfrac{2x^4}{4}+\dfrac{6x^2}{2}+5x+C$

$y = \dfrac{x^4}{2}+3x^2+5x+C$

72. $\dfrac{dy}{dx} = x^2+\dfrac{5x^4}{8}$

$y = \int\left(x^2+\dfrac{5x^4}{8}\right)dx = \dfrac{x^3}{3}+\dfrac{x^5}{8}+C$

73. $\dfrac{dy}{dx} = \dfrac{3x+1}{y}$

$y\,dy = (3x+1)dx$

$\int y\,dy = \int(3x+1)dx$

$\dfrac{y^2}{2} = \dfrac{3x^2}{2}+x+C$

$y^2 = 3x^2+2x+C$

74. $\dfrac{dy}{dx} = \dfrac{e^x+x}{y-1}$

$(y-1)dy = \left(e^x+x\right)dx$

$\dfrac{y^2}{2}-y = e^x+\dfrac{x^2}{2}+C$

75. $\dfrac{dy}{dx} = 5\left(e^{-x}-1\right)$; $y = 17$ when $x = 0$

$dy = 5\left(e^{-x}-1\right)dx \Rightarrow \int dy = \int 5\left(e^{-x}-1\right)dx \Rightarrow$

$y = 5\left[\dfrac{e^{-x}}{-1}-x\right]+C$

The general solution is $y = -5e^{-x}-5x+C$.

Let $x = 0$ and $y = 17$.

$17 = -5e^0-5(0)+C \Rightarrow 17 = -5+C \Rightarrow C = 22$

The particular solution is $y = -5e^{-x}-5x+22$.

76. $\dfrac{dy}{dx} = \dfrac{x}{x^2-3}+7$; $y = 52$ when $x = 2$

$y = \int\left(\dfrac{x}{x^2-3}+7\right)dx = \dfrac{1}{2}\int\left(\dfrac{2x}{x^2-3}+7\right)dx$

The general solution is

$y = \dfrac{1}{2}\ln\left|x^2-3\right|+7x+C$.

Let $x = 2$ and $y = 52$.

$52 = \dfrac{1}{2}\ln|4-3|+14+C \Rightarrow 38 = \dfrac{1}{2}\ln 1+C \Rightarrow$

$C = 38$

The particular solution is

$y = \dfrac{1}{2}\ln\left|x^2-3\right|+7x+38$.

77. $(5-2x)y = \dfrac{dy}{dx}$; $y = 2$ when $x = 0$

$(5-2x)dx = \dfrac{dy}{y} \Rightarrow \int(5-2x)dx = \int\dfrac{dy}{y}$

The general solution is $5x-x^2 = \ln|y|+C$.

Let $x = 0$ and $y = 2$.

$0 = \ln 2+C \Rightarrow C = -\ln 2 \Rightarrow$

$5x-x^2 = \ln|y|-\ln 2 = \ln\left|\dfrac{y}{2}\right| \Rightarrow$

$e^{5x-x^2} = \left|\dfrac{y}{2}\right| \Rightarrow 2e^{5x-x^2} = |y|$

The particular solution is $y = 2e^{5x-x^2}$.

78. $\sqrt{x}\dfrac{dy}{dx} = xy$; $y = 4$ when $x = 1$

Write $\sqrt{x}$ as $x^{1/2}$ and separate the variables.

$y = \dfrac{x^{1/2}}{x}\dfrac{dy}{dx} \Rightarrow \dfrac{1}{y}dy = x^{1/2}dx \Rightarrow$

$\displaystyle\int\dfrac{1}{y}dy = \int x^{1/2}dx \Rightarrow \ln|y| = \dfrac{2}{3}x^{3/2} + C$

Let $x = 1$ and $y = 4$.

$\ln 4 = \dfrac{2}{3}(1)^{3/2} + C \Rightarrow C = \ln 4 - \dfrac{2}{3} \approx .7196$

$\ln|y| \approx \dfrac{2}{3}x^{3/2} + .7196 \Rightarrow |y| \approx e^{.7196}e^{\frac{2x^{3/2}}{3}} \Rightarrow$

$y \approx \pm 2.054e^{\frac{2x^{3/2}}{3}}$

Since $y = 4$ when $x = 1$, we must use the positive

solution: $y = 2.054e^{\frac{2x^{3/2}}{3}}$

79. a. $\dfrac{dy}{dt} = .0264y$

b. $\dfrac{dy}{dt} = .0264y \Rightarrow \dfrac{dy}{y} = .0264dt \Rightarrow$

$\displaystyle\int\dfrac{dy}{y} = \int .0264dt \Rightarrow \ln|y| = .0264t + C \Rightarrow$

$y = e^{.0264t+C} = e^{.0264t}e^C \Rightarrow y = Me^{.0264t}$

Since $y = 172.3$ when $t = 0$, we have

$y = 172.3e^{.0264t}$.

c. $y = 172.3e^{.0264(7)} \approx 207.27$

In 2007, the CPI was about 207.27.

80. a. $\dfrac{dy}{dx} = 2.4e^{.1x} \Rightarrow dy = 2.4e^{.1x}dx \Rightarrow$

$\displaystyle\int dy = \int 2.4e^{.1x}dx \Rightarrow$

$y = \dfrac{2.4}{.1}e^{.1x} + C = 24e^{.1x} + C$

When $x = 0$, $y = 0$.

$0 = 24e^{.1(0)} + C \Rightarrow C \approx -24$

When $x = 8$, $y = 24e^{.1(8)} - 24 \approx 29.4$.

When 800 books are sold, profits are about
$2940.

b. When $x = 12$, $y = 24e^{.1(12)} - 24 \approx 55.68$.

When 1200 books are sold, profits are about
$5568.

81. $\dfrac{dy}{dx} = .2(125 - y) \Rightarrow \dfrac{dy}{125 - y} = .2\,dx \Rightarrow$

$\displaystyle\int\dfrac{dy}{125 - y} = \int .2\,dx \Rightarrow$

$-\ln|125 - y| = .2x + M \Rightarrow$

$\ln|125 - y| = -.2x + M \Rightarrow$

$|125 - y| = e^{-.2x+C} \Rightarrow |125 - y| = e^{-.2x}\cdot e^C \Rightarrow$

$|125 - y| = Me^{-.2x} \Rightarrow 125 - y = \pm Me^{-.2x} \Rightarrow$

$125 \pm Me^{-.2x} = y \Rightarrow y = 125 + Me^{-.2x}$

When $x = 0$, $y = 20$.

$20 = 125 + Me^0 \Rightarrow M = -105$

$y = 125 - 105e^{-.2x}$

a. Find the value of y when $x = 10$.

$y = 125 - 105e^{-2} \Rightarrow y \approx 110.79$

In 10 days the worker will produce about
111 items.

b. Theoretically, $125 - 105e^{-.2x}$ will never
equal 125, but, for practical purposes, for
large values of x it will approximately equal
125, so the worker can produce
approximately 125 items a day.

82. $\dfrac{dT}{dt} = k(T - T_F) \Rightarrow \dfrac{dT}{T - T_F} = k\,dt \Rightarrow$

$\displaystyle\int\dfrac{dT}{T - T_F} = \int k\,dt \Rightarrow \ln|T - T_F| = kt + C \Rightarrow$

$|T - T_F| = e^{kt+C} \Rightarrow |T - T_F| = e^{kt}\cdot e^C \Rightarrow$

$T - T_F = \pm e^C e^{kt} \Rightarrow T = T_F + Me^{kt}$

$T_F = 300°$, $T(0) = 40°$, and $T(1) = 150°$

$T = 300 + Me^{kt} \Rightarrow 40 = 300 + Me^0 \Rightarrow$

$M = -260$

$T = 300 - 260e^{kt} \Rightarrow 150 = 300 - 260e^{k\cdot 1} \Rightarrow$

$-150 = -260e^k \Rightarrow e^k = \dfrac{150}{260} \Rightarrow$

$k = \ln\left(\dfrac{150}{260}\right) \approx -.550 \Rightarrow T = 300 - 260e^{-.550t}$

a. After 2.5 hours, the temperature is

$T = 300 - 260e^{-.550(2.5)} \approx 234.26$

The temperature is approximately 234°.

b. If $T = 250$,

$$250 = 300 - 260e^{-.550t} \Rightarrow$$

$$-50 = -260e^{-.550t} \Rightarrow e^{-.550t} = \frac{50}{260} \Rightarrow$$

$$-.550t = \ln\left(\frac{5}{26}\right) \Rightarrow t = \frac{\ln\left(\frac{5}{26}\right)}{-.550} \approx 2.998$$

The temperature will be 250° in approximately 3 hours.

83. First note that $\dfrac{dy}{dx}$ is the ratio of $\dfrac{dy}{dt}$ to $\dfrac{dx}{dt}$.

Then the variables may be separated and an integration may be performed.

$$\frac{dy}{dx} = \frac{\frac{dy}{dt}}{\frac{dx}{dt}} = \frac{-.3y + .4xy}{.2x - .5xy}$$

$$\frac{dy}{dx} = \frac{(-.3 + .4x)y}{(.2 - .5y)x}$$

$$(.2 - .5y)\frac{dy}{y} = (-.3 + .4x)\frac{dx}{x}$$

$$\int \frac{.2 - .5y}{y}\,dy = \int \frac{-.3 + .4x}{x}\,dx$$

$$\int\left(\frac{.2}{y} - .5\right)dy = \int\left(-\frac{.3}{x} + .4\right)dx$$

$$.2\int\frac{dy}{y} - .5\int dy = -.3\int\frac{dx}{x} + .4\int dx$$

$$.2\ln|y| - .5y = -.3\ln|x| + .4x + C$$

This is an equation that relates x to y (we can assume $x > 0$ and $y > 0$). Both growth rates being zero means that $\dfrac{dx}{dt} = 0$ and $\dfrac{dy}{dt} = 0$. The system would then become

$$0 = .2x - .5xy$$

$$0 = -.3y + .4xy$$

Note that $x = 0$ and $y = 0$ is a solution to this system of equations, but this solution would mean that both species were absent. Suppose neither is 0 and see if there are other solutions. The first equation can be written as

$$.5xy = .2x \Rightarrow y = \frac{.2x}{.5x} = \frac{2}{5} \ (x \neq 0)$$

The second equation can be written as

$$.3y = .4xy \Rightarrow x = \frac{.3y}{.4y} = \frac{3}{4} \ (y \neq 0)$$

So both growth rates are 0 when $x = \dfrac{3}{4}$ unit and

$$y = \frac{2}{5} \text{ unit.}$$

Case 13 Bounded Population Growth

1. Verify $\dfrac{1}{P(C - P)} = \dfrac{\frac{1}{C}}{P} + \dfrac{\frac{1}{C}}{C - P}$

$$\frac{\frac{1}{C}}{P} + \frac{\frac{1}{C}}{C - P} = \frac{\left(\frac{1}{C}\right)(C - P)}{P(C - P)} + \frac{\left(\frac{1}{C}\right)P}{P(C - P)}$$

$$= \frac{\left(\frac{1}{C}\right)(C - P) + \left(\frac{1}{C}\right)P}{P(C - P)}$$

$$= \frac{\left(\frac{1}{C}\right)C - \left(\frac{1}{C}\right)P + \left(\frac{1}{C}\right)P}{P(C - P)}$$

$$= \frac{1 - 0}{P(C - P)} = \frac{1}{P(C - P)}$$

2. $P(0) = 10$

$$\frac{1}{200}\ln(P) - \frac{1}{200}\ln(200 - P) = .005t + K \Rightarrow$$

$$\ln P - \ln(200 - P) = t + 200K \Rightarrow$$

$$e^{\ln(P) - \ln(200 - P)} = e^{t + 200K} \Rightarrow$$

$$\frac{P}{200 - P} = e^{200K}e^{t} \Rightarrow$$

$$P = 200e^{200K}e^{t} - Pe^{200K}e^{t} \Rightarrow$$

$$P + P(e^{200K}e^{t}) = 200e^{200K}e^{t} \Rightarrow$$

$$P = \frac{200e^{200K}e^{t}}{1 + e^{200K}e^{t}} \Rightarrow P = \frac{200e^{t}}{e^{-200K} + e^{t}}$$

Let $A = e^{-200K}$. Then $P(t) = \dfrac{200e^{t}}{A + e^{t}}$

Since $P(0) = \dfrac{200}{A + 1}$ must equal 10, it follows that

$A = 19$ and $P(t) = \dfrac{200e^{t}}{e^{t} + 19}$.

3. $P(0) = 6$, $C = 20$, $k = .0011$

$$\frac{dP}{dt} = .0011P(20 - P)$$

$$\int \frac{1}{P(20 - P)}\, dP = \int .0011\, dt \Rightarrow$$

$$\frac{1}{20}\ln(P) - \frac{1}{20}\ln(20 - P) = .0011t + K$$

$$\ln(P) - \ln(20 - P) = .022t + 20K$$

Performing similar calculations as in Exercise 2,

$$P = \frac{20e^{.022t}}{e^{-20K} + e^{.022t}}$$

$$P(0) = \frac{20}{e^{-20K} + 1} \Rightarrow 6 = \frac{20}{e^{-20K} + 1} \Rightarrow$$

$$e^{-20K} + 1 = \frac{10}{3} \Rightarrow e^{-20K} = \frac{7}{3} \Rightarrow$$

$$-20K = \ln\frac{7}{3} \Rightarrow K = \frac{\ln\frac{7}{3}}{-20}$$

Rewrite P with the value of K.

$$P = \frac{20e^{.022t}}{\frac{7}{3} + e^{.022t}} \Rightarrow P(t) = \frac{60e^{.022t}}{7 + 3e^{.022t}}$$

Now find t when $P(t) = 18$.

$$18 = \frac{60e^{.022t}}{7 + 3e^{.022t}} \Rightarrow 18(7 + 3e^{.022t}) = 60e^{.022t} \Rightarrow$$

$$126 + 54e^{.022t} = 60e^{.022t} \Rightarrow 126 = 6e^{.022t} \Rightarrow$$

$$21 = e^{.022t} \Rightarrow \ln 21 = .022t \Rightarrow$$

$$t = \frac{\ln 21}{.022} \approx 138.387$$

The population will reach 18 billion after about 138 years.

4. No, $P(t)$ will approach but never reach 20 billion. This model was chosen because it agreed with the expected behavior of a bounded population. The population was bounded by assigning the carrying capacity as 20 billion.

Chapter 14 Multivariate Calculus

Section 14.1 Functions of Several Variables

1. $f(x, y) = 5x + 2y - 4$

$f(2, -1) = 5(2) + 2(-1) - 4$
$= 10 - 2 - 4 = 4$

$f(-4, 1) = 5(-4) + 2(1) - 4$
$= -20 + 2 - 4 = -22$

$f(-2, -3) = 5(-2) + 2(-3) - 4$
$= -10 - 6 - 4 = -20$

$f(0, 8) = 5(0) + 2(8) - 4$
$= 0 + 16 - 4 = 12$

3. $f(x, y) = \sqrt{y^2 + 2x^2}$

$f(2, -1) = \sqrt{(-1)^2 + 2(2)^2} = \sqrt{1 + 8}$
$= \sqrt{9} = 3$

$f(-4, 1) = \sqrt{(1)^2 + 2(-4)^2} = \sqrt{1 + 32}$
$= \sqrt{33}$

$f(-2, -3) = \sqrt{(-3)^2 + 2(-2)^2} = \sqrt{9 + 8}$
$= \sqrt{17}$

$f(0, 8) = \sqrt{(8)^2 + 2(0)^2} = \sqrt{64} = 8$

5. a. $g(-2, 4) = -(-2)^2 - 4(-2)(4) + (4)^3$
$= -4 + 32 + 64 = 92$

b. $g(-1, -2) = -(-1)^2 - 4(-1)(-2) + (-2)^3$
$= -1 - 8 - 8 = -17$

c. $g(-2, 3) = -(-2)^2 - 4(-2)(3) + (3)^3$
$= -4 + 24 + 27 = 47$

d. $g(5, 1) = -(5)^2 - 4(5)(1) + (1)^3$
$= -25 - 20 + 1 = -44$

7. $f(x, y) = x^2 + y^2$

The domain is the set of all ordered pairs (x, y) such that x and y are real numbers.

9. $B(h, w) = \dfrac{w}{h^2}$

The domain is the set of all ordered pairs (h, w) such that w is a real number and h is a nonzero real number.

11. $h(u, v, w) = \left| \dfrac{u + v}{v - w} \right|$

The domain is the set of all ordered triples (u, v, w) such that u, v, and w are real numbers, and $v \neq w$.

13. $P(l, w) = 2l + 2w$

15. Refer to sections 5.3 and 5.4 in the text.

$$S(R, i, n) = R \left[\dfrac{(1 + i)^n - 1}{i} \right]$$

17. The xy-, xz- and yz- traces of a graph are the curves that result when a surface that is the graph of a function of two variables, $z = f(x, y)$, is cut by the xy-, xz- and yz- planes, respectively.

19. $3x + 2y + z = 12$
Let $x = 0$ and $y = 0$. Then $z = 12$. The point $(0, 0, 12)$ is on the graph. Let $x = 0$ and $z = 0$. Then $y = 6$. The point $(0, 6, 0)$ is on the graph. Let $y = 0$ and $z = 0$. Then $x = 4$. The point $(4, 0, 0)$ is on the graph. The graph is a plane. We sketch the portion in the first octant through these three points.

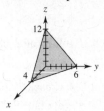

21. $x + y = 4$
We let $x = 0$ and find $y = 4$. The point $(0, 4, 0)$ is on the graph. We let $y = 0$ and find $x = 4$. The point $(4, 0, 0)$ is on the graph. Since there is no z-intercept, the plane is parallel to the z-axis. We sketch the portion in the first octant through these two points.

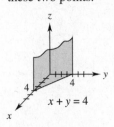

23. $z = 4$

The z-intercept is $(0, 0, 4)$. Since there is no x-intercept or y-intercept, the plane is parallel to the xy-plane. We sketch the portion in the first octant through the point $(0, 0, 4)$.

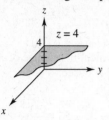

25. $3x + 2y + z = 18$

Value of z	Equation of level curve
$z = 0$	$3x + 2y = 18$
$z = 2$	$3x + 2y = 16$
$z = 4$	$3x + 2y = 14$

We graph the line $3x + 2y = 18$ in the plane $z = 0$. We graph the line $3x + 2y = 16$ in the plane $z = 2$. We graph the line $3x + 2y = 14$ in the plane $z = 4$.

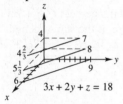

27. $y^2 - x = -z$

Value of z	Equation of level curve
$z = 0$	$x = y^2$
$z = 2$	$x = y^2 + 2$
$z = 4$	$x = y^2 + 4$

We graph the parabola $x = y^2$ (for $y > 0$) in the plane $z = 0$. We graph the parabola $x = y^2 + 2$ (for $y \geq 0$) in the plane $z = 2$. We graph the parabola $x = y^2 + 4$ (for $y \geq 0$) in the plane $z = 4$.

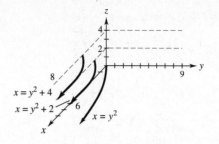

29. $z = x^7 y^3$ and $z = 500 = x^7 y^3 = 500$

$$y^3 = \frac{500}{x^7} = y^{3/10} = \frac{500}{x^{7/10}}$$

$$y = \left(\frac{500}{x^{7/10}}\right)^{10/3} = \frac{500^{10/3}}{x^{7/3}} \approx \frac{9.9 \times 10^8}{x^{7/3}} \approx \frac{10^9}{x^{7/3}}$$

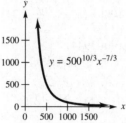

31. $z = x^7 y^3$

If x is doubled

$$z = (2x)^7 y^3 = \left(2^7\right) x^7 y^3 \approx 1.6 x^7 y^3.$$

The effect of doubling x is to multiply z by $2^7 \approx 1.6$.

If y is doubled,

$$z = x^7 (2y)^3 = (2)^3 x^7 y^3 \approx 1.2 x^7 y^3.$$

The effect of doubling y is to multiply z by $2^3 \approx 1.2$.

If both x and y are doubled,

$$z = (2x)^7 (2y)^3 = (2)^7 x^7 (2)^3 y^3$$
$$= 2^{7+3} x^7 y^3 = 2x^7 y^3$$

The effect of doubling both x and y is to double z.

33. a. If $x = 32$ and $y = 1$, then

$$P(32,1) = 100\left(\frac{3}{5}(32)^{-2/5} + \frac{2}{5}(1)^{-2/5}\right)^{-5}$$

$$= 100\left(\frac{3}{5}\left(\frac{1}{4}\right) + \frac{2}{5}(1)\right)^{-5}$$

$$= 100\left(\frac{11}{20}\right)^{-5} = 1986.9482$$

When 32 units of labor and 1 unit of capital are provided, production is about 1987 precision cameras.

b. If $x = 1$ and $y = 32$, then

$$P(1,32) = 100\left(\frac{3}{5}(1)^{-2/5} + \frac{2}{5}(32)^{-2/5}\right)^{-5}$$

$$= 100\left(\frac{3}{5}(1) + \frac{2}{5}\left(\frac{1}{4}\right)\right)^{-5}$$

$$= 100\left(\frac{3}{5} + \frac{1}{10}\right)^{-5} = 100\left(\frac{7}{10}\right)^{-5}$$

$$= 594.9902$$

When 1 unit of labor and 32 units of capital are provided, production is about 595 precision cameras.

c. If $x = 32$ and $y = 243$, then

$$P(32,243) = 100\left(\frac{3}{5}(32)^{-2/5} + \frac{2}{5}(243)^{-2/5}\right)^{-5}$$

$$= 100(.1944)^{-5}$$

$$\approx 359,768 \text{ precision cameras.}$$

35. $f(d,w) = 15 + .4(d + w)$

a. If $d = 85$ and $w = 60$,

$f(85,60) = 15 + .4(85 + 60) = 73$

A temperature humidity index of 73 would feel comfortable.

b. If $d = 85$ and $w = 70$,

$f(85,70) = 15 + .4(85 + 70) = 77$

A temperature humidity index of 77 would feel uncomfortable.

b. $\dfrac{\partial z}{\partial y} = 0 - 4x^2 + 18y = -4x^2 + 18y$

c. $f_x(2, 3) = 24(2)^2 - 8(2)(3) = 48$

d. $f_y(1, -2) = -4(1)^2 + 18(-2) = -40$

37. $A = .202 W^{.425} H^{.725}$

a. If $W = 72$ and $H = 1.78$,

$A = .202(72)^{.425}(1.78)^{.725} = 1.8892 \text{ m}^2$

b. If $W = 65$ and $H = 1.40$,

$A = .202(65)^{.425}(1.40)^{.725} = 1.5198 \text{ m}^2$

c. If $W = 70$ and $H = 1.60$,

$A = .202(70)^{.425}(1.60)^{.725} = 1.7278 \text{ m}^2$

d. Answers vary

39. $M(n, i, t) = \dfrac{(1 + .05)^{25}(1 - .33) + .33}{\left[1 + (1 - .33).05\right]^{25}} \approx 1.14$

Since $M > 1$, the IRA grows faster.

41. a. Solve for v. For a ferret,

$$2.56 = \frac{v^2}{(9.81)(.09)} \Rightarrow 2.56 = \frac{v^2}{.8829} \Rightarrow$$

$$2.260224 = v^2 \Rightarrow 1.5 \approx v$$

The velocity at which this change occurs for a ferret is about 1.5 m/sec.

For a rhinoceros,

$$2.56 = \frac{v^2}{(9.81)(1.2)} \Rightarrow 2.56 = \frac{v^2}{11.772} \Rightarrow$$

$$30.13632 = v^2 \Rightarrow 5.5 \approx v$$

The velocity at which this change occurs for a rhinoceros is about 5.5 m/sec.

b. Solve for v.

$$.025 = \frac{v^2}{(9.81)(4)} \Rightarrow .025 = \frac{v^2}{39.24} \Rightarrow$$

$$.981 = v^2 \Rightarrow 1 \approx v$$

The velocity at which the sauropods were traveling was about 1 m/sec.

43. The distance around the parcel is $2H + 2W$. The length is L. Therefore, if the sum of length and girth is given by f, $f(L, W, H) = L + 2H + 2W$.

45. $z = x^2 + y^2$

Find the traces.

xy-trace: $0 = x^2 + y^2$

xz-trace: $z = x^2$

yz-trace: $z = y^2$

These traces match with graph (c), the correct graph.

47. $x^2 - y^2 = z$

xy-trace: $x^2 - y^2 = 0$

yz-trace: $y^2 = -z$

xz-trace: $x^2 = z$

These traces match with graph (e), the correct graph.

49. $\dfrac{x^2}{16} + \dfrac{y^2}{25} + \dfrac{z^2}{4} = 1$

xy-trace: $\dfrac{x^2}{16} + \dfrac{y^2}{25} = 1$

yz-trace: $\dfrac{y^2}{25} + \dfrac{z^2}{4} = 1$

xz-trace: $\dfrac{x^2}{16} + \dfrac{z^2}{4} = 1$

These traces match with graph (b), the correct graph.

Section 14.2 Partial Derivatives

1. $z = f(x, y) = 8x^3 - 4x^2 y + 9y^2$

a. $\dfrac{\partial z}{\partial x} = 24x^2 - 8xy + 0 = 24x^2 - 8xy$

3. $f(x, y) = 6y^3 - 4xy + 5$
$f_x = 0 - 4y = -4y$
$f_y = 18y^2 - 4x$
$f_x(2, -1) = -4(-1) = 4$
$f_y(-4, 3) = 18(3^2) - 4(-4) = 178$

5. $f(x, y) = e^{2x+y}$
$f_x = e^{2x+y}(2) = 2e^{2x+y}$
$f_y = e^{2x+y}(1) = e^{2x+y}$
$f_x(2, -1) = 2e^{2(2)+(-1)} = 2e^3$
$f_y(-4, 3) = e^{2(-4)+3} = e^{-5}$

7. $f(x, y) = \dfrac{-2}{e^{x+2y}} = -2e^{-x-2y}$

$f_x = -2(-1)e^{-x-2y} = 2e^{-x-2y} = \dfrac{2}{e^{x+2y}}$

$f_y = -2(-2)e^{-x-2y} = 4e^{-x-2y} = \dfrac{4}{e^{x+2y}}$

$f_x(2, -1) = \dfrac{2}{e^{2+2(-1)}} = 2$

$f_y(-4, 3) = \dfrac{4}{e^{-4+2(3)}} = \dfrac{4}{e^2}$

9. $f(x, y) = \dfrac{x + 3y^2}{x^2 + y^3}$

Use the quotient rule to find $f_x(x, y)$ and $f_y(x, y)$.

$f_x = \dfrac{\left(x^2 + y^3\right)(1) - \left(x + 3y^2\right)(2x)}{\left(x^2 + y^3\right)^2}$

$\quad = \dfrac{x^2 + y^3 - 2x^2 - 6xy^2}{\left(x^2 + y^3\right)^2} = \dfrac{y^3 - x^2 - 6xy^2}{\left(x^2 + y^3\right)^2}$

$f_y = \dfrac{\left(x^2 + y^3\right)(6y) - \left(x + 3y^2\right)\left(3y^2\right)}{\left(x^2 + y^3\right)^2}$

$\quad = \dfrac{6x^2 y + 6y^4 - 3xy^2 - 9y^4}{\left(x^2 + y^3\right)^2}$

$\quad = \dfrac{6x^2 y - 3xy^2 - 3y^4}{\left(x^2 + y^3\right)^2}$

$f_x(2, -1) = \dfrac{(-1)^3 - 2^2 - 6(2)(-1)^2}{\left[(2)^2 + (-1)^3\right]^2} = -\dfrac{17}{9}$

$f_y(-4, 3) = \dfrac{6(-4)^2(3) - 3(-4)(3)^2 - 3(3)^4}{\left[(-4)^2 + (3)^3\right]^2}$

$\quad = \dfrac{153}{1849} \approx .083$

11. $f(x, y) = \ln\left|1 + 5x^3 y^2\right|$

$f_x = \dfrac{15x^2 y^2}{1 + 5x^3 y^2}$

$f_y = \dfrac{10x^3 y}{1 + 5x^3 y^2}$

$f_x(2, -1) = \dfrac{15(2)^2(-1)^2}{1 + 5(2)^3(-1)^2} = \dfrac{60}{41}$

$f_y(-4, 3) = \dfrac{10(-4)^3(3)}{1 + 5(-4)^3(3)^2} = \dfrac{1920}{2879}$

13. $f(x, y) = x^2 e^{2xy}$
$f_x = x^2 \cdot 2ye^{2xy} + 2xe^{2xy} = 2x^2 ye^{2xy} + 2xe^{2xy}$
$f_y = x^2 \cdot 2xe^{2xy} = 2x^3 e^{2xy}$
$f_x(2, -1) = 2(2)^2(-1)e^{2(2)(-1)} + 2(2)e^{2(2)(-1)}$
$\quad = -8e^{-4} + 4e^{-4} = -4e^{-4}$
$f_y(-4, 3) = 2(-4)^3 e^{2(-4)(3)} = -128e^{-24}$

15. $f(x, y) = 4x^2 y^2 - 16x^2 + 4y$

$f_x = 8xy^2 - 32x$

$f_y = 8x^2 y + 4$

$f_{xx} = 8y^2 - 32$

$f_{xy} = 16xy$

$f_{yy} = 8x^2$

$f_{yx} = 16xy$

17. $h(x, y) = -3y^2 - 4x^2 y^2 + 7xy^2$

$h_x = 0 - 8xy^2 + 7y^2 = -8xy^2 + 7y^2$

$h_y = -6y - 8x^2 y + 14xy$

$h_{xx} = -8y^2 + 0 = -8y^2$

$h_{xy} = -16xy + 14y$

$h_{yy} = -6 - 8x^2 + 14x$

$h_{yx} = 0 - 16xy + 14y = -16xy + 14y$

19. $r(x, y) = \dfrac{6y}{x+y}$

$r_x = \dfrac{(x+y)(0) - (6y)(1)}{(x+y)^2} = -\dfrac{6y}{(x+y)^2}$

$r_y = \dfrac{(x+y)(6) - (6y)(1)}{(x+y)^2} = \dfrac{6x}{(x+y)^2}$

$r_{xx} = \dfrac{(x+y)^2(0) - 2(x+y)(-6y)}{(x+y)^4} = \dfrac{12y}{(x+y)^3}$

$r_{xy} = \dfrac{(x+y)^2(-6) - (-6y)(2)(x+y)}{(x+y)^4}$

$\quad = \dfrac{-6(x+y)^2 + 12y(x+y)}{(x+y)^4}$

$\quad = \dfrac{-6(x+y) + 12y}{(x+y)^3} = \dfrac{6y - 6x}{(x+y)^3}$

$r_{yy} = \dfrac{(x+y)^2(0) - 6x(2)(x+y)}{(x+y)^4} = -\dfrac{12x}{(x+y)^3}$

$r_{yx} = \dfrac{(x+y)^2(6) - (6x)(2)(x+y)}{(x+y)^4}$

$\quad = \dfrac{6(x+y)^2 - 6x(2)(x+y)}{(x+y)^4} = \dfrac{6(x+y) - 12x}{(x+y)^3}$

$\quad = \dfrac{6y - 6x}{(x+y)^3}$

21. $z = 4xe^y$

$z_x = 4e^y, \quad z_y = 4xe^y$

$z_{xx} = 0, \quad z_{yy} = 4xe^y$

$z_{xy} = 4e^y, \quad z_{yx} = 4e^y$

23. $r = \ln|2x + y|$

$r_x = \dfrac{2}{2x + y}$

$r_y = \dfrac{1}{2x + y}$

$r_{xx} = \dfrac{(2x+y)(0) - 2(2)}{(2x+y)^2} = -\dfrac{4}{(2x+y)^2}$

$r_{xy} = \dfrac{(2x+y)(0) - 2(1)}{(2x+y)^2} = -\dfrac{2}{(2x+y)^2}$

$r_{yy} = \dfrac{(2x+y)(0) - 1(1)}{(2x+y)^2} = -\dfrac{1}{(2x+y)^2}$

$r_{yx} = \dfrac{(2x+y)(0) - 1(2)}{(2x+y)^2} = -\dfrac{2}{(2x+y)^2}$

25. $z = x\ln(xy) = x(\ln x + \ln y) = x\ln x + x\ln y$

$z_x = x\left(\dfrac{1}{x}\right) + 1(\ln x) + \ln y = 1 + \ln x + \ln y$

$z_y = 0 + x\left(\dfrac{1}{y}\right) = \dfrac{x}{y}$

$z_{xy} = 0 + 0 + \dfrac{1}{y} = \dfrac{1}{y}$

$z_{yy} = -\dfrac{x}{y^2}$

27. $f(x, y) = x\ln(xy)$

From exercise 25, $f_{xy} = \dfrac{1}{y}$ and $f_{yy} = -\dfrac{x}{y^2}$.

Therefore, $f_{xy}(1, 2) = \dfrac{1}{2}$ and

$f_{yy}(3, 1) = -\dfrac{3}{1^2} = -3.$

29. $f(x, y) = 6x^2 + 6y^2 + 6xy + 36x - 5$

$f_x(x, y) = 12x + 0 + 6y + 36$

$\quad = 12x + 6y + 36$

$f_y(x, y) = 0 + 12y + 6x + 0 - 0$

$\quad = 12y + 6x$

(continued on next page)

(*continued from page 454*)

Let $f_x(x, y) = 0$ and $f_y(x, y) = 0$. Solve the resulting system of equations.

$12x + 6y + 36 = 0$ (1)

$\quad 6x + 12y = 0$ (2)

To solve the system by the elimination method, multiply equation (2) by –2 and add the result to the first equation.

$\quad 12x + 6y = -36$

$\underline{-12x - 24y = 0}$

$\qquad\qquad -18y = -36 \Rightarrow y = 2$

Substitute 2 for y in equation (2).

$6x + 12(2) = 0 \Rightarrow x = -4$

Thus, $x = -4$ and $y = 2$.

31. $f(x, y) = 9xy - x^3 - y^3 - 6$

$f_x(x, y) = 9y - 3x^2 - 0 - 0 = 9y - 3x^2$

$f_y(x, y) = 9x - 0 - 3y^2 - 0 = 9x - 3y^2$

If $f_x(x, y) = 0$ and $f_y(x, y) = 0$, we have the nonlinear system of equations

$9y - 3x^2 = 0$

$9x - 3y^2 = 0.$

Divide each equation by 3.

$3y - x^2 = 0$ (1)

$3x - y^2 = 0$ (2)

We will solve this system by the substitution method, which may be used to solve nonlinear as well as linear systems. First, solve equation (1) for y.

$3y = x^2 \Rightarrow y = \dfrac{x^2}{3}$

Now substitute $\dfrac{x^2}{3}$ for y in equation (2) and solve the resulting equation for x.

$3x - \left(\dfrac{x^2}{3}\right)^2 = 0 \Rightarrow 3x - \dfrac{x^4}{9} = 0 \Rightarrow$

$27x - x^4 = 0 \Rightarrow x\left(27 - x^3\right) = 0 \Rightarrow x = 0, 3$

If $x = 0$, $y = \dfrac{0^2}{3} = 0$.

If $x = 3$, $y = \dfrac{3^2}{3} = 3$.

Thus, there are two solutions: $x = 0$, $y = 0$ and $x = 3$, $y = 3$.

33. $f(x, y, z) = x^2 + yz + z^4$

$f_x = 2x$

$f_y = z$

$f_z = y + 4z^3$

$f_{yz} = 1$

$f_y(3, 4, -2) = -2$

$f_{yz}(1, -1, 0) = 1$

35. $f(x, y, z) = \dfrac{6x - 5y}{4z + 5}$

$f_x = \dfrac{6}{4z + 5}$

$f_y = -\dfrac{5}{4z + 5}$

$f_z = -\dfrac{4(6x - 5y)}{(4z + 5)^2}$

$f_{yz} = \dfrac{20}{(4z + 5)^2}$

37. $f(x, y, z) = \ln\left(x^2 - 5xz^2 + y^4\right)$

$f_x = \dfrac{2x - 5z^2}{x^2 - 5xz^2 + y^4}$

$f_y = \dfrac{4y^3}{x^2 - 5xz^2 + y^4}$

$f_z = -\dfrac{10xz}{x^2 - 5xz^2 + y^4}$

$f_{yz} = \dfrac{4y^3(10zx)}{\left(x^2 - 5xz^2 + y^4\right)^2} = \dfrac{40xy^3z}{\left(x^2 - 5xz^2 + y^4\right)^2}$

39. A function with three independent variables has three partial derivatives. For a function $f(x, y, z)$, the 3 partial derivatives are f_x, f_y and f_z. In the second-order partial derivatives, each of the two subscripts may be x, y, or z, so there will be $3 \cdot 3 = 9$ possibilities. For a function $f(x, y, z)$, the 9 second-order partial derivatives are f_{xx}, f_{xy}, f_{xz}, f_{yx}, f_{yy}, f_{yz}, f_{zx}, f_{zy} and f_{zz}.

41. $M(x, y) = 40x^2 + 30y^2 - 10xy + 30$ where x is the cost of electronic chips and y is the cost of labor.

a. $M_y(x, y) = 60y - 10x$
$M_y(4, 2) = 60(2) - 10(4) = 80$
The manufacturing costs are $80.

b. $M_x(x, y) = 80x - 10y$
$M_x(3, 6) = 80(3) - 10(6) = 180$
The manufacturing costs are $180.

c. $\left(\dfrac{\partial M}{\partial x}\right) = 80x - 10y$
$\left(\dfrac{\partial M}{\partial x}\right)(2, 5) = 80(2) - 10(5) = 110$
The manufacturing costs are $110.

d. $\left(\dfrac{\partial M}{\partial y}\right) = 60y - 10x$
$\left(\dfrac{\partial M}{\partial y}\right)(6, 7) = 60(7) - 10(6) = 360$
The manufacturing costs are $360.

43. $f(p, i) = 99p - .5pi - .0025p^2$

a. If $p = 19,400$ and $i = 8$, then
$f(19400, 8)$
$= 99(19,400) - .5(19,400)(8)$
$\qquad\qquad\qquad - .0025(19,400)^2$
$= 1,920,600 - 77,600 - 940,900$
$= 902,100$
The weekly sales are $902,100.

b. $f(p, i) = 99p - .5pi - .0025p^2$
$f_p(p, i) = 99 - .5i - .005p$
$f_i(p, i) = -.5p$

$f_p(p, i)$ is the rate at which weekly sales are changing per unit change in price while the interest rate remains constant.
$f_i(p, i)$ is the rate at which weekly sales are changing per unit change in interest rate while price remains constant.

c. $f_i(19400, 8) = -.5(19400) = -9700$
Changing the interest rate from 8% to 9% would decrease weekly sales by $9700.

45. $f(m, v) = 25.92m^{.68} + \dfrac{3.62m^{.75}}{v}$

a. $f(300, 10) = 25.92(300)^{.68} + \dfrac{3.62(300)^{.75}}{10}$
$= 1253.36 + 26.09 = 1279.45 \approx 1279$
The energy expended is 1279 kcal per hour.

b. $f_m(m, v)$
$= 25.92(.68)m^{-.32} + \dfrac{3.62}{v}(.75m^{-.25})$
$= 17.6256m^{-.32} + \dfrac{2.715m^{-.25}}{v}$
$f_m(300, 10)$
$= 17.625(300)^{-.32} + \dfrac{2.715(300)^{-.25}}{10}$
$= 2.84 + .07 = 2.91$
The instantaneous rate of change of energy for a 300-kg animal traveling at 10 km per hour is about 2.91 kcal per hour per g. In other words, if the animal gains 1 g in mass, it would need to expend 2.91 kcal to walk or run 1 km.

47. $C(x, y) = 600x + 120y + 4000$

a. In $C(x, y)$, 4000 represents the fixed costs, that is, the cost when no items have been produced.

b. $\dfrac{\partial C}{\partial x} = \$600, \quad \dfrac{\partial C}{\partial y} = \$120$

$\dfrac{\partial C}{\partial x}$ is the marginal cost of producing 1 unit of radios when the production of flashlights remains constant. $\dfrac{\partial C}{\partial y}$ is the marginal cost of producing 1 unit of flashlights when the production of radios remains constant.

49. $A(M, H) = .202M^{.425}H^{.725}$

a. $A_M(M, H) = .202(.425)M^{-.575}H^{.725}$
$= .08585M^{-.575}H^{.725}$
$A_M(72, 1.8) = .08585(72)^{-.575}(1.8)^{.725}$
$= .0112\,\text{m}^2/\text{g}$

b. $A_H(M, H) = .202M^{.425}(.725)H^{-.275}$
$A_H(70, 1.6) = .14645M^{.425}H^{-.275}$
$= .14645(70)^{.425}(1.6)^{-.275}$
$= .783\,\text{m}^2/\text{m}$

51. $f(x, y) = \left[\frac{1}{3}x^{-1/3} + \frac{2}{3}y^{-1/3}\right]^{-3}$

a. $f(27, 64) = \left[\frac{1}{3}(27)^{-1/3} + \frac{2}{3}(64)^{-1/3}\right]^{-3}$

$\qquad = \left[\frac{1}{3}\left(\frac{1}{3}\right) + \frac{2}{3}\left(\frac{1}{4}\right)\right]^{-3}$

$\qquad = \left(\frac{1}{9} + \frac{1}{6}\right)^{-3} = \left(\frac{5}{18}\right)^{-3} = 46.656$

4665.6 units are produced when 27 units of labor and 64 units of capital are utilized.

b. $f_x = -3\left[\frac{1}{3}x^{-1/3} + \frac{2}{3}y^{-1/3}\right]^{-4}\left(-\frac{1}{9}x^{-4/3}\right)$

$f_x(27, 64)$

$\qquad = -3\left[\frac{1}{3}(27)^{-1/3} + \frac{2}{3}(64)^{-1/3}\right]^{-4}$

$\qquad\qquad \cdot \left(-\frac{1}{9}\right)(27)^{-4/3}$

$\qquad = -3\left(\frac{5}{18}\right)^{-4}\left(-\frac{1}{9}\right)\left(\frac{1}{81}\right) = \frac{432}{625}$

$f_x(27, 64) = .6912$, which represents the rate at which production is changing when labor changes by 1 unit from 27 to 28 and capital remains constant.

$f_y = -3\left[\frac{1}{3}x^{-1/3} + \frac{2}{3}y^{-1/3}\right]^{-4}\left(-\frac{2}{9}y^{-4/3}\right)$

$f_y(27, 64)$

$\qquad = -3\left[\frac{1}{3}(27)^{-1/3} + \frac{2}{3}(64)^{-1/3}\right]^{-4}$

$\qquad\qquad \cdot\left(-\frac{2}{9}\right)(64)^{-4/3}$

$\qquad = -3\left(\frac{5}{18}\right)^{-4}\left(-\frac{2}{9}\right)\left(\frac{1}{256}\right) = \frac{2187}{5000} = .4374$

which represents the rate at which production is changing when capital changes by 1 unit from 64 to 65 and labor remains constant.

c. If labor increases by 1 unit then production would increase by .6912(100) or about 69 units. (See part (b) of this solution.)

53. $z = x^7 y^3$, where x is labor, y is capital.

Marginal productivity of labor is

$\frac{\partial z}{\partial x} = .7x^{-.3}y^3 = \frac{.7y^3}{x^3}.$

Marginal productivity of capital is

$\frac{\partial z}{\partial y} = .3x^7 y^{-.7} = \frac{.3x^7}{y^7}.$

55. $C(a, b, v) = \frac{b}{a-v} = b(a-v)^{-1}$

a. $C(160, 200, 125) = \frac{200}{160-125} \approx 5.71$

b. $C(180, 260, 142) = \frac{260}{180-142} \approx 6.84$

c. $\frac{\partial C}{\partial b} = \frac{1}{a-v}$

d. $\frac{\partial C}{\partial v} = -b(a-v)^{-2}(-1) = \frac{b}{(a-v)^2}$

57. $p = f(s, n, a) = .078a + 4(sn)^{1/2}$

a. If $s = 8$, $n = 6$, and $a = 450$, then

$p = f(8, 6, 450) = .078(450) + 4(8 \cdot 6)^{1/2}$

$\qquad = 35.1 + 4(48)^{1/2} \approx 62.8$

The probability of passing the course is 62.8%.

b. If $s = 3$, $n = 3$, and $a = 320$, then

$p = f(3, 3, 320) = .078(320) + 4(3 \cdot 3)^{1/2}$

$\qquad = 24.96 + 4(9)^{1/2} = 36.96$

c. $f_n(s, n, a) = 4\left(s^{1/2}\right)\left(\frac{1}{2}n^{-1/2}\right) = 2s^{1/2}n^{-1/2}$

$f_n(3, 3, 320) = 2\left(3^{1/2}\right)\left(3^{-1/2}\right) = 2\left(3^0\right) = 2$

The rate of change of probability per additional semester of high school where placement score and SAT score remain constant is 2%.

$f_a(s, n, a) = 0.078 \Rightarrow f_a(3, 3, 320) = 0.078$

The rate of change of probability per unit change of SAT score when placement score and number of semesters remain constant is 0.078%.

59. $F = \dfrac{mgR^2}{r^2}$

    a. $F_m = \dfrac{gR^2}{r^2}$

        This represents the rate of change in force per unit change in mass.

        $F_r = \dfrac{-2mgR^2}{r^3}$

        This represents the rate of change in force per unit change in distance.

    b. Since all variables represent nonnegative

        values, $F_m = \dfrac{gR^2}{r^2} > 0$ and

        $F_r = \dfrac{-2mgR^2}{r^3} < 0.$

61. a. Let $w = 220$ and $h = 74$.

        $B = \dfrac{703(220)}{(74)^2} \approx 28.24$

    b. $\dfrac{\partial B}{\partial w} = \dfrac{703}{h^2}$

        $\dfrac{\partial B}{\partial h} = -2(703w)h^{-3} = \dfrac{-1406w}{h^3}$

Section 14.3 Extrema of Functions of Several Variables

1. Answers vary.

3. $f(x, y) = 2x^2 + 4xy + 6y^2 - 8x - 10$
$f_x = 4x + 4y + 0 - 8 - 0 = 4x + 4y - 8$
$f_y = 0 + 4x + 12y - 0 - 0 = 4x + 12y$

Set each of these two partial derivatives equal to zero, forming a system of linear equations.
$4x + 4y - 8 = 0$ (1)
$4x + 12y \quad = 0$ (2)

To solve this system by the elimination method, rewrite equation (1) in the form of $ax + by = c$ and multiply equation (2) by -1; then add the results.

$\quad 4x + \ 4y = 8$ (1)
$\underline{-4x - 12y = 0}$ (2)
$\quad\quad -8y = 8 \ \Rightarrow y = -1$

Substitute -1 for y in equation (2).
$4x + 12(-1) = 0 \Rightarrow 4x = 12 \Rightarrow x = 3$
The critical point is $(3, -1)$.

$f_{xx} = 4,\ f_{yy} = 12,$ and $f_{xy} = 4$

$M = f_{xx}(3, -1) \cdot f_{yy}(3, -1) - \left[f_{xy}(3, -1) \right]^2$
$\quad = 4 \cdot 12 - 4^2 = 32$

Since $M > 0$ and $f_{xx}(3, -1) > 0$, we have a local minimum at $(3, -1)$.

$f(3, -1) = 2(3)^2 + 4(3)(-1) + 6(-1)^2 - 8(3) - 10$
$\quad\quad\quad = 18 - 12 + 6 - 24 - 10 = -22$

There is a local minimum of -22 at $(3, -1)$.

5. $f(x, y) = x^2 - xy + y^2 + 2x + 2y + 6$
$f_x = 2x - y + 0 + 2 + 0 + 0 = 2x - y + 2$
$f_y = 0 - x + 2y + 0 + 2 + 0 = -x + 2y + 2$

Set $f_x = 0$ and $f_y = 0$. Solve the resulting system of linear equations by the elimination method.

$\begin{matrix} 2x - \ y + 2 = 0 \\ -x + 2y + 2 = 0 \end{matrix} \Rightarrow \begin{matrix} 2x - \ y + 2 = 0 \\ -2x + 4y + 4 = 0 \end{matrix} \Rightarrow$

$3y + 6 = 0 \Rightarrow 3y = -6 \Rightarrow y = -2$

$-x + 2(-2) + 2 = 0 \Rightarrow x = -2$

The critical point is $(-2, -2)$.

$f_{xx} = 2,\ f_{yy} = 2,$ and $f_{xy} = -1$

$M = f_{xx}(-2, -2) \cdot f_{yy}(-2, -2) - \left[f_{xy}(-2, -2) \right]^2$
$\quad = 2 \cdot 2 - (-1)^2 = 3$

Since $M > 0$ and $f_{xx}(-2, -2) > 0$, we have a local minimum at $(-2, -2)$.

$f(-2, -2) = (-2)^2 - (-2)(-2)$
$\quad\quad\quad\quad\quad + (-2)^2 + 2(-2) + 2(-2) + 6$
$\quad\quad\quad = 4 - 4 + 4 - 4 - 4 + 6 = 2$

There is a local minimum of 2 at $(-2, -2)$.

7. $f(x, y) = 3xy + 6y - 5x$
$f_x = 3y - 5;\ f_y = 3x + 6$

Solve the system of equations.
$\begin{matrix} 3y - 5 = 0 \\ 3x + 6 = 0 \end{matrix} \Rightarrow y = \dfrac{5}{3},\ x = -2$

The critical point is $\left(-2, \dfrac{5}{3} \right)$.

$f_{xx} = 0,\ f_{yy} = 0,$ and $f_{xy} = 3$

$M = f_{xx}\left(-2, \dfrac{5}{3} \right) \cdot f_{yy}\left(-2, \dfrac{5}{3} \right) - \left[f_{xy}\left(-2, \dfrac{5}{3} \right) \right]^2$

$\quad = 0 \cdot 0 - 3^2 = -9$

Since $M < 0$, we have a saddle point at $\left(-2, \dfrac{5}{3} \right)$

9. $f(x, y) = 4xy - 10x^2 - 4y^2 + 8x + 8y + 9$

$f_x = 4y - 20x + 8$

$f_y = 4x - 8y + 8$

Solve the system of equations.

$4y - 20x + 8 = 0$

$4x - 8y + 8 = 0$ $\Rightarrow$

$4y - 20x + 8 = 0$ $\;(1)$

$\underline{-4y + 2x + 4 = 0}$ $\;(2)$

$-18x + 12 = 0 \;\Rightarrow\; x = \dfrac{2}{3}$

$4y - 20\left(\dfrac{2}{3}\right) + 8 = 0 \Rightarrow y = \dfrac{4}{3}$

The critical point is $\left(\dfrac{2}{3}, \dfrac{4}{3}\right)$.

$f_{xx} = -20, \; f_{yy} = -8, \; f_{xy} = 4$

For $\left(\dfrac{2}{3}, \dfrac{4}{3}\right)$, $M = (-20)(-8) - 16 = 144 > 0$.

Since $f_{xx} < 0$, a local maximum occurs at

$\left(\dfrac{2}{3}, \dfrac{4}{3}\right)$.

$f\left(\dfrac{2}{3}, \dfrac{4}{3}\right) = 4\left(\dfrac{2}{3}\right)\left(\dfrac{4}{3}\right) - 10\left(\dfrac{2}{3}\right)^2 - 4\left(\dfrac{4}{3}\right)^2$

$\qquad\qquad + 8\left(\dfrac{2}{3}\right) + 8\left(\dfrac{4}{3}\right) + 9 = 17$

There is a local maximum of 17 at $\left(\dfrac{2}{3}, \dfrac{4}{3}\right)$.

11. $f(x, y) = x^2 + xy - 2x - 2y + 2$

$f_x = 2x + y - 2$

$f_y = x - 2$

Solve the system of equations.

$2x + y - 2 \;= 0$

$\underline{x - 2 \;= 0}$

$x = 2$

$2(2) + y - 2 = 0 \Rightarrow y = -2$

This critical point is $(2, -2)$.

$f_{xx} = 2, \; f_{yy} = 0, \; f_{xy} = 1$

For $(2, -2)$, $M = 2(0) - 1^2 = -1 < 0$.

There is a saddle point at $(2, -2)$.

13. $f(x, y) = x^2 - y^2 - 2x + 4y - 7$

$f_x = 2x - 2$

$f_y = -2y + 4$

If $2x - 2 = 0$, $x = 1$.

If $-2y + 4 = 0$, $y = 2$.

The critical point is $(1, 2)$.

$f_{xx} = 2, \; f_{yy} = -2, \; f_{xy} = 0$

For $(1, 2)$, $M = -4 - 0 < 0$.

There is a saddle point at $(1, 2)$.

15. $f(x, y) = 2x^3 + 2y^2 - 12xy + 15$

$f_x = 6x^2 + 0 - 12y + 0 = 6x^2 - 12y$

$f_y = 0 + 4y - 12x + 0 = 4y - 12x$

Setting $f_x = 0$ and $f_y = 0$, we obtain the

following nonlinear system.

$6x^2 - 12y = 0 \;\;(1)$

$4y - 12x = 0 \;\;(2)$

We will solve the system by the substitution method. Simplify equation (1) by dividing both sides by 6. Solve equation (2) for y.

$x^2 - 2y = 0 \quad (3)$

$\qquad y = 3x \quad (4)$

Now substitute $3x$ for y in equation (3) and solve the resulting equation for x.

$x^2 - 2(3x) = 0 \Rightarrow x(x - 6) = 0 \Rightarrow x = 0, \, 6$

If $x = 0$, $y = 0$. If $x = 6$, $y = 18$.

There are two critical points: $(0, 0)$ and $(6, 18)$.

$f_{xx} = 12x, \; f_{yy} = 4, \; \text{and} f_{xy} = -12$

At $(0, 0)$,

$M = f_{xx}(0, 0) \cdot f_{yy}(0, 0) - \left[f_{xy}(0, 0) \right]^2$

$\quad = 0(4) - (-12)^2 = -144$

Since $M < 0$, there is a saddle point at $(0, 0)$.

At $(6, 18)$,

$M = f_{xx}(6, 18) \cdot f_{yy}(6, 18) - \left[f_{xy}(6, 18) \right]^2$

$\quad = 72(4) - (-12)^2 = 144$

Since $M > 0$ and $f_{xx}(6, 18) > 0$, there is a local minimum at $(6, 18)$.

$f(6, 18) = 2(6)^3 + 2(18)^2 - 12(6)(18) + 15 = -201$

There is a local minimum of -201 at $(6, 18)$.

17. $f(x, y) = x^2 + 4y^3 - 6xy - 1$

$f_x = 2x - 6y$

$f_y = 12y^2 - 6x$

As in Exercise 15, setting $f_x = 0$ and $f_y = 0$

yields a nonlinear system of equations which we can solve by the substitution method.

$$\begin{matrix} 2x - 6y = 0 \\ 12y^2 - 6x = 0 \end{matrix} \Rightarrow \begin{matrix} x = 3y \\ 2y^2 - x = 0 \end{matrix} \Rightarrow$$

$2y^2 - 3y = 0 \Rightarrow y(2y - 3) = 0 \Rightarrow y = 0, \dfrac{3}{2}$

If $y = 0$, $x = 0$. If $y = \dfrac{3}{2}$, $x = \dfrac{9}{2}$.

There are two critical points, $(0, 0)$ and $\left(\dfrac{9}{2}, \dfrac{3}{2}\right)$.

$f_{xx} = 2$, $f_{yy} = 24y$, and $f_{xy} = -6$

At $(0, 0)$,

$M = f_{xx}(0, 0) \cdot f_{yy}(0, 0) - \left[f_{xy}(0, 0)\right]^2$

$\quad = 2(0) - (-6)^2 = -36$

Since $M < 0$, there is a saddle point at $(0, 0)$.

At $\left(\dfrac{9}{2}, \dfrac{3}{2}\right)$,

$M = f_{xx}\left(\dfrac{9}{2}, \dfrac{3}{2}\right) \cdot f_{yy}\left(\dfrac{9}{2}, \dfrac{3}{2}\right) - \left[f_{xy}\left(\dfrac{9}{2}, \dfrac{3}{2}\right)\right]^2$

$\quad = 2\left(24 \cdot \dfrac{3}{2}\right) - (-6)^2 = 180$

Since $M > 0$ and $f_{xx}\left(\dfrac{9}{2}, \dfrac{3}{2}\right) > 0$, there is a local

minimum at $\left(\dfrac{9}{2}, \dfrac{3}{2}\right)$.

$f\left(\dfrac{9}{2}, \dfrac{3}{2}\right) = \left(\dfrac{9}{2}\right)^2 + 4\left(\dfrac{3}{2}\right)^3 - 6\left(\dfrac{9}{2}\right)\left(\dfrac{3}{2}\right) - 1$

$\quad = -\dfrac{31}{4}$

There is a local minimum of $-\dfrac{31}{4}$ at $\left(\dfrac{9}{2}, \dfrac{3}{2}\right)$.

19. $f(x, y) = e^{xy}$

$f_x = ye^{xy}$, $f_y = xe^{xy}$

Solve the system of equations.

$$\begin{matrix} ye^{xy} = 0 \\ xe^{xy} = 0 \end{matrix} \Rightarrow ye^{xy} = xe^{xy} \Rightarrow x = y = 0$$

The critical point is $(0, 0)$.

$f_{xx} = y^2 e^{xy}$

$f_{yy} = x^2 e^{xy}$

$f_{xy} = e^{xy} + xye^{xy}$

For $(0, 0)$,

$M = 0 \cdot 0 - \left(e^0\right)^2 = -1 < 0$.

There is a saddle point at $(0, 0)$.

21. $z = f(x, y) = -3xy + x^3 - y^3 + \dfrac{1}{8}$

$f_x = -3y + 3x^2$

$f_y = -3x - 3y^2$

Solve the system $f_x = 0$, $f_y = 0$.

$-3y + 3x^2 = 0$

$-3x - 3y^2 = 0$

Solve the first equation for y.

$-3y + 3x^2 = 0 \Rightarrow 3y = 3x^2 \Rightarrow y = x^2$

Substitute into the second equation and solve for x.

$-3x - 3\left(x^2\right)^2 = 0 \Rightarrow -3x - 3x^4 = 0 \Rightarrow$

$-3x\left(1 + x^3\right) = 0 \Rightarrow x = 0, -1$

Then, $y = 0$ or $y = 1$.

The critical points are $(0, 0)$ and $(-1, 1)$.

$f_{xx} = 6x$, $f_{yy} = -6y$, $f_{xy} = -3$

For $(0, 0)$,

$M = f_{xx}(0, 0) \cdot f_{yy}(0, 0) - \left[f_{xy}(0, 0)\right]^2$

$\quad = 0(0) - (-3)^2 = -9 < 0$.

So $(0, 0)$ is a saddle point.

For $(-1, 1)$,

$M = f_{xx}(-1, 1) \cdot f_{yy}(-1, 1) - \left[f_{xy}(-1, 1)\right]^2$

$\quad = (-6)(-6) - [-3]^2 = 36 - 9 = 27 > 0$

So $(-1, 1)$ is a local maximum.

$f(-1, 1) = -3(-1)(1) + (-1)^3 - (1)^3 + \dfrac{1}{8} = \dfrac{9}{8}$

There is a relative maximum of $\dfrac{9}{8}$ or $1\dfrac{1}{8}$

at $(-1, 1)$. This is graph (a).

23. $z = f(x, y) = y^4 - 2y^2 + x^2 - \dfrac{17}{16}$

$f_x = 2x$

$f_y = 4y^3 - 4y$

Solve the system $f_x = 0$, $f_y = 0$.

$2x = 0$ (1)

$4y^3 - 4y = 0$ (2)

Equation (1) gives $2x = 0 \Rightarrow x = 0$.

Equation (2) gives

$4y(y^2 - 1) = 0 \Rightarrow y = 0, 1, -1$.

The critical points are $(0, 0)$, $(0, 1)$, and $(0, -1)$.

$f_{xx} = 2$, $f_{yy} = 12y^2 - 4$, $f_{xy} = 0$

For $(0, 0)$, $M = (2)(-4) - 0^2 = -8 < 0$.

There is a saddle point at $(0, 0)$.

For $(0, 1)$, $M = (2)(8) - 0^2 = 16 > 0$.

$f_{xx} = 2 > 0$, so $(0, 1)$ is a local minimum.

For $(0, -1)$,

$M = (2)(8) - 0^2 = 16 > 0$

$f_{xx} = 2 > 0$

So $(0, -1)$ is a local minimum.

$f(0, 1) = (1)^4 - 2(1)^2 + (0)^2 - \dfrac{17}{16} = -\dfrac{33}{16}$

$f(0, -1) = (-1)^4 - 2(-1)^2 + (0)^2 - \dfrac{17}{16} = -\dfrac{33}{16}$

A local minimum of $-\dfrac{33}{16}$ or $-2\dfrac{1}{16}$ occurs at $(0, 1)$ and $(0, -1)$. This is graph (b).

25. $f(x, y) = -x^4 + y^4 + 2x^2 - 2y^2 + \dfrac{1}{16}$

$f_x = -4x^3 + 4x$

$f_y = 4y^3 - 4y$

Solve $f_x = 0$, $f_y = 0$.

$-4x^3 + 4x = 0$ (1)

$\underline{4y^3 - 4y = 0}$ (2)

$-4x^3 + 4x = 0 \Rightarrow -4x(x^2 - 1) = 0 \Rightarrow x = 0, \pm 1$

$4y^3 - 4y = 0 \Rightarrow 4y(y^2 - 1) = 0 \Rightarrow y = 0, \pm 1$

Because any of the 3 x-values can be paired with any of the 3 y-values, there are 9 critical points. The critical points are $(0, 0)$, $(0, -1)$, $(0, 1)$, $(-1, 0)$, $(-1, -1)$, $(-1, 1)$, $(1, 0)$, $(1, -1)$, and $(1, 1)$.

$f_{xx} = -12x^2 + 4$, $f_{yy} = 12y^2 - 4$, $f_{xy} = 0$

For $(0, 0)$,

$M = (4)(-4) - 0 = -16 < 0$: Saddle point

For $(0, -1)$,

$M = (4)(8) - 0 = 32 > 0$ and $f_{xx} = 4$: Local minimum

For $(0, 1)$,

$M = (4)(8) - 0 = 32 > 0$ and $f_{xx} = 4$: Local minimum

For $(-1, 0)$,

$M = (-8)(-4) - 0 = 32 > 0$ and $f_{xx} = -8$: Local maximum

For $(-1, -1)$, $M = (-8)(8) - 0 = -64 < 0$: Saddle point

For $(-1, 1)$, $M = (-8)(8) - 0 = -64 < 0$: Saddle point

For $(1, 0)$, $M = (-8)(-4) - 0 = 32 > 0$ and $f_{xx} = -8$: Local maximum

For $(1, -1)$, $M = (-8)(8) - 0 = -64 < 0$: Saddle point

For $(1, 1)$, $M = (-8)(8) - 0 = -64 < 0$: Saddle point

The saddle points are at $(0, 0)$, $(-1, -1)$, $(-1, 1)$, $(1, -1)$, and $(1, 1)$.

The local maxima are at $(1, 0)$ and $(-1, 0)$.

The local minima are at $(0, 1)$ and $(0, -1)$.

$f(1, 0) = -(1)^4 + (0)^4 + 2(1)^2 - 2(0)^2 + \dfrac{1}{16} = \dfrac{17}{16}$

$f(-1, 0) = -(-1)^4 + (0)^4 + 2(-1)^2 - 2(0)^2 + \dfrac{1}{16}$

$= \dfrac{17}{16}$

A local maximum of $\dfrac{17}{16}$ occurs at $(1, 0)$ and $(-1, 0)$.

$f(0, 1) = -(0)^4 + (1)^4 + 2(0)^2 - 2(1)^2 + \dfrac{1}{16} = -\dfrac{15}{16}$

$f(0, -1) = -(0)^4 + (-1)^4 + 2(0)^2 - 2(-1)^2 + \dfrac{1}{16}$

$= -\dfrac{15}{16}$

A local minimum of $-\dfrac{15}{16}$ occurs at $(0, 1)$ and $(0, -1)$. This is graph (e).

27. $P(x, y) = 1000 + 24x - x^2 + 80y - y^2$

$P_x = 24 - 2x$

$P_y = 80 - 2y$

Solve the system

$24 - 2x = 0$

$80 - 2y = 0$ $\Rightarrow x = 12, \ y = 40$

Thus, there is a critical point at $(12, 40)$.

$P_{xx} = -2, \ P_{yy} = -2, \ P_{xy} = 0$

For $(12, 40)$, $M = -2(-2) - 0^2 = 4 > 0$.

Since $P_{xx} < 0$, $x = 12$ and $y = 40$ maximize the profit. Therefore,

$P(12, 40) = 1000 + 24(12) - (12)^2$
$\qquad\qquad + 80(40) - (40)^2 = 2744$

is the maximum profit.

29. $P(x, y) = 800 - 2x^3 + 12xy - y^2$

$P_x = 0 - 6x^2 + 12y - 0 = -6x^2 + 12y$

$P_y = 0 - 0 + 12x - 2y = 12x - 2y$

Set $P_x = 0$ and $P_y = 0$. Solve the resulting nonlinear system by the substitution method.

$-6x^2 + 12y = 0$
$12x - 2y = 0$ $\Rightarrow$ $x^2 - 2y = 0$
$y = 6x$ $\Rightarrow$

$x^2 - 2(6x) = 0 \Rightarrow x(x - 12) = 0 \Rightarrow x = 0, 12$

If $x = 0$, $y = 0$. If $x = 12$, $y = 72$.

The critical points are $(0, 0)$ and $(12, 72)$. We ignore $(0, 0)$ since it makes no sense in the problem.

$P_{xx} = -12x, \ P_{yy} = -2$, and $P_{xy} = 12$

At $(12, 72)$,

$M = P_{xx}(12, 72) \cdot P_{yy}(12, 72) - \left[P_{xy}(12, 72)\right]^2$

$\quad = (-144)(-2) - (12)^2 = 144$

Since $M > 0$ and $P_{xx}(12, 72) < 0$, there is a local maximum at $(12, 72)$.

$P(12, 72) = 800 - 2(12)^3 + 12(12)(72) - (72)^2$
$\qquad\qquad = 2528$

The profit function gives the profit in thousands of dollars. The maximum profit of $2,528,000 occurs when the cost of a unit of chips is $12 and the cost of a unit of labor is $72.

31. $C(x, y) = 2x^2 + 3y^2 - 2xy + 2x - 126y + 3800$

$C_x = 4x - 2y + 2; \ C_y = 6y - 2x - 126$

$4x - 2y + 2 = 0$ (1) $\quad 2x - y + 1 = 0$
$6y - 2x - 126 = 0$ (2) $\Rightarrow -2x + 6y - 126 = 0$ $\Rightarrow$

$5y - 125 = 0 \Rightarrow y = 25$

Substitute 25 for y in equation (1).

$4x - 2(25) + 2 = 0 \Rightarrow x = 12$

Critical point $(12, 25)$.

$C_{xx} = 4, \ C_{yy} = 6, \ C_{xy} = -2$

For $(12, 25)$, $M = 4(6) - 4 = 20 > 0$.

Since $C_{xx} > 0$, 12 units of electrical tape and 25 units of packing tape should be produced to yield a minimum cost.

$C(12, 25) = 2(12)^2 + 3(25)^2 - 2(12 \cdot 25)$
$\qquad\qquad + 2(12) - 126(25) + 3800$

$\qquad\qquad = 2237$

The minimum cost is 2237.

33. The volume is $V = xyz = 27$, so $z = \dfrac{27}{xy}$.

The surface area is

$S = 2xy + 2yz + 2xz$

$\quad = 2xy + 2y\left(\dfrac{27}{xy}\right) + 2x\left(\dfrac{27}{xy}\right)$

$\quad = 2xy + \dfrac{54}{x} + \dfrac{54}{y}$

$S_x = 2y - \dfrac{54}{x^2}, \ S_y = 2x - \dfrac{54}{y^2}$

Let $S_x = 0$ and $S_y = 0$ to obtain the following system.

$2y - \dfrac{54}{x^2} = 0$ (1)

$2x - \dfrac{54}{y^2} = 0$ (2)

Solve equation (1) for y.

$2y = \dfrac{54}{x^2} \Rightarrow y = \dfrac{27}{x^2}$

Substitute this expression for y into equation (2).

$2x - \dfrac{54}{\left(\dfrac{27}{x^2}\right)^2} = 0 \Rightarrow x = \dfrac{27}{\dfrac{(27)^2}{x^4}} \Rightarrow 27x = x^4 \Rightarrow$

$0 = x^4 - 27x \Rightarrow 0 = x\left(x^3 - 27\right) \Rightarrow x = 0, 3$

(*continued on next page*)

(continued from page 462)

If $x = 0$, $y = \dfrac{27}{0}$, which is undefined.

(If $x = 0$, there would be no box, so this is not relevant.)

If $x = 3$, $y = \dfrac{27}{x^2} = \dfrac{27}{3^2} = \dfrac{27}{9} = 3$.

Thus, the critical point is (3, 3).

$S_{xx} = \dfrac{108}{x^3}$, $S_{yy} = \dfrac{108}{y^3}$, $S_{xy} = 2$

$M = (4)(4) - 2^2 = 12 > 0$; $S_{xx} = \dfrac{108}{27} = 4 > 0$

A minimum surface area will occur when

$x = 3$, $y = 3$, and $z = \dfrac{27}{3(3)} = 3$.

The dimensions of the box will be
3 m by 3 m by 3 m.

35. $V = LWH$ and $2W + 2H + L = 108$
$L = 108 - 2W - 2H$
$V(H, W) = (108 - 2W - 2H)WH$
$\qquad = 108WH - 2W^2H - 2WH^2$

$V_H = 108W - 2W^2 - 4HW$

$V_W = 108H - 4WH - 2H^2$

$\left.\begin{array}{l} 108W - 2W^2 - 4HW = 0 \\ 108H - 4WH - 2H^2 = 0 \end{array}\right\} \Rightarrow$

$\left.\begin{array}{l} 2W(54 - W - 2H) = 0 \\ 2H(54 - 2W - H) = 0 \end{array}\right\} \Rightarrow$

$2W = 0$ or $54 - W - 2H = 0$

and

$2H = 0$ or $54 - 2W - H = 0$

Clearly $W = 0$ or $H = 0$ makes no sense in the problem, so we have the following system.

$W + 2H = 54$

$2W + H = 54$

Solve this system by the elimination method.

$-2W - 4H = -108$

$\underline{2W + H = 54}$

$\qquad -3H = -54 \Rightarrow H = 18$

If $H = 18$, $2W + 18 = 54$ and $W = 18$. The only critical point we need consider is (18, 18).

$V_{HH} = -4W$, $V_{WW} = -4H$

$V_{HW} = 108 - 4W - 4H$

$M = V_{HH}(18, 18) \cdot V_{WW}(18, 18) - \left[V_{HW}(18, 18)\right]^2$

$\quad = (-72)(-72) - [108 - 4(18) - 4(18)]^2 = 3888$

Since $M > 0$ and $V_{HH}(18, 18) < 0$, we have a local maximum at (18, 18).

$L = 108 - 2(18) - 2(18) = 36$

The dimensions of the box with maximum volume are 18 inches by 18 inches by 36 inches.

37. $V = LWH$ and $W + H + L = 45$
$L = 45 - W - H$
$V(H, W) = (45 - W - H)WH$
$\qquad = 45WH - W^2H - WH^2$

$V_H = 45W - W^2 - 2WH$

$V_W = 45H - 2WH - H^2$

$\left.\begin{array}{l} 45W - W^2 - 2WH = 0 \\ 45H - 2WH - H^2 = 0 \end{array}\right\} \Rightarrow$

$\left.\begin{array}{l} W(45 - W - 2H) = 0 \\ H(45 - 2W - H) = 0 \end{array}\right\} \Rightarrow$

$W = 0$ or $45 - W - 2H = 0$

and

$H = 0$ or $45 - 2W - H = 0$

Clearly $W = 0$ or $H = 0$ makes no sense in the problem, so we have the following system.

$W + 2H = 45$

$2W + H = 45$

Solve this system by the elimination method.

$-2W - 4H = -90$

$\underline{2W + H = 45}$

$\qquad -3H = -45 \Rightarrow H = 15$

If $H = 15$, $2W + 15 = 45$ and $W = 15$. The only critical point we need consider is (15, 15).

$V_{HH} = -2W$, $V_{WW} = -2H$

$V_{HW} = 45 - 2W - 2H$

$M = V_{HH}(15, 15) \cdot V_{WW}(15, 15) - \left[V_{HW}(15, 15)\right]^2$

$\quad = (-30)(-30) - [45 - 2(15) - 2(15)]^2 = 675$

Since $M > 0$ and $V_{HH}(15, 15) < 0$, we have a local maximum at (15, 15).

$L = 45 - 15 - 15 = 15$

The dimensions of the box with maximum volume are 15 inches by 15 inches by 15 inches.

39. $P(x, y) = R(x, y) - C(x, y)$

$\qquad = (2xy + 2y + 12) - \left(2x^2 + y^2\right)$

$\qquad = 2xy + 2y + 12 - 2x^2 - y^2$

$P_x(x, y) = 2y + 0 + 0 - 4x - 0$

$\qquad = 2y - 4x$

$P_y(x, y) = 2x + 2 + 0 - 0 - 2y$

$\qquad = 2x + 2 - 2y$

$\left.\begin{array}{l} 2y - 4x = 0 \quad (1) \\ 2x + 2 - 2y = 0 \quad (2) \end{array}\right\} \Rightarrow$

$\qquad y = 2x \quad (3)$

$\qquad x - y = -1 \quad (4)$

Substitute $2x$ for y in equation (4).

$x - 2x = -1 \Rightarrow x = 1$

If $x = 1$, $y = 2$.

The critical point is (1, 2).

$P_{xx} = -4$, $P_{yy} = -2$, and $P_{xy} = 2$

$M = P_{xx}(1, 2) \cdot P_{yy}(1, 2) - \left[P_{xy}(1, 2)\right]^2$

$\qquad = (-4)(-2) - (2)^2 = 4$

Since $M > 0$ and $P_{xx}(1, 2) < 0$, we have a local maximum at (1, 2).

$P(1, 2) = 2(1)(2) + 2(2) + 12 - 2(1)^2 - (2)^2 = 14$

A maximum profit of $1400 occurs when 1000 tons of grade A and 2000 tons of grade B ore are used.

41. $P(x, y) = 36xy - x^3 - 8y^3$

$P_x(x, y) = 36y - 3x^2$

$P_y(x, y) = 36x - 24y^2$

$\left.\begin{array}{l} 36y - 3x^2 = 0 \\ 36x - 24y^2 = 0 \end{array}\right\} \Rightarrow \begin{array}{l} y = \dfrac{x^2}{12} \quad (1) \\ 3x - 2y^2 = 0 \quad (2) \end{array}$

Substitute $\dfrac{x^2}{12}$ for y in equation (2).

$3x - 2\left(\dfrac{x^2}{12}\right)^2 = 0 \Rightarrow 3x - 2\left(\dfrac{x^4}{144}\right) = 0 \Rightarrow$

$3x - \dfrac{x^4}{72} = 0 \Rightarrow x\left(3 - \dfrac{x^3}{72}\right) = 0 \Rightarrow x = 0, 6$

Disregard $x = 0$ since it makes no sense in the problem. If $x = 6$, then $y = \dfrac{6^2}{12} = \dfrac{36}{12} = 3$.

Consider the critical point $(6, 3)$.

$P_{xx}(x, y) = -6x \Rightarrow P_{xx}(6, 3) = -6(6) = -36$

$P_{yy}(x, y) = -48y \Rightarrow P_{yy}(6, 3) = -48(3) = -144$

$P_{xy}(x, y) = 36 \Rightarrow P_{xy}(6, 3) = 36$

$M = (-36) \cdot (-144) - [36]^2 = 3888$

Since $M > 0$ and $P_{xx} < 0$, there is a local maximum at (6, 3).

$P(6, 3) = 36(6)(3) - (6)^3 - 8(3)^3 = 216$.

Profit is a maximum of $216,000 when 6 tons of steel and 3 tons of aluminum are used.

43. $E(t, T) = 436.16 - 10.57t - 5.46T$

$\qquad\qquad -.02t^2 + .02T^2 + .08Tt$

$E_t = -10.57 - .04t + .08T$

$E_T = -5.46 + .04T + .08t$

a. $E(0, 0) = 436.16$ kJ/mol

b. $E(10, 180)$

$\qquad = 436.16 - 10.57(10) - 5.46(180)$

$\qquad\quad -.02(10)^2 + .02(180)^2 + .08(180)(10)$

$\qquad = 137.66$ kJ/mol

c. First, solve the system $E_t = 0$, $E_T = 0$.

$\left.\begin{array}{l} -.04t + .08T = 10.57 \\ .08t + .04T = 5.46 \end{array}\right\} \Rightarrow$

$\qquad -.08t + .16T = 21.14$

$\qquad \underline{.08t + .04T = 5.46}$

$\qquad\qquad\quad .2T = 26.6 \Rightarrow T = 133$

$-.04t + .08(133) = 10.57 \Rightarrow -.04t = -.07 \Rightarrow$

$t = 1.75$

Then (1.75, 133) is a critical point.

$E_{tt} = -.04$, $E_{TT} = .04$, $E_{tT} = .08$

$M = (-.04)(.04) - (.08)^2 = -.008$

Since $M < 0$, (1.75, 133) is a saddle point.

Section 14.4 Lagrange Multipliers

1. Maximize $f(x, y) = 2xy$ subject to $x + y = 20$.
The Lagrange function is:
$$F(x, y, \lambda) = f(x, y) - \lambda \cdot g(x, y)$$
$$= 2xy - \lambda x - \lambda y + 20\lambda$$
Compute the partial derivatives:
$$F_x(x, y, \lambda) = 2y - \lambda$$
$$F_y(x, y, \lambda) = 2x - \lambda$$
$$F_\lambda(x, y, \lambda) = -x - y + 20$$
Set each partial derivative to zero to obtain the system:
$$2y - \lambda = 0$$
$$2x - \lambda = 0$$
$$-x - y + 20 = 0$$
Then solve the system:
$$\left. \begin{array}{c} 2y - \lambda = 0 \\ 2x - \lambda = 0 \end{array} \right\} \Rightarrow \left. \begin{array}{c} \lambda = 2y \\ \lambda = 2x \end{array} \right\} \Rightarrow \begin{array}{c} 2x = 2y \\ x = y \end{array}$$
$$-x - y + 20 = 0$$
$$-x - x + 20 = 0 \Rightarrow -2x = -20 \Rightarrow x = 10$$
Since $y = x$, $y = 10$ and $\lambda = 2x = 2 \cdot 10 = 20$.
So, the maximum value of $f(x, y) = 2xy$ subject to $x + y = 20$ occurs when $x = 10$ and when $y = 10$. The maximum value is
$$f(10, 10) = 2 \cdot 10 \cdot 10 = 200.$$

3. Maximize $f(x, y) = xy^2$ subject to $x + 4y = 15$.
The Lagrange function is:
$$F(x, y, \lambda) = f(x, y) - \lambda \cdot g(x, y)$$
$$= xy^2 - \lambda x - 4\lambda y + 15\lambda$$
Compute the partial derivatives:
$$F_x(x, y, \lambda) = y^2 - \lambda$$
$$F_y(x, y, \lambda) = 2xy - 4\lambda$$
$$F_\lambda(x, y, \lambda) = -x - 4y + 15$$
Set each partial derivative to zero to obtain the system:
$$y^2 - \lambda = 0$$
$$2xy - 4\lambda = 0$$
$$-x - 4y + 15 = 0$$
Then solve the system:
$$\left. \begin{array}{c} y^2 - \lambda = 0 \\ 2xy - 4\lambda = 0 \end{array} \right\} \Rightarrow \left. \begin{array}{c} y^2 = \lambda \\ 2xy = 4\lambda \end{array} \right\} \Rightarrow \left. \begin{array}{c} 4y^2 = 4\lambda \\ 2xy = 4\lambda \end{array} \right\} \Rightarrow$$
$$4y^2 - 2xy = 0 \Rightarrow 4y^2 = 2xy \Rightarrow 2y = x$$
$$-x - 4y + 15 = 0 \Rightarrow -2y - 4y + 15 = 0 \Rightarrow$$
$$-6y = -15 \Rightarrow y = \frac{5}{2}$$

Since $2y = x$, $x = 5$ and $\lambda = y^2 = \left(\frac{5}{2}\right)^2 = \frac{25}{4}$.
So, the maximum value of $f(x, y) = xy^2$ subject to $x + 4y = 15$ occurs when $x = 5$ and $y = \frac{5}{2}$.
The maximum value is
$$f\left(5, \frac{5}{2}\right) = 5\left(\frac{5}{2}\right)^2 = \frac{125}{4}.$$

5. Minimize $f(x, y) = x^2 + 2y^2 - xy$ subject to $x + y = 8$. The Lagrange function is:
$$F(x, y, \lambda) = f(x, y) - \lambda \cdot g(x, y)$$
$$= x^2 + 2y^2 - xy - x\lambda - y\lambda + 8\lambda$$
Compute the partial derivatives:
$$F_x(x, y, \lambda) = 2x - y - \lambda$$
$$F_y(x, y, \lambda) = 4y - x - \lambda$$
$$F_\lambda(x, y, \lambda) = -x - y + 8$$
Set each partial derivative to zero to obtain the system:
$$2x - y - \lambda = 0$$
$$4y - x - \lambda = 0$$
$$-x - y + 8 = 0$$
Then solve the system:
$$\left. \begin{array}{c} \lambda = 2x - y \\ \lambda = 4y - x \end{array} \right\} \Rightarrow 2x - y = 4y - x \Rightarrow 3x = 5y \Rightarrow$$
$$y = \frac{3}{5}x$$
$$-x - \frac{3}{5}x + 8 = 0 \Rightarrow -\frac{8}{5}x + 8 = 0 \Rightarrow x = 5$$
Since $y = \frac{3}{5}x$, $y = \frac{3}{5}(5) = 3$. So, the minimum value of $f(x, y) = x^2 + 2y^2 - xy$ subject to $x + y = 8$ occurs when $x = 5$ and when $y = 3$.
The maximum value is
$$f(5, 3) = 5^2 + 2(3)^2 - (5)(3) = 28.$$

7. Maximize $f(x, y) = x^2 - 10y^2$ subject to $x - y = 18$. The Lagrange function is:

$$F(x, y, \lambda) = f(x, y) - \lambda \cdot g(x, y)$$
$$= x^2 - 10y^2 - x\lambda + y\lambda + 18\lambda$$

Compute the partial derivatives:

$$F_x(x, y, \lambda) = 2x - \lambda$$
$$F_y(x, y, \lambda) = -20y + \lambda$$
$$F_\lambda(x, y, \lambda) = -x + y + 18$$

Set each partial derivative to zero to obtain the system:

$$2x - \lambda = 0$$
$$-20y + \lambda = 0$$
$$-x + y + 18 = 0$$

Then solve the system:

$$\left.\begin{array}{c} \lambda = 2x \\ \lambda = 20y \end{array}\right\} \Rightarrow \begin{array}{c} 2x = 20y \\ x = 10y \end{array}$$

$$-(10y) + y + 18 = 0 \Rightarrow -9y + 18 = 0 \Rightarrow y = 2$$

Since $x = 10y$, $x = 10(2) = 20$. So, the maximum value of $f(x, y) = x^2 - 10y^2$ subject to $x - y = 18$ occurs when $x = 20$ and when $y = 2$. The maximum value is

$$f(20, 2) = 20^2 - 10(2)^2 = 360.$$

9. Maximize $f(x, y, z) = xyz^2$ subject to $x + y + z = 6$. The Lagrange function is:

$$F(x, y, z, \lambda) = f(x, y, z) - \lambda \cdot g(x, y, z)$$
$$= xyz^2 - x\lambda - y\lambda - z\lambda + 6\lambda$$

Compute the partial derivatives:

$$F_x(x, y, z, \lambda) = yz^2 - \lambda$$
$$F_y(x, y, z, \lambda) = xz^2 - \lambda$$
$$F_z(x, y, z, \lambda) = 2xyz - \lambda$$
$$F_\lambda(x, y, z, \lambda) = -x - y - z + 6$$

Set each partial derivative to zero to obtain the system:

$$yz^2 - \lambda = 0$$
$$xz^2 - \lambda = 0$$
$$2xyz - \lambda = 0$$
$$-x - y - z + 6 = 0$$

Then solve the system:

$$\left.\begin{array}{c} \lambda = yz^2 \\ \lambda = xz^2 \\ \lambda = 2xyz \end{array}\right\} \Rightarrow \begin{array}{c} yz^2 = xz^2 \Rightarrow y = x \\ xz^2 = 2xyz \Rightarrow z = 2y \end{array}$$

$$-y - y - 2y + 6 = 0 \Rightarrow -4y = -6 \Rightarrow y = \frac{3}{2}$$

Since $x = y$, $x = \frac{3}{2}$ and since $z = 2y$,

$$z = 2\left(\frac{3}{2}\right) = 3.$$ So, the maximum value of

$f(x, y, z) = xyz^2$ subject to $x + y + z = 6$ occurs

when $x = \frac{3}{2}$, $y = \frac{3}{2}$, and when $z = 3$. The

maximum value is

$$f\left(\frac{3}{2}, \frac{3}{2}, 3\right) = \left(\frac{3}{2}\right)\left(\frac{3}{2}\right)(3)^2 = \frac{81}{4}.$$

11. Maximize $f(x, y) = 2xy^2$ subject to $x + y = 21$. The Lagrange function is:

$$F(x, y, \lambda) = f(x, y) - \lambda \cdot g(x, y)$$
$$= 2xy^2 - \lambda x - \lambda y + 21\lambda$$

Compute the partial derivatives:

$$F_x(x, y, \lambda) = 2y^2 - \lambda$$
$$F_y(x, y, \lambda) = 4xy - \lambda$$
$$F_\lambda(x, y, \lambda) = -x - y + 21$$

Set each partial derivative to zero to obtain the system:

$$2y^2 - \lambda = 0$$
$$4xy - \lambda = 0$$
$$-x - y + 21 = 0$$

Then solve the system:

$$\left.\begin{array}{c} \lambda = 2y^2 \\ \lambda = 4xy \end{array}\right\} \Rightarrow 2y^2 = 4xy \Rightarrow y = 2x$$

$$-x - (2x) + 21 = 0 \Rightarrow -3x = -21 \Rightarrow x = 7$$

Since $y = 2x$, $y = 2(7) = 14$. So, the maximum value of $f(x, y) = 2xy^2$ subject to $x + y = 21$ occurs when $x = 7$ and $y = 14$. The maximum value is $f(7, 14) = 2(7)(14)^2 = 2744$.

13. Maximize $f(x, y, z) = xyz$ subject to

$x + y + z = 102$. The Lagrange function is:

$$F(x, y, z, \lambda) = f(x, y, z) - \lambda \cdot g(x, y, z)$$
$$= xyz - \lambda x - \lambda y - \lambda z + 102\lambda$$

Compute the partial derivatives:

$F_x(x, y, z, \lambda) = yz - \lambda$
$F_y(x, y, z, \lambda) = xz - \lambda$
$F_z(x, y, z, \lambda) = xy - \lambda$
$F_\lambda(x, y, z, \lambda) = -x - y - z + 102$

Set each partial derivative to zero to obtain the system:

$yz - \lambda = 0$
$xz - \lambda = 0$
$xy - \lambda = 0$
$-x - y - z + 102 = 0$

Then solve the system:

$$\left.\begin{array}{l} \lambda = yz \\ \lambda = xz \\ \lambda = xy \end{array}\right\} \Rightarrow \begin{array}{l} x = y \\ y = z \end{array}$$

$x = y = z$

$-x - x - x + 102 = 0 \Rightarrow x = 34$

Since $x = y = z$, $y = 34$ and $z = 34$.

So, the maximum value of $f(x, y) = xyz$ subject to $x + y + z = 102$ occurs when $x = 34$, $y = 34$, and $z = 34$. The maximum value is

$f(34, 34, 34) = (34)(34)(34) = 39,304$.

15. Answers vary.

17. Answers vary.

$F(x, y, \lambda) = f(x, y) - \lambda \cdot g(x, y)$
$F_x(x, y, \lambda) = f_x(x, y) - \lambda \cdot g_x(x, y)$
$F_y(x, y, \lambda) = f_y(x, y) - \lambda \cdot g_y(x, y)$
$0 = f_x(x, y) - \lambda \cdot g_x(x, y) \Rightarrow$
$\qquad f_x(x, y) = \lambda g_x(x, y)$
$0 = f_y(x, y) - \lambda \cdot g_y(x, y) \Rightarrow$
$\qquad f_y(x, y) = \lambda g_y(x, y)$
$F_\lambda(x, y, \lambda) = f_\lambda(x, y) - \left[\lambda \cdot g_\lambda(x, y) + g(x, y)\right]$
$0 = 0 - \left[\lambda \cdot 0 + g(x, y)\right] \Rightarrow g(x, y) = 0$

19. Let x = length of the ends and y = length of the side opposite the building. Maximize

$A(x, y) = xy$ subject to $2(8x) + 6y = 1200$. The Lagrange function is:

$$F(x, y, \lambda) = f(x, y) - \lambda \cdot g(x, y)$$
$$= xy - 16\lambda x - 6\lambda y + 1200\lambda$$

Compute the partial derivatives:

$F_x(x, y, \lambda) = y - 16\lambda$
$F_y(x, y, \lambda) = x - 6\lambda$
$F_\lambda(x, y, \lambda) = -16x - 6y + 1200$

Set each partial derivative to zero to obtain the system:

$y - 16\lambda = 0$
$x - 6\lambda = 0$
$-16x - 6y + 1200 = 0$

Then solve the system:

$$\left.\begin{array}{l} 16\lambda = y \\ 6\lambda = x \end{array}\right\} \Rightarrow \left.\begin{array}{l} \lambda = \dfrac{y}{16} \\ \lambda = \dfrac{x}{6} \end{array}\right\} \Rightarrow \dfrac{y}{16} = \dfrac{x}{6} \Rightarrow x = \dfrac{6y}{16} = \dfrac{3}{8}y$$

$-16\left(\dfrac{3y}{8}\right) - 6y + 1200 = 0 \Rightarrow$

$-6y - 6y + 1200 = 0 \Rightarrow -12y = -1200 \Rightarrow y = 100$

Since $x = \dfrac{3}{8}y$, $x = \dfrac{300}{8} = 37.5$. So, the

maximum area is produced when $x = 37.5$ feet and $y = 100$ feet. The area is

$A(37.5, 100) = 37.5 \cdot 100 = 3750 \text{ ft}^2$.

21. Maximize $P(x, y) = -x^2 - y^2 + 4x + 8y$ subject to $x + y = 6$. The Lagrange function is:

$$F(x, y, \lambda) = f(x, y) - \lambda \cdot g(x, y)$$
$$= -x^2 - y^2 + 4x + 8y - \lambda x - \lambda y + 6\lambda$$

Compute the partial derivatives:

$F_x(x, y, \lambda) = -2x + 4 - \lambda$
$F_y(x, y, \lambda) = -2y + 8 - \lambda$
$F_\lambda(x, y, \lambda) = -x - y + 6$

Set each partial derivative to zero to obtain the system:

$-2x + 4 - \lambda = 0 \quad (1)$
$-2y + 8 - \lambda = 0 \quad (2)$
$-x - y + 6 = 0 \quad (3)$

Then solve the system:

$$\left.\begin{array}{l} \lambda = -2x + 4 \\ \lambda = -2y + 8 \end{array}\right\} \Rightarrow -2x + 4 = -2y + 8 \Rightarrow$$

$-2x = -2y + 4 \Rightarrow x = y - 2$

Using equation (3), we have

$-x - y + 6 = 0 \Rightarrow -(y - 2) - y + 6 = 0 \Rightarrow y = 4$

Since $x = y - 2$, $x = 4 - 2 = 2$. Profit is maximized when 2 automobile radiators and 4 generator radiators are sold. Maximum profit is

$P(2, 4) = -(2)^2 - (4)^2 + 4(2) + 8(4) = \$20$.

23. Maximize $f(x,y) = 12x^{3/4}y^{1/4}$ subject to $100x + 180y = 25,200$. The Lagrange function is: $F(x,y,\lambda) = f(x,y) - \lambda \cdot g(x,y)$
$$= 12x^{3/4}y^{1/4} - 100\lambda x$$
$$- 180\lambda y + 25,200\lambda$$

Compute the partial derivatives:
$$F_x(x,y,\lambda) = 9x^{-1/4}y^{1/4} - 100\lambda$$
$$F_y(x,y,\lambda) = 3x^{3/4}y^{-3/4} - 180\lambda$$
$$F_\lambda(x,y,\lambda) = -100x - 180y + 25,200$$

Set each partial derivative to zero to obtain the system:
$$9x^{-1/4}y^{1/4} - 100\lambda = 0$$
$$3x^{3/4}y^{-3/4} - 180\lambda = 0$$
$$-100x - 180y + 25,200 = 0$$

Then solve the system:
$$\left.\begin{array}{l} \lambda = \dfrac{9x^{-1/4}y^{1/4}}{100} \\[2mm] \lambda = \dfrac{x^{3/4}y^{-3/4}}{60} \end{array}\right\} \Rightarrow \dfrac{9x^{-1/4}y^{1/4}}{100} = \dfrac{x^{3/4}y^{-3/4}}{60} \Rightarrow$$
$$\dfrac{9y^{1/4}}{100x^{1/4}} = \dfrac{x^{3/4}}{60y^{3/4}} \Rightarrow 540y = 100x \Rightarrow x = 5.4y$$
$$-100(5.4y) - 180y + 25,200 = 0 \Rightarrow$$
$$-540y - 180y + 25,200 = 0 \Rightarrow$$
$$-720y = -25,200 \Rightarrow y = 35$$

Since $x = 5.4y$, $x = 5.4(35) = 189$. Maximum production occurs when 189 units of labor and when 35 units of capital are expended.

25. Let x = length of the ends and y = length of the side opposite. Maximize $A(x,y) = xy$ subject to $2x + y = 600$. The Lagrange function is:
$$F(x,y,\lambda) = f(x,y) - \lambda \cdot g(x,y)$$
$$= xy - 2\lambda x - \lambda y + 600\lambda$$

Compute the partial derivatives:
$$F_x(x,y,\lambda) = y - 2\lambda$$
$$F_y(x,y,\lambda) = x - \lambda$$
$$F_\lambda(x,y,\lambda) = -2x - y + 600$$

Set each partial derivative to zero to obtain the system:
$$y - 2\lambda = 0$$
$$x - \lambda = 0$$
$$-2x - y + 600 = 0$$

Then solve the system:
$$\left.\begin{array}{l} \lambda = \dfrac{y}{2} \\[2mm] \lambda = x \\[2mm] -2x - y + 600 = 0 \end{array}\right\} \Rightarrow x = \dfrac{y}{2}$$
$$-2\left(\dfrac{1}{2}y\right) - y + 600 = 0 \Rightarrow -2y = -600 \Rightarrow$$
$$y = 300$$

Since $x = \dfrac{1}{2}y$, $x = \dfrac{1}{2}(300) = 150$. The maximum area is enclosed by the fence occurs when the two ends are 150 meters each and the single opposite side is 300 meters, resulting in an area of $A(150, 300) = 45,000$ m^2.

27. Let h = height of cylinder and r = radius. Minimize the surface area,
$$S(h,r) = 2\pi r^2 + 2hr\pi \text{ subject to}$$
$$V(h,r) = h\pi r^2 = 25. \text{ The Lagrange function is:}$$
$$F(h,r,\lambda) = f(h,r) - \lambda \cdot g(h,r)$$
$$= 2\pi r^2 + 2\pi hr - \pi\lambda hr^2 + 25\lambda$$

Compute the partial derivatives:
$$F_h(h,r,\lambda) = 2\pi r - \pi r^2 \lambda$$
$$F_r(h,r,\lambda) = 4\pi r + 2\pi h - 2\pi hr\lambda$$
$$F_\lambda(h,r,\lambda) = -\pi hr^2 + 25$$

Set each partial derivative to zero to obtain the system:
$$2\pi r - \pi r^2 \lambda = 0$$
$$4\pi r + 2\pi h - 2\pi hr\lambda = 0$$
$$-\pi hr^2 + 25 = 0$$

Then solve the system:
$$\left.\begin{array}{l} \pi r^2 \lambda = 2\pi r \\[2mm] \lambda = \dfrac{2}{r} \\[2mm] 2\pi hr\lambda = 4\pi r + 2\pi h \\[2mm] hr\lambda = 2r + h \\[2mm] \lambda = \dfrac{2r+h}{hr} \end{array}\right\} \Rightarrow \dfrac{2}{r} = \dfrac{2r+h}{hr} \Rightarrow$$
$$2h = 2r + h \Rightarrow h = 2r$$
$$-\pi(2r)r^2 + 25 = 0 \Rightarrow -2\pi r^3 = -25 \Rightarrow$$
$$r^3 = \dfrac{25}{2\pi} \Rightarrow r = \sqrt[3]{\dfrac{25}{2\pi}} \approx 1.58$$

Since $h = 2r$, $h = 2(1.58) = 3.17$. The cylindrical can holding 25 in^3 has a minimum surface area if the radius is 1.58 inches and the height is 3.17 inches.

29. Let l = length, w = width, and h = height of the box. Minimize the surface area,

$S(l,w,h) = 2lh + 2wh + 2lw$, subject to the

constraint, $V(l,w,h) = lwh = 185$. The Lagrange function is:

$F(l,w,h,\lambda) = f(l,w,h) - \lambda \cdot g(l,w,h)$
$= 2lh + 2wh + 2lw - \lambda lwh + 185\lambda$

Compute the partial derivatives:

$F_l(l,w,h,\lambda) = 2h + 2w - \lambda wh$
$F_w(l,w,h,\lambda) = 2h + 2l - \lambda lh$
$F_h(l,w,h,\lambda) = 2l + 2w - \lambda lw$
$F_\lambda(l,w,h,\lambda) = -lwh + 185$

Set each partial derivative to zero to obtain the system:

$2h + 2w - \lambda wh = 0$
$2h + 2l - \lambda lh = 0$
$2l + 2w - \lambda lw = 0$
$-lwh + 185 = 0$

Then solve the system:

$\left.\begin{array}{l}\lambda = \dfrac{2h+2w}{wh} \\[6pt] \lambda = \dfrac{2h+2l}{lh} \\[6pt] \lambda = \dfrac{2l+2w}{lw}\end{array}\right\} \Rightarrow \left.\begin{array}{l}\dfrac{2h+2w}{wh} = \dfrac{2h+2l}{lh} \\[6pt] \dfrac{2h+2l}{lh} = \dfrac{2l+2w}{lw}\end{array}\right\} \Rightarrow$

$\left.\begin{array}{l}2hl + 2wl = 2hw + 2lw \Rightarrow w = l \\ 2hw + 2lw = 2lh + 2wh \Rightarrow w = h\end{array}\right\} \Rightarrow w = l = h$

$-www = -185 \Rightarrow w^3 = 185 \Rightarrow w = \sqrt[3]{185} \approx 5.70$

$w = l = h \approx 5.70$

A box having a volume of 185 in^3 has minimum surface area when it has dimensions 5.70 in × 5.70 in × 5.70 in.

31. Let l = length, w = width, and h = height of the acquarium. Minimize the surface area,

$S(l,w,h) = 2lh + 2wh + lw$, subject to the

constraint, $V(l,w,h) = lwh = 32$. The Lagrange function is:

$F(l,w,h,\lambda) = f(l,w,h) - \lambda \cdot g(l,w,h)$
$= 2lh + 2wh + lw - \lambda lwh + 32\lambda$

Compute the partial derivatives:

$F_l(l,w,h,\lambda) = 2h + w - \lambda wh$
$F_w(l,w,h,\lambda) = 2h + l - \lambda lh$
$F_h(l,w,h,\lambda) = 2l + 2w - \lambda lw$
$F_\lambda(l,w,h,\lambda) = -lwh + 32$

Set each partial derivative to zero to obtain the system:

$2h + w - \lambda wh = 0$
$2h + l - \lambda lh = 0$
$2l + 2w - \lambda lw = 0$
$-lwh + 32 = 0$

Then solve the system:

$\left.\begin{array}{l}\lambda = \dfrac{2h+2w}{wh} \\[6pt] \lambda = \dfrac{2h+l}{lh} \\[6pt] \lambda = \dfrac{2l+2w}{lw}\end{array}\right\} \Rightarrow \left.\begin{array}{l}\dfrac{2h+w}{wh} = \dfrac{2h+l}{lh} \\[6pt] \dfrac{2h+l}{lh} = \dfrac{2l+2w}{lw}\end{array}\right\} \Rightarrow$

$\left.\begin{array}{l}2hl + wl = 2hw + lw \Rightarrow w = l \\ 2hw + lw = 2lh + 2wh \Rightarrow w = 2h\end{array}\right\} \Rightarrow l = w = 2h$

$-(2h)(2h)h = -32 \Rightarrow 4h^3 = 32 \Rightarrow$

$h^3 = 8 \Rightarrow h = 2$

Since $w = 2h$, $w = 2(2) = 4$, and $l = 4$. The minimum surface area of the aquarium occurs when the dimensions are 4 ft × 4 ft × 2 ft.

33. a. $F(r,s,t,\lambda) = P(r,s,t) - \lambda \cdot (r + s + t - \alpha)$
$= rs(1-t) + (1-r)st + r(1-s)t$
$\qquad + rst - \lambda \cdot (r + s + t - \alpha)$
$= rs - 2rst + st + rt - \lambda r$
$\qquad - \lambda s - \lambda t + \lambda \alpha$

b. Assume $\alpha = .75$

$F(r,s,t,\lambda)$
$= rs - 2rst + st + rt - \lambda r - \lambda s - \lambda t + .75\lambda$

Partial derivatives of F:

$F_r(r,s,t,\lambda) = s - 2st + t - \lambda$
$F_s(r,s,t,\lambda) = r - 2rt + t - \lambda$
$F_t(r,s,t,\lambda) = -2rs + s + r - \lambda$
$F_\lambda(r,s,t,\lambda) = -r - s - t + .75$

Resulting system:

$s - 2st + t - \lambda = 0$
$r - 2rt + t - \lambda = 0$
$-2rs + s + r - \lambda = 0$
$-r - s - t + .75 = 0$

Solving the system:

$\left.\begin{array}{l}\lambda = s - 2st + t \\ \lambda = r - 2rt + t \\ \lambda = -2rs + s + r\end{array}\right\} \Rightarrow$

$s - 2st + t = r - 2rt + t \Rightarrow$
$s(1-2t) = r(1-2t) \Rightarrow r = s$

(continued on next page)

(continued from page 469)

$$r - 2rt + t = -2rs + s + r \Rightarrow$$
$$t(1 - 2r) = s(1 - 2r) \Rightarrow s = t$$
$$-r - r - r = -.75 \Rightarrow -3r = -.75 \Rightarrow r = .25$$

When $\alpha = .75$, the probability of convicting is maximized when $r = s = t = .25$.

c. If $\alpha = 3$, then $-3r = -3$; so,
$r = s = t = 1.0$.

Chapter 14 Review Exercises

1. $f(x, y) = 6y^2 - 5xy + 2x$
$$f(-3, 1) = 6(1)^2 - 5(-3)(1) + 2(-3) = 15$$
$$f(5, -2) = 6(-2)^2 - 5(5)(-2) + 2(5) = 84$$

2. $f(x, y) = -3x + 2x^2y^2 + 5y$
$$f(-3, 1) = -3(-3) + 2(-3)^2(1)^2 + 5(1) = 32$$
$$f(5, -2) = -3(5) + 2(5)^2(-2)^2 + 5(-2) = 175$$

3. $f(x, y) = \dfrac{2x - 4}{x + 3y}$
$$f(-3, 1) = \frac{2(-3) - 4}{(-3) + 3(1)} = \frac{-10}{0}, \text{ undefined}$$
$$f(5, -2) = \frac{2(5) - 4}{5 + 3(-2)} = \frac{6}{-1} = -6$$

4. $f(x, y) = x\sqrt{x^2 + y^2}$
$$f(-3, 1) = -3\sqrt{(-3)^2 + 1^2} = -3\sqrt{10}$$
$$f(5, -2) = 5\sqrt{5^2 + (-2)^2} = 5\sqrt{29}$$

5. Answers vary. Sample answer:
The graph of $2x + y + 4z = 12$ is a plane passing through $(6, 0, 0)$, $(0, 12, 0)$, and $(0, 0, 3)$.

6. Answers vary. Sample answer:
The graph of $y = 2$ is a plane parallel to the xz-plane passing through $(0, 2, 0)$.

7. $x + 2y + 4z = 4$
Let $x = 0$ and $y = 0$. Then $z = 1$. The point $(0, 0, 1)$ is on the graph. Let $x = 0$ and $z = 0$. Then $y = 2$. The point $(0, 2, 0)$ is on the graph. Let $y = 0$ and $z = 0$. Then $x = 4$. The point $(4, 0, 0)$ is on the graph. The graph is a plane. We sketch the portion in the first octant through these three points.

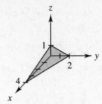

8. $3x + 2y = 6$
We let $x = 0$ and find $y = 3$. The point $(0, 3, 0)$ is on the graph. We let $y = 0$ and find $x = 2$. The point $(2, 0, 0)$ is on the graph. Since there is no z-intercept, the plane is parallel to the z-axis. We sketch the portion in the first octant through these two points.

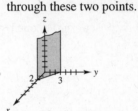

9. $4x + 5y = 20$
We let $x = 0$ and find $y = 4$. The point $(0, 4, 0)$ is on the graph. We let $y = 0$ and find $x = 5$. The point $(5, 0, 0)$ is on the graph. Since there is no z-intercept, the plane is parallel to the z-axis. We sketch the portion in the first octant through these two points.

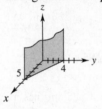

10. $x = 6$
The x-intercept is $(6, 0, 0)$. Since there is no y-intercept or z-intercept, the plane is parallel to the yz-plane. We sketch the portion in the first octant through the point $(6, 0, 0)$.

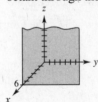

11. $z = f(x,\ y) = -2x^2 + 5xy + y^2$

    a. $\dfrac{\partial z}{\partial x} = -4x + 5y$

    b. $\dfrac{\partial z}{\partial y} = 5x + 2y$

       $\dfrac{\partial z}{\partial y}(-1,\ 4) = 5(-1) + 2(4) = 3$

    c. $f_{xy} = 5$

       $f_{xy}(2,\ -1) = 5$

12. $z = f(x,\ y) = \dfrac{2y + x^2}{3y - x}$

Use the quotient rule to find the partial derivatives.

    a. $\dfrac{\partial z}{\partial y} = \dfrac{(3y - x)(2) - \left(2y + x^2\right)(3)}{(3y - x)^2}$

       $= \dfrac{-2x - 3x^2}{(3y - x)^2}$

    b. $\dfrac{\partial z}{\partial x} = \dfrac{(3y - x)(2x) - \left(2y + x^2\right)(-1)}{(3y - x)^2}$

       $= \dfrac{6xy - x^2 + 2y}{(3y - x)^2}$

       $\dfrac{\partial z}{\partial x}(0,\ 2) = \dfrac{6(0)(2) - (0)^2 + 2(2)}{[3(2) - 0]^2} = \dfrac{4}{36} = \dfrac{1}{9}$

    c. $f_y = \left(-2x - 3x^2\right)(3y - x)^{-2}$

       $f_{yy} = -2\left(-2x - 3x^2\right)(3y - x)^{-3}(3)$

       $= \dfrac{6\left(2x + 3x^2\right)}{(3y - x)^3}$

       $f_{yy}(-1,\ 0) = \dfrac{6\left[2(-1) + 3(-1)^2\right]}{[3(0) - (-1)]^3} = \dfrac{6(1)}{1} = 6$

13. Answers vary.

14. $f(x,\ y) = 3y - 7x^2y^3$

    $f_x = 0 - 14xy^3 = -14xy^3$

    $f_y = 3 - 7 \cdot 3x^2y^2 = 3 - 21x^2y^2$

15. $f(x,\ y) = 4x^3y + 12xy^3$

    $f_x = 4 \cdot 3x^2y + 12 \cdot 1 \cdot y^3 = 12x^2y + 12y^3$

    $f_y = 4x^3 \cdot 1 + 12 \cdot 3xy^2 = 4x^3 + 36xy^2$

16. $f(x,\ y) = \sqrt{3x^2 + 2y^2} = \left(3x^2 + 2y^2\right)^{\frac{1}{2}}$

    $f_x = \dfrac{1}{2}\left(3x^2 + 2y^2\right)^{-\frac{1}{2}} \cdot 6x = \dfrac{3x}{\sqrt{3x^2 + 2y^2}}$

    $f_y = \dfrac{1}{2}\left(3x^2 + 2y^2\right)^{-\frac{1}{2}} \cdot 4y = \dfrac{2y}{\sqrt{3x^2 + 2y^2}}$

17. $f(x,\ y) = \dfrac{3x^2 - 2y^2}{x^2 + 4y^2}$

    $f_x = \dfrac{\left(x^2 + 4y^2\right)(6x) - \left(3x^2 - 2y^2\right)(2x)}{\left(x^2 + 4y^2\right)^2}$

    $= \dfrac{6x^3 + 24xy^2 - 6x^3 + 4xy^2}{\left(x^2 + 4y^2\right)^2} = \dfrac{28xy^2}{\left(x^2 + 4y^2\right)^2}$

    $f_y = \dfrac{\left(x^2 + 4y^2\right)(-4y) - \left(3x^2 - 2y^2\right)(8y)}{\left(x^2 + 4y^2\right)^2}$

    $= \dfrac{-4x^2y - 16y^3 - 24x^2y + 16y^3}{\left(x^2 + 4y^2\right)^2}$

    $= -\dfrac{28x^2y}{\left(x^2 + 4y^2\right)^2}$

18. $f(x,\ y) = x^3e^{3y}$

    $f_x = 3x^2e^{3y}$

    $f_y = x^3 \cdot 3e^{3y} = 3x^3e^{3y}$

19. $f(x,\ y) = (y + 1)^2 e^{2x + y}$

    $f_x = (y + 1)^2 \cdot 2e^{2x + y} = 2(y + 1)^2 e^{2x + y}$

    $f_y = (y + 1)^2 \cdot 1 \cdot e^{2x + y} + 2(y + 1)^1 \cdot 1 \cdot e^{2x + y}$

    $= (y + 1)^2 e^{2x + y} + 2(y + 1)e^{2x + y}$

    $= (y + 1)e^{2x + y}[(y + 1) + 2]$

    $= (y + 1)(y + 3)e^{2x + y}$

20. $f(x, y) = \ln\left|x^2 - 4y^3\right|$

$$f_x = \frac{2x}{x^2 - 4y^3}$$

$$f_y = \frac{-12y^2}{x^2 - 4y^3}$$

21. $f(x, y) = \ln\left|1 + x^3 y^2\right|$

$$f_x = \frac{0 + 3x^2 y^2}{1 + x^3 y^2} = \frac{3x^2 y^2}{1 + x^3 y^2}$$

$$f_y = \frac{0 + 2x^3 y}{1 + x^3 y^2} = \frac{2x^3 y}{1 + x^3 y^2}$$

22. Answers vary.

23. $f(x, y) = 4x^3 y^2 - 8xy$

$$f_x = 12x^2 y^2 - 8y$$

$$f_{xx} = 24xy^2$$

$$f_{xy} = 24x^2 y - 8$$

24. $f(x, y) = \frac{2x + y}{x - 2y}$

$$f_x = \frac{(x - 2y) \cdot 2 - (2x + y) \cdot 1}{(x - 2y)^2} = -\frac{5y}{(x - 2y)^2}$$

$$f_{xx} = \frac{(x - 2y)^2(0) - (-5y)(2)(x - 2y)}{(x - 2y)^4}$$

$$= \frac{10y}{(x - 2y)^3}$$

$$f_{xy} = \frac{(x - 2y)^2(-5) - (-5y)(-4)(x - 2y)}{(x - 2y)^4}$$

$$= \frac{(x - 2y)\left[(x - 2y)(-5) - (-5y)(-4)\right]}{(x - 2y)^4}$$

$$= \frac{-5x - 10y}{(x - 2y)^3}$$

25. $f(x, y) = -6xy^3 + 2x^2 y$

$$f_x = -6y^3 + 4xy$$

$$f_{xx} = 4y$$

$$f_{xy} = -18y^2 + 4x$$

26. $f(x, y) = \frac{3x + y}{x - 1}$

$$f_x = \frac{(x - 1) \cdot 3 - (3x + y) \cdot 1}{(x - 1)^2} = \frac{-3 - y}{(x - 1)^2}$$

$$= (-3 - y)(x - 1)^{-2}$$

$$f_{xx} = -2(-3 - y)(x - 1)^{-3} = \frac{2(3 + y)}{(x - 1)^3}$$

$$f_{xy} = \frac{-1}{(x - 1)^2}$$

27. $f(x, y) = x^2 e^y$

$$f_x = 2xe^y$$

$$f_{xx} = 2e^y, \quad f_{xy} = 2xe^y$$

28. $f(x, y) = ye^{x^2}$

$$f_x = 2xye^{x^2}$$

$$f_{xx} = 2xy \cdot 2xe^{x^2} + e^{x^2} \cdot 2y = 2ye^{x^2}\left(2x^2 + 1\right)$$

$$f_{xy} = 2xe^{x^2}$$

29. $f(x, y) = \ln\left(2 - x^2 y\right)$

$$f_x = \frac{1}{2 - x^2 y} \cdot (-2xy) = \frac{2xy}{x^2 y - 2}$$

$$f_{xx} = \frac{\left(x^2 y - 2\right) 2y - 2xy(2xy)}{\left(x^2 y - 2\right)^2}$$

$$= \frac{2y\left[\left(x^2 y - 2\right) - 2x^2 y\right]}{\left(x^2 y - 2\right)^2} = \frac{2y\left(-x^2 y - 2\right)}{\left(x^2 y - 2\right)^2}$$

$$= \frac{-2x^2 y^2 - 4y}{\left(x^2 y - 2\right)^2}$$

$$f_{xy} = \frac{\left(x^2 y - 2\right) 2x - (2xy)x^2}{\left(x^2 y - 2\right)^2}$$

$$= \frac{2x\left[\left(x^2 y - 2\right) - x^2 y\right]}{\left(x^2 y - 2\right)^2} = \frac{2x(-2)}{\left(x^2 y - 2\right)^2}$$

$$= -\frac{4x}{\left(x^2 y - 2\right)^2}$$

30. $f(x, y) = \ln\left(1 + 3xy^2\right)$

$$f_x = \frac{1}{1 + 3xy^2} \cdot 3y^2 = \frac{3y^2}{1 + 3xy^2} = 3y^2\left(1 + 3xy^2\right)^{-1}$$

$$f_{xx} = 3y^2 \cdot \left(-3y^2\right)\left(1 + 3xy^2\right)^{-2} = \frac{-9y^4}{\left(1 + 3xy^2\right)^2}$$

$$f_{xy} = \frac{\left(1 + 3xy^2\right) \cdot 6y - 3y^2(6xy)}{\left(1 + 3xy^2\right)^2} = \frac{6y}{\left(1 + 3xy^2\right)^2}$$

31. $C(x, y) = 4x^2 + 5y^2 - 4xy + 50$

    a. $C(10, 8) = 4(10)^2 + 5(8)^2 - 4(10)(8) + 50$
             $= 450$

        The cost when 10 hours of labor and 8 gallons of paint are used is $450.

    b. $C(12, 10) = 4(12)^2 + 5(10)^2 - 4(12)(10) + 50$
               $= 646$

        The cost when 12 hours of labor and 10 gallons of paint are used is $646.

    c. $C(14, 14) = 4(14)^2 + 5(14)^2 - 4(14)(14) + 50$
               $= 1030$

        The cost when 14 hours of labor and 14 gallons of paint are used is $1030.

32. Total area = area bottom and top + area of ends + area of sides, so
$$F(L, W, H) = 2LW + 2WH + 2LH.$$

33. $c(x, y) = 2x + y^2 + 2xy + 25$

    a. $\dfrac{\partial c}{\partial x} = 2 + 2y$

        $\dfrac{\partial c}{\partial x}(160, 4) = 2 + 2(4) = 10$

    b. $\dfrac{\partial c}{\partial y} = 2y + 2x$

        $\dfrac{\partial c}{\partial y}(350, 10) = 2(10) + 2(350) = 720$

34. $z = x^7 y^3$

    a. $\dfrac{\partial z}{\partial x} = .7x^{7-1}y^3 = \dfrac{.7y^3}{x^3}$

    b. $\dfrac{\partial z}{\partial y} = .3x^7 y^{3-1} = \dfrac{.3x^7}{y^7}$

35. $z = f(x, y) = x^2 + 2y^2 - 4y$
    $f_x = 2x$
    $f_y = 4y - 4$

Set each of these two partial derivatives equal to zero forming a system of linear equations.

$$\left.\begin{array}{r} 2x = 0 \\ 4y - 4 = 0 \end{array}\right\} \Rightarrow x = 0, \ y = 1$$

Thus, the critical point is $(0, 1)$.

$$f_{xx} = 2, \ f_{yy} = 4, \ f_{xy} = 0$$

$$M = f_{xx}(0, 1) \cdot f_{yy}(0, 1) - \left[f_{xy}(0, 1)\right]^2$$
$$= 2 \cdot 4 - 0 = 8$$

Since $M > 0$ and $f_{xx}(0, 1) > 0$, we have a local minimum at $(0, 1)$.

$$z = f(0, 1) = 0^2 + 2(1)^2 - 4(1) = -2$$

There is a local minimum of –2 at $(0, 1)$.

36. $z = f(x, y) = x^2 + y^2 + 9x - 8y + 1$
    $f_x = 2x + 9$
    $f_y = 2y - 8$

Set each of these two partial derivatives equal to zero forming a system of linear equations.

$$\left.\begin{array}{r} 2x + 9 = 0 \\ 2y - 8 = 0 \end{array}\right\} \Rightarrow x = -\left(\frac{9}{2}\right), \ y = 4$$

Thus, the critical point is $\left(-\dfrac{9}{2}, 4\right)$.

$$f_{xx} = 2, \ f_{yy} = 2, \ f_{xy} = 0$$

$$M = f_{xx}\left(-\frac{9}{2}, 4\right) \cdot f_{yy}\left(-\frac{9}{2}, 4\right) - \left[f_{xy}\left(-\frac{9}{2}, 4\right)\right]^2$$
$$= 2 \cdot 2 - 0 = 4$$

Since $M > 0$ and $f_{xx}\left(-\dfrac{9}{2}, 4\right) > 0$, we have a

local minimum at $\left(-\dfrac{9}{2}, 4\right)$.

$$z = f\left(-\frac{9}{2}, 4\right)$$
$$= \left(-\frac{9}{2}\right)^2 + (4)^2 + 9\left(-\frac{9}{2}\right) - 8(4) + 1 = -\frac{141}{4}$$

There is a local minimum of $-\dfrac{141}{4}$ at $\left(-\dfrac{9}{2}, 4\right)$.

37. $f(x,y) = x^2 + 5xy - 10x + 3y^2 - 12y$

$f_x = 2x + 5y - 10$

$f_y = 5x + 6y - 12$

Set each of these two partial derivatives equal to zero forming a system of linear equations.

$$\left.\begin{array}{r}2x + 5y - 10 = 0 \\ 5x + 6y - 12 = 0\end{array}\right\} \Rightarrow \left.\begin{array}{r}12x + 30y - 60 = 0 \\ -25x - 30y + 60 = 0\end{array}\right\} \Rightarrow$$

$13x = 0 \Rightarrow x = 0$

$2(0) + 5y - 10 = 0 \Rightarrow y = 2$

The critical point is $(0,2)$.

$f_{xx} = 2, \ f_{yy} = 6, \ f_{xy} = 5$

$M = f_{xx}(0,2) \cdot f_{yy}(0,2) - \left[f_{xy}(0,2)\right]^2$

$\quad = 2 \cdot 6 - (5)^2 = 12 - 25 = -13$

Since $M < 0$, we have a saddle point at $(0,2)$.

38. $z = f(x,y) = x^3 - 8y^2 + 6xy + 4$

$f_x = 3x^2 + 6y$

$f_y = -16y + 6x$

Set each of these two partial derivatives equal to zero forming a system of linear equations.

$$\left.\begin{array}{r}3x^2 + 6y = 0 \\ -16y + 6x = 0\end{array}\right\} \Rightarrow -16y + 6x = 0 \Rightarrow x = \frac{8}{3}y$$

$3\left(\frac{8}{3}y\right)^2 + 6y = 0 \Rightarrow \frac{64}{3}y^2 + 6y = 0 \Rightarrow$

$y\left(\frac{64}{3}y + 6\right) = 0 \Rightarrow y = 0 \text{ or } y = -\frac{9}{32}$

If $y = 0$, then $x = 0$.

If $y = -\dfrac{9}{32}$, then $x = -\dfrac{3}{4}$.

There are two critical points: $(0,0)$ and

$\left(-\dfrac{3}{4}, -\dfrac{9}{32}\right)$.

$f_{xx} = 6x, \ f_{yy} = -16, \ f_{xy} = 6$

At $(0,0)$

$M = f_{xx}(0,0) \cdot f_{yy}(0,0) - \left[f_{xy}(0,0)\right]^2$

$\quad = (6)(0) \cdot (-16) - (6)^2 = -36$

Since $M < 0$ there is a saddle point at $(0,0)$.

At $\left(-\dfrac{3}{4}, -\dfrac{9}{32}\right)$

$M = f_{xx}\left(-\dfrac{3}{4}, -\dfrac{9}{32}\right) \cdot f_{yy}\left(-\dfrac{3}{4}, -\dfrac{9}{32}\right)$

$\qquad - \left[f_{xy}\left(-\dfrac{3}{4}, -\dfrac{9}{32}\right)\right]^2$

$\quad = (6)\left(-\dfrac{3}{4}\right) \cdot (-16) - (6)^2 = 72 - 36 = 36$

Since $M > 0$ and $f_{xx}\left(-\dfrac{3}{4}, -\dfrac{9}{32}\right) < 0$, there is a

local maximum at $\left(-\dfrac{3}{4}, -\dfrac{9}{32}\right)$.

$f\left(-\dfrac{3}{4}, -\dfrac{9}{32}\right)$

$= \left(-\dfrac{3}{4}\right)^3 - 8\left(-\dfrac{9}{32}\right)^2 + 6\left(-\dfrac{3}{4}\right)\left(-\dfrac{9}{32}\right) + 4$

$= \dfrac{539}{128} \approx 4.21$

There is a local maximum of 4.21 at

$\left(-\dfrac{3}{4}, -\dfrac{9}{32}\right)$.

39. $z = f(x,y) = x^3 + y^2 + 2xy - 4x - 3y - 2$

$f_x = 3x^2 + 2y - 4$

$f_y = 2y + 2x - 3$

Set each of these two partial derivatives equal to zero forming a system of nonlinear equations.

$$\left.\begin{array}{r}3x^2 + 2y - 4 = 0 \\ 2y + 2x - 3 = 0\end{array}\right\} \Rightarrow 2y = 3 - 2x$$

Substitute $3 - 2x$ for $2y$ in the first equation.

$3x^2 + 3 - 2x - 4 = 0 \Rightarrow 3x^2 - 2x - 1 = 0 \Rightarrow$

$(3x + 1)(x - 1) = 0 \Rightarrow x = -\dfrac{1}{3}, \ 1$

If $x = -\dfrac{1}{3}$, $2y = 3 - 2\left(-\dfrac{1}{3}\right) = \dfrac{11}{3} \Rightarrow y = \dfrac{11}{6}$.

If $x = 1$, $2y = 3 - 2(1) = 1 \Rightarrow y = \dfrac{1}{2}$.

There are two critical points: $\left(-\dfrac{1}{3}, \dfrac{11}{6}\right)$ and

$\left(1, \dfrac{1}{2}\right)$.

(continued on next page)

(*continued from page 474*)

$$f_{xx} = 6x, \ f_{yy} = 2, \ f_{xy} = 2$$

At $\left(-\dfrac{1}{3}, \dfrac{11}{6}\right)$

$$M = f_{xx}\left(-\frac{1}{3}, \frac{11}{6}\right) \cdot f_{yy}\left(-\frac{1}{3}, \frac{11}{6}\right)$$
$$-\left[f_{xy}\left(-\frac{1}{3}, \frac{11}{6}\right)\right]^2$$
$$= (-2)(2) - (2)^2 = -4 - 4 = -8$$

Since $M < 0$ there is a saddle point at

$\left(-\dfrac{1}{3}, \dfrac{11}{6}\right)$.

At $\left(1, \dfrac{1}{2}\right)$

$$M = f_{xx}\left(1, \frac{1}{2}\right) \cdot f_{yy}\left(1, \frac{1}{2}\right) - \left[f_{xy}\left(1, \frac{1}{2}\right)\right]^2$$
$$= 6(1) \cdot 2 - (2)^2 = 12 - 4 = 8$$

$$z = f\left(1, \frac{1}{2}\right)$$
$$= (1)^3 + \left(\frac{1}{2}\right)^2 + 2(1)\left(\frac{1}{2}\right) - 4(1) - 3\left(\frac{1}{2}\right) - 2$$
$$= -\frac{21}{4} \approx -5.25$$

Since $M > 0$ and $f_{xx}\left(1, \dfrac{1}{2}\right) = 6 > 0$, there is a

local minimum of -5.25 at $\left(1, \dfrac{1}{2}\right)$.

40. $z = f(x, y) = 7x^2 + y^2 - 3x + 6y - 5xy$
$f_x = 14x - 3 - 5y$
$f_y = 2y + 6 - 5x$

Set each of these two partial derivatives equal to zero forming a system of linear equations.

$$\left.\begin{array}{r} 14x - 3 - 5y = 0 \\ 2y + 6 - 5x = 0 \end{array}\right\} \Rightarrow \left.\begin{array}{r} 14x - 5y = 3 \\ -5x + 2y = -6 \end{array}\right\} \Rightarrow$$

$$\left.\begin{array}{r} 28x - 10y = 6 \\ -25x + 10y = -30 \end{array}\right\} \Rightarrow 3x = -24 \Rightarrow x = -8$$

$$2y + 6 - 5(-8) = 0 \Rightarrow y = -23$$

The critical point is $(-8, -23)$.

$f_{xx} = 14, \ f_{yy} = 2, \ f_{xy} = -5$

$$M = f_{xx}(-8, -23) \cdot f_{yy}(-8, -23)$$
$$-\left[f_{xy}(-8, -23)\right]^2$$
$$= (14) \cdot (2) - (-5)^2 = 3$$

Since $M > 0$ and $f_{xx}(-8, -23) > 0$, we have a

local minimum at $(-8, -23)$.

$$z = f(-8, -23)$$
$$= 7(-8)^2 + (-23)^2 - 3(-8) + 6(-23)$$
$$\qquad\qquad -5(-8)(-23)$$
$$= -57$$

There is a local minimum of -57 at $(-8, -23)$.

41. a. $c(x, y) = x^2 + 5y^2 + 4xy - 70x - 164y + 1800$

$c_x = 2x + 4y - 70$

$c_y = 10y + 4x - 164$

$$\left.\begin{array}{r} 2x + 4y - 70 = 0 \\ 4x + 10y - 164 = 0 \end{array}\right\} \Rightarrow \left.\begin{array}{r} -4x - 8y + 140 = 0 \\ 4x + 10y - 164 = 0 \end{array}\right\} \Rightarrow$$

$$2y - 24 = 0 \Rightarrow y = 12$$

$$2x + 4(12) - 70 = 0 \Rightarrow x = 11$$

$$c_{xx} = 2, \ c_{yy} = 10, \ c_{xy} = 4$$

At (11, 12), $M = 2 \cdot 10 - 4^2 = 4 > 0$ and

$c_{xx} > 0$.

A local minimum occurs at (11, 12).

b. The minimum cost is

$c(11, 12)$

$$= (11)^2 + 5(12)^2 + 4(11)(12) - 70(11)$$
$$\qquad\qquad -164(12) + 1800$$
$$= 121 + 720 + 528 - 770 - 1968 + 1800$$
$$= \$431$$

42. $P(x, y)$
$$= .01\left(-x^2 + 3xy + 160x - 5y^2 + 200y + 2600\right)$$

a. $x + y = 280 \Rightarrow y = 280 - x$

$$P(x, y) = .01[-x^2 + 3x(280 - x) + 160x$$
$$-5(280 - x)^2 + 200(280 - x)$$
$$+ 2600]$$

$$= .01[-x^2 + 840x - 3x^2 + 160x$$
$$-5\left(78,400 - 560x + x^2\right)$$
$$+ 56,000 - 200x + 2600]$$

$$= .01\left(-x^2 + 840x - 3x^2 + 160x\right.$$
$$-392,000 + 2800x - 5x^2$$
$$\left. + 56,000 - 200x + 2600\right)$$

$$P(x, y) = .01\left(-9x^2 + 3600x - 333,400\right)$$

$$P'(x, y) = .01(-18x + 3600)$$

(*continued on next page*)

(continued from page 475)

$.01(-18x + 3600) = 0 \Rightarrow x = 200$

If $x < 200$, $P' > 0$. If $x > 200$, $P' < 0$.
There is a maximum when $x = 200$.
If $x = 200$, $y = 280 - 200 = 80$.

$$P(200, 80) = .01\left[-(200)^2 + 3(200)(80) \right.$$
$$+ 160(200) - 5(80)^2$$
$$+ 200(80) + 2600 \right]$$
$$= 266$$

$200 spent on fertilizer and $80 spent on seed will produce a maximum profit of $266 per acre.

b. $P_x = .01(-2x + 3y + 160)$
$P_y = .01(3x - 10y + 200)$

$\left. \begin{array}{l} -2x + 3y + 160 = 0 \\ 3x - 10y + 200 = 0 \end{array} \right\} \Rightarrow$

$\left. \begin{array}{l} -2x + 3y = -160 \\ 3x - 10y = -200 \end{array} \right\} \Rightarrow \left. \begin{array}{l} -6x + 9y = -480 \\ 6x - 20y = -400 \end{array} \right\} \Rightarrow$

$-11y = -880 \Rightarrow y = 80$
$-2x + 3(80) = -160 \Rightarrow -2x = -400 \Rightarrow$
$x = 200$
The critical point is (200, 80).

$P_{xx} = .01(-2) = -.02$

$P_{yy} = .01(-10) = -.1$

$P_{xy} = .01(3) = .03$

At (200, 80),

$M = P_{xx}(200, 80) \cdot P_{yy}(200, 80)$
$$- \left[P_{xy}(200, 80) \right]^2$$
$$= (-.02)(-.1) - (.03)^2 = .0011$$

Since $M > 0$ and $P_{xx}(200, 80) < 0$, there is a local maximum at (200, 80). This is the same as in part (a).

c. In (a), we saw that $P(200, 80) = 266$.
$200 spent on fertilizer and $80 spent on seed will produce a maximum profit of $266 per acre.

43. Minimize $f(x, y) = x^2 + y^2$ subject to
$x = y + 2$. The Lagrange function is:
$$F(x, y, \lambda) = f(x, y) - \lambda \cdot g(x, y)$$
$$= x^2 + y^2 - \lambda x + \lambda y + 2\lambda$$

Compute the partial derivatives:
$F_x(x, y, \lambda) = 2x - \lambda$
$F_y(x, y, \lambda) = 2y + \lambda$
$F_\lambda(x, y, \lambda) = -x + y + 2$

Set each partial derivative to zero to obtain the system:
$2x - \lambda = 0$
$2y + \lambda = 0$
$-x + y + 2 = 0$

Then solve the system:
$\left. \begin{array}{l} \lambda = 2x \\ \lambda = -2y \end{array} \right\} \Rightarrow 2x = -2y \Rightarrow x = -y$

$-(-y) + y + 2 = 0 \Rightarrow 2y = -2 \Rightarrow y = -1$

Since $x = -y$, $x = -(-1) = 1$. So, the minimum value of $f(x, y) = x^2 + y^2$ subject to $x = y + 2$ occurs when $x = 1$ and $y = -1$. The minimum value is $f(1, -1) = (1)^2 + (-1)^2 = 2$.

44. Minimize and maximize $f(x, y) = x^2 y$ subject to $x + y = 4$. The Lagrange function is:
$$F(x, y, \lambda) = f(x, y) - \lambda \cdot g(x, y)$$
$$= x^2 y - \lambda x - \lambda y + 4\lambda$$

Compute the partial derivatives:
$F_x(x, y, \lambda) = 2xy - \lambda$
$F_y(x, y, \lambda) = x^2 - \lambda$
$F_\lambda(x, y, \lambda) = -x - y + 4$

Set each partial derivative to zero to obtain the system:
$2xy - \lambda = 0$
$x^2 - \lambda = 0$
$-x - y + 4 = 0$

Then solve the system:
$\left. \begin{array}{l} \lambda = 2xy \\ \lambda = x^2 \end{array} \right\} \Rightarrow 2xy = x^2 \Rightarrow 2y = x.$

If $x = 2y$, then $-(2y) - y + 4 = 0 \Rightarrow y = \dfrac{4}{3}$.

Then $x = \dfrac{8}{3}$.

So, the maximum value of $f(x, y) = x^2 y$ subject to $x + y = 4$ occurs when $x = \dfrac{8}{3}$ and when $y = \dfrac{4}{3}$. The maximum value is

$$f\left(\frac{8}{3}, \frac{4}{3} \right) = \left(\frac{8}{3} \right)^2 \left(\frac{4}{3} \right) = \frac{256}{27}.$$

45. Maximize $f(x, y) = x^2 y$ subject to $x + y = 80$.

The Lagrange function is:
$$F(x, y, \lambda) = f(x, y) - \lambda \cdot g(x, y)$$
$$= x^2 y - \lambda x - \lambda y + 80\lambda$$

Compute the partial derivatives:
$$F_x(x, y, \lambda) = 2xy - \lambda$$
$$F_y(x, y, \lambda) = x^2 - \lambda$$
$$F_\lambda(x, y, \lambda) = -x - y + 80$$

Set each partial derivative to zero to obtain the system:
$$2xy - \lambda = 0$$
$$x^2 - \lambda = 0$$
$$-x - y + 80 = 0$$

Then solve the system:
$$\left. \begin{array}{r} \lambda = 2xy \\ \lambda = x^2 \end{array} \right\} \Rightarrow 2xy = x^2 \Rightarrow 2y = x$$

$$-2y - y + 80 = 0 \Rightarrow -3y = -80 \Rightarrow y = \frac{80}{3}$$

Since $x = 2y$, $x = 2\left(\dfrac{80}{3}\right) = \dfrac{160}{3}$. So, the

maximum value of $f(x, y) = x^2 y$ subject to

$x + y = 80$ occurs when $x = \dfrac{160}{3}$ and when

$y = \dfrac{80}{3}$. The maximum value is

$$f\left(\frac{160}{3}, \frac{80}{3}\right) = \frac{2{,}048{,}000}{27} \approx 75{,}852.$$

46. Maximize $f(x, y) = xy^2$ subject to $x + y = 50$.

The Lagrange function is:
$$F(x, y, \lambda) = f(x, y) - \lambda \cdot g(x, y)$$
$$= xy^2 - \lambda x - \lambda y + 50\lambda$$

Compute the partial derivatives:
$$F_x(x, y, \lambda) = y^2 - \lambda$$
$$F_y(x, y, \lambda) = 2xy - \lambda$$
$$F_\lambda(x, y, \lambda) = -x - y + 50$$

Set each partial derivative to zero to obtain the system:
$$y^2 - \lambda = 0$$
$$2xy - \lambda = 0$$
$$-x - y + 50 = 0$$

Then solve the system:
$$\left. \begin{array}{r} \lambda = 2xy \\ \lambda = y^2 \end{array} \right\} \Rightarrow 2xy = y^2 \Rightarrow y = 2x$$

$$-x - (2x) + 50 = 0 \Rightarrow -3x + 50 = 0 \Rightarrow x = \frac{50}{3}$$

Since $y = 2x$, $y = 2\left(\dfrac{50}{3}\right) = \dfrac{100}{3}$. So, the

maximum value of $f(x, y) = xy^2$ subject to

$x + y = 50$ occurs when $y = \dfrac{100}{3}$ and when

$x = \dfrac{50}{3}$. The maximum value is

$$f\left(\frac{50}{3}, \frac{100}{3}\right) = \left(\frac{50}{3}\right)\left(\frac{100}{3}\right)^2 = \frac{500{,}000}{27}.$$

47. Answers vary

48. Maximize $P(x,y) = .01\left(-x^2 + 3xy + 160x - 5y^2 + 200y + 2600\right)$ subject to $x + y = 280$

The Lagrange function is:

$$F(x,y,\lambda) = f(x,y) - \lambda \cdot g(x,y) = .01\left(-x^2 + 3xy + 160x - 5y^2 + 200y + 2600\right) - \lambda x - \lambda y + 280\lambda$$

$$= -.01x^2 + .03xy + 1.6x - .05y^2 + 2y + 26 - \lambda x - \lambda y + 280\lambda$$

Compute the partial derivatives: $F_x(x,y,\lambda) = -.02x + .03y + 1.6 - \lambda$

$$F_y(x,y,\lambda) = .03x - .1y + 2 - \lambda$$

$$F_\lambda(x,y,\lambda) = -x - y + 280$$

Set each partial derivatives to zero to obtain the system:

$$-.02x + .03y + 1.6 - \lambda = 0$$
$$.03x - .1y + 2 - \lambda = 0$$
$$-x - y + 280 = 0$$

Solving the system using the elimination method (or using technology), we obtain $x = 200$ and $y = 80$.

$$P(200,80) = .01(-(200)^2 + 3(200)(80) + 160(200) - 5(80)^2 + 200(80) + 2600) = 266$$

$P(x,y)$ has a maximum of \$266 when \$200 is spent on fertilizer and when \$80 is spent on hybrid seed.

Case 14 Global Warming and the Method of Least Squares

1. $f_m(m,b) = -101,033 + 22,825,000m + 11,700b$
$f_b(m,b) = -51.78 + 11,700m + 6b$
$f_{mm}(m,b) = 22,825,000$
$f_{bb}(m,b) = 6$
$f_{mb}(m,b) = 11,700$
$f_{mm}f_{bb} - [f_{mb}]^2 = 60,000$

At $m = .0062$, $b = -3.46$, we have $f_m = f_b = 0$, $f_{mm} > 0$, and $f_{mm}f_{bb} - [f_{mb}]^2 > 0$, so f has a minimum at $(.0062, -3.46)$.

2.

$$m = \frac{3(1900 \cdot 8.47 + 1950 \cdot 8.33 + 2000 \cdot 9.09) - (1900 + 1950 + 2000)(8.47 + 8.33 \cdot 9.09)}{3(1900^2 + 1950^2 + 2000^2) - (1900 + 1950 + 2000)^2} = .0062$$

$$b = \frac{8.47 + 8.33 + 9.09}{3} - .0062\left(\frac{1900 + 1950 + 2000}{3}\right) = -3.46$$

3. Using summation notation,

$$m = \frac{5\sum xy - \sum x \sum y}{5\sum x^2 - \left(\sum x\right)^2} = .00532$$

$$b = \frac{\sum y}{5} - m\frac{\sum x}{5} = -1.794$$

Answers vary. The slope .00532 of the least-squares line still suggests a gradual rise in the neighborhood of .005 – .006 degrees per year, but the pattern of data points suggests that the global land temperature fell during the first half of the century, and then rose during the second half at something like .015°C/year.

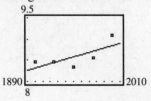